Deeper Conceptual Understanding

Successful students have the ability to apply their statistical ideas and knowledge to new concepts and real-world situations. MyLab Statistics offers a library of over 1,000 conceptual-based questions to help students internalize concepts, make interpretations, and think critically about statistics.

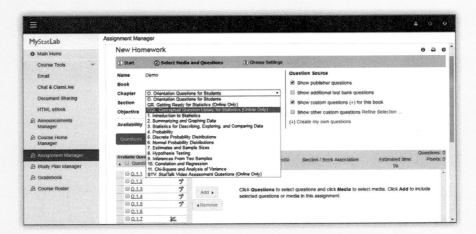

Real-World Connections

Students are motivated to succeed when they're engaged in the learning experience and understand the relevance and power of statistics. StatTalk Videos in MyLab Statistics demonstrate important statistical concepts through interesting stories and real-life events. Hosted by fun-loving statistician Andrew Vickers, this video series includes assignable questions built in MyLab Statistics and an instructor's guide.

pearson.com/mylab/statistics

SEVENTH EDITION

ANNOTATED INSTRUCTOR'S EDITION

Elementary Statistics
PICTURING THE WORLD

Ron Larson
The Pennsylvania State University
The Behrend College

Betsy Farber
Bucks County Community College

Director, Portfolio Management: *Deirdre Lynch*
Senior Courseware Portfolio Manager: *Patrick Barbera*
Editorial Assistant: *Morgan Danna*
Content Producer: *Tamela Ambush*
Managing Producer: *Karen Wernholm*
Media Producer: *Audra Walsh*
Manager, Courseware QA: *Mary Durnwald*
Manager, Content Development: *Robert Carroll*
Product Marketing Manager: *Emily Ockay*
Field Marketing Manager: *Andrew Noble*
Marketing Assistants: *Shannon McCormack, Erin Rush*
Senior Author Support/Technology Specialist: *Joe Vetere*
Manager, Rights and Permissions: *Gina Cheselka*
Manufacturing Buyer: *Carol Melville, LSC Communications*
Text Design, Cover Design: *Cenveo Publisher Services*
Production Coordination, Composition, and Illustrations: *Larson Texts, Inc.*

Library of Congress Cataloging-in-Publication Data

Names: Larson, Ron, 1941- author. | Farber, Elizabeth, author.
Title: Elementary statistics : picturing the world / Ron Larson, The
   Pennsylvania State University, The Behrend College, Betsy Farber, Bucks
   County Community College.
Description: 7th edition. | Boston, MA : Pearson Education, Inc., 2017.
Identifiers: LCCN 2017040872| ISBN 9780134683416 | ISBN 9780134683584
Subjects: LCSH: Statistics--Textbooks. | Mathematical statistics--Textbooks.
Classification: LCC QA276.12 .L373 2017 | DDC 519.5--dc23
LC record available at *https://lccn.loc.gov/2017040872*

1 17

2 17

 **Pearson**

ISBN-10: 0-13-468358-7 (Annotated Instructor's Edition)
ISBN-13: 978-0-13-468358-4
ISBN 10: 0-13-468341-2
ISBN 13: 978-0-13-468341-6

CONTENTS

PART 1 DESCRIPTIVE STATISTICS

 1 Introduction to Statistics

2 Descriptive Statistics *38*

PART 2 PROBABILITY AND PROBABILITY DISTRIBUTIONS

5 Normal Probability Distributions *232*

PART 3 STATISTICAL INFERENCE

6 Confidence Intervals *296*

Hypothesis Testing with One Sample *346*

Hypothesis Testing with Two Samples *416*

11 Nonparametric Tests (Web Only)*

 Where You've Been  Where You're Going

Appendices

PREFACE

Welcome to *Elementary Statistics: Picturing the World*, Seventh Edition. You will find that this textbook is written with a balance of rigor and simplicity. It combines step-by-step instruction, real-life examples and exercises, carefully developed features, and technology that makes statistics accessible to all.

I am grateful for the overwhelming acceptance of the first six editions. It is gratifying to know that my vision of combining theory, pedagogy, and design to exemplify how statistics is used to picture and describe the world has helped students learn about statistics and make informed decisions.

What's New in this Edition

The goal of the Seventh Edition was a thorough update of the key features, examples, and exercises:

Examples This edition has 213 examples, over 60% of which are new or revised. Also, several of the examples now show an alternate solution or a check using technology.

Technology Examples In addition to showing screen displays from Minitab®, Excel®, and the TI-84 Plus, this edition also shows screen displays from StatCrunch®.

Try It Yourself Over 40% of the 213 Try It Yourself exercises are new or revised.

Picturing the World Over 50% of these are new or revised.

Tech Tips New to this edition are technology tips that appear in most sections. These tips show how to use Minitab, Excel, the TI-84 Plus, or StatCrunch to solve a problem.

Exercises Over 40% of the more than 2300 exercises are new or revised.

Extensive Chapter Feature Updates Over 60% of the following key features are new or revised, making this edition fresh and relevant to today's students:

- Where You've Been and Where You're Going
- Uses and Abuses: Statistics in the Real World
- Real Statistics–Real Decisions: Putting it all together
- Chapter Technology Project

Revised Content Here is a summary of the content changes.

- **Section 1.1** now has more discussion about populations and samples, how to identify them, and their relationships to parameters and statistics. Also, the Venn Diagrams have been redrawn to use clearer labeling to help students distinguish between a population and a sample.
- **In Section 1.3,** the figure depicting systemic sampling has been redrawn to more clearly depict the sampling process.
- **Section 2.1** now has more discussion of class widths and open-ended classes. Also, a figure showing a histogram and its corresponding frequency polygon was added after Example 4.
- **In Section 2.4,** Example 9 was rewritten to explain the use of an open-ended class.
- **Section 2.5** now has a Study Tip discussing outliers and modified box-and-whisker plots. On pages 124 and 125,

students are shown how to create modified box-and-whisker plots using technology.

- **In Section 3.1,** the solutions to the examples were rewritten to explain why a formula was chosen to find a probability.
- **In Chapter 5,** in addition to using a table, examples were revised and Tech Tips were added to show how to find areas or probabilities using technology.
- **In Chapter 6,** in addition to using a table, examples were revised and Tech Tips were added to show how to find critical values using technology. Also, the exercises in this chapter were revised to ask more conceptual questions.
- **Section 6.2** now has more explanation about why the t-distribution is needed when σ is unknown. Also, the flowchart on page 314 was revised to illustrate when it is not possible to use the normal distribution or the t-distribution to construct a confidence interval.
- **In Chapters 7–9,** in addition to using a table, examples were revised and Tech Tips were added to show how to find P-values and critical values using technology.
- **Section 8.2** now shows the formula for the number of degrees of freedom for the t-test often used by technology.
- **In Section 9.1,** the requirements to use a correlation coefficient r to make an inference about a population have been revised.

Features of the Seventh Edition
Guiding Student Learning

Where You've Been and Where You're Going Each chapter begins with a two-page visual description of a real-life problem. *Where You've Been* connects the chapter to topics learned in earlier chapters. *Where You're Going* gives students an overview of the chapter.

What You Should Learn Each section is organized by learning objectives, presented in everyday language in *What You Should Learn*. The same objectives are then used as subsection titles throughout the section.

Definitions and Formulas are clearly presented in easy-to-locate boxes. They are often followed by **Guidelines**, which explain *In Words* and *In Symbols* how to apply the formula or understand the definition.

Margin Features help reinforce understanding:

- **Study Tips** show how to read a table, interpret a result, help drive home an important interpretation, or connect different concepts.
- **Tech Tips** show how to use Minitab, Excel, the TI-84 Plus, or StatCrunch to solve a problem.
- **Picturing the World** is a "mini case study" in each section that illustrates the important concept or concepts of the section. Each Picturing the World concludes with a question and can be used for general class discussion or group work. The answers to these questions are included in the *Annotated Instructor's Edition.*

Examples and Exercises

Examples Every concept in the text is clearly illustrated with one or more step-by-step examples. Most examples have an interpretation step that shows the student how the solution may be interpreted within the real-life context of the example and promotes critical thinking and writing skills. Each example, which is numbered and titled for easy reference, is followed by a similar exercise called **Try It Yourself** so students can immediately practice the skill learned. The answers to these exercises are in the back of the book, and the worked-out solutions are in the *Student's Solutions Manual*.

Technology Examples Many sections contain an example that shows how technology can be used to calculate formulas, perform tests, or display data. Screen displays from Minitab, Excel, the TI-84 Plus, and StatCrunch are shown. Additional screen displays are presented at the ends of selected chapters, and detailed instructions are given in separate technology manuals available with the book.

Exercises The exercises give students practice in performing calculations, making decisions, providing explanations, and applying results to a real-life setting. The section exercises are divided into three parts:

- **Building Basic Skills and Vocabulary** are short answer, true or false, and vocabulary exercises carefully written to nurture student understanding.
- **Using and Interpreting Concepts** are skill or word problems that move from basic skill development to more challenging and interpretive problems.
- **Extending Concepts** go beyond the material presented in the section. They tend to be more challenging and are not required as prerequisites for subsequent sections.

Technology Answers Answers in the back of the book are found using calculations by hand and by tables. Answers found using technology (usually the TI-84 Plus) are also included when there are discrepancies due to rounding.

Review and Assessment

Chapter Summary Each chapter concludes with a Chapter Summary that answers the question *What did you learn?* The objectives listed are correlated to Examples in the section as well as to the Review Exercises.

Chapter Review Exercises A set of Review Exercises follows each Chapter Summary. The order of the exercises follows the chapter organization. Answers to all odd-numbered exercises are given in the back of the book.

Chapter Quizzes Each chapter has a Chapter Quiz. The answers to all quiz questions are provided in the back of the book. For additional help, see the step-by-step video solutions available in MyLab Statistics.

Chapter Tests Each chapter has a Chapter Test. The questions are in random order. The answers to all test questions are provided in the *Annotated Instructor's Edition*.

Cumulative Review There is a Cumulative Review after Chapters 2, 5, 8, and 10. Exercises in the Cumulative Review are in random order and may incorporate multiple ideas. Answers to all odd-numbered exercises are given in the back of the book.

Statistics in the Real World

Uses and Abuses: Statistics in the Real World Each chapter discusses how statistical techniques should be used, while cautioning students about common abuses. The discussion includes ethics, where appropriate. Exercises help students apply their knowledge.

Applet Activities Selected sections contain activities that encourage interactive investigation of concepts in the lesson with exercises that ask students to draw conclusions. The applets are available in MyLab Statistics and at **www.pearson.com/math-stats-resources.**

Chapter Case Study Each chapter has a full-page Case Study featuring actual data from a real-world context and questions that illustrate the important concepts of the chapter.

Real Statistics – Real Decisions: Putting it all together This feature encourages students to think critically and make informed decisions about real-world data. Exercises guide students from interpretation to drawing of conclusions.

Chapter Technology Project Each chapter has a Technology project using Minitab, Excel, and the TI-84 Plus that gives students insight into how technology is used to handle large data sets or real-life questions.

Continued Strong Pedagogy from the Sixth Edition

Versatile Course Coverage The table of contents was developed to give instructors many options. For instance, the *Extending Concepts* exercises, applet activities, Real Statistics— Real Decisions, and Uses and Abuses provide sufficient content for the text to be used in a two-semester course. More commonly, I expect the text to be used in a three-credit semester course or a four-credit semester course that includes a lab component. In such cases, instructors will have to pare down the text's 46 sections.

Graphical Approach As with most introductory statistics texts, this text begins the descriptive statistics chapter (Chapter 2) with a discussion of different ways to display data graphically. A difference between this text and many others is that **it continues to incorporate the graphical display of data throughout the text.** For example, see the use of stem-and-leaf plots to display data on page 387. This emphasis on graphical displays is beneficial to all students, especially those utilizing visual learning strategies.

Balanced Approach The text strikes a **balance among computation, decision making, and conceptual understanding.** I have provided many Examples, Exercises, and Try It Yourself exercises that go beyond mere computation.

Variety of Real-Life Applications I have chosen real-life applications that are representative of the majors of students taking introductory statistics courses. I want statistics to come alive and appear relevant to students so they understand the importance of and rationale for studying statistics. I wanted the applications to be **authentic**—but they also need to be **accessible.** See the Index of Applications on page xvi.

Data Sets and Source Lines The data sets in the book were chosen for interest, variety, and their ability to illustrate concepts. Most of the **250-plus data sets** contain real data with

source lines. The remaining data sets contain simulated data that are representative of real-life situations. All data sets containing 20 or more entries are available in a variety of formats in MyLab™ Statistics or at **www.pearson.com/math-stats-resources.** In the exercise sets, the data sets that are available electronically are indicated by the icon.

Flexible Technology Although most formulas in the book are illustrated with "hand" calculations, I assume that most students have access to some form of technology, such as Minitab, Excel, StatCrunch, or the TI-84 Plus. Because technology varies widely, the text is flexible. **It can be used in courses with no more technology than a scientific calculator—or it can be used in courses that require sophisticated technology tools.** Whatever your use of technology, I am sure you agree with me that the goal of the course is not computation. Rather, it is to help students gain an understanding of the basic concepts and uses of statistics.

Prerequisites Algebraic manipulations are kept to a minimum—often I display informal versions of formulas using words in place of or in addition to variables.

Choice of Tables My experience has shown that students find a **cumulative distribution function** (CDF) table easier to use than a "0-to-z" table. Using the CDF table to find the area under the standard normal curve is a topic of Section 5.1 on pages 237–241. Because some teachers prefer to use the "0-to-z" table, an alternative presentation of this topic is provided in Appendix A.

Page Layout Statistics instruction is more accessible when it is carefully formatted on each page with a consistent open layout. This text is the first college-level statistics book to be written so that, when possible, its features are not split from one page to the next. Although this process requires extra planning, the result is a presentation that is clean and clear.

Meeting the Standards

MAA, AMATYC, NCTM Standards This text answers the call for a **student-friendly text that emphasizes the uses of statistics.** My goal is not to produce statisticians but to produce informed consumers of statistical reports. For this reason, I have included exercises that require students to interpret results, provide written explanations, find patterns, and make decisions.

GAISE Recommendations Funded by the American Statistical Association, the Guidelines for Assessment and Instruction in Statistics Education (GAISE) Project developed six recommendations for teaching introductory statistics in a college course. These recommendations are:

- Emphasize statistical literacy and develop statistical thinking.
- Use real data.
- Stress conceptual understanding rather than mere knowledge of procedures.
- Foster active learning in the classroom.
- Use technology for developing conceptual understanding and analyzing data.
- Use assessments to improve and evaluate student learning.

The examples, exercises, and features in this text embrace all of these recommendations.

Technology Resources
MyLab Statistics Online Course (access code required)

Used by nearly one million students a year, MyLab Statistics is the world's leading online program for teaching and learning statistics. MyLab Statistics delivers assessment, tutorials, and multimedia resources that provide engaging and personalized experiences for each student, so learning can happen in any environment.

Personalized Learning Not every student learns the same way or at the same rate. Personalized learning in the MyLab gives instructors the flexibility to incorporate the approach that best suits the needs of their course and students.

- Based on their performance on a quiz or test, **personalized homework** allows students to focus on just the topics they have not yet mastered.
- With **Companion Study Plan Assignments** you can assign the Study Plan as a prerequisite to a test or quiz, guiding students through the concepts they need to master.

Preparedness Preparedness is one of the biggest challenges in statistics courses. Pearson offers a variety of content and course options to support students with just-in-time remediation and key-concept review as needed.

- **Redesign-Ready Course Options** Many new course models have emerged in recent years, as institutions "redesign" to help improve retention and results. At Pearson, we're focused on tailoring solutions to support your plans and programs.
- **Getting Ready for Statistics Questions** This question library contains more than 450 exercises that cover the relevant developmental math topics for a given section. These can be made available to students for extra practice or assigned as a prerequisite to other assignments.

Conceptual Understanding Successful students have the ability to apply their statistical ideas and knowledge to new concepts and real-world situations. Providing frequent opportunities for data analysis and interpretation helps students develop the 21st century skills that they need in order to be successful in the classroom and workplace.

- **Conceptual Question Library** There are 1,000 questions in the Assignment Manager that require students to apply their statistical understanding.
- **Modern statistics is practiced with technology,** and MyLab Statistics makes learning and using software programs seamless and intuitive. Instructors can copy data sets from the text and MyLab Statistics exercises directly into software such as StatCrunch or Excel®. Students can also access instructional support tools including tutorial videos, Study Cards, and manuals for a variety of statistical software programs including StatCrunch, Excel, Minitab®, JMP®, R, SPSS, and TI 83/84 calculators.

Motivation Students are motivated to succeed when they are engaged in the learning experience and understand the relevance and power of statistics.

- **Exercises with Immediate Feedback** Homework and practice exercises in MyLab Statistics regenerate algorithmically to give students unlimited opportunity for

practice and mastery. Instructors can choose from the many exercises available for the author's approach—or even choose additional exercises from other MyLab Statistics courses. Most exercises include learning aids, such as guided solutions, sample problems, extra help at point-of-use, and immediate feedback when students enter incorrect answers.

- Instructors can create, import, and manage online homework assignments, quizzes, and tests—or start with sample assignments—all of which are automatically graded, allowing instructors to spend less time grading, and more time teaching.

Data & Analytics MyLab Statistics provides resources to help instructors assess and improve student results. A comprehensive gradebook with enhanced reporting functionality makes it easier for instructors to manage courses efficiently.

- **Reporting Dashboard** Instructors can view, analyze, and report learning outcomes, gaining the information they need to keep our students on track. Available via the Gradebook and fully mobile-ready, the Reporting Dashboard presents student performance data at the class, section, and program levels in an accessible, visual manner. Its finegrain reports allow instructors and administrators to compare performance across different courses, across individual sections and within each course.
- **Item Analysis** Instructors can track class-wide understanding of particular exercises in order to refine your class lectures or adjust the course/department syllabus. Just-in-time teaching has never been easier.

Accessibility Pearson works continuously to ensure our products are as accessible as possible to all students. We are working toward achieving WCAG 2.0 Level AA and Section 508 standards, as expressed in the Pearson Guidelines for Accessible Educational Web Media, **www.pearson.com/mylab/statistics/accessibility.**

StatCrunch

Integrated directly into MyLab Statistics, StatCrunch® is powerful web-based statistical software that allows users to perform complex analyses, share data sets, and generate compelling reports of their data.

- **Collect** Users can upload their own data to StatCrunch or search a large library of publicly shared data sets, spanning almost any topic of interest. A Featured Data page houses the best data sets, making it easy for instructors to use current data in their course. Data sets from the text and from online homework exercises can also be accessed and analyzed in StatCrunch. An online survey tool allows users to quickly collect data via web-based surveys.
- **Crunch** A full range of numerical and graphical methods allow users to analyze and gain insights from any data set. Interactive graphics help users understand statistical concepts, and are available for export to enrich reports with visual representations of data.
- **Communicate** Reporting options help users create a wide variety of visually-appealing representations of their data.

StatCrunch is integrated into MyLab Statistics, but it is also available by itself to qualified adopters. StatCrunch is also now available on your smartphone or tablet when you visit **www.statcrunch.com** from the device's browser. For more information, visit our website at **www.statcrunch.com,** or contact your Pearson representative.

MathXL Online Course (access code required)

Part of the world's leading collection of online homework, tutorial, and assessment products, MathXL® delivers assessment and tutorial resources that provide engaging and personalized experiences for each student. Each course is developed to accompany Pearson's best-selling content, authored by thought leaders across the math curriculum, and can be easily customized to fit any course format.

With MathXL, instructors can:

- Create, edit, and assign online homework and tests using algorithmically generated exercises correlated at the objective level to the textbook.
- Create and assign their own online exercises and import TestGen tests for added flexibility.
- Maintain records of all student work tracked in MathXL's online gradebook.

With MathXL, students can:

- Take chapter tests in MathXL and receive personalized study plans and/or personalized homework assignments based on their test results.
- Use the study plan and/or the homework to link directly to tutorial exercises for the objectives they need to study.
- Access supplemental animations and video clips directly from selected exercises.

MathXL is available to qualified adopters. For more information, visit our website at **www.pearson.com/mathxl,** or contact your Pearson representative.

Minitab and Minitab Express

Minitab and Minitab Express™ make learning statistics easy and provide students with a skill-set that is in demand in today's data driven workforce. Bundling Minitab software with educational materials ensures students have access to the software they need in the classroom, around campus, and at home. And having 12-month access to Minitab and Minitab Express ensures students can use the software for the duration of their course. ISBN 13: 978-0-13-445640-9 ISBN 10: 0-13-445640-8 (access card only; not sold as stand alone)

JMP Student Edition

JMP® Student Edition is an easy-to-use, streamlined version of JMP desktop statistical discovery software from SAS Institute, Inc. and is available for bundling with the text. ISBN-13: 978-0-13-467979-2 ISBN-10: 0-13-467979-2

XLSTAT

XLSTAT™ is an Excel add-in that enhances the analytical capabilities of Excel. XLSTAT is used by leading businesses and universities around the world. It is available to bundle with this text. For more information, go to **www.pearsonhighered.com/xlstat.** ISBN-13: 978-0-321-75932-0; ISBN-10: 0-321-75932-X

Resources for Success

P Pearson
MyLab

MyLab Statistics Online Course for *Elementary Statistics: Picturing the World*, 7e (access code required)

MyLab™ Statistics is available to accompany Pearson's market-leading text offerings. To give students a consistent tone, voice, and teaching method, each text's flavor and approach is tightly integrated throughout the accompanying MyLab Statistics course, making learning the material as seamless as possible. MyLab Statistics for *Elementary Statistics* includes the following new features, in addition to the resources listed on the previous page.

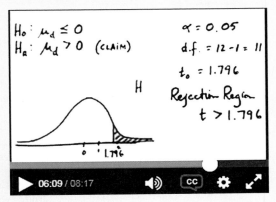

UPDATED! Video Program
Chapter Review Exercises come to life with new review videos that help students understand key chapter concepts. Section Lecture Videos work through examples and elaborate on key objectives.

NEW! StatCrunch Question Library
This library of questions provides opportunities for students to analyze and interpret data sets in StatCrunch. Instructors can assign individual questions from the library by topic or they can assign questions from the same data set as a longer assignment that spans multiple learning objectives.

NEW! Integrated Review Course
Designed for just-in-time prerequisite review or for co-requisite courses, the Integrated Review version of the MyLab Statistics course provides pre-made, assignable skill-review quizzes and personalized homework assignments that are integrated throughout the regular statistics course content.

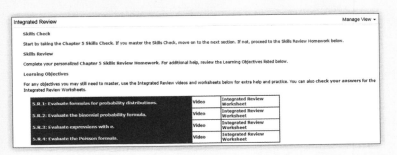

pearson.com/mylab/statistics

Resources for Success

Instructor Resources

Annotated Instructor's Edition

Includes suggested activities, additional ways to present material, common pitfalls, and other helpful teaching tips. All answers to the section and review exercises are provided in the margins next to the exercise. (ISBN-13: 978-0-13-468358-4; ISBN-10: 0-13-468358-7)

Instructor's Solutions Manual (downloadable)

Includes complete solutions to all of the exercises (including exercises in *Try it Yourself, Case Study, Technology, Uses and Abuses,* and *Real Statistics— Real Decisions* sections). It can be downloaded from within MyLab Statistics or from Pearson's online catalog, *www.pearson.com/us/higher-education.*

PowerPoint Lecture Slides (downloadable)

Classroom presentation slides feature key concepts, examples, and definitions from this text, along with notes with suggestions for presenting the material in class. They can be downloaded from within MyLab Statistics or from Pearson's online catalog, *www.pearson.com/us/higher-education.*

TestGen

TestGen® (www.pearson.com/testgen) enables instructors to build, edit, print, and administer tests using a computerized bank of questions developed to cover all the objectives of the text. TestGen is algorithmically based, allowing instructors to create multiple but equivalent versions of the same question or test with the click of a button. Instructors can also modify test bank questions or add new questions. The software and test bank are available for download from Pearson's online catalog, *www.pearson.com/us/higher-education.* The questions are also assignable in MyLab Statistics.

Learning Catalytics

Now included in all MyLab Statistics courses, this student response tool uses students' smartphones, tablets, or laptops to engage them in more interactive tasks and thinking during lecture. Learning Catalytics™ fosters student engagement and peer-to-peer learning with real-time analytics. Access pre-built exercises created specifically for statistics.

Student Resources

Video Resources

A comprehensive set of videos tied to the textbook contain short video clips with solutions to *Try It Yourself* exercises, Chapter Quiz Prep Videos, and Section Lecture Videos. Also, StatTalk Videos, hosted by fun-loving statistician Andrew Vickers, demonstrate important statistical concepts through interesting stories and real-life events. StatTalk Videos include assessment questions and an instructor's guide.

Student's Solutions Manual (softcover and downloadable)

This manual includes complete worked-out solutions to all of the *Try It Yourself* exercises, the odd-numbered exercises, and all of the Chapter Quiz exercises. This manual is available in print and can be downloaded from MyLab Statistics. (ISBN-13: 978-0-13-468361-4; ISBN-10: 0-13-468361-7)

Technology Manuals for Elementary Statistics (downloadable)

Technology-specific manuals for Graphing Calculator, Excel®, and Minitab® include tutorial instruction and worked-out examples from the book. Each manual can be downloaded from within MyLab Statistics.

pearson.com/mylab/statistics

ACKNOWLEDGMENTS

I owe a debt of gratitude to the many reviewers who helped me shape and refine *Elementary Statistics: Picturing the World,* Seventh Edition.

Reviewers of the Current Edition

Karen Benway, University of Vermont
B.K. Brinkley, Tidewater Community College
Christine Curtis, Hillsborough Community College–Dale Mabry
Carrie Elledge, San Juan College
Jason Malozzi, Lower Columbia College
Cynthia McGinnis, Northwest Florida State College
Larry Musolino, Pennsylvania State University
Cyndi Roemer, Union County College
Jean Rowley, American Public University and DeVry University
Heidi Webb, Horry Georgetown Technical College

Reviewers of the Previous Editions

Rosalie Abraham, Florida Community College at Jacksonville
Ahmed Adala, Metropolitan Community College
Olcay Akman, College of Charleston
Polly Amstutz, University of Nebraska, Kearney
John J. Avioli, Christopher Newport University
David P. Benzel, Montgomery College
John Bernard, University of Texas—Pan American
G. Andy Chang, Youngstown State University
Keith J. Craswell, Western Washington University
Carol Curtis, Fresno City College
Dawn Dabney, Northeast State Community College
Cara DeLong, Fayetteville Technical Community College
Ginger Dewey, York Technical College
David DiMarco, Neumann College
Gary Egan, Monroe Community College
Charles Ehler, Anne Arundel Community College
Harold W. Ellingsen, Jr., SUNY—Potsdam
Michael Eurgubian, Santa Rosa Jr. College
Jill Fanter, Walters State Community College
Patricia Foard, South Plains College
Douglas Frank, Indiana University of Pennsylvania
Frieda Ganter, California State University
David Gilbert, Santa Barbara City College
Donna Gorton, Butler Community College
Larry Green, Lake Tahoe Community College
Sonja Hensler, St. Petersburg Jr. College
Sandeep Holay, Southeast Community College, Lincoln Campus
Lloyd Jaisingh, Morehead State
Nancy Johnson, Manatee Community College

Martin Jones, College of Charleston
David Kay, Moorpark College
Mohammad Kazemi, University of North Carolina—Charlotte
Jane Keller, Metropolitan Community College
Susan Kellicut, Seminole Community College
Hyune-Ju Kim, Syracuse University
Rita Kolb, Cantonsville Community College
Rowan Lindley, Westchester Community College
Jeffrey Linek, St. Petersburg Jr. College
Benny Lo, DeVry University, Fremont
Diane Long, College of DuPage
Austin Lovenstein, Pulaski Technical College
Rhonda Magel, North Dakota State University
Mike McGann, Ventura Community College
Vicki McMillian, Ocean County College
Lynn Meslinsky, Erie Community College
Lyn A. Noble, Florida Community College at Jacksonville—South Campus
Julie Norton, California State University—Hayward
Lynn Onken, San Juan College
Lindsay Packer, College of Charleston
Nishant Patel, Northwest Florida State
Jack Plaggemeyer, Little Big Horn College
Eric Preibisius, Cuyamaca Community College
Melonie Rasmussen, Pierce College
Neal Rogness, Grand Valley State University
Elisabeth Schuster, Benedictine University
Jean Sells, Sacred Heart University
John Seppala, Valdosta State University
Carole Shapero, Oakton Community College
Abdullah Shuaibi, Harry S. Truman College
Aileen Solomon, Trident Technical College
Sandra L. Spain, Thomas Nelson Community College
Michelle Strager-McCarney, Penn State—Erie, The Behrend College
Jennifer Strehler, Oakton Community College
Deborah Swiderski, Macomb Community College
William J. Thistleton, SUNY—Institute of Technology, Utica
Millicent Thomas, Northwest University
Agnes Tuska, California State University—Fresno
Clark Vangilder, DeVry University
Ting-Xiu Wang, Oakton Community
Dex Whittinghall, Rowan University
Cathleen Zucco-Teveloff, Rider University

Many thanks to Betsy Farber for her significant contributions to previous editions of the text. Sadly, Betsy passed away in 2013.

I would also like to thank the staff of Larson Texts, Inc., who assisted with the production of the book. On a personal level, I am grateful to my spouse, Deanna Gilbert Larson, for her love, patience, and support. Also, a special thanks goes to R. Scott O'Neil.

I have worked hard to make this text a clean, clear, and enjoyable one from which to teach and learn statistics. Despite my best efforts to ensure accuracy and ease of use, many users will undoubtedly have suggestions for improvement. I welcome your suggestions.

Ron Larson, odx@psu.edu

INDEX OF APPLICATIONS

CHAPTER 1

Introduction to Statistics

For the first 10 months of 2016, construction completions of privately-owned housing units in the U.S. was greatest in the south.

You are already familiar with many of the practices of statistics, such as taking surveys, collecting data, and describing populations. What you may not know is that collecting accurate statistical data is often difficult and costly. Consider, for instance, the monumental task of counting and describing the entire population of the United States. If you were in charge of such a census, how would you do it? How would you ensure that your results are accurate? These and many more concerns are the responsibility of the United States Census Bureau, which conducts the census every decade.

In Chapter 1, you will be introduced to the basic concepts and goals of statistics. For instance, statistics were used to construct the figures below, which show the numbers, by region in the U.S., of construction completions of privately-owned housing units for October of 2016 and for the first 10 months of 2016, as numbers in thousands and as percents of the total.

For the 2010 Census, the Census Bureau sent short forms to every household. Short forms ask all members of every household such things as their gender, age, race, and ethnicity. Previously, a long form, which covered additional topics, was sent to about 17% of the population. But for the first time since 1940, the long form was replaced by the American Community Survey, which surveys more than 3.5 million households a year throughout the decade. These households form a sample. In this course, you will learn how the data collected from a sample are used to infer characteristics about the entire population.

**Housing Units Completed
in the U.S. (October 2016)**

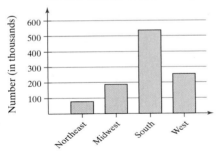

**Housing Units Completed
in the U.S. (October 2016)**

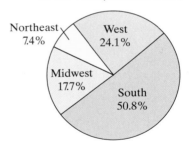

**Housing Units
Completed in the U.S.
(January–October 2016)**

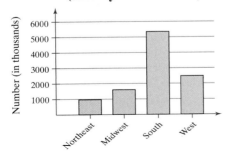

**Housing Units
Completed in the U.S.
(January–October 2016)**

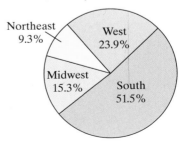

1.1   An Overview of Statistics

A Definition of Statistics ■ Data Sets ■ Branches of Statistics

A Definition of Statistics

Almost every day you are exposed to statistics. For instance, consider the next two statements.

- According to a survey, more than 7 in 10 Americans say a nursing career is a prestigious occupation. *(Source: The Harris Poll)*

- "Social media consumes kids today as well, as more score their first social media accounts at an average age of 11.4 years old." *(Source: Influence Central's 2016 Digital Trends Study)*

By learning the concepts in this text, you will gain the tools to become an informed consumer, understand statistical studies, conduct statistical research, and sharpen your critical thinking skills.

Many statistics are presented graphically. For instance, consider the figure shown below.

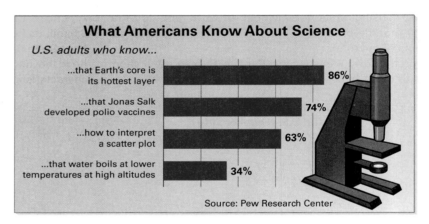

The information in the figure is based on the collection of **data.** In this instance, the data are based on the results of a science quiz given to 3278 U.S. adults.

DEFINITION

Data consist of information coming from observations, counts, measurements, or responses.

The use of statistics dates back to census taking in ancient Babylonia, Egypt, and later in the Roman Empire, when data were collected about matters concerning the state, such as births and deaths. In fact, the word *statistics* is derived from the Latin word *status*, meaning "state." The modern practice of statistics involves more than counting births and deaths, as you can see in the next definition.

DEFINITION

Statistics is the science of collecting, organizing, analyzing, and interpreting data in order to make decisions.

Data Sets

There are two types of data sets you will use when studying statistics. These data sets are called **populations** and **samples.**

DEFINITION

A **population** is the collection of *all* outcomes, responses, measurements, or counts that are of interest. A **sample** is a subset, or part, of a population.

A sample is used to gain information about a population. For instance, to estimate the unemployment rate for the *population* of the United States, the U.S. Bureau of Labor uses a *sample* of about 60,000 households.

A sample should be representative of a population so that sample data can be used to draw conclusions about that population. Sample data must be collected using an appropriate method, such as *random sampling*. When sample data are collected using an *inappropriate* method, the data cannot be used to draw conclusions about the population. (You will learn more about random sampling and data collection in Section 1.3.)

EXAMPLE 1

Identifying Data Sets

In a survey, 834 employees in the United States were asked whether they thought their jobs were highly stressful. Of the 834 respondents, 517 said yes. Identify the population and the sample. Describe the sample data set. *(Source: CareerCast Job Stress Report)*

SOLUTION

The population consists of the responses of all employees in the United States. The sample consists of the responses of the 834 employees in the survey. In the Venn diagram below, notice that the sample is a subset of the responses of all employees in the United States. Also, the sample data set consists of 517 people who said yes and 317 who said no.

Responses of All Employees (population)

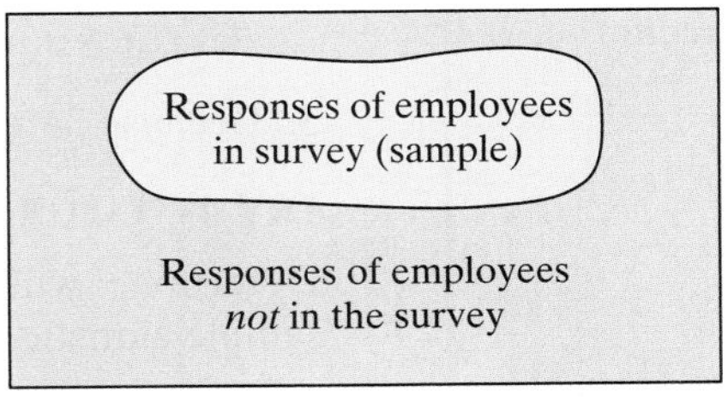

Responses of employees in survey (sample)

Responses of employees *not* in the survey

TRY IT YOURSELF 1

In a survey of 1501 ninth to twelfth graders in the United States, 1215 said "leaders today are more concerned with their own agenda than with achieving the overall goals of the organization they serve." Identify the population and the sample. Describe the sample data set. *(Source: National 4-H Council)*

Answer: Page A31

Whether a data set is a population or a sample usually depends on the context of the real-life situation. For instance, in Example 1, the population is the set of responses of all employees in the United States. Depending on the purpose of the survey, the population could have been the set of responses of all employees who live in California or who work in the healthcare industry.

Study Tip

A *census* consists of data from an entire population. But, unless a population is small, it is usually impractical to obtain all the population data. In most studies, information must be obtained from a random sample.

Note to Instructor

Point out to students that Venn diagrams show only which sets are subsets of other sets. They do not show relative sizes of the sets. For instance, the Venn diagram at the right should not be interpreted to mean that the sample size is roughly one-fourth of the population size.

Study Tip

To remember the terms parameter and statistic, try using the mnemonic device of matching the first letters in *population parameter* and the first letters in *sample statistic.*

Picturing the World

How accurate is the count of the U.S. population taken each decade by the Census Bureau? According to estimates, the net undercount of the U.S. population by the 1940 census was 5.4%. The accuracy of the census has improved greatly since then. The net undercount in the 2010 census was −0.01%. (This means that the 2010 census overcounted the U.S. population by 0.01%, which is about 36,000 people.) (Source: U.S. Census Bureau)

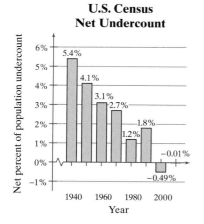

U.S. Census Net Undercount

What are some difficulties in collecting population data?

It is difficult to collect comprehensive data from large populations.

Two important terms that are used throughout this course are **parameter** and **statistic.**

DEFINITION

A **parameter** is a numerical description of a *population* characteristic.

A **statistic** is a numerical description of a *sample* characteristic.

It is important to note that a sample statistic can differ from sample to sample, whereas a population parameter is constant for a population. For instance, consider the survey in Example 1. The results showed that 517 of 834 employees surveyed think their jobs are highly stressful. Another sample may have a different number of employees that say their jobs are highly stressful. For the population, however, the number of employees who think that their jobs are highly stressful does not change.

EXAMPLE 2

Distinguishing Between a Parameter and a Statistic

Determine whether each number describes a population parameter or a sample statistic. Explain your reasoning.

1. A survey of several hundred collegiate student-athletes in the United States found that, during the season of their sport, the average time spent on athletics by student-athletes is 50 hours per week. *(Source: Penn Schoen Berland)*

2. The freshman class at a university has an average SAT math score of 514.

3. In a random check of several hundred retail stores, the Food and Drug Administration found that 34% of the stores were not storing fish at the proper temperature.

SOLUTION

1. Because the average of 50 hours per week is based on a subset of the population, it is a sample statistic.

2. Because the average SAT math score of 514 is based on the entire freshman class, it is a population parameter.

3. Because 34% is based on a subset of the population, it is a sample statistic.

TRY IT YOURSELF 2

Determine whether each number describes a population parameter or a sample statistic. Explain your reasoning.

a. Last year, a small company spent a total of $5,150,694 on employees' salaries.

b. In the United States, a survey of a few thousand adults with hearing loss found that 43% have difficulty remembering conversations. *(Source: The Harris Poll)*

Answer: Page A31

In this course, you will see how the use of statistics can help you make informed decisions that affect your life. Consider the census that the U.S. government takes every decade. When taking the census, the Census Bureau attempts to contact everyone living in the United States. Although it is impossible to count everyone, it is important that the census be as accurate as it can be because public officials make many decisions based on the census information. Data collected in the census will determine how to assign congressional seats and how to distribute public funds.

Branches of Statistics

The study of statistics has two major branches: **descriptive statistics** and **inferential statistics.**

DEFINITION

Descriptive statistics is the branch of statistics that involves the organization, summarization, and display of data.

Inferential statistics is the branch of statistics that involves using a sample to draw conclusions about a population. A basic tool in the study of inferential statistics is probability. (You will learn more about probability in Chapter 3.)

EXAMPLE 3

Descriptive and Inferential Statistics

For each study, identify the population and the sample. Then determine which part of the study represents the descriptive branch of statistics. What conclusions might be drawn from the study using inferential statistics?

1. A study of 2560 U.S. adults found that of adults not using the Internet, 23% are from households earning less than $30,000 annually, as shown in the figure at the left. *(Source: Pew Research Center)*

2. A study of 300 Wall Street analysts found that the percentage who incorrectly forecasted high-tech earnings in a recent year was 44%. *(Adapted from Bloomberg News)*

SOLUTION

1. The population consists of the responses of all U.S. adults, and the sample consists of the responses of the 2560 U.S. adults in the study. The part of this study that represents the descriptive branch of statistics involves the statement "23% [of U.S. adults not using the Internet] are from households earning less than $30,000 annually." Also, the figure represents the descriptive branch of statistics. A possible inference drawn from the study is that lower-income households cannot afford access to the Internet.

2. The population consists of the high-tech earnings forecasts of all Wall Street analysts, and the sample consists of the forecasts of the 300 Wall Street analysts in the study. The part of this study that represents the descriptive branch of statistics involves the statement "the percentage [of Wall Street analysts] who incorrectly forecasted high-tech earnings in a recent year was 44%." A possible inference drawn from the study is that the stock market is difficult to forecast, even for professionals.

TRY IT YOURSELF 3

A study of 1000 U.S. adults found that when they have a question about their medication, three out of four adults will consult with their physician or pharmacist and only 8% visit a medication-specific website. *(Source: Finn Futures™ Health poll)*

a. Identify the population and the sample.
b. Determine which part of the study represents the descriptive branch of statistics.
c. What conclusions might be drawn from the study using inferential statistics?

Answer: Page A31

Not Online
U.S. adults who do not use the Internet by household income

23%
12%
6%
3%

Less than $30,000 | $30,000 to $49,999 | $50,000 to $74,999 | $75,000 or more
Household income

Note to Instructor

In Example 3, ask students to suggest other inferences that might be drawn. This would be a good time to tell students that one can never be 100% sure about inferences.

Study Tip

Throughout this course you will see applications of both branches of statistics. A major theme in this course will be how to use sample statistics to make inferences about unknown population parameters.

1.1 EXERCISES

For Extra Help: MyLab Statistics

Answers (left column)

1. A sample is a subset of a population.

2. Is is usually impractical (too expensive and/or time-consuming) to obtain all the population data.

3. A parameter is a numerical description of a population characteristic. A statistic is a numerical description of a sample characteristic.

4. The two main branches of statistics are descriptive statistics and inferential statistics.

5. False. A statistic is a numerical description of a sample characteristic.

6. True   7. True

8. False. Inferential statistics involves using a sample to draw conclusions about a population.

9. False. A population is the collection of *all* outcomes, responses, measurements, or counts that are of interest.

10. False. A sample statistic can differ from sample to sample.

11. Population, because it is a collection of the salaries of each member of a Major League Baseball team.

12. Population, because it is a collection of the energy collected from all the solar panels on a photo voltaic power plant.

13. Sample, because the collection of the 300 people is a subset of the population of 13,000 people in the auditorium.

14. Population, because it is a collection of the revenue of all the stores at the shopping mall.

15. See Odd Answers, page A40.

16. See Selected Answers, page A89.

17. See Odd Answers, page A40.

18. See Selected Answers, page A89.

19. See Odd Answers, page A40.

20. See Selected Answers, page A89.

21. See Odd Answers, page A40.

22. See Selected Answers, page A89.

Building Basic Skills and Vocabulary

1. How is a sample related to a population?

2. Why is a sample used more often than a population?

3. What is the difference between a parameter and a statistic?

4. What are the two main branches of statistics?

True or False? *In Exercises 5–10, determine whether the statement is true or false. If it is false, rewrite it as a true statement.*

5. A statistic is a numerical description of a population characteristic.

6. A sample is a subset of a population.

7. It is impossible to obtain all the census data about the U.S. population.

8. Inferential statistics involves using a population to draw a conclusion about a corresponding sample.

9. A population is the collection of some outcomes, responses, measurements, or counts that are of interest.

10. A sample statistic will not change from sample to sample.

Classifying a Data Set *In Exercises 11–20, determine whether the data set is a population or a sample. Explain your reasoning.*

11. The salary of each member of a Major League Baseball team

12. The amount of energy collected from every solar panel on a photovoltaic power plant

13. A survey of 300 people from an auditorium with 13,000 people

14. The annual revenue of each store in a shopping mall

15. The triglyceride levels of 10 patients in a clinic with 50 patients

16. The number of wireless devices in each U.S. household

17. The final score of each gamer in a tournament

18. The age of every fourth person entering a grocery store

19. The political party of every U.S. senator

20. The air contamination levels at 20 locations near a factory

Graphical Analysis *In Exercises 21–24, use the Venn diagram to identify the population and the sample.*

21. **Parties of Registered Voters**

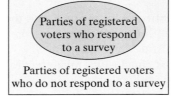

22. **Student Donations at a Food Drive**

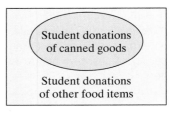

23. Population: Ages of adults in the United States who own automobiles

    Sample: Ages of adults in the United States who own Honda automobiles

24. Population: Incomes of home owners in Massachusetts

    Sample: Incomes of home owners in Massachusetts with mortgages

25. Population: Collections of the responses of all U.S. adults

    Sample: Collection of the responses of the 1020 U.S. adults surveyed

    Sample data set: 42% of adults who said they trust their political leaders and 58% who said they did not

26. Population: Collection of fetal tobacco exposure of all infants

    Sample: Collection of the fetal tobacco exposure of 203 infants

    Sample data set: Infants with fetal tobacco exposure and their focused attention levels

27. Population: Collection of the influenza immunization status of all adults in the United States

    Sample: Collection of the influenza immunization status of the 3301 U.S. adults surveyed

    Sample data set: 39% of U.S. adults who received an influenza vaccine and 61% who did not

28. Population: Collection of the responses of travelers with pets in the world

    Sample: Collection of the responses of the 1100 travelers surveyed with pets

    Sample data set: 53% of respondents with pets who said they travel with their pets and 47% who said they did not

29. See Odd Answers, page A40.

30. See Selected Answers, page A89.

31. See Odd Answers, page A40.

32. See Selected Answers, page A89.

33. See Odd Answers, page A40.

34. See Selected Answers, page A89.

35. See Odd Answers, page A40.

36. See Selected Answers, page A89.

37. See Odd Answers, page A40.

23. **Ages of Adults in the United States Who Own Automobiles**

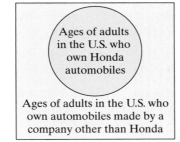

Ages of adults in the U.S. who own automobiles made by a company other than Honda

24. **Incomes of Home Owners in Massachusetts**

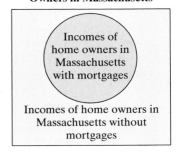

Incomes of home owners in Massachusetts without mortgages

Using and Interpreting Concepts

Identifying Data Sets *In Exercises 25–34, identify the population and the sample. Describe the sample data set.*

25. A survey of 1020 U.S. adults found that 42% trust their political leaders. *(Source: Gallup)*

26. A study of 203 infants was conducted to find a link between fetal tobacco exposure and focused attention in infancy. *(Source: Infant Behavior and Development)*

27. A survey of 3301 U.S. adults found that 39% received an influenza vaccine for a recent flu season. *(Source: U.S. Centers for Disease Control and Prevention)*

28. A survey of 1100 travelers worldwide found that 53% of respondents with pets travel with their pets.

29. A survey of 159 U.S. law firms found that the average hourly billing rate for partners was $604. *(Source: The National Law Journal)*

30. A survey of 496 students at a high school found that 95% planned on going to college.

31. A survey of 1029 U.S. adults found that 23% of those suffering with chronic pain had been diagnosed with a sleep disorder. *(Source: National Sleep Foundation)*

32. A survey of 1254 preowned automobile shoppers found that 5% bought extended warranties.

33. To gather information about starting salaries at companies listed in the Standard & Poor's 500, a researcher contacts 54 of the 500 companies.

34. A survey of 1060 parents of 13- to 17-year-olds found that 636 of the 1060 parents have checked their teen's social media profile. *(Source: Pew Research Center)*

Distinguishing Between a Parameter and a Statistic *In Exercises 35–42, determine whether the number describes a population parameter or a sample statistic. Explain your reasoning.*

35. The average salary for 45 of a consulting firm's 300 engineers is $72,000.

36. A survey of 1058 college board members found that 56.3% think that college completion is a major priority or the most important priority for their board. *(Source: Association of Governing Boards of Universities and Colleges)*

37. Sixty-two of the 97 passengers aboard the Hindenburg airship survived its explosion.

38. Population parameter. The value 62% is a numerical description of the total number of governors.

39. Sample statistic. The value 7% is a numerical description of a sample of computer users.

40. Population parameter. The value 87% is a numerical description of the total number of voters.

41. Sample statistic. The value 80% is a numerical description of a sample of U.S. adults.

42. Population parameter. The score 20.6 is a numerical description of the ACT scores for all graduates.

43. The statement "23% of those suffering with chronic pain had been diagnosed with a sleep disorder" is an example of descriptive statistics. Using inferential statistics, you may conclude that an association exists between chronic pain and sleep disorders.

44. The statement "5% bought extended warranties" is an example of descriptive statistics. Using inferential statistics, you may conclude that most pre-owned automobile shoppers do not buy extended warranties.

45–46. Answers will vary.

47. The inference may incorrectly imply that exercise increases a person's cognitive ability. The study shows a slower decline in cognitive ability, not an increase.

48. The inference may incorrectly imply that obesity trends will continue in future years. Even though the obesity rates have been increasing, that does not mean the rates will continue to increase for eternity.

49. (a) The sample is the results on the standardized test by the participants in the study.

(b) The population is the collection of all the results of the standardized test.

(c) The statement "the closer that participants were to an optimal sleep duration target, the better they performed on a standardized test" is an example of descriptive statistics.

(d) See Odd Answers, page A40.

38. In January 2016, 62% of the governors of the 50 states in the United States were Republicans. *(Source: National Governors Association)*

39. In a survey of 400 computer users, 7% said their computers had malfunctions that needed to be repaired by service technicians.

40. Voter registration records show that 87% of all voters in a county are registered as Democrats.

41. A survey of 2008 U.S. adults found that 80% think that the militant group known as ISIS is a major threat to the well-being of the United States. *(Source: Pew Research Center)*

42. In a recent year, the average math score on the ACT for all graduates was 20.6. *(Source: ACT, Inc.)*

43. Descriptive and Inferential Statistics   Which part of the survey described in Exercise 31 represents the descriptive branch of statistics? What conclusions might be drawn from the survey using inferential statistics?

44. Descriptive and Inferential Statistics   Which part of the survey described in Exercise 32 represents the descriptive branch of statistics? What conclusions might be drawn from the survey using inferential statistics?

Extending Concepts

45. Identifying Data Sets in Articles   Find an article that describes a survey.

(a) Identify the sample used in the survey.

(b) What is the population?

(c) Make an inference about the population based on the results of the survey.

46. Writing   Write an essay about the importance of statistics for one of the following.

• A study on the effectiveness of a new drug

• An analysis of a manufacturing process

• Drawing conclusions about voter opinions using surveys

47. Exercise and Cognitive Ability   A study of 876 senior citizens shows that participants who exercise regularly exhibit less of a decline in cognitive ability than those who barely exercise at all. From this study, a researcher infers that your cognitive ability increases the more your exercise. What is wrong with this type of reasoning? *(Source: Neurology)*

48. Increase in Obesity Rates   A study shows that the obesity rate among adolescents has steadily increased since 1988. From this study, a researcher infers that this trend will continue in future years. What is wrong with this type of reasoning? *(Source: Journal of the American Medical Association)*

49. Sleep and Student Achievement   A study shows the closer that participants were to an optimal sleep duration target, the better they performed on a standardized test. *(Source: Eastern Economics Journal)*

(a) Identify the sample used in the study.

(b) What is the population?

(c) Which part of the study represents the descriptive branch of statistics?

(d) Make an inference about the population based on the results of the study.

1.2 Data Classification

What You Should Learn

▶ How to distinguish between qualitative data and quantitative data

▶ How to classify data with respect to the four levels of measurement: nominal, ordinal, interval, and ratio

Types of Data ■ Levels of Measurement

Types of Data

When conducting a study, it is important to know the kind of data involved. The type of data you are working with will determine which statistical procedures can be used. In this section, you will learn how to classify data by type and by level of measurement. Data sets can consist of two types of data: **qualitative data** and **quantitative data.**

DEFINITION

Qualitative data consist of attributes, labels, or nonnumerical entries.

Quantitative data consist of numbers that are measurements or counts.

EXAMPLE 1

Classifying Data by Type

The table shows sports-related head injuries treated in U.S. emergency rooms during a recent five-year span for several sports. Which data are qualitative data and which are quantitative data? Explain your reasoning. *(Source: BMC Emergency Medicine)*

Sports-Related Head Injuries Treated in U.S. Emergency Rooms

Sport	Head injuries treated
Basketball	131,930
Baseball	83,522
Football	220,258
Gymnastics	33,265
Hockey	41,450
Soccer	98,710
Softball	41,216
Swimming	44,815
Volleyball	13,848

SOLUTION

The information shown in the table can be separated into two data sets. One data set contains the names of sports, and the other contains the numbers of head injuries treated. The names are nonnumerical entries, so these are qualitative data. The numbers of head injuries treated are numerical entries, so these are quantitative data.

TRY IT YOURSELF 1

The populations of several U.S. cities are shown in the table. Which data are qualitative data and which are quantitative data? Explain your reasoning. *(Source: U.S. Census Bureau)*

Answer: Page A31

City	Population
Baltimore, MD	621,849
Chicago, IL	2,720,546
Glendale, AZ	240,126
Miami, FL	441,003
Portland, OR	632,309
San Francisco, CA	864,816

Levels of Measurement

Another characteristic of data is its level of measurement. The level of measurement determines which statistical calculations are meaningful. The four levels of measurement, in order from lowest to highest, are **nominal, ordinal, interval,** and **ratio.**

DEFINITION

Data at the **nominal level of measurement** are qualitative only. Data at this level are categorized using names, labels, or qualities. No mathematical computations can be made at this level.

Data at the **ordinal level of measurement** are qualitative or quantitative. Data at this level can be arranged in order, or ranked, but differences between data entries are not meaningful.

When numbers are at the nominal level of measurement, they simply represent a label. Examples of numbers used as labels include Social Security numbers and numbers on sports jerseys. For instance, it would not make sense to add the numbers on the players' jerseys for the Chicago Bears.

EXAMPLE 2

Classifying Data by Level

For each data set, determine whether the data are at the nominal level or at the ordinal level. Explain your reasoning. *(Source: U.S. Bureau of Labor Statistics)*

1.

Top five U.S. occupations with the most job growth (projected 2024)
1. Personal care aides
2. Registered nurses
3. Home health aides
4. Combined food preparation and serving workers, including fast food
5. Retail salespersons

2.

Movie genres
Action
Adventure
Comedy
Drama
Horror

SOLUTION

1. This data set lists the ranks of the five fastest-growing occupations in the U.S. over the next few years. The data set consists of the ranks 1, 2, 3, 4, and 5. Because the ranks can be listed in order, these data are at the ordinal level. Note that the difference between a rank of 1 and 5 has no mathematical meaning.

2. This data set consists of the names of movie genres. No mathematical computations can be made with the names, and the names cannot be ranked, so these data are at the nominal level.

TRY IT YOURSELF 2

For each data set, determine whether the data are at the nominal level or at the ordinal level. Explain your reasoning.

1. The final standings for the Pacific Division of the National Basketball Association
2. A collection of phone numbers   *Answer: Page A31*

Picturing the World

For more than 25 years, the Harris Poll has conducted an annual study to determine the strongest brands, based on consumer response, in several industries. A recent study determined the top five health nonprofit brands, as shown in the table. (Source: Harris Poll)

Top five health nonprofit brands
1. St Jude Children's Research Hospital
2. Shriners Hospital for Children
3. Make-A-Wish
4. The Jimmy Fund
5. American Cancer Society

In this list, what is the level of measurement?

The data are at the ordinal level of measurement.

The two highest levels of measurement consist of quantitative data only.

DEFINITION

Data at the **interval level of measurement** can be ordered, and meaningful differences between data entries can be calculated. At the interval level, a zero entry simply represents a position on a scale; the entry is not an inherent zero.

Data at the **ratio level of measurement** are similar to data at the interval level, with the added property that a zero entry is an inherent zero. A ratio of two data entries can be formed so that one data entry can be meaningfully expressed as a multiple of another.

An *inherent zero* is a zero that implies "none." For instance, the amount of money you have in a savings account could be zero dollars. In this case, the zero represents no money; it is an inherent zero. On the other hand, a temperature of 0°C does not represent a condition in which no heat is present. The 0°C temperature is simply a position on the Celsius scale; it is not an inherent zero.

To distinguish between data at the interval level and at the ratio level, determine whether the expression "twice as much" has any meaning in the context of the data. For instance, $2 is twice as much as $1, so these data are at the ratio level. On the other hand, 2°C is not twice as warm as 1°C, so these data are at the interval level.

EXAMPLE 3

Classifying Data by Level

Two data sets are shown at the left. Which data set consists of data at the interval level? Which data set consists of data at the ratio level? Explain your reasoning. *(Source: Major League Baseball)*

SOLUTION

Both of these data sets contain quantitative data. Consider the dates of the Yankees' World Series victories. It makes sense to find differences between specific dates. For instance, the time between the Yankees' first and last World Series victories is

$$2009 - 1923 = 86 \text{ years.}$$

But it does not make sense to say that one year is a multiple of another. So, these data are at the interval level. However, using the home run totals, you can find differences *and* write ratios. For instance, Boston hit 23 more home runs than Cleveland hit because $208 - 185 = 23$ home runs. Also, Baltimore hit about 1.5 times as many home runs as Chicago hit because

$$\frac{253}{168} \approx 1.5.$$

So, these data are at the ratio level.

TRY IT YOURSELF 3

For each data set, determine whether the data are at the interval level or at the ratio level. Explain your reasoning.

1. The body temperatures (in degrees Fahrenheit) of an athlete during an exercise session
2. The heart rates (in beats per minute) of an athlete during an exercise session

Answer: Page A31

New York Yankees' World Series victories (years)

1923, 1927, 1928, 1932, 1936, 1937, 1938, 1939, 1941, 1943, 1947, 1949, 1950, 1951, 1952, 1953, 1956, 1958, 1961, 1962, 1977, 1978, 1996, 1998, 1999, 2000, 2009

2016 American League home run totals (by team)

Team	Total
Baltimore	253
Boston	208
Chicago	168
Cleveland	185
Detroit	211
Houston	198
Kansas City	147
Los Angeles	156
Minnesota	200
New York	183
Oakland	169
Seattle	223
Tampa Bay	216
Texas	215
Toronto	221

The tables below summarize which operations are meaningful at each of the four levels of measurement. When identifying a data set's level of measurement, use the highest level that applies.

Level of measurement	Put data in categories	Arrange data in order	Subtract data entries	Determine whether one data entry is a multiple of another
Nominal	Yes	No	No	No
Ordinal	Yes	Yes	No	No
Interval	Yes	Yes	Yes	No
Ratio	Yes	Yes	Yes	Yes

Summary of Four Levels of Measurement

	Example of a data set	Meaningful calculations
Nominal level (Qualitative data)	*Types of Shows Televised by a Network* Comedy Documentaries Drama Cooking Reality Shows Soap Operas Sports Talk Shows	*Put in a category.* For instance, a show televised by the network could be put into one of the eight categories shown.
Ordinal level (Qualitative or quantitative data)	*Motion Picture Association of America Ratings Description* G General Audiences PG Parental Guidance Suggested PG-13 Parents Strongly Cautioned R Restricted NC-17 No One 17 and Under Admitted	Put in a category and *put in order.* For instance, a PG rating has a stronger restriction than a G rating.
Interval level (Quantitative data)	*Average Monthly Temperatures (in degrees Fahrenheit) for Denver, CO* Jan 30.7 Jul 74.2 Feb 32.5 Aug 72.5 Mar 40.4 Sep 63.4 Apr 47.4 Oct 50.9 May 57.1 Nov 38.3 Jun 67.4 Dec 30.0 *(Source: National Climatic Data Center)*	Put in a category, put in order, and *find differences between data entries.* For instance, $72.5 - 63.4 = 9.1°F$. So, August is 9.1°F warmer than September.
Ratio level (Quantitative data)	*Average Monthly Precipitation (in inches) for Orlando, FL* Jan 2.35 Jul 7.27 Feb 2.38 Aug 7.13 Mar 3.77 Sep 6.06 Apr 2.68 Oct 3.31 May 3.45 Nov 2.17 Jun 7.58 Dec 2.58 *(Source: National Climatic Data Center)*	Put in a category, put in order, find differences between data entries, and *find ratios of data entries.* For instance, $$\frac{7.58}{3.77} \approx 2.$$ So, there is about twice as much precipitation in June as in March.

1.2 EXERCISES

Answers (left column):

1. Nominal and ordinal

2. Ordinal, interval, and ratio

3. False. Data at the ordinal level can be qualitative or quantitative.

4. False. For data at the interval level, you can calculate meaningful differences between data entries. You cannot calculate meaningful differences at the nominal or ordinal levels.

5. False. More types of calculations can be performed with data at the interval level than with data at the nominal level.

6. False. Data at the ratio level can be placed in a meaningful order.

7. Quantitative, because dog weights are numerical measurements.

8. Quantitative, because carrying capacities are numerical measurements.

9. Qualitative, because hair colors are attributes.

10. Qualitative, because student ID numbers are labels.

11. Quantitative, because infant heights are numerical measurements.

12. Qualitative, because mammal species are labels.

13. Qualitative, because the poll responses are attributes.

14. Quantitative, because wait times are numerical measurements.

15. Interval. Data can be ordered and meaningful differences can be calculated, but it does not make sense to say that one year is a multiple of another.

16. Ordinal. Data can be arranged in order, but the differences between data entries are not meaningful.

Building Basic Skills and Vocabulary

1. Name each level of measurement for which data can be qualitative.

2. Name each level of measurement for which data can be quantitative.

True or False? *In Exercises 3–6, determine whether the statement is true or false. If it is false, rewrite it as a true statement.*

3. Data at the ordinal level are quantitative only.

4. For data at the interval level, you cannot calculate meaningful differences between data entries.

5. More types of calculations can be performed with data at the nominal level than with data at the interval level.

6. Data at the ratio level cannot be put in order.

Using and Interpreting Concepts

Classifying Data by Type *In Exercises 7–14, determine whether the data are qualitative or quantitative. Explain your reasoning.*

7. Weights of dogs at an animal rescue facility

8. Carrying capacities of flatbed trucks

9. Hair colors of classmates

10. Student ID numbers

11. Heights of infants in a maternity ward

12. Species of mammals in a rain forest

13. Responses on an opinion poll

14. Wait times at a the Department of Motor Vehicles

Classifying Data By Level *In Exercises 15–20, determine the level of measurement of the data set. Explain your reasoning.*

15. **Comedy Series** The years that a television show on ABC won the Emmy for best comedy series are listed. *(Source: Academy of Television Arts and Sciences)*

| 1955 | 1979 | 1980 | 1981 | 1982 | 1988 |
| 2010 | 2011 | 2012 | 2013 | 2014 | |

16. **Business Schools** The top ten business schools in the United States for a recent year according to Forbes are listed. *(Source: Forbes Media LLC)*

1. Stanford
2. Harvard
3. Northwestern (Kellogg)
4. Columbia
5. Dartmouth (Tuck)
6. Chicago (Booth)
7. Pennsylvania (Wharton)
8. UC Berkeley (Haas)
9. MIT (Sloan)
10. Cornell (Johnson)

17. Nominal. No mathematical computations can be made, and data are categorized using numbers.

18. Ratio. A ratio of two data values can be formed, so one data value can be expressed as a multiple of another.

19. Ordinal. Data can be arranged in order, but the differences between data entries are not meaningful.

20. Interval. Data can be ordered and meaningful differences can be calculated, but it does not make sense to say that one time is a multiple of another.

21. Horizontal: Nominal
Vertical: Ratio

22. Horizontal: Ordinal
Vertical: Ratio

23. Horizontal: Nominal
Vertical: Ratio

24. Horizontal: Interval
Vertical: Ratio

17. **Flight Departures**   The flight numbers of 21 departing flights from Chicago O'Hare International Airport on an afternoon in October of 2016 are listed. *(Source: Chicago O'Hare International Airport)*

1785	5159	4509	1575	6827	3486	7676
1989	522	6868	1893	3133	3337	3266
3458	334	6320	8385	3112	2110	7664

18. **Songs**   The lengths (in seconds) of songs on an album are listed.

228	233	268	265	252
335	103	338	252	371
586	290	532	282	

19. **Best Sellers List**   The top ten fiction books on The New York Times Best Sellers List on October 9, 2016, are listed. *(Source: The New York Times)*

1. The Girl on the Train
2. Home
3. The Kept Woman
4. Magic Binds
5. Commonwealth
6. The Light Between Oceans
7. Immortal Nights
8. A Man Called Ove
9. Thrice the Brinded Cat Hath Mew'd
10. The Woman in Cabin 10

20. **Cell Phone**   The times of the day when a person checks his or her cell phone are listed.

8:28 A.M.	9:30 A.M.	9:43 A.M.	10:18 A.M.
11:25 A.M.	11:46 A.M.	12:27 P.M.	2:18 P.M.
2:26 P.M.	2:49 P.M.	3:05 P.M.	4:18 P.M.
5:28 P.M.	5:57 P.M.	8:17 P.M.	

Graphical Analysis   *In Exercises 21–24, determine the level of measurement of the data listed on the horizontal and vertical axes in the figure.*

21.   **What is the Format of the Books You Read?**

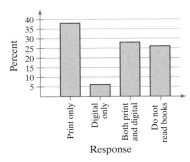

(Source: Pew Research Center)

22.   **How Many Vacations Are You Planning to Take This Summer?**

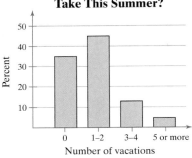

(Source: The Harris Poll)

23.   **Gender Profile of the 114th Congress**

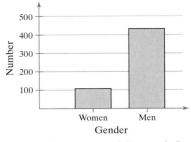

(Source: Congressional Research Service)

24.   **Motor Vehicle Fatalities by Year**

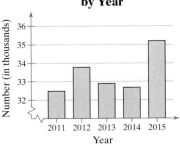

(Source: National Highway Traffic Safety Administration)

25. (a) Interval (b) Nominal
(c) Ratio (d) Ordinal

26. (a) Interval (b) Nominal
(c) Interval (d) Ratio

27. Qualitative. Ordinal. Data can
be arranged in order, but the
differences between data entries
make no sense.

28. Qualitative. Nominal. No
mathematical computations
can be made, and data are
categorized by political party.

29. Qualitative. Nominal. No
mathematical computations
can be made and data are
categorized by region.

30. Quantitative. Interval. Data can
be ordered and meaningful
differences can be calculated,
but it does not make sense to
say that one score is a multiple
of another.

31. Qualitative. Ordinal. Data can
be arranged in order, but the
differences between data entries
are not meaningful.

32. Quantitative. Ratio. A ratio
of two data entries can be
formed, so one data entry can
be expressed as a multiple of
another.

33. An inherent zero is a zero
that implies "none." Answers
will vary.

34. Answers will vary.

25. The items below appear on a physician's intake form. Determine the level of measurement of the data for each category.
(a) Temperature (b) Allergies
(c) Weight (d) Pain level (scale of 0 to 10)

26. The items below appear on an employment application. Determine the level of measurement of the data for each category.
(a) Highest grade level completed (b) Gender
(c) Year of college graduation (d) Number of years at last job

Classifying Data by Type and Level *In Exercises 27–32, determine whether the data are qualitative or quantitative, and determine the level of measurement of the data set.*

27. Football The top ten teams in the final college football poll released in January 2017 are listed. *(Source: Associated Press)*

1. Clemson **6.** Ohio State
2. Alabama **7.** Penn State
3. USC **8.** Florida State
4. Washington **9.** Wisconsin
5. Oklahoma **10.** Michigan

28. Politics The three political parties in the 114th Congress are listed.
Republican Democrat Independent

29. Top Salespeople The regions representing the top salespeople in a corporation for the past six years are listed.

Southeast Northwest
Northeast Southeast
Southwest Southwest

30. Diving The scores for the gold medal winning diver in the men's 10-meter platform event from the 2016 Summer Olympics are listed. *(Source: International Olympic Committee)*

91.80 91.00 88.20
97.20 99.90 91.80

31. Concert Tours The top ten highest grossing worldwide concert tours for 2016 are listed. *(Source: Pollstar)*

1. Bruce Springsteen & the E Street Band **6.** Justin Bieber
2. Beyoncé **7.** Paul McCartney
3. Coldplay **8.** Garth Brooks
4. Guns N' Roses **9.** The Rolling Stones
5. Adele **10.** Celine Dion

32. Numbers of Performances The numbers of performances for the 10 longest-running Broadway shows at the end of the 2016 season are listed. *(Source: The Broadway League)*

11,782 8107 7705 7485 6680
6137 5959 5758 5461 5238

Extending Concepts

33. Writing What is an inherent zero? Describe three examples of data sets that have inherent zeros and three that do not.

34. Describe two examples of data sets for each of the four levels of measurement. Justify your answer.

Reputations of Companies in the U.S.

For more than 50 years, The Harris Poll has conducted surveys using a representative sample of people in the United States. The surveys have been used to represent the opinions of people in the United States on many subjects, such as health, politics, the U.S. economy, and sports.

Since 1999, The Harris Poll has conducted an annual survey to measure the reputations of the most visible companies in the United States, as perceived by U.S. adults. The Harris Poll uses a sample of about 23,000 U.S. adults for the survey. The survey respondents rate companies according to 20 attributes that are classified into six categories: (1) social responsibility, (2) vision and leadership, (3) financial performance, (4) products and services, (5) emotional appeal, and (6) workplace environment. This information is used to determine the reputation of a company as Excellent, Very Good, Good, Fair, Poor, Very Poor, or Critical. The reputations (along with some additional information) of 10 companies are shown in the table.

All U.S. Adults

U.S. adults in The Harris Poll sample (about 23,000 U.S. adults)

U.S. adults *not* in The Harris Poll sample (about 242.8 million U.S. adults)

Reputations of 10 Companies in the U.S.

Company Name	Year Company Formed	Reputation	Industry	Number of Employees
Amazon.com	1994	Excellent	Retail	230,800
Apple, Inc.	1977	Excellent	Computers and peripherals	116,000
Netflix, Inc.	1999	Very Good	Internet television	4,700
The Kraft Heinz Co.	2015	Very Good	Food products	41,000
Facebook, Inc.	2004	Good	Internet	17,048
Ford Motor Co.	1903	Good	Automotive	201,000
Chipotle Mexican Grill, Inc.	1993	Fair	Restaurant	64,570
Comcast Corp.	1963	Poor	Cable television	136,000
Exxon Mobil Corp.	1999	Poor	Petroleum (integrated)	71,100
Wells Fargo & Co.	1998	Critical	Banking	265,000

(Source: The Harris Poll; Amazon.com; Apple, Inc.; Netflix, Inc.; The Kraft Heinz Co.; Facebook, Inc.; Ford Motor Co.; Chipotle Mexican Grill, Inc.; Comcast Corp.; Exxon Mobil Corp.; Wells Fargo & Co.)

EXERCISES

1. **Sampling Percent** What percentage of the total number of U.S. adults did The Harris Poll sample for its survey? (Assume the total number of U.S. adults is 242.8 million.)

2. **Nominal Level of Measurement** Identify any column in the table with data at the nominal level.

3. **Ordinal Level of Measurement** Identify any column in the table with data at the ordinal level. Describe two ways that the data can be ordered.

4. **Interval Level of Measurement** Identify any column in the table with data at the interval level. How can these data be ordered?

5. **Ratio Level of Measurement** Identify any column in the table with data at the ratio level.

6. **Inferences** What decisions can be made on the basis of The Harris Poll survey that measures the reputations of the most visible companies in the United States?

1.3 Data Collection and Experimental Design

Design of a Statistical Study ■ Data Collection ■ Experimental Design ■ Sampling Techniques

Design of a Statistical Study

The goal of every statistical study is to collect data and then use the data to make a decision. Any decision you make using the results of a statistical study is only as good as the process used to obtain the data. When the process is flawed, the resulting decision is questionable.

Although you may never have to develop a statistical study, it is likely that you will have to interpret the results of one. Before interpreting the results of a study, however, you should determine whether the results are reliable. In other words, you should be familiar with how to design a statistical study.

GUIDELINES

Designing a Statistical Study
1. Identify the variable(s) of interest (the focus) and the population of the study.
2. Develop a detailed plan for collecting data. If you use a sample, make sure the sample is representative of the population.
3. Collect the data.
4. Describe the data, using descriptive statistics techniques.
5. Interpret the data and make decisions about the population using inferential statistics.
6. Identify any possible errors.

A statistical study can usually be categorized as an observational study or an experiment. In an **observational study,** a researcher does not influence the responses. In an **experiment,** a researcher deliberately applies a treatment before observing the responses. Here is a brief summary of these types of studies.

- In an **observational study,** a researcher observes and measures characteristics of interest of part of a population but does not change existing conditions. For instance, an observational study was conducted in which researchers measured the amount of time people spent doing various activities, such as paid work, childcare, and socializing. *(Source: U.S. Bureau of Labor Statistics)*

- In performing an **experiment,** a **treatment** is applied to part of a population, called a **treatment group,** and responses are observed. Another part of the population may be used as a **control group,** in which no treatment is applied. (The subjects in both groups are called **experimental units.**) In many cases, subjects in the control group are given a **placebo,** which is a harmless, fake treatment that is made to look like the real treatment. The responses of both groups can then be compared and studied. In most cases, it is a good idea to use the same number of subjects for each group. For instance, an experiment was performed in which overweight subjects in a treatment group were given the artificial sweetener sucralose to drink while a control group drank water. After performing a glucose test, researchers concluded that "sucralose affects the glycemic and insulin responses" in overweight people who do not normally consume artificial sweeteners. *(Source: Diabetes Care)*

EXAMPLE 1

Distinguishing Between an Observational Study and an Experiment

Determine whether each study is an observational study or an experiment.

1. Researchers study the effect of vitamin D_3 supplementation among patients with antibody deficiency or frequent respiratory tract infections. To perform the study, 70 patients receive 4000 IU of vitamin D_3 daily for a year. Another group of 70 patients receive a placebo daily for one year. *(Source: British Medical Journal)*

2. Researchers conduct a study to determine how confident Americans are in the U.S. economy. To perform the study, researchers call 3040 U.S. adults and ask them to rate current U.S. economic conditions and whether the U.S. economy is getting better or worse. *(Source: Gallup)*

SOLUTION

1. Because the study applies a treatment (vitamin D_3) to the subjects, the study is an experiment.

2. Because the study does not attempt to influence the responses of the subjects (there is no treatment), the study is an observational study.

TRY IT YOURSELF 1

The Pennsylvania Game Commission conducted a study to count the number of elk in Pennsylvania. The commission captured and released 636 elk, which included 350 adult cows, 125 calves, 110 branched bulls, and 51 spikes. Is this study an observational study or an experiment? *(Source: Pennsylvania Game Commission)*

Answer: Page A31

Data Collection

There are several ways to collect data. Often, the focus of the study dictates the best way to collect data. Here is a brief summary of two methods of data collection.

Note to Instructor

Students will be given the opportunity to use simulations when they study probability. Refer to page 187 to see an interesting example of a simulation.

- A **simulation** is the use of a mathematical or physical model to reproduce the conditions of a situation or process. Collecting data often involves the use of computers. Simulations allow you to study situations that are impractical or even dangerous to create in real life, and often they save time and money. For instance, automobile manufacturers use simulations with dummies to study the effects of crashes on humans. Throughout this course, you will have the opportunity to use applets that simulate statistical processes on a computer.

- A **survey** is an investigation of one or more characteristics of a population. Most often, surveys are carried out on *people* by asking them questions. The most common types of surveys are done by interview, Internet, phone, or mail. In designing a survey, it is important to word the questions so that they do not lead to biased results, which are not representative of a population. For instance, a survey is conducted on a sample of female physicians to determine whether the primary reason for their career choice is financial stability. In designing the survey, it would be acceptable to make a list of reasons and ask each individual in the sample to select her first choice.

Experimental Design

To produce meaningful unbiased results, experiments should be carefully designed and executed. It is important to know what steps should be taken to make the results of an experiment valid. Three key elements of a well-designed experiment are *control*, *randomization*, and *replication*.

Because experimental results can be ruined by a variety of factors, being able to control these influential factors is important. One such factor is a **confounding variable.**

> ### DEFINITION
>
> A **confounding variable** occurs when an experimenter cannot tell the difference between the effects of different factors on the variable.

Study Tip

The *Hawthorne effect* occurs in an experiment when subjects change their behavior simply because they know they are participating in an experiment.

For instance, to attract more customers, a coffee shop owner experiments by remodeling the shop using bright colors. At the same time, a shopping mall nearby has its grand opening. If business at the coffee shop increases, it cannot be determined whether it is because of the new colors or the new shopping mall. The effects of the colors and the shopping mall have been confounded.

Another factor that can affect experimental results is the *placebo effect*. The **placebo effect** occurs when a subject reacts favorably to a placebo when in fact the subject has been given a fake treatment. To help control or minimize the placebo effect, a technique called **blinding** can be used.

> ### DEFINITION
>
> **Blinding** is a technique where the subjects do not know whether they are receiving a treatment or a placebo. In a **double-blind experiment,** neither the experimenter nor the subjects know whether the subjects are receiving a treatment or a placebo. The experimenter is informed after all the data have been collected. This type of experimental design is preferred by researchers.

One challenge for experimenters is assigning subjects to groups so the groups have similar characteristics (such as age, height, weight, and so on). When treatment and control groups are similar, experimenters can conclude that any differences between groups is due to the treatment. To form groups with similar characteristics, experimenters use **randomization.**

> ### DEFINITION
>
> **Randomization** is a process of randomly assigning subjects to different treatment groups.

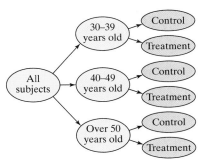

Randomized Block Design

In a **completely randomized design,** subjects are assigned to different treatment groups through random selection. In some experiments, it may be necessary for the experimenter to use **blocks,** which are groups of subjects with similar characteristics. A commonly used experimental design is a **randomized block design.** To use a randomized block design, the experimenter divides the subjects with similar characteristics into blocks, and then, within each block, randomly assign subjects to treatment groups. For instance, an experimenter who is testing the effects of a new weight loss drink may first divide the subjects into age categories such as 30–39 years old, 40–49 years old, and over 50 years old, and then, within each age group, randomly assign subjects to either the treatment group or the control group (see figure at the left).

Another type of experimental design is a **matched-pairs design,** where subjects are paired up according to a similarity. One subject in each pair is randomly selected to receive one treatment while the other subject receives a different treatment. For instance, two subjects may be paired up because of their age, geographical location, or a particular physical characteristic.

Sample size, which is the number of subjects in a study, is another important part of experimental design. To improve the validity of experimental results, **replication** is required.

Study Tip

The *validity* of an experiment refers to the accuracy and reliability of the experimental results. The results of a valid experiment are more likely to be accepted in the scientific community.

DEFINITION

Replication is the repetition of an experiment under the same or similar conditions.

For instance, suppose an experiment is designed to test a vaccine against a strain of influenza. In the experiment, 10,000 people are given the vaccine and another 10,000 people are given a placebo. Because of the sample size, the effectiveness of the vaccine would most likely be observed. But, if the subjects in the experiment are not selected so that the two groups are similar (according to age and gender), the results are of less value.

EXAMPLE 2

Analyzing an Experimental Design

A company wants to test the effectiveness of a new gum developed to help people quit smoking. Identify a potential problem with each experimental design and suggest a way to improve it.

1. The company identifies ten adults who are heavy smokers. Five of the subjects are given the new gum and the other five subjects are given a placebo. After two months, the subjects are evaluated and it is found that the five subjects using the new gum have quit smoking.

2. The company identifies one thousand adults who are heavy smokers. The subjects are divided into blocks according to gender. Females are given the new gum and males are given the placebo. After two months, a significant number of the female subjects have quit smoking.

SOLUTION

1. The sample size being used is not large enough to validate the results of the experiment. The experiment must be replicated to improve the validity.

2. The groups are not similar. The new gum may have a greater effect on women than on men, or vice versa. The subjects can be divided into blocks according to gender, but then, within each block, they should be randomly assigned to be in the treatment group or in the control group.

TRY IT YOURSELF 2

The company in Example 2 identifies 240 adults who are heavy smokers. The subjects are randomly assigned to be in a gum treatment group or in a control group. Each subject is also given a DVD featuring the dangers of smoking. After four months, most of the subjects in the treatment group have quit smoking. Identify a potential problem with the experimental design and suggest a way to improve it.

Answer: Page A31

Sampling Techniques

A **census** is a count or measure of an *entire* population. Taking a census provides complete information, but it is often costly and difficult to perform. A **sampling** is a count or measure of *part* of a population and is more commonly used in statistical studies. To collect unbiased data, a researcher must ensure that the sample is representative of the population. Appropriate sampling techniques must be used to ensure that inferences about the population are valid. Remember that when a study is done with faulty data, the results are questionable. Even with the best methods of sampling, a **sampling error** may occur. A sampling error is the difference between the results of a sample and those of the population. When you learn about inferential statistics, you will learn techniques of controlling sampling errors.

A **random sample** is one in which every member of the population has an equal chance of being selected. A **simple random sample** is a sample in which every possible sample of the same size has the same chance of being selected. One way to collect a simple random sample is to assign a different number to each member of the population and then use a random number table like Table 1 in Appendix B. Responses, counts, or measures for members of the population whose numbers correspond to those generated using the table would be in the sample. Calculators and computer software programs are also used to generate random numbers (see page 36).

Table 1—Random Numbers

92630	78240	19267	95457	53497	23894	37708	79862
79445	78735	71549	44843	26104	67318	00701	34986
59654	71966	27386	50004	05358	94031	29281	18544
31524	49587	76612	39789	13537	48086	59483	60680
06348	76938	90379	51392	55887	71015	09209	79157

Portion of Table 1 found in Appendix B

Consider a study of the number of people who live in West Ridge County. To use a simple random sample to count the number of people who live in West Ridge County households, you could assign a different number to each household, use a technology tool or table of random numbers to generate a sample of numbers, and then count the number of people living in each selected household.

EXAMPLE 3

Using a Simple Random Sample

There are 731 students currently enrolled in a statistics course at your school. You wish to form a sample of eight students to answer some survey questions. Select the students who will belong to the simple random sample.

SOLUTION

Assign numbers 1 to 731 to the students in the course. In the table of random numbers, choose a starting place at random and read the digits in groups of three (because 731 is a three-digit number). For instance, if you started in the third row of the table at the beginning of the second column, you would group the numbers as follows:

719|66   2|738|6   50|004| 053|58   9|403|1   29|281| 185|44

Ignoring numbers greater than 731, the first eight numbers are 719, 662, 650, 4, 53, 589, 403, and 129. The students assigned these numbers will make up the sample. To find the sample using a TI-84 Plus, follow the instructions shown at the left.

Study Tip

A *biased sample* is one that is not representative of the population from which it is drawn. For instance, a sample consisting of only 18- to 22-year-old U.S. college students would not be representative of the entire 18- to 22-year-old population in the United States.

1.3 To explore this topic further, see **Activity 1.3** on page 27.

Tech Tip

You can use technology such as Minitab, Excel, StatCrunch, or the TI-84 Plus to generate random numbers. (Detailed instructions for using Minitab, Excel, and the TI-84 Plus are shown in the technology manuals that accompany this text.) For instance, here are instructions for using the random integer generator on a TI-84 Plus for Example 3.

MATH

Choose the PRB menu.

5: randInt(

1 , 7 3 1 , 8)

ENTER

```
randInt(1,731,8)
{537 33 249 728…
```

Continuing to press ENTER will generate more random samples of 8 integers.

TRY IT YOURSELF 3

A company employs 79 people. Choose a simple random sample of five to survey.

Answer: Page A31

When you choose members of a sample, you should decide whether it is acceptable to have the same population member selected more than once. If it is acceptable, then the sampling process is said to be *with replacement*. If it is not acceptable, then the sampling process is said to be *without replacement*.

There are several other commonly used sampling techniques. Each has advantages and disadvantages.

- *Stratified Sample*   When it is important for the sample to have members from each segment of the population, you should use a stratified sample. Depending on the focus of the study, members of the population are divided into two or more subsets, called *strata*, that share a similar characteristic such as age, gender, ethnicity, or even political preference. A sample is then randomly selected from each of the strata. Using a stratified sample ensures that each segment of the population is represented. For instance, to collect a stratified sample of the number of people who live in West Ridge County households, you could divide the households into socioeconomic levels and then randomly select households from each level. In using a stratified sample, care must be taken to ensure that all strata are sampled in proportion to their actual percentages of occurrence in the population. For instance, if 40% of the people in West Ridge County belong to the low-income group, then the proportion of the sample should have 40% from this group.

Group 1:   Group 2:   Group 3:
Low income   Middle income   High income

Stratified Sampling

Study Tip

Be sure you understand that stratified sampling randomly selects a *sample of members* from *all* strata. Cluster sampling uses *all members* from a randomly selected sample of *clusters* (but not all, so some clusters will not be part of the sample). For instance, in the figure for "Stratified Sampling" at the right, a *sample of households* in West Ridge County is randomly selected from *all* three income groups. In the figure for "Cluster Sampling," *all households* in a randomly selected *cluster* (Zone 1) are used. (Notice that the other zones are not part of the sample.)

- *Cluster Sample*   When the population falls into naturally occurring subgroups, each having similar characteristics, a cluster sample may be the most appropriate. To select a cluster sample, divide the population into groups, called *clusters*, and select all of the members in one or more (but not all) of the clusters. Examples of clusters could be different sections of the same course or different branches of a bank. For instance, to collect a cluster sample of the number of people who live in West Ridge County households, divide the households into groups according to zip codes, then select all the households in one or more, but not all, zip codes and count the number of people living in each household. In using a cluster sample, care must be taken to ensure that all clusters have similar characteristics. For instance, if one of the zip code clusters has a greater proportion of high-income people, the data might not be representative of the population.

Zip Code Zones in West Ridge County

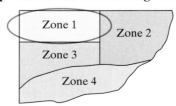

Cluster Sampling

Picturing the World

The research firm Gallup conducts many polls (or surveys) regarding the president, Congress, and political and nonpolitical issues. A commonly cited Gallup poll is the public approval rating of the president. For instance, the approval ratings for President Barack Obama for selected months in 2016 are shown in the figure. (Each rating is from the poll conducted at the end of the indicated month.)

President's Approval Ratings, 2016

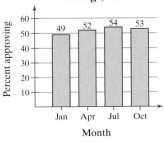

Discuss some ways that Gallup could select a biased sample to conduct a poll. How could Gallup select a sample that is unbiased?

Gallup may have surveyed only people who like President Obama. An unbiased sample could be obtained from random sampling by phone.

- **Systematic Sample**   A systematic sample is a sample in which each member of the population is assigned a number. The members of the population are ordered in some way, a starting number is randomly selected, and then sample members are selected at regular intervals from the starting number. (For instance, every 3rd, 5th, or 100th member is selected.) For instance, to collect a systematic sample of the number of people who live in West Ridge County households, you could assign a different number to each household, randomly choose a starting number, select every 100th household, and count the number of people living in each. An advantage of systematic sampling is that it is easy to use. In the case of any regularly occurring pattern in the data, however, this type of sampling should be avoided.

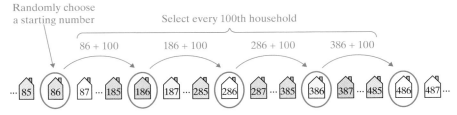

Systematic Sampling

A type of sample that often leads to biased studies (so it is not recommended) is a **convenience sample.** A convenience sample consists only of members of the population that are easy to get.

EXAMPLE 4

Identifying Sampling Techniques

You are doing a study to determine the opinions of students at your school regarding stem cell research. Identify the sampling technique you are using when you select the samples listed. Discuss potential sources of bias (if any).

1. You divide the student population with respect to majors and randomly select and question some students in each major.

2. You assign each student a number and generate random numbers. You then question each student whose number is randomly selected.

3. You select students who are in your biology class.

SOLUTION

1. Because students are divided into strata (majors) and a sample is selected from each major, this is a stratified sample.

2. Each sample of the same size has an equal chance of being selected and each student has an equal chance of being selected, so this is a simple random sample.

3. Because the sample is taken from students that are readily available, this is a convenience sample. The sample may be biased because biology students may be more familiar with stem cell research than other students and may have stronger opinions.

TRY IT YOURSELF 4

You want to determine the opinions of students regarding stem cell research. Identify the sampling technique you are using when you select these samples.

1. You select a class at random and question each student in the class.

2. You assign each student a number and, after choosing a starting number, question every 25th student.   *Answer: Page A31*

1.3 EXERCISES

Building Basic Skills and Vocabulary

1. In an experiment, a treatment is applied to part of a population and responses are observed. In an observational study, a researcher measures characteristics of interest of a part of a population but does not change existing conditions.

2. A census includes the entire population. A sampling includes only a portion of the population.

3. In a random sample, every member of the population has an equal chance of being selected. In a simple random sample, every possible sample of the same size has an equal chance of being selected.

4. Replication is the repetition of an experiment under the same or similar conditions. Replication is important because it enhances the validity of the results.

5. False. A placebo is a fake treatment.

6. False. A double-blind experiment is used to decrease the placebo effect.

7. False. Using stratified sampling guarantees that members of each group within a population will be sampled.

8. False. A convenience sample is not representative of a population.

9. False. A systematic sample is selected by ordering a population in some way and then selecting members of the population at regular intervals.

10. True

11. Observational study. The study does not apply a treatment to the adults.

12. Experiment. The study applies a treatment (intensive program to lower systolic blood pressure) to the subjects.

13. See Odd Answers, page A41.

14. See Selected Answers, page A89.

15–18. Answers will vary.

1. What is the difference between an observational study and an experiment?

2. What is the difference between a census and a sampling?

3. What is the difference between a random sample and a simple random sample?

4. What is replication in an experiment? Why is replication important?

True or False?   *In Exercises 5–10, determine whether the statement is true or false. If it is false, rewrite it as a true statement.*

5. A placebo is an actual treatment.

6. A double-blind experiment is used to increase the placebo effect.

7. Using a systematic sample guarantees that members of each group within a population will be sampled.

8. A convenience sample is always representative of a population.

9. The method for selecting a stratified sample is to order a population in some way and then select members of the population at regular intervals.

10. To select a cluster sample, divide a population into groups and then select all of the members in at least one (but not all) of the groups.

Distinguishing Between an Observational Study and an Experiment
In Exercises 11–14, determine whether the study is an observational study or an experiment. Explain.

11. In a survey of 1033 U.S. adults, 51% said U.S. presidents should release all medical information that might affect their ability to serve. *(Source: Gallup)*

12. Researchers demonstrated that adults using an intensive program to lower systolic blood pressure to less than 120 millimeters of mercury reduce the risk of death from all causes by 27%. *(Source: American Heart Association)*

13. To study the effects of social media on teenagers' brains, researchers showed a few dozen teenagers photographs that had varying numbers of "likes" while scanning the reactions in their brains. *(Source: NPR)*

14. In a study designed to research the effect of music on driving habits, 1000 motorists ages 17–25 years old were asked whether the music they listened to influenced their driving. *(Source: More Th>n)*

15. Random Number Table   Use the sixth row of Table 1 in Appendix B to generate 12 random numbers between 1 and 99.

16. Random Number Table   Use the tenth row of Table 1 in Appendix B to generate 10 random numbers between 1 and 920.

Random Numbers   *In Exercises 17 and 18, use technology to generate the random numbers.*

17. Fifteen numbers between 1 and 150

18. Nineteen numbers between 1 and 1000

Using and Interpreting Concepts

19. (a) The experimental units are the 500 females ages 25 to 45 years old who suffer from migraine headaches. The treatment is the new drug used to treat migraine headaches.

(b) A problem with the design is that the sample is not representative of the entire population because only females ages 25 to 45 were used. To increase validity, use a stratified sample.

(c) For the experiment to be double-blind, neither the subjects nor the company would know whether the subjects are receiving the drug or the placebo.

20. See Selected Answers, page A89.

21. *Sample answer:* Treatment group: Jake, Maria, Lucy, Adam, Bridget, Vanessa, Rick, Dan, and Mary. Control group: Mike, Ron, Carlos, Steve, Susan, Kate, Pete, Judy, and Connie.

A random number table was used.

22. Answers will vary.

23. Simple random sampling is used because each employee has an equal chance of being contacted, and all samples of 300 people have an equal chance of being selected. A possible source of bias is that the random sample may contain a much greater percentage of employees from one department than from others.

24. Convenience sampling is used because the students are chosen due to their convenience of location. Bias may enter into the sample because the students sampled may not be representative of the population of students.

25. Cluster sampling is used because the disaster area is divided into grids, and 30 grids are then entirely selected. A possible source of bias is that certain grids may have been much more severely damaged than others.

19. **Allergy Drug** A pharmaceutical company wants to test the effectiveness of a new drug used to treat migraine headaches. The company identifies 500 females ages 25 to 45 years old who suffer from migraine headaches. The subjects are randomly assigned into two groups. One group is given the drug and the other is given a placebo that looks exactly like the drug. After three months, the subjects' symptoms are studied and compared.

(a) Identify the experimental units and treatments used in this experiment.

(b) Identify a potential problem with the experimental design being used and suggest a way to improve it.

(c) How could this experiment be designed to be double-blind?

20. **Dietary Supplement** Researchers in Germany tested the effect of a dietary supplement designed to control metabolism in patients with type 2 diabetes. Thirty-one patients with type 2 diabetes completed the study. The patients were assigned at random either the supplement or a placebo for 12 weeks. After a subsequent "wash-out" period of 12 weeks, the patients were assigned the other product. At the conclusion of the study, the patients' glycated hemoglobin, fasting blood glucose, and fructosamine levels were checked, as well as their lipid parameters. *(Source: Food and Nutrition Research)*

(a) Identify the experimental units and treatments used in this experiment.

(b) Identify a potential problem with the experimental design being used and suggest a way to improve it.

(c) The experiment is described as a placebo-controlled, double-blind study. Explain what this means.

(d) How could blocking be used in designing this experiment?

21. **Sleep Deprivation** A researcher wants to study the effects of sleep deprivation on motor skills. Eighteen people volunteer for the experiment: Jake, Maria, Mike, Lucy, Ron, Adam, Bridget, Carlos, Steve, Susan, Vanessa, Rick, Dan, Kate, Pete, Judy, Mary, and Connie. Use a random number generator to choose nine subjects for the treatment group. The other nine subjects will go into the control group. List the subjects in each group. Tell which method you used to generate the random numbers.

22. **Using a Simple Random Sample** Volunteers for an experiment are numbered from 1 to 90. The volunteers are to be randomly assigned to two different treatment groups. Use a random number generator different from the one you used in Exercise 21 to choose 45 subjects for the treatment group. The other 45 subjects will go into the control group. List the subjects, according to number, in each group. Tell which method you used to generate the random numbers.

Identifying Sampling Techniques *In Exercises 23–28, identify the sampling technique used, and discuss potential sources of bias (if any). Explain.*

23. Selecting employees at random from an employee directory, researchers contact 300 people and ask what obstacles (such as computer problems) keep them from accomplishing tasks at work.

24. Questioning university students as they leave a fraternity party, a researcher asks 463 students about their study habits.

25. After a hurricane, a disaster area is divided into 200 equal grids. Thirty of the grids are selected, and every occupied household in the grid is interviewed to help focus relief efforts on what residents require the most.

26. Systematic sampling is used because every tenth person entering the shopping mall is sampled. It is possible for bias to enter into the sample if, for some reason, there is a regular pattern to the people entering the shopping mall.

27. Stratified sampling is used because a sample is taken from each one-acre subplot.

28. Simple random sampling is used because each telephone number has an equal chance of being dialed, and all samples of 1012 phone numbers have an equal chance of being selected. The sample may be biased because telephone sampling only samples individuals who have telephones, who are available, and who are willing to respond.

29. Census, because it is relatively easy to obtain the ages of the 115 residents.

30. Sampling, because the population of subscribers is too large for their most popular movie types to be easily recorded. Random sampling would be advised because it would be easy to select subscribers randomly and then record their most popular movie types.

31. The question is biased because it already suggests that eating whole-grain foods improves your health. The question could be rewritten as "How does eating whole-grain foods affect your health?"

32. The question is biased because it already suggests that text messaging while driving increases the risk of a crash. The question could be rewritten as "Does text messaging while driving affect the risk of a crash?"

33. The survey question is unbiased.

34. See Selected Answers, page A89.

35. Answers will vary.

36. Answers will vary.

37. See Odd Answers, page A41.

38. See Selected Answers, page A89.

---

26. Every tenth person entering a mall is asked to name his or her favorite store.

27. Soybeans are planted on a 48-acre field. The field is divided into one-acre subplots. A sample is taken from each subplot to estimate the harvest.

28. From calls made with randomly generated telephone numbers, 1012 respondents are asked if they rent or own their residences.

Choosing Between a Census and a Sampling   *In Exercises 29 and 30, determine whether you would take a census or use a sampling. If you would use a sampling, determine which sampling technique you would use. Explain.*

29. The average age of the 115 residents of a retirement community

30. The most popular type of movie among 100,000 online movie rental subscribers

Recognizing a Biased Question   *In Exercises 31–34, determine whether the survey question is biased. If the question is biased, suggest a better wording.*

31. Why does eating whole-grain foods improve your health?

32. Why does text messaging while driving increase the risk of a crash?

33. How much do you exercise during an average week?

34. How does the media influence the opinions of voters?

Extending Concepts

35. **Analyzing a Study**   Find an article or a news story that describes a statistical study.
(a) Identify the population and the sample.
(b) Classify the data as qualitative or quantitative. Determine the level of measurement.
(c) Is the study an observational study or an experiment? If it is an experiment, identify the treatment.
(d) Identify the sampling technique used to collect the data.

36. **Designing and Analyzing a Study**   Design a study for some subject that is of interest to you. Answer parts (a)–(d) of Exercise 35 for this study.

37. **Open and Closed Questions**   Two types of survey questions are open questions and closed questions. An open question allows for any kind of response; a closed question allows for only a fixed response. An open question and a closed question with its possible choices are given below. List an advantage and a disadvantage of each question.

Open Question   What can be done to get students to eat healthier foods?

Closed Question   How would you get students to eat healthier foods?
   1. Mandatory nutrition course
   2. Offer only healthy foods in the cafeteria and remove unhealthy foods
   3. Offer more healthy foods in the cafeteria and raise the prices on unhealthy foods

38. **Natural Experiments**   Observational studies are sometimes referred to as *natural experiments*. Explain, in your own words, what this means.

1.3 ACTIVITY

Random Numbers

APPLET

You can find the interactive applet for this activity within MyLab Statistics or at *www.pearsonhighered.com/ mathstatsresources.*

The *random numbers* applet is designed to allow you to generate random numbers from a range of values. You can specify integer values for the minimum value, maximum value, and the number of samples in the appropriate fields. You should not use decimal points when filling in the fields. When SAMPLE is clicked, the applet generates random values, which are displayed as a list in the text field.

Minimum value: ▢
Maximum value: ▢
Number of samples: ▢

Sample

EXPLORE

Step 1 Specify a minimum value.
Step 2 Specify a maximum value.
Step 3 Specify the number of samples.
Step 4 Click SAMPLE to generate a list of random values.

DRAW CONCLUSIONS

APPLET

1. Specify the minimum, maximum, and number of samples to be 1, 20, and 8, respectively, as shown. Run the applet. Continue generating lists until you obtain one that shows that the random sample is taken with replacement. Write down this list. How do you know that the list is a random sample taken with replacement?

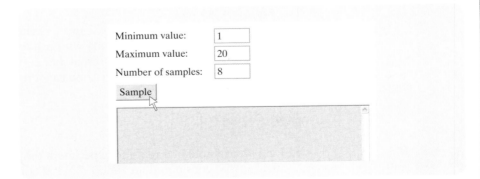

2. Use the applet to repeat Example 3 on page 21. What values did you use for the minimum, maximum, and number of samples? Which method do you prefer? Explain.

Uses

An experiment studied 321 women with advanced breast cancer. All of the women had been previously treated with other drugs, but the cancer had stopped responding to the medications. The women were then given the opportunity to take a new drug combined with a chemotherapy drug.

The subjects were divided into two groups, one that took the new drug combined with a chemotherapy drug, and one that took only the chemotherapy drug. After three years, results showed that the new drug in combination with the chemotherapy drug delayed the progression of cancer in the subjects. The results were so significant that the study was stopped, and the new drug was offered to all women in the study. The Food and Drug Administration has since approved use of the new drug in conjunction with a chemotherapy drug.

Abuses

For four years, one hundred eighty thousand teenagers in Norway were used as subjects to test a new vaccine against the deadly bacteria *meningococcus b*. A brochure describing the possible effects of the vaccine stated, "it is unlikely to expect serious complications," while information provided to the Norwegian Parliament stated, "serious side effects can not be excluded." The vaccine trial had some disastrous results: More than 500 side effects were reported, with some considered serious, and several of the subjects developed serious neurological diseases. The results showed that the vaccine was providing immunity in only 57% of the cases. This result was not sufficient for the vaccine to be added to Norway's vaccination program. Compensations have since been paid to the vaccine victims.

Ethics

Experiments help us further understand the world that surrounds us. But, in some cases, they can do more harm than good. In the Norwegian experiments, several ethical questions arise. Was the Norwegian experiment unethical if the best interests of the subjects were neglected? When should the experiment have been stopped? Should it have been conducted at all? When serious side effects are not reported and are withheld from subjects, there is no ethical question here, it is just wrong.

On the other hand, the breast cancer researchers would not want to deny the new drug to a group of patients with a life-threatening disease. But again, questions arise. How long must a researcher continue an experiment that shows better-than-expected results? How soon can a researcher conclude a drug is safe for the subjects involved?

EXERCISES

1. Find an example of a real-life experiment other than the one described above that may be considered an "abuse." What could have been done to avoid the outcome of the experiment?

2. *Stopping an Experiment* In your opinion, what are some problems that may arise when clinical trials of a new experimental drug or vaccine are stopped early and then the drug or vaccine is distributed to other subjects or patients?

1 Chapter Summary

What Did You Learn?	Example(s)	Review Exercises
Section 1.1		
▶ How to distinguish between a population and a sample	1	1–4
▶ How to distinguish between a parameter and a statistic	2	5–8
▶ How to distinguish between descriptive statistics and inferential statistics	3	9, 10
Section 1.2		
▶ How to distinguish between qualitative data and quantitative data	1	11–14
▶ How to classify data with respect to the four levels of measurement: nominal, ordinal, interval, and ratio	2, 3	15–18

Level of measurement	Put data in categories	Arrange data in order	Subtract data entries	Determine whether one data entry is a multiple of another
Nominal	Yes	No	No	No
Ordinal	Yes	Yes	No	No
Interval	Yes	Yes	Yes	No
Ratio	Yes	Yes	Yes	Yes

	Example(s)	Review Exercises
Section 1.3		
▶ How to design a statistical study and how to distinguish between an observational study and an experiment	1	19, 20
▶ How to design an experiment	2	21, 22
▶ How to create a sample using random sampling, simple random sampling, stratified sampling, cluster sampling, and systematic sampling and how to identify a biased sample	3, 4	23–29

Sampling Techniques

Random: A sample in which every member of a population has an equal chance of being selected.

Simple random: A sample in which every possible sample of the same size has the same chance of being selected from a population.

Stratified: Members of a population are divided into two or more subsets, called strata, that share a similar characteristic. A *sample* is then randomly selected from *each* of the strata. Using a stratified sample ensures that each segment of the population is represented.

Cluster: The population is divided into groups (or clusters) and *all of the members in one or more* (but not all) of the clusters are selected. To avoid a biased sample, care must be taken to ensure that all clusters have similar characteristics.

Systematic: Each member of a population is assigned a number. The members of the population are ordered in some way, a starting number is randomly selected, and then sample members are selected at regular intervals from the starting number. (For instance, every 3rd, 5th, or 100th member is selected.)

Review Exercises

1. See Odd Answers, page A41.

2. See Selected Answers, page A89.

3. See Odd Answers, page A42.

4. See Selected Answers, page A89.

5. See Odd Answers, page A42.

6. See Selected Answers, page A89.

7. Population parameter. The 10 students minoring in physics is a numerical description of all math majors at a university.

8. Sample statistic. The value 30% is a numerical description of a sample of U.S. workers.

9. The statement "62% would encourage a child to pursue a career as a video game developer or designer" is an example of descriptive statistics. An inference drawn from the sample is that a majority of people encourage children to pursue a career as a video game developer or designer.

10. The statement "48% have visited a public library or a bookmobile over a recent span of 12 months" is an example of descriptive statistics. An inference drawn from the sample is that about half of U.S. children and adults ages 16 years and older have visited a public library or a bookmobile over a recent span of 12 months.

11. Quantitative, because ages are numerical measurements.

12. Qualitative, because zip codes are labels for customers.

13. Quantitative, because revenues are numerical measurements.

14. Qualitative, because marital statuses are attributes.

15. Interval. The data can be ordered and meaningful differences can be calculated, but it does not make sense to say that 84 degrees is 1.05 times as hot as 80 degrees.

16. Ordinal. The data are qualitative and could be arranged in order of car size.

Section 1.1

In Exercises 1–4, identify the population and the sample. Describe the sample data set.

1. A survey of 4787 U.S. adults found that 15% use ride-hailing applications. *(Source: Pew Research Center)*

2. Eighty-three doctors working in the St. Louis area were surveyed concerning their opinions of health care reform.

3. A survey of 2223 U.S. adults found that 62% would encourage a child to pursue a career as a video game developer or designer. *(Source: The Harris Poll)*

4. A survey of 1601 U.S. children and adults ages 16 years and older found that 48% have visited a public library or a bookmobile over a recent span of 12 months. *(Source: Pew Research Center)*

In Exercises 5–8, determine whether the number describes a population parameter or a sample statistic. Explain your reasoning.

5. In 2016, the National Science Foundation announced $22.7 million in infrastructure-strengthening investments. *(Source: National Science Foundation)*

6. In a survey of 1000 likely U.S. voters, 29% trust media fact-checking of candidates' comments. *(Source: Rasmussen Reports)*

7. In a study of math majors at a university, 10 students minored in physics.

8. Thirty percent of a sample of 521 U.S. workers say that they worry about having their benefits reduced. *(Source: Gallup)*

9. Which part of the survey described in Exercise 3 represents the descriptive branch of statistics? Make an inference based on the results of the survey.

10. Which part of the survey described in Exercise 4 represents the descriptive branch of statistics? Make an inference based on the results of the survey.

Section 1.2

In Exercises 11–14, determine whether the data are qualitative or quantitative. Explain your reasoning.

11. The ages of a sample of 350 employees of a software company

12. The zip codes of a sample of 200 customers at a sporting goods store

13. The revenues of the companies on the Fortune 500 list

14. The marital statuses of all engineers at an electric utility

In Exercises 15–18, determine the level of measurement of the data set. Explain.

15. The daily high temperatures (in degrees Fahrenheit) for Sacramento, California, for a week in September are listed. *(Source: National Climatic Data Center)*

    90   80   76   84   91   94   97

16. The vehicle size classes for a sample of sedans are listed.

    Minicompact   Subcompact   Compact   Mid-size   Large

17. Nominal. The data are qualitative and cannot be arranged in a meaningful order.

18. Ratio. The data are quantitative, and it makes sense to say that $53.2 million is 1.12 times as much as $47.5 million.

19. Experiment. The study applies a treatment (drug to treat hypertension in patients with obstructive sleep apnea) to the subjects.

20. Observational study. The study does not attempt to influence the responses of the subjects and there is no treatment.

21. *Sample answer:* The subjects could be split into male and female and then be randomly assigned to each of the five treatment groups.

22. *Sample answer:* Number the volunteers and then use a random number generator to assign subjects randomly to one of the treatment groups or the control group.

23. Simple random sampling is used because random telephone numbers were generated and called. A potential source of bias is that telephone sampling only samples individuals who have telephones, who are available, and who are willing to respond.

24. Convenience sampling is used because the student sampled a convenient group of friends. The study may be biased toward the opinions of the student's friends.

25. Cluster sampling is used because each district is considered a cluster and every pregnant woman in a selected district is surveyed. A potential source of bias is that the selected districts may not be representative of the entire area.

26. Systematic sampling is used because every third vehicle is checked. A potential source of bias is that the street the law enforcement officials are using may be near a bar.

27. See Odd Answers, page A42.

28. See Selected Answers, page A89.

29. See Odd Answers, page A42.

17. The four departments of a printing company are listed.

Administration   Sales   Production   Billing

18. The total compensations (in millions of dollars) of the ten highest-paid CEOs at U.S. public companies are listed. *(Source: Equilar, Inc.)*

94.6   56.4   54.1   53.2   53.2   51.6   47.5   43.5   39.2   37.0

Section 1.3

In Exercises 19 and 20, determine whether the study is an observational study or an experiment. Explain.

19. Researchers conduct a study to determine whether a drug used to treat hypertension in patients with obstructive sleep apnea works better when taken in the morning or in the evening. To perform the study, 78 patients are given one pill to take in the morning and one pill to take in the evening (one containing the drug and the other a placebo). After 6 weeks, researchers collected blood pressure information on the patients. *(Source: American Thoracic Society)*

20. Researchers conduct a study to determine the effect of coffee consumption on the development of multiple sclerosis. To perform the study, researchers asked 4408 adults in Sweden and 2331 adults in the United States how many cups of coffee they drink per day. *(Source: American Association for the Advancement of Science)*

In Exercises 21 and 22, two hundred students volunteer for an experiment to test the effects of sleep deprivation on memory recall. The students will be placed in one of five different treatment groups, including the control group.

21. Explain how you could design an experiment so that it uses a randomized block design.

22. Explain how you could design an experiment so that it uses a completely randomized design.

In Exercises 23–28, identify the sampling technique used, and discuss potential sources of bias (if any). Explain.

23. Using random digit dialing, researchers ask 1201 U.S. adults whether enough is being done to fight opioid addiction. *(Source: Kaiser Family Foundation)*

24. A student asks 18 friends to participate in a psychology experiment.

25. A study in a town in northwest Ethiopia designed to determine prevalence and predictors of depression among pregnant women randomly selects four districts of the town, then interviews all pregnant women in these districts. *(Source: Public Library of Science)*

26. Law enforcement officials stop and check the driver of every third vehicle for blood alcohol content.

27. Twenty-five students are randomly selected from each grade level at a high school and surveyed about their study habits.

28. A journalist interviews 154 people waiting at an airport baggage claim and asks them how safe they feel during air travel.

29. You want to know the favorite spring break destination among 15,000 students at a university. Determine whether you would take a census or use a sampling. If you would use a sampling, determine which sampling technique you would use. Explain your reasoning.

Chapter Quiz

1. Population: Collection of the school performance of all Korean adolescents

Sample: Collection of the school performance of the 359,264 Korean adolescents in the study

2. (a) Sample statistic. The value 52% is a numerical description of a sample of U.S. adults.

(b) Population parameter. The 90% of members that approved the contract of the new president is a numerical description of all Board of Trustees members.

(c) Sample statistic. The value 25% is a numerical description of a sample of small business owners.

3. (a) Qualitative, because debit card personal identification numbers are labels and it does not make sense to find differences between numbers.

(b) Quantitative, because final scores are numerical measurements.

4. (a) Ordinal, because badge numbers can be ordered and often indicate seniority of service, but no meaningful mathematical computation can be performed.

(b) Ratio, because one data entry can be expressed as a multiple of another.

(c) Ordinal, because data can be arranged in order, but the differences between data entries make no sense.

(d) Interval, because meaningful differences between entries can be calculated but a zero entry is not an inherent zero.

5. See Odd Answers, page A42.

6. See Odd Answers, page A42.

7. See Odd Answers, page A42.

8. See Odd Answers, page A42.

Take this quiz as you would take a quiz in class. After you are done, check your work against the answers given in the back of the book.

1. A study of the dietary habits of 359,264 Korean adolescents was conducted to find a link between dietary habits and school performance. Identify the population and the sample in the study. *(Source: Wolters Kluwer Health, Inc.)*

2. Determine whether each number describes a population parameter or a sample statistic. Explain your reasoning.

(a) A survey of 1000 U.S. adults found that 52% think that the introduction of driverless cars will make roads less safe. *(Source: Rasmussen Reports)*

(b) At a college, 90% of the members of the Board of Trustees approved the contract of the new president.

(c) A survey of 727 small business owners found that 25% reported job openings they could not fill. *(Source: National Federation of Independent Business)*

3. Determine whether the data are qualitative or quantitative. Explain.

(a) A list of debit card personal identification numbers

(b) The final scores on a video game

4. Determine the level of measurement of the data set. Explain your reasoning.

(a) A list of badge numbers of police officers at a precinct

(b) The horsepowers of racing car engines

(c) The top 10 grossing films released in a year

(d) The years of birth for the runners in the Boston marathon

5. Determine whether the study is an observational study or an experiment. Explain.

(a) Researchers conduct a study to determine whether body mass index (BMI) influences mortality. To conduct the study, researchers obtained the BMIs of 3,951,455 people. *(Source: Elsevier, Ltd.)*

(b) Researchers conduct a study to determine whether taking a multivitamin daily affects cognitive health among men as they age. To perform the study, researchers studied 5947 male physicians ages 65 years or older and had one group take a multivitamin daily and had another group take a placebo daily. *(Source: American College of Physicians)*

6. An experiment is performed to test the effects of a new drug on high blood pressure. The experimenter identifies 320 people ages 35–50 years old with high blood pressure for participation in the experiment. The subjects are divided into equal groups according to age. Within each group, subjects are then randomly selected to be in either the treatment group or the control group. What type of experimental design is being used for this experiment?

7. Identify the sampling technique used in each study. Explain your reasoning.

(a) A journalist asks people at a campground about air pollution.

(b) For quality assurance, every tenth machine part is selected from an assembly line and measured for accuracy.

(c) A study on attitudes about smoking is conducted at a college. The students are divided by class (freshman, sophomore, junior, and senior). Then a random sample is selected from each class and interviewed.

8. Which technique used in Exercise 7 could lead to a biased study? Explain.

1 Chapter Test

Answer column (left)

1. (a) Sampling, because the population of New Jersey is too large for the most popular type of investment to be easily recorded. Random sampling would be advised because it would be easy to select people from New Jersey randomly and then record their most popular type of investment.

   (b) Census, because the population is small and it is relatively easy to obtain the ages of the 30 employees.

2. (a) Sample statistic. The value of 72% is a numerical description of a sample of U.S. adults ages 18 years and older.

   (b) Population parameter. The average evidence-based reading and writing score of 543 is a numerical description of all test takers in a recent year.

3. (a) Stratified sampling is used because the high school students are divided into strata (male and female), and a sample is selected from each stratum.

   (b) Simple random sampling is used because each customer has an equal chance of being contacted, and all samples of 625 customers have an equal chance of being selected.

   (c) Convenience sampling is used because a sample is taken from members of a population that are readily available. The sample may be biased because the teachers at the school may not be representative of the population of teachers.

4. See Selected Answers, page A89.

5. See Selected Answers, page A89.

6. See Selected Answers, page A90.

Test column (right)

Take this test as you would take a test in class.

1. Determine whether you would take a census or use a sampling. If you would use a sampling, determine which sampling technique you would use. Explain.

   (a) The most popular type of investment among investors in New Jersey

   (b) The average age of the 30 employees of a company

2. Determine whether each number describes a population parameter or a sample statistic. Explain.

   (a) A survey of 1003 U.S. adults ages 18 years and older found that 72% own a smartphone. *(Source: Pew Research Center)*

   (b) In a recent year, the average evidence-based reading and writing score on the SAT was 543. *(Source: The College Board)*

3. Identify the sampling technique used, and discuss potential sources of bias (if any). Explain.

   (a) Chosen at random, 200 male and 200 female high school students are asked about their plans after high school.

   (b) Chosen at random, 625 customers at an electronics store are contacted and asked their opinions of the service they received.

   (c) Questioning teachers as they leave a faculty lounge, a researcher asks 45 of them about their teaching styles.

4. Determine whether the data are qualitative or quantitative, and determine the level of measurement of the data set. Explain your reasoning.

   (a) The numbers of employees at fast-food restaurants in a city are listed.

20	11	6	31	17	23	12	18	40	22
13	8	18	14	37	32	25	27	25	18

   (b) The grade point averages (GPAs) for a class of students are listed.

3.6	3.2	2.0	3.8	3.0	3.5	1.7	3.2
2.2	4.0	2.5	1.9	2.8	3.6	2.5	3.7

5. Determine whether the survey question is biased. If the question is biased, suggest a better wording.

   (a) How many hours of sleep do you get on a normal night?

   (b) Do you agree that the town's ban on skateboarding in parks is unfair?

6. Researchers surveyed 19,183 U.S. physicians, asking for the information below. *(Source: Medscape from WebMD)*

   location (region of the U.S.)   income (dollars)
   employment status (private practice or an employee)
   benefits received (health insurance, liability coverage, etc.)
   specialty (cardiology, family medicine, radiology, etc.)
   time spent seeing patients per week (hours)

   (a) Identify the population and the sample.

   (b) Is the data collected qualitative, quantitative, or both? Explain your reasoning.

   (c) Determine the level of measurement for each item above.

   (d) Determine whether the study is an observational study or an experiment. Explain.

You are a researcher for a professional research firm. Your firm has won a contract to conduct a study for a technology publication. The editors of the publication would like to know their readers' thoughts on using smartphones for making and receiving payments, for redeeming coupons, and as tickets to events. They would also like to know whether people are interested in using smartphones as digital wallets that store data from their drivers' licenses, health insurance cards, and other cards.

The editors have given you their readership database and 20 questions they would like to ask (two sample questions from a previous study are given at the right). You know that it is too expensive to contact all of the readers, so you need to determine a way to contact a representative sample of the entire readership population.

EXERCISES

1. How Would You Do It?

(a) What sampling technique would you use to select the sample for the study? Why?

(b) Will the technique you chose in part (a) give you a sample that is representative of the population?

(c) Describe the method for collecting data.

(d) Identify possible flaws or biases in your study.

2. Data Classification

(a) What type of data do you expect to collect: qualitative, quantitative, or both? Why?

(b) At what levels of measurement do you think the data in the study will be? Why?

(c) Will the data collected for the study represent a population or a sample?

(d) Will the numerical descriptions of the data be parameters or statistics?

3. How They Did It

When The Harris Poll did a similar study, they used an Internet survey.

(a) Describe some possible errors in collecting data by Internet surveys.

(b) Compare your method for collecting data in Exercise 1 to this method.

When do you think smartphone payments will replace payment card transactions for a majority of purchases?

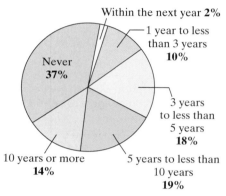

(Source: The Harris Poll)

How interested are you in being able to use your smartphone to make payments, rather than using cash or payment cards?

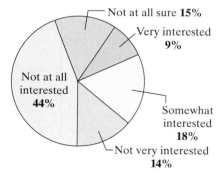

(Source: The Harris Poll)

HISTORY OF STATISTICS-TIMELINE

17TH CENTURY

John Graunt (1620–1674)

Studied records of deaths in London in the early 1600s. The first to make extensive statistical observations from massive amounts of data (Chapter 2), his work laid the foundation for modern statistics.

Blaise Pascal (1623–1662)
Pierre de Fermat (1601–1665)

Pascal and Fermat corresponded about basic probability problems (Chapter 3)—especially those dealing with gaming and gambling.

18TH CENTURY

Pierre Laplace (1749–1827)

▶ **Carl Friedrich Gauss (1777–1855)**

Studied probability (Chapter 3) and is credited with putting probability on a sure mathematical footing.

Studied regression and the method of least squares (Chapter 9) through astronomy. In his honor, the normal distribution (Chapter 5) is sometimes called the Gaussian distribution.

19TH CENTURY

Lambert Quetelet (1796–1874)

◀ **Florence Nightingale (1820–1910)**

Francis Galton (1822–1911)

Used descriptive statistics (Chapter 2) to analyze crime and mortality data and studied census techniques. Described normal distributions (Chapter 5) in connection with human traits such as height.

A nurse during the Crimean War, she was one of the first to advocate the importance of sanitation in hospitals. One of the first statisticians to use descriptive statistics (Chapter 2) as a way to argue for social change and credited with having developed the Coxcomb chart.

Used regression and correlation (Chapter 9) to study genetic variation in humans. He is credited with the discovery of the Central Limit Theorem (Chapter 5).

20TH CENTURY

Karl Pearson (1857–1936)

Studied natural selection using correlation (Chapter 9). Formed first academic department of statistics and helped develop chi-square analysis (Chapter 6).

William Gosset (1876–1937)

Studied process of brewing and developed *t*-test to correct problems connected with small sample sizes (Chapter 6).

Charles Spearman (1863–1945)

British psychologist who was one of the first to develop intelligence testing using factor analysis (Chapter 10).

Ronald Fisher (1890–1962)

Studied biology and natural selection and developed ANOVA (Chapter 10), stressed the importance of experimental design (Chapter 1), and was the first to identify the null and alternative hypotheses (Chapter 7).

20TH CENTURY (later)

Frank Wilcoxon (1892–1965)

◀ **John Tukey (1915–2000)**

David Kendall (1918–2007)

Biochemist who used statistics to study plant pathology. He introduced two-sample tests (Chapter 8), which led the way to the development of nonparametric statistics.

Worked at Princeton during World War II. Introduced exploratory data analysis techniques such as stem-and-leaf plots (Chapter 2). Also, worked at Bell Laboratories and is best known for his work in inferential statistics (Chapters 6–11).

Worked at Princeton and Cambridge. Was a leading authority on applied probability and data analysis (Chapters 2 and 3).

Using Technology in Statistics

With large data sets, you will find that calculators or computer software programs can help perform calculations and create graphics. These calculations can be performed on many calculators and statistical software programs, such as Minitab, Excel, and the TI-84 Plus.

The following example shows a sample generated by each of these three technologies to generate a list of random numbers. This list of random numbers can be used to select sample members or perform simulations.

EXAMPLE

Generating a List of Random Numbers

A quality control department inspects a random sample of 15 of the 167 cars that are assembled at an auto plant. How should the cars be chosen?

SOLUTION

One way to choose the sample is to first number the cars from 1 to 167. Then you can use technology to form a list of random numbers from 1 to 167. Each of the technology tools shown requires different steps to generate the list. Each, however, does require that you identify the minimum value as 1 and the maximum value as 167. Check your user's manual for specific instructions.

MINITAB	
↓	C1
1	167
2	11
3	74
4	160
5	18
6	70
7	80
8	56
9	37
10	6
11	82
12	126
13	98
14	104
15	137

EXCEL	
	A
1	41
2	16
3	91
4	58
5	151
6	36
7	96
8	154
9	2
10	113
11	157
12	103
13	64
14	135
15	90

TI-84 PLUS

```
randInt (1, 167, 15)
{17  42  152  59  5  116
125  64  122  55  58  60
82  152  105}
```

Recall that when you generate a list of random numbers, you should decide whether it is acceptable to have numbers that repeat. If it is acceptable, then the sampling process is said to be with replacement. If it is not acceptable, then the sampling process is said to be without replacement.

With each of the three technology tools shown on page 36, you have the capability of sorting the list so that the numbers appear in order. Sorting helps you see whether any of the numbers in the list repeat. If it is not acceptable to have repeats, you should specify that the tool generate more random numbers than you need.

EXERCISES

1. The SEC (Securities and Exchange Commission) is investigating a financial services company. The company being investigated has 86 brokers. The SEC decides to review the records for a random sample of 10 brokers. Describe how this investigation could be done. Then use technology to generate a list of 10 random numbers from 1 to 86 and order the list.

2. A quality control department is testing 25 smartphones from a shipment of 300 smartphones. Describe how this test could be done. Then use technology to generate a list of 25 random numbers from 1 to 300 and order the list.

3. Consider the population of ten digits: 0, 1, 2, 3, 4, 5, 6, 7, 8, and 9. Select three random samples of five digits from this list. Find the average of each sample. Compare your results with the average of the entire population. Comment on your results. (Hint: To find the average, sum the data entries and divide the sum by the number of entries.)

4. Consider the population of 41 whole numbers from 0 to 40. What is the average of these numbers? Select three random samples of seven numbers from this list. Find the average of each sample. Compare your results with the average of the entire population. Comment on your results. (Hint: To find the average, sum the data entries and divide the sum by the number of entries.)

5. Use random numbers to simulate rolling a six-sided die 60 times. How many times did you obtain each number from 1 to 6? Are the results what you expected?

6. You rolled a six-sided die 60 times and got the following tally.

   20 ones
   20 twos
   15 threes
   3 fours
   2 fives
   0 sixes

   Does this seem like a reasonable result? What inference might you draw from the result?

7. Use random numbers to simulate tossing a coin 100 times. Let 0 represent heads, and let 1 represent tails. How many times did you obtain each number? Are the results what you expected?

8. You tossed a coin 100 times and got 77 heads and 23 tails. Does this seem like a reasonable result? What inference might you draw from the result?

9. A political analyst would like to survey a sample of the registered voters in a county. The county has 47 election districts. How could the analyst use random numbers to obtain a cluster sample?

Extended solutions are given in the technology manuals that accompany this text. Technical instruction is provided for Minitab, Excel, and the TI-84 Plus.

CHAPTER 2

Descriptive Statistics

Since the 1966 season, the National Football League has determined its champion in the Super Bowl. The winning team receives the Lombardi Trophy.

Where You've Been

In Chapter 1, you learned that there are many ways to collect data. Usually, researchers must work with sample data in order to analyze populations, but occasionally it is possible to collect all the data for a given population. For instance, the data at the right represents the points scored by the winning teams in the first 51 Super Bowls. *(Source: NFL.com)*

35, 33, 16, 23, 16, 24, 14, 24, 16, 21, 32, 27, 35, 31, 27, 26, 27, 38, 38, 46, 39, 42, 20, 55, 20, 37, 52, 30, 49, 27, 35, 31, 34, 23, 34, 20, 48, 32, 24, 21, 29, 17, 27, 31, 31, 21, 34, 43, 28, 24, 34

Where You're Going

In Chapter 2, you will learn ways to organize and describe data sets. The goal is to make the data easier to understand by describing trends, averages, and variations. For instance, in the raw data showing the points scored by the winning teams in the first 51 Super Bowls, it is not easy to see any patterns or special characteristics. Here are some ways you can organize and describe the data.

Make a frequency distribution.

Class	Frequency, f
14–19	5
20–25	12
26–31	13
32–37	11
38–43	5
44–49	3
50–55	2

Draw a histogram.

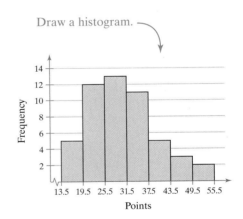

$$\text{Mean} = \frac{35 + 33 + 16 + 23 + 16 + \cdots + 43 + 28 + 24 + 34}{51}$$

$$= \frac{1541}{51}$$

$$\approx 30.2 \text{ points} \qquad \text{Find an average.}$$

$$\text{Range} = 55 - 14$$

$$= 41 \text{ points} \qquad \text{Find how the data vary.}$$

2.1 | Frequency Distributions and Their Graphs

What You Should Learn

▶ How to construct a frequency distribution, including limits, midpoints, relative frequencies, cumulative frequencies, and boundaries

▶ How to construct frequency histograms, frequency polygons, relative frequency histograms, and ogives

Frequency Distributions ▪ Graphs of Frequency Distributions

Frequency Distributions

There are many ways to organize and describe a data set. Important characteristics to look for when organizing and describing a data set are its **center,** its **variability** (or spread), and its **shape.** Measures of center and shapes of distributions are covered in Section 2.3. Measures of variability are covered in Section 2.4.

When a data set has many entries, it can be difficult to see patterns. In this section, you will learn how to organize data sets by grouping the data into **intervals** called **classes** and forming a **frequency distribution.** You will also learn how to use frequency distributions to construct graphs.

DEFINITION

A **frequency distribution** is a table that shows **classes** or **intervals** of data entries with a count of the number of entries in each class. The **frequency** f of a class is the number of data entries in the class.

Example of a Frequency Distribution

Class	Frequency, f
1–5	5
6–10	8
11–15	6
16–20	8
21–25	5
26–30	4

In the frequency distribution shown at the left, there are six classes. The frequencies for each of the six classes are 5, 8, 6, 8, 5, and 4. Each class has a **lower class limit,** which is the least number that can belong to the class, and an **upper class limit,** which is the greatest number that can belong to the class. In the frequency distribution shown, the lower class limits are 1, 6, 11, 16, 21, and 26, and the upper class limits are 5, 10, 15, 20, 25, and 30. The **class width** is the distance between lower (or upper) limits of consecutive classes. For instance, the class width in the frequency distribution shown is $6 - 1 = 5$. Notice that the classes do not overlap.

The difference between the maximum and minimum data entries is called the **range.** In the frequency table shown, suppose the maximum data entry is 29, and the minimum data entry is 1. The range then is $29 - 1 = 28$. You will learn more about the range of a data set in Section 2.4.

Study Tip

In general, the frequency distributions shown in this text will use the minimum data entry for the lower limit of the first class. Sometimes it may be more convenient to choose a lower limit that is slightly less than the minimum data entry. The frequency distribution produced will vary slightly.

GUIDELINES

Constructing a Frequency Distribution from a Data Set

1. Decide on the number of classes to include in the frequency distribution. The number of classes should be between 5 and 20; otherwise, it may be difficult to detect any patterns.

2. Find the class width as follows. Determine the range of the data, divide the range by the number of classes, and *round up to the next convenient number.*

3. Find the class limits. You can use the minimum data entry as the lower limit of the first class. To find the remaining lower limits, add the class width to the lower limit of the preceding class. Then find the upper limit of the first class. Remember that classes cannot overlap. Find the remaining upper class limits.

4. Make a tally mark for each data entry in the row of the appropriate class.

5. Count the tally marks to find the total frequency f for each class.

Study Tip

If you obtain a whole number when calculating the class width of a frequency distribution, use the next whole number as the class width. Doing this ensures that you will have enough space in your frequency distribution for all the data entries.

Lower limit	Upper limit
155	190
191	226
227	262
263	298
299	334
335	370
371	406

Study Tip

The uppercase Greek letter sigma (Σ) is used throughout statistics to indicate a summation of values.

EXAMPLE 1

Constructing a Frequency Distribution from a Data Set

The data set lists the out-of-pocket prescription medicine expenses (in dollars) for 30 U.S. adults in a recent year. Construct a frequency distribution that has seven classes. *(Adapted from: Health, United States, 2015)*

200	239	155	252	384	165	296	405	303	400
307	241	256	315	330	317	352	266	276	345
238	306	290	271	345	312	293	195	168	342

SOLUTION

1. The number of classes (7) is stated in the problem.

2. The minimum data entry is 155 and the maximum data entry is 405, so the range is $405 - 155 = 250$. Divide the range by the number of classes and round up to find the class width.

$$\text{Class width} = \frac{250}{7} \qquad \frac{\text{Range}}{\text{Number of classes}}$$

$$\approx 35.71 \qquad \text{Round up to the next convenient number, 36.}$$

3. The minimum data entry is a convenient lower limit for the first class. To find the lower limits of the remaining six classes, add the class width of 36 to the lower limit of each previous class. So, the lower limits of the other classes are $155 + 36 = 191$, $191 + 36 = 227$, and so on. The upper limit of the first class is 190, which is one less than the lower limit of the second class. The upper limits of the other classes are $190 + 36 = 226$, $226 + 36 = 262$, and so on. The lower and upper limits for all seven classes are shown at the left.

4. Make a tally mark for each data entry in the appropriate class. For instance, the data entry 168 is in the 155–190 class, so make a tally mark in that class. Continue until you have made a tally mark for each of the 30 data entries.

5. The number of tally marks for a class is the frequency of that class.

The frequency distribution is shown below. The first class, 155–190, has three tally marks. So, the frequency of this class is 3. Notice that the sum of the frequencies is 30, which is the number of entries in the data set. The sum is denoted by Σf where Σ is the uppercase Greek letter **sigma.**

**Frequency Distribution for Out-of-Pocket
Prescription Medicine Expenses (in dollars)**

Expenses →   ← Number of adults

Class	Tally	Frequency, f
155–190	|||	3
191–226	||	2
227–262	ℋℋ	5
263–298	ℋℋ		6
299–334	ℋℋ ||	7
335–370	||||	4
371–406	|||	3
		$\Sigma f = 30$

Check that the sum of the frequencies equals the number in the sample.

TRY IT YOURSELF 1

Construct a frequency distribution using the points scored by the 51 winning teams listed on page 39. Use six classes. *Answer: Page A31*

Population of Iowa

Ages	Frequency
0–9	399,859
10–19	424,850
20–29	412,354
30–39	387,363
40–49	368,620
50–59	421,726
60–69	356,124
70–79	203,053
80 and older	143,699

The last class, 80 and older, is open-ended.

(Source: U.S. Census Bureau)

Note in Example 1 that the classes do not overlap, so each of the original data entries belongs to exactly one class. Also, the classes are of equal width. In general, all classes in a frequency distribution have the same width. However, this may not always be possible because a class can be *open-ended*. For instance, the frequency distribution for the population of Iowa shown at the left has an open-ended class, "80 and older."

After constructing a standard frequency distribution such as the one in Example 1, you can include several additional features that will help provide a better understanding of the data. These features (the **midpoint, relative frequency,** and **cumulative frequency** of each class) can be included as additional columns in your table.

> ### DEFINITION
>
> The **midpoint** of a class is the sum of the lower and upper limits of the class divided by two. The midpoint is sometimes called the *class mark*.
>
> $$\text{Midpoint} = \frac{(\text{Lower class limit}) + (\text{Upper class limit})}{2}$$
>
> The **relative frequency** of a class is the portion, or percentage, of the data that falls in that class. To find the relative frequency of a class, divide the frequency f by the sample size n.
>
> $$\text{Relative frequency} = \frac{\text{Class frequency}}{\text{Sample size}} = \frac{f}{n} \qquad \text{Note that } n = \Sigma f.$$
>
> The **cumulative frequency** of a class is the sum of the frequencies of that class and all previous classes. The cumulative frequency of the last class is equal to the sample size n.

You can use the formula shown above to find the midpoint of each class, or after finding the first midpoint, you can find the remaining midpoints by adding the class width to the previous midpoint. For instance, the midpoint of the first class in Example 1 is

$$\text{Midpoint} = \frac{155 + 190}{2} = 172.5. \qquad \text{Midpoint of first class.}$$

Using the class width of 36, the remaining midpoints are

$172.5 + 36 = 208.5$ Midpoint of second class.

$208.5 + 36 = 244.5$ Midpoint of third class.

$244.5 + 36 = 280.5$ Midpoint of fourth class.

and so on.

You can write the relative frequency as a fraction, decimal, or percent. The sum of the relative frequencies of all the classes should be equal to 1, or 100%. Due to rounding, the sum may be slightly less than or greater than 1. So, values such as 0.99 and 1.01 are sufficient.

EXAMPLE 2

Finding Midpoints, Relative Frequencies, and Cumulative Frequencies

Using the frequency distribution constructed in Example 1, find the midpoint, relative frequency, and cumulative frequency of each class. Describe any patterns.

SOLUTION

The midpoints, relative frequencies, and cumulative frequencies of the first five classes are calculated as follows.

Class	f	Midpoint	Relative frequency	Cumulative frequency
155–190	3	$\dfrac{155 + 190}{2} = 172.5$	$\dfrac{3}{30} = 0.1$	3
191–226	2	$\dfrac{191 + 226}{2} = 208.5$	$\dfrac{2}{30} \approx 0.07$	3 + 2 = 5
227–262	5	$\dfrac{227 + 262}{2} = 244.5$	$\dfrac{5}{30} \approx 0.17$	5 + 5 = 10
263–298	6	$\dfrac{263 + 298}{2} = 280.5$	$\dfrac{6}{30} = 0.2$	10 + 6 = 16
299–334	7	$\dfrac{299 + 334}{2} = 316.5$	$\dfrac{7}{30} \approx 0.23$	16 + 7 = 23

The remaining midpoints, relative frequencies, and cumulative frequencies are shown in the expanded frequency distribution below.

Frequency Distribution for Out-of-Pocket Prescription Medicine Expenses (in dollars)

Expenses

Number of adults

Portion of adults

Class	Frequency, f	Midpoint	Relative frequency	Cumulative frequency
155–190	3	172.5	0.1	3
191–226	2	208.5	0.07	5
227–262	5	244.5	0.17	10
263–298	6	280.5	0.2	16
299–334	7	316.5	0.23	23
335–370	4	352.5	0.13	27
371–406	3	388.5	0.1	30
	$\Sigma f = 30$		$\Sigma\dfrac{f}{n} = 1$	

Interpretation There are several patterns in the data set. For instance, the most common range for the expenses is $299 to $334. Also, about half of the expenses are less than $299.

TRY IT YOURSELF 2

Using the frequency distribution constructed in Try It Yourself 1, find the midpoint, relative frequency, and cumulative frequency of each class. Describe any patterns.

Answer: Page A31

Graphs of Frequency Distributions

Sometimes it is easier to discover patterns in a data set by looking at a graph of the frequency distribution. One such graph is a **frequency histogram.**

DEFINITION

A **frequency histogram** uses bars to represent the frequency distribution of a data set. A histogram has the following properties.

1. The horizontal scale is quantitative and measures the data entries.
2. The vertical scale measures the frequencies of the classes.
3. Consecutive bars must touch.

Because consecutive bars of a histogram must touch, bars must begin and end at class boundaries instead of class limits. **Class boundaries** are the numbers that separate classes *without* forming gaps between them. For data that are integers, subtract 0.5 from each lower limit to find the lower class boundaries. To find the upper class boundaries, add 0.5 to each upper limit. The upper boundary of a class will equal the lower boundary of the next higher class.

EXAMPLE 3

Constructing a Frequency Histogram

Draw a frequency histogram for the frequency distribution in Example 2. Describe any patterns.

SOLUTION

First, find the class boundaries. Because the data entries are integers, subtract 0.5 from each lower limit to find the lower class boundaries and add 0.5 to each upper limit to find the upper class boundaries. So, the lower and upper boundaries of the first class are as follows.

First class lower boundary = 155 − 0.5 = 154.5
First class upper boundary = 190 + 0.5 = 190.5

The boundaries of the remaining classes are shown in the table at the left. To construct the histogram, choose possible frequency values for the vertical scale. You can mark the horizontal scale either at the midpoints or at the class boundaries. Both histograms are shown below.

Class	Class boundaries	Frequency, f
155–190	154.5–190.5	3
191–226	190.5–226.5	2
227–262	226.5–262.5	5
263–298	262.5–298.5	6
299–334	298.5–334.5	7
335–370	334.5–370.5	4
371–406	370.5–406.5	3

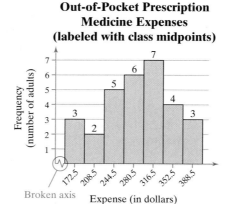

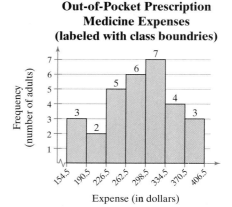

Study Tip

It is customary in bar graphs to have spaces between the bars, whereas with histograms, it is customary that the bars have no spaces between them.

Interpretation From either histogram, you can see that two-thirds of the adults are paying more than $262.50 for out-of-pocket prescription medicine expenses.

TRY IT YOURSELF 3

Use the frequency distribution from Try It Yourself 2 to construct a frequency histogram that represents the points scored by the 51 winning teams listed on page 39. Describe any patterns. *Answer: Page A32*

Another way to graph a frequency distribution is to use a frequency polygon. A **frequency polygon** is a line graph that emphasizes the continuous change in frequencies.

EXAMPLE 4

Constructing a Frequency Polygon

Draw a frequency polygon for the frequency distribution in Example 2. Describe any patterns.

SOLUTION

To construct the frequency polygon, use the same horizontal and vertical scales that were used in the histogram labeled with class midpoints in Example 3. Then plot points that represent the midpoint and frequency of each class and connect the points in order from left to right with line segments. Because the graph should begin and end on the horizontal axis, extend the left side to one class width before the first class midpoint and extend the right side to one class width after the last class midpoint.

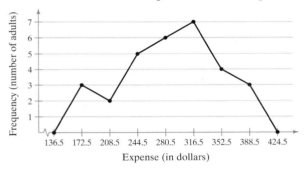

Out-of-Pocket Prescription Medicine Expenses

You can check your answer using technology, as shown below.

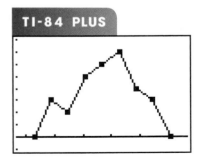

TI-84 PLUS

Interpretation   You can see that the frequency of adults increases up to an expense of $316.50 and then the frequency decreases.

TRY IT YOURSELF 4

Use the frequency distribution from Try It Yourself 2 to construct a frequency polygon that represents the points scored by the 51 winning teams listed on page 39. Describe any patterns. *Answer: Page A32*

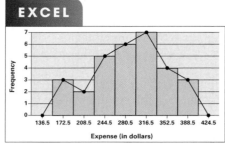

A histogram and its corresponding frequency polygon are often drawn together, as shown at the left using Excel. To do this by hand, first, construct the frequency polygon by choosing appropriate horizontal and vertical scales. The horizontal scale should consist of the class midpoints, and the vertical scale should consist of appropriate frequency values. Then plot the points that represent the midpoint and frequency of each class. After connecting the points with line segments, finish by drawing the bars for the histogram.

A **relative frequency histogram** has the same shape and the same horizontal scale as the corresponding frequency histogram. The difference is that the vertical scale measures the *relative* frequencies, not frequencies.

EXAMPLE 5

Constructing a Relative Frequency Histogram

Draw a relative frequency histogram for the frequency distribution in Example 2.

SOLUTION

The relative frequency histogram is shown. Notice that the shape of the histogram is the same as the shape of the frequency histogram constructed in Example 3. The only difference is that the vertical scale measures the relative frequencies.

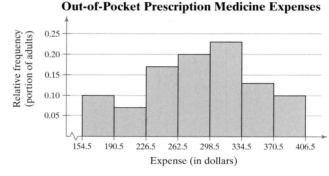

Out-of-Pocket Prescription Medicine Expenses

Interpretation   From this graph, you can quickly see that 0.2, or 20%, of the adults have expenses between $262.50 and $298.50, which is not immediately obvious from the frequency histogram in Example 3.

TRY IT YOURSELF 5

Use the frequency distribution in Try It Yourself 2 to construct a relative frequency histogram that represents the points scored by the 51 winning teams listed on page 39.

Answer: Page A32

To describe the number of data entries that are less than or equal to a certain value, construct a **cumulative frequency graph.**

DEFINITION

A **cumulative frequency graph,** or **ogive** (pronounced ō′jīve), is a line graph that displays the cumulative frequency of each class at its upper class boundary. The upper boundaries are marked on the horizontal axis, and the cumulative frequencies are marked on the vertical axis.

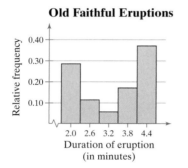

Picturing the World

Old Faithful, a geyser at Yellowstone National Park, erupts on a regular basis. The time spans of a sample of eruptions are shown in the relative frequency histogram. (Source: Yellowstone National Park)

Old Faithful Eruptions

About 50% of the eruptions last less than how many minutes?

About 50% of the eruptions last less than 3.5 minutes.

GUIDELINES

Constructing an Ogive (Cumulative Frequency Graph)

1. Construct a frequency distribution that includes cumulative frequencies as one of the columns.

2. Specify the horizontal and vertical scales. The horizontal scale consists of upper class boundaries, and the vertical scale measures cumulative frequencies.

3. Plot points that represent the upper class boundaries and their corresponding cumulative frequencies.

4. Connect the points in order from left to right with line segments.

5. The graph should start at the lower boundary of the first class (cumulative frequency is 0) and should end at the upper boundary of the last class (cumulative frequency is equal to the sample size).

EXAMPLE 6

Constructing an Ogive

Draw an ogive for the frequency distribution in Example 2.

SOLUTION

Using the cumulative frequencies, you can construct the ogive shown. The upper class boundaries, frequencies, and cumulative frequencies are shown in the table. Notice that the graph starts at 154.5, where the cumulative frequency is 0, and the graph ends at 406.5, where the cumulative frequency is 30.

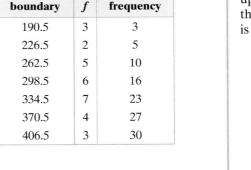

Upper class boundary	f	Cumulative frequency
190.5	3	3
226.5	2	5
262.5	5	10
298.5	6	16
334.5	7	23
370.5	4	27
406.5	3	30

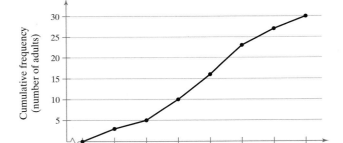

Interpretation   From the ogive, you can see that 10 adults had expenses of $262.50 or less. Also, the greatest increase in cumulative frequency occurs between $298.50 and $334.50 because the line segment is steepest between these two class boundaries.

TRY IT YOURSELF 6

Use the frequency distribution from Try It Yourself 2 to construct an ogive that represents the points scored by the 51 winning teams listed on page 39.

Answer: Page A32

Another type of ogive uses percent as the vertical axis instead of frequency (see Example 5 in Section 2.5).

If you have access to technology such as Minitab, Excel, StatCrunch, or the TI-84 Plus, you can use it to draw the graphs discussed in this section.

EXAMPLE 7

Using Technology to Construct Histograms

Use technology to construct a histogram for the frequency distribution in Example 2.

SOLUTION

Using the instructions for a TI-84 Plus shown in the Tech Tip at the left, you can draw a histogram similar to the one below on the left. To investigate the graph, you can use the *trace* feature. After pressing TRACE , the midpoint and the frequency of the first class are displayed, as shown in the figure on the right. Use the right and left arrow keys to move through each bar.

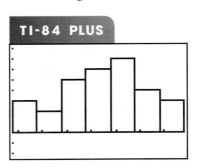

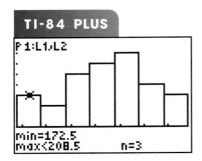

Histograms made using Minitab, Excel, and StatCrunch are shown below.

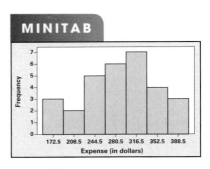

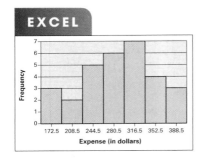

TRY IT YOURSELF 7

Use technology and the frequency distribution from Try It Yourself 2 to construct a frequency histogram that represents the points scored by the 51 winning teams listed on page 39.

Answer: Page A32

Tech Tip

You can use technology such as Minitab, Excel, StatCrunch, or the TI-84 Plus to create a histogram. (Detailed instructions for using Minitab, Excel, and the TI-84 Plus are shown in the technology manuals that accompany this text.) For instance, here are instructions for creating a histogram on a TI-84 Plus.

STAT ENTER

Enter midpoints in L1.
Enter frequencies in L2.

2nd STAT PLOT

Turn on Plot 1.
Highlight Histogram.

Xlist: L1

Freq: L2

ZOOM 9

2.1 EXERCISES

Building Basic Skills and Vocabulary

1. What are some benefits of representing data sets using frequency distributions? What are some benefits of using graphs of frequency distributions?

2. Why should the number of classes in a frequency distribution be between 5 and 20?

3. What is the difference between class limits and class boundaries?

4. What is the difference between relative frequency and cumulative frequency?

5. After constructing an expanded frequency distribution, what should the sum of the relative frequencies be? Explain.

6. What is the difference between a frequency polygon and an ogive?

True or False? *In Exercises 7–10, determine whether the statement is true or false. If it is false, rewrite it as a true statement.*

7. In a frequency distribution, the class width is the distance between the lower and upper limits of a class.

8. The midpoint of a class is the sum of its lower and upper limits divided by two.

9. An ogive is a graph that displays relative frequencies.

10. Class boundaries ensure that consecutive bars of a histogram touch.

In Exercises 11–14, use the minimum and maximum data entries and the number of classes to find the class width, the lower class limits, and the upper class limits.

11. min = 9, max = 64, 7 classes   **12.** min = 12, max = 88, 6 classes

13. min = 17, max = 135, 8 classes   **14.** min = 54, max = 247, 10 classes

Reading a Frequency Distribution *In Exercises 15 and 16, use the frequency distribution to find the (a) class width, (b) class midpoints, and (c) class boundaries.*

15.   Travel Time to Work (in minutes)

Class	Frequency, f
0–10	188
11–21	372
22–32	264
33–43	205
44–54	83
55–65	76
66–76	32

16.   Toledo, OH, Average Normal Temperatures (°F)

Class	Frequency, f
25–32	86
33–40	39
41–48	41
49–56	48
57–64	43
65–72	68
73–80	40

17. Use the frequency distribution in Exercise 15 to construct an expanded frequency distribution, as shown in Example 2.

18. Use the frequency distribution in Exercise 16 to construct an expanded frequency distribution, as shown in Example 2.

Answers (sidebar)

1. Organizing the data into a frequency distribution may make patterns within the data more evident. Sometimes it is easier to identify patterns of a data set by looking at a graph of the frequency distribution.

2. If there are too few or too many classes, it may be difficult to detect patterns because the data are too condensed or too spread out.

3. Class limits determine which numbers can belong to each class. Class boundaries are the numbers that separate classes without forming gaps between them.

4. Relative frequency of a class is the portion, or percentage, of the data that falls in that class. Cumulative frequency of a class is the sum of the frequencies of that class and all previous classes.

5. The sum of the relative frequencies must be 1 or 100% because it is the sum of all portions or percentages of the data.

6. A frequency polygon displays frequencies or relative frequencies whereas an ogive displays cumulative frequencies.

7. False. Class width is the difference between lower or upper limits of consecutive classes.

8. True

9. False. An ogive is a graph that displays cumulative frequencies.

10. True

11. See Odd Answers, page A43.

12. See Selected Answers, page A90.

13. See Odd Answers, page A43.

14. See Selected Answers, page A90.

15. See Odd Answers, page A43.

16. See Selected Answers, page A90.

17. See Odd Answers, page A43.

18. See Selected Answers, page A90.

19. (a) 7

   (b) Greatest frequency: about 300
   Least frequency: about 10

   (c) 10

   (d) *Sample answer:* About half
   of the employee salaries
   are between $50,000 and
   $69,000.

20. (a) 6

   (b) Greatest frequency: 37
   Least frequency: 1

   (c) 53

   (d) *Sample answer:* The heights
   of most roller coasters are
   less than 231 feet.

21. Class with greatest frequency:
   506–510

   Classes with least frequency:
   474–478

22. Class with greatest frequency:
   3.5–4.5 miles

   Class with least frequency:
   0.5–1.5 miles

23. (a) Class with greatest
   relative frequency:
   35–36 centimeters

   Class with least
   relative frequency:
   39–40 centimeters

   (b) Greatest relative frequency
   ≈ 0.25

   Least relative frequency
   ≈ 0.01

   (c) *Sample answer:* From
   the graph, 0.25 or 25%
   of females have a fibula
   length between 35 and 36
   centimeters.

24. (a) Class with greatest relative
   frequency: 11–12 minutes

   Class with least relative
   frequency: 14–15 minutes

   (b) Greatest relative frequency
   ≈ 38%

   Least relative frequency
   ≈ 4%

   (c) *Sample answer:* From the
   graph, about 0.75 or 75% of
   campus security response
   times are between 11 and
   13 minutes.

Graphical Analysis *In Exercises 19 and 20, use the frequency histogram to*

(a) *determine the number of classes.*

(b) *estimate the greatest and least frequencies.*

(c) *determine the class width.*

(d) *describe any patterns with the data.*

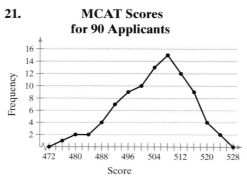

Graphical Analysis *In Exercises 21 and 22, use the frequency polygon to identify the class with the greatest, and the class with the least, frequency.*

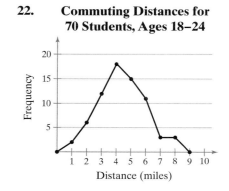

Graphical Analysis *In Exercises 23 and 24, use the relative frequency histogram to*

(a) *identify the class with the greatest, and the class with the least, relative frequency.*

(b) *approximate the greatest and least relative frequencies.*

(c) *describe any patterns with the data.*

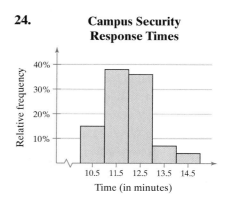

25. (a) 75
    (b) 158.5–201.5 pounds
26. (a) 77
    (b) 68–70 inches
27. (a) 47
    (b) 287.5 pounds
    (c) 40
    (d) 6
28. (a) 71
    (b) 68 inches
    (c) 56
    (d) 30
29. See Odd Answers, page A43.
30. See Selected Answers, page A90.

Graphical Analysis   *In Exercises 25 and 26, use the ogive to approximate*

(a) the number in the sample.

(b) the location of the greatest increase in frequency.

25.

Black Bears

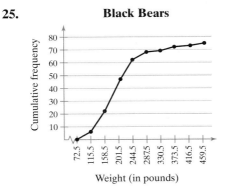

Weight (in pounds)

26.

Adult Males

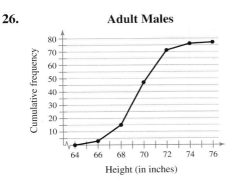

Height (in inches)

27. Use the ogive in Exercise 25 to approximate
    (a) the cumulative frequency for a weight of 201.5 pounds.
    (b) the weight for which the cumulative frequency is 68.
    (c) the number of black bears that weigh between 158.5 pounds and 244.5 pounds.
    (d) the number of black bears that weigh more than 330.5 pounds.

28. Use the ogive in Exercise 26 to approximate
    (a) the cumulative frequency for a height of 72 inches.
    (b) the height for which the cumulative frequency is 15.
    (c) the number of adult males that are between 68 and 72 inches tall.
    (d) the number of adult males that are taller than 70 inches.

Using and Interpreting Concepts

Constructing a Frequency Distribution   *In Exercises 29 and 30, construct a frequency distribution for the data set using the indicated number of classes. In the table, include the midpoints, relative frequencies, and cumulative frequencies. Which class has the greatest class frequency and which has the least class frequency?*

 29. Music Blog Reading Times
Number of classes: 5
Data set: Times (in minutes) spent reading a music blog in a day

7	39	13	25	8	22	0	2	18	2	30	7
35	12	15	8	6	5	29	0	11	39	16	15

 30. Textbook Spending
Number of classes: 6
Data set: Amounts (in dollars) spent on textbooks for a semester

91	472	279	249	530	376	188	341	266	199
142	273	189	130	489	266	248	101	375	486
190	398	188	269	43	30	127	354	84	319

---

indicates that the data set for this exercise is available within MyStatLab or at *www.pearsonhighered.com/mathstatsresources.*

31. See Odd Answers, page A43.

32. See Selected Answers, page A90.

33. See Odd Answers, page A44.

34. See Selected Answers, page A91.

35. See Odd Answers, page A44.

36. See Selected Answers, page A91.

Constructing a Frequency Distribution and a Frequency Histogram

In Exercises 31–34, construct a frequency distribution and a frequency histogram for the data set using the indicated number of classes. Describe any patterns.

31. Sales

Number of classes: 6

Data set: July sales (in dollars) for 21 sales representatives at a company

2114	2468	7119	1876	4105	3183	1932
1355	4278	2000	1077	5835	1512	1697
2478	3981	1643	1858	1500	4608	1000

32. Pepper Pungencies

Number of classes: 5

Data set: Pungencies (in thousands of Scoville units) of 24 tabasco peppers

35	51	44	42	37	38	36	39	44	43	40	40
32	39	41	38	42	39	40	46	37	35	41	39

33. Reaction Times

Number of classes: 8

Data set: Reaction times (in milliseconds) of 30 adult females to an auditory stimulus

507	389	305	291	336	310	514	442	373	428
387	454	323	441	388	426	411	382	320	450
309	416	359	388	307	337	469	351	422	413

34. Finishing Times

Number of classes: 8

Data set: Finishing times (in seconds) of 21 participants in a 10K race

3449	2645	3255	3712	4183	3896	3760
5008	3983	2855	3789	3176	2923	2281
2574	2252	4223	2588	2243	2837	3292

Constructing a Frequency Distribution and a Frequency Polygon

In Exercises 35 and 36, construct a frequency distribution and a frequency polygon for the data set using the indicated number of classes. Describe any patterns.

35. Ages of the Presidents

Numbers of classes: 6

Data set: Ages of the U.S. presidents at Inauguration *(Source: The White House)*

57	61	57	57	58	57	61	54	68	51	49	64	50	48	65
52	56	46	54	49	51	47	55	55	54	42	51	56	55	51
54	51	60	62	43	55	56	61	52	69	64	46	54	47	70

36. Declaration of Independence

Number of classes: 5

Data set: Numbers of children of those who signed the Declaration of Independence *(Source: The U.S. National Archives & Records Administration)*

5	2	12	18	7	4	10	8	16	3	3	7	3	1
2	7	13	0	8	3	7	5	2	6	0	6	7	9
0	11	9	10	7	8	13	5	8	3	5	0	3	13
3	15	5	6	3	2	5	2	0	3	7	12	4	1

37. See Odd Answers, page A44.

38. See Selected Answers, page A91.

39. See Odd Answers, page A44.

40. See Selected Answers, page A91.

41. See Odd Answers, page A45.

42. See Selected Answers, page A92.

Constructing a Frequency Distribution and a Relative Frequency Histogram

In Exercises 37–40, construct a frequency distribution and a relative frequency histogram for the data set using five classes. Which class has the greatest relative frequency and which has the least relative frequency?

37. Taste Test

Data set: Ratings from 1 (lowest) to 10 (highest) provided by 36 people after taste-testing a new flavor of protein bar

```
2   6   9   2   9   9   6   10  5
8   7   6   5   10  1   4   9   3
4   5   3   6   5   2   4   9   2
9   3   3   6   5   1   9   4   2
```

38. Years of Service

Data set: Years of service of 28 Ohio state government employees

```
13  8   10  9   10  9   13
11  10  11  7   9   14  13
11  12  8   15  13  10  9
11  10  12  14  9   15  19
```

39. Fijian Banded Iguanas

Data set: Lengths (in centimeters) of 28 adult Fijian banded iguanas

```
68  65  70  61  60  60  69
61  64  74  64  62  70  70
63  75  74  71  70  66  72
64  67  66  70  73  72  70
```

40. Triglyceride Levels

Data set: Triglyceride levels (in milligrams per deciliter of blood) of 28 patients

```
209  140  155  170  265  138  180
295  250  320  270  225  215  390
420  462  150  200  400  295  240
200  190  145  160  175  195  223
```

Constructing a Cumulative Frequency Distribution and an Ogive

In Exercises 41 and 42, construct a cumulative frequency distribution and an ogive for the data set using six classes. Then describe the location of the greatest increase in frequency.

41. Retirement Ages

Data set: Retirement ages of 35 English professors

```
72  62  55  61  53  62  65
66  69  55  66  63  67  69
55  65  67  57  67  68  73
75  65  54  71  57  52  58
58  71  72  67  63  65  61
```

42. Saturated Fat Intakes

Data set: Daily saturated fat intakes (in grams) of 28 people

```
18  12  14  19  20  26  12
17  19  13  8   20  25  16
13  14  22  16  11  13  17
14  15  11  13  15  23  7
```

43. See Odd Answers, page A45.

44. See Selected Answers, page A92.

45. (a)

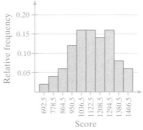
Daily Withdrawals

(b) 16.7%, because the sum of the relative frequencies for the last three classes is 0.167.

(c) $9700, because the sum of the relative frequencies for the last two classes is 0.10.

46. (a)

SAT Scores

(b) 64%; The portion of the scores greater than or equal to 1070 is 0.64.

(c) A score of 908 or above, because the sum of the relative frequencies of the class starting with 908 and all classes with higher scores is 0.88.

47. See Odd Answers, page A45.

In Exercises 43 and 44, use the data set and the indicated number of classes to construct (a) an expanded frequency distribution, (b) a frequency histogram, (c) a frequency polygon, (d) a relative frequency histogram, and (e) an ogive.

 43. Pulse Rates
Number of classes: 6
Data set: Pulse rates of all students in a class

68	105	95	80	90	100	75	70	84	98	102	70
65	88	90	75	78	94	110	120	95	80	76	108

 44. Hospitals
Number of classes: 8
Data set: Number of hospitals in each of the 50 states *(Source: American Hospital Directory)*

12	92	50	75	345	54	34	8	216	114
14	40	17	141	98	58	75	108	78	52
21	105	56	88	68	16	106	10	29	14
73	37	26	197	148	99	38	176	12	63
26	109	382	36	95	7	62	75	36	14

Extending Concepts

 45. What Would You Do? You work at a bank and are asked to recommend the amount of cash to put in an ATM each day. You do not want to put in too much (which would cause security concerns) or too little (which may create customer irritation). Here are the daily withdrawals (in hundreds of dollars) for 30 days.

72	84	61	76	104	76	86	92	80	88	98	76	97	82	84
67	70	81	82	89	74	73	86	81	85	78	82	80	91	83

(a) Construct a relative frequency histogram for the data. Use 8 classes.

(b) If you put $9000 in the ATM each day, what percent of the days in a month should you expect to run out of cash? Explain.

(c) If you are willing to run out of cash on 10% of the days, how much cash should you put in the ATM each day? Explain.

 46. What Would You Do? The admissions department for a college is asked to recommend the minimum SAT scores that the college will accept for full-time students. Here are the SAT scores of 50 applicants.

1170	1000	910	870	1070	1290	920	1470	1080	1180
770	900	1120	1070	1370	1160	970	930	1240	1270
1250	1330	1010	1010	1410	1130	1210	1240	960	820
650	1010	1190	1500	1400	1270	1310	1050	950	1150
1450	1290	1310	1100	1330	1410	840	1040	1090	1080

(a) Construct a relative frequency histogram for the data. Use 10 classes.

(b) If you set the minimum score at 1070, what percent of the applicants will meet this requirement? Explain.

(c) If you want to accept the top 88% of the applicants, what should the minimum score be? Explain.

 47. Writing Use the data set listed and technology to create frequency histograms with 5, 10, and 20 classes. Which graph displays the data best? Explain.

2	7	3	2	11	3	15	8	4	9	10	13	9
7	11	10	1	2	12	5	6	4	2	9	15	14

2.2 More Graphs and Displays

What You Should Learn

▶ How to graph and interpret quantitative data sets using stem-and-leaf plots and dot plots

▶ How to graph and interpret qualitative data sets using pie charts and Pareto charts

▶ How to graph and interpret paired data sets using scatter plots and time series charts

Graphing Quantitative Data Sets ■ Graphing Qualitative Data Sets ■ Graphing Paired Data Sets

Graphing Quantitative Data Sets

In Section 2.1, you learned several ways to display quantitative data graphically. In this section, you will learn more ways to display quantitative data, beginning with **stem-and-leaf plots.** Stem-and-leaf plots are examples of **exploratory data analysis (EDA),** which was developed by John Tukey in 1977.

In a stem-and-leaf plot, each number is separated into a **stem** (for instance, the entry's leftmost digits) and a **leaf** (for instance, the rightmost digit). You should have as many leaves as there are entries in the original data set and the leaves should be single digits. A stem-and-leaf plot is similar to a histogram but has the advantage that the graph still contains the original data. Another advantage of a stem-and-leaf plot is that it provides an easy way to sort data.

Number of Text Messages Sent				
76	49	102	58	88
122	76	89	67	80
66	80	78	69	56
76	115	99	72	19
41	86	48	52	28
26	29	33	26	20
33	24	43	16	39
29	32	29	29	40
23	33	30	41	33
38	34	53	30	149

EXAMPLE 1

Constructing a Stem-and-Leaf Plot

The data set at the left lists the numbers of text messages sent in one day by 50 cell phone users. Display the data in a stem-and-leaf plot. Describe any patterns. *(Adapted from Pew Research)*

SOLUTION

Because the data entries go from a low of 16 to a high of 149, you should use stem values from 1 to 14. To construct the plot, list these stems to the left of a vertical line. For each data entry, list a leaf to the right of its stem. For instance, the entry 102 has a stem of 10 and a leaf of 2. Make the plot with the leaves in increasing order from left to right. Be sure to include a key.

Number of Text Messages Sent

```
 1 | 6 9                      Key: 10|2 = 102
 2 | 0 3 4 6 6 8 9 9 9 9
 3 | 0 0 2 3 3 3 3 4 8 9
 4 | 0 1 1 3 8 9
 5 | 2 3 6 8
 6 | 6 7 9
 7 | 2 6 6 6 8
 8 | 0 0 6 8 9
 9 | 9
10 | 2
11 | 5
12 | 2
13 |
14 | 9
```

Interpretation From the display, you can see that more than 50% of the cell phone users sent between 20 and 50 text messages.

Study Tip

It is important to include a key for a stem-and-leaf plot to identify the data entries. This is done by showing an entry represented by a stem and one leaf.

Tech Tip

You can use technology such as Minitab, StatCrunch, or Excel (with the XLSTAT add-in) to construct a stem-and-leaf plot.

For instance, a StatCrunch stem-and-leaf plot for the data in Example 1 is shown below.

STATCRUNCH

Variable: Number of text messages sent

Decimal point is 1 digit(s) to the right of the colon.
Leaf unit = 1

```
 1 : 69
 2 : 0346689999
 3 : 0023333489
 4 : 011389
 5 : 2368
 6 : 679
 7 : 26668
 8 : 00689
 9 : 9
10 : 2
11 : 5
12 : 2
13 :
14 : 9
```

Study Tip

You can use stem-and-leaf plots to identify unusual data entries called *outliers*. In Examples 1 and 2, the data entry 149 is an outlier. You will learn more about outliers in Section 2.3.

TRY IT YOURSELF 1

Use a stem-and-leaf plot to organize the points scored by the 51 winning teams listed on page 39. Describe any patterns. *Answer: Page A32*

EXAMPLE 2

Constructing Variations of Stem-and-Leaf Plots

Organize the data set in Example 1 using a stem-and-leaf plot that has two rows for each stem. Describe any patterns.

SOLUTION

Use the stem-and-leaf plot from Example 1, except now list each stem twice. Use the leaves 0, 1, 2, 3, and 4 in the first stem row and the leaves 5, 6, 7, 8, and 9 in the second stem row. The revised stem-and-leaf plot is shown. Notice that by using two rows per stem, you obtain a more detailed picture of the data.

Number of Text Messages Sent

```
 1 |              Key: 10|2 = 102
 1 | 6 9
 2 | 0 3 4
 2 | 6 6 8 9 9 9
 3 | 0 0 2 3 3 3 3 4
 3 | 8 9
 4 | 0 1 1 3
 4 | 8 9
 5 | 2 3
 5 | 6 8
 6 |
 6 | 6 7 9
 7 | 2
 7 | 6 6 6 8
 8 | 0 0
 8 | 6 8 9
 9 |
 9 | 9
10 | 2
10 |
11 |
11 | 5
12 | 2
12 |
13 |
13 |
14 |
14 | 9
```

Interpretation From the display, you can see that most of the cell phone users sent between 20 and 80 text messages.

TRY IT YOURSELF 2

Using two rows for each stem, revise the stem-and-leaf plot you constructed in Try It Yourself 1. Describe any patterns. *Answer: Page A32*

You can also use a dot plot to graph quantitative data. In a **dot plot,** each data entry is plotted, using a point, above a horizontal axis. Like a stem-and-leaf plot, a dot plot allows you to see how data are distributed, to determine specific data entries, and to identify unusual data entries.

EXAMPLE 3

Constructing a Dot Plot

Use a dot plot to organize the data set in Example 1. Describe any patterns.

Number of Text Messages Sent									
76	49	102	58	88	122	76	89	67	80
66	80	78	69	56	76	115	99	72	19
41	86	48	52	28	26	29	33	26	20
33	24	43	16	39	29	32	29	29	40
23	33	30	41	33	38	34	53	30	149

SOLUTION

So that each data entry is included in the dot plot, the horizontal axis should include numbers between 15 and 150. To represent a data entry, plot a point above the entry's position on the axis. When an entry is repeated, plot another point above the previous point.

Number of Text Messages Sent

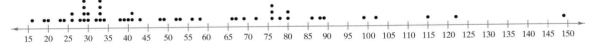

Interpretation From the dot plot, you can see that most entries occur between 20 and 80 and only 4 people sent more than 100 text messages. You can also see that 149 is an unusual data entry.

TRY IT YOURSELF 3

Use a dot plot to organize the points scored by the 51 winning teams listed on page 39. Describe any patterns.

Answer: Page A32

Technology can be used to construct dot plots. For instance, Minitab and StatCrunch dot plots for the text messaging data are shown below.

MINITAB

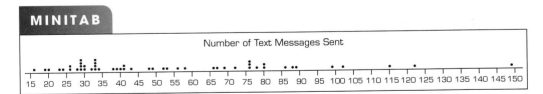

Number of Text Messages Sent

STATCRUNCH

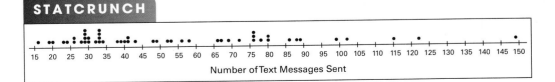

Number of Text Messages Sent

Graphing Qualitative Data Sets

Pie charts provide a convenient way to present qualitative data graphically as percents of a whole. A **pie chart** is a circle that is divided into sectors that represent categories. The area of each sector is proportional to the frequency of each category. In most cases, you will be interpreting a pie chart or constructing one using technology. Example 4 shows how to construct a pie chart by hand.

EXAMPLE 4

Constructing a Pie Chart

The numbers of earned degrees conferred (in thousands) in 2014 are shown in the table at the right. Use a pie chart to organize the data. *(Source: U.S. National Center for Education Statistics)*

Earned Degrees Conferred in 2014

Type of degree	Number (in thousands)
Associate's	1003
Bachelor's	1870
Master's	754
Doctoral	178

SOLUTION

Begin by finding the relative frequency, or percent, of each category. Then construct the pie chart using the central angle that corresponds to each category. To find the central angle, multiply 360° by the category's relative frequency. For instance, the central angle for associate's degrees is $360°(0.264) \approx 95°$.

Type of degree	f	Relative frequency	Angle
Associate's	1003	0.264	95°
Bachelor's	1870	0.491	177°
Master's	754	0.198	71°
Doctoral	178	0.047	17°

Earned Degrees Conferred in 2014

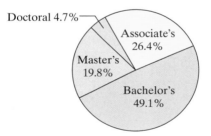

Interpretation From the pie chart, you can see that almost one-half of the degrees conferred in 2014 were bachelor's degrees.

TRY IT YOURSELF 4

The numbers of earned degrees conferred (in thousands) in 1990 are shown in the table. Use a pie chart to organize the data. Compare the 1990 data with the 2014 data. *(Source: U.S. National Center for Education Statistics)*

Earned Degrees Conferred in 1990

Type of degree	Number (in thousands)
Associate's	455
Bachelor's	1051
Master's	330
Doctoral	104

Answer: Page A32

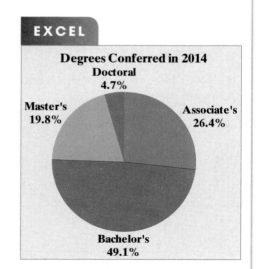

You can use technology to construct a pie chart. For instance, an Excel pie chart for the degrees conferred in 2014 is shown at the left.

Another way to graph qualitative data is to use a Pareto chart. A **Pareto chart** is a vertical bar graph in which the height of each bar represents frequency or relative frequency. The bars are positioned in order of decreasing height, with the tallest bar positioned at the left. Such positioning helps highlight important data and is used frequently in business.

EXAMPLE 5

Constructing a Pareto Chart

In 2014, these were the leading causes of death in the United States.

> Accidents: 136,053
>
> Cancer: 591,699
>
> Chronic lower respiratory disease: 147,101
>
> Heart disease: 614,348
>
> Stroke (cerebrovascular diseases): 133,103

Use a Pareto chart to organize the data. What was the leading cause of death in the United States in 2014? *(Source: Health, United States, 2015, Table 19)*

SOLUTION

Using frequencies for the vertical axis, you can construct the Pareto chart as shown.

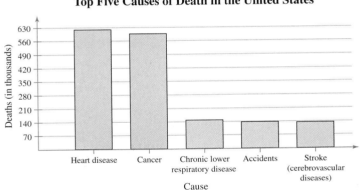

Top Five Causes of Death in the United States

Interpretation From the Pareto chart, you can see that the leading cause of death in the United States in 2014 was from heart disease. Also, heart disease and cancer caused more deaths than the other three causes combined.

TRY IT YOURSELF 5

Every year, the Better Business Bureau (BBB) receives complaints from customers. Here are some complaints the BBB received in a recent year.

> 16,281 complaints about auto dealers (used cars)
>
> 8384 complaints about insurance companies
>
> 3634 complaints about mortgage brokers
>
> 19,277 complaints about collection agencies
>
> 6985 complaints about travel agencies and bureaus

Use a Pareto chart to organize the data. Which industry is the greatest cause of complaints? *(Source: Council of Better Business Bureaus)*

Answer: Page A32

Picturing the World

According to data from the U.S. Bureau of Labor Statistics, earnings increase as educational attainment rises. The average weekly earnings data by educational attainment are shown in the Pareto chart. (Source: Based on U.S. Bureau of Labor Statistics)

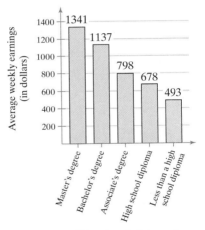

Average Weekly Earnings by Educational Attainment

The average worker with an associate's degree makes how much more in a year (52 weeks) than the average worker with a high school diploma?

In 52 weeks, the average worker with an associate's degree makes $41,496, and the average worker with a high school diploma makes $35,256. So, the worker with an associate's degree makes $6240 more than the worker with a high school diploma.

Graphing Paired Data Sets

When each entry in one data set corresponds to one entry in a second data set, the sets are called **paired data sets.** For instance, a data set contains the costs of an item and a second data set contains sales amounts for the item at each cost. Because each cost corresponds to a sales amount, the data sets are paired. One way to graph paired data sets is to use a **scatter plot,** where the ordered pairs are graphed as points in a coordinate plane. A scatter plot is used to show the relationship between two quantitative variables.

EXAMPLE 6

Interpreting a Scatter Plot

The British statistician Ronald Fisher (see page 35) introduced a famous data set called Fisher's Iris data set. This data set describes various physical characteristics, such as petal length and petal width (in millimeters), for three species of iris. In the scatter plot shown, the petal lengths form the first data set and the petal widths form the second data set. As the petal length increases, what tends to happen to the petal width? *(Source: Fisher, R. A., 1936)*

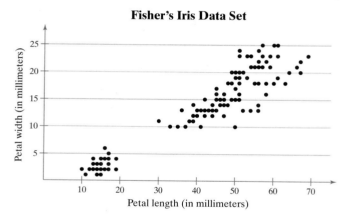

Fisher's Iris Data Set

SOLUTION

The horizontal axis represents the petal length, and the vertical axis represents the petal width. Each point in the scatter plot represents the petal length and petal width of one flower.

Interpretation   From the scatter plot, you can see that as the petal length increases, the petal width also tends to increase.

TRY IT YOURSELF 6

The lengths of employment and the salaries of 10 employees are listed in the table below. Graph the data using a scatter plot. Describe any trends.

Length of employment (in years)	5	4	8	4	2
Salary (in dollars)	32,000	32,500	40,000	27,350	25,000

Length of employment (in years)	10	7	6	9	3
Salary (in dollars)	43,000	41,650	39,225	45,100	28,000

Answer: Page A33

You will learn more about scatter plots and how to analyze them in Chapter 9.

A data set that is composed of quantitative entries taken at regular intervals over a period of time is called a **time series.** For instance, the amount of precipitation measured each day for one month is a time series. You can use a **time series chart** to graph a time series.

EXAMPLE 7

See Minitab and TI-84 Plus steps on pages 124 and 125.

Constructing a Time Series Chart

The table lists the number of motor vehicle thefts (in millions) and burglaries (in millions) in the United States for the years 2005 through 2015. Construct a time series chart for the number of motor vehicle thefts. Describe any trends. *(Source: Federal Bureau of Investigation, Crime in the United States)*

Year	Motor vehicle thefts (in millions)	Burglaries (in millions)
2005	1.24	2.16
2006	1.20	2.19
2007	1.10	2.19
2008	0.96	2.23
2009	0.80	2.20
2010	0.74	2.17
2011	0.72	2.19
2012	0.72	2.11
2013	0.70	1.93
2014	0.69	1.71
2015	0.71	1.58

SOLUTION

Let the horizontal axis represent the years and let the vertical axis represent the number of motor vehicle thefts (in millions). Then plot the paired data and connect them with line segments

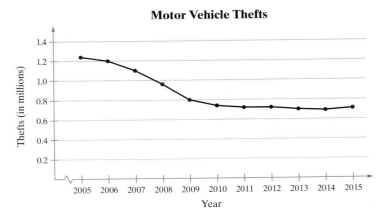

Motor Vehicle Thefts

Interpretation The time series chart shows that the number of motor vehicle thefts decreased until 2011 and then remained about the same through 2015.

TRY IT YOURSELF 7

Use the table in Example 7 to construct a time series chart for the number of burglaries for the years 2005 through 2015. Describe any trends.

Answer: Page A33

2.2 EXERCISES

1. Quantitative: stem-and-leaf plot, dot plot, histogram, scatter plot, time series chart

Qualitative: pie chart, Pareto chart

2. Unlike the histogram, the stem-and-leaf plot still contains the original data values. However, some data are difficult to organize in a stem-and-leaf plot.

3. Both the stem-and-leaf plot and the dot plot allow you to see how data are distributed, to determine specific data entries, and to identify unusual data values.

4. In a Pareto chart, the height of each bar represents frequency or relative frequency and the bars are positioned in order of decreasing height with the tallest bar positioned at the left.

5. b **6.** d **7.** a **8.** c

9. 27, 32, 41, 43, 43, 44, 47, 47, 48, 50, 51, 51, 52, 53, 53, 53, 54, 54, 54, 54, 55, 56, 56, 58, 59, 68, 68, 68, 73, 78, 78, 85

Max: 85; Min: 27

10. 12.9, 13.3, 13.6, 13.7, 13.7, 14.1, 14.1, 14.1, 14.1, 14.3, 14.4, 14.4, 14.6, 14.9, 14.9, 15.0, 15.0, 15.0, 15.1, 15.2, 15.4, 15.6, 15.7, 15.8, 15.8, 15.8, 15.9, 16.1, 16.6, 16.7

Max: 16.7; Min: 12.9

11. 13, 13, 14, 14, 14, 15, 15, 15, 15, 15, 16, 17, 17, 18, 19

Max: 19; Min: 13

12. 214, 214, 214, 216, 216, 217, 218, 218, 220, 221, 223, 224, 225, 225, 227, 228, 228, 228, 228, 230, 230, 231, 235, 237, 239

Max: 239; Min: 214

Building Basic Skills and Vocabulary

1. Name some ways to display quantitative data graphically. Name some ways to display qualitative data graphically.

2. What is an advantage of using a stem-and-leaf plot instead of a histogram? What is a disadvantage?

3. In terms of displaying data, how is a stem-and-leaf plot similar to a dot plot?

4. How is a Pareto chart different from a standard vertical bar graph?

Putting Graphs in Context *In Exercises 5–8, match the plot with the description of the sample.*

5.
```
0 | 8          Key: 0|8 = 0.8
1 | 5 6 8
2 | 1 3 4 5
3 | 0 9
4 | 0 0
```

6.
```
6 | 7 8          Key: 6|7 = 67
7 | 4 5 5 8 8 8
8 | 1 3 5 5 8 8 9
9 | 0 0 0 2 4
```

7.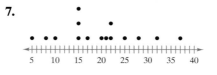

8.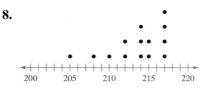

(a) Times (in minutes) it takes a sample of employees to drive to work

(b) Grade point averages of a sample of students with finance majors

(c) Top speeds (in miles per hour) of a sample of high-performance sports cars

(d) Ages (in years) of a sample of residents of a retirement home

Graphical Analysis *In Exercises 9–12, use the stem-and-leaf plot or dot plot to list the actual data entries. What is the maximum data entry? What is the minimum data entry?*

9.
```
2 | 7                          Key: 2|7 = 27
3 | 2
4 | 1 3 3 4 7 7 8
5 | 0 1 1 2 3 3 3 4 4 4 5 6 6 8 9
6 | 8 8 8
7 | 3 8 8
8 | 5
```

10.
```
12 |                Key: 12|9 = 12.9
12 | 9
13 | 3
13 | 6 7 7
14 | 1 1 1 1 3 4 4
14 | 6 9 9
15 | 0 0 0 1 2 4
15 | 6 7 8 8 8 9
16 | 1
16 | 6 7
```

11.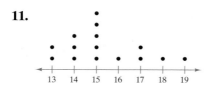

12.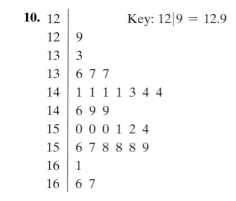

13. *Sample answer:* Facebook has the most users, and Pinterest has the least. Tumblr and Instagram have about the same number of users.

14. *Sample answer:* The year 2010 had the most motor vehicle thefts and 2013 had the least. Motor vehicle thefts decreased the most between 2011 and 2012.

15. *Sample answer:* The Texter is the least popular driver. The Left-Lane Hog is tolerated more than the Tailgater. The Speedster and the Drifter have the same popularity.

16. *Sample answer:* Food is the most costly aspect of pet care and live animal purchases is the least. The amounts spent on veterinarian care and supplies/OTC medicine are about the same.

17. **Exam Scores**

```
6 | 7 8        Key: 6|7 = 67
7 | 3 5 5 6 9
8 | 0 0 2 3 5 5 7 7 8
9 | 0 1 1 1 2 4 5 5
```

Sample answer: Most grades for the biology midterm were in the 80s or 90s.

18. **Hours Worked by Nurses**

```
2 | 4              Key: 2|4 = 24
3 | 0 2 2 3 5 5 6 6 6 6 8 8 9
4 | 0 0 0 0 0 0 0 0 8
5 | 0
```

Sample answer: Most nurses work from 30 to 40 hours per week.

19. See Odd Answers, page A46.

20. See Selected Answers, page A92.

Tomato prices (in dollars per pound)				
1.71	1.60	1.83	1.64	2.07
2.08	1.54	1.78	1.82	1.91
1.57	1.64	1.74	1.87	1.61
2.13	1.63	1.79	2.07	1.68
1.97	1.61	1.93	1.98	1.66
2.11	1.77	1.89	1.86	1.78

TABLE FOR EXERCISE 20

Using and Interpreting Concepts

Graphical Analysis   *In Exercises 13–16, give three observations that can be made from the graph.*

13.
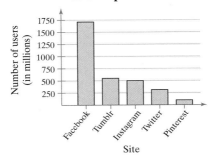
Monthly Active Users on 5 Social Networking Sites as of September 2016

(*Source: Statista*)

14.
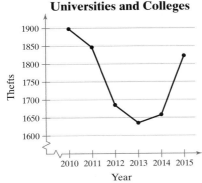
Motor Vehicle Thefts at U.S. Universities and Colleges

(*Source: Federal Bureau of Investigation*)

15.
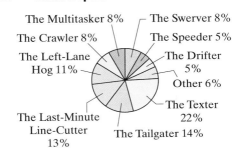
Least Popular American Drivers

The Multitasker 8%   The Swerver 8%
The Crawler 8%   The Speeder 5%
The Left-Lane Hog 11%   The Drifter 5%
Other 6%
The Texter 22%
The Last-Minute Line-Cutter 13%   The Tailgater 14%

(*Source: Expedia*)

16.
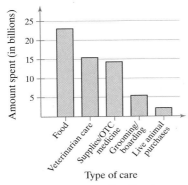
Amount Spent on Pet Care

(*Source: American Pet Products Association*)

Graphing Data Sets   *In Exercises 17–32, organize the data using the indicated type of graph. Describe any patterns.*

 17. **Exam Scores**   Use a stem-and-leaf plot to display the data, which represent the scores of a biology class on a midterm exam.

```
75  85  90  80  87  67  82  88  95  91  73  80
83  92  94  68  75  91  79  95  87  76  91  85
```

 18. **Nursing**   Use a stem-and-leaf plot to display the data, which represent the numbers of hours 24 nurses work per week.

```
40  40  35  48  38  40  36  50  32  36  40  35
30  24  40  36  40  36  40  39  33  40  32  38
```

 19. **Ice Thickness**   Use a stem-and-leaf plot to display the data, which represent the thicknesses (in centimeters) of ice measured at 20 different locations on a frozen lake.

```
5.8  6.4  6.9  7.2  5.1  4.9  4.3  5.8  7.0  6.8
8.1  7.5  7.2  6.9  5.8  7.2  8.0  7.0  6.9  5.9
```

 20. **Tomato Prices**   Use a stem-and-leaf plot to display the data shown in the table at the left, which represent the monthly average prices (in dollars per pound) charged by 30 retail outlets for fresh tomatoes.

21. Incomes (in millions) of Highest Paid Athletes

```
3 | 3 4 4 4        Key: 3|3 = 33
3 | 5 6 7 7 8 8 8
4 | 1 2 3 4 4
4 | 5 5 5 6
5 | 0 3 3 3
5 | 6 6
6 |
6 | 8
7 |
7 | 7
8 | 1
8 | 8
```

Sample answer: Most of the highest-paid athletes have an income of $33 million to $56 million.

22. See Selected Answers, page A92.

23. See Odd Answers, page A46.

24.

Housefly Life Spans

Life span (in days)

Sample answer: The life span of a housefly tends to be from 6 to 14 days.

25. See Odd Answers, page A46.

26. See Selected Answers, page A92.

27. See Odd Answers, page A46.

28. See Selected Answers, page A92.

29. See Odd Answers, page A46.

Hours	Hourly wage
33	17.16
37	14.98
34	15.79
40	16.71
35	16.80
33	16.51
40	18.65
33	17.05
28	15.54
45	15.33
37	16.57
28	15.17

TABLE FOR EXERCISE 29

21. Highest-Paid Athletes Use a stem-and-leaf plot that has two rows for each stem to display the data, which represent the incomes (in millions) of the top 30 highest-paid athletes. *(Source: Forbes Media LLC)*

37	36	44	50	35	56	56	68	45	33	46	37	77	45	34
44	43	81	53	38	53	38	34	38	88	34	53	41	42	45

22. Electoral Votes Use a stem-and-leaf plot that has two rows for each stem to display the data, which represent the numbers of electoral votes for each of the 50 states. *(Source: U.S. Census Bureau)*

9	3	11	6	55	9	7	3	29	16
4	4	20	11	6	6	8	8	4	10
11	16	10	6	10	3	5	6	4	14
5	29	15	3	18	7	7	20	4	9
3	11	38	6	3	13	12	5	10	3

23. Systolic Blood Pressures Use a dot plot to display the data, which represent the systolic blood pressures (in millimeters of mercury) of 24 patients at a doctor's office.

120	135	140	145	130	150	120	170	145	125	130	110
160	180	200	150	200	135	140	120	130	170	165	140

24. Life Spans of Houseflies Use a dot plot to display the data, which represent the life spans (in days) of 30 houseflies.

9	9	4	11	10	5	13	9	7	11	6	8	14	10	6
10	10	7	14	11	7	8	6	13	10	14	14	8	13	10

25. Student Loans Use a pie chart to display the data, which represent the numbers of student loan borrowers (in millions) by balance owed in the fourth quarter of 2015. *(Source: Federal Reserve Bank of New York)*

$1 to $10,000	16.7	$10,001 to $25,000	12.4
$25,001 to $50,000	8.3	$50,001+	6.7

26. New York City Marathon Use a pie chart to display the data, which represent the number of men's New York City Marathon winners from each country through 2016. *(Source: New York Road Runners)*

United States	15	Tanzania	1	Great Britain	1
Italy	4	Kenya	12	Brazil	2
Ethiopia	2	Mexico	4	New Zealand	1
South Africa	2	Morocco	1	Eritrea	1

27. Olympics The medal counts for five countries at the 2016 Summer Olympics include Germany (42 medals), Great Britain (67 medals), the United States (121 medals), Russia (56 medals), and China (70 medals). Use a Pareto chart to display the data. *(Source: International Olympic Committee)*

28. Vehicle Costs The average owning and operating costs for four types of vehicles in the United States in 2016 include small sedans ($6579), medium sedans ($8604), SUVs ($10,255), and minivans ($9262). Use a Pareto chart to display the data. *(Source: American Automobile Association)*

29. Hourly Wages Use a scatter plot to display the data shown in the table at the left. The data represent the numbers of hours worked and the hourly wages (in dollars) of 12 production workers.

Number of students per teacher	Average teacher's salary
16.6	66.4
22.8	45.4
24.3	72.5
16.2	50.9
15.6	55.2
20.6	56.7
13.2	77.0
22.2	59.8
23.0	46.0
14.1	45.4

TABLE FOR EXERCISE 30

30. **Salaries**   Use a scatter plot to display the data shown in the table at the left. The data represent the numbers of students per teacher and the average teacher salaries (in thousands of dollars) of 10 school districts.

31. Engineering Degrees   Use a time series chart to display the data shown in the table. The data represent the numbers of bachelor's degrees in engineering (in thousands) conferred in the U.S. *(Source: American Society for Engineering Education)*

Year	2008	2009	2010	2011	2012	2013	2014	2015
Degrees	74.2	74.4	78.3	83.0	88.2	93.4	99.2	106.7

32. Construction   Use a time series chart to display the data shown in the table. The data represent the percentages of the U.S. gross domestic product (GDP) that come from the construction sector. *(Source: U.S. Bureau of Economic Analysis)*

Year	2004	2005	2006	2007	2008	2009
Percent	4.8%	5.0%	5.0%	4.9%	4.4%	4.0%

Year	2010	2011	2012	2013	2014	2015
Percent	3.6%	3.5%	3.6%	3.7%	3.9%	4.1%

30.

Sample answer: It appears that there is no relation between a teacher's average salary and the number of students per teacher.

31.

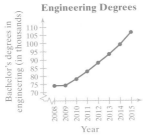

Sample answer: The number of bachelor's degrees in engineering conferred in the U.S. has increased from 2008 to 2015.

32. See Selected Answers, page A93.

33. See Odd Answers, page A46.

34 See Selected Answers, page A93.

35. See Odd Answers, page A47.

36. See Selected Answers, page A93.

33. Basketball   Display the data below in a stem-and-leaf plot. Describe the differences in how the dot plot and the stem-and-leaf plot show patterns in the data.

Heights of Players on a College Basketball Team

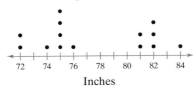

Inches

34. Phone Screen Sizes   Display the data below in a dot plot. Describe the differences in how the stem-and-leaf plot and the dot plot show patterns in the data.

Phone Screen Sizes (in inches)

5	0 0 Key: 5|0 = 5.0
5	5 5 5 6 7 8 8 9
6	0 0 0 1 2 3 4 4
6	5 5 6 8 8 9
7	0
7	

35. Favorite Season   Display the data below in a Pareto chart. Describe the differences in how the pie chart and the Pareto chart show patterns in the data. *(Source: Ipsos Public Affairs)*

Favorite Season of U.S. Adults Ages 18 and Older

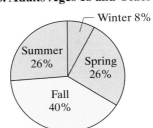

36. Favorite Day of the Week   Display the data below in a pie chart. Describe the differences in how the Pareto chart and the pie chart show patterns in the data.

Favorite Day of the Week

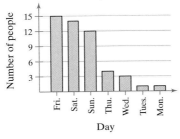

37. (a) The graph is misleading because the large gap from 0 to 90 makes it appear that the sales for the 3rd quarter are disproportionately larger than the other quarters.

  (b) See Odd Answers, page A47.

38. (a) The graph is misleading because the vertical axis has no break. The percent of middle schoolers that responded "yes" appears three times as large as either of the others when the difference is only 10%.

  (b) See Selected Answers, page A93.

39. See Odd Answers, page A47.

Law Firm A		Law Firm B
5 0	9	0 3
8 5 2 2 2	10	5 7
9 9 7 0 0	11	0 0 5
1 1	12	0 3 3 5
	13	2 2 5 9
	14	1 3 3 3 9
	15	5 5 5 6
	16	4 9 9
9 9 5 1 0	17	1 2 5
5 5 5 2 1	18	9
9 9 8 7 5	19	0
3	20	

Key: 5|19|0 = $195,000 for Law Firm A and $190,000 for Law Firm B

FIGURE FOR EXERCISE 41

3:00 P.M. Class				8:00 P.M. Class			
40	60	73	77	19	18	20	29
51	68	68	35	39	43	71	56
68	53	64	75	44	44	18	19
76	69	59	55	19	18	18	20
38	57	68	84	25	29	25	22
75	62	73	75	31	24	24	23
85	77			19	19	18	28
				20	31		

TABLE FOR EXERCISE 42

40. See Selected Answers, page A93.

41. See Odd Answers, page A47.

42. See Selected Answers, page A93.

Extending Concepts

A Misleading Graph?   *A misleading graph is not drawn appropriately, which can misrepresent data and lead to false conclusions. In Exercises 37–40, (a) explain why the graph is misleading, and (b) redraw the graph so that it is not misleading.*

37.

Sales for Company A

38.

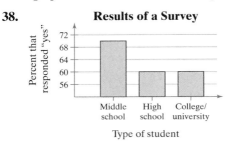

Results of a Survey

39.

Sales for Company B

40.

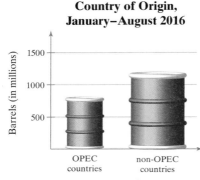

U.S. Crude Oil Imports by Country of Origin, January–August 2016

(Source: U.S. Energy Information Administration)

41. Law Firm Salaries   A **back-to-back stem-and-leaf plot** compares two data sets by using the same stems for each data set. Leaves for the first data set are on one side while leaves for the second data set are on the other side. The back-to-back stem-and-leaf plot at the left shows the salaries (in thousands of dollars) of all lawyers at two small law firms.

  (a) What are the lowest and highest salaries at Law Firm A? at Law Firm B? How many lawyers are in each firm?

  (b) Compare the distribution of salaries at each law firm. What do you notice?

42. Yoga Classes   The data sets at the left show the ages of all participants in two yoga classes.

  (a) Make a back-to-back stem-and-leaf plot as described in Exercise 41 to display the data.

  (b) What are the lowest and highest ages of participants in the 3:00 P.M. class? in the 8:00 P.M. class? How many participants are in each class?

  (c) Compare the distribution of ages in each class. What observation(s) can you make?

43. Choosing an Appropriate Display   Use technology to create (a) a stem-and-leaf plot, (b) a dot plot, (c) a pie chart, (d) a frequency histogram, and (e) an ogive for the data. Which graph displays the data best? Explain.

64	46	40	55	70	31	47	44	55	63
49	49	26	72	64	55	44	71	45	72

2.3   Measures of Central Tendency

What You Should Learn

► How to find the mean, median, and mode of a population and of a sample

► How to find a weighted mean of a data set, and how to estimate the sample mean of grouped data

► How to describe the shape of a distribution as symmetric, uniform, or skewed, and how to compare the mean and median for each

Mean, Median, and Mode ■ Weighted Mean and Mean of Grouped Data ■ The Shapes of Distributions

Mean, Median, and Mode

In Sections 2.1 and 2.2, you learned about the graphical representations of quantitative data. In Sections 2.3 and 2.4, you will learn how to supplement graphical representations with numerical statistics that describe the center and variability of a data set.

A **measure of central tendency** is a value that represents a typical, or central, entry of a data set. The three most commonly used measures of central tendency are the **mean,** the **median,** and the **mode.**

DEFINITION

The **mean** of a data set is the sum of the data entries divided by the number of entries. To find the mean of a data set, use one of these formulas.

$$\text{Population Mean: } \mu = \frac{\Sigma x}{N}$$

$$\text{Sample Mean: } \bar{x} = \frac{\Sigma x}{n}$$

The lowercase Greek letter μ (pronounced mu) represents the population mean and $\bar{x}$ (read as "x bar") represents the sample mean. Note that N represents the number of entries in a *population* and n represents the number of entries in a *sample.* Recall that the uppercase Greek letter sigma (Σ) indicates a summation of values.

EXAMPLE 1

Finding a Sample Mean

The weights (in pounds) for a sample of adults before starting a weight-loss study are listed. What is the mean weight of the adults?

274   235   223   268   290   285   235

SOLUTION   The sum of the weights is

$\Sigma x = 274 + 235 + 223 + 268 + 290 + 285 + 235 = 1810.$

There are 7 adults in the sample, so $n = 7$. To find the mean weight, divide the sum of the weights by the number of adults in the sample.

$$\bar{x} = \frac{\Sigma x}{n} = \frac{1810}{7} \approx 258.6.$$

Round the last calculation to one more decimal place than the original data.

So, the mean weight of the adults is about 258.6 pounds.

TRY IT YOURSELF 1

Find the mean of the points scored by the 51 winning teams listed on page 39.

Answer: Page A33

Study Tip

Notice that the mean in Example 1 has one more decimal place than the original set of data entries. When the mean needs to be rounded, this *round-off rule* will be used in the text. Another important *round-off rule* is that rounding should not be done until the last calculation.

DEFINITION

The **median** of a data set is the value that lies in the middle of the data when the data set is ordered. The median measures the center of an ordered data set by dividing it into two equal parts. When the data set has an odd number of entries, the median is the middle data entry. When the data set has an even number of entries, the median is the mean of the two middle data entries.

Tech Tip

You can use technology such as Minitab, Excel, StatCrunch, or the TI-84 Plus to find the mean and median of a data set. For instance, to find the mean and median of the weights listed in Example 1 on a TI-84 Plus, enter the data in L1. Next, press 2nd LIST and from the MATH menu choose *mean*. Then press 2nd LIST and from the MATH menu choose *median*.

TI-84 PLUS

```
mean(L1)
        258.5714286
median(L1)
             268
```

EXAMPLE 2

Finding the Median

Find the median of the weights listed in Example 1.

SOLUTION To find the median weight, first order the data.

223   235   235   268   274   285   290

Because there are seven entries (an odd number), the median is the middle, or fourth, entry. So, the median weight is 268 pounds.

TRY IT YOURSELF 2

Find the median of the points scored by the 51 winning teams listed on page 39.

Answer: Page A33

In a data set, the number of data entries above the median is the same as the number below the median. For instance, in Example 2, three of the weights are below 268 pounds and three are above 268 pounds.

EXAMPLE 3

Finding the Median

In Example 2, the adult weighing 285 pounds decides to not participate in the study. What is the median weight of the remaining adults?

SOLUTION The remaining weights, in order, are

223   235   235   268   274   290.

Because there are six entries (an even number), the median is the mean of the two middle entries.

$$\text{Median} = \frac{235 + 268}{2} = 251.5$$

So, the median weight of the remaining adults is 251.5 pounds. You can check your answer using technology, as shown below using Excel.

EXCEL

	A	B	C
1	MEDIAN(223,235,235,268,274,290)		
2			251.5

TRY IT YOURSELF 3

The points scored by the winning teams in the Super Bowls for the National Football League's 2001 through 2016 seasons are listed. Find the median.

20   48   32   24   21   29   17   27

31   31   21   34   43   28   24   34

Answer: Page A33

DEFINITION

The **mode** of a data set is the data entry that occurs with the greatest frequency. A data set can have one mode, more than one mode, or no mode. When no entry is repeated, the data set has no mode. When two entries occur with the same greatest frequency, each entry is a mode and the data set is called **bimodal.**

EXAMPLE 4

Finding the Mode

Find the mode of the weights listed in Example 1.

SOLUTION

To find the mode, first order the data.

223   235   235   268   274   285   290

From the ordered data, you can see that the entry 235 occurs twice, whereas the other data entries occur only once. So, the mode of the weights is 235 pounds.

TRY IT YOURSELF 4

Find the mode of the points scored by the 51 winning teams listed on page 39.

Answer: Page A33

EXAMPLE 5

Finding the Mode

At a political debate, a sample of audience members were asked to name the political party to which they belonged. Their responses are shown in the table. What is the mode of the responses?

Political party	Frequency, f
Democrat	46
Republican	34
Independent	39
Other/don't know	5

SOLUTION

The response occurring with the greatest frequency is Democrat. So, the mode is Democrat.

Interpretation   In this sample, there were more Democrats than people of any other single affiliation.

TRY IT YOURSELF 5

In a survey, 1534 adults were asked, "How much do you, personally, care about the issue of global climate change?" Of those surveyed, 550 said "a great deal," 578 said "some," 274 said "not too much," 119 said "not at all," and 13 did not provide an answer. What is the mode of the responses? *(Adapted from Pew Research Center)*

Answer: Page A33

The mode is the only measure of central tendency that can be used to describe data at the nominal level of measurement. But when working with quantitative data, the mode is rarely used.

Although the mean, the median, and the mode each describe a typical entry of a data set, there are advantages and disadvantages of using each. The mean is a reliable measure because it takes into account every entry of a data set. The mean can be greatly affected, however, when the data set contains **outliers.**

Ages in a class						
20	20	20	20	20	20	21
21	21	21	22	22	22	23
23	23	23	24	24	65	

DEFINITION

An **outlier** is a data entry that is far removed from the other entries in the data set. (You will learn a formal way for determining an outlier in Section 2.5.)

While some outliers are valid data, other outliers may occur due to data-recording errors. A data set can have one or more outliers, causing **gaps** in a distribution. Conclusions that are drawn from a data set that contains outliers may be flawed.

EXAMPLE 6

Comparing the Mean, the Median, and the Mode

The table at the left shows the sample ages of students in a class. Find the mean, median, and mode of the ages. Are there any outliers? Which measure of central tendency best describes a typical entry of this data set?

SOLUTION

From the histogram below, it appears that the data entry 65 is an outlier because it is far removed from the other ages in the class.

Mean: $\bar{x} = \dfrac{\Sigma x}{n} = \dfrac{475}{20} \approx 23.8$ years

Median: Median $= \dfrac{21 + 22}{2} = 21.5$ years

Mode: The entry occurring with the greatest frequency is 20 years.

Interpretation   The mean takes every entry into account but is influenced by the outlier of 65. The median also takes every entry into account, and it is not affected by the outlier. In this case the mode exists, but it does not appear to represent a typical entry. Sometimes a graphical comparison can help you decide which measure of central tendency best represents a data set. The histogram shows the distribution of the data and the locations of the mean, the median, and the mode. In this case, it appears that the median best describes the data set.

Ages of Students in a Class

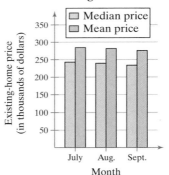

TRY IT YOURSELF 6

Remove the data entry 65 from the data set in Example 6. Then rework the example. How does the absence of this outlier change each of the measures?

Answer: Page A33

Weighted Mean and Mean of Grouped Data

Sometimes data sets contain entries that have a greater effect on the mean than do other entries. To find the mean of such a data set, you must find the **weighted mean.**

DEFINITION

A **weighted mean** is the mean of a data set whose entries have varying weights. The weighted mean is given by

$$\bar{x} = \frac{\Sigma xw}{\Sigma w} \qquad \frac{\text{Sum of the products of the entries and the weights}}{\text{Sum of the weights}}$$

where w is the weight of each entry x.

Tech Tip

You can use technology such as Minitab, Excel, StatCrunch, or the TI-84 Plus to find the weighted mean. For instance, to find the weighted mean in Example 7 on a TI-84 Plus, enter the points in L1 and the credit hours in L2. Then, use the *1-Var Stats* feature with L1 as the list and L2 as the frequency list to calculate the mean (and other statistics), as shown below.

TI-84 PLUS

```
            1-Var Stats
x̄=2.5  ◄────── Mean
Σx=40
Σx²=112
Sx=.894427191
σx=.8660254038
↓n=16
```

EXAMPLE 7

Finding a Weighted Mean

Your grades from last semester are in the table. The grading system assigns points as follows: A = 4, B = 3, C = 2, D = 1, F = 0. Determine your grade point average (weighted mean).

Final Grade	Credit Hours
C	3
C	4
D	1
A	3
C	2
B	3

SOLUTION

Let x be the points assigned to the letter grade and w be the credit hours. You can organize the points and hours in a table.

Points, x	Credit hours, w	xw
2	3	6
2	4	8
1	1	1
4	3	12
2	2	4
3	3	9
	$\Sigma w = 16$	$\Sigma(x \cdot w) = 40$

$$\bar{x} = \frac{\Sigma xw}{\Sigma w} = \frac{40}{16} = 2.5$$

Last semester, your grade point average was 2.5.

TRY IT YOURSELF 7

In Example 7, your grade in the two-credit course is changed to a B. What is your new weighted mean?

Answer: Page A33

For data presented in a frequency distribution, you can estimate the mean as shown in the next definition.

Study Tip

For a frequency distribution that represents a population, the mean of the frequency distribution is estimated by

$$\mu = \frac{\Sigma xf}{N}$$

where $N = \Sigma f$.

DEFINITION

The **mean of a frequency distribution** for a sample is estimated by

$$\bar{x} = \frac{\Sigma xf}{n} \qquad \text{Note that } n = \Sigma f.$$

where x and f are the midpoint and frequency of each class, respectively.

GUIDELINES

Finding the Mean of a Frequency Distribution

In Words	In Symbols
1. Find the midpoint of each class.	$x = \dfrac{(\text{Lower limit}) + (\text{Upper limit})}{2}$
2. Find the sum of the products of the midpoints and the frequencies.	Σxf
3. Find the sum of the frequencies.	$n = \Sigma f$
4. Find the mean of the frequency distribution.	$\bar{x} = \dfrac{\Sigma xf}{n}$

EXAMPLE 8

Finding the Mean of a Frequency Distribution

The frequency distribution at the left shows the out-of-pocket prescription medicine expenses (in dollars) for 30 U.S. adults in a recent year. Use the frequency distribution to estimate the mean expense. Using the sample mean formula from page 67 with the original data set (see Example 1 in Section 2.1), the mean expense is $285.50. Compare this with the estimated mean.

Class midpoint, x	Frequency, f	xf
172.5	3	517.5
208.5	2	417.0
244.5	5	1222.5
280.5	6	1683.0
316.5	7	2215.5
352.5	4	1410.0
388.5	3	1165.5
	$n = 30$	$\Sigma = 8631$

SOLUTION

$$\bar{x} = \frac{\Sigma xf}{n}$$

$$= \frac{8631}{30}$$

$$= 287.7$$

Interpretation The mean expense is $287.70. This value is an estimate because it is based on class midpoints instead of the original data set. Although it is not substantially different, the mean of $285.50 found using the original data set is a more accurate result.

TRY IT YOURSELF 8

Use a frequency distribution to estimate the mean of the points scored by the 51 winning teams listed on page 39. (See Try It Yourself 2 on page 43.) Using the population mean formula from page 67 with the original data set, the mean is about 30.2 points. Compare this with the estimated mean.

Answer: Page A33

The Shapes of Distributions

A graph reveals several characteristics of a frequency distribution. One such characteristic is the shape of the distribution.

2.3 To explore this topic further, see **Activity 2.3** on page 81.

> **DEFINITION**
>
> A frequency distribution is **symmetric** when a vertical line can be drawn through the middle of a graph of the distribution and the resulting halves are approximately mirror images.
>
> A frequency distribution is **uniform** (or **rectangular**) when all entries, or classes, in the distribution have equal or approximately equal frequencies. A uniform distribution is also symmetric.
>
> A frequency distribution is skewed when the "tail" of the graph elongates more to one side than to the other. A distribution is **skewed left (negatively skewed)** when its tail extends to the left. A distribution is **skewed right (positively skewed)** when its tail extends to the right.

When a distribution is symmetric and unimodal, the mean, median, and mode are equal. When a distribution is skewed left, the mean is less than the median and the median is usually less than the mode. When a distribution is skewed right, the mean is greater than the median and the median is usually greater than the mode. Examples of these commonly occurring distributions are shown.

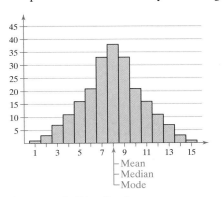

Symmetric Distribution

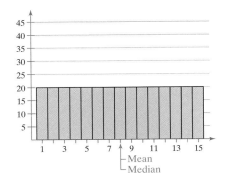

Uniform Distribution

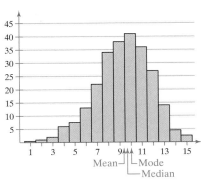

Skewed Left Distribution

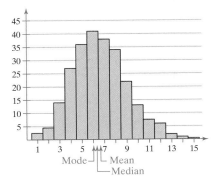

Skewed Right Distribution

The mean will always fall in the direction in which the distribution is skewed. For instance, when a distribution is skewed left, the mean is to the left of the median.

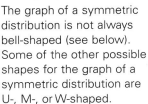

Study Tip

The graph of a symmetric distribution is not always bell-shaped (see below). Some of the other possible shapes for the graph of a symmetric distribution are U-, M-, or W-shaped.

Study Tip

Be aware that there are many different shapes of distributions. In some cases, the shape cannot be classified as symmetric, uniform, or skewed. A distribution can have several gaps caused by outliers or *clusters* of data. Clusters may occur when several types of data entries are used in a data set. For instance, a data set of gas mileages for trucks (which get low gas mileage) and hybrid cars (which get high gas mileage) would have two clusters.

2.3 EXERCISES

For Extra Help: MyLab Statistics

1. True
2. False. Every quantitative data set has a median.
3. True   4. True
5. *Sample answer:* 1, 2, 2, 2, 3
6. *Sample answer:* 2, 4, 5, 5, 6, 8
7. *Sample answer:* 2, 5, 7, 9, 35
8. *Sample answer:* 1, 2, 3, 3, 3, 4, 5
9. The shape of the distribution is skewed right because the bars have a "tail" to the right.
10. The shape of the distribution is symmetric because a vertical line can be drawn down the middle, creating two halves that look approximately the same.
11. The shape of the distribution is uniform because the bars are approximately the same height.
12. The shape of the distribution is skewed left because the bars have a "tail" to the left.
13. (11), because the distribution of values ranges from 1 to 12 and has (approximately) equal frequencies.
14. (9), because the distribution has values in the thousands and is skewed right due to the few vehicles that have much higher mileages than the majority of the vehicles.
15. (12), because the distribution has a maximum value of 90 and is skewed left due to a few students scoring much lower than the majority of the students.
16. (10), because the distribution is approximately symmetric and the weights range from 80 to 160 pounds.

Building Basic Skills and Vocabulary

True or False? *In Exercises 1–4, determine whether the statement is true or false. If it is false, rewrite it as a true statement.*

1. The mean is the measure of central tendency most likely to be affected by an outlier.

2. Some quantitative data sets do not have medians.

3. A data set can have the same mean, median, and mode.

4. When each data class has the same frequency, the distribution is symmetric.

Constructing Data Sets *In Exercises 5–8, construct the described data set. The entries in the data set cannot all be the same.*

5. Median and mode are the same.

6. Mean and mode are the same.

7. Mean is *not* representative of a typical number in the data set.

8. Mean, median, and mode are the same.

Graphical Analysis *In Exercises 9–12, determine whether the approximate shape of the distribution in the histogram is symmetric, uniform, skewed left, skewed right, or none of these. Justify your answer.*

9.
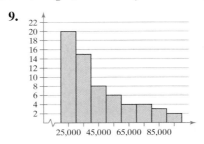

10.
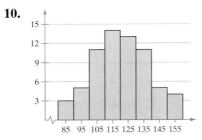

11.
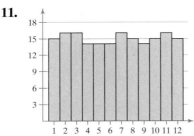

12.
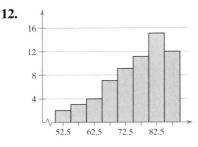

Matching *In Exercises 13–16, match the distribution with one of the graphs in Exercises 9–12. Justify your decision.*

13. The frequency distribution of 180 rolls of a dodecagon (a 12-sided die)

14. The frequency distribution of mileages of service vehicles at a business where a few vehicles have much higher mileages than the majority of vehicles

15. The frequency distribution of scores on a 90-point test where a few students scored much lower than the majority of students

16. The frequency distribution of weights for a sample of seventh-grade boys

17. $\bar{x} \approx 14.9$; median = 15; mode = 16

18. $\bar{x} \approx 172.1$; median = 172; mode = 169; The mode does not represent the center of the data because 169 is the smallest number in the data set.

19. $\bar{x} \approx 902.3$; median = 788; mode = none; The mode cannot be found because no data entry is repeated. The mean does not represent the center of data because it is influenced by the outliers of 1242 and 1462.

20. $\bar{x} = 52.3$; median = 59.5; mode = 38, 63

21. $\bar{x} \approx 49.8$; median = 50.5; mode = 51

22. $\bar{x} \approx 200.4$; median = 186; mode = none; The mode cannot be found because no data entry is repeated.

23. $\bar{x} \approx 7.4$; median = 6; mode = 6

24. $\bar{x} \approx 59.1$; median = 49; mode = 80, 125; The modes do not represent the center of the data set because they are large values compared to the rest of the data.

25. $\bar{x} \approx 14.3$; median = 9; mode = none; The mode cannot be found because no data entry is repeated. The mean does not represent the center of the data set because it is influenced by the outlier of 42.

26. $\bar{x} \approx 11.8$; median = 12; mode = 10

Using and Interpreting Concepts

Finding and Discussing the Mean, Median, and Mode *In Exercises 17–34, find the mean, the median, and the mode of the data, if possible. If any measure cannot be found or does not represent the center of the data, explain why.*

17. College Credits The numbers of credits being taken by a sample of 14 full-time college students for a semester

12 14 16 15 13 14 15
18 16 16 12 16 15 17

18. LSAT Scores The Law School Admission Test (LSAT) scores for a sample of seven students accepted into a law school

174 172 169 176 169 170 175

19. Word Counts The lengths (in words) of seven articles from *The New York Post* *(Source: The New York Post)*

650 1242 788 1462 662 709 803

20. Representatives The ages of the new members of the House of Representatives in the 115th Congress from Florida as of January 3, 2017 *(Source: C-SPAN)*

60 34 38 59 68 63 63 36 64 38

21. Tuition The 2016–2017 tuition and fees (in thousands of dollars) for the top 14 universities in the U.S. *(Source: U.S. News & World Report)*

45 47 52 49 55 48 48
51 51 50 51 48 51 51

22. Cholesterol The cholesterol levels of a sample of 10 female employees

154 240 171 188 235 203 184 173 181 275

23. Ports of Entry The maximum numbers of passenger vehicle lanes at 16 Canadian border ports of entry *(Source: U.S. Customs and Border Protection)*

8 6 10 3 6 11 17 2
2 6 1 10 3 19 10 5

24. Power Failures The durations (in minutes) of power failures at a residence in the last 10 years

18 26 45 75 125 80 33
40 44 49 89 80 96 125
12 61 31 63 103 28 19

25. Treatment of Depression The numbers of patients who responded to various combinations of electroconvulsive therapy, medication, and cognitive-behavioral therapy to treat acute depression over different time periods *(Source: Adapted from Bipolar Network News)*

42 15 8 9 13 6 7

26. Number One Songs The numbers of weeks the 33 longest leading Hot 100 songs remained at number 1 as of November 19, 2016 *(Source: Billboard)*

10 10 13 14 11 14 16 14 11 11 14
13 12 10 11 10 10 12 12 14 10 10
10 12 14 10 12 10 14 12 10 10 12

How purchases are made	Frequency, f
Research online and in store, buy in store	1173
Search and buy online	2238
Search and buy in store	1066
Research online and in store, buy online	853

TABLE FOR EXERCISE 27

Small Businesses

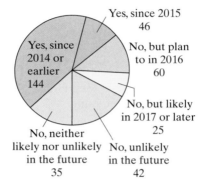

FIGURE FOR EXERCISE 30

27. $\overline{x}$ is not possible; median is not possible; mode = "Search and buy online"; The mean and median cannot be found because the data are at the nominal level of measurement.

28. $\overline{x}$ is not possible; median is not possible; mode = "Mental health," "Education"; The mean and median cannot be found because the data are at the nominal level of measurement.

29. $\overline{x}$ is not possible; median is not possible; mode = "Junior"; The mean and median cannot be found because the data are at the nominal level of measurement.

30. $\overline{x}$ is not possible; median is not possible; mode = "Yes, since 2014 or earlier"; The mean and median cannot be found because the data are at the nominal level of measurement.

31. See Odd Answers, page A48.

32. See Selected Answers, page A93.

33. See Odd Answers, page A48.

34. See Selected Answers, page A93.

35. See Odd Answers, page A48.

36. See Selected Answers, page A93.

27. Online Shopping The responses of a sample of 5330 shoppers who were asked how their purchases are made are shown in the table at the left. *(Adapted from UPS)*

28. Criminal Justice The responses of a sample of 34 young adult United Kingdom males in custodial sentences who were asked what is affected by such sentences *(Adapted from User Voice)*

    Mental health: 8
    Trust: 3
    Education: 8
    Personal development: 5
    Family: 3
    Future opportunities: 3
    Other: 4

29. Class Level The class levels of 25 students in a physics course

    Freshman: 2   Junior: 10
    Sophomore: 5   Senior: 8

30. Small Business Websites The pie chart at the left shows the responses of a sample of 352 small-business owners who were asked whether their business has a website. *(Source: Clutch)*

31.
    Weights (in pounds) of Packages on a Delivery Truck

```
0 | 5 8                    Key: 3|0 = 30
1 | 0 1 3 6
2 | 1 3 3 3 6 7 7
3 | 0 1 2 4 4 4 5 7 8
4 | 3 4 5 6 9
5 | 2
```

32.
    Grade Point Averages of Students in a Class

```
0 | 8                     Key: 0|8 = 0.8
1 | 5 6 8
2 | 1 3 4 5
3 | 0 9
4 | 0 0
```

33.
    Times (in minutes) It Takes Employees to Drive to Work

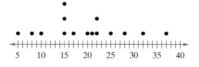

34.
    Prices (in dollars) of Flights from Chicago to Alanta

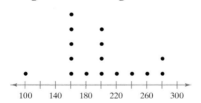

Graphical Analysis *In Exercises 35 and 36, identify any clusters, gaps, or outliers.*

35.
    Model Year 2017 Ethanol Flexible Fuel Vehicles

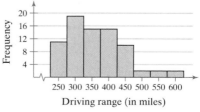

(Source: United States Environmental Protection Agency)

36.
    Model Year 2017 Hybrid Electric Cars

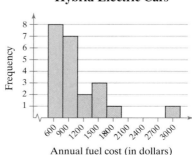

(Source: Based on United States Environmental Protection Agency)

37. Mode, because the data are at the nominal level of measurement.

38. Mean, because the distribution is symmetric and there are no outliers.

39. Mean, because the distribution is symmetric and there are no outliers.

40. Median, because there is an outlier.

41. 90.5   42. 93.5

43. $612.73   44. $449.21

In Exercises 37–40, without performing any calculations, determine which measure of central tendency best represents the graphed data. Explain your reasoning.

37.

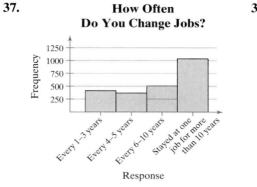

How Often Do You Change Jobs?

(*Source: Jobvite*)

38.

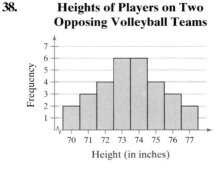

Heights of Players on Two Opposing Volleyball Teams

39.

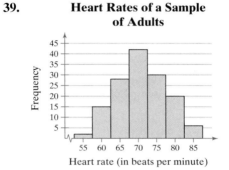

Heart Rates of a Sample of Adults

40.

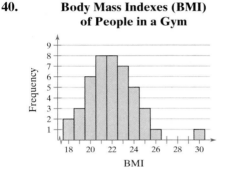

Body Mass Indexes (BMI) of People in a Gym

Finding a Weighted Mean   *In Exercises 41–46, find the weighted mean of the data.*

41. Final Grade   The scores and their percents of the final grade for a statistics student are shown below. What is the student's mean score?

	Score	Percent of final grade
Homework	85	5%
Quizzes	80	35%
Project/Speech	100	35%
Final exam	93	25%

42. Final Grade   The scores and their percents of the final grade for an archaeology student are shown below. What is the student's mean score?

	Score	Percent of final grade
Quizzes	100	20%
Midterm exam	89	30%
Student lecture	100	10%
Final exam	92	40%

43. Account Balance   For the month of April, a checking account has a balance of $523 for 24 days, $2415 for 2 days, and $250 for 4 days. What is the account's mean daily balance for April?

44. Credit Card Balance   For the month of October, a credit card has a balance of $115.63 for 12 days, $637.19 for 6 days, $1225.06 for 7 days, $0 for 2 days, and $34.88 for 4 days. What is the account's mean daily balance for October?

45. 84   **46.** 2.8

47. 87   **48.** 3.0

49. 36.2 miles per gallon

50. 29 miles per gallon

51. 42.3 years old

52. 74.5 thousand

53.

Class	Frequency, f	Midpoint
127–161	7	144
162–196	6	179
197–231	3	214
232–266	3	249
267–301	1	284

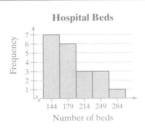

Positively skewed

45. Scores   The mean scores for students in a statistics course (by major) are shown below. What is the mean score for the class?

9 engineering majors: 85   5 math majors: 90   13 business majors: 81

46. Grades   A student receives the grades shown below, with an A worth 4 points, a B worth 3 points, a C worth 2 points, and a D worth 1 point. What is the student's grade point average?

A in 1 four-credit class   C in 1 three-credit class
B in 2 three-credit classes   D in 1 two-credit class

47. Final Grade   In Exercise 41, an error was made in grading your final exam. Instead of getting 93, you scored 85. What is your new weighted mean?

48. Grades   In Exercise 46, one of the student's B grades gets changed to an A. What is the student's new grade point average?

Finding the Mean of a Frequency Distribution   *In Exercises 49–52, approximate the mean of the frequency distribution.*

49. Fuel Economy   The gas mileages (in miles per gallon) for 30 small cars

Gas Mileage (in miles per gallon)	Frequency
29–33	11
34–38	12
39–43	2
44–48	5

50. Fuel Economy   The gas mileages (in miles per gallon) for 24 family sedans

Gas Mileage (in miles per gallon)	Frequency
22–27	16
28–33	2
34–39	2
40–45	4

51. Ages   The ages (in years) of the residents of a small town in 2016

Age (in years)	Frequency
0–9	78
10–19	97
20–29	54
30–39	63
40–49	69
50–59	86
60–69	73
70–79	53
80–89	43
90–99	15

52. Populations   The populations (in thousands) of the parishes of Louisiana in 2015 *(Source: U.S. Census Bureau)*

Population (in thousands)	Frequency
0–49	41
50–99	9
100–149	6
150–199	2
200–249	1
250–299	2
300–349	0
350–399	1
400–449	2

Identifying the Shape of a Distribution   *In Exercises 53–56, construct a frequency distribution and a frequency histogram for the data set using the indicated number of classes. Describe the shape of the histogram as symmetric, uniform, negatively skewed, positively skewed, or none of these.*

 53. Hospital Beds
Number of classes: 5
Data set: The number of beds in a sample of 20 hospitals

167   162   127   130   180   160   167   221   145   137
194   207   150   254   262   244   297   137   204   180

54. See Selected Answers, page A93.

55. See Odd Answers, page A48.

56. See Selected Answers, page A94.

57. (a) $\bar{x} \approx 1518.2$,
median = 1520.5

(b) $\bar{x} \approx 1521.2$,
median = 1522.5

(c) Mean

58. (a) $\bar{x} \approx 46.76$, median = 22.5

(b) $\bar{x} \approx 27.91$, median = 21.7;
Mean

(c) $\bar{x} = 44.68$, median = 21.7;
Mean

59. The data are skewed right.

A = mode, because it is the data entry that occurred most often.

B = median, because the median is to the left of the mean in a skewed right distribution.

C = mean, because the mean is to the right of the median in a skewed right distribution.

U.S. trade deficits (in billions of dollars)	
China: 367.2	Germany: 74.8
Japan: 68.9	Mexico: 60.7
Vietnam: 30.9	Ireland: 30.4
South Korea: 28.3	Italy: 28.0
India: 23.3	Malaysia: 21.7
France: 17.7	Thailand: 17.4
Canada: 15.5	Taiwan: 15.0
Indonesia: 12.5	Israel: 10.9
Russian Federation: 9.3	
Switzerland: 9.2	

TABLE FOR EXERCISE 58

60. The data are skewed left.

A = mean, because the mean is to the left of the median in a skewed left distribution.

B = median, because the median is to the right of the mean in a skewed left distribution.

C = mode, because it is the data entry that occurred most often.

54. Emergency Room
Number of classes: 6
Data set: The numbers of patients visiting an emergency room per day over a two-week period

256 317 237 182 382 106 162
112 162 264 104 194 236 227

55. Heights of Males
Number of classes: 5
Data set: The heights (to the nearest inch) of 30 males

67 76 69 68 72 68 65 63 75 69
66 72 67 66 69 73 64 62 71 73
68 72 71 65 69 66 74 72 68 69

56. Six-Sided Die
Number of classes: 6
Data set: The results of rolling a six-sided die 30 times

1 4 6 1 5 3 2 5 4 6
1 2 4 3 5 6 3 2 1 1
5 6 2 4 4 3 1 6 2 4

57. Protein Powder During a quality assurance check, the actual contents (in grams) of six containers of protein powder were recorded as 1525, 1526, 1502, 1516, 1529, and 1511.

(a) Find the mean and the median of the contents.

(b) The third value was incorrectly measured and is actually 1520. Find the mean and the median of the contents again.

(c) Which measure of central tendency, the mean or the median, was affected more by the data entry error?

58. U.S. Trade Deficits The table at the left shows the U.S. trade deficits (in billions of dollars) with 18 countries in 2015. *(Source: U.S. Department of Commerce)*

(a) Find the mean and the median of the trade deficits.

(b) Find the mean and the median without the Chinese trade deficit. Which measure of central tendency, the mean or the median, was affected more by the elimination of the Chinese trade deficit?

(c) The Austrian trade deficit was $7.3 billion. Find the mean and the median with the Austrian trade deficit added to the original data set. Which measure of central tendency was affected more?

Graphical Analysis *In Exercises 59 and 60, the letters A, B, and C are marked on the horizontal axis. Describe the shape of the data. Then determine which is the mean, which is the median, and which is the mode. Justify your answers.*

59. **Sick Days Used by Employees**

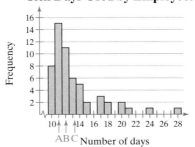

60. **Hourly Wages of Employees**

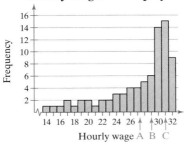

61. Increase one of the three-credit B classes to an A. The three-credit class is weighted more than the two-credit classes, so it will have a greater effect on the grade point average.

62. (a) $\bar{x} = 358$, median $= 375$

(b) $\bar{x} = 1074$, median $= 1125$

(c) The mean and median in part (b) are three times the mean and median in part (a).

(d) If you multiply the mean and median of the original data set by 36, you will get the mean and median of the data set in inches.

63. (a) Mean, because Car A has the highest mean of the three.

Car

	A	B	C
Run 1	28	31	29
Run 2	32	29	32
Run 3	28	31	28
Run 4	30	29	32
Run 5	34	31	30

TABLE FOR EXERCISE 63

(b) Median, because Car B has the highest median of the three.

(c) Mode, because Car C has the highest mode of the three.

64. Car A, because the midrange is the largest.

65. See Odd Answers, page A48.

Test scores

44	51	11	90	76	36	64	37
43	72	53	62	36	74	51	72
37	28	38	61	47	63	36	41
22	37	51	46	85	13		

TABLE FOR EXERCISE 65

66. (a) About 49.2

(b) $\bar{x} \approx 49.2$; median $= 46.5$; mode $= 36, 37, 51$; midrange $= 50.5$

(c) Using the trimmed mean eliminates potential outliers that could affect the mean of the entries.

Extending Concepts

61. Writing In an academic year, a student receives the grades shown below, with an A worth 4 points, a B worth 3 points, and a C worth 2 points.

A in 2 four-credit classes and 3 three-credit classes
B in 2 three-credit classes and 2 two-credit classes
C in 1 two-credit class

The student can increase one of the Bs or Cs by one letter grade. Which one should the student choose? Explain your reasoning.

62. Golf The distances (in yards) for nine holes of a golf course are listed.

336   393   408   522   147   504   177   375   360

(a) Find the mean and the median of the data.

(b) Convert the distances to feet. Then rework part (a).

(c) Compare the measures you found in part (b) with those found in part (a). What do you notice?

(d) Use your results from part (c) to explain how to quickly find the mean and the median of the original data set when the distances are converted to inches.

63. Data Analysis A consumer testing service obtained the gas mileages (in miles per gallon) shown in the table at the left in five test runs performed with three types of compact cars.

(a) The manufacturer of Car A wants to advertise that its car performed best in this test. Which measure of central tendency—mean, median, or mode—should be used for its claim? Explain your reasoning.

(b) The manufacturer of Car B wants to advertise that its car performed best in this test. Which measure of central tendency—mean, median, or mode—should be used for its claim? Explain your reasoning.

(c) The manufacturer of Car C wants to advertise that its car performed best in this test. Which measure of central tendency—mean, median, or mode—should be used for its claim? Explain your reasoning.

64. Midrange Another measure of central tendency, which is rarely used, is the **midrange.** It can be found by using the formula

$$\text{Midrange} = \frac{(\text{Maximum data entry}) + (\text{Minimum data entry})}{2}.$$

Which of the manufacturers in Exercise 63 would prefer to use the midrange statistic in their ads? Explain your reasoning.

65. Data Analysis Students in an experimental psychology class did research on depression as a sign of stress. A test was administered to a sample of 30 students. The scores are shown in the table at the left.

(a) Find the mean and the median of the data.

(b) Draw a stem-and-leaf plot for the data using one row per stem. Locate the mean and the median on the display.

(c) Describe the shape of the distribution.

66. Trimmed Mean To find the 10% **trimmed mean** of a data set, order the data, delete the lowest 10% of the entries and the highest 10% of the entries, and find the mean of the remaining entries.

(a) Find the 10% trimmed mean for the data in Exercise 65.

(b) Compare the four measures of central tendency, including the midrange.

(c) What is the benefit of using a trimmed mean versus using a mean found using all data entries? Explain your reasoning.

2.3 ACTIVITY

Mean Versus Median

APPLET

You can find the interactive applet for this activity within **MyLab Statistics** or at *www.pearsonhighered.com/ mathstatsresources.*

The *mean versus median* applet is designed to allow you to investigate interactively the mean and the median as measures of the center of a data set. Points can be added to the plot by clicking the mouse above the horizontal axis. The mean of the points is shown as a green arrow and the median is shown as a red arrow. When the two values are the same, a single yellow arrow is displayed. Numeric values for the mean and the median are shown above the plot. Points on the plot can be removed by clicking on the point and then dragging the point into the trash can. All of the points on the plot can be removed by simply clicking inside the trash can. The range of values for the horizontal axis can be specified by inputting lower and upper limits and then clicking UPDATE.

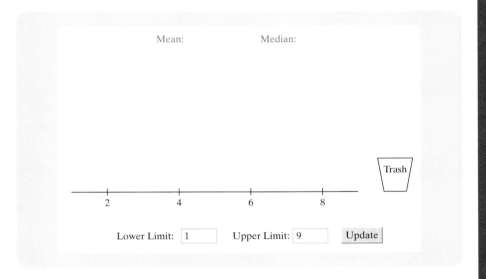

EXPLORE

Step 1 Specify a lower limit.
Step 2 Specify an upper limit.
Step 3 Add 15 points to the plot.
Step 4 Remove all of the points from the plot.

DRAW CONCLUSIONS

APPLET

1. Specify the lower limit to be 1 and the upper limit to be 50. Add at least 10 points that range from 20 to 40 so that the mean and the median are the same. What is the shape of the distribution? What happens at first to the mean and the median when you add a few points that are less than 10? What happens over time as you continue to add points that are less than 10?

2. Specify the lower limit to be 0 and the upper limit to be 0.75. Place 10 points on the plot. Then change the upper limit to 25. Add 10 more points that are greater than 20 to the plot. Can the mean be any one of the points that were plotted? Can the median be any one of the points that were plotted? Explain.

2.4 Measures of Variation

What You Should Learn

▶ How to find the range of a data set

▶ How to find the variance and standard deviation of a population and of a sample

▶ How to use the Empirical Rule and Chebychev's Theorem to interpret standard deviation

▶ How to estimate the sample standard deviation for grouped data

▶ How to use the coefficient of variation to compare variation in different data sets

Range ■ Variance and Standard Deviation ■ Interpreting Standard Deviation ■ Standard Deviation for Grouped Data ■ Coefficient of Variation

Range

In this section, you will learn different ways to measure the variation (or spread) of a data set. The simplest measure is the **range** of the set.

DEFINITION

The **range** of a data set is the difference between the maximum and minimum data entries in the set. To find the range, the data must be quantitative.

$$\text{Range} = (\text{Maximum data entry}) - (\text{Minimum data entry})$$

EXAMPLE 1

Finding the Range of a Data Set

Two corporations each hired 10 graduates. The starting salaries for each graduate are shown. Find the range of the starting salaries for Corporation A.

Starting Salaries for Corporation A (in thousands of dollars)

Salary	41	38	39	45	47	41	44	41	37	42

Starting Salaries for Corporation B (in thousands of dollars)

Salary	40	23	41	50	49	32	41	29	52	58

SOLUTION

Ordering the data helps to find the least and greatest salaries.

37   38   39   41   41   41   42   44   45   47

Minimum ⟶   ⟵ Maximum

$$\text{Range} = (\text{Maximum salary}) - (\text{Minimum salary})$$
$$= 47 - 37$$
$$= 10$$

So, the range of the starting salaries for Corporation A is 10, or $10,000.

TRY IT YOURSELF 1

Find the range of the starting salaries for Corporation B. Compare the result to the one in Example 1.   *Answer: Page A33*

Corporation A

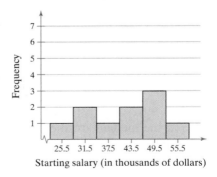

Corporation B

Both data sets in Example 1 have a mean of 41.5, or $41,500, a median of 41, or $41,000, and a mode of 41, or $41,000. And yet the two sets differ significantly. The difference is that the entries in the second set have greater variation. As you can see in the figures at the left, the starting salaries for Corporation B are more spread out than those for Corporation A.

Variance and Standard Deviation

As a measure of variation, the range has the advantage of being easy to compute. Its disadvantage, however, is that it uses only two entries from the data set. Two measures of variation that use all the entries in a data set are the *variance* and the *standard deviation*. Before you learn about these measures of variation, you need to know what is meant by the **deviation** of an entry in a data set.

Note to Instructor

Remind students of the reason for the difference between the symbols μ and $\bar{x}$.

> **DEFINITION**
>
> The **deviation** of an entry x in a population data set is the difference between the entry and the mean μ of the data set.
>
> Deviation of $x = x - \mu$

Consider the starting salaries for Corporation A in Example 1. The mean starting salary is $\mu = 415/10 = 41.5$, or \$41,500. The table at the left lists the deviation of each salary from the mean. For instance, the deviation of 41 is $41 - 41.5 = -0.5$. Notice that the sum of the deviations is 0. In fact, the sum of the deviations for *any* data set is 0. So, it does not make sense to find the average of the deviations. To overcome this problem, take the square of each deviation. The sum of the squares of the deviations, or **sum of squares,** is denoted by SS_x. In a population data set, the average of the squares of the deviations is the **population variance.**

Deviations of Starting Salaries for Corporation A

Salary (in 1000s of dollars) x	Deviation (in 1000s of dollars) $x - \mu$
41	−0.5
38	−3.5
39	−2.5
45	3.5
47	5.5
41	−0.5
44	2.5
41	−0.5
37	−4.5
42	0.5
$\Sigma x = 415$	$\Sigma(x - \mu) = 0$

The sum of the deviations is 0.

> **DEFINITION**
>
> The **population variance** of a population data set of N entries is
>
> $$\text{Population variance} = \sigma^2 = \frac{\Sigma(x - \mu)^2}{N}.$$
>
> The symbol σ is the lowercase Greek letter sigma.

As a measure of variation, one disadvantage with the variance is that its units are different from the data set. For instance, the variance for the starting salaries (in thousands of dollars) in Example 1 is measured in "square thousands of dollars." To overcome this problem, take the square root of the variance to get the **standard deviation.**

> **DEFINITION**
>
> The **population standard deviation** of a population data set of N entries is the square root of the population variance.
>
> $$\text{Population standard deviation} = \sigma = \sqrt{\sigma^2} = \sqrt{\frac{\Sigma(x - \mu)^2}{N}}$$

Here are some observations about the standard deviation.

- The standard deviation measures the variation of the data set about the mean and has the same units of measure as the data set.

- The standard deviation is always greater than or equal to 0. When $\sigma = 0$, the data set has no variation and all entries have the same value.

- As the entries get farther from the mean (that is, more spread out), the value of σ increases.

Note to Instructor

The formulas used here are derived from the definition of the population variance and standard deviation because we feel they are easier to remember than the alternative formula. If you prefer to use the alternative formula, we have included it on page 98 in Exercise 51.

To find the variance and standard deviation of a population data set, use these guidelines.

GUIDELINES

Finding the Population Variance and Standard Deviation

In Words	In Symbols
1. Find the mean of the population data set.	$\mu = \dfrac{\Sigma x}{N}$
2. Find the deviation of each entry.	$x - \mu$
3. Square each deviation.	$(x - \mu)^2$
4. Add to get the sum of squares.	$SS_x = \Sigma(x - \mu)^2$
5. Divide by N to get the population variance.	$\sigma^2 = \dfrac{\Sigma(x - \mu)^2}{N}$
6. Find the square root of the variance to get the population standard deviation.	$\sigma = \sqrt{\dfrac{\Sigma(x - \mu)^2}{N}}$

Sum of Squares of Starting Salaries for Corporation A

Salary x	Deviation $x - \mu$	Squares $(x - \mu)^2$
41	−0.5	0.25
38	−3.5	12.25
39	−2.5	6.25
45	3.5	12.25
47	5.5	30.25
41	−0.5	0.25
44	2.5	6.25
41	−0.5	0.25
37	−4.5	20.25
42	0.5	0.25
$\Sigma x = 415$		$SS_x = 88.5$

EXAMPLE 2

Finding the Population Variance and Standard Deviation

Find the population variance and standard deviation of the starting salaries for Corporation A listed in Example 1.

SOLUTION

For this data set, $N = 10$ and $\Sigma x = 415$. The mean is

$$\mu = \frac{415}{10} = 41.5. \qquad \text{Mean}$$

The table at the left summarizes the steps used to find SS_x. Because

$$SS_x = 88.5 \qquad \text{Sum of squares}$$

you can find the variance and standard deviation as shown.

$$\sigma^2 = \frac{88.5}{10} \approx 8.9 \qquad \text{Round to one more decimal place than the original data.}$$

$$\sigma = \sqrt{\frac{88.5}{10}} \approx 3.0 \qquad \text{Round to one more decimal place than the original data.}$$

So, the population variance is about 8.9, and the population standard deviation is about 3.0, or $3000.

TRY IT YOURSELF 2

Find the population variance and standard deviation of the starting salaries for Corporation B in Example 1. *Answer: Page A33*

Study Tip

Notice that the variance and standard deviation in Example 2 have one more decimal place than the original set of data entries. This is the same *round-off rule* that was used to calculate the mean.

The formulas shown on the next page for the sample variance s^2 and sample standard deviation s of a sample data set differ slightly from those of a population. For instance, to find s, the formula uses $\bar{x}$. Also, SS_x is divided by $n - 1$. Why divide by one less than the number of entries? In many cases, a statistic is calculated to estimate the corresponding parameter, such as using $\bar{x}$ to estimate μ. Statistical theory has shown that the best estimates of σ^2 and σ are obtained when dividing SS_x by $n - 1$ in the formulas for s^2 and s.

The **sample variance** and **sample standard deviation** of a sample data set of n entries are listed below.

$$\text{Sample variance} = s^2 = \frac{\Sigma(x - \bar{x})^2}{n - 1}$$

$$\text{Sample standard deviation} = s = \sqrt{s^2} = \sqrt{\frac{\Sigma(x - \bar{x})^2}{n - 1}}$$

Symbols in Variance and Standard Deviation Formulas

	Population	Sample
Variance	σ^2	s^2
Standard deviation	σ	s
Mean	μ	$\bar{x}$
Number of entries	N	n
Deviation	$x - \mu$	$x - \bar{x}$
Sum of squares	$\Sigma(x - \mu)^2$	$\Sigma(x - \bar{x})^2$

GUIDELINES

Finding the Sample Variance and Standard Deviation

In Words / In Symbols

1. Find the mean of the sample data set. $\bar{x} = \frac{\Sigma x}{N}$
2. Find the deviation of each entry. $x - \bar{x}$
3. Square each deviation. $(x - \bar{x})^2$
4. Add to get the sum of squares. $SS_x = \Sigma(x - \bar{x})^2$
5. Divide by $n - 1$ to get the sample variance. $s^2 = \frac{\Sigma(x - \bar{x})^2}{n - 1}$
6. Find the square root of the variance to get the sample standard deviation. $s = \sqrt{\frac{\Sigma(x - \bar{x})^2}{n - 1}}$

EXAMPLE 3

See Minitab and TI-84 Plus steps on pages 124 and 125.

Finding the Sample Variance and Standard Deviation

In a study of high school football players that suffered concussions, researchers placed the players in two groups. Players that recovered from their concussions in 14 days or less were placed in Group 1. Those that took more than 14 days were placed in Group 2. The recovery times (in days) for Group 1 are listed below. Find the sample variance and standard deviation of the recovery times. *(Adapted from The American Journal of Sports Medicine)*

4 7 6 7 9 5 8 10 9 8 7 10

SOLUTION

For this data set, $n = 12$ and $\Sigma x = 90$. The mean is $\bar{x} = 90/12 = 7.5$. To calculate s^2 and s, note that $n - 1 = 12 - 1 = 11$.

Time x	Deviation $x - \bar{x}$	Squares $(x - \bar{x})^2$
4	−3.5	12.25
7	−0.5	0.25
6	−1.5	2.25
7	−0.5	0.25
9	1.5	2.25
5	−2.5	6.25
8	0.5	0.25
10	2.5	6.25
9	1.5	2.25
8	0.5	0.25
7	−0.5	0.25
10	2.5	6.25
$\Sigma x = 90$		$SS_x = 39$

$SS_x = 39$ — Sum of squares (see table at left)

$s^2 = \frac{39}{11} \approx 3.5$ — Sample variance (divide SS_x by $n - 1$)

$s = \sqrt{\frac{39}{11}} \approx 1.9$ — Sample standard deviation

So, the sample variance is about 3.5, and the sample standard deviation is about 1.9 days.

TRY IT YOURSELF 3

Refer to the study in Example 3. The recovery times (in days) for Group 2 are listed below. Find the sample variance and standard deviation of the recovery times.

43 57 18 45 47 33 49 24

Answer: Page A33

EXAMPLE 4

Using Technology to Find the Standard Deviation

Sample office rental rates (in dollars per square foot per year) for Los Angeles are shown in the table at the left. Use technology to find the mean rental rate and the sample standard deviation. *(Adapted from LoopNet.com)*

Office rental rates		
51	30	15
47	14	87
33	11	35
74	42	51
24	40	26
36	22	40
41	35	36
42	29	24

SOLUTION

Minitab, Excel, and the TI-84 Plus each have features that calculate the means and the standard deviations of data sets. Try using this technology to find the mean and the standard deviation of the office rental rates. From the displays, you can see that

$$\bar{x} \approx 36.9 \text{ and } s \approx 17.4.$$

MINITAB

Descriptive Statistics: Office Rental Rates

Variable	N	Mean	SE Mean	StDev	Minimum
Rental Rates	24	36.88	3.55	17.39	11.00

Variable	Q1	Median	Q3	Maximum
Rental Rates	24.50	35.50	42.00	87.00

EXCEL

	A	B
1	Mean	36.875
2	Standard Error	3.550011
3	Median	35.5
4	Mode	51
5	Standard Deviation	17.39143
6	Sample Variance	302.462
7	Kurtosis	2.354212
8	Skewness	1.214477
9	Range	76
10	Minimum	11
11	Maximum	87
12	Sum	885
13	Count	24

TI-84 PLUS

1-Var Stats
$\bar{x}=36.875$
$\Sigma x=885$
$\Sigma x^2=39591$
$Sx=17.39143342$
$\sigma x=17.02525697$
$\downarrow n=24$

Sample Mean
Sample Standard Deviation

Note to Instructor

The standard deviations reported by Minitab and Excel represent sample standard deviations. The TI-84 Plus also reports σ, the population standard deviation. Ask students to compare the values of s and σ derived from the same data.

TRY IT YOURSELF 4

Sample office rental rates (in dollars per square foot per year) for Dallas are listed. Use technology to find the mean rental rate and the sample standard deviation. *(Adapted from LoopNet.com)*

18 27 21 14 20 20 24 11

16 7 12 22 10 15 21 34

23 13 38 16 18 30 15 30

Answer: Page A33

Interpreting Standard Deviation

When interpreting the standard deviation, remember that it is a measure of the typical amount an entry deviates from the mean. The more the entries are spread out, the greater the standard deviation.

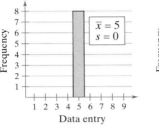 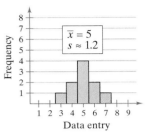 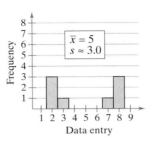

EXAMPLE 5

Estimating Standard Deviation

Without calculating, estimate the population standard deviation of each data set.

1. 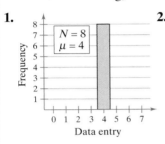 **2.** 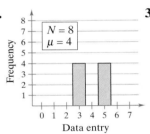 **3.**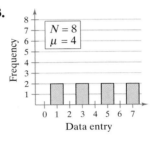

SOLUTION

1. Each of the eight entries is 4. The deviation of each entry is 0, so

$$\sigma = 0. \qquad \text{Standard deviation}$$

2. Each of the eight entries has a deviation of ± 1. So, the population standard deviation should be 1. By calculating, you can see that

$$\sigma = 1. \qquad \text{Standard deviation}$$

3. Each of the eight entries has a deviation of ± 1 or ± 3. So, the population standard deviation should be about 2. By calculating, you can see that σ is greater than 2, with

$$\sigma \approx 2.2. \qquad \text{Standard deviation}$$

TRY IT YOURSELF 5

Write a data set that has 10 entries, a mean of 10, and a population standard deviation that is approximately 3. (There are many correct answers.)

Answer: Page A33

Data entries that lie more than two standard deviations from the mean are considered unusual, while those that lie more than three standard deviations from the mean are very unusual. Unusual and very unusual entries have a greater influence on the standard deviation than entries closer to the mean. This happens because the deviations are squared. Consider the data entries from Example 5, part 3 (see table at the left). The squares of the deviations of the entries farther from the mean (1 and 7) have a greater influence on the value of the standard deviation than those closer to the mean (3 and 5).

Study Tip

You can use standard deviation to compare variation in data sets that use the same units of measure and have means that are about the same. For instance, in the data sets with $\bar{x} = 5$ shown at the right, the data set with $s \approx 3.0$ is more spread out than the other data sets. Not all data sets, however, use the same units of measure or have approximately equal means. To compare variation in these data sets, use the *coefficient of variation,* which is discussed later in this section.

2.4 To explore this topic further, see **Activity 2.4** on page 100.

Entry x	Deviation $x - \mu$	Squares $(x - \mu)^2$
1	−3	9
3	−1	1
5	1	1
7	3	9

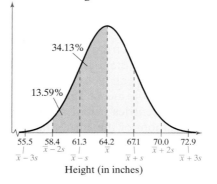

Many real-life data sets have distributions that are approximately symmetric and bell-shaped (see figure below). For instance, the distributions of men's and women's heights in the United States are approximately symmetric and bell-shaped (see the figures at the left and bottom left). Later in the text, you will study bell-shaped distributions in greater detail. For now, however, the **Empirical Rule** can help you see how valuable the standard deviation can be as a measure of variation.

Bell-Shaped Distribution

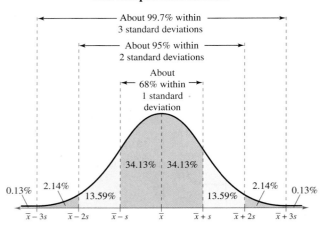

Empirical Rule (or 68–95–99.7 Rule)

For data sets with distributions that are approximately symmetric and bell-shaped (see figure above), the standard deviation has these characteristics.

1. About 68% of the data lie within one standard deviation of the mean.

2. About 95% of the data lie within two standard deviations of the mean.

3. About 99.7% of the data lie within three standard deviations of the mean.

EXAMPLE 6

Using the Empirical Rule

In a survey conducted by the National Center for Health Statistics, the sample mean height of women in the United States (ages 20–29) was 64.2 inches, with a sample standard deviation of 2.9 inches. Estimate the percent of women whose heights are between 58.4 inches and 64.2 inches. (Adapted from National Center for Health Statistics)

SOLUTION

The distribution of women's heights is shown at the left. Because the distribution is bell-shaped, you can use the Empirical Rule. The mean height is 64.2, so when you subtract two standard deviations from the mean height, you get

$$\bar{x} - 2s = 64.2 - 2(2.9) = 58.4.$$

Because 58.4 is two standard deviations below the mean height, the percent of the heights between 58.4 and 64.2 inches is about $13.59\% + 34.13\% = 47.72\%$.

Interpretation   So, about 47.72% of women are between 58.4 and 64.2 inches tall.

TRY IT YOURSELF 6

Estimate the percent of women ages 20–29 whose heights are between 64.2 inches and 67.1 inches.   *Answer: Page A33*

The Empirical Rule applies only to (symmetric) bell-shaped distributions. What if the distribution is not bell-shaped, or what if the shape of the distribution is not known? The next theorem gives an inequality statement that applies to *all* distributions. It is named after the Russian statistician Pafnuti Chebychev (1821–1894).

Chebychev's Theorem

The portion of any data set lying within k standard deviations ($k > 1$) of the mean is at least

$$1 - \frac{1}{k^2}.$$

- $k = 2$: In any data set, at least $1 - \frac{1}{2^2} = \frac{3}{4}$, or 75%, of the data lie within 2 standard deviations of the mean.

- $k = 3$: In any data set, at least $1 - \frac{1}{3^2} = \frac{8}{9}$, or about 88.9%, of the data lie within 3 standard deviations of the mean.

EXAMPLE 7

Using Chebychev's Theorem

The age distributions for Georgia and Iowa are shown in the histograms. Apply Chebychev's Theorem to the data for Georgia using $k = 2$. What can you conclude? Is an age of 100 unusual for a Georgia resident? Explain.
(Source: Based on U.S. Census Bureau)

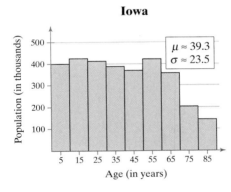

SOLUTION

The histogram on the left shows Georgia's age distribution. Moving two standard deviations to the left of the mean puts you below 0, because $\mu - 2\sigma \approx 37.3 - 2(22.3) = -7.3$. Moving two standard deviations to the right of the mean puts you at

$$\mu + 2\sigma \approx 37.3 + 2(22.3) = 81.9.$$

By Chebychev's Theorem, you can say that at least 75% of the population of Georgia is between 0 and 81.9 years old. Also, because $100 > 81.9$, an age of 100 lies more than two standard deviations from the mean. So, this age is unusual.

TRY IT YOURSELF 7

Apply Chebychev's Theorem to the data for Iowa using $k = 2$. What can you conclude? Is an age of 80 unusual for an Iowa resident? Explain.

Answer: Page A33

Study Tip

In Example 7, Chebychev's Theorem gives you an inequality statement that says at least 75% of the population of Georgia is under the age of 81.9. This is a true statement, but it is not nearly as strong a statement as could be made from reading the histogram.

In general, Chebychev's Theorem gives the minimum percent of data entries that fall within the given number of standard deviations of the mean. Depending on the distribution, there is probably a higher percent of data falling in the given range.

Study Tip

Remember that formulas for grouped data require you to multiply by the frequencies.

Standard Deviation for Grouped Data

In Section 2.1, you learned that large data sets are usually best represented by frequency distributions. The formula for the sample standard deviation for a frequency distribution is

$$\text{Sample standard deviation} = s = \sqrt{\frac{\Sigma(x - \bar{x})^2 f}{n - 1}}$$

where $n = \Sigma f$ is the number of entries in the data set.

EXAMPLE 8

Finding the Standard Deviation for Grouped Data

You collect a random sample of the number of children per household in a region. The results are listed below. Find the sample mean and the sample standard deviation of the data set.

$$
\begin{array}{cccccccccccccccccc}
1 & 3 & 1 & 1 & 1 & 1 & 2 & 2 & 1 & 0 & 1 & 1 & 0 & 0 & 0 & 1 & 5 \\
0 & 3 & 6 & 3 & 0 & 3 & 1 & 1 & 1 & 1 & 6 & 0 & 1 & 3 & 6 & 6 & 1 \\
2 & 2 & 3 & 0 & 1 & 1 & 4 & 1 & 1 & 2 & 2 & 0 & 3 & 0 & 2 & 4 \\
\end{array}
$$

SOLUTION

These data could be treated as 50 individual entries, and you could use the formulas for mean and standard deviation. Because there are so many repeated numbers, however, it is easier to use a frequency distribution.

x	f	xf		$x - \bar{x}$	$(x - \bar{x})^2$	$(x - \bar{x})^2 f$
0	10	0		−1.82	3.3124	33.1240
1	19	19		−0.82	0.6724	12.7756
2	7	14		0.18	0.0324	0.2268
3	7	21		1.18	1.3924	9.7468
4	2	8		2.18	4.7524	9.5048
5	1	5		3.18	10.1124	10.1124
6	4	24		4.18	17.4724	69.8896
	$\Sigma = 50$	$\Sigma = 91$				$\Sigma = 145.38$

$$\bar{x} = \frac{\Sigma xf}{n} = \frac{91}{50} = 1.82 \approx 1.8 \qquad \text{Sample mean}$$

Use the sum of squares to find the sample standard deviation.

$$s = \sqrt{\frac{\Sigma(x - \bar{x})^2 f}{n - 1}} = \sqrt{\frac{145.38}{49}} \approx 1.7 \qquad \text{Sample standard deviation}$$

So, the sample mean is about 1.8 children, and the sample standard deviation is about 1.7 children.

TRY IT YOURSELF 8

Change three of the 6's in the data set to 4's. How does this change affect the sample mean and sample standard deviation?

Answer: Page A33

When a frequency distribution has classes, you can estimate the sample mean and the sample standard deviation by using the midpoint of each class.

EXAMPLE 9

Using Midpoints of Classes

The figure below shows the results of a survey in which 1000 adults were asked how much they spend in preparation for personal travel each year. Make a frequency distribution for the data. Then use the table to estimate the sample mean and the sample standard deviation of the data set. *(Adapted from Travel Industry Association of America)*

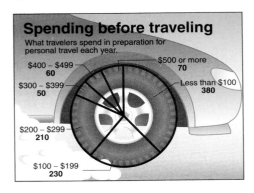

Spending before traveling
What travelers spend in preparation for personal travel each year.

$400 – $499: 60
$300 – $399: 50
$200 – $299: 210
$100 – $199: 230
$500 or more: 70
Less than $100: 380

SOLUTION

Begin by using a frequency distribution to organize the data. Because the class of $500 or more is open-ended, you must choose a value to represent the midpoint, such as 599.5.

Class	x	f	xf	$x - \bar{x}$	$(x - \bar{x})^2$	$(x - \bar{x})^2 f$
0–99	49.5	380	18,810	−142.5	20,306.25	7,716,375.0
100–199	149.5	230	34,385	−42.5	1,806.25	415,437.5
200–299	249.5	210	52,395	57.5	3,306.25	694,312.5
300–399	349.5	50	17,475	157.5	24,806.25	1,240,312.5
400–499	449.5	60	26,970	257.5	66,306.25	3,978,375.0
500+	599.5	70	41,965	407.5	166,056.25	11,623,937.5
		$\Sigma = 1000$	$\Sigma = 192,000$			$\Sigma = 25,668,750.0$

$$\bar{x} = \frac{\Sigma xf}{n} = \frac{192,000}{1000} = 192 \qquad \text{Sample mean}$$

Use the sum of squares to find the sample standard deviation.

$$s = \sqrt{\frac{\Sigma (x - \bar{x})^2 f}{n - 1}} = \sqrt{\frac{25,668,750}{999}} \approx 160.3 \qquad \text{Sample standard deviation}$$

So, an estimate for the sample mean is $192 per year, and an estimate for the sample standard deviation is $160.30 per year.

TRY IT YOURSELF 9

In the frequency distribution in Example 9, 599.5 was chosen as the midpoint for the class of $500 or more. How does the sample mean and standard deviation change when the midpoint of this class is 650?

Answer: Page A33

Coefficient of Variation

To compare variation in different data sets, you can use standard deviation when the data sets use the same units of measure and have means that are about the same. For data sets with different units of measure or different means, use the **coefficient of variation.**

DEFINITION

The **coefficient of variation (CV)** of a data set describes the standard deviation as a percent of the mean.

$$\text{Population: } CV = \frac{\sigma}{\mu} \cdot 100\% \qquad \text{Sample: } CV = \frac{s}{\bar{x}} \cdot 100\%$$

Note that the coefficient of variation measures the variation of a data set relative to the mean of the data.

EXAMPLE 10

Comparing Variation in Different Data Sets

The table below shows the population heights (in inches) and weights (in pounds) of the members of a basketball team. Find the coefficient of variation for the heights and the weights. Then compare the results.

Heights and Weights of a Basketball Team

Heights	72	74	68	76	74	69	72	79	70	69	77	73
Weights	180	168	225	201	189	192	197	162	174	171	185	210

SOLUTION

The mean height is $\mu \approx 72.8$ inches with a standard deviation of $\sigma \approx 3.3$ inches. The coefficient of variation for the heights is

$$CV_{\text{height}} = \frac{\sigma}{\mu} \cdot 100\%$$

$$= \frac{3.3}{72.8} \cdot 100\%$$

$$\approx 4.5\%.$$

The mean weight is $\mu \approx 187.8$ pounds with a standard deviation of $\sigma \approx 17.7$ pounds. The coefficient of variation for the weights is

$$CV_{\text{weight}} = \frac{\sigma}{\mu} \cdot 100\%$$

$$= \frac{17.7}{187.8} \cdot 100\%$$

$$\approx 9.4\%.$$

Interpretation The weights (9.4%) are more variable than the heights (4.5%).

TRY IT YOURSELF 10

Find the coefficient of variation for the office rental rates in Los Angeles (see Example 4) and for those in Dallas (see Try It Yourself 4). Then compare the results.

Answer: Page A33

2.4 EXERCISES

For Extra Help: **MyLab Statistics**

Building Basic Skills and Vocabulary

1. Explain how to find the range of a data set. What is an advantage of using the range as a measure of variation? What is a disadvantage?

2. Explain how to find the deviation of an entry in a data set. What is the sum of all the deviations in any data set?

3. Why is the standard deviation used more frequently than the variance?

4. Explain the relationship between variance and standard deviation. Can either of these measures be negative? Explain.

5. Describe the difference between the calculation of population standard deviation and that of sample standard deviation.

6. Given a data set, how do you know whether to calculate σ or s?

7. Discuss the similarities and the differences between the Empirical Rule and Chebychev's Theorem.

8. What must you know about a data set before you can use the Empirical Rule?

Using and Interpreting Concepts

Finding the Range of a Data Set *In Exercises 9 and 10, find the range of the data set represented by the graph.*

9.   **Median Annual Income by State**

10.

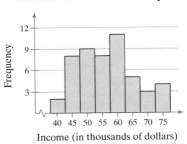

(Source: U.S. Census Bureau)

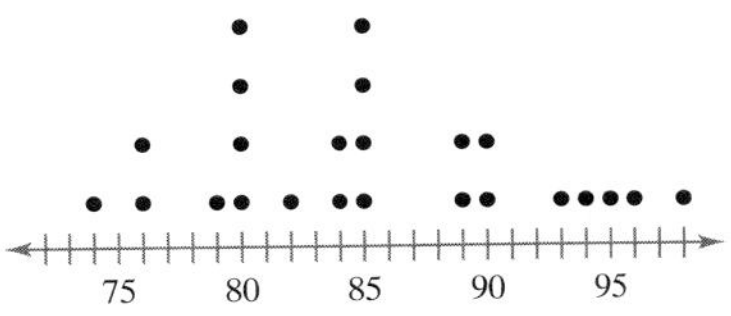

11. Archaeology The depths (in inches) at which 10 artifacts are found are listed.

20.7   24.8   30.5   26.2   36.0   34.3   30.3   29.5   27.0   38.5

(a) Find the range of the data set.

(b) Change 38.5 to 60.5 and find the range of the new data set.

12. In Exercise 11, compare your answer to part (a) with your answer to part (b). How do outliers affect the range of a data set?

Finding Population Statistics *In Exercises 13 and 14, find the range, mean, variance, and standard deviation of the population data set.*

13. Drunk Driving The numbers of alcohol-impaired crash fatalities (in thousands) per year from 2005 through 2015 *(Source: National Highway Traffic Safety Administration)*

14   13   13   12   11   10   10   10   10   10   10

Answers (left column)

1. The range is the difference between the maximum and minimum values of a data set. The advantage of the range is that it is easy to calculate. The disadvantage is that it uses only two entries from the data set.

2. A deviation $(x - \mu)$ is the difference between an entry x and the mean of the data μ. The sum of the deviations is always zero.

3. The units of variance are squared. Its units are meaningless (example: dollars^2). The units of standard deviation are the same as the data.

4. The standard deviation is the positive square root of the variance. The standard deviation and variance can never be negative because squared deviations can never be negative.

5. When calculating the population standard deviation, you divide the sum of the squared deviations by N, then take the square root of that value. When calculating the sample standard deviation, you divide the sum of the squared deviations by $n - 1$, then take the square root of that value.

6. When given a data set, you would have to determine if it represented the population or if it was a sample taken from the population. If the data are a population, then σ is calculated. If the data are a sample, then s is calculated.

7. See Odd Answers, page A49.

8. You must know that the distribution is approximately symmetric and bell-shaped.

9. Approximately 35, or $35,000

10. 24

11. (a) 17.8   (b) 39.8

12. Changing the maximum value of the data set greatly affects the range.

13. Range = 4; $\mu \approx 11.2$; $\sigma^2 \approx 2.1$; $\sigma \approx 1.5$

14. Range $= 7869.91$; $\mu = 2251.1$;
$\sigma^2 \approx 5{,}125{,}758.4$; $\sigma \approx 2264.0$

15. Range $= 6$; $\bar{x} = 19$; $s^2 \approx 3.5$;
$s \approx 1.9$

16. Range $= 35$; $\bar{x} \approx 281.0$;
$s^2 \approx 107.7$; $s \approx 10.4$

17. The data set in (a) has a
standard deviation of 2.4
and the data set in (b) has a
standard deviation of 5 because
the data in (b) have more
variability.

18. The data set in (a) has a
standard deviation of 24 and the
data set in (b) has a standard
deviation of 16 because the data
in (a) have more variability.

19. Company B; An offer of $43,000
is two standard deviations
from the mean of Company A's
starting salaries, which makes
it unlikely. The same offer is
within one standard deviation
of the mean of Company B's
starting salaries, which makes
the offer likely.

14. Density   The densities (in kilograms per cubic meter) of the ten most
abundant elements by weight in Earth's crust

| 1.4 | 2330 | 2700 | 7870 | 1500 |
| 970 | 900 | 1740 | 4500 | 0.09 |

Finding Sample Statistics   *In Exercises 15 and 16, find the range, mean,
variance, and standard deviation of the sample data set.*

 15. Ages of Students   The ages (in years) of a random sample of students
in a campus dining hall

| 19 | 20 | 17 | 19 | 17 | 21 | 23 | 21 | 17 | 17 |
| 19 | 19 | 17 | 20 | 23 | 18 | 18 | 18 | 18 | 19 |

 16. Pregnancy Durations   The durations (in days) of pregnancies for a
random sample of mothers

277	291	295	280	268	278	291
277	282	279	296	285	269	293
267	281	286	269	264	299	275

17. Estimating Standard Deviation   Both data sets shown in the histograms
have a mean of 50. One has a standard deviation of 2.4, and the other has
a standard deviation of 5. By looking at the histograms, which is which?
Explain your reasoning.

(a) 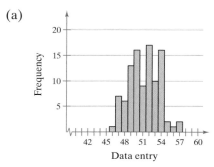   (b)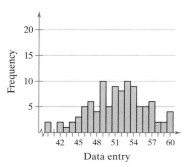

18. Estimating Standard Deviation   Both data sets shown in the stem-and-leaf
plots have a mean of 165. One has a standard deviation of 16, and the other
has a standard deviation of 24. By looking at the stem-and-leaf plots, which
is which? Explain your reasoning.

(a)
```
12 | 8  9          Key: 12|8 = 128
13 | 5  5  8
14 | 1  2
15 | 0  0  6  7
16 | 4  5  9
17 | 1  3  6  8
18 | 0  8  9
19 | 6
20 | 3  5  7
```

(b)
```
12 |                 Key: 13|1 = 131
13 | 1
14 | 2  3  5
15 | 0  4  5  6  8
16 | 1  1  2  3  3  3
17 | 1  5  8  8
18 | 2  3  4  5
19 | 0  2
20 |
```

19. Salary Offers   You are applying for jobs at two companies. Company A
offers starting salaries with $\mu = \$41{,}000$ and $\sigma = \$1000$. Company B offers
starting salaries with $\mu = \$41{,}000$ and $\sigma = \$5000$. From which company
are you more likely to get an offer of $43,000 or more? Explain your
reasoning.

20. Company C; An offer of $62,000 is three standard deviations from the mean of Company D's starting salaries, which makes it unlikely. The same offer is within two standard deviations of the mean of Company C's starting salaries, which makes the offer somewhat likely.

21. (a) Greatest sample standard deviation: (ii)

Data set (ii) has more entries that are farther away from the mean.

Least sample standard deviation: (iii)

Data set (iii) has more entries that are close to the mean.

(b) The three data sets have the same mean, median, and mode, but have a different standard deviation.

(c) Estimates will vary; (i) $s \approx 1.1$; (ii) $s \approx 1.3$; (iii) $s \approx 0.8$

22. (a) Greatest sample standard deviation: (ii)

Data set (ii) has more entries that are farther away from the mean.

Least sample standard deviation: (iii)

Data set (iii) has more entries that are close to the mean.

(b) The three data sets have the same mean and median, but have a different mode and standard deviation.

(c) Estimates will vary; (i) $s \approx 1.6$; (ii) $s \approx 2.9$; (iii) $s \approx 0.8$

23. See Odd Answers, page A49.

24. See Selected Answers, page A94.

25. *Sample answer:* 3, 3, 3, 7, 7, 7

26. *Sample answer:* 3, 3, 3, 3, 9, 9, 9, 9

27. *Sample answer:* 9, 9, 9, 9, 9, 9, 9

28. *Sample answer:* 5, 5, 5, 9, 9, 9

20. Salary Offers You are applying for jobs at two companies. Company C offers starting salaries with $\mu = \$59,000$ and $\sigma = \$1500$. Company D offers starting salaries with $\mu = \$59,000$ and $\sigma = \$1000$. From which company are you more likely to get an offer of $62,000 or more? Explain your reasoning.

Graphical Analysis *In Exercises 21–24, you are asked to compare three data sets. (a) Without calculating, determine which data set has the greatest sample standard deviation and which has the least sample standard deviation. Explain your reasoning. (b) How are the data sets the same? How do they differ? (c) Estimate the sample standard deviations. Then determine how close each of your estimates is by finding the sample standard deviations.*

21. (i) (ii) (iii)

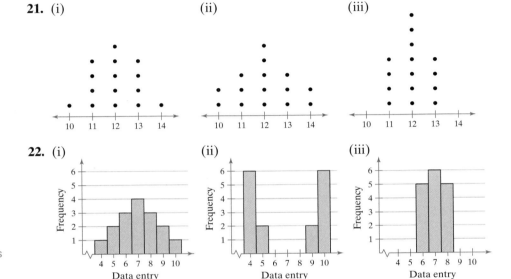

22. (i) (ii) (iii)

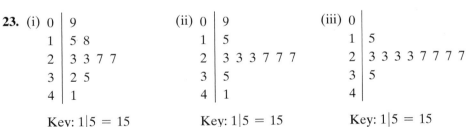

23. (i)

```
0 | 9
1 | 5 8
2 | 3 3 7 7
3 | 2 5
4 | 1
```
Key: 1|5 = 15

(ii)

```
0 | 9
1 | 5
2 | 3 3 3 7 7 7
3 | 5
4 | 1
```
Key: 1|5 = 15

(iii)

```
0 |
1 | 5
2 | 3 3 3 3 7 7 7 7
3 | 5
4 |
```
Key: 1|5 = 15

24. (i) (ii) (iii)

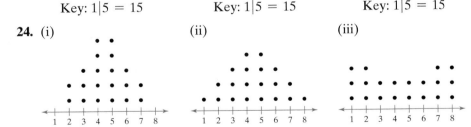

Constructing Data Sets *In Exercises 25–28, construct a data set that has the given statistics.*

25. $N = 6$
$\mu = 5$
$\sigma \approx 2$

26. $N = 8$
$\mu = 6$
$\sigma \approx 3$

27. $n = 7$
$\bar{x} = 9$
$s = 0$

28. $n = 6$
$\bar{x} = 7$
$s \approx 2$

29. 68%

30. $54 and $86

31. (a) 51   (b) 17

32. (a) 38   (b) 19

33. 78, 76, and 82 are unusual; 82 is very unusual because it is more than 3 standard deviations from the mean.

34. $52 and $98 are unusual; $98 is very unusual because it is more than 3 standard deviations from the mean.

35. 30

36. 75

37. At least 93.75% of the exam scores are from 70 to 94.

38. At least 75% of the runs per game scored by the Chicago Cubs during the 2016 World Series are from 0 to 10 (note that −2.86 and 10.58 do not make sense in the context of the data).

Using the Empirical Rule   *In Exercises 29–34, use the Empirical Rule.*

29. The mean speed of a sample of vehicles along a stretch of highway is 67 miles per hour, with a standard deviation of 4 miles per hour. Estimate the percent of vehicles whose speeds are between 63 miles per hour and 71 miles per hour. (Assume the data set has a bell-shaped distribution.)

30. The mean monthly utility bill for a sample of households in a city is $70, with a standard deviation of $8. Between what two values do about 95% of the data lie? (Assume the data set has a bell-shaped distribution.)

31. Use the sample statistics from Exercise 29 and assume the number of vehicles in the sample is 75.
    (a) Estimate the number of vehicles whose speeds are between 63 miles per hour and 71 miles per hour.
    (b) In a sample of 25 additional vehicles, about how many vehicles would you expect to have speeds between 63 miles per hour and 71 miles per hour?

32. Use the sample statistics from Exercise 30 and assume the number of households in the sample is 40.
    (a) Estimate the number of households whose monthly utility bills are between $54 and $86.
    (b) In a sample of 20 additional households, about how many households would you expect to have monthly utility bills between $54 and $86?

33. The speeds for eight vehicles are listed. Using the sample statistics from Exercise 29, determine which of the data entries are unusual. Are any of the data entries very unusual? Explain your reasoning.

    70, 78, 62, 71, 65, 76, 82, 64

34. The monthly utility bills for eight households are listed. Using the sample statistics from Exercise 30, determine which of the data entries are unusual. Are any of the data entries very unusual? Explain your reasoning.

    $65, $52, $63, $83, $77, $98, $84, $70

35. **Using Chebychev's Theorem**   You are conducting a survey on the number of pets per household in your region. From a sample with $n = 40$, the mean number of pets per household is 2 pets and the standard deviation is 1 pet. Using Chebychev's Theorem, determine at least how many of the households have 0 to 4 pets.

36. **Using Chebychev's Theorem**   Old Faithful is a famous geyser at Yellowstone National Park. From a sample with $n = 100$, the mean interval between Old Faithful's eruptions is 101.56 minutes and the standard deviation is 42.69 minutes. Using Chebychev's Theorem, determine at least how many of the intervals lasted between 16.18 minutes and 186.94 minutes. *(Adapted from Geyser Times)*

37. **Using Chebychev's Theorem**   The mean score on a Statistics exam is 82 points, with a standard deviation of 3 points. Apply Chebychev's Theorem to the data using $k = 4$. Interpret the results.

38. **Using Chebychev's Theorem**   The mean number of runs per game scored by the Chicago Cubs during the 2016 World Series was 3.86 runs, with a standard deviation of 3.36 runs. Apply Chebychev's Theorem to the data using $k = 2$. Interpret the results. *(Adapted from Major League Baseball)*

39. See Odd Answers, page A49.

40. See Selected Answers, page A94.

41. See Odd Answers, page A49.

42. See Selected Answers, page A94.

43. See Odd Answers, page A49.

44. See Selected Answers, page A94.

Finding the Sample Mean and Standard Deviation for Grouped Data *In Exercises 39 and 40, make a frequency distribution for the data. Then use the table to find the sample mean and the sample standard deviation of the data set.*

39. 3 3 5 3 8 0 3 9 6 6 7 1 6 3 2 6 9 1 8 5 0 2 3 4 9
    5 8 1 9 7 6 9 6 7 0 6 3 8 6 8 7 3 8 9 3 7 2 4 4 1

40. 1 1 1 0 0 0 0 0 1 0 1 0 0 1 0 1 0 1 1 0 0 0 0 1 0 0 1
    1 1 0 0 1 1 0 0 0 0 0 1 0 1 1 1 0 0 0 0 1 1 0 0 0

Estimating the Sample Mean and Standard Deviation for Grouped Data *In Exercises 41–44, make a frequency distribution for the data. Then use the table to estimate the sample mean and the sample standard deviation of the data set.*

41. **College Expenses**   The distribution of the tuitions, fees, and room and board charges of a random sample of public 4-year degree-granting postsecondary institutions is shown in the pie chart. Use $26,249.50 as the midpoint for "$25,000 or more."

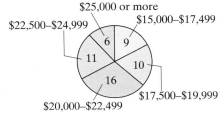

FIGURE FOR EXERCISE 41

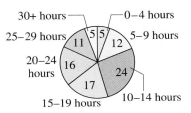

FIGURE FOR EXERCISE 42

42. **Weekly Study Hours**   The distribution of the numbers of hours that a random sample of college students study per week is shown in the pie chart. Use 32 as the midpoint for "30+ hours."

43. **Teaching Load**   The numbers of courses taught per semester by a random sample of university professors are shown in the histogram.

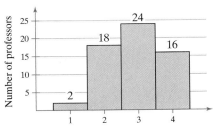

44. **Amounts of Caffeine**   The amounts of caffeine in a sample of five-ounce servings of brewed coffee are shown in the histogram.

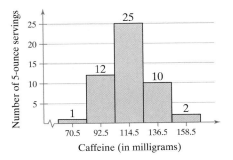

45. $CV_{\text{Denver}} \approx 9.7\%$
$CV_{\text{LA}} \approx 8.8\%$

Salaries for entry level architects are more variable in Denver than in Los Angeles.

46. $CV_{\text{Raleigh}} \approx 12.4\%$
$CV_{\text{Wichita}} \approx 15.8\%$

Salaries for entry level software engineers are more variable in Wichita than in Raleigh.

47. $CV_{\text{ages}} \approx 13.3\%$
$CV_{\text{heights}} \approx 3.5\%$

Ages are more variable than heights for all members of the 2016 Women's U.S. Olympic swimming team.

48. $CV_{\text{ages}} \approx 12.7\%$
$CV_{\text{weight classes}} \approx 28.4\%$

Weight classes are more variable than ages for all members of the 2016 Men's U.S. Olympic wrestling team.

49. $CV_{\text{males}} \approx 20.7\%$
$CV_{\text{females}} \approx 17.6\%$

SAT scores are more variable for males than for females.

50. $CV_{\text{males}} \approx 22.7\%$
$CV_{\text{females}} \approx 25.3\%$

Grade point averages are more variable for females than for males.

51. (a) Answers will vary.

(b) $s \approx 1.9$

(c) They are the same.

Comparing Variation in Different Data Sets   *In Exercises 45–50, find the coefficient of variation for each of the two data sets. Then compare the results.*

 45. Annual Salaries   Sample annual salaries (in thousands of dollars) for entry level architects in Denver, CO, and Los Angeles, CA, are listed.

Denver	45.8	46.4	44.4	40.7	51.5	39.5
	44.2	53.1	44.8	51.6	41.3	49.0
Los Angeles	56.7	50.6	56.0	48.5	55.7	55.6
	47.6	56.3	48.1	46.3	51.9	61.2

46. Annual Salaries   Sample annual salaries (in thousands of dollars) for entry level software engineers in Raleigh, NC, and Wichita, KS, are listed.

Raleigh	63.7	68.4	59.3	50.7	59.8	56.3	73.7	66.6	52.5
Wichita	54.8	71.1	52.2	74.8	50.3	65.8	75.1	73.9	59.3

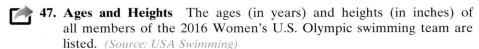 **47. Ages and Heights**   The ages (in years) and heights (in inches) of all members of the 2016 Women's U.S. Olympic swimming team are listed.   *(Source: USA Swimming)*

Ages	24	24	19	23	22	21	21	24	19	19	19
	24	25	21	20	26	21	21	28	30	19	21
Heights	70	70	68	66	69	67	74	68	69	71	71
	68	67	70	75	73	74	69	73	73	70	71

48. Ages and Weights   The ages (in years) and weight classes (in kilograms) of all members of the 2016 Men's U.S. Olympic wrestling team are listed.   *(Source: U.S. Olympic Committee)*

Ages	24	29	26	29	29	28	21	30	27	20
Weight Classes	59	75	85	130	57	74	86	125	65	97

49. SAT Scores   Sample SAT scores for eight males and eight females are listed.

Males	1010	1170	1410	920	1320	1100	690	1140
Females	1190	1010	1000	1300	1470	1250	840	1060

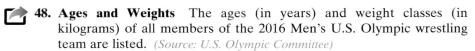 **50. Grade Point Averages**   Sample grade point averages for ten male students and ten female students are listed.

Males	2.4	3.7	3.8	3.9	2.8	2.6	3.6	3.3	4.0	1.9
Females	2.8	3.7	2.1	3.9	3.6	4.0	2.0	3.9	3.7	2.3

Extending Concepts

51. Alternative Formula   You used $SS_x = \Sigma(x - \bar{x})^2$ when calculating variance and standard deviation. An alternative formula that is sometimes more convenient for hand calculations is

$$SS_x = \Sigma x^2 - \frac{(\Sigma x)^2}{n}.$$

You can find the sample variance by dividing the sum of squares by $n - 1$ and the sample standard deviation by finding the square root of the sample variance.

(a) Show how to obtain the alternative formula.

(b) Use the alternative formula to calculate the sample standard deviation for the data set in Exercise 15.

(c) Compare your result with the sample standard deviation obtained in Exercise 15.

52. (a) 1.35; The mean absolute
deviation is less than the
sample standard deviation.

(b) 8.4; The mean absolute
deviation is less than the
sample standard deviation.

53. (a) $\bar{x} \approx 42.1$; $s \approx 5.6$

(b) $\bar{x} \approx 44.3$; $s \approx 5.9$

(c) 3.5, 3, 3, 4, 4, 2.75, 4.25, 3.25,
3.25, 3.5, 3.25, 3.75, 3.5, 4.17
$\bar{x} \approx 3.5$; $s \approx 0.47$

(d) When each entry is
multiplied by a constant k,
the new sample mean is
$k \cdot x$, and the new sample
standard deviation is $k \cdot s$.

54. (a) $\bar{x} \approx 41.2$, $s \approx 6.0$

(b) $\bar{x} \approx 42.2$. $s \approx 6.0$

(c) $\bar{x} \approx 39.2$, $s \approx 6.0$

(d) Adding a constant k to, or
subtracting it from, each
entry makes the new sample
mean $\bar{x} + k$, or $\bar{x} - k$,
with the sample standard
deviation being unaffected.

55. (a) $P \approx -2.61$

The data are skewed left.

(b) $P \approx 4.12$

The data are skewed right.

(c) $P = 0$

The data are symmetric.

(d) $P = 1$

The data are skewed right.

(e) $P = -3$

The data are skewed left.

56. 10

Set $1 - \dfrac{1}{k^2} = 0.99$ and solve for k.

52. Mean Absolute Deviation Another useful measure of variation for a data set is the **mean absolute deviation (MAD).** It is calculated by the formula

$$MAD = \frac{\Sigma |x - \bar{x}|}{n}.$$

(a) Find the mean absolute deviation of the data set in Exercise 15. Compare your result with the sample standard deviation obtained in Exercise 15.

(b) Find the mean absolute deviation of the data set in Exercise 16. Compare your result with the sample standard deviation obtained in Exercise 16.

53. Scaling Data Sample annual salaries (in thousands of dollars) for employees at a company are listed.

42 36 48 51 39 39 42
36 48 33 39 42 45 50

(a) Find the sample mean and the sample standard deviation.

(b) Each employee in the sample receives a 5% raise. Find the sample mean and the sample standard deviation for the revised data set.

(c) Find each monthly salary. Then find the sample mean and the sample standard deviation for the monthly salaries.

(d) What can you conclude from the results of (a), (b), and (c)?

54. Shifting Data Sample annual salaries (in thousands of dollars) for employees at a company are listed.

40 35 49 53 38 39 40
37 49 34 38 43 47 35

(a) Find the sample mean and the sample standard deviation.

(b) Each employee in the sample receives a $1000 raise. Find the sample mean and the sample standard deviation for the revised data set.

(c) Each employee in the sample takes a pay cut of $2000 from their original salary. Find the sample mean and the sample standard deviation for the revised data set.

(d) What can you conclude from the results of (a), (b), and (c)?

55. Pearson's Index of Skewness The English statistician Karl Pearson (1857–1936) introduced a formula for the skewness of a distribution.

$$P = \frac{3(\bar{x} - \text{median})}{s} \qquad \text{Pearson's index of skewness}$$

Most distributions have an index of skewness between -3 and 3. When $P > 0$, the data are skewed right. When $P < 0$, the data are skewed left. When $P = 0$, the data are symmetric. Calculate the coefficient of skewness for each distribution. Describe the shape of each.

(a) $\bar{x} = 17$, $s = 2.3$, median $= 19$

(b) $\bar{x} = 32$, $s = 5.1$, median $= 25$

(c) $\bar{x} = 9.2$, $s = 1.8$, median $= 9.2$

(d) $\bar{x} = 42$, $s = 6.0$, median $= 40$

(e) $x = 155$, $s = 20.0$, median $= 175$

56. Chebychev's Theorem At least 99% of the data in any data set lie within how many standard deviations of the mean? Explain how you obtained your answer.

APPLET

You can find the interactive applet for this activity within **MyLab Statistics** or at *www.pearsonhighered.com/ mathstatsresources.*

The *standard deviation* applet is designed to allow you to investigate interactively the standard deviation as a measure of spread for a data set. Points can be added to the plot by clicking the mouse above the horizontal axis. The mean of the points is shown as a green arrow. A numeric value for the standard deviation is shown above the plot. Points on the plot can be removed by clicking on the point and then dragging the point into the trash can. All of the points on the plot can be removed by simply clicking inside the trash can. The range of values for the horizontal axis can be specified by inputting lower and upper limits and then clicking UPDATE.

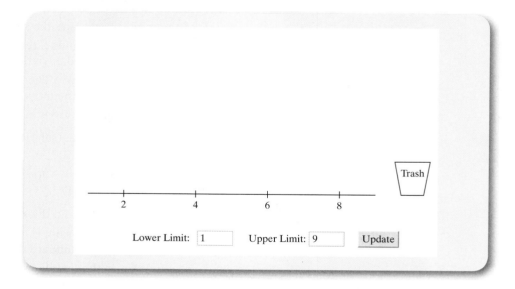

EXPLORE

Step 1 Specify a lower limit.
Step 2 Specify an upper limit.
Step 3 Add 15 points to the plot.
Step 4 Remove all of the points from the plot.

DRAW CONCLUSIONS

APPLET

1. Specify the lower limit to be 10 and the upper limit to be 20. Plot 10 points that have a mean of about 15 and a standard deviation of about 3. Write the estimates of the values of the points. Plot a point with a value of 15. What happens to the mean and standard deviation? Plot a point with a value of 20. What happens to the mean and standard deviation?

2. Specify the lower limit to be 30 and the upper limit to be 40. How can you plot eight points so that the points have the greatest possible standard deviation? Use the applet to plot the set of points and then use the formula for standard deviation to confirm the value given in the applet. How can you plot eight points so that the points have the least possible standard deviation? Explain.

CASE STUDY

Business Size

The numbers of employees at businesses can vary. A business can have anywhere from a single employee to more than 1000 employees. The data shown below are the numbers of manufacturing businesses for nine states in a recent year. *(Source: U.S. Census Bureau)*

State	Number of manufacturing businesses
California	38,293
Illinois	13,531
Indiana	8,036
Michigan	12,361
New York	16,076
Ohio	14,208
Pennsylvania	13,684
Texas	19,681
Wisconsin	8,858

Number of Manufacturing Businesses by Number of Employees

State	1–4	5–9	10–19	20–49	50–99	100–249	250–499	500–999	1000+
California	15,320	7,074	5,862	5,494	2,276	1,609	433	144	81
Illinois	4,683	2,234	2,103	2,165	1,123	852	241	96	34
Indiana	2,225	1,319	1,276	1,403	797	640	229	99	48
Michigan	4,055	2,103	2,008	2,044	974	800	254	76	47
New York	7,048	2,810	2,342	2,134	885	587	171	75	24
Ohio	4,274	2,469	2,281	2,495	1,233	982	311	112	51
Pennsylvania	4,505	2,292	2,185	2,335	1,125	860	268	79	35
Texas	7,019	3,409	2,994	3,078	1,501	1,114	358	145	63
Wisconsin	2,657	1,372	1,342	1,520	889	725	227	97	29

EXERCISES

Use the information given in the above tables.

1. **Employees** Which state has the greatest number of manufacturing employees? Explain your reasoning.

2. **Mean Business Size** Estimate the mean number of employees at a manufacturing business for each state. Use 1500 as the midpoint for "1000+."

3. **Employees** Which state has the greatest number of employees per manufacturing business? Explain your reasoning.

4. **Standard Deviation** Estimate the standard deviation for the number of employees at a manufacturing business for each state. Use 1500 as the midpoint for "1000+."

5. **Standard Deviation** Which state has the greatest standard deviation? Explain your reasoning.

6. **Distribution** Describe the distribution of the number of employees at manufacturing businesses for each state.

2.5 | Measures of Position

Quartiles ▪ Percentiles and Other Fractiles ▪ The Standard Score

Quartiles

In this section, you will learn how to use fractiles to specify the position of a data entry within a data set. **Fractiles** are numbers that partition, or divide, an ordered data set into equal parts (each part has the same number of data entries). For instance, the median is a fractile because it divides an ordered data set into two equal parts.

DEFINITION

The three **quartiles,** Q_1, Q_2, and Q_3, divide an ordered data set into four equal parts. About one-quarter of the data fall on or below the **first quartile** Q_1. About one-half of the data fall on or below the **second quartile** Q_2 (the second quartile is the same as the median of the data set). About three-quarters of the data fall on or below the **third quartile** Q_3.

EXAMPLE 1

Finding the Quartiles of a Data Set

Each year in the U.S., automobile commuters waste fuel due to traffic congestion. The amounts (in gallons per year) of fuel wasted by commuters in the 15 largest U.S. urban areas are listed. (Large urban areas have populations of at least 3 million.) Find the first, second, and third quartiles of the data set. What do you observe? *(Source: Based on 2015 Urban Mobility Scorecard)*

20   30   29   22   25   29   25   24   35   23   25   11   33   28   35

SOLUTION

First, order the data set and find the median Q_2. The first quartile Q_1 is the median of the data entries to the left of Q_2. The third quartile Q_3 is the median of the data entries to the right of Q_2.

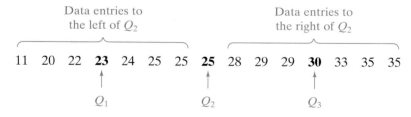

Interpretation   In about one-quarter of the large urban areas, auto commuters waste 23 gallons of fuel or less, about one-half waste 25 gallons or less, and about three-quarters waste 30 gallons or less.

TRY IT YOURSELF 1

Find the first, second, and third quartiles for the points scored by the 51 winning teams using the data set listed on page 39. What do you observe?

Answer: Page A33

EXAMPLE 2

Using Technology to Find Quartiles

The tuition costs (in thousands of dollars) for 25 liberal arts colleges are listed. Use technology to find the first, second, and third quartiles. What do you observe? *(Source: U.S. News & World Report)*

50	52	51	49	52	51	25	41	47	36	30	44	40
35	40	45	34	33	23	34	27	16	18	18	35	

SOLUTION

Minitab and the TI-84 Plus each have features that calculate quartiles. Try using this technology to find the first, second, and third quartiles of the tuition data. From the displays, you can see that

$$Q_1 = 28.5, \ Q_2 = 36, \text{ and } Q_3 = 48.$$

MINITAB

Descriptive Statistics: Tuition

Variable	N	Mean	SE Mean	StDev	Minimum
Tuition	25	37.04	2.27	11.36	16.00

Variable	Q1	Median	Q3	Maximum
Tuition	28.50	36.00	48.00	52.00

TI-84 PLUS

1-Var Stats

↑n=25
minX=16
Q_1=28.5
Med=36
Q_3=48
maxX=52

STATCRUNCH

Summary statistics:

Column	Q1	Median	Q3
Tuition	30	36	47

Interpretation About one-quarter of these colleges charge tuition of $28,500 or less; about one-half charge $36,000 or less; and about three-quarters charge $48,000 or less.

TRY IT YOURSELF 2

The tuition costs (in thousands of dollars) for 25 universities are listed. Use technology to find the first, second, and third quartiles. What do you observe? *(Source: U.S. News & World Report)*

44	30	38	23	20	29	19	44	29	17	45	39	29
18	43	45	39	24	44	26	34	20	35	30	36	

Answer: Page A33

The median (the second quartile) is a measure of central tendency based on position. A measure of variation that is based on position is the **interquartile range.** The interquartile range tells you the spread of the middle half of the data, as shown in the next definition.

Tech Tip

Note that you may get results that differ slightly when comparing results obtained by different technology tools. For instance, in Example 2, the first quartile, as determined by Minitab and the TI-84 Plus, is 28.5, whereas the result using Excel is 30 (see below).

EXCEL

	A	B
1	50	
2	52	Quartile.inc(A1:A25,1)
3	51	30
4	49	
5	52	Quartile.inc(A1:A25,2)
6	51	36
7	25	
8	41	Quartile.inc(A1:A25,3)
9	47	47
10	36	
11	30	
12	44	
13	40	
14	35	
15	40	
16	45	
17	34	
18	33	
19	23	
20	34	
21	27	
22	16	
23	18	
24	18	
25	35	

DEFINITION

The **interquartile range (IQR)** of a data set is a measure of variation that gives the range of the middle portion (about half) of the data. The IQR is the difference between the third and first quartiles.

$$\text{IQR} = Q_3 - Q_1$$

In Section 2.3, an outlier was described as a data entry that is far removed from the other entries in the data set. One way to identify outliers is to use the interquartile range.

GUIDELINES

Using the Interquartile Range to Identify Outliers

1. Find the first (Q_1) and third (Q_3) quartiles of the data set.
2. Find the interquartile range: $\text{IQR} = Q_3 - Q_1$.
3. Multiply IQR by 1.5: $1.5(\text{IQR})$.
4. Subtract $1.5(\text{IQR})$ from Q_1. Any data entry less than $Q_1 - 1.5(\text{IQR})$ is an outlier.
5. Add $1.5(\text{IQR})$ to Q_3. Any data entry greater than $Q_3 + 1.5(\text{IQR})$ is an outlier.

EXAMPLE 3

Using the Interquartile Range to Identify an Outlier

Find the interquartile range of the data set in Example 1. Are there any outliers?

SOLUTION

From Example 1, you know that $Q_1 = 23$ and $Q_3 = 30$. So, the interquartile range is $\text{IQR} = Q_3 - Q_1 = 30 - 23 = 7$. To identify any outliers, first note that $1.5(\text{IQR}) = 1.5(7) = 10.5$. There is a data entry, 11, that is less than

$$Q_1 - 1.5(\text{IQR}) = 23 - 10.5 \qquad \text{Subtract } 1.5(\text{IQR}) \text{ from } Q_1.$$
$$= 12.5 \qquad \text{A data entry less than 12.5 is an outlier.}$$

but there are no data entries greater than

$$Q_3 + 1.5(\text{IQR}) = 30 + 10.5 \qquad \text{Add } 1.5(\text{IQR}) \text{ from } Q_3.$$
$$= 40.5. \qquad \text{A data entry greater than 40.5 is an outlier.}$$

So, 11 is an outlier.

Interpretation   In large urban areas, the amount of fuel wasted by auto commuters in the middle of the data set varies by at most 10.5 gallons. Notice that the outlier, 11, does not affect the IQR.

TRY IT YOURSELF 3

Find the interquartile range for the points scored by the 51 winning teams listed on page 39. Are there any outliers?

Answer: Page A33

Another important application of quartiles is to represent data sets using box-and-whisker plots. A **box-and-whisker plot** (or **boxplot**) is an exploratory data analysis tool that highlights the important features of a data set. To graph a box-and-whisker plot, you must know the values shown at the top of the next page.

Picturing the World

Since 1970, there have been 2845 fatalities in the United States attributed to lightning strikes. The box-and-whisker plot summarizes the fatalities for each year since 1970. (Source: National Weather Service)

Lightning Fatalities

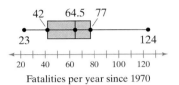

Fatalities per year since 1970

About how many fatalities are represented by the right whisker? There were 27 lightning fatalities in 2015. Into what quartile does this number of fatalities fall?

About 711 fatalities are represented by the right whisker. The 27 lightning fatalities in 2015 falls into the first quartile.

Study Tip

For data sets that have outliers, you can represent them graphically using a *modified box-and-whisker plot*. A *modified box-and-whisker plot* is a box-and-whisker plot that uses symbols (such as an asterisk or a point) to indicate outliers. The horizontal line of a modified box-and-whisker plot extends as far as the minimum data entry that is not an outlier and the maximum data entry that is not an outlier. For instance, on pages 124 and 125, Minitab and the TI-84 Plus were used to draw modified box-and-whisker plots that represent the data set in Example 1. Compare these results with the one in Example 4.

1. The minimum entry
2. The first quartile Q_1
3. The median Q_2
4. The third quartile Q_3
5. The maximum entry

These five numbers are called the **five-number summary** of the data set.

GUIDELINES

Drawing a Box-and-Whisker Plot

1. Find the five-number summary of the data set.
2. Construct a horizontal scale that spans the range of the data.
3. Plot the five numbers above the horizontal scale.
4. Draw a box above the horizontal scale from Q_1 to Q_3 and draw a vertical line in the box at Q_2.
5. Draw whiskers from the box to the minimum and maximum entries.

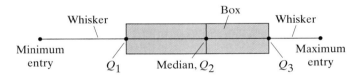

EXAMPLE 4

See Minitab and TI-84 Plus steps on pages 124 and 125.

Drawing a Box-and-Whisker Plot

Draw a box-and-whisker plot that represents the data set in Example 1. What do you observe?

SOLUTION Here is the five-number summary of the data set.

$$\text{Minimum} = 11 \quad Q_1 = 23 \quad Q_2 = 25 \quad Q_3 = 30 \quad \text{Maximum} = 35$$

Using these five numbers, you can construct the box-and-whisker plot shown.

Gallons of Fuel Wasted Per Year

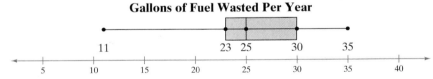

Interpretation The box represents about half of the data, which means about 50% of the data entries are between 23 and 30. The left whisker represents about one-quarter of the data, so about 25% of the data entries are less than 23. The right whisker represents about one-quarter of the data, so about 25% of the data entries are greater than 30. Also, the length of the left whisker is much longer than the right one. This indicates that the data set has a possible outlier to the left. (You already know from Example 3 that the data entry of 11 is an outlier).

TRY IT YOURSELF 4

Draw a box-and-whisker plot that represents the points scored by the 51 winning teams listed on page 39. What do you observe?

Answer: Page A33

You can use a box-and-whisker plot to determine the shape of a distribution. Notice that the box-and-whisker plot in Example 4 represents a distribution that is skewed left.

Percentiles and Other Fractiles

In addition to using quartiles to specify a measure of position, you can also use percentiles and deciles. Here is a summary of these common fractiles.

Fractiles	Summary	Symbols
Quartiles	Divide a data set into 4 equal parts.	Q_1, Q_2, Q_3
Deciles	Divide a data set into 10 equal parts.	$D_1, D_2, D_3, \ldots, D_9$
Percentiles	Divide a data set into 100 equal parts.	$P_1, P_2, P_3, \ldots, P_{99}$

Study Tip

Notice that the 25th percentile is the same as Q_1; the 50th percentile is the same as Q_2, or the median; and the 75th percentile is the same as Q_3.

Percentiles are often used in education and health-related fields to indicate how one individual compares with others in a group. Percentiles can also be used to identify unusually high or unusually low values. For instance, children's growth measurements are often expressed in percentiles. Measurements in the 95th percentile and above are unusually high, while those in the 5th percentile and below are unusually low.

Study Tip

Be sure you understand what a percentile means. For instance, the weight of a six-month-old infant is at the 78th percentile. This means the infant weighs the same as or more than 78% of all six-month-old infants. It does not mean that the infant weighs 78% of some ideal weight.

EXAMPLE 5

Interpreting Percentiles

The ogive at the right represents the cumulative frequency distribution for SAT scores of college-bound students in a recent year. What score represents the 80th percentile? *(Source: The College Board)*

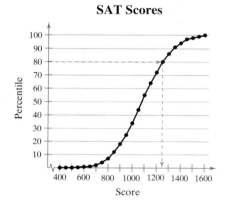

SAT Scores

SOLUTION

From the ogive, you can see that the 80th percentile corresponds to a score of 1250.

Interpretation This means that approximately 80% of the students had an SAT score of 1250 or less.

TRY IT YOURSELF 5

The points scored by the 51 winning teams in the Super Bowl (see page 39) are represented in the ogive at the left. What score represents the 10th percentile? How should you interpret this?

Answer: Page A33

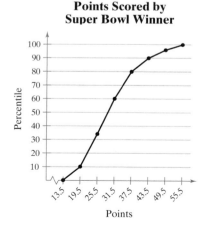

Points Scored by Super Bowl Winner

In Example 5, you used an ogive to approximate a data entry that corresponds to a percentile. You can also use an ogive to approximate a percentile that corresponds to a data entry. Another way to find a percentile is to use a formula.

DEFINITION

To find the **percentile that corresponds to a specific data entry x,** use the formula

$$\text{Percentile of } x = \frac{\text{number of data entries less than } x}{\text{total number of data entries}} \cdot 100$$

and then round to the nearest whole number.

EXAMPLE 6

Finding a Percentile

For the data set in Example 2, find the percentile that corresponds to $34,000.

SOLUTION

Recall that the tuition costs are in thousands of dollars, so $34,000 is the data entry 34. Begin by ordering the data.

$$16 \quad 18 \quad 18 \quad 23 \quad 25 \quad 27 \quad 30 \quad 33 \quad 34 \quad 34 \quad 35 \quad 35 \quad 36$$
$$40 \quad 40 \quad 41 \quad 44 \quad 45 \quad 47 \quad 49 \quad 50 \quad 51 \quad 51 \quad 52 \quad 52$$

There are 8 data entries less than 34 and the total number of data entries is 25.

$$\text{Percentile of } 34 = \frac{\text{number of data entries less than } 34}{\text{total number of entries}} \cdot 100$$

$$= \frac{8}{25} \cdot 100$$

$$= 32$$

The tuition cost of $34,000 corresponds to the 32nd percentile.

Interpretation   The tuition cost of $34,000 is greater than 32% of the other tuition costs.

TRY IT YOURSELF 6

For the data set in Try It Yourself 2, find the percentile that corresponds to $26,000, which is the data entry 26.

Answer: Page A33

The Standard Score

When you know the mean and standard deviation of a data set, you can measure the position of an entry in the data set with a **standard score,** or **z-score.**

DEFINITION

> The **standard score,** or **z-score,** represents the number of standard deviations a value x lies from the mean μ. To find the z-score for a value, use the formula
>
> $$z = \frac{\text{Value} - \text{Mean}}{\text{Standard deviation}} = \frac{x - \mu}{\sigma}.$$

A z-score can be negative, positive, or zero. When z is negative, the corresponding x-value is less than the mean. When z is positive, the corresponding x-value is greater than the mean. For $z = 0$, the corresponding x-value is equal to the mean. A z-score can be used to identify an unusual value of a data set that is approximately bell-shaped.

When a distribution is approximately bell-shaped, you know from the Empirical Rule that about 95% of the data lie within 2 standard deviations of the mean. So, when this distribution's values are transformed to z-scores, about 95% of the z-scores should fall between -2 and 2. A z-score outside of this range will occur about 5% of the time and would be considered unusual. So, according to the Empirical Rule, a z-score less than -3 or greater than 3 would be very unusual, with such a score occurring about 0.3% of the time.

Very unusual scores
Unusual scores
Usual scores

$-3 \quad -2 \quad -1 \quad 0 \quad 1 \quad 2 \quad 3$

z–score

EXAMPLE 7

Finding z-Scores

The mean speed of vehicles along a stretch of highway is 56 miles per hour with a standard deviation of 4 miles per hour. You measure the speeds of three cars traveling along this stretch of highway as 62 miles per hour, 47 miles per hour, and 56 miles per hour. Find the z-score that corresponds to each speed. Assume the distribution of the speeds is approximately bell-shaped.

SOLUTION   The z-score that corresponds to each speed is calculated below.

$x = 62$ mph $x = 47$ mph $x = 56$ mph

$$z = \frac{62 - 56}{4} = 1.5 \qquad z = \frac{47 - 56}{4} = -2.25 \qquad z = \frac{56 - 56}{4} = 0$$

Interpretation   From the z-scores, you can conclude that a speed of 62 miles per hour is 1.5 standard deviations above the mean; a speed of 47 miles per hour is 2.25 standard deviations below the mean; and a speed of 56 miles per hour is equal to the mean. The car traveling 47 miles per hour is said to be traveling unusually slow, because its speed corresponds to a z-score of -2.25.

TRY IT YOURSELF 7

The monthly utility bills in a city have a mean of $70 and a standard deviation of $8. Find the z-scores that correspond to utility bills of $60, $71, and $92. Assume the distribution of the utility bills is approximately bell-shaped.

Answer: Page A33

EXAMPLE 8

Comparing z-Scores from Different Data Sets

The table shows the mean heights and standard deviations for a population of men and a population of women. Compare the z-scores for a 6-foot-tall man and a 6-foot-tall woman. Assume the distributions of the heights are approximately bell-shaped.

Men's heights	Women's heights
$\mu = 69.9$ in.	$\mu = 64.3$ in.
$\sigma = 3.0$ in.	$\sigma = 2.6$ in.

SOLUTION   Note that 6 feet $=$ 72 inches. Find the z-score for each height.

z-score for 6-foot-tall man   **z-score for 6-foot-tall woman**

$$z = \frac{x - \mu}{\sigma} = \frac{72 - 69.9}{3.0} = 0.7 \qquad z = \frac{x - \mu}{\sigma} = \frac{72 - 64.3}{2.6} \approx 3.0$$

Interpretation   The z-score for the 6-foot-tall man is within 1 standard deviation of the mean (69.9 inches). This is among the typical heights for a man. The z-score for the 6-foot-tall woman is about 3 standard deviations from the mean (64.3 inches). This is an unusual height for a woman.

TRY IT YOURSELF 8

Use the information in Example 8 to compare the z-scores for a 5-foot-tall man and a 5-foot-tall woman.

Answer: Page A33

2.5 EXERCISES

Building Basic Skills and Vocabulary

1. The length of a guest lecturer's talk represents the third quartile for talks in a guest lecture series. Make an observation about the length of the talk.

2. A motorcycle's fuel efficiency represents the ninth decile of vehicles in its class. Make an observation about the motorcycle's fuel efficiency.

3. A student's score on the Fundamentals of Engineering exam is in the 89th percentile. Make an observation about the student's exam score.

4. A student's IQ score is in the 91st percentile on the Weschler Adult Intelligence Scale. Make an observation about the student's IQ score.

5. Explain how to identify outliers using the interquartile range.

6. Describe the relationship between quartiles and percentiles.

True or False? *In Exercises 7–10, determine whether the statement is true or false. If it is false, rewrite it as a true statement.*

7. About one-quarter of a data set falls below Q_1.

8. The second quartile is the mean of an ordered data set.

9. An outlier is any number above Q_3 or below Q_1.

10. It is impossible to have a z-score of 0.

Using and Interpreting Concepts

Finding Quartiles, Interquartile Range, and Outliers *In Exercises 11 and 12, (a) find the quartiles, (b) find the interquartile range, and (c) identify any outliers.*

11. 56 63 51 60 57 60 60 54 63 59 80 63 60 62 65

12. 22 25 22 24 20 24 19 22 29 21
21 20 23 25 23 23 21 25 23 22

Graphical Analysis *In Exercises 13 and 14, use the box-and-whisker plot to identify the five-number summary.*

13.

14.

Drawing a Box-and-Whisker Plot *In Exercises 15–18, (a) find the five-number summary, and (b) draw a box-and-whisker plot that represents the data set.*

15. 39 36 30 27 26 24 28 35 39 60 50 41 35 32 51

16. 171 176 182 150 178 180 173 170 174 178 181 180

17. 4 7 7 5 2 9 7 6 8 5 8 4 1 5 2 8 7 6 6 9

18. 2 7 1 3 1 2 8 9 9 2 5 4 7 3 7 5 4
2 3 5 9 5 6 3 9 3 4 9 8 8 2 3 9 5

Answers (left column)

1. The talk is longer in length than 75% of the lectures in the series.

2. The motorcycle's fuel efficiency is higher than 90% of the other vehicles in its class.

3. The student scored higher than 89% of the students who took the Fundamentals of Engineering exam.

4. The student has a higher IQ score than 91% of the students in the same age group.

5. The interquartile range of a data set can be used to identify outliers because data entries that are greater than $Q_3 + 1.5(\text{IQR})$ or less than $Q_1 - 1.5(\text{IQR})$ are considered outliers.

6. Quartiles are special cases of percentiles. Q_1 is the 25th percentile, Q_2 is the 50th percentile, and Q_3 is the 75th percentile.

7. True

8. False. The second quartile is the median of an ordered data set.

9. False; An outlier is any number above $Q_3 + 1.5(\text{IQR})$ or below $Q_1 - 1.5(\text{IQR})$.

10. False. It is possible to have a z-score of 0 when the x-value equals the mean.

11. (a) $Q_1 = 57$, $Q_2 = 60$, $Q_3 = 63$ (b) IQR $= 6$ (c) 80

12. (a) $Q_1 = 21$, $Q_2 = 22.5$, $Q_3 = 24$ (b) IQR $= 3$ (c) 29

13. Min $= 0$, $Q_1 = 2$, $Q_2 = 5$, $Q_3 = 8$, Max $= 10$

14. Min $= 500$, $Q_1 = 580$, $Q_2 = 605$, $Q_3 = 630$, Max $= 720$

15. See Odd Answers, page A50.

16. See Selected Answers, page A94.

17. See Odd Answers, page A50.

18. See Selected Answers, page A95.

19. None. The data are not skewed or symmetric.

20. Skewed right. Most of the data lie to the left on the box plot.

21. Skewed left. Most of the data lie to the right on the box plot.

22. Symmetric. The data are evenly spaced to the left and to the right of the median.

23.
Studying
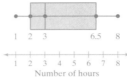

Number of hours

24.
Vacation Days

Number of days

25.
Commuting Distances

Distance (in miles)

26.
Consulting Firm Employees
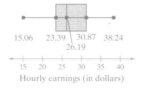
Hourly earnings (in dollars)

27. (a) 6.5 hours
   (b) About 50%
   (c) About 25%

28. (a) $26.19 per hour
   (b) About 75%
   (c) About 75%
   (d) About 25%

Graphical Analysis *In Exercises 19–22, use the box-and-whisker plot to determine whether the shape of the distribution represented is symmetric, skewed left, skewed right, or none of these. Justify your answer.*

19.

20.
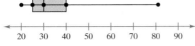

21.
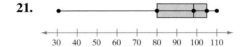

22.
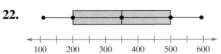

Using Technology to Find Quartiles and Draw Graphs *In Exercises 23–26, use technology to draw a box-and-whisker plot that represents the data set.*

 23. **Studying** The numbers of hours spent studying per day by a sample of 28 students

2	8	7	2	3	3	3	2	2	7	8	3	5	1
1	2	6	1	5	7	3	8	5	3	3	7	6	2

24. **Vacation Days** The numbers of vacation days used by a sample of 20 employees in a recent year

3	9	2	1	7	5	3	2	2	6
4	0	10	0	3	5	7	8	6	5

25. **Commuting Distances** The commuting distances (in miles) of a sample of 30 employees

7	6	7	5	2	1	1	2	3	8
15	24	3	8	9	19	12	17	45	4
4	3	11	26	10	4	21	1	5	12

26. **Hourly Earnings** The hourly earnings (in dollars) of a sample of 21 employees at a consulting firm

25.89	27.09	31.76	28.28	26.19	27.43	24.06
25.61	22.56	29.76	18.01	23.66	38.24	37.27
32.70	31.12	25.87	15.06	23.12	30.62	19.85

27. **Studying** Refer to the data set in Exercise 23 and the box-and-whisker plot you drew that represents the data set.
   (a) About 75% of the students studied no more than how many hours per day?
   (b) What percent of the students studied more than 3 hours per day?
   (c) You randomly select one student from the sample. What is the likelihood that the student studied less than 2 hours per day? Write your answer as a percent.

28. **Hourly Earnings** Refer to the data set in Exercise 26 and the box-and-whisker plot you drew that represents the data set.
   (a) About 50% of the employees made less than what amount per hour?
   (b) What percent of the employees made more than $23.39 per hour?
   (c) What percent of the employees made between $23.39 and $38.24 per hour?
   (d) You randomly select one employee from the sample. What is the likelihood that the employee made more than $30.87 per hour? Write your answer as a percent.

29. About 158; About 70% of quantitative reasoning scores on the Graduate Record Examination are less than 158.

30. About 150; About 40% of quantitative reasoning scores on the Graduate Record Examination are less than 150.

31. About 8th percentile; About 8% of quantitative reasoning scores on the Graduate Record Examination are less that 140.

32. About 97th percentile; About 97% of quantitative reasoning scores on the Graduate Record Examination are less than 170.

33. 10th percentile

34. 70th percentile

35. 57, 57, 61, 61, 65, 66

36. 28, 35, 38, 40, 41, 41, 42

37. See Odd Answers, page A50.

38. About 8 minutes; About 50% of wait times are less than 8 minutes.

39. About 85th percentile

40. About 4.5 minutes to 16 minutes

41. See Odd Answers, page A50.

42. See Selected Answers, page A95.

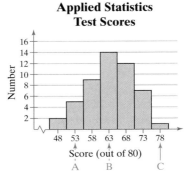

FIGURE FOR EXERCISE 41

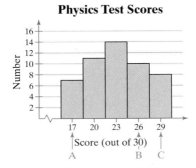

FIGURE FOR EXERCISE 42

Interpreting Percentiles   *In Exercises 29–32, use the ogive, which represents the cumulative frequency distribution for quantitative reasoning scores on the Graduate Record Examination in a recent range of years.* (*Adapted from Educational Testing Service*)

29. What score represents the 70th percentile? How should you interpret this?

30. Which score represents the 40th percentile? How should you interpret this?

31. What percentile is a score of 140? How should you interpret this?

32. What percentile is a score of 170? How should you interpret this?

Finding a Percentile   *In Exercises 33–36, use the data set, which represents the ages of 30 executives.*

43 57 65 47 57 41 56 53 61 54
56 50 66 56 50 61 47 40 50 43
54 41 48 45 28 35 38 43 42 44

33. Find the percentile that corresponds to an age of 40 years old.

34. Find the percentile that corresponds to an age of 56 years old.

35. Which ages are above the 75th percentile?

36. Which ages are below the 25th percentile?

Finding and Interpreting Percentiles   *In Exercises 37–40, use the data set, which represents wait times (in minutes) for various services at a state's Department of Motor Vehicles locations.*

6 10 1 22 23 10 6 7 2 1 6 6 2 4 14 15 16 4
19 3 19 26 5 3 4 7 6 10 9 10 20 18 3 20 10 13
14 11 14 17 4 27 4 8 4 3 26 18 21 1 3 3 5 5

37. Draw an ogive to show corresponding percentiles for the data.

38. Which wait time represents the 50th percentile? How would you interpret this?

39. Find the percentile that corresponds to a wait time of 20 minutes.

40. Which wait times are between the 25th and 75th percentiles?

Graphical Analysis   *In Exercises 41 and 42, the midpoints A, B, and C are marked on the histogram at the left. Match them with the indicated z-scores. Which z-scores, if any, would be considered unusual?*

41. $z = 0$, $z = 2.14$, $z = -1.43$

42. $z = 0.77$, $z = 1.54$, $z = -1.54$

43. Not unusual; The z-score is 0.94, so the age of 31 is about 0.94 standard deviation above the mean.

44. Not unusual; The z-score is −1.18, so the age of 24 is about 1.18 standard deviations below the mean.

45. Not unusual; The z-score is −0.27, so the age of 27 is about 0.27 standard deviation below the mean.

46. Unusual; The z-score is 2.45, so the age of 36 is about 2.45 standard deviations above the mean.

47. Unusual; The z-score is −2.39, so the age of 20 is about 2.39 standard deviations below the mean.

48. Not unusual; The z-score is 0.03, so the age of 28 is about 0.03 standard deviation above the mean.

49. (a) For 34,000, $z \approx -0.44$; For 37,000, $z \approx 0.89$; For 30,000, $z \approx -2.22$

   The tire with a life span of 30,000 miles has an unusually short life span.

   (b) For 30,500, about 2.5th percentile; For 37,250, about 84th percentile; For 35,000, about 50th percentile

50. (a) For 34, $z = 0.25$; For 30, $z = -0.75$; For 42, $z = 2.25$

   The life span of 42 days is unusual.

   (b) For 29, about 16th percentile; For 41, about 98th percentile; For 25, about 2.5th percentile

51. Robert Duvall: $z \approx 1.07$; Jack Nicholson: $z \approx -0.32$; The age of Robert Duvall was about 1 standard deviation above the mean age of Best Actor winners, and the age of Jack Nicholson was less than 1 standard deviation below the mean age of Best Supporting Actor winners. Neither actor's age is unusual.

52. See Selected Answers, page A95.

53. See Odd Answers, page A50.

54. See Selected Answers, page A95.

Finding z-Scores *The distribution of the ages of the winners of the Tour de France from 1903 to 2016 is approximately bell-shaped. The mean age is 27.9 years, with a standard deviation of 3.3 years. In Exercises 43–48, use the corresponding z-score to determine whether the age is unusual. Explain your reasoning.* (*Source: Le Tour de France*)

	Winner	Year	Age
43.	Christopher Froome	2016	31
44.	Jan Ullrich	1997	24
45.	Antonin Magne	1931	27
46.	Firmin Lambot	1922	36
47.	Henri Cornet	1904	20
48.	Christopher Froome	2013	28

49. Life Spans of Tires A brand of automobile tire has a mean life span of 35,000 miles, with a standard deviation of 2250 miles. Assume the life spans of the tires have a bell-shaped distribution.

(a) The life spans of three randomly selected tires are 34,000 miles, 37,000 miles, and 30,000 miles. Find the z-score that corresponds to each life span. Determine whether any of these life spans are unusual.

(b) The life spans of three randomly selected tires are 30,500 miles, 37,250 miles, and 35,000 miles. Using the Empirical Rule, find the percentile that corresponds to each life span.

50. Life Spans of Fruit Flies The life spans of a species of fruit fly have a bell-shaped distribution, with a mean of 33 days and a standard deviation of 4 days.

(a) The life spans of three randomly selected fruit flies are 34 days, 30 days, and 42 days. Find the z-score that corresponds to each life span. Determine whether any of these life spans are unusual.

(b) The life spans of three randomly selected fruit flies are 29 days, 41 days, and 25 days. Using the Empirical Rule, find the percentile that corresponds to each life span.

Comparing z-Scores from Different Data Sets *The table shows population statistics for the ages of Best Actor and Best Supporting Actor winners at the Academy Awards from 1929 to 2016. The distributions of the ages are approximately bell-shaped. In Exercises 51–54, compare the z-scores for the actors.*

Best actor	Best supporting actor
$\mu \approx 43.7$ yr	$\mu \approx 50.4$ yr
$\sigma \approx 8.7$ yr	$\sigma \approx 13.8$ yr

51. Best Actor 1984: Robert Duvall, Age: 53
Best Supporting Actor 1984: Jack Nicholson, Age: 46

52. Best Actor 2005: Jamie Foxx, Age: 37
Best Supporting Actor 2005: Morgan Freeman, Age: 67

53. Best Actor 1970: John Wayne, Age: 62
Best Supporting Actor 1970: Gig Young, Age: 56

54. Best Actor 1982: Henry Fonda, Age: 76
Best Supporting Actor 1982: John Gielgud, Age: 77

55. 5

56. 33.75

57. (a) The distribution of Concert 1 is symmetric. The distribution of Concert 2 is skewed right. Concert 1 has less variation.

(b) Concert 2 is more likely to have outliers because it has more variation.

(c) Concert 1, because 68% of the data should be between ±16.3 of the mean.

(d) No, you do not know the number of songs played at either concert or the actual lengths of the songs.

58.

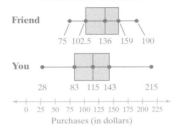

The shape of your distribution is symmetric, and the shape of your friend's distribution is slightly skewed right.

59. (a) 24, 2

(b)

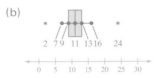

60. (a) 62, 95

(b)

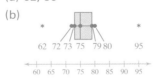

61. (a) 1

(b)

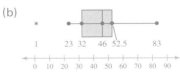

62. (a) 90

(b)

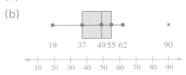

63. Answers will vary.

Extending Concepts

Midquartile *Another measure of position is called the* **midquartile.** *You can find the midquartile of a data set by using the formula below.*

$$\text{Midquartile} = \frac{Q_1 + Q_3}{2}$$

In Exercises 55 and 56, find the midquartile of the data set.

55. 5 7 1 2 3 10 8 7 5 3

56. 23 36 47 33 34 40 39 24 32 22 38 41

57. Song Lengths   Side-by-side box-and-whisker plots can be used to compare two or more different data sets. Each box-and-whisker plot is drawn on the same number line to compare the data sets more easily. The lengths (in seconds) of songs played at two different concerts are shown.

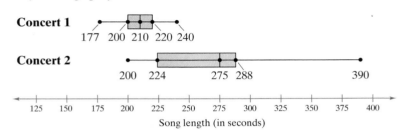

(a) Describe the shape of each distribution. Which concert has less variation in song lengths?

(b) Which distribution is more likely to have outliers? Explain.

(c) Which concert do you think has a standard deviation of 16.3? Explain.

(d) Can you determine which concert lasted longer? Explain.

58. Credit Card Purchases The credit card purchases (rounded to the nearest dollar) over the last three months for you and a friend are listed.

You	60	95	102	110	130	130	162	200	215	120	124	28
	58	40	102	105	141	160	130	210	145	90	46	76
Friend	100	125	132	90	85	75	140	160	180	190	160	105
	145	150	151	82	78	115	170	158	140	130	165	125

Use technology to draw side-by-side box-and-whisker plots that represent the data sets. Then describe the shapes of the distributions.

Modified Box-and-Whisker Plot *In Exercises 59–62, (a) identify any outliers and (b) draw a modified box-and-whisker plot that represents the data set. Use asterisks (*) to identify outliers.*

59. 16 9 11 12 8 10 12 13 11 10 24 9 2 15 7

60. 75 78 80 75 62 72 74 75 80 95 76 72

61. 47 29 59 83 46 1 46 23 52 53 35 37 49

62. 36 38 47 50 53 54 19 27 30 47 48 50 56 60 90 62

63. Project Find a real-life data set and use the techniques of Chapter 2, including graphs and numerical quantities, to discuss the center, variation, and shape of the data set. Describe any patterns.

Statistics in the Real World

Uses

Descriptive statistics help you see trends or patterns in a set of raw data. A good description of a data set consists of (1) a measure of the center of the data, (2) a measure of the variability (or spread) of the data, and (3) the shape (or distribution) of the data. When you read reports, news items, or advertisements prepared by other people, you are rarely given the raw data used for a study. Instead, you see graphs, measures of central tendency, and measures of variability. To be a discerning reader, you need to understand the terms and techniques of descriptive statistics.

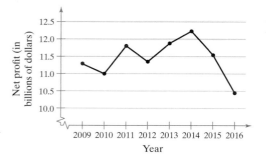

Procter & Gamble's Net Profit

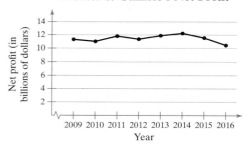

Procter & Gamble's Net Profit

Abuses

Knowing how statistics are calculated can help you analyze questionable statistics. For instance, you are interviewing for a sales position and the company reports that the average yearly commission earned by the five people in its sales force is $60,000. This is a misleading statement if it is based on four commissions of $25,000 and one of $200,000. The median would more accurately describe the yearly commission, but the company used the mean because it is a greater amount.

Statistical graphs can also be misleading. Compare the two time series charts at the left, which show the net profits for the Procter & Gamble Corporation from 2009 through 2016. The data are the same for each chart. The first time series chart, however, has a cropped vertical axis, which makes it appear that the net profit decreased greatly from 2009 to 2010, from 2011 to 2012, and from 2014 to 2016, and increased greatly from 2010 to 2011 and from 2012 to 2014. In the second time series chart, the scale on the vertical axis begins at zero. This time series chart correctly shows that the net profit changed modestly during this time period. *(Source: Procter & Gamble Corporation)*

Ethics

Mark Twain helped popularize the saying, "There are three kinds of lies: lies, damned lies, and statistics." In short, even the most accurate statistics can be used to support studies or statements that are incorrect. Unscrupulous people can use misleading statistics to "prove" their point. Being informed about how statistics are calculated and questioning the data are ways to avoid being misled.

EXERCISES

1. Use the Internet or some other resource to find an example of a graph that might lead to incorrect conclusions.

2. You are publishing an article that discusses how drinking red wine can help prevent heart disease. Because drinking red wine might help people at risk for heart disease, you include a graph that exaggerates the effects of drinking red wine and preventing heart disease. Do you think it is ethical to publish this graph? Explain.

2 Chapter Summary

What Did You Learn?	Example(s)	Review Exercises
Section 2.1		
▶ How to construct a frequency distribution including limits, midpoints, relative frequencies, cumulative frequencies, and boundaries	1, 2	1
▶ How to construct frequency histograms, frequency polygons, relative frequency histograms, and ogives	3–7	2–6
Section 2.2		
▶ How to graph and interpret quantitative data sets using stem-and-leaf plots and dot plots	1–3	7, 8
▶ How to graph and interpret qualitative data sets using pie charts and Pareto charts	4, 5	9, 10
▶ How to graph and interpret paired data sets using scatter plots and time series charts	6, 7	11, 12
Section 2.3		
▶ How to find the mean, median, and mode of a population and of a sample	1–6	13, 14
▶ How to find a weighted mean of a data set, and how to estimate the sample mean of grouped data	7, 8	15–18
▶ How to describe the shape of a distribution as symmetric, uniform, or skewed, and how to compare the mean and median for each		19–24
Section 2.4		
▶ How to find the range of a data set, and how to find the variance and standard deviation of a population and of a sample	1–4	25–28
▶ How to use the Empirical Rule and Chebychev's Theorem to interpret standard deviation	5–7	29–32
▶ How to estimate the sample standard deviation for grouped data	8, 9	33, 34
▶ How to use the coefficient of variation to compare variation in different data sets	10	35, 36
Section 2.5		
▶ How to find the first, second, and third quartiles of a data set, how to find the interquartile range of a data set, and how to represent a data set graphically using a box-and-whisker plot	1–4	37–42
▶ How to interpret other fractiles such as percentiles, and how to find percentiles for a specific data entry	5, 6	43, 44
▶ How to find and interpret the standard score (z-score)	7, 8	45–48

2 Review Exercises

1. See Odd Answers, page A51.

2. See Selected Answers, page A95.

3.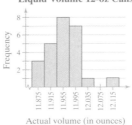
Liquid Volume 12-oz Cans

Volumes (in ounces)

11.95	11.91	11.86	11.94	12.00
11.93	12.00	11.94	12.10	11.95
11.99	11.94	11.89	12.01	11.99
11.94	11.92	11.98	11.88	11.94
11.98	11.92	11.95	11.93	12.04

TABLE FOR EXERCISES 3 AND 4

4.
Liquid Volume 12-oz Cans

5. See Odd Answers, page A51.

6. Rooms Reserved

7. See Odd Answers, page A51.

8. See Selected Answers, page A95.

9. See Odd Answers, page A51.

10. See Selected Answers, page A95.

Section 2.1

In Exercises 1 and 2, use the data set, which represents the overall average class sizes for 20 national universities. (Adapted from Public University Honors)

| 37 | 34 | 42 | 44 | 39 | 40 | 41 | 51 | 49 | 31 |
| 55 | 26 | 31 | 40 | 30 | 27 | 36 | 43 | 49 | 35 |

1. Construct a frequency distribution for the data set using five classes. Include class limits, midpoints, boundaries, frequencies, relative frequencies, and cumulative frequencies.

2. Construct a relative frequency histogram using the frequency distribution in Exercise 1. Then determine which class has the greatest relative frequency and which has the least relative frequency.

In Exercises 3 and 4, use the data set shown in the table at the left, which represents the actual liquid volumes (in ounces) in 25 twelve-ounce cans.

3. Construct a frequency histogram for the data set using seven classes.

4. Construct a relative frequency histogram for the data set using seven classes.

In Exercises 5 and 6, use the data set, which represents the numbers of rooms reserved during one night's business at a sample of hotels.

153	104	118	166	89	104	100	79	93	96	116
94	140	84	81	96	108	111	87	126	101	111
122	108	126	93	108	87	103	95	129	93	124

5. Construct a frequency distribution for the data set with six classes and draw a frequency polygon.

6. Construct an ogive for the data set using six classes.

Section 2.2

In Exercises 7 and 8, use the data set, which represents the pollution indices for 24 U.S. cities. (Adapted from Numbeo)

| 22 | 41 | 46 | 50 | 38 | 57 | 65 | 49 | 33 | 28 | 53 | 32 |
| 41 | 23 | 38 | 65 | 28 | 36 | 63 | 54 | 39 | 43 | 56 | 39 |

7. Use a stem-and-leaf plot to display the data set. Describe any patterns.

8. Use a dot plot to display the data set. Describe any patterns.

In Exercises 9 and 10, use the data set, which represents the results of a survey that asked U.S. full-time university and college students about their activities and time use on an average weekday. (Source: Bureau of Labor Statistics)

Response	Sleeping	Leisure and Sports	Working	Educational Activities	Other
Time (in hours)	8.8	4.0	2.3	3.5	5.4

9. Use a pie chart to display the data set. Describe any patterns.

10. Use a Pareto chart to display the data set. Describe any patterns.

11.

Heights of Buildings

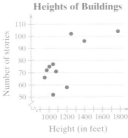

Sample answer: The number of stories appears to increase with height.

12.

U.S. Unemployment Rate

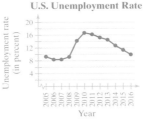

Sample answer: The real unemployment rate varied by a couple of percentage points from 2005 to 2008, then increased dramatically from 2008 to 2010, and then decreased from 2011 to 2016.

13. $\bar{x} = 29.5$; median $= 29.5$; mode $= 29.5$

14. $\bar{x}$ is not possible; median is not possible; mode $=$ "$250–999"; The mean and median cannot be found because the data are at the nominal level of measurement.

15. 82.1

16. 88.8

17. 38.4

18. About 2.1

19. Skewed right

20. Skewed right

11. The heights (in feet) and the numbers of stories of the ten tallest buildings in New York City are listed. Use a scatter plot to display the data. Describe any patterns. *(Source: Emporis)*

Height (in feet)	1776	1398	1250	1200	1079	1046	1046	1005	975	952
Stories	104	96	102	58	71	77	52	75	72	66

12. The U.S. real unemployment rates over a 12-year period are listed. Use a time series chart to display the data. Describe any patterns. *(Source: U.S. Bureau of Labor Statistics)*

Year	2005	2006	2007	2008	2009	2010
Rate	9.3%	8.4%	8.4%	9.2%	14.2%	16.7%

Year	2011	2012	2013	2014	2015	2016
Rate	16.2%	15.2%	14.5%	12.7%	11.3%	9.9%

Section 2.3

In Exercises 13 and 14, find the mean, the median, and the mode of the data, if possible. If any measure cannot be found or does not represent the center of the data, explain why.

13. The vertical jumps (in inches) of a sample of 10 college basketball players at the 2016 NBA Draft Combine *(Source: DraftExpress)*

   33.0   35.5   37.5   31.0   28.0   29.5   21.0   26.0   24.0   29.5

14. The responses of 1019 adults who were asked how much money they think they will spend on Christmas gifts in a recent year *(Adapted from Gallup)*

   $1000 or more: 306 $250–999: 336   Less than $250: 234
   Not sure: 51   None/do not celebrate Christmas: 92

15. For the six test scores 78, 72, 86, 91, 87, and 80, the first 5 test scores are 15% of the final grade and the last test score is 25% of the final grade. Find the weighted mean of the test scores.

16. For the four test scores 96, 85, 91, and 86, the first 3 test scores are 20% of the final grade, and the last test score is 40% of the final grade. Find the weighted mean of the test scores.

17. Estimate the mean of the frequency distribution you made in Exercise 1.

18. The frequency distribution shows the numbers of magazine subscriptions per household for a sample of 60 households. Find the mean number of subscriptions per household.

Number of magazines	0	1	2	3	4	5	6
Frequency	13	9	19	8	5	2	4

19. Describe the shape of the distribution for the histogram you made in Exercise 3 as symmetric, uniform, skewed left, skewed right, or none of these.

20. Describe the shape of the distribution for the histogram you made in Exercise 4 as symmetric, uniform, skewed left, skewed right, or none of these.

21. Skewed right

22. Skewed left

23. Mean; When a distribution is skewed right, the mean is to the right of the median.

24. Median; When a distribution is skewed left, the mean is to the left of the median.

25. Range = 14; $\mu \approx 6.9$;
$\sigma^2 \approx 21.1$; $\sigma \approx 4.6$

26. Range = 27; $\mu \approx 69.25$;
$\sigma^2 \approx 86.19$; $\sigma \approx 9.28$

27. Range = \$2044; $\bar{x} \approx$ \$6266.81;
$s^2 \approx 455{,}944.30$; $s \approx$ \$675.24

28. Range = \$26,414;
$\bar{x} =$ \$54,127.50;
$s^2 \approx 82{,}778{,}148.86$;
$s \approx$ \$9098.25

29. \$75 and \$145

30. 68%

31. 30 customers

32. 101 flights

In Exercises 21 and 22, determine whether the approximate shape of the distribution in the histogram is symmetric, uniform, skewed left, skewed right, or none of these.

21.

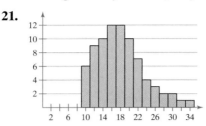

22.

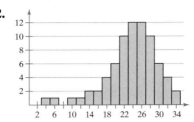

23. For the histogram in Exercise 21, which is greater, the mean or the median? Explain your reasoning.

24. For the histogram in Exercise 22, which is greater, the mean or the median? Explain your reasoning.

Section 2.4

In Exercises 25 and 26, find the range, mean, variance, and standard deviation of the population data set.

25. The mileages (in thousands of miles) for a rental car company's fleet.

     4   2   9   12   15   3   6   8   1   4   14   12   3   3

26. The ages of the Supreme Court justices as of December 22, 2016 *(Source: Supreme Court of the United States)*

     61   80   68   83   78   66   62   56

In Exercises 27 and 28, find the range, mean, variance, and standard deviation of the sample data set.

27. Dormitory room charges (in dollars) for one school year for a random sample of four-year universities

     5816   6045   5612   6341   6106   7361   6320   6265
     7220   7439   5395   6908   5561   5710   5538   6632

28. Salaries (in dollars) of a random sample of teachers

     62,222   56,719   50,259   45,120   47,692   45,985   53,489   71,534

In Exercises 29 and 30, use the Empirical Rule.

29. The mean charge for electricity for a sample of households was \$110.00 per month, with a standard deviation of \$17.50 per month. Between what two values do 95% of the data lie? (Assume the data set has a bell-shaped distribution.)

30. The mean charge for satellite television for a sample of households was \$87.50 per month, with a standard deviation of \$14.50 per month. Estimate the percent of satellite television charges between \$73.00 and \$102.00. (Assume the data set has a bell-shaped distribution.)

31. The mean sale per customer for 40 customers at a gas station is \$32.00, with a standard deviation of \$4.00. Using Chebychev's Theorem, determine at least how many of the customers spent between \$24.00 and \$40.00.

32. The mean duration of the 135 space shuttle flights was about 9.9 days, and the standard deviation was about 3.8 days. Using Chebychev's Theorem, determine at least how many of the flights lasted between 2.3 days and 17.5 days. *(Source: NASA)*

33. $\bar{x} \approx 2.5$; $s \approx 1.2$
34. $\bar{x} \approx 2.4$; $s \approx 1.7$
35. $CV_{\text{freshmen}} \approx 41.3\%$
    $CV_{\text{seniors}} \approx 24.2\%$
    Grade point averages are more variable for freshmen than seniors.
36. $CV_{\text{age}} \approx 21.6\%$
    $CV_{\text{experience}} \approx 38.3\%$
    Years of experience are more variable than ages of the lawyers.
37. Min = 16, Q_1 = 25, Q_2 = 35, Q_3 = 56, Max = 136
38. 31 miles per gallon
39. **Model 2017 Vehicle Fuel Economies**

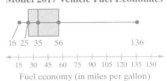

16 25 35 56   136

15 30 45 60 75 90 105 120 135 150
Fuel economy (in miles per gallon)
40. 20
41. 7 inches
42. **Weights of Football Players**

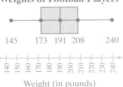

145   173 191 208   240

140 150 160 170 180 190 200 210 220 230 240
Weight (in pounds)

The distribution of the weights is symmetric.
43. 35%
44. 84th percentile
45. Not unusual; The z-score is 1.97, so a towing capacity of 16,500 pounds is about 1.97 standard deviations above the mean.
46. Unusual; The z-score is −2.67, so a towing capacity of 5500 pounds is about 2.67 standard deviations below the mean.
47. Unusual; The z-score is 2.60, so a towing capacity of 18,000 pounds is about 2.60 standard deviations above the mean.
48. Not unusual; The z-score is −0.22, so a towing capacity of 11,300 pounds is about 0.22 standard deviation below the mean.

33. From a random sample of households, the numbers of televisions are listed. Find the sample mean and the sample standard deviation of the data.

Number of televisions	0	1	2	3	4	5
Number of households	1	8	13	10	5	3

34. From a random sample of airplanes, the numbers of defects found in their fuselages are listed. Find the sample mean and the sample standard deviation of the data.

Number of defects	0	1	2	3	4	5	6
Number of airplanes	4	5	2	9	1	3	1

In Exercises 35 and 36, find the coefficient of variation for each of the two data sets. Then compare the results.

35. Sample grade point averages for freshmen and seniors are listed.

Freshmen	2.8	1.8	4.0	3.8	2.4	2.0	0.9	3.6	1.8
Seniors	2.3	3.3	1.8	4.0	3.1	2.7	3.9	2.6	2.9

36. The ages and years of experience for all lawyers at a firm are listed.

Ages	66	54	47	61	36	59	50	33
Years of experience	37	20	23	32	14	29	22	8

Section 2.5

In Exercises 37–40, use the data set, which represents the model 2017 vehicles with the highest fuel economies (in miles per gallon) in the most popular classes. (Source: U.S. Environmental Protection Agency)

35 35 112 34 124 35 107 46 136 56 58 119 50
41 25 25 22 16 16 52 22 22 22 34 30 30

37. Find the five-number summary of the data set.
38. Find the interquartile range of the data set.
39. Draw a box-and-whisker plot that represents the data set.
40. About how many vehicles fall on or below the third quartile?
41. Find the interquartile range of the data set from Exercise 13.
42. The weights (in pounds) of the defensive players on a high school football team are shown below. Draw a box-and-whisker plot that represents the data set and describe the shape of the distribution.

173 145 205 192 197 227 156 240 172
208 185 190 167 212 228 190 184 195

43. A student's test grade of 75 represents the 65th percentile of the grades. What percent of students scored higher than 75?
44. As of December 2016, there were 721 adult contemporary radio stations in the United States. One station finds that 115 stations have a larger daily audience than it has. What percentile does this station come closest to in the daily audience rankings? *(Source: Radio-Locator.com)*

The towing capacities (in pounds) of all the pickup trucks at a dealership have a bell-shaped distribution, with a mean of 11,830 pounds and a standard deviation of 2370 pounds. In Exercises 45–48, use the corresponding z-score to determine whether the towing capacity is unusual. Explain your reasoning.

45. 16,500 pounds   **46.** 5500 pounds   **47.** 18,000 pounds   **48.** 11,300 pounds

2 Chapter Quiz

1. See Odd Answers, page A52.

2. $\bar{x} \approx 126.1$; $s \approx 13.0$

3. (a)

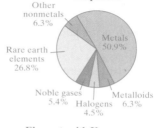

Elements with Known Properties

Other nonmetals 6.3%
Rare earth elements 26.8%
Metals 50.9%
Noble gases 5.4%
Halogens 4.5%
Metalloids 6.3%

(b)

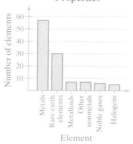

Elements with Known Properties

Number of elements

Metals
Rare earth elements
Metalloids
Other nonmetals
Noble gases
Halogens

Element

4. (a) $\bar{x} \approx 1016.4$; median = 1019; mode = 1100; The mean or median best describes a typical salary because there are no outliers.

(b) Range = 666; $s^2 \approx 47{,}120.9$; $s \approx 217.1$

(c) $CV \approx 21.4\%$

5. $150,000 and $210,000

6. (a) Unusual; The z-score is 3, so a new home price of $225,000 is about 3 standard deviations above the mean.

(b) Unusual; The z-score is −6.67, so a new home price of $80,000 is about 6.67 standard deviations below the mean.

(c) Not unusual; The z-score is 1.33, so a new home price of $200,000 is about 1.33 standard deviations above the mean.

(d) Unusual; The z-score is −2.2, so a new home price of $147,000 is about 2.2 standard deviations below the mean.

7. See Odd Answers, page A53.

Take this quiz as you would take a quiz in class. After you are done, check your work against the answers given in the back of the book.

 1. The data set represents the numbers of minutes a sample of 27 people exercise each week.

108	139	120	123	120	132	123	131	131
157	150	124	111	101	135	119	116	117
127	128	139	119	118	114	127	142	130

(a) Construct a frequency distribution for the data set using five classes. Include class limits, midpoints, boundaries, frequencies, relative frequencies, and cumulative frequencies.

(b) Display the data using a frequency histogram and a frequency polygon on the same axes.

(c) Display the data using a relative frequency histogram.

(d) Describe the shape of the distribution as symmetric, uniform, skewed left, skewed right, or none of these.

(e) Display the data using an ogive.

(f) Display the data using a stem-and-leaf plot. Use one line per stem.

(g) Display the data using a box-and-whisker plot.

2. Use frequency distribution formulas to approximate the sample mean and the sample standard deviation of the data set in Exercise 1.

3. The elements with known properties can be classified as metals (57 elements), metalloids (7 elements), halogens (5 elements), noble gases (6 elements), rare earth elements (30 elements), and other nonmetals (7 elements). Display the data using (a) a pie chart and (b) a Pareto chart.

4. Weekly salaries (in dollars) for a sample of construction workers are listed.

| 1100 | 720 | 1384 | 1124 | 1255 | 976 | 718 | 1316 |
| 749 | 1062 | 1248 | 891 | 969 | 790 | 860 | 1100 |

(a) Find the mean, median, and mode of the salaries. Which best describes a typical salary?

(b) Find the range, variance, and standard deviation of the data set.

(c) Find the coefficient of variation of the data set.

5. The mean price of new homes from a sample of houses is $180,000 with a standard deviation of $15,000. The data set has a bell-shaped distribution. Using the Empirical Rule, between what two prices do 95% of the houses fall?

6. Refer to the sample statistics from Exercise 5 and determine whether any of the house prices below are unusual. Explain your reasoning.

(a) $225,000   (b) $80,000   (c) $200,000   (d) $147,000

 7. The numbers of regular season wins for each Major League Baseball team in 2016 are listed. Display the data using a box-and-whisker plot.

(Source: Major League Baseball)

93	89	89	84	68	94	86	81	78	59
95	86	84	74	69	95	87	79	71	68
103	86	78	73	68	91	87	75	69	68

2 Chapter Test

1. (a) $\bar{x} \approx 80.3$; median = 83;
   mode = 87; The median best
   represents the center of the
   data.
   (b) Range = 36; $s^2 \approx 132.6$;
   $s \approx 11.5$
   (c) $CV \approx 14.3\%$
   (d) **Average Scores**

   | 6 | 3 7 8 | Key: 6\|3 = 63
7	2 3
8	1 5 7 7 8
9	4 9

2. See Selected Answers, page A95.

3. $\bar{x} \approx 112.7$; $s \approx 46.9$

4. 75th percentile

5. See Selected Answers, page A95.

6. **Students in a Statistics Class**

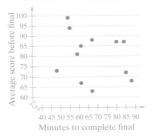

Sample answer: It appears that
there is no relation between
minutes to complete the final
exam and average score before
the final exam.

Certification	Number of albums
Diamond	6
Multi-Platinum	26
Platinum	42
Gold	48

TABLE FOR EXERCISE 5

7. (a)   **Ages of College Professors**

   box-and-whisker plot with values 28, 40, 55, 60, 68 and axis 25 30 35 40 45 50 55 60 65 70 Age

(b) 75%

8. (a) About 141
   (b) Unusual

Take this test as you would take a test in class.

1. The overall averages of 12 students in a statistics class prior to taking the final exam are listed.

   67   72   88   73   99   85   81   87   63   94   68   87

   (a) Find the mean, median, and mode of the data set. Which best represents the center of the data?
   (b) Find the range, variance, and standard deviation of the sample data set.
   (c) Find the coefficient of variation of the data set.
   (d) Display the data in a stem-and-leaf plot. Use one line per stem.

2. The data set represents the numbers of movies that a sample of 20 people watched in a year.

   121   148   94   142   170   88   221   106   18   67
   149   28   60   101   134   168   92   154   53   66

   (a) Construct a frequency distribution for the data set using six classes. Include class limits, midpoints, boundaries, frequencies, relative frequencies, and cumulative frequencies.
   (b) Display the data using a frequency histogram and a frequency polygon on the same axes.
   (c) Display the data using a relative frequency histogram.
   (d) Describe the shape of the distribution as symmetric, uniform, skewed left, skewed right, or none of these.
   (e) Display the data using an ogive.

3. Use frequency distribution formulas to estimate the sample mean and the sample standard deviation of the data set in Exercise 2.

4. For the data set in Exercise 2, find the percentile that corresponds to 149 movies watched in a year.

5. The table lists the numbers of albums by The Beatles that received sales certifications. Display the data using (a) a pie chart and (b) a Pareto chart. *(Source: Recording Industry Association of America)*

6. The numbers of minutes it took 12 students in a statistics class to complete the final exam are listed. Use a scatter plot to display this data set and the data set in Exercise 1. The data sets are in the same order. Describe any patterns.

   61   85   67   48   54   61   59   80   67   55   88   84

7. The data set represents the ages of 15 college professors.

   46   51   60   58   37   65   40   55   30   68   28   62   56   42   59

   (a) Display the data in a box-and-whisker plot.
   (b) About what percent of the professors are over the age of 40?

8. The mean gestational length of a sample of 208 horses is 343.7 days, with a standard deviation of 10.4 days. The data set has a bell-shaped distribution.

   (a) Estimate the number of gestational lengths between 333.3 and 354.1 days.
   (b) Determine whether a gestational length of 318.4 days feet is unusual.

You are a member of your local apartment association. The association represents rental housing owners and managers who operate residential rental property throughout the greater metropolitan area. Recently, the association has received several complaints from tenants in a particular area of the city who feel that their monthly rental fees are much higher compared to other parts of the city.

You want to investigate the rental fees. You gather the data shown in the table at the right. Area A represents the area of the city where tenants are unhappy about their monthly rents. The data represent the monthly rents paid by a random sample of tenants in Area A and three other areas of similar size. Assume all the apartments represented are approximately the same size with the same amenities.

The Monthly Rents (in dollars) Paid by 12 Randomly Selected Apartment Tenants in 4 Areas of Your City

Area A	Area B	Area C	Area D
1435	1265	1221	1044
1249	1074	931	1234
1097	917	893	970
970	1213	1317	827
1171	949	1034	898
1122	839	1061	914
1259	896	851	1387
1022	918	861	1166
1002	1056	911	1123
1187	1218	1148	1029
968	844	799	1131
1097	791	872	1047

EXERCISES

1. How Would You Do It?

(a) How would you investigate the complaints from renters who are unhappy about their monthly rents?

(b) Which statistical measure do you think would best represent the data sets for the four areas of the city?

(c) Calculate the measure from part (b) for each of the four areas.

2. Displaying the Data

(a) What type of graph would you choose to display the data? Explain your reasoning.

(b) Construct the graph from part (a).

(c) Based on your data displays, does it appear that the monthly rents in Area A are higher than the rents in the other areas of the city? Explain.

3. Measuring the Data

(a) What other statistical measures in this chapter could you use to analyze the monthly rent data?

(b) Calculate the measures from part (a).

(c) Compare the measures from part (b) with the graph you constructed in Exercise 2. Do the measurements support your conclusion in Exercise 2? Explain.

4. Discussing the Data

(a) Do you think the complaints in Area A are legitimate? How do you think they should be addressed?

(b) What reasons might you give as to why the rents vary among different areas of the city?

Highest Monthly Rents For Two-Bedroom Apartments

MEDIAN PER CITY

San Francisco, CA	$4570
New York, NY	$4310
Boston, MA	$3200
Jersey City, NJ	$3200
Washington, DC	$3000

(Source: Apartment List)

Parking Tickets

According to data from the city of Toronto, Ontario, Canada, there were more than 180,000 parking infractions in the city for December 2015, with fines totaling over 8,500,000 Canadian dollars.

The fines (in Canadian dollars) for a random sample of 105 parking infractions in Toronto, Ontario, Canada, for December 2015 are listed below. *(Source: City of Toronto)*

30	30	30	30	40	60	40
15	50	150	40	30	30	30
40	30	40	30	30	30	40
40	40	30	60	60	30	150
40	30	250	40	30	30	30
30	30	30	40	30	40	30
50	15	40	40	30	40	30
40	30	30	40	30	30	30
100	30	40	30	30	30	40
30	30	30	40	100	30	40
30	40	30	40	40	40	40
30	30	30	60	30	40	40
30	40	15	60	30	15	150
150	40	40	30	30	150	60
30	40	60	30	40	40	30

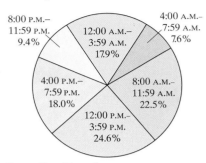

Parking Infractions by Time of Day

8:00 P.M.–11:59 P.M. 9.4%
12:00 A.M.–3:59 A.M. 17.9%
4:00 A.M.–7:59 A.M. 7.6%
4:00 P.M.–7:59 P.M. 18.0%
8:00 A.M.–11:59 A.M. 22.5%
12:00 P.M.–3:59 P.M. 24.6%

(Source: City of Toronto)

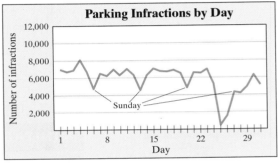

Parking Infractions by Day

(Source: City of Toronto)

The figures above show parking infractions in Toronto, Ontario, Canada, for December 2015 by time of day and by day.

EXERCISES

In Exercises 1–5, use technology. If possible, print your results.

1. Find the sample mean of the data.

2. Find the sample standard deviation of the data.

3. Find the five-number summary of the data.

4. Make a frequency distribution for the data. Use a class width of 15.

5. Draw a histogram for the data. Does the distribution appear to be bell-shaped?

6. What percent of the distribution lies within one standard deviation of the mean? Within two standard deviations of the mean? Within three standard deviations of the mean?

7. Do the results of Exercise 6 agree with the Empirical Rule? Explain.

8. Do the results of Exercise 6 agree with Chebychev's Theorem? Explain.

9. Use the frequency distribution in Exercise 4 to estimate the sample mean and sample standard deviation of the data. Do the formulas for grouped data give results that are as accurate as the individual entry formulas? Explain.

10. **Writing** Do you think the mean or the median better represents the data? Explain your reasoning.

Extended solutions are given in the technology manuals that accompany this text. Technical instruction is provided for Minitab, Excel, and the TI-84 Plus.

2 Using Technology to Determine Descriptive Statistics

Here are some Minitab and TI-84 Plus printouts for three examples in this chapter.

See Example 7, page 61.

Bar Chart...
Pie Chart...
Time Series Plot...
Area Graph...
Contour Plot...
3D Scatterplot...
3D Surface Plot...

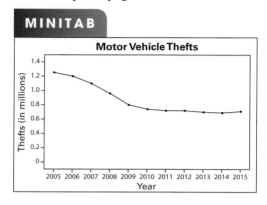

MINITAB

Motor Vehicle Thefts

See Example 3, page 85.

Display Descriptive Statistics...
Store Descriptive Statistics...
Graphical Summary...
1-Sample Z...
1-Sample t...
2-Sample t...
Paired t...

MINITAB

Descriptive Statistics: Recovery times

Variable	N	Mean	SE Mean	StDev	Minimum
Recovery times	12	7.500	0.544	1.883	4.000

Variable	Q1	Median	Q3	Maximum
Recovery times	6.250	7.500	9.000	10.000

See Example 4, page 105.

Empirical CDF...
Probability Distribution Plot ...
Boxplot...
Interval Plot...
Individual Value Plot...
Line Plot...

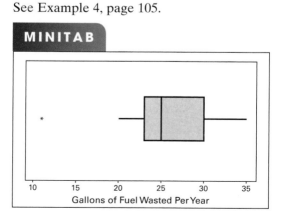

MINITAB

See Example 7, page 61.

TI-84 PLUS

STAT PLOTS
1: Plot1...Off
 L1 L2 .
2: Plot2...Off
 L1 L2 .
3: Plot3...Off
 L1 L2 .
4↓ PlotsOff

TI-84 PLUS

Plot1 Plot2 Plot3
On Off
Type:

Xlist: L1
Ylist: L2
Mark: ■ + .

TI-84 PLUS

ZOOM MEMORY
4↑ ZDecimal
5: ZSquare
6: ZStandard
7: ZTrig
8: ZInteger
9: ZoomStat
0↓ ZoomFit

TI-84 PLUS

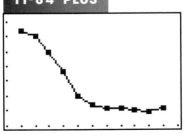

See Example 3, page 85.

TI-84 PLUS

EDIT CALC TESTS
1: 1-Var Stats
2: 2-Var Stats
3: Med-Med
4: LinReg(ax+b)
5: QuadReg
6: CubicReg
7↓ QuartReg

TI-84 PLUS

1-Var Stats
List:L1
FreqList:
Calculate

TI-84 PLUS

1-Var Stats
$\bar{x}=7.5$

$\Sigma x=90$

$\Sigma x^2=714$

Sx=1.882937743
σx=1.802775638

↓n=12

See Example 4, page 105.

TI-84 PLUS

STAT PLOTS
1: Plot1...Off
 L1 L2 .
2: Plot2...Off
 L1 L2 .
3: Plot3...Off
 L1 L2 .
4↓ PlotsOff

TI-84 PLUS

Plot1 Plot2 Plot3
On Off
Type:

Xlist: L1
Freq: 1
Mark: ■ + .

TI-84 PLUS

ZOOM MEMORY
4↑ ZDecimal
5: ZSquare
6: ZStandard
7: ZTrig
8: ZInteger
9: ZoomStat
0↓ ZoomFit

TI-84 PLUS

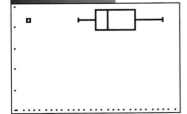

CHAPTERS 1&2
CUMULATIVE REVIEW

Section 1.3 *In Exercises 1 and 2, identify the sampling technique used, and discuss potential sources of bias (if any). Explain.*

1. For quality assurance, every fortieth toothbrush is taken from each of four assembly lines and tested to make sure the bristles stay in the toothbrush.

2. Using random digit dialing, researchers asked 1090 U.S. adults their level of education.

Section 2.2 **3.** In 2016, a worldwide study of workplace fraud found that initial detections of fraud resulted from a tip (39.1%), an internal audit (16.5%), management review (13.4%), detection by accident (5.6%), account reconciliation (5.5%), surveillance/monitoring (1.9%), confession (1.3%), or some other means (16.7%). Use a Pareto chart to organize the data. *(Source: Association of Certified Fraud Examiners)*

Section 1.1 *In Exercises 4 and 5, determine whether the number is a parameter or a statistic. Explain your reasoning.*

4. In 2016, the median annual salary of a marketing account executive was $68,232.

5. In a survey of 1002 U.S. adults, 88% said that fake news has caused a great deal of confusion or some confusion. *(Source: Pew Research Center)*

Sections 2.4 and 2.5 **6.** The mean annual salary for a sample of electrical engineers is $86,500, with a standard deviation of $1500. The data set has a bell-shaped distribution.

(a) Use the Empirical Rule to estimate the percent of electrical engineers whose annual salaries are between $83,500 and $89,500.

(b) The salaries of three randomly selected electrical engineers are $93,500, $85,600, and $82,750. Find the z-score that corresponds to each salary. Determine whether any of these salaries are unusual.

Section 1.1 *In Exercises 7 and 8, identify the population and the sample.*

7. A survey of 339 college and university admissions directors and enrollment officers found that 72% think their institution is losing potential applicants due to concerns about accumulating student loan debt. *(Source: Gallup)*

8. A survey of 67,901 Americans ages 12 years or older found that 1.6% had used pain relievers for nonmedical purposes. *(Source: Substance Abuse and Mental Health Services Administration)*

Section 1.3 *In Exercises 9 and 10, determine whether the study is an observational study or an experiment. Explain.*

9. To study the effect of using digital devices in the classroom on exam performance, researchers divided 726 undergraduate students into three groups, including a group that was allowed to use digital devices, a group that had restricted access to tablets, and a control group that was "technology-free." *(Source: Massachusetts Institute of Technology)*

10. In a study of 7847 children in grades 1 through 5, 15.5% have attention deficit hyperactivity disorder. *(Source: Gallup)*

In Exercises 11 and 12, determine whether the data are qualitative or quantitative, and determine the level of measurement of the data set.

11. The numbers of stolen bases during the 2016 season for Chicago Cubs players who stole at least one base are listed. *(Source: Major League Baseball)*

$$2 \quad 8 \quad 3 \quad 2 \quad 13 \quad 12 \quad 6 \quad 2 \quad 1 \quad 5 \quad 11 \quad 1$$

12. The six top-earning states in 2015 by median household income are listed. *(Source: U.S. Census Bureau)*

1. New Hampshire   2. Alaska   3. Maryland
4. Connecticut   5. Minnesota   6. New Jersey

13. The numbers of tornadoes by state in 2016 are listed. (a) Draw a box-and-whisker plot that represents the data set and (b) describe the shape of the distribution. *(Source: National Oceanic and Atmospheric Administration)*

87	0	3	23	7	45	0	0	48	27
0	1	50	40	46	99	32	31	2	2
2	15	44	67	23	4	47	0	2	2
3	1	16	32	31	55	4	9	0	3
16	11	90	3	0	12	6	6	11	1

14. Five test scores are shown below. The first 4 test scores are 15% of the final grade, and the last test score is 40% of the final grade. Find the weighted mean of the test scores.

$$85 \quad 92 \quad 84 \quad 89 \quad 91$$

15. Tail lengths (in feet) for a sample of American alligators are listed.

$$6.5 \quad 3.4 \quad 4.2 \quad 7.1 \quad 5.4 \quad 6.8 \quad 7.5 \quad 3.9 \quad 4.6$$

(a) Find the mean, median, and mode of the tail lengths. Which best describes a typical American alligator tail length? Explain your reasoning.

(b) Find the range, variance, and standard deviation of the data set.

16. A study shows that life expectancies for Americans have increased or remained stable every year for the past five years.

(a) Make an inference based on the results of the study.

(b) What is wrong with this type of reasoning?

In Exercises 17–19, use the data set, which represents the points scored by each player on the Montreal Canadiens in the 2015–2016 NHL season. (Source: National Hockey League)

17	10	0	19	2	18	9	5	1	29
5	26	0	12	20	10	56	40	2	6
0	2	0	0	2	44	1	2	19	64
7	16	54	0	4	12	51	2	0	26

17. Construct a frequency distribution for the data set using eight classes. Include class limits, midpoints, boundaries, frequencies, relative frequencies, and cumulative frequencies.

18. Describe the shape of the distribution.

19. Construct a relative frequency histogram using the frequency distribution in Exercise 17. Then determine which class has the greatest relative frequency and which has the least relative frequency.

CHAPTER 3

Probability

Nine-time Olympic gold medalist Carl Lewis tested positive for banned stimulant use during the 1988 Olympic trials, but the results were overturned and Lewis was allowed to compete.

Where You've Been

In Chapters 1 and 2, you learned how to collect and describe data. Once the data are collected and described, you can use the results to write summaries, draw conclusions, and make decisions. For instance, the International Olympic Committee has been testing Olympic athletes for the use of performance enhancing drugs (PEDs) since 1968. Like most lab tests, Olympic drug tests have a chance of producing false results. By collecting and analyzing data, you can determine the accuracy of these tests in order to minimize the chance of a false result.

Over 10,000 athletes competed in the 2016 Summer Olympics in Rio, Brazil. It is estimated that around 29% of Olympic athletes are guilty of using PEDs. An athlete is not penalized for PED use unless the International Olympic Committee is almost 100% sure the athlete is guilty. Given the percentage of guilty athletes and the accuracy of the drug test, how sure can they be?

Where You're Going

In Chapter 3, you will learn how to determine the probability of an event. For instance, you can find the probability $P(\text{guilty}|\text{fail})$ that an athlete is guilty of using PEDs, given that the athlete failed the drug test. This type of probability is a *conditional probability,* which will be discussed in Section 3.2.

Assume a drug test has been shown to be sensitive enough that an athlete using PEDs has a 90% chance of testing positive. In other words, the drug test has a *sensitivity* of 0.9. You can use this percentage and the estimated overall percentage of guilty athletes to calculate $P(\text{guilty}|\text{fail})$.

The table below shows the probabilities $P(\text{guilty}|\text{fail})$ for drug tests with different sensitivities.

Drug Test Sensitivity	$P(\text{guilty}|\text{fail})$
0.9	0.79
0.95	0.89
0.99	0.98
0.999	0.998

3.1 Basic Concepts of Probability and Counting

What You Should Learn

▶ How to identify the sample space of a probability experiment and how to identify simple events

▶ How to use the Fundamental Counting Principle to find the number of ways two or more events can occur

▶ How to distinguish among classical probability, empirical probability, and subjective probability

▶ How to find the probability of the complement of an event

▶ How to use a tree diagram and the Fundamental Counting Principle to find probabilities

Probability Experiments

When weather forecasters say that there is a 90% chance of rain or a physician says there is a 35% chance for a successful surgery, they are stating the likelihood, or *probability*, that a specific event will occur. Decisions such as "should you go golfing" or "should you proceed with surgery" are often based on these probabilities. In the preceding chapter, you learned about the role of the descriptive branch of statistics. The second branch, inferential statistics, has probability as its foundation, so it is necessary to learn about probability before proceeding.

> **DEFINITION**
>
> A **probability experiment** is an action, or trial, through which specific results (counts, measurements, or responses) are obtained. The result of a single trial in a probability experiment is an **outcome.** The set of all possible outcomes of a probability experiment is the **sample space.** An **event** is a subset of the sample space. It may consist of one or more outcomes.

Study Tip

Here is a simple example of the use of the terms *probability experiment, sample space, event,* and *outcome.*

Probability Experiment:
    Roll a six-sided die.
Sample Space:
    {1, 2, 3, 4, 5, 6}
Event:
    Roll an even number, {2, 4, 6}.
Outcome:
    Roll a 2, {2}.

Note to Instructor

Bring dice, a deck of cards, and some coins to class to illustrate examples. Differently colored gaming chips work well also.

EXAMPLE 1

Identifying the Sample Space of a Probability Experiment

A survey consists of asking people for their blood types (O, A, B, and AB), including whether they are Rh-positive or Rh-negative. Determine the number of outcomes and identify the sample space.

SOLUTION

There are four blood types: O, A, B, and AB. For each person, they are either Rh-positive or Rh-negative. A **tree diagram** gives a visual display of the outcomes of a probability experiment by using branches that originate from a starting point. It can be used to find the number of possible outcomes in a sample space as well as individual outcomes.

Tree Diagram for Blood Types

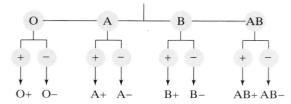

From the tree diagram, you can see that the sample space has eight possible outcomes, which are listed below.

{O+, O−, A+, A−, B+, B−, AB+, AB−}   Sample space

SURVEY

Does your favorite
team's win or loss
affect your mood?

Check one response:

☐ Yes
☐ No
☐ Not sure

Source: Rasmussen

**Diagram for Coin and
Die Experiment**

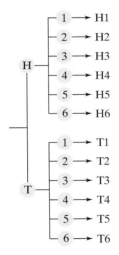

Note to Instructor

Ask students to use a tree diagram
to develop the sample space that
results from rolling two six-sided dice.
Emphasize that the outcome {3, 2}
is different from the outcome {2, 3}
for a statistician, but not for a player.
Point out that the event {3, 2} is a
simple event. The event "roll a sum
of 5" is not simple because it consists
of the four outcomes {1, 4}, {2, 3},
{3, 2}, and {4, 1}.

TRY IT YOURSELF 1

For each probability experiment, determine the number of outcomes and
identify the sample space.

1. A probability experiment consists of recording a response to the survey
statement at the left and the gender of the respondent.

2. A probability experiment consists of recording a response to the survey
statement at the left and the age (18–34, 35–49, 50 and older) of the
respondent.

3. A probability experiment consists of recording a response to the survey
statement at the left and the geographic location (Northeast, South,
Midwest, West) of the respondent.

Answer: Page A33

In the rest of this chapter, you will learn how to calculate the probability or
likelihood of an event. Events are often represented by uppercase letters, such as
A, B, and C. An event that consists of a single outcome is called a **simple event.**
For instance, consider a probability experiment that consists of tossing a coin and
then rolling a six-sided die, as shown in the tree diagram at the left. The event
"tossing heads and rolling a 3" is a simple event and can be represented as

$$A = \{H3\}.$$ Event A has one outcome, so it is a simple event.

In contrast, the event "tossing heads and rolling an even number" is not simple
because it consists of three possible outcomes and can be represented as

$$B = \{H2, H4, H6\}.$$ Event B has more than one outcome, so it is
not simple.

EXAMPLE 2

Identifying Simple Events

Determine the number of outcomes in each event. Then decide whether each
event is simple or not. Explain your reasoning.

1. For quality control, you randomly select a machine part from a batch that
has been manufactured that day. Event A is selecting a specific defective
machine part.

2. You roll a six-sided die. Event B is rolling at least a 4.

SOLUTION

1. Event A has only one outcome: choosing the specific defective machine
part. So, the event is a simple event.

2. Event B has three outcomes: rolling a 4, a 5, or a 6. Because the event has
more than one outcome, it is not simple.

TRY IT YOURSELF 2

You ask for a student's age at his or her last birthday. Determine the number
of outcomes in each event. Then decide whether each event is simple or not.
Explain your reasoning.

1. Event C: The student's age is between 18 and 23, inclusive.

2. Event D: The student's age is 20.

Answer: Page A34

The Fundamental Counting Principle

In some cases, an event can occur in so many different ways that it is not practical to write out all the outcomes. When this occurs, you can rely on the Fundamental Counting Principle. The **Fundamental Counting Principle** can be used to find the number of ways two or more events can occur in sequence.

The Fundamental Counting Principle

If one event can occur in m ways and a second event can occur in n ways, then the number of ways the two events can occur in sequence is $m \cdot n$. This rule can be extended to any number of events occurring in sequence.

In words, the number of ways that events can occur in sequence is found by multiplying the number of ways one event can occur by the number of ways the other event(s) can occur.

EXAMPLE 3

Using the Fundamental Counting Principle

You are purchasing a new car. The possible manufacturers, car sizes, and colors are listed in the table.

Manufacturer	Car size	Color
Ford	compact	white (W)
GM	midsize	red (R)
Honda		black (B)
		green (G)

How many different ways can you select one manufacturer, one car size, and one color? Use a tree diagram to check your result.

SOLUTION

There are three choices of manufacturers, two choices of car sizes, and four choices of colors. Using the Fundamental Counting Principle, you can determine that the number of ways to select one manufacturer, one car size, and one color is

$3 \cdot 2 \cdot 4 = 24$ ways.

Using a tree diagram, you can see why there are 24 options.

Tree Diagram for Car Selections

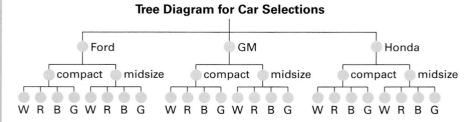

TRY IT YOURSELF 3

You add another manufacturer, Toyota, and another color, tan, to the choices in Example 3. How many different ways can you select one manufacturer, one car size, and one color? Use a tree diagram to check your result.

Answer: Page A34

EXAMPLE 4

Using the Fundamental Counting Principle

The access code for a car's security system consists of four digits. Each digit can be any number from 0 through 9.

Access Code

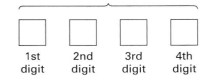

| 1st digit | 2nd digit | 3rd digit | 4th digit |

How many access codes are possible when

1. each digit can be used only once and not repeated?

2. each digit can be repeated?

3. each digit can be repeated but the first digit cannot be 0 or 1?

SOLUTION

1. Because each digit can be used only once, there are 10 choices for the first digit, 9 choices left for the second digit, 8 choices left for the third digit, and 7 choices left for the fourth digit. Using the Fundamental Counting Principle, you can conclude that there are

$$10 \cdot 9 \cdot 8 \cdot 7 = 5040$$

possible access codes.

2. Because each digit can be repeated, there are 10 choices for each of the four digits. So, there are

$$10 \cdot 10 \cdot 10 \cdot 10 = 10^4$$
$$= 10{,}000$$

possible access codes.

3. Because the first digit cannot be 0 or 1, there are 8 choices for the first digit. Then there are 10 choices for each of the other three digits. So, there are

$$8 \cdot 10 \cdot 10 \cdot 10 = 8000$$

possible access codes.

Remember that you can use technology to check your answers. For instance, at the left, a TI-84 Plus was used to check the results in Example 4.

TI-84 PLUS

```
10*9*8*7
               5040
10^4
              10000
8*10*10*10
               8000
```

TRY IT YOURSELF 4

How many license plates can you make when a license plate consists of

1. six (out of 26) alphabetical letters, each of which can be repeated?

2. six (out of 26) alphabetical letters, each of which cannot be repeated?

3. six (out of 26) alphabetical letters, each of which can be repeated but the first letter cannot be A, B, C, or D?

4. one digit (any number 1 through 9) and five (out of 26) alphabetical letters, each of which can be repeated?

Answer: Page A34

Types of Probability

The method you will use to calculate a probability depends on the type of probability. There are three types of probability: **classical probability, empirical probability,** and **subjective probability.** The probability that event E will occur is written as $P(E)$ and is read as "the probability of event E."

Study Tip

Probabilities can be written as fractions, decimals, or percents. In Example 5, the probabilities are written as reduced fractions and decimals, with decimals rounded to three places when possible. For very small probabilities, round to the first nonzero digit. For example, 0.0000271 would be 0.00003. In general, these *round-off rules* will be used throughout the text. (Note that some results may be rounded differently for accuracy.)

DEFINITION

Classical (or **theoretical**) **probability** is used when each outcome in a sample space is equally likely to occur. The classical probability for an event E is given by

$$P(E) = \frac{\text{Number of outcomes in event } E}{\text{Total number of outcomes in sample space}}.$$

EXAMPLE 5

Finding Classical Probabilities

You roll a six-sided die. Find the probability of each event.

1. Event A: rolling a 3

2. Event B: rolling a 7

3. Event C: rolling a number less than 5

SOLUTION

When a six-sided die is rolled, the sample space consists of six outcomes: $\{1, 2, 3, 4, 5, 6\}$. Because each outcome in the sample space is equally likely to occur, you can use the formula for classical probability.

1. There is one outcome in event $A = \{3\}$. So,

$$P(\text{rolling a 3}) = \frac{1}{6} \approx 0.167. \qquad \text{Round to three decimal places.}$$

The probability of rolling a 3 is $\frac{1}{6}$, or about 0.167.

2. Because 7 is not in the sample space, there are no outcomes in event B. So,

$$P(\text{rolling a 7}) = \frac{0}{6} = 0. \qquad \text{Event is not possible.}$$

The probability of rolling a 7 is 0, so it is not possible for the event to occur.

3. There are four outcomes in event $C = \{1, 2, 3, 4\}$. So,

$$P(\text{rolling a number less than 5}) = \frac{4}{6} = \frac{2}{3} \approx 0.667.$$

The probability of rolling a number less than 5 is $\frac{2}{3}$, or about 0.667.

TRY IT YOURSELF 5

You select a card from a standard deck of playing cards. Find the probability of each event.

1. Event D: Selecting the nine of clubs

2. Event E: Selecting a heart

3. Event F: Selecting a diamond, heart, club, or spade

Answer: Page A34

Standard Deck of Playing Cards

Hearts	Diamonds	Spades	Clubs
A ♥	A ♦	A ♠	A ♣
K ♥	K ♦	K ♠	K ♣
Q ♥	Q ♦	Q ♠	Q ♣
J ♥	J ♦	J ♠	J ♣
10 ♥	10 ♦	10 ♠	10 ♣
9 ♥	9 ♦	9 ♠	9 ♣
8 ♥	8 ♦	8 ♠	8 ♣
7 ♥	7 ♦	7 ♠	7 ♣
6 ♥	6 ♦	6 ♠	6 ♣
5 ♥	5 ♦	5 ♠	5 ♣
4 ♥	4 ♦	4 ♠	4 ♣
3 ♥	3 ♦	3 ♠	3 ♣
2 ♥	2 ♦	2 ♠	2 ♣

When an experiment is repeated many times, regular patterns are formed. These patterns make it possible to find empirical probability. Empirical probability can be used even when each outcome of an event is not equally likely to occur.

Picturing the World

It seems that no matter how strange an event is, somebody wants to know the probability that it will occur. The table below lists the probabilities that some intriguing events will happen. (Adapted from Life: The Odds)

Event	Probability
Being audited by the IRS	0.6%
Writing a *New York Times* best seller	0.005
Winning an Academy Award	0.00009
Having your identity stolen	0.5%
Spotting a UFO	0.0000003

Which of these events is most likely to occur? Least likely?

The event most likely to occur is being audited by the IRS. The event least likely to occur is spotting a UFO.

3.1 To explore this topic further, see **Activity 3.1** on page 146.

DEFINITION

Empirical (or **statistical**) **probability** is based on observations obtained from probability experiments. The empirical probability of an event E is the relative frequency of event E.

$$P(E) = \frac{\text{Frequency of event } E}{\text{Total frequency}}$$

$$= \frac{f}{n} \qquad \text{Note that } n = \Sigma f.$$

EXAMPLE 6

Finding Empirical Probabilities

A company is conducting an online survey of randomly selected U.S. adults to determine how they read books during the past year, if at all. So far, 1490 adults have been surveyed. The pie chart shows the results. (Note that digital books include ebooks as well as audio books.) What is the probability that the next adult surveyed read only print books during the last year? *(Pew Research Center, September 2016, "Book Reading 2016")*

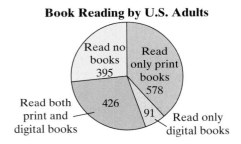

Book Reading by U.S. Adults

SOLUTION

Note that the responses are *not* equally likely to occur and are based on observations. So, you *cannot* use the formula for classical probability, but you can use the formula for empirical probability. The event is a response of "read only print books." The frequency of this event is 578. The total of the frequencies is

$$n = 578 + 91 + 426 + 395 \qquad \text{Add frequency of each response.}$$
$$= 1490. \qquad \text{Total frequency}$$

The empirical probability that the response of the next adult is "read only print books" is

$$P(\text{read only print books}) = \frac{578}{1490} \qquad \text{Find empirical probability.}$$

$$\approx 0.388. \qquad \text{Round to three decimal places.}$$

TRY IT YOURSELF 6

In Example 6, determine the probability that the next adult surveyed read only digital books during the last year.

Answer: Page A34

Ages	Frequency, f
18 to 22	156
23 to 35	312
36 to 49	254
50 to 64	195
65 and over	58
	$\Sigma f = 975$

EXAMPLE 7

Using a Frequency Distribution to Find Probabilities

A company is conducting a phone survey of randomly selected individuals to determine the ages of social networking site users. So far, 975 social networking site users have been surveyed. The frequency distribution at the right shows the results. What is the probability that the next user surveyed is 23 to 35 years old? *(Adapted from Pew Research Center)*

SOLUTION

Because the responses are *not* equally likely to occur and are based on observations, use the formula for empirical probability. The event is a response of "23 to 35 years old." The frequency of this event is 312. Because the total of the frequencies is 975, the empirical probability that the next user is 23 to 35 years old is

$$P(\text{age 23 to 35}) = \frac{312}{975} = 0.32.$$

TRY IT YOURSELF 7

Find the probability that the next user surveyed is 36 to 49 years old.

Answer: Page A34

As you increase the number of times a probability experiment is repeated, the empirical probability (relative frequency) of an event approaches the theoretical probability of the event. This is known as the **law of large numbers.**

Law of Large Numbers

As an experiment is repeated over and over, the empirical probability of an event approaches the theoretical (actual) probability of the event.

As an example of this law, suppose you want to determine the probability of tossing a head with a fair coin. You toss the coin 10 times and get 3 heads, so you obtain an empirical probability of $\frac{3}{10}$. Because you tossed the coin only a few times, your empirical probability is not representative of the theoretical probability, which is $\frac{1}{2}$. The law of large numbers tells you that the empirical probability after tossing the coin several thousand times will be very close to the theoretical or actual probability.

The scatter plot below shows the results of simulating a coin toss 150 times. Notice that, as the number of tosses increases, the probability of tossing a head gets closer and closer to the theoretical probability of 0.5.

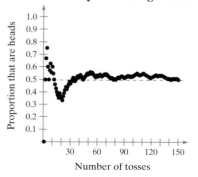

Probability of Tossing a Head

Proportion that are heads (vertical axis: 0.1 to 1.0)

Number of tosses (horizontal axis: 30, 60, 90, 120, 150)

The third type of probability is **subjective probability.** Subjective probabilities result from intuition, educated guesses, and estimates. For instance, given a patient's health and extent of injuries, a doctor may feel that the patient has a 90% chance of a full recovery. Or a business analyst may predict that the chance of the employees of a certain company going on strike is 0.25.

EXAMPLE 8

Classifying Types of Probability

Classify each statement as an example of classical probability, empirical probability, or subjective probability. Explain your reasoning.

1. The probability that you will get an A on your next test is 0.9.

2. The probability that a voter chosen at random will be younger than 35 years old is 0.3.

3. The probability of winning a 1000-ticket raffle with one ticket is $\frac{1}{1000}$.

SOLUTION

1. This probability is most likely based on an educated guess. It is an example of subjective probability.

2. This statement is most likely based on a survey of a sample of voters, so it is an example of empirical probability.

3. Because you know the number of outcomes and each is equally likely, this is an example of classical probability.

TRY IT YOURSELF 8

Based on previous counts, the probability of a salmon successfully passing through a dam on the Columbia River is 0.85. Is this statement an example of classical probability, empirical probability, or subjective probability? *(Source: Army Corps of Engineers)*

Answer: Page A34

A probability cannot be negative or greater than 1, as stated in the rule below.

Range of Probabilities Rule

The probability of an event E is between 0 and 1, inclusive. That is,

$$0 \le P(E) \le 1.$$

When the probability of an event is 1, the event is certain to occur. When the probability of an event is 0, the event is impossible. A probability of 0.5 indicates that an event has an even chance of occurring or not occurring.

The figure below shows the possible range of probabilities and their meanings.

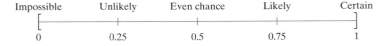

An event that occurs with a probability of 0.05 or less is typically considered unusual. Unusual events are highly unlikely to occur. Later in this course you will identify unusual events when studying inferential statistics.

Complementary Events

The sum of the probabilities of all outcomes in a sample space is 1 or 100%. An important result of this fact is that when you know the probability of an event E, you can find the probability of the **complement of event** E.

Sample Space

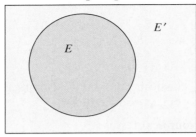

The area of the rectangle represents the total probability of the sample space (1 = 100%). The area of the circle represents the probability of event E, and the area outside the circle represents the probability of the complement of event E.

DEFINITION

The **complement of event** E is the set of all outcomes in a sample space that are not included in event E. The complement of event E is denoted by E' and is read as "E prime." The Venn diagram at the left illustrates the relationship between the sample space, event E, and its complement E'.

For instance, when you roll a die and let E be the event "the number is at least 5," the complement of E is the event "the number is less than 5." In symbols, $E = \{5, 6\}$ and $E' = \{1, 2, 3, 4\}$.

Using the definition of the complement of an event and the fact that the sum of the probabilities of all outcomes is 1, you can determine the formulas below.

$$P(E) + P(E') = 1$$
$$P(E) = 1 - P(E')$$
$$P(E') = 1 - P(E)$$

EXAMPLE 9

Finding the Probability of the Complement of an Event

The frequency distribution from Example 7 is shown below. Find the probability of randomly selecting a social networking site user who is not 23 to 35 years old.

Ages	Frequency, f
18 to 22	156
23 to 35	312
36 to 49	254
50 to 64	195
65 and over	58
	$\Sigma f = 975$

SOLUTION

From Example 7, you know that

$$P(\text{age 23 to 35}) = \frac{312}{975}$$
$$= 0.32.$$

So, the probability that a user is not 23 to 35 years old is

$$P(\text{age is not 23 to 35}) = 1 - \frac{312}{975} = \frac{663}{975} = 0.68.$$

TRY IT YOURSELF 9

Use the frequency distribution in Example 7 to find the probability of randomly selecting a user who is not 18 to 22 years old.

Answer: Page A34

Probability Applications

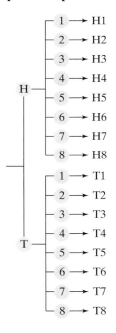

Tree Diagram for Coin and Spinner Experiment

EXAMPLE 10

Using a Tree Diagram

A probability experiment consists of tossing a coin and spinning the spinner shown at the left. The spinner is equally likely to land on each number. Use a tree diagram to find the probability of each event.

1. Event A: tossing a tail and spinning an odd number

2. Event B: tossing a head or spinning a number greater than 3

SOLUTION

From the tree diagram at the left, you can see that there are 16 outcomes. The outcomes are equally likely to occur, so use the formula for classical probability.

1. There are four outcomes in event $A = \{$T1, T3, T5, T7$\}$. So,

$$P(\text{tossing a tail and spinning an odd number}) = \frac{4}{16} = \frac{1}{4} = 0.25.$$

2. There are 13 outcomes in event $B = \{$H1, H2, H3, H4, H5, H6, H7, H8, T4, T5, T6, T7, T8$\}$. So,

$$P(\text{tossing a head or spinning a number greater than 3}) = \frac{13}{16} \approx 0.813.$$

TRY IT YOURSELF 10

Find the probability of tossing a tail and spinning a number less than 6.

Answer: Page A34

EXAMPLE 11

Using the Fundamental Counting Principle

Your college identification number consists of eight digits. Each digit can be 0 through 9 and each digit can be repeated. What is the probability of getting your college identification number when randomly generating eight digits?

SOLUTION

Because each digit can be repeated, there are 10 choices for each of the 8 digits. So, using the Fundamental Counting Principle, there are

$$10 \cdot 10 \cdot 10 \cdot 10 \cdot 10 \cdot 10 \cdot 10 \cdot 10 = 10^8 = 100{,}000{,}000$$

possible identification numbers. But only one of those numbers corresponds to your college identification number. So, the probability of randomly generating 8 digits and getting your college identification number is

$$\frac{1}{100{,}000{,}000}, \text{ or } 0.00000001.$$

TRY IT YOURSELF 11

Your college identification number consists of nine digits. The first two digits of the number will be the last two digits of the year you are scheduled to graduate. The other digits can be any number from 0 through 9, and each digit can be repeated. What is the probability of getting your college identification number when randomly generating the other seven digits?

Answer: Page A34

3.1 EXERCISES

For Extra Help: **MyLab Statistics**

Building Basic Skills and Vocabulary

1. What is the difference between an outcome and an event?

2. Determine whether each number could represent the probability of an event. Explain your reasoning.

     (a) $\frac{25}{25}$ (b) 333.3%   (c) 2.3   (d) -0.0004 (e) 0   (f) $\frac{320}{105}$

3. Explain why the statement is incorrect: *The probability of rain is 150%.*

4. When you use the Fundamental Counting Principle, what are you counting?

5. Describe the law of large numbers in your own words. Give an example.

6. List the three formulas that can be used to describe complementary events.

True or False? *In Exercises 7–10, determine whether the statement is true or false. If it is false, rewrite it as a true statement.*

7. You are taking a test that has true or false and multiple choice questions. The event "choosing false on a true or false question and choosing A or B on a multiple choice question" is a simple event.

8. You toss a fair coin nine times and it lands tails up each time. The probability it will land heads up on the tenth toss is greater than 0.5.

9. A probability of $\frac{1}{10}$ indicates an unusual event.

10. When an event is almost certain to happen, its complement will be an unusual event.

Matching Probabilities *In Exercises 11–16, match the event with its probability.*

(a) 0.95   (b) 0.005   (c) 0.25   (d) 0   (e) 0.375   (f) 0.5

11. A random number generator is used to select a number from 1 to 100. What is the probability of selecting the number 153?

12. A random number generator is used to select a number from 1 to 100. What is the probability of selecting an even number?

13. You randomly select a number from 0 to 9 and then randomly select a number from 0 to 19. What is the probability of selecting a 3 both times?

14. A game show contestant must randomly select a door. One door doubles her money while the other three doors leave her with no winnings. What is the probability she selects the door that doubles her money?

15. Five of the 100 digital video recorders (DVRs) in an inventory are known to be defective. What is the probability you randomly select a DVR that is not defective?

16. You toss a coin four times. What is the probability of tossing tails exactly half of the time?

Finding the Probability of the Complement of an Event
In Exercises 17–20, the probability that an event will happen is given. Find the probability that the event will not happen.

17. $P(E) = \frac{19}{23}$ 18. $P(E) = 0.55$ 19. $P(E) = 0.03$ 20. $P(E) = \frac{2}{7}$

---

1. An outcome is the result of a single trial in a probability experiment, whereas an event is a set of one or more outcomes.

2. (a) Could represent the probability of an event. The probability of an event occurring must be contained in the interval [0, 1] or [0%, 100%].

     (b) Could not represent the probability of an event. The probability of an event occurring cannot be greater than 100%.

     (c)–(f) See Selected Answers, page A96.

3. The probability of an event cannot exceed 100%.

4. The Fundamental Counting Principle counts the number of ways that two or more events can occur in sequence.

5. The law of large numbers states that as an experiment is repeated over and over, the probabilities found in the experiment will approach the actual probabilities of the event. Examples will vary.

6. $P(E) + P(E') = 1$
$P(E) = 1 - P(E')$
$P(E') = 1 - P(E)$

7. False. The event "choosing false on a true or false question and choosing A or B on a multiple choice question" is not simple because it consists of two possible outcomes and can be represented as $A = \{FA, FB\}$.

8. False. You toss a fair coin nine times and it lands tails up each time. The probability it will land heads up on the tenth toss is 0.5.

9. False. A probability of less than $\frac{1}{20} = 0.05$ indicates an unusual event.

10. True   11. d   12. f   13. b

14. c   15. a   16. e   17. $\frac{4}{23}$

18. 0.45   19. 0.97   20. $\frac{5}{7}$

21. 0.05   **22.** 0.87

23. $\frac{1}{4}$ **24.** $\frac{40}{61}$

25. {A, B, C, D, E, F, G, H, I, J, K, L, M, N, O, P, Q, R, S, T, U, V, W, X, Y, Z}; 26

26. {A, B, C, D, F}; 5

27. See Odd Answers, page A54.

28. {(Blue, Blonde), (Blue, Black), (Blue, Brown), (Blue, Red), (Blue, Other), (Brown, Blonde), (Brown, Black), (Brown, Brown), (Brown, Red), (Brown, Other), (Green, Blonde), (Green, Black), (Green, Brown), (Green, Red), (Green, Other), (Hazel, Blonde), (Hazel, Black), (Hazel, Brown), (Hazel, Red), (Hazel, Other), (Gray, Blonde), (Gray, Black), (Gray, Brown), (Gray, Red), (Gray, Other), (Other, Blonde), (Other, Black), (Other, Brown), (Other, Red), (Other, Other),}; 30

29.

{HH, HT, TH, TT}; 4

30. See Selected Answers, page A96.

31. See Odd Answers, page A54.

32. {1HH1, 1HH2, 1HH3, 1HT1, 1HT2, 1HT3, 1TH1, 1TH2, 1TH3, 1TT1, 1TT2, 1TT3, 2HH1, 2HH2, 2HH3, 2HT1, 2HT2, 2HT3, 2TH1, 2TH2, 2TH3, 2TT1, 2TT2, 2TT3, 3HH1, 3HH2, 3HH3, 3HT1, 3HT2, 3HT3, 3TH1, 3TH2, 3TH3, 3TT1, 3TT2, 3TT3, 4HH1, 4HH2, 4HH3, 4HT1, 4HT2, 4HT3, 4TH1, 4TH2, 4TH3, 4TT1, 4TT2, 4TT3, 5HH1, 5HH2, 5HH3, 5HT1, 5HT2, 5HT3, 5TH1, 5TH2, 5TH3, 5TT1, 5TT2, 5TT3, 6HH1, 6HH2, 6HH3, 6HT1, 6HT2, 6HT3, 6TH1, 6TH2, 6TH3, 6TT1, 6TT2, 6TT3}; 72

33. 1; Simple event because it is an event that consists of a single outcome.

34. 499; Not a simple event because it is an event that consists of more than a single outcome.

35. 13; Not a simple event because it is an event that consists of more than a single outcome.

36. 1; Simple event because it is an event that consists of a single outcome.

37. 576   **38.** 864

39. 4500   **40.** 64

Finding the Probability of an Event *In Exercises 21–24, the probability that an event will not happen is given. Find the probability that the event will happen.*

21. $P(E') = 0.95$ **22.** $P(E') = 0.13$ **23.** $P(E') = \frac{3}{4}$ **24.** $P(E') = \frac{21}{61}$

Using and Interpreting Concepts

Identifying the Sample Space of a Probability Experiment *In Exercises 25–32, identify the sample space of the probability experiment and determine the number of outcomes in the sample space. Draw a tree diagram when appropriate.*

25. Guessing the initial of a student's middle name

26. Guessing a student's letter grade (A, B, C, D, F) in a class

27. Drawing one card from a standard deck of cards

28. Identifying a person's eye color (blue, brown, green, hazel, gray, other) and hair color (blonde, black, brown, red, other).

29. Tossing two coins

30. Tossing three coins

31. Rolling a pair of six-sided dice

32. Rolling a six-sided die, tossing two coins, and then drawing one card from a hand of three cards

Identifying Simple Events *In Exercises 33–36, determine the number of outcomes in the event. Then decide whether the event is a simple event or not. Explain your reasoning.*

33. A spreadsheet is used to randomly generate a number from 1 to 2000. Event *A* is generating the number 253.

34. A spreadsheet is used to randomly generate a number from 1 to 4000. Event *B* is generating a number less than 500.

35. You randomly select one card from a standard deck of 52 playing cards. Event *A* is selecting a diamond.

36. You randomly select one card from a standard deck of 52 playing cards. Event B is selecting the ace of spades.

Using the Fundamental Counting Principle *In Exercises 37–40, use the Fundamental Counting Principle.*

37. **Menu** A restaurant offers a $15 dinner special that lets you choose from 6 appetizers, 12 entrées, and 8 desserts. How many different meals are available when you select an appetizer, an entrée, and a dessert?

38. **Tablet** A tablet has 4 choices for an operating system, 3 choices for a screen size, 4 choices for a processor, 6 choices for memory size, and 3 choices for a battery. How many ways can you customize the tablet?

39. **Realty** A realtor uses a lock box to store the keys to a house that is for sale. The access code for the lock box consists of four digits. The first digit cannot be zero and the last digit must be even. How many different codes are available?

40. **True or False Quiz** Assuming that no questions are left unanswered, in how many ways can a six-question true or false quiz be answered?

41. 0.083   **42.** 0.083   **43.** 0.667
44. 0.583   **45.** 0.333   **46.** 0.167
47. 0.712   **48.** 0.072   **49.** 0.216
50. 0.238   **51.** 0.344   **52.** 0.203

53. Empirical probability because company records were used to calculate the frequency of a washing machine breaking down.

54. Classical probability because each outcome in the sample space is equally likely to occur.

Response	Number of times, f
None	1584
One	205
Two	160
Three	71
Four or more	205

TABLE FOR EXERCISES 47 AND 48

Ages	Frequency, f (in millions)
18 to 29	48.9
30 to 44	53.9
45 to 64	78.1
65 and over	46.0

TABLE FOR EXERCISES 49–52

55. Subjective probability because it is most likely based on an educated guess.

56. Empirical probability because survey results were used to calculate the frequency of an adult favoring the ban.

57. See Odd Answers, page A54.

58. See Selected Answers, page A96.

59. 0.842   **60.** 0.824
61. 0.777   **62.** 0.843

Ages	Frequency, f
0–14	173
15–29	123
30–44	92
45–59	137
60–74	130
75 and over	122

TABLE FOR EXERCISES 59–62

Finding Classical Probabilities *In Exercises 41–46, a probability experiment consists of rolling a 12-sided die, numbered 1 to 12. Find the probability of the event.*

41. Event *A*: rolling a 2

42. Event *B*: rolling a 10

43. Event *C*: rolling a number greater than 4

44. Event *D*: rolling a number less than 8

45. Event *E*: rolling a number divisible by 3

46. Event *F*: rolling a number divisible by 5

Finding Empirical Probabilities *A polling organization is asking a sample of U.S. adults how many tattoos they have. The frequency distribution at the left shows the results. In Exercises 47 and 48, use the frequency distribution. (Adapted from The Harris Poll)*

47. What is the probability that the next person asked does not have a tattoo?

48. What is the probability that the next person asked has two tattoos?

Using a Frequency Distribution to Find Probabilities *In Exercises 49–52, use the frequency distribution at the left, which shows the number of voting-age American citizens (in millions) by age, to find the probability that a citizen chosen at random is in the age range. (Source: U.S. Census Bureau)*

49. 18 to 29 years old

50. 30 to 44 years old

51. 45 to 64 years old

52. 65 years old and older

Classifying Types of Probability *In Exercises 53–58, classify the statement as an example of classical probability, empirical probability, or subjective probability. Explain your reasoning.*

53. According to company records, the probability that a washing machine will need repairs during a six-year period is 0.10.

54. The probability of choosing 6 numbers from 1 to 40 that match the 6 numbers drawn by a state lottery is $1/3{,}838{,}380 \approx 0.00000026$.

55. An analyst feels that a certain stock's probability of decreasing in price over the next week is 0.75.

56. According to a survey, the probability that an adult chosen at random is in favor of a sprinkling ban is about 0.45.

57. The probability that a randomly selected number from 1 to 100 is divisible by 6 is 0.16.

58. You think that a football team's probability of winning its next game is about 0.80.

Finding the Probability of the Complement of an Event *The age distribution of the residents of Kadoka, South Dakota, is shown at the left. In Exercises 59–62, find the probability of the event. (Adapted from U.S. Census Bureau)*

59. Event *A*: randomly choosing a resident who is not 15 to 29 years old

60. Event *B*: randomly choosing a resident who is not 45 to 59 years old

61. Event *C*: randomly choosing a resident who is not 14 years old or younger

62. Event *D*: randomly choosing a resident who is not 75 years old or older

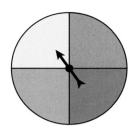

FIGURE FOR EXERCISES 63–66

63, 65, 67. See Odd Answers, page A54.

64, 66, 68. See Selected Answers, page A96.

69. 0.125   70. 0.125   71. 0.375

72. 0.875   73. 0.450   74. 0.404

75. 0.033   76. 0.253   77. 0.275

78. 0.044

79. See Odd Answers, page A54.

80. See Selected Answers, page A96.

2016 Presidential Election Voters from Virginia

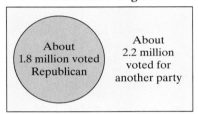

FIGURE FOR EXERCISE 73

All Registered Voters in Texas

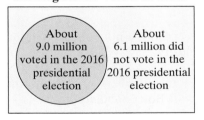

FIGURE FOR EXERCISE 74

Level of Education

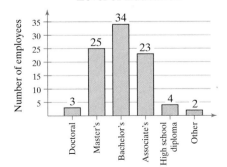

FIGURE FOR EXERCISES 75–78

Using a Tree Diagram *In Exercises 63–66, a probability experiment consists of rolling a six-sided die and spinning the spinner shown at the left. The spinner is equally likely to land on each color. Use a tree diagram to find the probability of the event. Then explain whether the event can be considered unusual.*

63. Event *A*: rolling a 5 and the spinner landing on blue

64. Event *B*: rolling an odd number and the spinner landing on green

65. Event *C*: rolling a number less than 6 and the spinner landing on yellow

66. Event *D*: not rolling a number less than 6 and the spinner landing on yellow

67. **Access Code** An access code consists of three digits. Each digit can be any number from 0 through 9, and each digit can be repeated.

   (a) What is the probability of randomly selecting the correct access code on the first try?

   (b) What is the probability of not selecting the correct access code on the first try?

68. **Access Code** An access code consists of six characters. For each character, any letter or number can be used, with the exceptions that the first character cannot be 0 and the last two characters must be odd numbers.

   (a) What is the probability of randomly selecting the correct access code on the first try?

   (b) What is the probability of not selecting the correct access code on the first try?

Wet or Dry? *You are planning a three-day trip to Seattle, Washington, in October. In Exercises 69–72, use the fact that on each day, it could either be sunny or rainy.*

69. What is the probability that it is sunny all three days?

70. What is the probability that it rains all three days?

71. What is the probability that it rains on exactly one day?

72. What is the probability that it rains on at least one day?

Graphical Analysis *In Exercises 73 and 74, use the diagram at the left.*

73. What is the probability that a voter from Virginia chosen at random voted Republican in the 2016 presidential election? *(Source: Virginia Department of Elections)*

74. What is the probability that a registered voter in Texas chosen at random did not vote in the 2016 presidential election? *(Source: Texas Secretary of State)*

Using a Bar Graph to Find Probabilities *In Exercises 75–78, use the bar graph at the left, which shows the highest level of education received by employees of a company. Find the probability that the highest level of education for an employee chosen at random is*

75. a doctorate.

76. an associate's degree.

77. a master's degree.

78. a high school diploma.

79. **Unusual Events** Can any of the events in Exercises 49–52 be considered unusual? Explain.

80. **Unusual Events** Can any of the events in Exercises 75–78 be considered unusual? Explain.

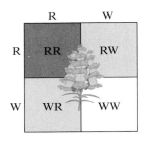

FIGURE FOR EXERCISE 81

81. (a) 0.5   (b) 0.25   (c) 0.25
82. 0.5   83. 0.808   84. 0.192
85. 0.103   86. 0.984
87. (a) 0.313   (b) 0.078
    (c) 0.031; This event is unusual
    because its probability is
    less than or equal to 0.05.

81. Genetics A *Punnett square* is a diagram that shows all possible gene combinations in a cross of parents whose genes are known. When two pink snapdragon flowers (RW) are crossed, there are four equally likely possible outcomes for the genetic makeup of the offspring: red (RR), pink (RW), pink (WR), and white (WW), as shown in the Punnett square at the left. When two pink snapdragons are crossed, what is the probability that the offspring will be (a) pink, (b) red, and (c) white?

82. Genetics There are six basic types of coloring in registered collies: sable (SSmm), tricolor (ssmm), trifactored sable (Ssmm), blue merle (ssMm), sable merle (SSMm), and trifactored sable merle (SsMm). The Punnett square below shows the possible coloring of the offspring of a trifactored sable merle collie and a trifactored sable collie. What is the probability that the offspring will have the same coloring as one of its parents?

Parents: Ssmm and SsMm

	SM	Sm	sM	sm
Sm	SSMm	SSmm	SsMm	Ssmm
Sm	SSMm	SSmm	SsMm	Ssmm
sm	SsMm	Ssmm	ssMm	ssmm
sm	SsMm	Ssmm	ssMm	ssmm

Using a Pie Chart to Find Probabilities *In Exercises 83–86, use the pie chart at the left, which shows the number of workers (in thousands) by industry for the United States.* (*Source: U.S. Bureau of Labor Statistics*)

83. Find the probability that a worker chosen at random is employed in the services industry.

84. Find the probability that a worker chosen at random is not employed in the services industry.

85. Find the probability that a worker chosen at random is employed in the manufacturing industry.

86. Find the probability that a worker chosen at random is not employed in the agriculture, forestry, fishing, and hunting industry.

87. College Football A stem-and-leaf plot for the numbers of touchdowns allowed by all 128 NCAA Division I Football Bowl Subdivision teams in the 2016–2017 season is shown. Find the probability that a team chosen at random allowed (a) at least 51 touchdowns, (b) between 20 and 30 touchdowns, inclusive, and (c) more than 63 touchdowns. Are any of these events unusual? Explain. (*Source: National Collegiate Athletic Association*)

Workers (in thousands) by Industry for the U.S.

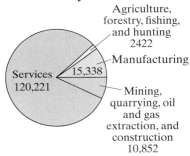

Agriculture, forestry, fishing, and hunting 2422

Manufacturing 15,338

Services 120,221

Mining, quarrying, oil and gas extraction, and construction 10,852

FIGURE FOR EXERCISES 83–86

```
1 | 6                                                                    Key: 1|6 = 16
2 | 1 2 2 3 4 7 8 8
3 | 0 0 1 3 3 3 3 4 4 4 5 5 5 6 6 7 7 7 7 7 7 7 7 8 8 8 9 9 9 9
4 | 0 0 0 0 0 1 1 1 2 2 2 2 3 3 3 3 3 4 4 4 4 5 5 5 6 6 6 6 7 7 7 8 8 8 8 8 9 9 9
5 | 0 0 0 0 0 1 1 2 2 2 3 3 3 3 3 4 4 4 4 4 5 5 5 6 6 6 8 8 9 9
6 | 0 0 0 1 1 1 2 2 2 3 3 3 7 7 8 8
```

88. (a) 0.25   (b) 0.5   (c) 0.5

89. The probability of randomly choosing a tea drinker who does not have a college degree

90. The probability of randomly choosing a smoker whose mother did not smoke

91. No. The odds of winning a prize are 1 : 6 (one winning cap and six losing caps). So, the statement should read, "one in seven game pieces wins a prize."

92. The first game; The probability of winning the second game is $\frac{1}{11} \approx 0.091$, which is less than $\frac{1}{10}$.

93. (a) 0.444   (b) 0.556

94. 13 : 39 = 1 : 3

95. 39 : 13 = 3 : 1

96. p = number of successful outcomes

q = number of unsuccessful outcomes

$P(A) =$

$\dfrac{\text{number of successful outcomes}}{\text{total number of outcomes}}$

$= \dfrac{p}{p + q}$

97. (a)

Sum	Probability
2	0.028
3	0.056
4	0.083
5	0.111
6	0.139
7	0.167
8	0.139
9	0.111
10	0.083
11	0.056
12	0.028

(b) Answers will vary.

(c) Answers will vary.

88. Individual Stock Price   An individual stock is selected at random from the portfolio represented by the box-and-whisker plot shown. Find the probability that the stock price is (a) less than $21, (b) between $21 and $50, and (c) $30 or more.

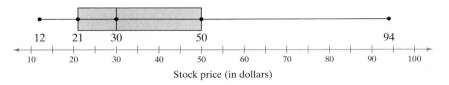

Stock price (in dollars)

Writing   *In Exercises 89 and 90, write a statement that represents the complement of the probability.*

89. The probability of randomly choosing a tea drinker who has a college degree (Assume that you are choosing from the population of all tea drinkers.)

90. The probability of randomly choosing a smoker whose mother also smoked (Assume that you are choosing from the population of all smokers.)

Extending Concepts

Odds   *The chances of winning are often written in terms of odds rather than probabilities. The **odds of winning** is the ratio of the number of successful outcomes to the number of unsuccessful outcomes. The **odds of losing** is the ratio of the number of unsuccessful outcomes to the number of successful outcomes. For example, when the number of successful outcomes is 2 and the number of unsuccessful outcomes is 3, the odds of winning are 2 : 3 (read "2 to 3"). In Exercises 91–96, use this information about odds.*

91. A beverage company puts game pieces under the caps of its drinks and claims that one in six game pieces wins a prize. The official rules of the contest state that the odds of winning a prize are 1 : 6. Is the claim "one in six game pieces wins a prize" correct? Explain your reasoning.

92. The probability of winning an instant prize game is $\frac{1}{10}$. The odds of winning a different instant prize game are 1 : 10. You want the best chance of winning. Which game should you play? Explain your reasoning.

93. The odds of an event occurring are 4 : 5. Find (a) the probability that the event will occur and (b) the probability that the event will not occur.

94. A card is picked at random from a standard deck of 52 playing cards. Find the odds that it is a spade.

95. A card is picked at random from a standard deck of 52 playing cards. Find the odds that it is not a spade.

96. The odds of winning an event A are $p : q$. Show that the probability of event A is given by $P(A) = \dfrac{p}{p + q}$.

97. Rolling a Pair of Dice   You roll a pair of six-sided dice and record the sum.

(a) List all of the possible sums and determine the probability of rolling each sum.

(b) Use technology to simulate rolling a pair of dice and record the sum 100 times. Make a tally of the 100 sums and use these results to list the probability of rolling each sum.

(c) Compare the probabilities in part (a) with the probabilities in part (b). Explain any similarities or differences.

APPLET

You can find the interactive applet for this activity within MyLab Statistics or at *www.pearsonhighered.com/ mathstatsresources.*

The *simulating the stock market* applet allows you to investigate the probability that the stock market will go up on any given day. The plot at the top left corner shows the probability associated with each outcome. In this case, the market has a 50% chance of going up on any given day. When SIMULATE is clicked, outcomes for *n* days are simulated. The results of the simulations are shown in the frequency plot. When the *animate* option is checked, the display will show each outcome dropping into the frequency plot as the simulation runs. The individual outcomes are shown in the text field at the far right of the applet. The center plot shows in red the cumulative proportion of times that the market went up. The green line in the plot reflects the theoretical probability of the market going up. As the experiment is conducted over and over, the cumulative proportion should converge to the theoretical probability.

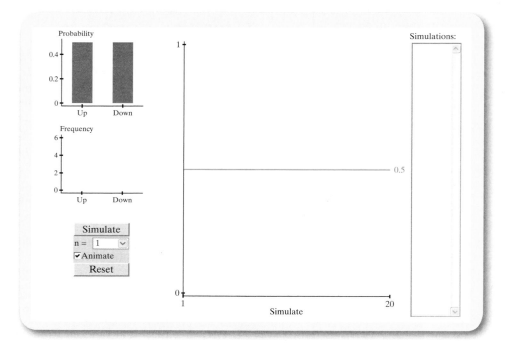

EXPLORE

Step 1 Specify a value for *n*.
Step 2 Click SIMULATE four times.
Step 3 Click RESET.
Step 4 Specify another value for *n*.
Step 5 Click SIMULATE.

DRAW CONCLUSIONS

APPLET

1. Run the simulation using $n = 1$ without clicking RESET. How many days did it take until there were three straight days on which the stock market went up? three straight days on which the stock market went down?

2. Run the applet to simulate the stock market activity over the next 35 business days. Find the empirical probability that the market goes up on day 36.

3.2 Conditional Probability and the Multiplication Rule

What You Should Learn

▶ How to find the probability of an event given that another event has occurred

▶ How to distinguish between independent and dependent events

▶ How to use the Multiplication Rule to find the probability of two or more events occurring in sequence and to find conditional probabilities

Conditional Probability ■ Independent and Dependent Events
■ The Multiplication Rule

Conditional Probability

In this section, you will learn how to find the probability that two events occur in sequence. Before you can find this probability, however, you must know how to find **conditional probabilities.**

DEFINITION

A **conditional probability** is the probability of an event occurring, given that another event has already occurred. The conditional probability of event B occurring, given that event A has occurred, is denoted by $P(B|A)$ and is read as "probability of B, given A."

Note to Instructor

Point out to students that the table in Example 1 (part 2) is an example of a contingency table (or two-way table). They will learn more about contingency tables in Section 10.2.

	Have you ever been offended by something on social media?		
Gender	**Yes**	**No**	**Total**
Female	619	549	1168
Male	532	576	1108
Total	1151	1125	2276

Sample Space

Gender	**Yes**
Female	619
Male	532
Total	1151

EXAMPLE 1

Finding Conditional Probabilities

1. Two cards are selected in sequence from a standard deck of 52 playing cards. Find the probability that the second card is a queen, given that the first card is a king. (Assume that the king is not replaced.)

2. The table at the left shows the results of a survey in which 2276 social media users were asked whether they have ever been offended by something (posts, comments, or photos) they saw on social media. Find the probability that a user is male, given that the user was offended by something on social media. *(Adapted from The Harris Poll)*

SOLUTION

1. Because the first card is a king and is not replaced, the remaining deck has 51 cards, 4 of which are queens. So,

$$P(B|A) = \frac{4}{51} \approx 0.078.$$

The probability that the second card is a queen, given that the first card is a king, is about 0.078.

2. There are 1151 users who said they were offended by something on social media. So, the sample space consists of these 1151 users, as shown at the left. Of these, 532 are males. So,

$$P(B|A) = \frac{532}{1151} \approx 0.462.$$

The probability that a user is male, given that the user was offended, is about 0.462.

TRY IT YOURSELF 1

Refer to the survey in the second part of Example 1. Find the probability that a user is female, given that the user was not offended by something on social media. *Answer: Page A34*

Picturing the World

Truman Collins, a probability and statistics enthusiast, wrote a program that finds the probability of landing on each square of a Monopoly® board during a game. Collins explored various scenarios, including the effects of the Chance and Community Chest cards and the various ways of landing in or getting out of jail. Interestingly, Collins discovered that the length of each jail term affects the probabilities. (Note that the probabilities are rounded to more than three decimal places so that it is easier to see how going to jail affects the probabilities.)

Monopoly square	Probability given short jail term	Probability given long jail term
Go	0.0310	0.0291
Chance	0.0087	0.0082
In Jail	0.0395	0.0946
Free Parking	0.0288	0.0283
Park Place	0.0219	0.0206
B&O RR	0.0307	0.0289
Water Works	0.0281	0.0265

Why do the probabilities depend on how long you stay in jail?

The probabilities depend on how long you stay in jail because the more turns you spend in jail, the fewer chances you have to land on any other spot.

Independent and Dependent Events

In some experiments, one event does not affect the probability of another. For instance, when you roll a die and toss a coin, the outcome of the roll of the die does not affect the probability of the coin landing heads up. These two events are *independent*. The question of the independence of two or more events is important to researchers in fields such as marketing, medicine, and psychology. You can use conditional probabilities to determine whether events are **independent.**

DEFINITION

Two events are **independent** when the occurrence of one of the events does not affect the probability of the occurrence of the other event. Two events A and B are independent when

$$P(B|A) = P(B) \qquad \text{Occurrence of } A \text{ does not affect probability of } B$$

or when

$$P(A|B) = P(A). \qquad \text{Occurrence of } B \text{ does not affect probability of } A$$

Events that are not independent are **dependent.**

To determine whether A and B are independent, first calculate $P(B)$, the probability of event B. Then calculate $P(B|A)$, the probability of B, given A. If the values are equal, then the events are independent. If $P(B) \neq P(B|A)$, then A and B are dependent events.

EXAMPLE 2

Classifying Events as Independent or Dependent

Determine whether the events are independent or dependent.

1. Selecting a king (A) from a standard deck of 52 playing cards, not replacing it, and then selecting a queen (B) from the deck

2. Tossing a coin and getting a head (A), and then rolling a six-sided die and obtaining a 6 (B)

3. Driving over 85 miles per hour (A), and then getting in a car accident (B)

SOLUTION

1. $P(B) = \frac{4}{52}$ and $P(B|A) = \frac{4}{51}$. The occurrence of A changes the probability of the occurrence of B, so the events are dependent.

2. $P(B) = \frac{1}{6}$ and $P(B|A) = \frac{1}{6}$. The occurrence of A does not change the probability of the occurrence of B, so the events are independent.

3. Driving over 85 miles per hour increases the chances of getting in an accident, so these events are dependent.

TRY IT YOURSELF 2

Determine whether the events are independent or dependent.

1. Smoking a pack of cigarettes per day (A) and developing emphysema, a chronic lung disease (B)

2. Tossing a coin and getting a head (A), and then tossing the coin again and getting a tail (B)

Answer: Page A34

The Multiplication Rule

To find the probability of two events occurring in sequence, you can use the **Multiplication Rule.**

Study Tip

In words, to use the Multiplication Rule,

1. find the probability that the first event occurs,
2. find the probability that the second event occurs given that the first event has occurred, and
3. multiply these two probabilities.

> ### The Multiplication Rule for the Probability of *A* and *B*
>
> The probability that two events *A* and *B* will occur in sequence is
>
> $$P(A \text{ and } B) = P(A) \cdot P(B|A).$$ Events *A* and *B* are dependent.
>
> If events *A* and *B* are independent, then the rule can be simplified to
>
> $$P(A \text{ and } B) = P(A) \cdot P(B).$$ Events *A* and *B* are independent.
>
> This simplified rule can be extended to any number of independent events.

EXAMPLE 3

Using the Multiplication Rule to Find Probabilities

1. Two cards are selected, without replacing the first card, from a standard deck of 52 playing cards. Find the probability of selecting a king and then selecting a queen.

2. A coin is tossed and a die is rolled. Find the probability of tossing a head and then rolling a 6.

SOLUTION

Study Tip

Recall from Section 3.1 that a probability of 0.05 or less is typically considered unusual. In the first part of Example 3, $0.006 < 0.05$. This means that selecting a king and then a queen (without replacement) from a standard deck is an unusual event.

1. Because the first card is not replaced, the events are dependent.

$$P(K \text{ and } Q) = P(K) \cdot P(Q|K)$$
$$= \frac{4}{52} \cdot \frac{4}{51}$$
$$= \frac{16}{2652}$$
$$\approx 0.006$$

So, the probability of selecting a king and then a queen without replacement is about 0.006.

2. The events are independent.

$$P(H \text{ and } 6) = P(H) \cdot P(6)$$
$$= \frac{1}{2} \cdot \frac{1}{6}$$
$$= \frac{1}{12}$$
$$\approx 0.083$$

So, the probability of tossing a head and then rolling a 6 is about 0.083.

TRY IT YOURSELF 3

1. The probability that a salmon swims successfully through a dam is 0.85. Find the probability that two salmon swim successfully through the dam.

2. Two cards are selected from a standard deck of 52 playing cards without replacement. Find the probability that they are both hearts.

Answer: Page A34

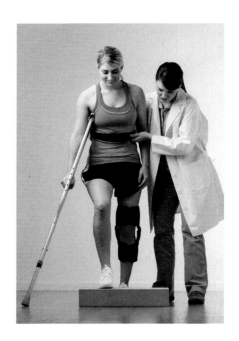

EXAMPLE 4

Using the Multiplication Rule to Find Probabilities

For anterior cruciate ligament (ACL) reconstructive surgery, the probability that the surgery is successful is 0.95. *(Source: The Orthopedic Center of St. Louis)*

1. Find the probability that three ACL surgeries are successful.

2. Find the probability that none of the three ACL surgeries are successful.

3. Find the probability that at least one of the three ACL surgeries is successful.

SOLUTION

1. The probability that each ACL surgery is successful is 0.95. The chance of success for one surgery is independent of the chances for the other surgeries.

$$P(\text{three surgeries are successful}) = (0.95)(0.95)(0.95) \approx 0.857$$

So, the probability that all three surgeries are successful is about 0.857.

2. Because the probability of success for one surgery is 0.95, the probability of failure for one surgery is $1 - 0.95 = 0.05$.

$$P(\text{none of the three are successful}) = (0.05)(0.05)(0.05) \approx 0.0001$$

So, the probability that none of the surgeries are successful is about 0.0001. Note that because 0.0001 is less than 0.05, this can be considered an unusual event.

3. The phrase "at least one" means one or more. The complement to the event "at least one is successful" is the event "none are successful." Use the complement found in part 2 to find the probability. (To avoid rounding in the second step below, use (0.05)(0.05)(0.05), not the rounded result.)

$$
\begin{aligned}
P(\text{at least one is successful}) &= 1 - P(\text{none are successful}) \\
&= 1 - (0.05)(0.05)(0.05) \\
&\approx 0.9999.
\end{aligned}
$$

So, the probability that at least one of the three surgeries is successful is about 0.9999. Note that this probability is not rounded to three decimal places because the result would be 1.000, which implies the event is certain. Even though it is highly likely that at least one of the three surgeries is successful, it is not a certain event.

TRY IT YOURSELF 4

The probability that a particular rotator cuff surgery is successful is 0.9. *(Source: The Orthopedic Center of St. Louis)*

1. Find the probability that three rotator cuff surgeries are successful.

2. Find the probability that none of the three rotator cuff surgeries are successful.

3. Find the probability that at least one of the three rotator cuff surgeries is successful.

Answer: Page A34

In Example 4, you were asked to find a probability using the phrase "at least one." Notice that it was easier to find the probability of its complement, "none," and then subtract the probability of its complement from 1. In general, this probability can be written as

$$P(\text{at least one occurrence of event } A) = 1 - P(\text{no occurrence of event } A).$$

EXAMPLE 5

Using the Multiplication Rule to Find Probabilities

In a recent year, there were 18,187 U.S. allopathic medical school seniors who applied to residency programs and submitted their residency program choices. Of these seniors, 17,057 were matched with residency positions, with about 79.2% getting one of their top three choices. Medical students rank the residency programs in their order of preference, and program directors in the United States rank the students. The term "match" refers to the process whereby a student's preference list and a program director's preference list overlap, resulting in the placement of the student in a residency position. *(Source: National Resident Matching Program)*

1. Find the probability that a randomly selected senior was matched with a residency position *and* it was one of the senior's top three choices.

2. Find the probability that a randomly selected senior who was matched with a residency position did *not* get matched with one of the senior's top three choices.

3. Would it be unusual for a randomly selected senior to be matched with a residency position *and* that it was one of the senior's top three choices?

SOLUTION

Let $A = \{$matched with residency position$\}$ and $B = \{$matched with one of top three choices$\}$. So,

$$P(A) = \frac{17{,}057}{18{,}187} \quad \text{and} \quad P(B|A) = 0.792.$$

1. The events are dependent.

$$P(A \text{ and } B) = P(A) \cdot P(B|A) = \left(\frac{17{,}057}{18{,}187}\right)(0.792) \approx 0.743$$

So, the probability that a randomly selected senior was matched with one of the senior's top three choices is about 0.743.

2. To find this probability, use the complement.

$$P(B'|A) = 1 - P(B|A) = 1 - 0.792 = 0.208$$

So, the probability that a randomly selected senior was matched with a residency position that was not one of the senior's top three choices is 0.208.

3. It is not unusual because the probability of a senior being matched with a residency position that was one of the senior's top three choices is about 0.743, which is greater than 0.05. In fact, with a probability of 0.743, this event is *likely* to happen.

TRY IT YOURSELF 5

In a jury selection pool, 65% of the people are female. Of these 65%, one out of four works in a health field.

1. Find the probability that a randomly selected person from the jury pool is female and works in a health field. Is this event unusual?

2. Find the probability that a randomly selected person from the jury pool is female and does not work in a health field. Is this event unusual?

Answer: Page A34

3.2 EXERCISES

Answers (left column)

1. Two events are independent when the occurrence of one of the events does not affect the probability of the occurrence of the other event, whereas two events are dependent when the occurrence of one of the events does affect the probability of the occurrence of the other event.

2. (a) *Sample answer:* Roll a die twice. The outcome of the second roll is independent of the outcome of the first roll.

   (b) *Sample answer:* Draw two cards (without replacement) from a standard deck of 52 playing cards. The outcome of the second draw is dependent on the outcome of the first draw.

3. The notation $P(B|A)$ means the probability of event B occurring, given that event A has occurred.

4. The complement of "at least one" is "none." So, the probability of getting at least one item is equal to $1 - P$(none of the items).

5. False. If two events are independent, then $P(A|B) = P(A)$.

6. False. If events A and B are independent, then $P(A \text{ and } B) = P(A) \cdot P(B)$.

7. (a) 0.526   (b) 0.159

8. (a) 0.464   (b) 0.552

Building Basic Skills and Vocabulary

1. What is the difference between independent and dependent events?

2. Give an example of
   (a) two events that are independent.
   (b) two events that are dependent.

3. What does the notation $P(B|A)$ mean?

4. Explain how to use the complement to find the probability of getting at least one item of a particular type.

True or False? *In Exercises 5 and 6, determine whether the statement is true or false. If it is false, rewrite it as a true statement.*

5. If two events are independent, then $P(A|B) = P(B)$.

6. If events A and B are dependent, then $P(A \text{ and } B) = P(A) \cdot P(B)$.

Using and Interpreting Concepts

Finding Conditional Probabilities *In Exercises 7 and 8, use the table to find each conditional probability.*

7. **Business Degrees** The table shows the numbers of male and female students in the United States who received bachelor's degrees in business in a recent year. *(Source: National Center for Educational Statistics)*

	Business degrees	Nonbusiness degrees	Total
Male	191,310	621,359	812,669
Female	172,489	909,776	1,082,265
Total	363,799	1,531,135	1,894,934

   (a) Find the probability that a randomly selected student is male, given that the student received a business degree.

   (b) Find the probability that a randomly selected student received a business degree, given that the student is female.

8. **Retirement Savings** The table shows the results of a survey in which 250 male and 250 female workers ages 25 to 64 were asked if they contribute to a retirement savings plan at work.

	Contribute	Do not contribute	Total
Male	116	134	250
Female	143	107	250
Total	259	241	500

   (a) Find the probability that a randomly selected worker contributes to a retirement savings plan at work, given that the worker is male.

   (b) Find the probability that a randomly selected worker is female, given that the worker contributes to a retirement savings plan at work.

Classifying Events as Independent or Dependent *In Exercises 9–14, determine whether the events are independent or dependent. Explain your reasoning.*

9. Selecting a king from a standard deck of 52 playing cards, replacing it, and then selecting a queen from the deck

10. A father having hazel eyes and a daughter having hazel eyes

11. Returning a rented movie after the due date and receiving a late fee

12. Not putting money in a parking meter and getting a parking ticket

13. Rolling a six-sided die and then rolling the die a second time so that the sum of the two rolls is five

14. A ball is selected from a bin of balls numbered from 1 through 52. It is replaced, and then a second numbered ball is selected from the bin.

Classifying Events Based on Studies *In Exercises 15–18, identify the two events described in the study. Do the results indicate that the events are independent or dependent? Explain your reasoning.*

15. A study found that people who suffer from obstructive sleep apnea are at increased risk of having heart disease. *(Source: American Sleep Association)*

16. Certain components in coffee have been found to cause the body to produce higher amounts of acid, which can irritate already existing stomach ulcers. But, coffee does not cause stomach ulcers. *(Source: Top 10 Home Remedies)*

17. A study found that there is no relationship between playing violent video games and aggressive or bullying behavior in teenagers. *(Source: Stetson University)*

18. According to researchers, high engagement with mobile technology for escapism is linked to depression and anxiety in college-age students *(Source: University of Illinois)*

Using the Multiplication Rule *In Exercises 19–32, use the Multiplication Rule.*

19. Cards Two cards are selected from a standard deck of 52 playing cards. The first card is not replaced before the second card is selected. Find the probability of selecting a heart and then selecting a club.

20. Coin and Die A coin is tossed and a die is rolled. Find the probability of tossing a tail and then rolling a number greater than 2.

21. BRCA1 Gene Research has shown that approximately 1 woman in 600 carries a mutation of the BRCA1 gene. About 60% of women with this mutation develop breast cancer. Find the probability that a randomly selected woman will carry the mutation of the BRCA1 gene and will develop breast cancer. *(Adapted from Susan G. Komen)*

Sample Space: Women

---

9. Independent. The outcome of the first draw does not affect the outcome of the second draw.

10. Dependent. The outcome of a father having hazel eyes affects the outcome of a daughter having hazel eyes.

11. Dependent. The outcome of returning a movie after its due date affects the outcome of receiving a late fee.

12. Dependent. The outcome of not putting money in a parking meter affects the outcome of getting a parking ticket.

13. Dependent. The sum of the rolls depends on which numbers came up on the first and second rolls.

14. Independent. The outcome of the first selection does not affect the outcome of the second selection.

15. Events: obstructive sleep apnea, heart disease; Dependent. People with obstructive sleep apnea are more likely to have heart disease.

16. Events: certain components in coffee, stomach ulcers; Independent. Coffee only irritates existing stomach ulcers.

17. Events: playing violent video games, aggressive or bullying behavior; Independent. Playing violent video games does not cause aggressive or bullying behavior in teens.

18. Events: high engagement with mobile technology for escapism, depression, and anxiety; Dependent. High engagement with mobile technology for escapism may cause depression and anxiety in college-age students.

19. 0.063

20. 0.333

21. 0.001

22. 0.067
23. (a) 0.040   (b) 0.640   (c) 0.360
24. (a) 0.008   (b) 0.239   (c) 0.761
25. (a) 0.022   (b) 0.722   (c) 0.278
    (d) The event in part (a) is unusual because its probability is less than or equal to 0.05.
26. (a) 0.002   (b) 0.658   (c) 0.998
    (d) The event in part (a) is unusual because its probability is less than or equal to 0.05.

22. **Pickup Trucks** In a survey, 510 U.S. adults were asked whether they drive a pickup truck and whether they drive a Ford. The results showed that three in ten adults surveyed drive a Ford. Of the adults surveyed that drive Fords, two in nine drive a pickup truck. Find the probability that a randomly selected adult drives a Ford and drives a pickup truck.

Sample Space: U.S. Adults

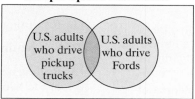

23. **Celebrities as Role Models** In a sample of 1000 U.S. adults, 200 think that most Hollywood celebrities are good role models. Two U.S. adults are selected at random without replacement. *(Adapted from Rasmussen Reports)*

(a) Find the probability that both adults think that most Hollywood celebrities are good role models.

(b) Find the probability that neither adult thinks that most Hollywood celebrities are good role models.

(c) Find the probability that at least one of the two adults thinks that most Hollywood celebrities are good role models.

24. **Knowing a Murder Victim** In a sample of 1000 U.S. adults, 300 said they know a murder victim. Four U.S. adults are selected at random without replacement. *(Adapted from Rasmussen Reports)*

(a) Find the probability that all four adults know a murder victim.

(b) Find the probability that none of the four adults knows a murder victim.

(c) Find the probability that at least one of the four adults knows a murder victim.

25. **Best President** In a sample of 1446 U.S. registered voters, 217 said that John Kennedy was the best president since World War II. Two registered voters are selected at random without replacement. *(Adapted from Quinnipiac University)*

(a) Find the probability that both registered voters say that John Kennedy was the best president since World War II.

(b) Find the probability that neither registered voter says that John Kennedy was the best president since World War II.

(c) Find the probability that at least one of the two registered voters says that John Kennedy was the best president since World War II.

(d) Which of the events can be considered unusual? Explain.

26. **Worst President** In a sample of 1446 U.S. registered voters, 188 said that Richard Nixon was the worst president since World War II. Three registered voters are selected at random without replacement. *(Adapted from Quinnipiac University)*

(a) Find the probability that all three registered voters say that Richard Nixon was the worst president since World War II.

(b) Find the probability that none of the three registered voters say that Richard Nixon was the worst president since World War II.

(c) Find the probability that at most two of the three registered voters say that Richard Nixon was the worst president since World War II.

(d) Which of the events can be considered unusual? Explain.

27. (a) 0.011   (b) 0.022   (c) 0.978
    (d) The events in parts (a) and
        (b) are unusual because their
        probabilities are less than or
        equal to 0.05.
28. (a) 0.00000001
    (b) 0.961   (c) 0.039
    (d) The events in parts (a) and
        (c) are unusual because their
        probabilities are less than or
        equal to 0.05.
29. (a) 0.007   (b) 0.589
    (c) Yes, this is unusual because
        the probability is less than or
        equal to 0.05.
30. (a) 0.135   (b) 0.725
    (c) No, this is not unusual
        because the probability is
        not less than or equal to
        0.05.
31. 0.32
32. 0.3

27. Blood Types   The probability that an African American person in the United States has type O+ blood is 47%. Six unrelated African American people in the United States are selected at random. *(Source: American National Red Cross)*

(a) Find the probability that all six have type O+ blood.

(b) Find the probability that none of the six have type O+ blood.

(c) Find the probability that at least one of the six has type O+ blood.

(d) Which of the events can be considered unusual? Explain.

28. Blood Types   The probability that a Caucasian person in the United States has type AB− blood is 1%. Four unrelated Caucasian people in the United States are selected at random. *(Source: American National Red Cross)*

(a) Find the probability that all four have type AB− blood.

(b) Find the probability that none of the four have type AB− blood.

(c) Find the probability that at least one of the four has type AB− blood.

(d) Which of the events can be considered unusual? Explain.

29. In Vitro Fertilization   In a recent year, about 1.6% of all infants born in the U.S. were conceived through in vitro fertilization (IVF). Of the IVF deliveries, about 41.1% resulted in multiple births. *(Source: American Society for Reproductive Medicine)*

(a) Find the probability that a randomly selected infant was conceived through IVF *and* was part of a multiple birth.

(b) Find the probability that a randomly selected infant conceived through IVF was *not* part of a multiple birth.

(c) Would it be unusual for a randomly selected infant to have been conceived through IVF and to have been part of a multiple birth? Explain.

30. Lottery Tickets   According to a survey, 49% of U.S. adults have purchased a state lottery ticket in the past 12 months. Of these 49%, about 27.5% have annual incomes less than $36,000. *(Adapted from Gallup)*

(a) Find the probability that a randomly selected U.S. adult purchased a state lottery ticket in the past 12 months *and* has an annual income less than $36,000.

(b) Find the probability that a randomly selected U.S. adult who purchased a state lottery ticket in the past 12 months has an annual income greater than or equal to $36,000.

(c) Would it be unusual for a randomly selected U.S. adult to have purchased a state lottery ticket in the past 12 months and to have an annual income less than $36,000? Explain.

31. Digital Content in Schools   According to a study, 80% of K–12 schools or districts in the United States use digital content such as ebooks, audio books, and digital textbooks. Of these 80%, 4 out of 10 use digital content as part of their curriculum. Find the probability that a randomly selected school or district uses digital content and uses it as part of their curriculum. *(Source: School Library Journal)*

32. Surviving Surgery   A doctor gives a patient a 60% chance of surviving bypass surgery after a heart attack. If the patient survives the surgery, then the patient has a 50% chance that the heart heals. Find the probability that the patient survives surgery and the heart heals.

33. 0.444
34. 0.4
35. 0.167
36. 0.797
37. 0.792
38. 0.321
39. (a) 0.074   (b) 0.999
40. (a) 0.462   (b) 0.538
    (c) Answers will vary.
41. 0.954
42. 0.933

Extending Concepts

*According to **Bayes' Theorem,** the probability of event A, given that event B has occurred, is*

$$P(A \mid B) = \frac{P(A) \cdot P(B \mid A)}{P(A) \cdot P(B \mid A) + P(A') \cdot P(B \mid A')}.$$

In Exercises 33–38, use Bayes' Theorem to find $P(A \mid B)$.

33. $P(A) = \frac{2}{3}$, $P(A') = \frac{1}{3}$, $P(B \mid A) = \frac{1}{5}$, and $P(B \mid A') = \frac{1}{2}$

34. $P(A) = \frac{3}{8}$, $P(A') = \frac{5}{8}$, $P(B \mid A) = \frac{2}{3}$, and $P(B \mid A') = \frac{3}{5}$

35. $P(A) = 0.25$, $P(A') = 0.75$, $P(B \mid A) = 0.3$, and $P(B \mid A') = 0.5$

36. $P(A) = 0.62$, $P(A') = 0.38$, $P(B \mid A) = 0.41$, and $P(B \mid A') = 0.17$

37. $P(A) = 73\%$, $P(A') = 17\%$, $P(B \mid A) = 46\%$, and $P(B \mid A') = 52\%$

38. $P(A) = 12\%$, $P(A') = 88\%$, $P(B \mid A) = 66\%$, and $P(B \mid A') = 19\%$

39. **Reliability of Testing**   A virus infects one in every 200 people. A test used to detect the virus in a person is positive 80% of the time when the person has the virus and 5% of the time when the person does not have the virus. (This 5% result is called a *false positive*.) Let *A* be the event "the person is infected" and *B* be the event "the person tests positive."

    (a) Using Bayes' Theorem, when a person tests positive, determine the probability that the person is infected.

    (b) Using Bayes' Theorem, when a person tests negative, determine the probability that the person is *not* infected.

40. **Birthday Problem**   You are in a class that has 24 students. You want to find the probability that at least two of the students have the same birthday.

    (a) Find the probability that each student has a different birthday.

    (b) Use the result of part (a) to find the probability that at least two students have the same birthday.

    (c) Use technology to simulate the "Birthday Problem" by generating 24 random numbers from 1 to 365. Repeat the simulation 10 times. How many times did you get at least two people with the same birthday?

The Multiplication Rule and Conditional Probability   *By rewriting the formula for the Multiplication Rule, you can write a formula for finding conditional probabilities. The conditional probability of event B occurring, given that event A has occurred, is*

$$P(B \mid A) = \frac{P(A \text{ and } B)}{P(A)}.$$

In Exercises 41 and 42, use the information below.

- *The probability that an airplane flight departs on time is 0.89.*

- *The probability that a flight arrives on time is 0.87.*

- *The probability that a flight departs and arrives on time is 0.83.*

41. Find the probability that a flight departed on time given that it arrives on time.

42. Find the probability that a flight arrives on time given that it departed on time.

3.3 The Addition Rule

What You Should Learn

▶ How to determine whether two events are mutually exclusive

▶ How to use the Addition Rule to find the probability of two events

Mutually Exclusive Events ■ The Addition Rule ■ A Summary of Probability

Mutually Exclusive Events

In Section 3.2, you learned how to find the probability of two events, A and B, occurring in sequence. Such probabilities are denoted by $P(A \text{ and } B)$. In this section, you will learn how to find the probability that at least one of two events will occur. Probabilities such as these are denoted by $P(A \text{ or } B)$ and depend on whether the events are **mutually exclusive.**

DEFINITION

Two events A and B are **mutually exclusive** when A and B cannot occur at the same time. That is, A and B have no outcomes in common.

The Venn diagrams show the relationship between events that are mutually exclusive and events that are not mutually exclusive. Note that when events A and B are mutually exclusive, they have no outcomes in common, so $P(A \text{ and } B) = 0$.

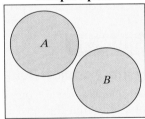

Sample Space

A and B are mutually exclusive.

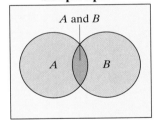

Sample Space

A and B are not mutually exclusive.

Study Tip

In probability and statistics, the word *or* is mostly used as an "inclusive or" rather than an "exclusive or." For instance, there are three ways for "event *A* or *B*" to occur.

(1) *A* occurs and *B* does not occur.

(2) *B* occurs and *A* does not occur.

(3) *A* and *B* both occur.

EXAMPLE 1

Recognizing Mutually Exclusive Events

Determine whether the events are mutually exclusive. Explain your reasoning.

1. Event A: Roll a 3 on a die.
   Event B: Roll a 4 on a die.

2. Event A: Randomly select a male student.
   Event B: Randomly select a nursing major.

3. Event A: Randomly select a blood donor with type O blood.
   Event B: Randomly select a female blood donor.

SOLUTION

1. Event A has one outcome, a 3. Event B also has one outcome, a 4. These outcomes cannot occur at the same time, so the events are mutually exclusive.

2. Because the student can be a male nursing major, the events are not mutually exclusive.

3. Because the donor can be a female with type O blood, the events are not mutually exclusive.

TRY IT YOURSELF 1

Determine whether the events are mutually exclusive. Explain your reasoning.

1. Event *A*: Randomly select a jack from a standard deck of 52 playing cards.
   Event *B*: Randomly select a face card from a standard deck of 52 playing cards.

2. Event *A*: Randomly select a vehicle that is a Ford.
   Event *B*: Randomly select a vehicle that is a Toyota.

Answer: Page A34

3.3   To explore this topic further, see **Activity 3.3** on page 166.

The Addition Rule

The Addition Rule for the Probability of *A* or *B*

The probability that events *A* or *B* will occur, $P(A \text{ or } B)$, is given by

$$P(A \text{ or } B) = P(A) + P(B) - P(A \text{ and } B).$$

If events *A* and *B* are mutually exclusive, then the rule can be simplified to

$$P(A \text{ or } B) = P(A) + P(B).$$ Events *A* and *B* are mutually exclusive.

This simplified rule can be extended to any number of mutually exclusive events.

In words, to find the probability that one event or the other will occur, add the individual probabilities of each event and subtract the probability that they both occur. As shown in the Venn diagram at the left, subtracting $P(A \text{ and } B)$ avoids double counting the probability of outcomes that occur in both *A* and *B*.

Outcomes here are double counted by $P(A) + P(B)$

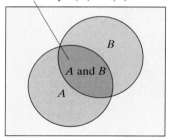

EXAMPLE 2

Using the Addition Rule to Find Probabilities

1. You select a card from a standard deck of 52 playing cards. Find the probability that the card is a 4 or an ace.

2. You roll a die. Find the probability of rolling a number less than 3 or rolling an odd number.

SOLUTION

1. A card that is a 4 cannot be an ace. So, the events are mutually exclusive, as shown in the Venn diagram. The probability of selecting a 4 or an ace is

$$P(4 \text{ or ace}) = P(4) + P(\text{ace}) = \frac{4}{52} + \frac{4}{52} = \frac{8}{52} = \frac{2}{13} \approx 0.154.$$

Deck of 52 Cards

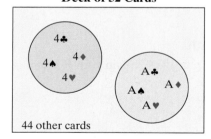

44 other cards

2. The events are not mutually exclusive because 1 is an outcome of both events, as shown in the Venn diagram. So, the probability of rolling a number less than 3 or an odd number is

$$P(\text{less than 3 or odd}) = P(\text{less than 3}) + P(\text{odd}) - P(\text{less than 3 and odd})$$
$$= \frac{2}{6} + \frac{3}{6} - \frac{1}{6}$$
$$= \frac{4}{6}$$
$$= \frac{2}{3}$$
$$\approx 0.667.$$

Roll a Die

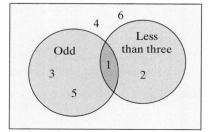

TRY IT YOURSELF 2

1. A die is rolled. Find the probability of rolling a 6 or an odd number.

2. A card is selected from a standard deck of 52 playing cards. Find the probability that the card is a face card or a heart.

Answer: Page A34

EXAMPLE 3

Finding Probabilities of Mutually Exclusive Events

The frequency distribution shows volumes of sales (in dollars) and the number of months in which a sales representative reached each sales level during the past three years. Using this sales pattern, find the probability that the sales representative will sell between $75,000 and $124,999 next month.

Sales volume (in dollars)	Months
0–24,999	3
25,000–49,999	5
50,000–74,999	6
75,000–99,999	7
100,000–124,999	9
125,000–149,999	2
150,000–174,999	3
175,000–199,999	1

SOLUTION

To solve this problem, define events A and B as

$A = \{\text{monthly sales between } \$75{,}000 \text{ and } \$99{,}999\}$

and

$B = \{\text{monthly sales between } \$100{,}000 \text{ and } \$124{,}999\}.$

The events are mutually exclusive, as shown in the Venn diagram.

Monthly Sales Volume

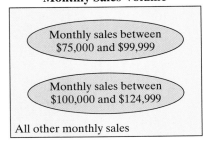

Because events A and B are mutually exclusive, the probability that the sales representative will sell between $75,000 and $124,999 next month is

$$P(A \text{ or } B) = P(A) + P(B) = \frac{7}{36} + \frac{9}{36} = \frac{16}{36} = \frac{4}{9} \approx 0.444.$$

TRY IT YOURSELF 3

Find the probability that the sales representative will sell between $0 and $49,999.

Answer: Page A34

Picturing the World

A survey of 1520 U.S. adults ages 18 and older asked them whether they had a smartphone, a tablet computer, or a home broadband subscription. Overall, 39% said they have all three; 28% said they have two of the three; 17% said they have one of the three; and 16% said they have none of them, as shown in the pie chart. (Source: Pew Research)

Do you have a smartphone, tablet computer, or a home broadband subscription?

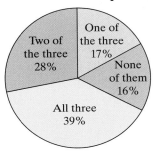

A U.S. adult is selected at random. What is the probability that when asked whether the adult has a smartphone, a tablet computer, or a home broadband, the response will be "none of them" or "one of the three?"

The events "none of them" and "one of the three" are mutually exclusive. For an adult selected at random, the probability that the response of the adult will be "none of them" or "one of the three" is 33%.

Note to Instructor

Point out to students that the notations $P(A \cap B)$ and $P(A \cup B)$ are sometimes used instead of $P(A$ and $B)$ and $P(A$ or $B)$, respectively. The symbol $\cap$ represents the intersection of two sets and the symbol $\cup$ represents the union of two sets. So, the Addition Rule can be written as $P(A \cup B) = P(A) + P(B) - P(A \cap B)$.

EXAMPLE 4

Using the Addition Rule to Find Probabilities

A blood bank catalogs the types of blood, including whether it is Rh-positive or Rh-negative, given by donors during the last five days. The number of donors who gave each blood type is shown in the table.

1. Find the probability that a donor selected at random has type O or type A blood.

2. Find the probability that a donor selected at random has type B blood or is Rh-negative.

		Blood type				
		O	A	B	AB	Total
Rh-factor	Positive	156	139	37	12	344
	Negative	28	25	8	4	65
	Total	184	164	45	16	409

SOLUTION

1. Because a donor cannot have type O blood and type A blood, these events are mutually exclusive. So, using the Addition Rule, the probability that a randomly chosen donor has type O or type A blood is

$$P(\text{type O or type A}) = P(\text{type O}) + P(\text{type A})$$
$$= \frac{184}{409} + \frac{164}{409}$$
$$= \frac{348}{409}$$
$$\approx 0.851.$$

2. Because a donor can have type B blood and be Rh-negative, these events are not mutually exclusive. So, using the Addition Rule, the probability that a randomly chosen donor has type B blood or is Rh-negative is

$$P(\text{type B or Rh-neg}) = P(\text{type B}) + P(\text{Rh-neg}) - P(\text{type B and Rh-neg})$$
$$= \frac{45}{409} + \frac{65}{409} - \frac{8}{409}$$
$$= \frac{102}{409}$$
$$\approx 0.249.$$

TRY IT YOURSELF 4

1. Find the probability that a donor selected at random has type B or type AB blood.

2. Find the probability that a donor selected at random does not have type O or type A blood.

3. Find the probability that a donor selected at random has type O blood or is Rh-positive.

4. Find the probability that a donor selected at random has type A blood or is Rh-negative.

Answer: Page A34

A Summary of Probability

Type of probability and probability rules	In words	In symbols
Classical Probability	The number of outcomes in the sample space is known and each outcome is equally likely to occur.	$P(E) = \dfrac{\text{Number of outcomes in event } E}{\text{Number of outcomes in sample space}}$
Empirical Probability	The frequency of each outcome in the sample space is estimated from experimentation.	$P(E) = \dfrac{\text{Frequency of event } E}{\text{Total frequency}} = \dfrac{f}{n}$
Range of Probabilities Rule	The probability of an event is between 0 and 1, inclusive.	$0 \le P(E) \le 1$
Complementary Events	The complement of event E is the set of all outcomes in a sample space that are not included in E, and is denoted by E'.	$P(E') = 1 - P(E)$
Multiplication Rule	The Multiplication Rule is used to find the probability of two events occurring in sequence.	$P(A \text{ and } B) = P(A) \cdot P(B\|A)$ Dependent events $P(A \text{ and } B) = P(A) \cdot P(B)$ Independent events
Addition Rule	The Addition Rule is used to find the probability of at least one of two events occurring.	$P(A \text{ or } B) = P(A) + P(B) - P(A \text{ and } B)$ $P(A \text{ or } B) = P(A) + P(B)$ Mutually exclusive events

EXAMPLE 5

Combining Rules to Find Probabilities

Use the figure at the right to find the probability that a randomly selected draft pick is not a running back or a wide receiver.

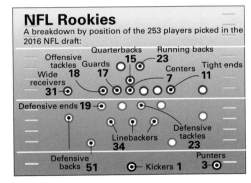

NFL Rookies
A breakdown by position of the 253 players picked in the 2016 NFL draft:
(Source: National Football League)

SOLUTION

Define events A and B.

　　A: Draft pick is a running back.
　　B: Draft pick is a wide receiver.

These events are mutually exclusive, so the probability that the draft pick is a running back or wide receiver is

$$P(A \text{ or } B) = P(A) + P(B) = \frac{23}{253} + \frac{31}{253} = \frac{54}{253}.$$

By taking the complement of $P(A \text{ or } B)$, you can determine that the probability of randomly selecting a draft pick who is not a running back or wide receiver is

$$1 - P(A \text{ or } B) = 1 - \frac{54}{253} = \frac{199}{253} \approx 0.787.$$

TRY IT YOURSELF 5

Find the probability that a randomly selected draft pick is not a linebacker or a quarterback.

Answer: Page A34

3.3 EXERCISES

1. $P(A \text{ and } B) = 0$ because A and B cannot occur at the same time.

2. (a) *Sample answer:* Toss coin once: $A = \{head\}$ and $B = \{tail\}$
   (b) *Sample answer:* Draw one card: $A = \{ace\}$ and $B = \{spade\}$

3. True

4. False. Two events being independent does not imply they are mutually exclusive.

5. False. The probability that event A or event B will occur is
   $P(A \text{ or } B) =$
   $\quad P(A) + P(B) - P(A \text{ and } B)$.

6. True

7. Not mutually exclusive. A presidential candidate can lose the popular vote and win the election.

8. Mutually exclusive. A movie cannot have two ratings.

9. Not mutually exclusive. A psychology major can be male and 20 years old.

10. Mutually exclusive. A student cannot have a birthday in both months.

11. Mutually exclusive. A voter cannot be both a Republican and a Democrat.

12. Not mutually exclusive. A member of Congress can be a male senator.

13. 0.625

14. 0.760

Building Basic Skills and Vocabulary

1. When two events are mutually exclusive, why is $P(A \text{ and } B) = 0$?

2. Give an example of (a) two events that are mutually exclusive and (b) two events that are not mutually exclusive.

True or False? *In Exercises 3–6, determine whether the statement is true or false. If it is false, explain why.*

3. When two events are mutually exclusive, they have no outcomes in common.

4. When two events are independent, they are also mutually exclusive.

5. The probability that event A or event B will occur is
$$P(A \text{ or } B) = P(A) + P(B) + P(A \text{ and } B).$$

6. If events A and B are mutually exclusive, then
$$P(A \text{ or } B) = P(A) + P(B).$$

Graphical Analysis *In Exercises 7 and 8, determine whether the events shown in the Venn diagram are mutually exclusive. Explain your reasoning.*

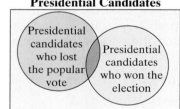

7. **Sample Space: Presidential Candidates**
   Presidential candidates who lost the popular vote / Presidential candidates who won the election

8. **Sample Space: Movies**
   Movies that are rated R / Movies that are rated PG-13

Using and Interpreting Concepts

Recognizing Mutually Exclusive Events *In Exercises 9–12, determine whether the events are mutually exclusive. Explain your reasoning.*

9. Event A: Randomly select a male psychology major.
   Event B: Randomly select a psychology major who is 20 years old.

10. Event A: Randomly select a student with a birthday in April.
    Event B: Randomly select a student with a birthday in May.

11. Event A: Randomly select a voter who is a registered Republican.
    Event B: Randomly select a voter who is a registered Democrat.

12. Event A: Randomly select a member of the U.S. Congress.
    Event B: Randomly select a male U.S. Senator.

13. **Students** A physics class has 40 students. Of these, 12 students are physics majors and 16 students are female. Of the physics majors, three are female. Find the probability that a randomly selected student is female or a physics major.

14. **Conference** A teaching conference has an attendance of 6855 people. Of these, 3120 are college professors and 3595 are male. Of the college professors, 1505 are male. Find the probability that a randomly selected attendee is male or a college professor.

15. 0.126
16. 0.997
17. (a) 0.308   (b) 0.538
    (c) 0.308
18. (a) 0.5   (b) 0.667
    (c) 0.833
19. (a) 0.06   (b) 0.426
    (c) 0.81   (d) 0.201
20. (a) 0.603   (b) 0.397
    (c) 0.16   (d) 0.187

15. **Carton Defects** Of the cartons produced by a company, 5% have a puncture, 8% have a smashed corner, and 0.4% have both a puncture and a smashed corner. Find the probability that a randomly selected carton has a puncture or a smashed corner.

16. **Can Defects** Of the cans produced by a company, 96% do not have a puncture, 93% do not have a smashed edge, and 89.3% do not have a puncture and do not have a smashed edge. Find the probability that a randomly selected can does not have a puncture or a smashed edge.

17. **Selecting a Card** A card is selected at random from a standard deck of 52 playing cards. Find the probability of each event.
    (a) Randomly selecting a club or a 3
    (b) Randomly selecting a red suit or a king
    (c) Randomly selecting a 9 or a face card

18. **Rolling a Die** You roll a die. Find the probability of each event.
    (a) Rolling a 5 or a number greater than 3
    (b) Rolling a 2 or an odd number
    (c) Rolling a number less than 4 or an even number

19. **U.S. Age Distribution** The estimated percent distribution of the U.S. population for 2025 is shown in the pie chart. Find the probability of each event. *(Source: U.S. Census Bureau)*
    (a) Randomly selecting someone who is under 5 years old
    (b) Randomly selecting someone who is 45 years or over
    (c) Randomly selecting someone who is not 65 years or over
    (d) Randomly selecting someone who is between 20 and 34 years old

U.S. Age Distribution

75 years or over 8.3%
65–74 years 10.7%
45–64 years 23.6%
35–44 years 13.2%
25–34 years 13.7%
20–24 years 6.4%
15–19 years 6.1%
5–14 years 11.9%
Under 5 years 6.0%

FIGURE FOR EXERCISE 19

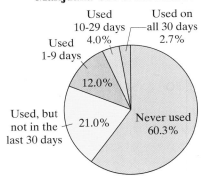

Marijuana Use in the Last 30 Days

Used 10-29 days 4.0%
Used on all 30 days 2.7%
Used 1-9 days 12.0%
Never used 60.3%
Used, but not in the last 30 days 21.0%

FIGURE FOR EXERCISE 20

20. **Marijuana Use** The percent of college students' marijuana use for a sample of 95,761 students is shown in the pie chart. Find the probability of each event. *(Source: American College Health Association)*
    (a) Randomly selecting a student who never used marijuana
    (b) Randomly selecting a student who used marijuana
    (c) Randomly selecting a student who used marijuana between 1 and 29 of the last 30 days
    (d) Randomly selecting a student who used marijuana on at least 1 of the last 30 days

How Would You Grade the Media for the Way They Conducted Themselves in the 2016 Presidential Campaign?

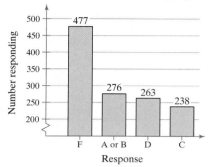

FIGURE FOR EXERCISE 21

How Important Is the Brexit Story to You Personally?

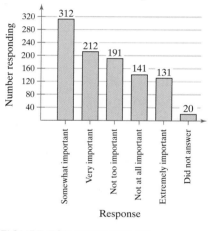

FIGURE FOR EXERCISE 22

21. (a) 0.780   (b) 0.410
    (c) 0.590   (d) 0.400
22. (a) 0.310   (b) 0.140
    (c) 0.330   (d) 0.341
23. (a) 0.520   (b) 0.899
    (c) 0.909
24. (a) 0.708   (b) 0.830
    (c) 0.960

21. **Media Conduct** The responses of 1254 voters to a survey about the way the media conducted themselves in the 2016 presidential campaign are shown in the Pareto chart. Find the probability of each event. *(Adapted from Pew Research Center)*

   (a) Randomly selecting a person from the sample who did not give the media an A or a B

   (b) Randomly selecting a person from the sample who gave the media a grade better than a D

   (c) Randomly selecting a person from the sample who gave the media a D or an F

   (d) Randomly selecting a person from the sample who gave the media a C or a D

22. **Brexit** The responses of 1007 American adults to a survey question about the story of Britons' vote to leave the European Union are shown in the Pareto chart. Find the probability of each event. *(Adapted from GfK Public Affairs and Corporate Communications)*

   (a) Randomly selecting an adult who thinks the story is somewhat important

   (b) Randomly selecting an adult who thinks the story is not at all important

   (c) Randomly selecting an adult who thinks the story is not too important or not at all important

   (d) Randomly selecting an adult who thinks the story is extremely important or very important

23. **Business Degrees** The table shows the numbers of male and female students in the U.S. who received bachelor's degrees in business in a recent year. A student is selected at random. Find the probability of each event. *(Source: National Center for Educational Statistics)*

	Business degrees	Nonbusiness degrees	Total
Males	191,310	621,359	812,669
Females	172,489	909,776	1,082,265
Total	363,799	1,531,135	1,894,934

   (a) The student is male or received a business degree.

   (b) The student is female or received a nonbusiness degree.

   (c) The student is not female or received a nonbusiness degree.

24. **Education Tax** The table shows the results of a survey that asked 506 Maine adults whether they favored or opposed a tax to fund education. A person is selected at random from the sample. Find the probability of each event. *(Adapted from Portland Press Herald)*

	Support	Oppose	Unsure	Total
Males	128	99	20	247
Females	173	65	21	259
Total	301	164	41	506

   (a) The person opposes the tax or is female.

   (b) The person supports the tax or is male.

   (c) The person is not unsure or is female.

25. (a) 0.589   (b) 0.762
    (c) 0.461   (d) 0.922
26. (a) 0.475   (b) 0.595
    (c) 0.662   (d) 0.752
27. 0.63
28. 0.55
29. If events *A*, *B*, and *C* are not mutually exclusive, then *P*(*A* and *B* and *C*) must be added because *P*(*A*) + *P*(*B*) + *P*(*C*) counts the intersection of all three events three times and −*P*(*A* and *B*) − *P*(*A* and *C*) −*P*(*B* and *C*) subtracts the intersection of all three events three times. So, if *P*(*A* and *B* and *C*) is not added at the end, then it will not be counted.
30. No; If two events *A* and *B* are independent, then *P*(*A* and *B*) = *P*(*A*) · *P*(*B*). If two events are mutually exclusive, then *P*(*A* and *B*) = 0. The only scenario when two events can be independent and mutually exclusive is when *P*(*A*) = 0 or *P*(*B*) = 0.

25. **Charity**   The table shows the results of a survey that asked 2850 people whether they were involved in any type of charity work. A person is selected at random from the sample. Find the probability of each event.

	Frequently	Occasionally	Not at all	Total
Males	221	456	795	1472
Females	207	430	741	1378
Total	428	886	1536	2850

(a) The person is male or frequently involved in charity work.
(b) The person is female or not involved in charity work at all.
(c) The person is frequently or occasionally involved in charity work.
(d) The person is female or not frequently involved in charity work.

26. **Eye Survey**   The table shows the results of a survey that asked 3203 people whether they wore contacts or glasses. A person is selected at random from the sample. Find the probability of each event.

	Only contacts	Only glasses	Both	Neither	Total
Males	64	841	177	456	1538
Females	189	427	368	681	1665
Total	253	1268	545	1137	3203

(a) The person wears only contacts or only glasses.
(b) The person is male or wears both contacts and glasses.
(c) The person is female or wears neither contacts nor glasses.
(d) The person is male or does not wear glasses.

Extending Concepts

Addition Rule for Three Events   *The Addition Rule for the probability that event A or B or C will occur, P(A or B or C), is given by*

$$P(A \text{ or } B \text{ or } C) = P(A) + P(B) + P(C) - P(A \text{ and } B) - P(A \text{ and } C)$$
$$- P(B \text{ and } C) + P(A \text{ and } B \text{ and } C).$$

In the Venn diagram shown at the left, P(A or B or C) is represented by the blue areas. In Exercises 27 and 28, find P(A or B or C).

27. *P*(*A*) = 0.40, *P*(*B*) = 0.10, *P*(*C*) = 0.50, *P*(*A* and *B*) = 0.05, *P*(*A* and *C*) = 0.25, *P*(*B* and *C*) = 0.10, *P*(*A* and *B* and *C*) = 0.03

28. *P*(*A*) = 0.38, *P*(*B*) = 0.26, *P*(*C*) = 0.14, *P*(*A* and *B*) = 0.12, *P*(*A* and *C*) = 0.03, *P*(*B* and *C*) = 0.09, *P*(*A* and *B* and *C*) = 0.01

29. Explain, in your own words, why in the Addition Rule for *P*(*A* or *B* or *C*), *P*(*A* and *B* and *C*) is added at the end of the formula.

30. **Writing**   Can two events with nonzero probabilities be both independent and mutually exclusive? Explain your reasoning.

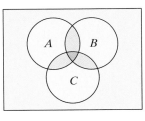

FIGURE FOR EXERCISES 27 AND 28

You can find the interactive applet for this activity within **MyLab Statistics** or at *www.pearsonhighered.com/ mathstatsresources.*

The *simulating the probability of rolling a 3 or 4* applet allows you to investigate the probability of rolling a 3 or 4 on a fair die. The plot at the top left corner shows the probability associated with each outcome of a die roll. When ROLL is clicked, *n* simulations of the experiment of rolling a die are performed. The results of the simulations are shown in the frequency plot. When the *animate* option is checked, the display will show each outcome dropping into the frequency plot as the simulation runs. The individual outcomes are shown in the text field at the far right of the applet. The center plot shows in blue the cumulative proportion of times that an event of rolling a 3 or 4 occurs. The green line in the plot reflects the theoretical probability of rolling a 3 or 4. As the experiment is conducted over and over, the cumulative proportion should converge to the theoretical probability.

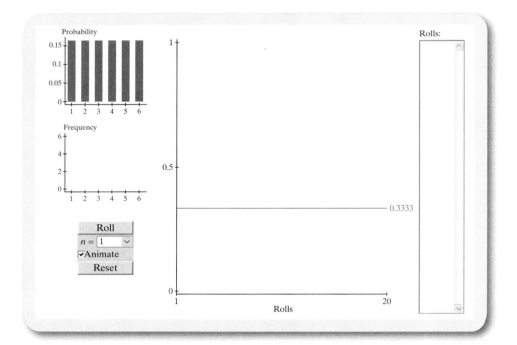

EXPLORE

Step 1 Specify a value for *n*.
Step 2 Click ROLL four times.
Step 3 Click RESET.
Step 4 Specify another value for *n*.
Step 5 Click ROLL.

DRAW CONCLUSIONS

1. Run the simulation using each value of *n* one time. Clear the results after each trial. Compare the cumulative proportion of "rolls" that result in a 3 or 4 for each trial with the theoretical probability of rolling a 3 or 4.

2. You want to modify the applet so you can find the probability of rolling a number less than 4. Describe the placement of the green line.

United States Congress

Congress is made up of the House of Representatives and the Senate. Members of the House of Representatives serve two-year terms and represent a district in a state. The number of representatives for each state is determined by population. States with larger populations have more representatives than states with smaller populations. The total number of representatives is set by law at 435. Members of the Senate serve six-year terms and represent a state. Each state has 2 senators, for a total of 100 senators. The tables show the makeup of the 115th Congress by gender and political party as of January 3, 2017.

House of Representatives

		Political party			
		Republican	Democrat	Independent	Total
Gender	Male	218	129	0	347
	Female	23	65	0	88
	Total	241	194	0	435

Senate

		Political party			
		Republican	Democrat	Independent	Total
Gender	Male	47	30	2	79
	Female	5	16	0	21
	Total	52	46	2	100

EXERCISES

1. Find the probability that a randomly selected representative is female. Find the probability that a randomly selected senator is female.

2. Compare the probabilities from Exercise 1.

3. A representative is selected at random. Find the probability of each event.

(a) The representative is male.

(b) The representative is a Republican.

(c) The representative is male given that the representative is a Republican.

(d) The representative is female and a Democrat.

4. Among members of the House of Representatives, are the events "being female" and "being a Democrat" independent or dependent events? Explain.

5. A senator is selected at random. Find the probability of each event.

(a) The senator is male.

(b) The senator is not a Democrat.

(c) The senator is female or a Republican.

(d) The senator is male or a Democrat.

6. Among members of the Senate, are the events "being female" and "being an Independent" mutually exclusive? Explain.

7. Using the same row and column headings as the tables above, create a combined table for Congress.

8. A member of Congress is selected at random. Use the table from Exercise 7 to find the probability of each event.

(a) The member is Independent.

(b) The member is female and a Republican.

(c) The member is male or a Democrat.

3.4 Additional Topics in Probability and Counting

What You Should Learn

▶ How to find the number of ways a group of objects can be arranged in order

▶ How to find the number of ways to choose several objects from a group without regard to order

▶ How to use counting principles to find probabilities

Permutations ■ Combinations ■ Applications of Counting Principles

Permutations

In Section 3.1, you learned that the Fundamental Counting Principle is used to find the number of ways two or more events can occur in sequence. An important application of the Fundamental Counting Principle is determining the number of ways that *n* objects can be arranged in order. An ordering of *n* objects is called a **permutation.**

DEFINITION

A **permutation** is an ordered arrangement of objects. The number of different permutations of *n* distinct objects is $n!$.

The expression ***n*!** is read as ***n* factorial.** If *n* is a positive integer, then $n!$ is defined as follows.

$$n! = n \cdot (n-1) \cdot (n-2) \cdot (n-3) \cdots 3 \cdot 2 \cdot 1$$

As a special case, $0! = 1$. Here are several other values of $n!$.

$$1! = 1 \qquad 2! = 2 \cdot 1 = 2 \qquad 3! = 3 \cdot 2 \cdot 1 = 6 \qquad 4! = 4 \cdot 3 \cdot 2 \cdot 1 = 24$$

Study Tip

Notice that small values of *n* can produce very large values of *n*!. For instance, $10! = 3,628,800$. Be sure you know how to use the factorial key on your calculator.

EXAMPLE 1

Finding the Number of Permutations of *n* Objects

The objective of a 9 × 9 Sudoku number puzzle is to fill the grid so that each row, each column, and each 3 × 3 grid contain the digits 1 through 9. How many different ways can the first row of a blank 9 × 9 Sudoku grid be filled?

SOLUTION

The number of permutations is

$$9! = 9 \cdot 8 \cdot 7 \cdot 6 \cdot 5 \cdot 4 \cdot 3 \cdot 2 \cdot 1 = 362,880.$$

So, there are 362,880 different ways the first row can be filled.

TRY IT YOURSELF 1

The Big 12 is a collegiate athletic conference with 10 schools: Baylor, Iowa State, Kansas, Kansas State, Oklahoma, Oklahoma State, TCU, Texas, Texas Tech, and West Virginia. How many different final standings are possible for the Big 12's football teams?

Answer: Page A34

You may want to choose some of the objects in a group and put them in order. Such an ordering is called a **permutation of *n* objects taken *r* at a time.**

Sudoku Number Puzzle

6	7	1				2	4	9
8			7		2			1
2				6				3
	5		6		3		2	
		8				7		
	1		8		4		6	
9				1				6
1			5		9			7
5	8	7				9	1	2

Permutations of *n* Objects Taken *r* at a Time

The number of permutations of *n* distinct objects taken *r* at a time is

$$_nP_r = \frac{n!}{(n-r)!}, \text{ where } r \le n.$$

Tech Tip

You can use technology such as Minitab, Excel, StatCrunch, or the TI-84 Plus to find the number of permutations of n objects taken r at a time. For instance, here is how to find $_nP_r$ in Example 2 on a TI-84 Plus.

Enter the total number of objects, $n = 10$.

| MATH |

Choose the PRB menu.

    2: nPr

Enter the number of objects taken, $r = 4$.

| ENTER |

```
TI-84 PLUS
10 nPr 4
                5040
```

EXAMPLE 2

Finding $_nP_r$

Find the number of ways of forming four-digit codes in which no digit is repeated.

SOLUTION

To form a four-digit code with no repeating digits, you need to select 4 digits from a group of 10, so $n = 10$ and $r = 4$.

$$
\begin{aligned}
_nP_r &= {}_{10}P_4 \\
&= \frac{10!}{(10-4)!} \\
&= \frac{10!}{6!} \\
&= \frac{10 \cdot 9 \cdot 8 \cdot 7 \cdot 6!}{6!} \\
&= 5040
\end{aligned}
$$

So, there are 5040 possible four-digit codes that do not have repeating digits.

TRY IT YOURSELF 2

A psychologist shows a list of eight activities to a subject in an experiment. How many ways can the subject pick a first, second, and third activity?

Answer: Page A34

EXAMPLE 3

Finding $_nP_r$

Each year, 33 race cars start the Indianapolis 500. How many ways can the cars finish first, second, and third?

SOLUTION

You need to select three race cars from a group of 33, so

    $n = 33$ and $r = 3$.

Because the order is important, the number of ways the cars can finish first, second, and third is

$$
_nP_r = {}_{33}P_3 = \frac{33!}{(33-3)!} = \frac{33!}{30!} = \frac{33 \cdot 32 \cdot 31 \cdot 30!}{30!} = 32{,}736.
$$

TRY IT YOURSELF 3

The board of directors of a company has 12 members. One member is the president, another is the vice president, another is the secretary, and another is the treasurer. How many ways can these positions be assigned?

Answer: Page A34

In Example 3, note that the Fundamental Counting Principle can be used to obtain the same result. There are 33 choices for first place, 32 choices for second place, and 31 choices for third place. So, there are

    $33 \cdot 32 \cdot 31 \cdot = 32{,}736$

ways the cars can finish first, second, and third.

You may want to order a group of n objects in which some of the objects are the same. For instance, consider the group of letters

AAAABBC.

This group has four A's, two B's, and one C. How many ways can you order such a group? Using the formula for $_nP_r$, you might conclude that there are

$$_7P_7 = 7! = 5040$$

possible orders. However, because some of the objects are the same, not all of these permutations are *distinguishable*. How many distinguishable permutations are possible? The answer can be found using the formula for the number of **distinguishable permutations.**

Distinguishable Permutations

The number of **distinguishable permutations** of n objects, where n_1 are of one type, n_2 are of another type, and so on, is

$$\frac{n!}{n_1! \cdot n_2! \cdot n_3! \cdots n_k!}$$

where

$$n_1 + n_2 + n_3 + \cdots + n_k = n.$$

Using the formula for distinguishable permutations, you can determine that the number of distinguishable permutations of the letters AAAABBC is

$$\frac{7!}{4! \cdot 2! \cdot 1!} = \frac{7 \cdot 6 \cdot 5}{2}$$

$$= 105 \text{ distinguishable permutations.}$$

EXAMPLE 4

Finding the Number of Distinguishable Permutations

A building contractor is planning to develop a subdivision. The subdivision is to consist of 6 one-story houses, 4 two-story houses, and 2 split-level houses. In how many distinguishable ways can the houses be arranged?

SOLUTION

There are to be 12 houses in the subdivision, 6 of which are of one type (one-story), 4 of another type (two-story), and 2 of a third type (split-level). So, there are

$$\frac{12!}{6! \cdot 4! \cdot 2!} = \frac{12 \cdot 11 \cdot 10 \cdot 9 \cdot 8 \cdot 7 \cdot 6!}{6! \cdot 4! \cdot 2!}$$

$$= 13,860 \text{ distinguishable ways.}$$

You can check your answer using technology, as shown at the left on a TI-84 Plus.

Interpretation There are 13,860 distinguishable ways to arrange the houses in the subdivision.

TI-84 PLUS

```
12!/(6!4!2!)
              13860
```

TRY IT YOURSELF 4

The contractor wants to plant six oak trees, nine maple trees, and five poplar trees along the subdivision street. The trees are to be spaced evenly. In how many distinguishable ways can they be planted?

Answer: Page A34

Combinations

A state park manages five beaches labeled A, B, C, D, and E. Due to budget constraints, new restrooms will be built at only three beaches. There are 10 ways for the state to select the three beaches.

ABC, ABD, ABE, ACD, ACE, ADE, BCD, BCE, BDE, CDE

In each selection, order does not matter (ABC is the same as BAC). The number of ways to choose *r* objects from *n* objects without regard to order is called the number of **combinations of *n* objects taken *r* at a time.**

Combinations of *n* Objects Taken *r* at a Time

The number of combinations of *r* objects selected from a group of *n* objects *without regard to order* is

$$_nC_r = \frac{n!}{(n-r)!\,r!}$$

where $r \le n$.

You can think of a combination of *n* objects chosen *r* at a time as a permutation of *n* objects in which the *r* selected objects are alike and the remaining $n - r$ (not selected) objects are alike.

Tech Tip

You can use technology such as Minitab, Excel, StatCrunch, or the TI-84 Plus to find the number of combinations of *n* objects taken *r* at a time. For instance, here is how to find $_nC_r$ in Example 5 on a TI-84 Plus.

Enter the total number of objects, $n = 16$.

MATH

Choose the PRB menu.

3: nCr

Enter the number of objects taken, $r = 4$.

ENTER

```
TI-84 PLUS
16 nCr 4
              1820
```

EXAMPLE 5

Finding the Number of Combinations

A state's department of transportation plans to develop a new section of interstate highway and receives 16 bids for the project. The state plans to hire four of the bidding companies. How many different combinations of four companies can be selected from the 16 bidding companies?

SOLUTION

The state is selecting four companies from a group of 16, so

$$n = 16 \text{ and } r = 4.$$

Because order is not important, there are

$$
\begin{aligned}
_nC_r &= {}_{16}C_4 \\
&= \frac{16!}{(16-4)!\,4!} \\
&= \frac{16!}{12!\,4!} \\
&= \frac{16 \cdot 15 \cdot 14 \cdot 13 \cdot 12!}{12! \cdot 4!} \\
&= 1820 \text{ different combinations.}
\end{aligned}
$$

Interpretation There are 1820 different combinations of four companies that can be selected from the 16 bidding companies.

TRY IT YOURSELF 5

The manager of an accounting department wants to form a three-person advisory committee from the 20 employees in the department. In how many ways can the manager form this committee?

Answer: Page A34

Study Tip

To solve a problem using a counting principle, be sure you choose the appropriate counting principle. To help you do this, consider these questions.

- *Are there two or more separate events?* Fundamental Counting Principle
- *Is the order of the objects important?* Permutation
- *Are the chosen objects from a larger group of objects in which order is not important?* Combination

Note that some problems may require you to use more than one counting principle (see Example 8).

Applications of Counting Principles

The table summarizes the counting principles.

Principle	Description	Formula
Fundamental Counting Principle	If one event can occur in m ways and a second event can occur in n ways, then the number of ways the two events can occur in sequence is $m \cdot n$.	$m \cdot n$
Permutations	The number of permutations of n distinct objects	$n!$
	The number of permutations of n distinct objects taken r at a time, where $r \le n$	$_nP_r = \dfrac{n!}{(n-r)!}$
	The number of distinguishable permutations of n objects where n_1 are of one type, n_2 are of another type, and so on, and $n_1 + n_2 + n_3 + \cdots + n_k = n$	$\dfrac{n!}{n_1! \cdot n_2! \cdots n_k!}$
Combinations	The number of combinations of r objects selected from a group of n objects without regard to order, where $r \le n$	$_nC_r = \dfrac{n!}{(n-r)!r!}$

EXAMPLE 6

Finding Probabilities

A student advisory board consists of 17 members. Three members will be chosen to serve as the board's chair, secretary, and webmaster. Each member is equally likely to serve in any of the positions. What is the probability of randomly selecting the three members who will be chosen for the board?

SOLUTION

Note that order is important because the positions (chair, secretary, and webmaster) are distinct objects. There is one favorable outcome and there are

$$_{17}P_3 = \frac{17!}{(17-3)!} = \frac{17!}{14!} = \frac{17 \cdot 16 \cdot 15 \cdot 14!}{14!} = 17 \cdot 16 \cdot 15 = 4080$$

ways the three positions can be filled. So, the probability of correctly selecting the three members who hold each position is

$$P(\text{selecting the three members}) = \frac{1}{4080} \approx 0.0002.$$

You can check your answer using technology. For instance, using Excel's PERMUT command, you can find the probability of selecting the three members, as shown at the left.

TRY IT YOURSELF 6

A student advisory board consists of 20 members. Two members will be chosen to serve as the board's chair and secretary. Each member is equally likely to serve in either of the positions. What is the probability of randomly selecting the two members who will be chosen for the board?

Answer: Page A34

EXCEL

	A
1	PERMUT(17,3)
2	4080
3	1/A2
4	0.000245098

Picturing the World

One of the largest lottery jackpots ever, $656 million, was won in the Mega Millions lottery. When the jackpot was won, five different numbers were chosen from 1 to 56 and one number, the Mega Ball, was chosen from 1 to 46. The winning numbers are shown below.

2   4   23

38   46   23

↑
Mega Ball

In 2013, the lottery changed its rules. Now, a player chooses five different numbers from 1 to 75 and one number from 1 to 15. A player wins the jackpot by matching all six winning numbers in a drawing.

You purchase one ticket in the Mega Millions lottery. Find the probability of winning the jackpot using the old rules and the new rules. Which set of rules provides you with a better chance of winning the jackpot? How likely is your chance of winning?

The probability of winning the jackpot using the old rules is

$$\frac{1}{_{56}C_5 \cdot 46} = \frac{1}{175{,}711{,}536}$$
$$\approx 0.000000006.$$

The probability of winning the jackpot using the new rules is

$$\frac{1}{_{75}C_5 \cdot 15} = \frac{1}{258{,}890{,}850}$$
$$\approx 0.000000004.$$

The probability of winning was slightly higher using the old rules. Either way, the probability of winning is highly unlikely.

EXAMPLE 7

Finding Probabilities

Find the probability of being dealt 5 diamonds from a standard deck of 52 playing cards.

SOLUTION

In a standard deck of playing cards, 13 cards are diamonds. Note that it does not matter what order the cards are selected. The possible number of ways of choosing 5 diamonds out of 13 is $_{13}C_5$. The number of possible five-card hands is $_{52}C_5$. So, the probability of being dealt 5 diamonds is

$$P(5 \text{ diamonds}) = \frac{_{13}C_5}{_{52}C_5}$$
$$= \frac{1287}{2{,}598{,}960}$$
$$\approx 0.0005.$$

TRY IT YOURSELF 7

Find the probability of being dealt 5 diamonds from a standard deck of playing cards that also includes two jokers. In this case, the joker is considered to be a wild card that can be used to represent any card in the deck.

Answer: Page A34

EXAMPLE 8

Finding Probabilities

A food manufacturer is analyzing a sample of 400 corn kernels for the presence of a toxin. In this sample, three kernels have dangerously high levels of the toxin. Four kernels are randomly selected from the sample. What is the probability that exactly one kernel contains a dangerously high level of the toxin?

SOLUTION

Note that it does not matter what order the kernels are selected. The possible number of ways of choosing one toxic kernel out of three toxic kernels is $_3C_1$. The possible number of ways of choosing 3 nontoxic kernels from 397 nontoxic kernels is $_{397}C_3$. So, using the Fundamental Counting Principle, the number of ways of choosing one toxic kernel and three nontoxic kernels is

$$_3C_1 \cdot {}_{397}C_3 = 3 \cdot 10{,}349{,}790 = 31{,}049{,}370.$$

The number of possible ways of choosing 4 kernels from 400 kernels is

$$_{400}C_4 = 1{,}050{,}739{,}900.$$

So, the probability of selecting exactly 1 toxic kernel is

$$P(1 \text{ toxic kernel}) = \frac{_3C_1 \cdot {}_{397}C_3}{_{400}C_4}$$
$$= \frac{31{,}049{,}370}{1{,}050{,}739{,}900}$$
$$\approx 0.030.$$

TRY IT YOURSELF 8

A jury consists of five men and seven women. Three jury members are selected at random for an interview. Find the probability that all three are men.

Answer: Page A34

3.4 EXERCISES

1. The number of ordered arrangements of *n* objects taken *r* at a time.

   Sample answer: An example of a permutation is the number of seating arrangements of you and three of your friends.

2. The number of ways to select *r* of the *n* objects without regard to order.

   Sample answer: An example of a combination is the number of selections of different playoff teams from a volleyball tournament.

3. False. A permutation is an ordered arrangement of objects.

4. True

5. True

6. True

7. 15,120

8. 2184

9. 56

10. 203,490

11. 0.076

12. 0.035

13. 0.462

14. 0.018

15. Permutation. The order of the 16 floats in line matters.

16. Combination. The order of the committee members does not matter.

17. Combination. The order does not matter because the position of one captain is the same as the other.

18. Permutation. The order of the letters matters in the password.

19. 5040

20. 40,320

21. 720

22. 3,628,800

Building Basic Skills and Vocabulary

1. When you calculate the number of permutations of *n* distinct objects taken *r* at a time, what are you counting? Give an example.

2. When you calculate the number of combinations of *r* objects taken from a group of *n* objects, what are you counting? Give an example.

True or False? *In Exercises 3–6, determine whether the statement is true or false. If it is false, rewrite it as a true statement.*

3. A combination is an ordered arrangement of objects.

4. The number of different ordered arrangements of *n* distinct objects is *n*!.

5. When you divide the number of permutations of 11 objects taken 3 at a time by 3!, you will get the number of combinations of 11 objects taken 3 at a time.

6. $_7C_5 = _7C_2$

In Exercises 7–14, perform the indicated calculation.

7. $_9P_5$

8. $_{14}P_3$

9. $_8C_3$

10. $_{21}C_8$

11. $\dfrac{_8C_4}{_{12}C_6}$

12. $\dfrac{_{10}C_7}{_{14}C_7}$

13. $\dfrac{_3P_2}{_{13}P_1}$

14. $\dfrac{_7P_3}{_{12}P_4}$

In Exercises 15–18, determine whether the situation involves permutations, combinations, or neither. Explain your reasoning.

15. The number of ways 16 floats can line up in a row for a parade

16. The number of ways a four-member committee can be chosen from 10 people

17. The number of ways 2 captains can be chosen from 28 players on a lacrosse team

18. The number of four-letter passwords that can be created when no letter can be repeated

Using and Interpreting Concepts

19. **Video Games**   You have seven different video games. How many different ways can you arrange the games side by side on a shelf?

20. **Skiing**   Eight people compete in a downhill ski race. Assuming that there are no ties, in how many different orders can the skiers finish?

21. **Security Code**   In how many ways can the letters A, B, C, D, E, and F be arranged for a six-letter security code?

22. **Starting Lineup**   The starting lineup for a softball team consists of 10 players. How many different batting orders are possible using the starting lineup?

23. 117,600
24. 524,160
25. 96,909,120
26. 1,771,560
27. 2,042,040
28. 11,085,360
29. 50,400
30. 56
31. 4845
32. 1,251,677,700
33. 9880
34. 20,358,520
35. 6240
36. 19,600
37. 86,296,950
38. 171,360

23. Footrace There are 50 runners in a race. How many ways can the runners finish first, second, and third?

24. Singing Competition There are 16 finalists in a singing competition. The top five singers receive prizes. How many ways can the singers finish first through fifth?

25. Playlist A DJ is preparing a playlist of 24 songs. How many different ways can the DJ choose the first six songs?

26. Archaeology Club An archaeology club has 38 members. How many different ways can the club select a president, vice president, treasurer, and secretary?

27. Blood Donors At a blood drive, 8 donors with type O+ blood, 6 donors with type A+ blood, and 3 donors with type B+ blood are in line. In how many distinguishable ways can the donors be in line?

28. Necklaces You are putting 9 pieces of blue beach glass, 3 pieces of red beach glass, and 7 pieces of green beach glass on a necklace. In how many distinguishable ways can the beach glass be put on the necklace?

29. Letters In how many distinguishable ways can the letters in the word *statistics* be written?

30. Computer Science A byte is a sequence of eight bits. A bit can be a 0 or a 1. In how many distinguishable ways can you have a byte with five 0's and three 1's?

31. Experimental Group In order to conduct an experiment, 4 subjects are randomly selected from a group of 20 subjects. How many different groups of four subjects are possible?

32. Jury Selection From a group of 36 people, a jury of 12 people is selected. In how many different ways can a jury of 12 people be selected?

33. Students A class has 40 students. In how many different ways can three students form a group to work on a class project? (Assume the order of the students is not important.)

34. Lottery Number Selection A lottery has 52 numbers. In how many different ways can 6 of the numbers be selected? (Assume that order of selection is not important.)

35. Menu A restaurant offers a dinner special that lets you choose from 10 entrées, 8 side dishes, and 13 desserts. You can choose one entrée, one side dish, and two desserts. How many different meals are possible?

36. Floral Arrangements A floral arrangement consists of 6 different colored roses, 3 different colored carnations, and 3 different colored daisies. You can choose from 8 different colors of roses, 6 different colors of carnations, and 7 different colors of daisies. How many different arrangements are possible?

37. Water Pollution An environmental agency is analyzing water samples from 80 lakes for pollution. Five of the lakes have dangerously high levels of dioxin. Six lakes are randomly selected from the sample. Using technology, how many ways could one polluted lake and five nonpolluted lakes be chosen?

38. Property Inspection A property inspector is visiting 24 properties. Six of the properties are one acre or less in size, and the rest are greater than one acre in size. Eight properties are randomly selected. Using technology, how many ways could three properties that are each one acre or less and five properties that are each larger than one acre be chosen?

39. 0.005
40. 0.05
41. 0.005
42. 0.012
43. (a) 0.016   (b) 0.385
44. (a) 0.015   (b) 0.070
45. 0.0009
46. 0.036
47. 0.242
48. 0.105
49. 0.0000015

39. Senate Committee   The U.S. Senate Committee on Homeland Security and Governmental Affairs has 15 members. Two members are chosen to serve as the committee chair and the ranking member. Each committee member is equally likely to serve in either of these positions. What is the probability of randomly selecting the chair and the ranking member? *(Source: U.S. Senate)*

40. University Committee   The University of California Health Services committee has five members. Two members are chosen to serve as the committee chair and vice chair. Each committee member is equally likely to serve in either of these positions. What is the probability of randomly selecting the chair and the vice chair? *(Source: University of California)*

41. Horse Race   A horse race has 12 entries. Assuming that there are no ties, what is the probability that the three horses owned by one person finish first, second, and third?

42. Pizza Toppings   A pizza shop offers nine toppings. No topping is used more than once. What is the probability that the toppings on a three-topping pizza are pepperoni, onions, and mushrooms?

43. Jukebox   You look over the songs on a jukebox and determine that you like 15 of the 56 songs.
 (a) What is the probability that you like the next three songs that are played? (Assume a song cannot be repeated.)
 (b) What is the probability that you do not like the next three songs that are played? (Assume a song cannot be repeated.)

44. Officers   The offices of president, vice president, secretary, and treasurer for an environmental club will be filled from a pool of 14 candidates. Six of the candidates are members of the debate team.
 (a) What is the probability that all of the offices are filled by members of the debate team?
 (b) What is the probability that none of the offices are filled by members of the debate team?

How Many of Your Closest Family and Friends Have Food Allergies or Intolerances?

FIGURE FOR EXERCISES 45–48

Food Allergies or Intolerances   *In Exercises 45–48, use the pie chart, which shows the results of a survey of 1500 U.S. adults who were asked how many of their closest family and friends have food allergies or intolerances. (Adapted from Pew Research Center)*

45. You choose 2 adults at random. What is the probability that both say most of their closest family and friends have food allergies or intolerances?

46. You choose 3 adults at random. What is the probability that all three say none of their closest family and friends have food allergies or intolerances?

47. You choose 6 adults at random. What is the probability that none of the six say some of their closest family and friends have food allergies or intolerances?

48. You choose 4 adults at random. What is the probability that none of the four say only a few of their closest family and friends have food allergies or intolerances?

49. Lottery   In a state lottery, you must correctly select 5 numbers (in any order) out of 40 to win the top prize. You purchase one lottery ticket. What is the probability that you will win the top prize?

50. (a) 2,535,650,040
    (b) 481,008,528
    (c) 0.190
    (d) Yes. The probability that no minorities are selected is small.
51. 0.166
52. 0.285
53. 0.070
54. 0.108
55. 0.933
56. 0.988
57. 0.086
58. 0.686
59. 0.066
60. 0.0002
61. 0.001
62. 0.021

50. **Committee** A company that has 200 employees chooses a committee of 5 to represent employee retirement issues. When the committee is formed, none of the 56 minority employees are selected.

    (a) Use technology to find the number of ways 5 employees can be chosen from 200.

    (b) Use technology to find the number of ways 5 employees can be chosen from 144 nonminorities.

    (c) What is the probability that the committee contains no minorities when the committee is chosen randomly (without bias)?

    (d) Does your answer to part (c) indicate that the committee selection is biased? Explain your reasoning.

Warehouse *In Exercises 51–54, a warehouse employs 24 workers on first shift, 17 workers on second shift, and 13 workers on third shift. Eight workers are chosen at random to be interviewed about the work environment.*

51. Find the probability of choosing five first-shift workers.

52. Find the probability of choosing three second-shift workers.

53. Find the probability of choosing four third-shift workers.

54. Find the probability of choosing two second-shift workers and two third-shift workers.

Extending Concepts

55. **Defective Units** A shipment of 10 microwave ovens contains 2 defective units. A restaurant buys three units. What is the probability of the restaurant buying at least two nondefective units?

56. **Defective Disks** A pack of 100 recordable DVDs contains 5 defective disks. You select four disks. What is the probability of selecting at least three nondefective disks?

57. **Employee Selection** Four sales representatives for a company are to be chosen at random to participate in a training program. The company has eight sales representatives, two in each of four regions. What is the probability that the four sales representatives chosen to participate in the training program will be from only two of the four regions?

58. **Employee Selection** In Exercise 57, what is the probability that the four sales representatives chosen to participate in the training program will be from only three of the four regions?

Cards *In Exercises 59–62, you are dealt a hand of five cards from a standard deck of 52 playing cards.*

59. Find the probability of being dealt two clubs and one of each of the other three suits.

60. Find the probability of being dealt four of a kind.

61. Find the probability of being dealt a full house (three of one kind and two of another kind).

62. Find the probability of being dealt three of a kind (the other two cards are different from each other).

Uses

Probability affects decisions when the weather is forecast, when medications are selected, and even when players are selected for professional sports teams. Although intuition is often used for determining probabilities, you will be better able to assess the likelihood of an event by applying the rules of probability.

For instance, you work for a real estate company and are asked to estimate the likelihood that a particular house will sell for a particular price within the next 90 days. You could use your intuition, but you could better assess the probability by looking at sales records for similar houses.

Abuses

One common abuse of probability is thinking that probabilities have "memories." For instance, the probability that a coin tossed eight times will land heads up every time is about 0.004. However, when seven heads have been tossed in a row, the probability that the eighth toss lands heads up is 0.5. Each toss is independent of all other tosses. The coin does not "remember" that it has already landed heads up seven times.

A famous instance of this abuse happened at a casino in Monte Carlo, Monaco, in 1913. After a roulette wheel landed on black 15 times in a row, people started rushing to bet on red, thinking that the wheel was bound to land on red soon. The wheel kept landing on black, and players doubled and tripled their bets, using the same reasoning. The wheel ended up landing on black a record 26 times in a row, costing players millions.

Ethics

A study by economists Daniel Chen, Tobias Moskowitz, and Kelly Shue found evidence that the gambler's fallacy occasionally leads baseball umpires, loan officers, and judges in refugee asylum courts to make mistakes. For instance, when loan officers have approved five loan applications in a row, they might think that six deserving loans in a row is unlikely and reject the sixth application based on a minor flaw when objectively it should be approved. The study concluded that up to 9% of loan decisions are influenced by this fallacy.

Similarly, when judges are reviewing a request for asylum, they might be more likely to deny the case if they approved the last two cases. The authors of the study estimated that as many as 2% of asylum cases may be affected. Although not as serious an injustice as the first two examples, the study also found that baseball umpires are about 1.5% less likely to call a pitch a strike when they called the previous pitch a strike. For decision makers such as judges to make ethical decisions, they must attempt to view each case as independent from previous cases.

EXERCISES

A "Daily Number" lottery has a three-digit number from 000 to 999. You buy one ticket each day. Your number is 389.

1. What is the probability of winning next Tuesday and Wednesday?

2. You won on Tuesday. What is the probability of winning on Wednesday?

3. You did not win on Tuesday. What is the probability of winning on Wednesday?

3 Chapter Summary

What Did You Learn?	Example(s)	Review Exercises
Section 3.1		
▶ How to identify the sample space of a probability experiment and how to identify simple events	1, 2	1–4
▶ How to use the Fundamental Counting Principle to find the number of ways two or more events can occur	3, 4	5, 6
▶ How to distinguish among classical probability, empirical probability, and subjective probability	5–8	7–12
▶ How to find the probability of the complement of an event and how to use a tree diagram and the Fundamental Counting Principle to find probabilities	9–11	13–16
Section 3.2		
▶ How to find the probability of an event given that another event has occurred	1	17, 18
▶ How to distinguish between independent and dependent events	2	19–22
▶ How to use the Multiplication Rule to find the probability of two or more events occurring in sequence and to find conditional probabilities	3–5	23, 24

$P(A \text{ and } B) = P(A) \cdot P(B|A)$ Events A and B are dependent.

$P(A \text{ and } B) = P(A) \cdot P(B)$ Events A and B are independent.

Section 3.3		
▶ How to determine whether two events are mutually exclusive	1	25, 26
▶ How to use the Addition Rule to find the probability of two events	2–5	27–40

$P(A \text{ or } B) = P(A) + P(B) - P(A \text{ and } B)$

$P(A \text{ or } B) = P(A) + P(B)$ Events A and B are mutually exclusive.

Section 3.4		
▶ How to find the number of ways a group of objects can be arranged in order and the number of ways to choose several objects from a group without regard to order	1–5	41–48

$_nP_r = \dfrac{n!}{(n-r)!}$ Permutations of n objects taken r at a time

$\dfrac{n!}{n_1! \cdot n_2! \cdot n_3! \cdots n_k!}$ Distinguishable permutations

$_nC_r = \dfrac{n!}{(n-r)!r!}$ Combinations of n objects taken r at a time

▶ How to use counting principles to find probabilities	6–8	49–53

3 Review Exercises

1. See Odd Answers, page A56.

2. See Selected Answers, page A96.

3. {January, February, March, April, May, June, July, August, September, October, November, December}; 3

4.

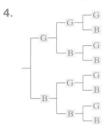

{GGG, GGB, GBG, GBB, BGG, BGB, BBG, BBB}; 3

5. 84

6. 175,760,000

7. Empirical probability because prior counts were used to calculate the frequency of a part being defective.

8. Classical probability because each outcome in the sample space is equally likely to occur.

9. Subjective probability because it is based on opinion.

10. Empirical probability because it is based on observations obtained from probability experiments.

11. Classical probability because all of the outcomes in the event and the sample space can be counted.

12. Empirical probability because it is based on observations obtained from probability experiments.

13. 0.258

14. 0.805

Section 3.1

In Exercises 1–4, identify the sample space of the probability experiment and determine the number of outcomes in the event. Draw a tree diagram when appropriate.

1. *Experiment:* Tossing four coins
   Event: Getting three heads

2. *Experiment:* Rolling 2 six-sided dice
   Event: Getting a sum of 4 or 5

3. *Experiment:* Choosing a month of the year
   Event: Choosing a month that begins with the letter J

4. *Experiment:* Guessing the gender(s) of the three children in a family
   Event: Guessing that the family has two boys

In Exercises 5 and 6, use the Fundamental Counting Principle.

5. A student must choose from 7 classes to take at 8:00 A.M., 4 classes to take at 9:00 A.M., and 3 classes to take at 10:00 A.M. How many ways can the student arrange the schedule?

6. The state of Virginia's license plates have three letters and four digits. Assuming that any letter or digit can be used, how many different license plates are possible?

In Exercises 7–12, classify the statement as an example of classical probability, empirical probability, or subjective probability. Explain your reasoning.

7. On the basis of prior counts, a quality control officer says there is a 0.05 probability that a randomly chosen part is defective.

8. The probability of randomly selecting five cards of the same suit from a standard deck of 52 playing cards is about 0.002.

9. The chance that Corporation A's stock price will fall today is 75%.

10. The probability that a person can roll his or her tongue is 70%.

11. The probability of rolling 2 six-sided dice and getting a sum of 9 is $\frac{1}{9}$.

12. The chance that a randomly selected person in the United States is between 17 and 23 years old is about 9.5%. *(Source: U.S. Census Bureau)*

In Exercises 13 and 14, use the table, which shows the numbers (in thousands) of bachelor's degrees for a recent year. (Source: National Center for Education Statistics)

Degree	Business	Health Professions	Social Sciences/ History	Psychology	Other
Percent	361	181	178	114	1006

13. Find the probability that a randomly selected degree will be in business or psychology.

14. Find the probability that a randomly selected degree will not be in health professions or social sciences/history.

15. 1.25×10^{-7}

16. 0.999999875

17. 0.555

18. 0.379

19. Independent. The outcomes of the first four coin tosses do not affect the outcome of the fifth coin toss.

20. Dependent. The outcome of selecting an ace affects the outcome of selecting a jack.

21. Dependent. The outcome of taking a driver's education course affects the outcome of passing the driver's license exam.

22. Dependent. The outcome of getting high grades affects the outcome of being awarded an academic scholarship.

23. 0.025; Yes, the event is unusual because its probability is less than or equal to 0.05.

24. 0.235; No, the event is not unusual because its probability is not less than or equal to 0.05.

25. Mutually exclusive. A jelly bean cannot be both completely red and completely yellow.

26. Not mutually exclusive. A person who loves cats can also own a dog.

Telephone Numbers *The telephone numbers for a region of Pennsylvania have an area code of 570. The next seven digits represent the local telephone numbers for that region. These cannot begin with a 0 or 1. In Exercises 15 and 16, assume your cousin lives within the given area code.*

15. What is the probability of randomly generating your cousin's telephone number on the first try?

16. What is the probability of not randomly generating your cousin's telephone number on the first try?

Section 3.2

In Exercises 17 and 18, use the table, which shows the numbers of students from American Bar Association approved law schools who took the Bar Examination for the first time in a recent year and the numbers of students who repeated the exam that year. (Source: National Conference of Bar Examiners)

	Passed	Failed	Total
First time	36,534	13,194	49,728
Repeat	6,454	10,581	17,035
Total	42,988	23,775	66,763

17. Find the probability that a student took the exam for the first time, given that the student failed.

18. Find the probability that a student passed, given that the student repeated the exam.

In Exercises 19–22, determine whether the events are independent or dependent. Explain your reasoning.

19. Tossing a coin four times, getting four heads, and tossing it a fifth time and getting a head

20. Selecting an ace from a standard deck of 52 playing cards, and then selecting a jack from the deck without replacing the ace.

21. Taking a driver's education course and passing the driver's license exam

22. Getting high grades and being awarded an academic scholarship

23. Your roommate has asked you to buy toothpaste and dental rinse, but your roommate did not tell you which brands to get. The store has eight brands of toothpaste and five brands of dental rinse. What is the probability that you will purchase the correct brands of both products? Is this an unusual event? Explain.

24. Your sock drawer has 18 folded pairs of socks, with 8 pairs of white, 6 pairs of black, and 4 pairs of blue. What is the probability, without looking in the drawer, that you will first select and remove a black pair, then select either a blue or a white pair? Is this an unusual event? Explain.

Section 3.3

In Exercises 25 and 26, determine whether the events are mutually exclusive. Explain your reasoning.

25. Event *A*: Randomly select a red jelly bean from a jar.
Event *B*: Randomly select a yellow jelly bean from the same jar.

26. Event *A*: Randomly select a person who loves cats.
Event *B*: Randomly select a person who owns a dog.

27. 0.9
28. 0.54
29. 0.538
30. 0.538
31. 0.583
32. 0.5
33. 0.576
34. 0.698

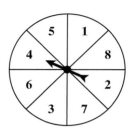

FIGURE FOR EXERCISE 32

Students in Public Schools

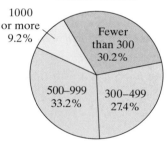

1000 or more 9.2%

Fewer than 300 30.2%

500–999 33.2%

300–499 27.4%

FIGURE FOR EXERCISES 33 AND 34

35. 0.584
36. 0.384
37. 0.568
38. 0.886

27. A random sample of 250 working adults found that 74% access the Internet at work, 88% access the Internet at home, and 72% access the Internet at both work and home. Find the probability that a person in this sample selected at random accesses the Internet at home or at work.

28. A sample of 6500 automobiles found that 1560 of the automobiles were black, 3120 of the automobiles were sedans, and 1170 of the automobiles were black sedans. Find the probability that a randomly chosen automobile from this sample is black or a sedan.

In Exercises 29–32, find the probability.

29. A card is randomly selected from a standard deck of 52 playing cards. Find the probability that the card is between 4 and 8, inclusive, or is a club.

30. A card is randomly selected from a standard deck of 52 playing cards. Find the probability that the card is red or a queen.

31. A 12-sided die, numbered 1 to 12, is rolled. Find the probability that the roll results in an odd number or a number less than 4.

32. The spinner shown at the left is spun. The spinner is equally likely to land on each number. Find the probability than the spinner lands on a multiple of 3 or a number greater than 5.

In Exercises 33 and 34, use the pie chart at the left, which shows the percent distribution of the number of students in U.S. public schools in a recent year. (Source: U.S. National Center for Education Statistics)

33. Find the probability of randomly selecting a school with fewer than 500 students.

34. Find the probability of randomly selecting a school with 300 or more students.

In Exercises 35–38, use the Pareto chart, which shows the results of a survey in which 3078 adults were asked with which social class they identify. (Adapted from Gallup)

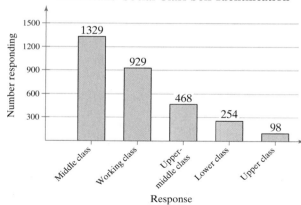

Americans' Social Class Self-Identification

35. Find the probability of randomly selecting an adult who identifies as middle or upper-middle class.

36. Find the probability of randomly selecting an adult who identifies as working or lower class.

37. Find the probability of randomly selecting an adult who does not identify as middle class.

38. Find the probability of randomly selecting an adult who does not identify as upper or lower class.

39. No; You do not know whether events A and B are mutually exclusive.

40. Yes; The Addition Rule is

$P(A \text{ or } B) =$
$P(A) + P(B) - P(A \text{ and } B).$

Substitute the given values for $P(A \text{ or } B)$ and $P(A) + P(B)$ and then solve for $P(A \text{ and } B).$

41. 110

42. 20,160

43. 35

44. 0.083

45. 2730

46. 120

47. 2380

48. 78

49. 0.000009

50. 0.000006

51. (a) 0.955   (b) 0.0000008
(c) 0.045   (d) 0.9999992

52. (a) 0.006   (b) 0.411
(c) 0.012   (d) 0.156

53. (a) 0.071   (b) 0.005
(c) 0.429   (d) 0.114

Letter grade	Number of students
A	8
B	10
C	12
D	6
F	4

TABLE FOR EXERCISE 52

39. You are given that $P(A) = 0.15$ and $P(B) = 0.40$. Do you have enough information to find $P(A \text{ or } B)$? Explain.

40. You are given that $P(A \text{ or } B) = 0.55$ and $P(A) + P(B) = 1$. Do you have enough information to find $P(A \text{ and } B)$? Explain.

Section 3.4

In Exercises 41–44, perform the indicated calculation.

41. $_{11}P_2$ **42.** $_8P_6$ **43.** $_7C_4$ **44.** $\dfrac{_5C_3}{_{10}C_3}$

In Exercises 45–48, use combinations and permutations.

45. Fifteen cyclists enter a race. How many ways can the cyclists finish first, second, and third?

46. Five players on a basketball team must each choose a player on the opposing team to defend. In how many ways can the players choose their defensive assignments?

47. A literary magazine editor must choose 4 short stories for this month's issue from 17 submissions. In how many ways can the editor choose this month's stories?

48. An employer must hire 2 people from a list of 13 applicants. In how many ways can the employer choose to hire the 2 people?

In Exercises 49–53, use counting principles to find the probability.

49. A full house consists of three of one kind and two of another kind. You are dealt a hand of five cards from a standard deck of 52 playing cards. Find the probability of being dealt a full house consisting of three kings and two queens.

50. A security code consists of three letters and one digit. The first letter cannot be A, B, or C. What is the probability of guessing the security code on the first try?

51. A shipment of 200 calculators contains 3 defective units. What is the probability that a sample of three calculators will have

(a) no defective calculators?

(b) all defective calculators?

(c) at least one defective calculator?

(d) at least one nondefective calculator?

52. A class of 40 students takes a statistics exam. The results are shown in the table at the left. Three students are selected at random. What is the probability that

(a) all three students received an A?

(b) all three students received a C or better?

(c) all three students received a D or an F?

(d) all three students received a B or a C?

53. A corporation has six male senior executives and four female senior executives. Four senior executives are chosen at random to attend a technology seminar. What is the probability of choosing

(a) four men?   (b) four women?

(c) two men and two women?   (d) one man and three women?

Chapter Quiz

Answers

1. 450,000
2. (a) 0.713   (b) 0.662
   (c) 0.778   (d) 0.937
   (e) 0.049   (f) 0.606
   (g) 0.346   (h) 0.515
3. The event in part (e) is unusual because its probability is less than or equal to 0.05.
4. Not mutually exclusive. A bowler can have the highest game in a 40-game tournament and still lose the tournament.

   Dependent. One event can affect the occurrence of the second event.
5. 657,720
6. (a) 2,481,115   (b) 1
   (c) 2,572,999
7. (a) 0.964
   (b) 0.0000004
   (c) 0.9999996

Take this quiz as you would take a quiz in class. After you are done, check your work against the answers given in the back of the book.

1. The access code for a warehouse's security system consists of six digits. The first digit cannot be 0 and the last digit must be even. How many access codes are possible?

2. The table shows the numbers (in thousands) of earned degrees by level in two different fields, conferred in the United States in a recent year. *(Source: U.S. National Center for Education Statistics)*

		Field		
		Natural sciences/ mathematics	**Computer science/ engineering**	**Total**
Level of degree	**Bachelor's**	154.9	164.3	319.2
	Master's	28.2	71.9	100.1
	Doctoral	16.0	12.1	28.1
	Total	199.1	248.3	447.4

   A person who earned a degree in the year is randomly selected. Find the probability of selecting someone who
   (a) earned a bachelor's degree.
   (b) earned a bachelor's degree, given that the degree is in computer science/ engineering.
   (c) earned a bachelor's degree, given that the degree is not in computer science/engineering.
   (d) earned a bachelor's degree or a master's degree.
   (e) earned a doctorate, given that the degree is in computer science/engineering.
   (f) earned a master's degree or the degree is in natural sciences/mathematics.
   (g) earned a bachelor's degree and the degree is in natural sciences/mathematics.
   (h) earned a degree in computer science/engineering, given that the person earned a bachelor's degree.

3. Which event(s) in Exercise 2 can be considered unusual? Explain.

4. Determine whether the events are mutually exclusive. Then determine whether the events are independent or dependent. Explain your reasoning.

   Event *A*: A bowler having the highest game in a 40-game tournament
   Event *B*: Losing the bowling tournament

5. From a pool of 30 candidates, the offices of president, vice president, secretary, and treasurer will be filled. In how many different ways can the offices be filled?

6. A shipment of 250 netbooks contains 3 defective units. Determine how many ways a vending company can buy three of these units and receive
   (a) no defective units.   (b) all defective units.   (c) at least one good unit.

7. In Exercise 6, find the probability of the vending company receiving
   (a) no defective units.   (b) all defective units.   (c) at least one good unit.

3 Chapter Test

Answers (left column):

1. 0.005

2. (a) 0.00000015
   (b) 0.99999985
   (c) 0.000008
   (d) Classical probability because each outcome in the sample space is equally likely to occur.

3. Mutually exclusive. The month of February does not have 30 days.

4. (a) 0.327 (b) 0.664
   (c) 0.334 (d) 0.679
   (e) 0.548 (f) 0.222

5. None are unusual because none of the events have a probability of 0.05 or less.

6. Dependent, because $P(B|A) \neq P(B)$ and $P(A|B) \neq P(A)$.

7. (a) 3360 (b) 50,450,400

Take this test as you would take a test in class.

1. Sixty-five runners compete in a 10k race. Your school has 12 runners in the race. What is the probability that three runners from your school place first, second, and third?

2. A security code consists of a person's first and last initials and four digits.
   (a) What is the probability of guessing a person's code on the first try?
   (b) What is the probability of not guessing a person's code on the first try?
   (c) You know a person's first name and that the last digit is odd. What is the probability of guessing this person's code on the first try?
   (d) Are the statements in parts (a)–(c) examples of classical probability, empirical probability, or subjective probability? Explain your reasoning.

3. Determine whether the events are mutually exclusive. Explain your reasoning.
   Event *A*: Randomly select a student born on the 30th of a month
   Event *B*: Randomly select a student with a birthday in February

4. The table shows the sixth, seventh, and eighth grade student enrollment levels (in thousands) in Minnesota and Ohio schools in a recent year. *(Source: U.S. National Center for Education Statistics)*

	Sixth grade	Seventh grade	Eighth grade	Total
Minnesota	61.8	63.6	62.9	188.3
Ohio	130.8	134.3	135.0	400.1
Total	192.6	197.9	197.9	588.4

   A student in one of the indicated grades and states is randomly selected. Find the probability of selecting a student who
   (a) is in sixth grade
   (b) is in sixth or seventh grade
   (c) is in eighth grade, given that the student is enrolled in Minnesota
   (d) is enrolled in Ohio, given that the student is in seventh grade
   (e) is in seventh grade or is enrolled in Minnesota
   (f) is in sixth grade and is enrolled in Ohio

5. Which event(s) in Exercise 4 can be considered unusual? Explain your reasoning.

6. A person is selected at random from the sample in Exercise 4. Are the events "the student is in sixth grade" and "the student is enrolled in Minnesota" independent or dependent? Explain your reasoning.

7. There are 16 students giving final presentations in your history course.
   (a) Three students present per day. How many presentation orders are possible for the first day?
   (b) Presentation subjects are based on the units of the course. Unit B is covered by three students, Unit C is covered by five students, and Units A and D are each covered by four students. How many presentation orders are possible when presentations on the same unit are indistinguishable from each other?

You work in the security department of a bank's website. To access their accounts, customers of the bank must create an 8-digit password. It is your job to determine the password requirements for these accounts. Security guidelines state that for the website to be secure, the probability that an 8-digit password is guessed on one try must be less than $\frac{1}{60^8}$, assuming all passwords are equally likely.

Your job is to use the probability techniques you have learned in this chapter to decide what requirements a customer must meet when choosing a password, including what sets of characters are allowed, so that the website is secure according to the security guidelines.

ACCOUNT REGISTRATION FORM
register here to access your account

Select your username:

Create an 8-digit password:

Verify Password:

EXERCISES

1. *How Would You Do It?*

(a) How would you investigate the question of what password requirements you should set to meet the security guidelines?

(b) What statistical methods taught in this chapter would you use?

2. *Answering the Question*

(a) What password requirements would you set? What characters would be allowed?

(b) Show that the probability that a password is guessed on one try is less than $\frac{1}{60^8}$, when the requirements in part (a) are used and all passwords are equally likely.

3. *Additional Security*

For additional security, each customer creates a 5-digit PIN (personal identification number). The table on the right shows the 10 most commonly chosen 5-digit PINs. From the table, you can see that more than a third of all 5-digit PINs could be guessed by trying these 10 numbers. To discourage customers from using predictable PINs, you consider prohibiting PINs that use the same digit more than once.

(a) How would this requirement affect the number of possible 5-digit PINs?

(b) Would you decide to prohibit PINs that use the same digit more than once? Explain.

Most Popular 5-Digit PINs

Rank	PIN	Percent
1	12345	22.80%
2	11111	4.48%
3	55555	1.77%
4	00000	1.26%
5	54321	1.20%
6	13579	1.11%
7	77777	0.62%
8	22222	0.45%
9	12321	0.41%
10	99999	0.40%

(Source: Datagenetics.com)

TECHNOLOGY

Simulation: Composing Mozart Variations with Dice

Wolfgang Mozart (1756–1791) composed a wide variety of musical pieces. In his Musical Dice Game, he wrote a minuet with an almost endless number of variations. Each minuet has 16 bars. In the eighth and sixteenth bars, the player has a choice of two musical phrases. In each of the other 14 bars, the player has a choice of 11 phrases.

To create a minuet, Mozart suggested that the player toss 2 six-sided dice 16 times. For the eighth and sixteenth bars, choose Option 1 when the dice total is odd and Option 2 when it is even. For each of the other 14 bars, subtract 1 from the dice total. The minuet shown is the result of the following sequence of numbers.

$$
\begin{array}{cccccccc}
5 & 7 & 1 & 6 & 4 & 10 & 5 & 1 \\
6 & 6 & 2 & 4 & 6 & 8 & 8 & 2
\end{array}
$$

EXERCISES

1. How many phrases did Mozart write to create the Musical Dice Game minuet? Explain.

2. How many possible variations are there in Mozart's Musical Dice Game minuet? Explain.

3. Use technology to randomly select a number from 1 to 11.

   (a) What is the theoretical probability of each number from 1 to 11 occurring?

   (b) Use this procedure to select 100 integers from 1 to 11. Tally your results and compare them with the probabilities in part (a).

4. What is the probability of randomly selecting option 6, 7, or 8 for the first bar? For all 14 bars? Find each probability using (a) theoretical probability and (b) the results of Exercise 3(b).

5. Use technology to randomly select two numbers from 1 to 6. Find the sum and subtract 1 to obtain a total.

   (a) What is the theoretical probability of each total from 1 to 11?

   (b) Use this procedure to select 100 totals from 1 to 11. Tally your results and compare them with the probabilities in part (a).

6. Repeat Exercise 4 using the results of Exercise 5.

Extended solutions are given in the technology manuals that accompany this text.
Technical instruction is provided for Minitab, Excel, and the TI-84 Plus.

Discrete Probability Distributions

The National Climatic Data Center (NCDC) is the world's largest active archive of weather data. NCDC archives weather data from the Coast Guard, Federal Aviation Administration, Military Services, the National Weather Service, and voluntary observers.

In Chapters 1 through 3, you learned how to collect and describe data and how to find the probability of an event. These skills are used in many different types of careers. For instance, data about climatic conditions are used to analyze and forecast the weather throughout the world. On a typical day, meteorologists use data from aircraft, National Weather Service cooperative observers, radar, remote sensing systems, satellites, ships, weather balloons, wind profilers, and a variety of other data-collection devices to forecast the weather. Even with this much data, meteorologists cannot forecast the weather with certainty. Instead, they assign probabilities to certain weather conditions. For instance, a meteorologist might determine that there is a 40% chance of rain (based on the relative frequency of rain under similar weather conditions).

In Chapter 4, you will learn how to create and use probability distributions. Knowing the shape, center, and variability of a probability distribution enables you to make decisions in inferential statistics. For example, you are a meteorologist working on a three-day forecast. Assuming that having rain on one day is independent of having rain on another day, you have determined that there is a 40% probability of rain (and a 60% probability of no rain) on each of the three days. What is the probability that it will rain on 0, 1, 2, or 3 of the days? To answer this, you can create a probability distribution for the possible outcomes.

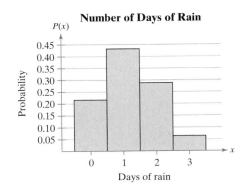

Day 1	Day 2	Day 3	Probability	Days of Rain
			$P(\text{☀},\text{☀},\text{☀}) = 0.216$	0
			$P(\text{☀},\text{☀},\text{💧}) = 0.144$	1
			$P(\text{☀},\text{💧},\text{☀}) = 0.144$	1
			$P(\text{☀},\text{💧},\text{💧}) = 0.096$	2
			$P(\text{💧},\text{☀},\text{☀}) = 0.144$	1
			$P(\text{💧},\text{☀},\text{💧}) = 0.096$	2
			$P(\text{💧},\text{💧},\text{☀}) = 0.096$	2
			$P(\text{💧},\text{💧},\text{💧}) = 0.064$	3

Using the *Addition Rule* with the probabilities in the tree diagram, you can determine the probabilities of having rain on various numbers of days. You can then use this information to construct and graph a probability distribution.

Probability Distribution

Days of rain	Tally	Probability
0	1	0.216
1	3	0.432
2	3	0.288
3	1	0.064

Number of Days of Rain

4.1 Probability Distributions

Random Variables ■ Discrete Probability Distributions ■ Mean, Variance, and Standard Deviation ■ Expected Value

Random Variables

The outcome of a probability experiment is often a count or a measure. When this occurs, the outcome is called a **random variable.**

> ### DEFINITION
>
> A **random variable** x represents a value associated with each outcome of a probability experiment.

The word *random* indicates that x is determined by chance. There are two types of random variables: **discrete** and **continuous.**

> ### DEFINITION
>
> A random variable is **discrete** when it has a finite or countable number of possible outcomes that can be listed.
>
> A random variable is **continuous** when it has an uncountable number of possible outcomes, represented by an interval on a number line.

In most applications, discrete random variables represent counted data, while continuous random variables represent measured data. For instance, consider the following example. You conduct a study of the number of calls a telemarketing firm makes in one day. The possible values of the random variable x are 0, 1, 2, 3, 4, and so on. Because the set of possible outcomes $\{0, 1, 2, 3, \dots\}$ can be listed, x is a discrete random variable. You can represent its values as points on a number line.

Number of Calls (Discrete)

x can be any whole number: 0, 1, 2, 3, . . .

A different way to conduct the study would be to measure the time (in hours) the telemarketing firm spends making calls in one day. Because the time spent making calls can be any number from 0 to 24 (including fractions and decimals), x is a continuous random variable. You can represent its values with an interval on a number line.

Hours Spent on Calls (Continuous)

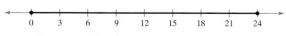

x can be any value between 0 and 24.

When a random variable is discrete, you can list the possible values the variable can assume. However, it is impossible to list all values for a continuous random variable.

Discrete Variables and Continuous Variables

Determine whether each random variable x is discrete or continuous. Explain your reasoning.

1. Let x represent the number of Fortune 500 companies that lost money in the previous year.

2. Let x represent the volume of gasoline in a 21-gallon tank.

SOLUTION

1. The number of companies that lost money in the previous year can be counted. The set of possible outcomes is

$$\{0, 1, 2, 3, \ldots, 500\}.$$

So, x is a *discrete* random variable.

2. The amount of gasoline in the tank can be any volume between 0 gallons and 21 gallons. So, x is a *continuous* random variable.

TRY IT YOURSELF 1

Determine whether each random variable x is discrete or continuous. Explain your reasoning.

1. Let x represent the speed of a rocket.

2. Let x represent the number of calves born on a farm in one year.

3. Let x represent the number of days of rain for the next three days (see page 189).

Answer: Page A34

Study Tip

Values of variables such as volume, age, height, and weight are sometimes rounded to the nearest whole number. These values represent measured data, however, so they are continuous random variables.

It is important that you can distinguish between discrete and continuous random variables because different statistical techniques are used to analyze each. The remainder of this chapter focuses on discrete random variables and their probability distributions. Your study of continuous probability distributions will begin in Chapter 5.

Discrete Probability Distributions

Each value of a discrete random variable can be assigned a probability. By listing each value of the random variable with its corresponding probability, you are forming a **discrete probability distribution.**

DEFINITION

A **discrete probability distribution** lists each possible value the random variable can assume, together with its probability. A discrete probability distribution must satisfy these conditions.

In Words	In Symbols
1. The probability of each value of the discrete random variable is between 0 and 1, inclusive.	$0 \leq P(x) \leq 1$
2. The sum of all the probabilities is 1.	$\Sigma P(x) = 1$

Because probabilities represent relative frequencies, a discrete probability distribution can be graphed with a relative frequency histogram.

Note to Instructor

It is important to establish the correspondence between the area under each region of the probability histogram and the probability that the random variable takes on a particular value. Ask students to find the area of each bar in the histogram shown in Example 2.

Frequency Distribution

Score, x	Frequency, f
1	24
2	33
3	42
4	30
5	21

Passive-Aggressive Traits

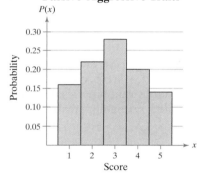

Frequency Distribution

Sales per day, x	Number of days, f
0	16
1	19
2	15
3	21
4	9
5	10
6	8
7	2

GUIDELINES

Constructing a Discrete Probability Distribution

Let x be a discrete random variable with possible outcomes $x_1, x_2, \ldots, x_n$.

1. Make a frequency distribution for the possible outcomes.
2. Find the sum of the frequencies.
3. Find the probability of each possible outcome by dividing its frequency by the sum of the frequencies.
4. Check that each probability is between 0 and 1, inclusive, and that the sum of all the probabilities is 1.

EXAMPLE 2

Constructing and Graphing a Discrete Probability Distribution

An industrial psychologist administered a personality inventory test for passive-aggressive traits to 150 employees. Each individual was given a whole number score from 1 to 5, where 1 is extremely passive and 5 is extremely aggressive. A score of 3 indicated neither trait. The results are shown at the left. Construct a probability distribution for the random variable x. Then graph the distribution using a histogram.

SOLUTION

Divide the frequency of each score by the total number of individuals in the study to find the probability for each value of the random variable.

$$P(1) = \frac{24}{150} = 0.16 \qquad P(2) = \frac{33}{150} = 0.22 \qquad P(3) = \frac{42}{150} = 0.28$$

$$P(4) = \frac{30}{150} = 0.20 \qquad P(5) = \frac{21}{150} = 0.14$$

The discrete probability distribution is shown in the table below.

x	1	2	3	4	5
$P(x)$	0.16	0.22	0.28	0.20	0.14

Note that the probability of each value of x is between 0 and 1, and the sum of the probabilities is 1. So, the distribution is a probability distribution. The graph of the distribution is shown in the histogram at the left. Because the width of each bar is one, the area of each bar is equal to the probability of a particular outcome. Also, the probability of an event corresponds to the sum of the areas of the outcomes included in the event. For instance, the probability of the event "having a score of 2 or 3" is equal to the sum of the areas of the second and third bars,

$$(1)(0.22) + (1)(0.28) = 0.22 + 0.28 = 0.50.$$

Interpretation   You can see that the distribution is approximately symmetric.

TRY IT YOURSELF 2

A company tracks the number of sales new employees make each day during a 100-day probationary period. The results for one new employee are shown at the left. Construct a probability distribution for the random variable x. Then graph the distribution using a histogram.

Answer: Page A34

EXAMPLE 3

Verifying a Probability Distribution

Verify that the distribution for the three-day forecast (see page 189) and the number of days of rain is a probability distribution.

Days of rain, x	0	1	2	3
Probability, $P(x)$	0.216	0.432	0.288	0.064

SOLUTION

If the distribution is a probability distribution, then (1) each probability is between 0 and 1, inclusive, and (2) the sum of all the probabilities equals 1.

1. Each probability is between 0 and 1.

2. $\Sigma P(x) = 0.216 + 0.432 + 0.288 + 0.064$
$= 1.$

Interpretation Because both conditions are met, the distribution is a probability distribution.

TRY IT YOURSELF 3

Verify that the distribution you constructed in Try It Yourself 2 is a probability distribution.

Answer: Page A34

EXAMPLE 4

Identifying Probability Distributions

Determine whether each distribution is a probability distribution. Explain your reasoning.

1.

x	5	6	7	8
$P(x)$	0.28	0.21	0.43	0.15

2.

x	1	2	3	4
$P(x)$	$\frac{1}{2}$	$\frac{1}{4}$	$\frac{5}{4}$	-1

SOLUTION

1. Each probability is between 0 and 1, but the sum of all the probabilities is 1.07, which is greater than 1. The sum of all the probabilities in a probability distribution always equals 1. So, this distribution is *not* a probability distribution.

2. The sum of all the probabilities is equal to 1, but $P(3)$ and $P(4)$ are not between 0 and 1. Probabilities can never be negative or greater than 1. So, this distribution is *not* a probability distribution.

TRY IT YOURSELF 4

Determine whether each distribution is a probability distribution. Explain your reasoning.

1.

x	5	6	7	8
$P(x)$	$\frac{1}{16}$	$\frac{5}{8}$	$\frac{1}{4}$	$\frac{1}{16}$

2.

x	1	2	3	4
$P(x)$	0.09	0.36	0.49	0.10

Answer: Page A34

Picturing the World

A study was conducted to determine how many credit cards people have. The results are shown in the histogram. (Adapted from American Association of Retired Persons)

How Many Credit Cards Do You Have?

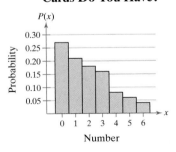

Estimate the probability that a randomly selected person has two or three credit cards.

The probability that a randomly selected person has two or three credit cards is about 0.35.

Mean, Variance, and Standard Deviation

You can measure the center of a probability distribution with its mean and measure the variability with its variance and standard deviation. The mean of a discrete random variable is defined as follows.

Mean of a Discrete Random Variable

The **mean** of a discrete random variable is given by

$$\mu = \Sigma x P(x).$$

Each value of x is multiplied by its corresponding probability and the products are added.

The mean of a random variable represents the "theoretical average" of a probability experiment and sometimes is not a possible outcome. If the experiment were performed many thousands of times, then the mean of all the outcomes would be close to the mean of the random variable.

Note to Instructor

In Example 5, point out that 2.94 is not a possible score for one person. When the scores of the 150 employees are added up and divided by 150, however, the result is 2.94.

EXAMPLE 5

Finding the Mean of a Probability Distribution

The probability distribution for the personality inventory test for passive-aggressive traits discussed in Example 2 is shown below. Find the mean score.

Score, x	1	2	3	4	5
Probability, $P(x)$	0.16	0.22	0.28	0.20	0.14

SOLUTION

Use a table to organize your work, as shown below.

x	$P(x)$	$xP(x)$
1	0.16	$1(0.16) = 0.16$
2	0.22	$2(0.22) = 0.44$
3	0.28	$3(0.28) = 0.84$
4	0.20	$4(0.20) = 0.80$
5	0.14	$5(0.14) = 0.70$
	$\Sigma P(x) = 1$	$\Sigma xP(x) = 2.94 \approx 2.9$ ◄—— Mean

From the table, you can see that the mean score is

$$\mu = 2.94 \approx 2.9. \qquad \text{Round to one decimal place.}$$

Note that the mean is rounded to one more decimal place than the possible values of the random variable x.

Interpretation Recall that a score of 3 represents an individual who exhibits neither passive nor aggressive traits and the mean is slightly less than 3. So, the mean personality trait is neither extremely passive nor extremely aggressive, but is slightly closer to passive.

Study Tip

Notice that the mean in Example 5 is rounded to one decimal place. This rounding was done because the mean of a probability distribution should be rounded to one more decimal place than was used for the random variable x. This *round-off rule* is also used for the variance and standard deviation of a probability distribution.

TRY IT YOURSELF 5

Find the mean of the probability distribution you constructed in Try It Yourself 2. What can you conclude?

Answer: Page A34

Although the mean of the random variable of a probability distribution describes a typical outcome, it gives no information about how the outcomes vary. To study the variation of the outcomes, you can use the variance and standard deviation of the random variable of a probability distribution.

Study Tip

An alternative formula for the variance of a probability distribution is

$$\sigma^2 = \left[\Sigma x^2 P(x)\right] - \mu^2.$$

Variance and Standard Deviation of a Discrete Random Variable

The **variance** of a discrete random variable is

$$\sigma^2 = \Sigma(x - \mu)^2 P(x).$$

The **standard deviation** is

$$\sigma = \sqrt{\sigma^2} = \sqrt{\Sigma(x - \mu)^2 P(x)}.$$

EXAMPLE 6

Finding the Variance and Standard Deviation

The probability distribution for the personality inventory test for passive-aggressive traits discussed in Example 2 is shown below. Find the variance and standard deviation of the probability distribution.

Score, x	1	2	3	4	5
Probability, $P(x)$	0.16	0.22	0.28	0.20	0.14

SOLUTION

To find the variance and standard deviation, note that from Example 5 the mean of the distribution before rounding is $\mu = 2.94$. (Use this value to avoid rounding until the last calculation.) Use a table to organize your work, as shown below.

x	$P(x)$	$x - \mu$	$(x - \mu)^2$	$(x - \mu)^2 P(x)$
1	0.16	−1.94	3.7636	0.602176
2	0.22	−0.94	0.8836	0.194392
3	0.28	0.06	0.0036	0.001008
4	0.20	1.06	1.1236	0.224720
5	0.14	2.06	4.2436	0.594104
	$\Sigma P(x) = 1$			$\Sigma(x - \mu)^2 P(x) = 1.6164$

Variance

So, the variance is

$$\sigma^2 = 1.6164 \approx 1.6$$

and the standard deviation is

$$\sigma = \sqrt{\sigma^2} = \sqrt{1.6164} \approx 1.3.$$

Interpretation Most of the data values differ from the mean by no more than 1.3.

TRY IT YOURSELF 6

Find the variance and standard deviation of the probability distribution constructed in Try It Yourself 2.

Answer: Page A34

Tech Tip

You can use technology such as Minitab, Excel, StatCrunch, or the TI-84 Plus to find the mean and standard deviation of a discrete random variable. For instance, to find the mean and standard deviation of the discrete random variable in Example 6 on a TI-84 Plus, enter the possible values of the discrete random variable x in L1. Next, enter the probabilities $P(x)$ in L2. Then, use the *1-Var Stats* feature with L1 as the list and L2 as the frequency list to calculate the mean and standard deviation (and other statistics), as shown below.

```
TI-84 PLUS
  1-Var Stats
x̄=2.94
Σx=2.94
Σx²=10.26
Sx=
σx=1.271377206
↓n=1
```

Expected Value

The mean of a random variable represents what you would expect to happen over thousands of trials. It is also called the **expected value.**

Note to Instructor

Tell students that expected value plays a role in decision theory.

DEFINITION

The **expected value** of a discrete random variable is equal to the mean of the random variable.

$$\text{Expected Value} = E(x) = \mu = \Sigma x P(x)$$

In most applications, an expected value of 0 has a practical interpretation. For instance, in games of chance, an expected value of 0 implies that a game is fair (an unlikely occurrence). In a profit and loss analysis, an expected value of 0 represents the break-even point.

Although probabilities can never be negative, the expected value of a random variable can be negative, as shown in the next example.

EXAMPLE 7

Finding an Expected Value

At a raffle, 1500 tickets are sold at $2 each for four prizes of $500, $250, $150, and $75. You buy one ticket. Find the expected value and interpret its meaning.

SOLUTION

To find the gain for each prize, subtract the price of the ticket from the prize. For instance, your gain for the $500 prize is

$$\$500 - \$2 = \$498$$

and your gain for the $250 prize is

$$\$250 - \$2 = \$248.$$

Write a probability distribution for the possible gains (or outcomes). Note that a gain represented by a negative number is a loss.

Gain, x	$498	$248	$148	$73	$-$2
Probability, $P(x)$	$\frac{1}{1500}$	$\frac{1}{1500}$	$\frac{1}{1500}$	$\frac{1}{1500}$	$\frac{1496}{1500}$

$-$2 represents a loss of $2

Then, using the probability distribution, you can find the expected value.

$$E(x) = \Sigma x P(x)$$
$$= \$498 \cdot \frac{1}{1500} + \$248 \cdot \frac{1}{1500} + \$148 \cdot \frac{1}{1500} + \$73 \cdot \frac{1}{1500} + (-\$2) \cdot \frac{1496}{1500}$$
$$= -\$1.35$$

Interpretation Because the expected value is negative, you can expect to lose an average of $1.35 for each ticket you buy.

TRY IT YOURSELF 7

At a raffle, 2000 tickets are sold at $5 each for five prizes of $2000, $1000, $500, $250, and $100. You buy one ticket. Find the expected value and interpret its meaning.

Answer: Page A34

4.1 EXERCISES

For Extra Help: **MyLab Statistics**

Building Basic Skills and Vocabulary

1. A random variable represents a value associated with each outcome of a probability experiment.

   Examples: Answers will vary.

2. A discrete probability distribution lists each possible value a random variable can assume, together with its probability.

   Condition 1: $0 \leq P(x) \leq 1$
   Condition 2: $\Sigma P(x) = 1$

3. No; The expected value may not be a possible value of x for one trial, but it represents the average value of x over a large number of trials.

4. The mean of a probability distribution represents the "theoretical average" of a probability experiment.

5. False. In most applications, discrete random variables represent counted data, while continuous random variables represent measured data.

6. True

7. False. The mean of the random variable of a probability distribution describes a typical outcome. The variance and standard deviation of the random variable of a probability distribution describe how the outcomes vary.

8. False. The expected value of a random variable can be positive, negative, or zero.

9. Discrete; Attendance is a random variable that is countable.

10. See Selected Answers, page A96.

11. See Odd Answers, page A57.

12. See Selected Answers, page A96.

13. See Odd Answers, page A57.

14. See Selected Answers, page A96.

15. See Odd Answers, page A57.

16. See Selected Answers, page A96.

17. See Odd Answers, page A57.

18. See Selected Answers, page A96.

1. What is a random variable? Give an example of a discrete random variable and a continuous random variable. Justify your answer.

2. What is a discrete probability distribution? What are the two conditions that a discrete probability distribution must satisfy?

3. Is the expected value of the probability distribution of a random variable always one of the possible values of x? Explain.

4. What does the mean of a probability distribution represent?

True or False? *In Exercises 5–8, determine whether the statement is true or false. If it is false, rewrite it as a true statement.*

5. In most applications, continuous random variables represent counted data, while discrete random variables represent measured data.

6. For a random variable x, the word *random* indicates that the value of x is determined by chance.

7. The mean of the random variable of a probability distribution describes how the outcomes vary.

8. The expected value of a random variable can never be negative.

Graphical Analysis *In Exercises 9–12, determine whether the graph on the number line represents a discrete random variable or a continuous random variable. Explain your reasoning.*

9. The attendance at concerts for a rock group

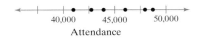
Attendance

10. The length of time student-athletes practice each week

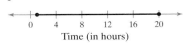
Time (in hours)

11. The distance a baseball travels after being hit

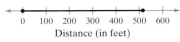
Distance (in feet)

12. The total annual arrests (in millions) in the United States *(Source: U.S. Department of Justice)*

Arrests

Using and Interpreting Concepts

Discrete Variables and Continuous Variables *In Exercises 13–18, determine whether the random variable x is discrete or continuous. Explain.*

13. Let x represent the number of cars in a university parking lot.

14. Let x represent the length of time it takes to complete an exam.

15. Let x represent the volume of blood drawn for a blood test.

16. Let x represent the number of tornadoes in the month of May in Oklahoma.

17. Let x represent the number of texts a student sends in one day.

18. Let x represent the snowfall (in inches) in Nome, Alaska, last winter.

19. (a)

x	P(x)
0	0.01
1	0.17
2	0.28
3	0.54

(b)

Televisions per Household

Skewed left

20. (a)

x	P(x)
0	0.031
1	0.063
2	0.151
3	0.297
4	0.219
5	0.156
6	0.083

(b)

Overtime

Approximately symmetric

21. (a) 0.45   **(b)** 0.82
(c) 0.99   **(d)** 0.46

22. (a) 0.214   **(b)** 0.245   **(c)** 0.755
(d) 0.511   **(e)** 0.761

23. Yes, because the probability is less than 0.05.

24. No, because the probability is greater than 0.05.

25. 0.34

26. 0.15

27. Yes

28. No, $\Sigma P(x) = 0.97$.

Constructing and Graphing Discrete Probability Distributions   *In Exercises 19 and 20, (a) construct a probability distribution, and (b) graph the probability distribution using a histogram and describe its shape.*

19. Televisions   The number of high-definition (HD) televisions per household in a small town

Televisions	0	1	2	3
Households	26	442	728	1404

20. Overtime Hours   The number of overtime hours worked in one week per employee

Overtime hours	0	1	2	3	4	5	6
Employees	6	12	29	57	42	30	16

21. Finding Probabilities   Use the probability distribution you made in Exercise 19 to find the probability of randomly selecting a household that has (a) one or two HD televisions, (b) two or more HD televisions, (c) between one and three HD televisions, inclusive, and (d) at most two HD televisions.

22. Finding Probabilities   Use the probability distribution you made in Exercise 20 to find the probability of randomly selecting an employee whose overtime is (a) one or two hours, (b) two hours or less, (c) between three and six hours, inclusive, (d) between one and three hours, inclusive, and (e) at most four hours.

23. Unusual Events   In Exercise 19, would it be unusual for a household to have no HD televisions? Explain your reasoning.

24. Unusual Events   In Exercise 20, would it be unusual for an employee to work two hours of overtime? Explain your reasoning.

Determining a Missing Probability   *In Exercises 25 and 26, determine the missing probability for the probability distribution.*

25.

x	0	1	2	3	4
P(x)	0.06	0.12	0.18	?	0.30

26.

x	0	1	2	3	4	5	6
P(x)	0.05	?	0.23	0.21	0.17	0.11	0.08

Identifying Probability Distributions   *In Exercises 27 and 28, determine whether the distribution is a probability distribution. If it is not a probability distribution, explain why.*

27.

x	0	1	2	3	4
P(x)	0.30	0.25	0.25	0.15	0.05

28.

x	0	1	2	3	4	5
P(x)	$\frac{3}{4}$	$\frac{1}{10}$	$\frac{1}{20}$	$\frac{1}{25}$	$\frac{1}{50}$	$\frac{1}{100}$

29. (a) $\mu \approx 0.5$; $\sigma^2 \approx 0.8$; $\sigma \approx 0.9$

   (b) The mean is 0.5, so the average number of dogs per household is about 0 or 1 dog. The standard deviation is 0.9, so most of the households differ from the mean by no more than about 1 dog.

30. (a) $\mu \approx 5.8$; $\sigma^2 \approx 1.4$; $\sigma \approx 1.2$

   (b) The mean is 5.8, so the average number of games played per World Series was about 6. The standard deviation is 1.2, so most of the World Series differed from the mean by no more than about 1 game.

31. (a) $\mu \approx 1.5$; $\sigma^2 \approx 1.5$; $\sigma \approx 1.2$

   (b) The mean is 1.5, so the average batch of 1000 machine parts has 1 or 2 defects. The standard deviation is 1.2, so most of the batches of 1000 differ from the mean by no more than about 1 defect.

32. (a) $\mu \approx 3.3$; $\sigma^2 \approx 3.4$; $\sigma \approx 1.8$

   (b) The mean is 3.3, so the average student is involved in about 3 extracurricular activities. The standard deviation is 1.8, so most of the students differ from the mean by no more than about 2 activities.

33. (a) $\mu \approx 2.0$; $\sigma^2 \approx 1.0$; $\sigma \approx 1.0$

   (b) The mean is 2.0, so the average hurricane that hits the U.S. mainland is a category 2 hurricane. The standard deviation is 1.0, so most of the hurricanes differ from the mean by no more than 1 category level.

34. (a) $\mu \approx 4.0$; $\sigma^2 \approx 1.2$; $\sigma \approx 1.1$

   (b) The mean is 4.0, so the average reviewer rating is 4. The standard deviation is 1.1, so most of the ratings differ from the mean by no more than about 1.

35. An expected value of 0 means that the money gained is equal to the money spent, representing the break-even point.

36. See Selected Answers, page A96.

Finding the Mean, Variance, and Standard Deviation   *In Exercises 29–34, (a) find the mean, variance, and standard deviation of the probability distribution, and (b) interpret the results.*

29. Dogs   The number of dogs per household in a neighborhood

Dogs	0	1	2	3	4	5
Probability	0.686	0.195	0.077	0.022	0.013	0.007

30. Baseball   The number of games played in each World Series from 1903 through 2016 *(Source: Adapted from Major League Baseball)*

Games played	4	5	6	7	8
Probability	0.188	0.223	0.214	0.348	0.027

31. Machine Parts   The number of defects per 1000 machine parts inspected

Defects	0	1	2	3	4	5
Probability	0.263	0.285	0.243	0.154	0.041	0.014

32. Extracurricular Activities   The number of school-related extracurricular activities per student

Activities	0	1	2	3	4	5	6	7
Probability	0.059	0.122	0.163	0.178	0.213	0.128	0.084	0.053

33. Hurricanes   The histogram shows the distribution of hurricanes that have hit the U.S. mainland from 1851 through 2015 by Saffir-Simpson category, where 1 is the weakest level and 5 is the strongest level. *(Source: National Oceanic & Atmospheric Administration)*

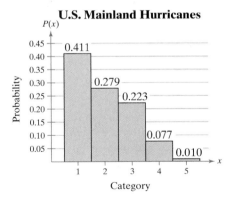

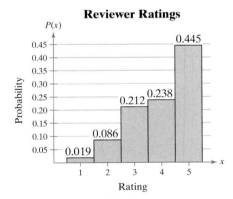

FIGURE FOR EXERCISE 33   FIGURE FOR EXERCISE 34

34. Reviewer Ratings   The histogram shows the reviewer ratings on a scale from 1 (lowest) to 5 (highest) of a recently published book.

35. Writing   The expected value of an accountant's profit and loss analysis is 0. Explain what this means.

36. Writing   In a game of chance, what is the relationship between a "fair bet" and its expected value? Explain.

37. −$0.05
38. −$7.78
39. $47,980
40. $4391.13
41. 1018; 30
42. 171.3

Finding an Expected Value *In Exercises 37 and 38, find the expected value* $E(x)$ *to the player for one play of the game. If x is the gain to a player in a game of chance, then* $E(x)$ *is usually negative. This value gives the average amount per game the player can expect to lose.*

37. In American roulette, the wheel has the 38 numbers, 00, 0, 1, 2, . . ., 34, 35, and 36, marked on equally spaced slots. If a player bets $1 on a number and wins, then the player keeps the dollar and receives an additional $35. Otherwise, the dollar is lost.

38. A high school basketball team is selling $10 raffle tickets as part of a fund-raising program. The first prize is a trip to the Bahamas valued at $5460, and the second prize is a weekend ski package valued at $496. The remaining 18 prizes are $100 gas cards. The number of tickets sold is 3500.

Extending Concepts

Linear Transformation of a Random Variable *In Exercises 39 and 40, use this information about linear transformations. For a random variable x, a new random variable y can be created by applying a **linear transformation** $y = a + bx$, where a and b are constants. If the random variable x has mean μ_x and standard deviation σ_x, then the mean, variance, and standard deviation of y are given by the formulas* $\mu_y = a + b\mu_x$, $\sigma_y^2 = b^2\sigma_x^2$, *and* $\sigma_y = |b|\sigma_x$.

39. The mean annual salary of employees at an office is originally $46,000. Each employee receives an annual bonus of $600 and a 3% raise (based on salary). What is the new mean annual salary (including the bonus and raise)?

40. The mean annual salary of employees at an office is originally $44,000 with a variance of 18,000,000. Each employee receives an annual bonus of $1000 and a 3.5% raise (based on salary). What is the standard deviation of the new salaries?

Independent and Dependent Random Variables *Two random variables x and y are **independent** when the value of x does not affect the value of y. When the variables are not independent, they are **dependent**. A new random variable can be formed by finding the sum or difference of random variables. If a random variable x has mean μ_x and a random variable y has mean μ_y, then the means of the sum and difference of the variables are given by the formulas below.*

$$\mu_{x+y} = \mu_x + \mu_y \qquad \mu_{x-y} = \mu_x - \mu_y$$

If random variables are independent, then the variance and standard deviation of the sum or difference of the random variables can be found. So, if a random variable x has variance σ_x^2 and a random variable y has variance σ_y^2, then the variances of the sum and difference of the variables are given by the formulas below. Note that the variance of the difference is the sum of the variances.

$$\sigma_{x+y}^2 = \sigma_x^2 + \sigma_y^2 \qquad \sigma_{x-y}^2 = \sigma_x^2 + \sigma_y^2$$

In Exercises 41 and 42, the distribution of SAT mathematics scores for college-bound male seniors in 2016 has a mean of 524 and a standard deviation of 126. The distribution of SAT mathematics scores for college-bound female seniors in 2016 has a mean of 494 and a standard deviation of 116. One male and one female are randomly selected. Assume their scores are independent. (Source: The College Board)

41. What is the average sum of their scores? What is the average difference of their scores?

42. What is the standard deviation of the difference of their scores?

4.2 Binomial Distributions

What You Should Learn

▶ How to determine whether a probability experiment is a binomial experiment

▶ How to find binomial probabilities using the binomial probability formula

▶ How to find binomial probabilities using technology, formulas, and a binomial probability table

▶ How to construct and graph a binomial distribution

▶ How to find the mean, variance, and standard deviation of a binomial probability distribution

Binomial Experiments ■ Binomial Probability Formula ■ Finding Binomial Probabilities ■ Graphing Binomial Distributions ■ Mean, Variance, and Standard Deviation

Binomial Experiments

There are many probability experiments for which the results of each trial can be reduced to two outcomes: success and failure. For instance, when a basketball player attempts a free throw, he or she either makes the basket or does not. Probability experiments such as these are called **binomial experiments.**

DEFINITION

A **binomial experiment** is a probability experiment that satisfies these conditions.

1. The experiment has a fixed number of trials, where each trial is independent of the other trials.
2. There are only two possible outcomes of interest for each trial. Each outcome can be classified as a success (S) or as a failure (F).
3. The probability of a success is the same for each trial.
4. The random variable x counts the number of successful trials.

Notation for Binomial Experiments

Symbol	Description
n	The number of trials
p	The probability of success in a single trial
q	The probability of failure in a single trial ($q = 1 - p$)
x	The random variable represents a count of the number of successes in n trials: $x = 0, 1, 2, 3, \ldots, n$.

Trial	Outcome	S or F?
1		F
2		S
3		F
4		F
5		S

There are two successful outcomes. So, $x = 2$.

In a binomial experiment, success does not imply something good occurred. For instance, in an experiment a survey asks 1012 people about identity theft. A success is a person who was a victim of identity theft.

Here is an example of a binomial experiment. From a standard deck of cards, you pick a card, note whether it is a club or not, and replace the card. You repeat the experiment five times, so $n = 5$. The outcomes of each trial can be classified in two categories: S = selecting a club and F = selecting another suit. The probabilities of success and failure are

$$p = \frac{1}{4} \quad \text{and} \quad q = 1 - \frac{1}{4} = \frac{3}{4}.$$

The random variable x represents the number of clubs selected in the five trials. So, the possible values of the random variable are $x = 0, 1, 2, 3, 4, 5$. For instance, if $x = 2$, then exactly two of the five cards are clubs and the other three are not clubs. An example of an experiment with $x = 2$ is shown at the left. Note that x is a discrete random variable because its possible values can be counted.

Picturing the World

A recent survey of 1520 U.S. adults was conducted to study the ways in which Americans use social media. One of the questions from the survey and the responses (either yes or no) are shown below. (Source: Pew Research)

Survey question: Do you ever use the Internet or a mobile app to use Facebook?

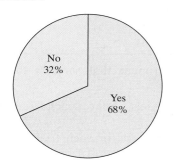

Why is this a binomial experiment? Identify the probability of success p. Identify the probability of failure q.

This is a binomial experiment because there are only two outcomes and the probability for success is the same for each trial, with $p = 0.68$ and $q = 0.32$.

EXAMPLE 1

Identifying and Understanding Binomial Experiments

Determine whether each experiment is a binomial experiment. If it is, specify the values of n, p, and q, and list the possible values of the random variable x. If it is not, explain why.

1. A certain surgical procedure has an 85% chance of success. A doctor performs the procedure on eight patients. The random variable represents the number of successful surgeries.

2. A jar contains five red marbles, nine blue marbles, and six green marbles. You randomly select three marbles from the jar, *without replacement*. The random variable represents the number of red marbles.

SOLUTION

1. The experiment is a binomial experiment because it satisfies the four conditions of a binomial experiment. In the experiment, each surgery represents one trial. There are eight surgeries, and each surgery is independent of the others. There are only two possible outcomes for each surgery—either the surgery is a success or it is a failure. Also, the probability of success for each surgery is 0.85. Finally, the random variable x represents the number of successful surgeries.

$n = 8$	Number of trials
$p = 0.85$	Probability of success
$q = 1 - 0.85$	
$\quad = 0.15$	Probability of failure
$x = 0, 1, 2, 3, 4, 5, 6, 7, 8$	Possible values of x

2. The experiment is not a binomial experiment because it does not satisfy all four conditions of a binomial experiment. In the experiment, each marble selection represents one trial, and selecting a red marble is a success. When the first marble is selected, the probability of success is 5/20. However, because the marble is not replaced, the probability of success for subsequent trials is no longer 5/20. So, the trials are not independent, and the probability of a success is not the same for each trial.

TRY IT YOURSELF 1

Determine whether the experiment is a binomial experiment. If it is, specify the values of n, p, and q, and list the possible values of the random variable x. If it is not, explain why.

> You take a multiple-choice quiz that consists of 10 questions. Each question has four possible answers, only one of which is correct. To complete the quiz, you randomly guess the answer to each question. The random variable represents the number of correct answers.

Answer: Page A34

For a random sample collected without replacement, such as in a survey, the events are dependent. However, you can treat this situation as a binomial experiment by treating the events as independent when the sample size is no more than 5% of the population. That is, $n \leq 0.05N$.

Binomial Probability Formula

There are several ways to find the probability of x successes in n trials of a binomial experiment. One way is to use a tree diagram and the Multiplication Rule. Another way is to use the **binomial probability formula.**

Study Tip

In the binomial probability formula, $_nC_x$ determines the number of ways of getting x successes in n trials, regardless of order.

$$_nC_x = \frac{n!}{(n-x)!x!}$$

Binomial Probability Formula

In a binomial experiment, the probability of exactly x successes in n trials is

$$P(x) = {_nC_x}p^xq^{n-x} = \frac{n!}{(n-x)!\,x!}p^xq^{n-x}.$$

Note that the number of failures is $n - x$.

EXAMPLE 2

Finding a Binomial Probability

Rotator cuff surgery has a 90% chance of success. The surgery is performed on three patients. Find the probability of the surgery being successful on exactly two patients. *(Source: The Orthopedic Center of St. Louis)*

SOLUTION

Method 1: Draw a tree diagram and use the Multiplication Rule.

1st Surgery	2nd Surgery	3rd Surgery	Outcome	Number of Successes	Probability
		S	SSS	3	$\frac{9}{10}\cdot\frac{9}{10}\cdot\frac{9}{10}=\frac{729}{1000}$
	S	F	SSF	2	$\frac{9}{10}\cdot\frac{9}{10}\cdot\frac{1}{10}=\frac{81}{1000}$
S		S	SFS	2	$\frac{9}{10}\cdot\frac{1}{10}\cdot\frac{9}{10}=\frac{81}{1000}$
	F	F	SFF	1	$\frac{9}{10}\cdot\frac{1}{10}\cdot\frac{1}{10}=\frac{9}{1000}$
		S	FSS	2	$\frac{1}{10}\cdot\frac{9}{10}\cdot\frac{9}{10}=\frac{81}{1000}$
	S	F	FSF	1	$\frac{1}{10}\cdot\frac{9}{10}\cdot\frac{1}{10}=\frac{9}{1000}$
F		S	FFS	1	$\frac{1}{10}\cdot\frac{1}{10}\cdot\frac{9}{10}=\frac{9}{1000}$
	F	F	FFF	0	$\frac{1}{10}\cdot\frac{1}{10}\cdot\frac{1}{10}=\frac{1}{1000}$

There are three outcomes that have exactly two successes, and each has a probability of $\frac{81}{1000}$. So, the probability of a successful surgery on exactly two patients is $3\left(\frac{81}{1000}\right) = 0.243$.

Method 2: Use the binomial probability formula.

In this binomial experiment, the values of n, p, q, and x are

$$n = 3, \quad p = \frac{9}{10}, \quad q = \frac{1}{10}, \quad \text{and} \quad x = 2.$$

The probability of exactly two successful surgeries is

$$P(2) = \frac{3!}{(3-2)!2!}\left(\frac{9}{10}\right)^2\left(\frac{1}{10}\right)^1 = 3\left(\frac{81}{100}\right)\left(\frac{1}{10}\right) = 3\left(\frac{81}{1000}\right) = 0.243.$$

Study Tip

Recall that $n!$ is read "n factorial" and represents the product of all integers from n to 1. For instance,

$$5! = 5\cdot4\cdot3\cdot2\cdot1$$
$$= 120.$$

TRY IT YOURSELF 2

A card is selected from a standard deck and replaced. This experiment is repeated a total of five times. Find the probability of selecting exactly three clubs.

Answer: Page A34

By listing the possible values of x with the corresponding probabilities, you can construct a **binomial probability distribution.**

EXAMPLE 3

Constructing a Binomial Distribution

In a survey, U.S. adults were asked to identify which social media platforms they use. The results are shown in the figure. Six adults who participated in the survey are randomly selected and asked whether they use the social media platform Facebook. Construct a binomial probability distribution for the number of adults who respond yes. *(Source: Pew Research)*

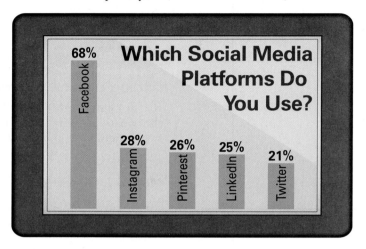

SOLUTION

From the figure, you can see that 68% of adults use the social media platform Facebook. So,

$$p = 0.68 \quad \text{and} \quad q = 0.32.$$

Because $n = 6$, the possible values of x are 0, 1, 2, 3, 4, 5, and 6. The probabilities of each value of x are

$$P(0) = {}_6C_0(0.68)^0(0.32)^6 = 1(0.68)^0(0.32)^6 \approx 0.001$$

$$P(1) = {}_6C_1(0.68)^1(0.32)^5 = 6(0.68)^1(0.32)^5 \approx 0.014$$

$$P(2) = {}_6C_2(0.68)^2(0.32)^4 = 15(0.68)^2(0.32)^4 \approx 0.073$$

$$P(3) = {}_6C_3(0.68)^3(0.32)^3 = 20(0.68)^3(0.32)^3 \approx 0.206$$

$$P(4) = {}_6C_4(0.68)^4(0.32)^2 = 15(0.68)^4(0.32)^2 \approx 0.328$$

$$P(5) = {}_6C_5(0.68)^5(0.32)^1 = 6(0.68)^5(0.32)^1 \approx 0.279$$

and

$$P(6) = {}_6C_6(0.68)^6(0.32)^0 = 1(0.68)^6(0.32)^0 \approx 0.099.$$

Notice in the table at the left that all the probabilities are between 0 and 1 and that the sum of the probabilities is 1.

TRY IT YOURSELF 3

Five adults who participated in the survey in Example 5 are randomly selected and asked whether they use the social media platform Instagram. Construct a binomial distribution for the number of adults who respond yes.

Answer: Page A35

Study Tip

When probabilities are rounded to a fixed number of decimal places, the sum of the probabilities may differ slightly from 1.

x	$P(x)$
0	0.001
1	0.014
2	0.073
3	0.206
4	0.328
5	0.279
6	0.099
	$\Sigma P(x) = 1$

Finding Binomial Probabilities

In Examples 2 and 3, you used the binomial probability formula to find the probabilities. A more efficient way to find binomial probabilities is to use technology. For instance, you can find binomial probabilities using Minitab, Excel, StatCrunch, and the TI-84 Plus.

Tech Tip

You can use technology such as Minitab, Excel, StatCrunch, or the TI-84 Plus to find a binomial probability. For instance, here are instructions for finding a binomial probability on a TI-84 Plus. From the DISTR menu, choose the *binompdf(* feature. Enter the values of *n*, *p*, and *x*. Then calculate the probability.

EXAMPLE 4

Finding a Binomial Probability Using Technology

A survey found that 26% of U.S. adults believe there is no difference between secured and unsecured wireless networks. (A secured network uses barriers, such as firewalls and passwords, to protect information; an unsecured network does not.) You randomly select 100 adults. What is the probability that exactly 35 adults believe there is no difference between secured and unsecured networks? Use technology to find the probability. *(Source: University of Phoenix)*

SOLUTION

Minitab, Excel, StatCrunch, and the TI-84 Plus each have features that allow you to find binomial probabilities. Try using these technologies. You should obtain results similar to these displays.

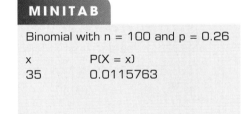

MINITAB

Binomial with n = 100 and p = 0.26

x	P(X = x)
35	0.0115763

STATCRUNCH

Binomial Distribution

n:100   p:0.26

P(X = 35) = 0.0115763

TI-84 PLUS

binompdf(100, .26, 35)
        .0115762984

EXCEL

	A	B	C	D
1	BINOM.DIST(35,100,0.26,FALSE)			
2				0.011576298

Study Tip

Recall that a probability of 0.05 or less is considered unusual.

Interpretation From these displays, you can see that the probability that exactly 35 adults believe there is no difference between secured and unsecured networks is about 0.012. Because 0.012 is less than 0.05, this can be considered an unusual event.

TRY IT YOURSELF 4

A survey found that 52% of U.S. adults associate professional football with negative moral values. You randomly select 150 adults. What is the probability that exactly 65 adults associate professional football with negative moral values? Use technology to find the probability. *(Source: The Harris Poll)*

Answer: Page A35

Note to Instructor

Point out that Example 4 gives the probability that exactly *x* successes will occur. Each of these technology tools has a cumulative distribution function (CDF) that allows an easy computation of the probability of "*x* or fewer" successes. (This subject is discussed after Try It Yourself 5 on the next page along with the TI-84 Plus.) This is a good time to point out that the CDF adds the areas for the given *x*-value and all those to its left.

EXAMPLE 5

Finding Binomial Probabilities Using Formulas

A survey found that 17% of U.S. adults say that Google News is a major source of news for them. You randomly select four adults and ask them whether Google News is a major source of news for them. Find the probability that (1) exactly two of them respond yes, (2) at least two of them respond yes, and (3) fewer than two of them respond yes. *(Source: Ipsos Public Affairs)*

SOLUTION

1. Using $n = 4$, $p = 0.17$, $q = 0.83$, and $x = 2$, the probability that exactly two adults will respond yes is

$$
\begin{aligned}
P(2) &= {}_4C_2(0.17)^2(0.83)^2 \\
&= 6(0.17)^2(0.83)^2 \\
&\approx 0.119.
\end{aligned}
$$

2. To find the probability that at least two adults will respond yes, find the sum of $P(2)$, $P(3)$, and $P(4)$. Begin by using the binomial probability formula to write an expression for each probability.

$$
\begin{aligned}
P(2) &= {}_4C_2(0.17)^2(0.83)^2 = 6(0.17)^2(0.83)^2 \\
P(3) &= {}_4C_3(0.17)^3(0.83)^1 = 4(0.17)^3(0.83)^1 \\
P(4) &= {}_4C_4(0.17)^4(0.83)^0 = 1(0.17)^4(0.83)^0
\end{aligned}
$$

So, the probability that at least two will respond yes is

$$
\begin{aligned}
P(x \geq 2) &= P(2) + P(3) + P(4) \\
&= 6(0.17)^2(0.83)^2 + 4(0.17)^3(0.83)^1 + (0.17)^4(0.83)^0 \\
&\approx 0.137.
\end{aligned}
$$

3. To find the probability that fewer than two adults will respond yes, find the sum of $P(0)$ and $P(1)$. Begin by using the binomial probability formula to write an expression for each probability.

$$
\begin{aligned}
P(0) &= {}_4C_0(0.17)^0(0.83)^4 = 1(0.17)^0(0.83)^4 \\
P(1) &= {}_4C_1(0.17)^1(0.83)^3 = 4(0.17)^1(0.83)^3
\end{aligned}
$$

So, the probability that fewer than two will respond yes is

$$
\begin{aligned}
P(x < 2) &= P(0) + P(1) \\
&= (0.17)^0(0.83)^4 + 4(0.17)^1(0.83)^3 \\
&\approx 0.863.
\end{aligned}
$$

TRY IT YOURSELF 5

The survey in Example 5 found that 27% of U.S. adults say that CNN is a major source of news for them. You randomly select five adults and ask them whether CNN is a major source of news for them. Find the probability that (1) exactly two of them respond yes, (2) at least two of them respond yes, and (3) fewer than two of them respond yes. *(Source: Ipsos Public Affairs)*

Answer: Page A35

You can use technology to check your answers. For instance, the TI-84 Plus screen at the left shows how to check parts 1 and 3 of Example 5. Note that the second entry uses the *binomial CDF* feature. A cumulative distribution function (CDF) computes the probability of "*x* or fewer" successes by adding the areas for the given *x*-value and all those to its left.

Study Tip

The complement of "*x* is at least 2" is "*x* is less than 2." So, another way to find the probability in part 3 of Example 5 is

$$
\begin{aligned}
P(x < 2) &= 1 - P(x \geq 2) \\
&\approx 1 - 0.137 \\
&= 0.863.
\end{aligned}
$$

Note to Instructor

For Try It Yourself 5, remind students to decide for each question whether it is easier to calculate a probability as given or to use the complement rule to get their answer.

TI-84 PLUS

```
binompdf(4, .17, 2)
               .11945526
binomcdf(4, .17, 1)
               .86339837
```

Finding binomial probabilities with the binomial probability formula can be a tedious process. To make this process easier, you can use a binomial probability table. Table 2 in Appendix B lists the binomial probabilities for selected values of n and p.

Note to Instructor

Discuss limitations on the use of the table. It can be used only for particular values of p and particular values of n up to 20. Later in the course, students will learn how to calculate probabilities for $n > 20$ using the normal approximation to the binomial.

EXAMPLE 6

Finding a Binomial Probability Using a Table

About 10% of workers (ages 16 years and older) in the United States commute to their jobs by carpooling. You randomly select eight workers. What is the probability that exactly four of them carpool to work? Use a table to find the probability. *(Source: American Community Survey)*

SOLUTION

A portion of Table 2 in Appendix B is shown here. Using the distribution for $n = 8$ and $p = 0.1$, you can find the probability that $x = 4$, as shown by the highlighted areas in the table.

		p												
n	***x***	**.01**	**.05**	**.10**	**.15**	**.20**	**.25**	**.30**	**.35**	**.40**	**.45**	**.50**	**.55**	**.60**
2	0	.980	.902	.810	.723	.640	.563	.490	.423	.360	.303	.250	.203	.160
	1	.020	.095	.180	.255	.320	.375	.420	.455	.480	.495	.500	.495	.480
	2	.000	.002	.010	.023	.040	.063	.090	.123	.160	.203	.250	.303	.360
3	0	.970	.857	.729	.614	.512	.422	.343	.275	.216	.166	.125	.091	.064
	1	.029	.135	.243	.325	.384	.422	.441	.444	.432	.408	.375	.334	.288
	2	.000	.007	.027	.057	.096	.141	.189	.239	.288	.334	.375	.408	.432
	3	.000	.000	.001	.003	.008	.016	.027	.043	.064	.091	.125	.166	.216
8	0	.923	.663	.430	.272	.168	.100	.058	.032	.017	.008	.004	.002	.001
	1	.075	.279	.383	.385	.336	.267	.198	.137	.090	.055	.031	.016	.008
	2	.003	.051	.149	.238	.294	.311	.296	.259	.209	.157	.109	.070	.041
	3	.000	.005	.033	.084	.147	.208	.254	.279	.279	.257	.219	.172	.124
	4	.000	.000	.005	.018	.046	.087	.136	.188	.232	.263	.273	.263	.232
	5	.000	.000	.000	.003	.009	.023	.047	.081	.124	.172	.219	.257	.279
	6	.000	.000	.000	.000	.001	.004	.010	.022	.041	.070	.109	.157	.209
	7	.000	.000	.000	.000	.000	.000	.001	.003	.008	.016	.031	.055	.090
	8	.000	.000	.000	.000	.000	.000	.000	.000	.001	.002	.004	.008	.017

According to the table, the probability is 0.005. You can check this result using technology. As shown at the right using Minitab, the probability is

0.0045927.

After rounding to three decimal places, the probability is 0.005, which is the same value found using the table.

MINITAB

Probability Density Function

Binomial with n = 8 and p = 0.1

x	P(X = x)
4	0.0045927

Interpretation So, the probability that exactly four of the eight workers carpool to work is 0.005. Because 0.005 is less than 0.05, this can be considered an unusual event.

4.2 To explore this topic further, see **Activity 4.2** on page 214.

TRY IT YOURSELF 6

About 5% of workers (ages 16 years and older) in the United States commute to their jobs by using public transportation (excluding taxicabs). You randomly select six workers. What is the probability that exactly two of them use public transportation to get to work? Use a table to find the probability. *(Source: American Community Survey)*

Answer: Page A35

Graphing Binomial Distributions

In Section 4.1, you learned how to graph discrete probability distributions. Because a binomial distribution is a discrete probability distribution, you can use the same process.

EXAMPLE 7

Graphing a Binomial Distribution

Sixty-two percent of cancer survivors are ages 65 years or older. You randomly select six cancer survivors and ask them whether they are 65 years of age or older. Construct a probability distribution for the random variable x. Then graph the distribution. *(Source: National Cancer Institute)*

SOLUTION

To construct the binomial distribution, find the probability for each value of x. Using $n = 6$, $p = 0.62$, and $q = 0.38$, you can obtain the following.

x	0	1	2	3	4	5	6
$P(x)$	0.003	0.029	0.120	0.262	0.320	0.209	0.057

Notice in the table that all the probabilities are between 0 and 1 and that the sum of the probabilities is 1. You can graph the probability distribution using a histogram as shown below.

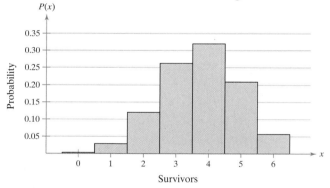

Cancer Survivors 65 Years of Age or Older

Interpretation   From the histogram, you can see that it would be unusual for none or only one of the survivors to be age 65 years or older because both probabilities are less than 0.05.

TRY IT YOURSELF 7

A recent study found that 28% of U.S. adults read an ebook in the last 12 months. You randomly select 4 adults and ask them whether they read an ebook in the last 12 months. Construct a probability distribution for the random variable x. Then graph the distribution. *(Source: Pew Research)*

Answer: Page A35

Notice in Example 7 that the histogram is skewed left. The graph of a binomial distribution with $p > 0.5$ is skewed left, whereas the graph of a binomial distribution with $p < 0.5$ is skewed right. The graph of a binomial distribution with $p = 0.5$ is symmetric.

Mean, Variance, and Standard Deviation

Although you can use the formulas you learned in Section 4.1 for mean, variance, and standard deviation of a discrete probability distribution, the properties of a binomial distribution enable you to use much simpler formulas.

Population Parameters of a Binomial Distribution

$$\text{Mean: } \mu = np$$
$$\text{Variance: } \sigma^2 = npq$$
$$\text{Standard deviation: } \sigma = \sqrt{npq}$$

EXAMPLE 8

Note to Instructor

It is useful to have students calculate the mean, variance, and standard deviation for a binomial experiment using the general formulas for discrete probability distributions given in Section 4.1. A quick example (and good review) would be to use the distribution for $n = 4$ and $p = 0.3$.

Point out that the formula $\mu = np$ gives the same result rounded to two decimal places.

X	P(x)	xP(x)
0	0.240	0
1	0.412	0.412
2	0.265	0.530
3	0.076	0.228
4	0.008	0.032
	$\Sigma \approx 1$	$\Sigma = 1.202$

Finding and Interpreting Mean, Variance, and Standard Deviation

In Pittsburgh, Pennsylvania, about 56% of the days in a year are cloudy. Find the mean, variance, and standard deviation for the number of cloudy days during the month of June. Interpret the results and determine any unusual values. *(Source: National Climatic Data Center)*

SOLUTION

There are 30 days in June. Using $n = 30$, $p = 0.56$, and $q = 0.44$, you can find the mean, variance, and standard deviation as shown below.

$$\mu = np$$
$$= 30 \cdot 0.56$$
$$= 16.8 \qquad \text{Mean}$$

$$\sigma^2 = npq$$
$$= 30 \cdot 0.56 \cdot 0.44$$
$$\approx 7.4 \qquad \text{Variance}$$

$$\sigma = \sqrt{npq}$$
$$= \sqrt{30 \cdot 0.56 \cdot 0.44}$$
$$\approx 2.7 \qquad \text{Standard deviation}$$

Interpretation   On average, there are 16.8 cloudy days during the month of June. The standard deviation is about 2.7 days. Values that are more than two standard deviations from the mean are considered unusual. Because

$$16.8 - 2(2.7) = 11.4$$

a June with 11 cloudy days or less would be unusual. Similarly, because

$$16.8 + 2(2.7) = 22.2$$

a June with 23 cloudy days or more would also be unusual.

TRY IT YOURSELF 8

In San Francisco, California, about 44% of the days in a year are clear. Find the mean, variance, and standard deviation for the number of clear days during the month of May. Interpret the results and determine any unusual events. *(Source: National Climatic Data Center)*

Answer: Page A35

4.2 EXERCISES

1. Each trial is independent of the other trials when the outcome of one trial does not affect the outcome of any of the other trials.

2. The random variable measures the number of successes in n trials.

3. c; Because the probability is greater than 0.5, the distribution is skewed left.

4. b; Because the probability is 0.5, the distribution is symmetric.

5. a; Because the probability is less than 0.5, the distribution is skewed right.

6. c; The histogram shows probabilities for 12 trials.

7. a; The histogram shows probabilities for 4 trials.

8. b; The histogram shows probabilities for 8 trials.

   As n increases, the distribution becomes more symmetric.

9. (3) 0, 1

   (4) 0, 5

   (5) 4, 5

10. (6) 0, 1, 2, 3, 4, 11, 12

   (7) 0

   (8) 0, 1, 2, 8

11. $\mu = 20$; $\sigma^2 = 12$; $\sigma \approx 3.5$

12. $\mu = 54.6$; $\sigma^2 \approx 19.1$; $\sigma \approx 4.4$

13. $\mu \approx 32.2$; $\sigma^2 \approx 23.9$; $\sigma \approx 4.9$

14. $\mu \approx 259.1$; $\sigma^2 \approx 46.6$; $\sigma \approx 6.8$

15. Binomial experiment

   Success: frequent gamer who plays video games on smartphone

   $n = 10$; $p = 0.36$; $q = 0.64$; $x = 0, 1, 2, 3, 4, 5, 6, 7, 8, 9, 10$

Building Basic Skills and Vocabulary

1. In a binomial experiment, what does it mean to say that each trial is independent of the other trials?

2. In a binomial experiment with n trials, what does the random variable measure?

Graphical Analysis *In Exercises 3–5, the histogram represents a binomial distribution with 5 trials. Match the histogram with the appropriate probability of success p. Explain your reasoning.*

(a) $p = 0.25$ (b) $p = 0.50$ (c) $p = 0.75$

3.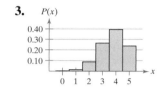

4.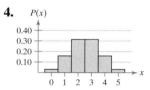

5.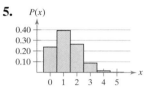

Graphical Analysis *In Exercises 6–8, the histogram represents a binomial distribution with probability of success p. Match the histogram with the appropriate number of trials n. Explain your reasoning. What happens as the value of n increases and p remains the same?*

(a) $n = 4$ (b) $n = 8$ (c) $n = 12$

6.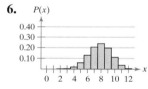

7.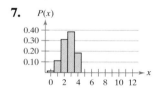

8.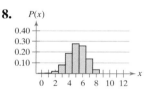

9. Identify the unusual values of x in each histogram in Exercises 3–5.

10. Identify the unusual values of x in each histogram in Exercises 6–8.

Mean, Variance, and Standard Deviation
In Exercises 11–14, find the mean, variance, and standard deviation of the binomial distribution with the given values of n and p.

11. $n = 50, p = 0.4$

12. $n = 84, p = 0.65$

13. $n = 124, p = 0.26$

14. $n = 316, p = 0.82$

Using and Interpreting Concepts

Identifying and Understanding Binomial Experiments *In Exercises 15–18, determine whether the experiment is a binomial experiment. If it is, identify a success, specify the values of n, p, and q, and list the possible values of the random variable x. If it is not a binomial experiment, explain why.*

15. Video Games A survey found that 36% of frequent gamers play video games on their smartphones. Ten frequent gamers are randomly selected. The random variable represents the number of frequent gamers who play video games on their smartphones. *(Source: Entertainment Software Association)*

16. Not a binomial experiment because the probability of a success is not the same for each trial.

17. Not a binomial experiment because the probability of a success is not the same for each trial.

18. Binomial experiment

Success: Woman age 18–33 who is a mother

$n = 8$; $p = 0.42$; $q = 0.58$; $x = 0, 1, 2, 3, 4, 5, 6, 7, 8$

19. (a) 0.019 (b) 0.272 (c) 0.905
20. (a) 0.227 (b) 0.891 (c) 0.939
21. (a) 0.150 (b) 0.759 (c) 0.712
22. (a) 0.006 (b) 0.387 (c) 0.862
23. (a) 0.221 (b) 0.247 (c) 0.753
24. (a) 0.113 (b) 0.859 (c) 0.141
25. (a) 0.089 (b) 0.017 (c) 0.106
26. (a) 0.199 (b) 0.318 (c) 0.511

16. Cards You draw five cards, one at a time, from a standard deck. You do not replace a card once it is drawn. The random variable represents the number of cards that are hearts.

17. Lottery A state lottery official randomly chooses 6 balls numbered from 1 through 40 without replacement. You choose six numbers and purchase a lottery ticket. The random variable represents the number of matches on your ticket to the numbers drawn in the lottery.

18. Women Who Are Mothers A survey found that 42% of women ages 18 to 33 are mothers. Eight women ages 18 to 33 are randomly selected. The random variable represents the number of women ages 18 to 33 who are mothers. *(Source: Pew Research Center)*

Finding Binomial Probabilities *In Exercises 19–26, find the indicated probabilities. If convenient, use technology or Table 2 in Appendix B.*

19. Newspapers Thirty-four percent of U.S. adults have very little confidence in newspapers. You randomly select eight U.S. adults. Find the probability that the number of U.S. adults who have very little confidence in newspapers is (a) exactly six, (b) at least four, and (c) less than five. *(Source: Gallup)*

20. Drone Use in Police Work Thirty-six percent of U.S. adults favor the use of unmanned drones by police agencies. You randomly select nine U.S. adults. Find the probability that the number of U.S. adults who favor the use of unmanned drones by police agencies is (a) exactly four, (b) at least two, and (c) less than six. *(Source: Rasmussen Reports)*

21. Flu Shots Fifty-six percent of U.S. adults say they intend to get a flu shot. You randomly select 10 U.S. adults. Find the probability that the number of U.S. adults who intend to get a flu shot is (a) exactly four, (b) at least five, and (c) less than seven. *(Source: Rasmussen Reports)*

22. Fast Food Eleven percent of U.S. adults eat fast food four to six times per week. You randomly select 12 U.S. adults. Find the probability that the number of U.S. adults who eat fast food four to six times per week is (a) exactly five, (b) at least two, and (c) less than three. *(Source: Statista)*

23. Consumer Electronics Forty percent of consumers prefer to purchase electronics online. You randomly select 11 consumers. Find the probability that the number of consumers who prefer to purchase electronics online is (a) exactly five, (b) more than five, and (c) at most five. *(Source: PwC)*

24. Grocery Shopping Twenty percent of consumers prefer to purchase groceries online. You randomly select 16 consumers. Find the probability that the number of consumers who prefer to purchase groceries online is (a) exactly one, (b) more than one, and (c) at most one. *(Source: PwC)*

25. Workplace Drug Testing Four percent of the U.S. workforce test positive for illicit drugs. You randomly select 14 workers. Find the probability that the number of workers who test positive for illicit drugs is (a) exactly two, (b) more than two, and (c) between two and five, inclusive. *(Source: Quest Diagnostics)*

26. Tax Holiday Forty-four percent of U.S. adults say they are more likely to make purchases during a sales tax holiday. You randomly select 15 adults. Find the probability that the number of adults who say they are more likely to make purchases during a sales tax holiday is (a) exactly seven, (b) more than seven, and (c) between seven and eleven, inclusive. *(Source: Rasmussen Reports)*

27. (a)

x	P(x)
0	0.008974
1	0.060355
2	0.173965
3	0.278572
4	0.267647
5	0.154291
6	0.049413
7	0.006782

(b) See Odd Answers, page A57.

(c) The values 0, 6, and 7 are unusual because their probabilities are less than 0.05.

28. See Selected Answers, page A96.

29. See Odd Answers, page A57

30. See Selected Answers, page A96

31. $\mu \approx 5.0$; $\sigma^2 \approx 1.4$; $\sigma \approx 1.2$; On average, 5 out of every 7 U.S. adults think that political correctness is a problem in America today. The standard deviation is 1.2, so most samples of 7 U.S. adults would differ from the mean by at most 1.2 U.S. adults.

32. $\mu = 2$; $\sigma^2 = 1$; $\sigma = 1$; On average, 2 out of every 4 adults are offended by how men portray women in rap and hip-hop music. The standard deviation is 1, so most samples of 4 adults would differ from the mean by at most 1 adult.

33. $\mu \approx 6.3$; $\sigma^2 \approx 1.3$; $\sigma \approx 1.2$; On average, 6.3 out of every 8 adults believe that life on other planets is possible. The standard deviation is 1.2, so most samples of 8 adults would differ from the mean by at most 1.2 adults.

34. $\mu = 1.8$; $\sigma^2 \approx 1.2$; $\sigma \approx 1.1$; On average, 1.8 out of every 5 likely U.S. voters believe that the federal government should get more involved in fighting local crime. The standard deviation is 1.1, so more samples of 5 likely U.S. voters would differ from the mean by at most 1.1 likely U.S. voters.

Constructing and Graphing Binomial Distributions
In Exercises 27–30, (a) construct a binomial distribution, (b) graph the binomial distribution using a histogram and describe its shape, and (c) identify any values of the random variable x that you would consider unusual. Explain your reasoning.

27. Working Mothers Forty-nine percent of working mothers do not have enough money to cover their health insurance deductibles. You randomly select seven working mothers and ask them whether they have enough money to cover their health insurance deductibles. The random variable represents the number of working mothers who do not have enough money to cover their health insurance deductibles. *(Source: Aflac)*

28. Workplace Cleanliness Fifty-seven percent of employees judge their peers by the cleanliness of their workspaces. You randomly select 10 employees and ask them whether they judge their peers by the cleanliness of their workspaces. The random variable represents the number of employees who judge their peers by the cleanliness of their workspaces. *(Source: Adecco)*

29. Living to Age 100 Seventy-seven percent of adults want to live to age 100. You randomly select five adults and ask them whether they want to live to age 100. The random variable represents the number of adults who want to live to age 100. *(Source: Standford Center on Longevity)*

30. Meal Programs Fifty-seven percent of school districts offer locally sourced fruits and vegetables in their meal programs. You randomly select eight school districts and ask them whether they offer locally sourced fruits and vegetables in their meal programs. The random variable represents the number of school districts that offer locally sourced fruits and vegetables in their meal programs. *(Source: School Nutrition Association)*

Finding and Interpreting Mean, Variance, and Standard Deviation
In Exercises 31–36, find the mean, variance, and standard deviation of the binomial distribution for the given random variable. Interpret the results.

31. Political Correctness Seventy-one percent of U.S. adults think that political correctness is a problem in America today. You randomly select seven U.S. adults and ask them whether they think that political correctness is a problem in America today. The random variable represents the number of U.S. adults who think that political correctness is a problem in America today. *(Source: Rasmussen Reports)*

32. Rap and Hip-Hop Music Fifty percent of adults are offended by how men portray women in rap and hip-hop music. You randomly select four adults and ask them whether they are offended by how men portray women in rap and hip-hop music. The random variable represents the number of adults who are offended by how men portray women in rap and hip-hop music. *(Source: Empower Women)*

33. Life on Other Planets Seventy-nine percent of U.S. adults believe that life on other planets is plausible. You randomly select eight U.S. adults and ask them whether they believe that life on other planets is plausible. The random variable represents the number of adults who believe that life on other planets is plausible. *(Source: Ipsos)*

34. Federal Involvement in Fighting Local Crime Thirty-six percent of likely U.S. voters think that the federal government should get more involved in fighting local crime. You randomly select five likely U.S. voters and ask them whether they think that the federal government should get more involved in fighting local crime. The random variable represents the number of likely U.S. voters who think that the federal government should get more involved in fighting local crime. *(Source: Rasmussen Reports)*

35. $\mu \approx 1.9;\ \sigma^2 \approx 1.3;\ \sigma \approx 1.1;$
On average, 1.9 out of every
6 U.S. employees who are late
for work blame oversleeping.
The standard deviation is 1.1,
so most samples of 6 U.S.
employees who are late would
differ from the mean by at most
1.1 U.S. employees.

36. $\mu \approx 0.5;\ \sigma^2 \approx 0.5;\ \sigma \approx 0.7;$
On average, 0.5 out of every
5 college graduates think that
Judge Judy serves on the
Supreme Court. The standard
deviation is 0.7, so most
samples of 5 college graduates
would differ from the mean by
at most 0.7 graduate.

37. 0.033

38. 0.002

39. (a) 0.107

(b) 0.107

(c) The results are the same.

35. Late for Work Thirty-two percent of U.S. employees who are late for work blame oversleeping. You randomly select six U.S. employees who are late for work and ask them whether they blame oversleeping. The random variable represents the number of U.S. employees who are late for work and blame oversleeping. *(Source: CareerBuilder)*

36. Supreme Court Ten percent of college graduates think that Judge Judy serves on the Supreme Court. You randomly select five college graduates and ask them whether they think that Judge Judy serves on the Supreme Court. The random variable represents the number of college graduates who think that Judge Judy serves on the Supreme Court. *(Source: CNN)*

Extending Concepts

Multinomial Experiments *In Exercises 37 and 38, use the information below.*

A **multinomial experiment** satisfies these conditions.

- The experiment has a fixed number of trials n, where each trial is independent of the other trials.

- Each trial has k possible mutually exclusive outcomes: $E_1, E_2, E_3, \ldots, E_k$.

- Each outcome has a fixed probability. So, $P(E_1) = p_1$, $P(E_2) = p_2$, $P(E_3) = p_3$, $\ldots$, $P(E_k) = p_k$. The sum of the probabilities for all outcomes is $p_1 + p_2 + p_3 + \cdots + p_k = 1$.

- The number of times E_1 occurs is x_1, the number of times E_2 occurs is x_2, the number of times E_3 occurs is x_3, and so on.

- The discrete random variable x counts the number of times x_1, x_2, x_3, $\ldots$, x_k that each outcome occurs in n independent trials where $x_1 + x_2 + x_3 + \cdots + x_k = n$. The probability that x will occur is

$$P(x) = \frac{n!}{x_1!x_2!x_3!\cdots x_k!}\, p_1^{x_1} p_2^{x_2} p_3^{x_3} \cdots p_k^{x_k}.$$

37. Genetics According to a theory in genetics, when tall and colorful plants are crossed with short and colorless plants, four types of plants will result: tall and colorful, tall and colorless, short and colorful, and short and colorless, with corresponding probabilities of $\frac{9}{16}$, $\frac{3}{16}$, $\frac{3}{16}$, and $\frac{1}{16}$. Ten plants are selected. Find the probability that 5 will be tall and colorful, 2 will be tall and colorless, 2 will be short and colorful, and 1 will be short and colorless.

38. Genetics Another proposed theory in genetics gives the corresponding probabilities for the four types of plants described in Exercise 37 as $\frac{5}{16}$, $\frac{4}{16}$, $\frac{1}{16}$, and $\frac{6}{16}$. Ten plants are selected. Find the probability that 5 will be tall and colorful, 2 will be tall and colorless, 2 will be short and colorful, and 1 will be short and colorless.

39. Manufacturing An assembly line produces 10,000 automobile parts. Twenty percent of the parts are defective. An inspector randomly selects 10 of the parts.

(a) Use the Multiplication Rule (discussed in Section 3.2) to find the probability that none of the selected parts are defective. (Note that the events are dependent.)

(b) Because the sample is only 0.1% of the population, treat the events as *independent* and use the binomial probability formula to approximate the probability that none of the selected parts are defective.

(c) Compare the results of parts (a) and (b).

The *binomial distribution* applet allows you to simulate values from a binomial distribution. You can specify the parameters for the binomial distribution (n and p) and the number of values to be simulated (N). When you click SIMULATE, N values from the specified binomial distribution will be plotted at the right. The frequency of each outcome is shown in the plot.

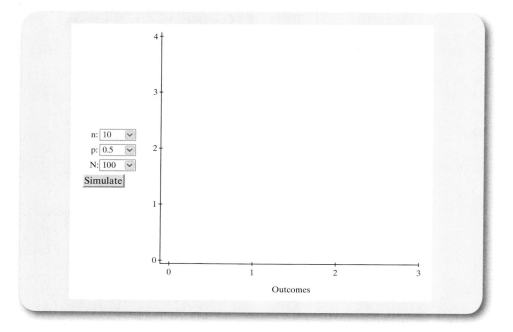

EXPLORE

Step 1 Specify a value of n. **Step 2** Specify a value of p.
Step 3 Specify a value of N. **Step 4** Click SIMULATE.

DRAW CONCLUSIONS

APPLET

1. During a presidential election year, 70% of a county's eligible voters cast a vote. Simulate selecting $n = 10$ eligible voters $N = 10$ times (for 10 communities in the county). Use the results to estimate the probability that the number who voted in this election is (a) exactly 5, (b) at least 8, and (c) at most 7.

2. During a non-presidential election year, 20% of the eligible voters in the same county as in Exercise 1 cast a vote. Simulate selecting $n = 10$ eligible voters $N = 10$ times (for 10 communities in the county). Use the results to estimate the probability that the number who voted in this election is (a) exactly 4, (b) at least 5, and (c) less than 4.

3. For the election in Exercise 1, simulate selecting $n = 10$ eligible voters $N = 100$ times. Estimate the probability that the number who voted in this election is exactly 5. Compare this result with the result in Exercise 1 part (a). Which of these is closer to the probability found using the binomial probability formula?

The official website of Major League Baseball, *MLB.com,* records detailed statistics about players and games.

During the 2016 regular season, Dustin Pedroia of the Boston Red Sox had a batting average of 0.318. The graphs below show the number of hits he had in games in which he had different numbers of at-bats.

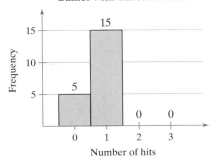

Games with Three At-Bats

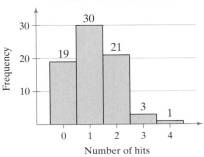

Games with Four At-Bats

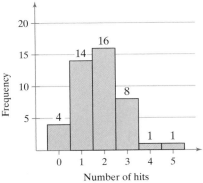

Games with Five At-Bats

EXERCISES

1. Construct a probability distribution for the number of hits in games with

   (a) 3 at-bats. (b) 4 at-bats. (c) 5 at-bats.

2. Construct binomial probability distributions for $p = 0.318$ and (a) $n = 3$, (b) $n = 4$, and (c) $n = 5$.

3. Compare your distributions from Exercise 1 and Exercise 2. Is a binomial distribution a good model for determining the numbers of hits in a baseball game for a given number of at-bats? Explain your reasoning and include a discussion of the four conditions for a binomial experiment.

4. During the 2016 regular season, Kris Bryant of the Chicago Cubs had 37 games with 3 at-bats. Of these games, he had 11 games with no hits, 18 games with one hit, 7 games with two hits, and 1 game with three hits.

   (a) Based on Pedroia's and Bryant's hits in games with 3 at-bats, which player do you think had the higher batting average?

   (b) Look up Bryant's 2016 regular season batting average. Was your expectation from part (a) correct? If not, propose a reason why.

4.3 More Discrete Probability Distributions

What You Should Learn

▶ How to find probabilities using the geometric distribution

▶ How to find probabilities using the Poisson distribution

The Geometric Distribution ■ The Poisson Distribution ■ Summary of Discrete Probability Distributions

The Geometric Distribution

Many actions in life are repeated until a success occurs. For instance, you might have to send an email several times before it is successfully sent. A situation such as this can be represented by a **geometric distribution.**

Tech Tip

You can use technology such as Minitab, Excel, StatCrunch, or the TI-84 Plus to find a geometric probability. For instance, here are instructions for finding a geometric probability on a TI-84 Plus. From the DISTR menu, choose the *geometpdf(* feature. Enter the values of *p* and *x*. Then calculate the probability.

Note to Instructor

If you choose, this section can be omitted. None of the material included here will be needed in later sections.

DEFINITION

A **geometric distribution** is a discrete probability distribution of a random variable x that satisfies these conditions.

1. A trial is repeated until a success occurs.
2. The repeated trials are independent of each other.
3. The probability of success p is the same for each trial.
4. The random variable x represents the number of the trial in which the first success occurs.

The probability that the first success will occur on trial number x is

$$P(x) = pq^{x-1}, \text{ where } q = 1 - p.$$

In other words, when the first success occurs on the third trial, the outcome is FFS, and the probability is $P(3) = q \cdot q \cdot p$, or $P(3) = p \cdot q^2$.

EXAMPLE 1

Using the Geometric Distribution

A study found that the smartphones made by a certain manufacturer had a failure rate of 43%. Four smartphones made by this manufacturer are selected at random. Find the probability that the fourth smartphone is the first one to have a failure. *(Source: Blancco Technology Group)*

SOLUTION

Using $p = 0.43$, $q = 0.57$, and $x = 4$, you have

$$P(4) = 0.43(0.57)^{4-1}$$
$$= 0.43(0.57)^3$$
$$\approx 0.080.$$

So, the probability that the fourth smartphone is the first one to have a failure is about 0.080. You can use technology to check this result. For instance, using a TI-84 Plus, you can find $P(4)$, as shown below.

```
TI-84 PLUS

geometpdf(.43,4)
              .07963299
```

TRY IT YOURSELF 1

The study in Example 1 found that the smartphones made by a second manufacturer had a failure rate of 14%. Six smartphones made by this manufacturer are selected at random. Find the probability that the sixth smartphone is the first one to have a failure. *(Source: Blancco Technology Group)*

Answer: Page A35

Even though theoretically a success may never occur, the geometric distribution is a discrete probability distribution because the values of x can be listed: 1, 2, 3, ... Notice that as x becomes larger, $P(x)$ gets closer to zero. For instance, in Example 1, the probability that the thirtieth smartphone is the first one to have a failure is

$$P(30) = 0.43(0.57)^{30-1} = 0.43(0.57)^{29} \approx 0.00000004$$

The Poisson Distribution

In a binomial experiment, you are interested in finding the probability of a specific number of successes in a given number of trials. Suppose instead that you want to know the probability that a specific number of occurrences takes place within a given unit of time, area, or volume. For instance, to determine the probability that an employee will take 15 sick days within a year, you can use the **Poisson distribution.**

Tech Tip

You can use technology such as Minitab, Excel, StatCrunch, or the TI-84 Plus to find a Poisson probability.

For instance, here are instructions for finding a Poisson probability on a TI-84 Plus. From the DISTR menu, choose the *poissonpdf(* feature. Enter the values of μ and x. (Note that the TI-84 Plus uses the Greek letter lambda, λ, in place of μ.) Then calculate the probability.

DEFINITION

The **Poisson distribution** is a discrete probability distribution of a random variable x that satisfies these conditions.

1. The experiment consists of counting the number of times x an event occurs in a given interval. The interval can be an interval of time, area, or volume.
2. The probability of the event occurring is the same for each interval.
3. The number of occurrences in one interval is independent of the number of occurrences in other intervals.

The probability of exactly x occurrences in an interval is

$$P(x) = \frac{\mu^x e^{-\mu}}{x!}$$

where e is an irrational number approximately equal to 2.71828 and μ is the mean number of occurrences per interval unit.

TI-84 PLUS

poissonpdf(3,4)
                .1680313557

EXAMPLE 2

Using the Poisson Distribution

The mean number of accidents per month at a certain intersection is three. What is the probability that in any given month four accidents will occur at this intersection?

SOLUTION

Using $x = 4$ and $\mu = 3$, the probability that 4 accidents will occur in any given month at the intersection is

$$P(4) \approx \frac{3^4(2.71828)^{-3}}{4!} \approx 0.168.$$

You can use technology to check this result. For instance, using a TI-84 Plus, you can find $P(4)$, as shown at the left.

In Example 2, you used a formula to determine a Poisson probability. You can also use a table to find Poisson probabilities. Table 3 in Appendix B lists the Poisson probabilities for selected values of x and μ. You can also use technology tools, such as Minitab, Excel, and the TI-84 Plus, to find Poisson probabilities.

EXAMPLE 3

Finding a Poisson Probability Using a Table

A population count shows that the average number of rabbits per acre living in a field is 3.6. Use a table to find the probability that seven rabbits are found on any given acre of the field.

SOLUTION

A portion of Table 3 in Appendix B is shown here. Using the distribution for $\mu = 3.6$ and $x = 7$, you can find the Poisson probability as shown by the highlighted areas in the table.

						μ	
x	3.1	3.2	3.3	3.4	3.5	3.6	3.7
0	.0450	.0408	.0369	.0334	.0302	.0273	.0247
1	.1397	.1304	.1217	.1135	.1057	.0984	.0915
2	.2165	.2087	.2008	.1929	.1850	.1771	.1692
3	.2237	.2226	.2209	.2186	.2158	.2125	.2087
4	.1734	.1781	.1823	.1858	.1888	.1912	.1931
5	.1075	.1140	.1203	.1264	.1322	.1377	.1429
6	.0555	.0608	.0662	.0716	.0771	.0826	.0881
7	.0246	.0278	.0312	.0348	.0385	(.0425)	.0466
8	.0095	.0111	.0129	.0148	.0169	.0191	.0215
9	.0033	.0040	.0047	.0056	.0066	.0076	.0089
10	.0010	.0013	.0016	.0019	.0023	.0028	.0033

According to the table, the probability is 0.0425. You can check this result using technology. As shown below using Excel, the probability is 0.042484. After rounding to four decimal places, the probability is 0.0425, which is the same value found using the table.

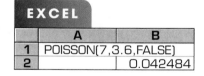

	A	B
1	POISSON(7,3.6,FALSE)	
2		0.042484

Interpretation So, the probability that seven rabbits are found on any given acre is 0.0425. Because 0.0425 is less than 0.05, this can be considered an unusual event.

Picturing the World

The first successful suspension bridge built in the United States, the Tacoma Narrows Bridge, spans the Tacoma Narrows in Washington State. The average occupancy of vehicles that travel across the bridge is 1.6. The probability distribution shown below represents the vehicle occupancy on the bridge during a five-day period. (Adapted from Washington State Department of Transportation)

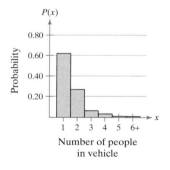

During the five-day period, what is the probability that a randomly selected vehicle has two occupants or fewer?

The probability that a randomly selected vehicle has two occupants or fewer is about 0.85.

Summary of Discrete Probability Distributions

The table summarizes the discrete probability distributions discussed in this chapter.

Distribution	Summary	Formulas
Binomial Distribution	A binomial experiment satisfies these conditions. 1. The experiment has a fixed number n of independent trials. 2. There are only two possible outcomes for each trial. Each outcome can be classified as a success or as a failure. 3. The probability of success p is the same for each trial. 4. The random variable x counts the number of successful trials. The parameters of a binomial distribution are n and p.	n = the number of trials x = the number of successes in n trials p = probability of success in a single trial q = probability of failure in a single trial $q = 1 - p$ The probability of exactly x successes in n trials is $$P(x) = {}_nC_x p^x q^{n-x}$$ $$= \frac{n!}{(n-x)!\,x!} p^x q^{n-x}.$$ $\mu = np$ $\sigma^2 = npq$ $\sigma = \sqrt{npq}$
Geometric Distribution	A geometric distribution is a discrete probability distribution of a random variable x that satisfies these conditions. 1. A trial is repeated until a success occurs. 2. The repeated trials are independent of each other. 3. The probability of success p is the same for each trial. 4. The random variable x represents the number of the trial in which the first success occurs. The parameter of a geometric distribution is p.	x = the number of the trial in which the first success occurs p = probability of success in a single trial q = probability of failure in a single trial $q = 1 - p$ The probability that the first success occurs on trial number x is $$P(x) = pq^{x-1}.$$
Poisson Distribution	The Poisson distribution is a discrete probability distribution of a random variable x that satisfies these conditions. 1. The experiment consists of counting the number of times x an event occurs over a specified interval of time, area, or volume. 2. The probability of the event occurring is the same for each interval. 3. The number of occurrences in one interval is independent of the number of occurrences in other intervals. The parameter of the Poisson distribution is μ.	x = the number of occurrences in the given interval μ = the mean number of occurrences in a given interval unit The probability of exactly x occurrences in an interval is $$P(x) = \frac{\mu^x e^{-\mu}}{x!}.$$

4.3 EXERCISES

1. 0.080
2. 0.45
3. 0.062
4. 0.028
5. 0.175
6. 0.089
7. 0.251
8. 0.042
9. In a binomial distribution, the value of x represents the number of successes in n trials. In a geometric distribution, the value of x represents the first trial that results in a success.
10. In a binomial distribution, the value of x represents the number of successes in n trials. In a Poisson distribution, the value of x represents the number of occurrences in an interval.
11. (a) 0.082   (b) 0.469   (c) 0.531
12. (a) 0.009; unusual
    (b) 0.030; unusual   (c) 0.904
13. (a) 0.036; unusual   (b) 0.053
    (c) 0.017; unusual
14. (a) 0.195   (b) 0.433   (c) 0.567
15. (a) 0.230   (b) 0.871   (c) 0.129
16. (a) 0.75   (b) 0.188   (c) 0.063

Building Basic Skills and Vocabulary

In Exercises 1–4, find the indicated probability using the geometric distribution.

1. Find $P(3)$ when $p = 0.65$.　　**2.** Find $P(1)$ when $p = 0.45$.

3. Find $P(5)$ when $p = 0.09$.　　**4.** Find $P(8)$ when $p = 0.28$.

In Exercises 5–8, find the indicated probability using the Poisson distribution.

5. Find $P(4)$ when $\mu = 5$.　　**6.** Find $P(3)$ when $\mu = 6$.

7. Find $P(2)$ when $\mu = 1.5$.　　**8.** Find $P(5)$ when $\mu = 9.8$.

9. In your own words, describe the difference between the value of x in a binomial distribution and in a geometric distribution.

10. In your own words, describe the difference between the value of x in a binomial distribution and in the Poisson distribution.

Using and Interpreting Concepts

Using a Distribution to Find Probabilities   *In Exercises 11–26, find the indicated probabilities using the geometric distribution, the Poisson distribution, or the binomial distribution. Then determine whether the events are unusual. If convenient, use a table or technology to find the probabilities.*

11. Telephone Sales   The probability that you will make a sale on any given telephone call is 0.19. Find the probability that you (a) make your first sale on the fifth call, (b) make your first sale on the first, second, or third call, and (c) do not make a sale on the first three calls.

12. Defective Parts   An auto parts seller finds that 1 in every 100 parts sold is defective. Find the probability that (a) the first defective part is the tenth part sold, (b) the first defective part is the first, second, or third part sold, and (c) none of the first 10 parts sold are defective.

13. Migrants   The mean number of international migrants gained per minute in the United States in a recent year was about two. Find the probability that the number of international migrants gained in any given minute is (a) exactly five, (b) at least five, and (c) more than five. *(Source: U.S. Census Bureau)*

14. Typographical Errors   A newspaper finds that the mean number of typographical errors per page is four. Find the probability that the number of typographical errors found on any given page is (a) exactly three, (b) at most three, and (c) more than three.

15. Pass Completions   Football player Ben Roethlesberger completes a pass 64.1% of the time. Find the probability that (a) the first pass he completes is the second pass, (b) the first pass he completes is the first or second pass, and (c) he does not complete his first two passes. *(Source: National Football League)*

16. Pilot Test   The probability that a student passes the written test for a private pilot license is 0.75. Find the probability that the student (a) passes on the first attempt, (b) passes on the second attempt, and (c) does not pass on the first or second attempt.

17. (a) 0.002; unusual
    (b) 0.006; unusual   (c) 0.980
18. (a) 0.105   (b) 0.578   (c) 0.316
19. (a) 0.311   (b) 0.493   (c) 0.507
20. (a) 0.066   (b) 0.990   (c) 0.077
21. (a) 0.273   (b) 0.615   (c) 0.868
22. (a) 0.058   (b) 0.766   (c) 0.512
23. (a) 0.322   (b) 0.513   (c) 0.809
24. (a) 0.328   (b) 0.913   (c) 0.901
25. (a) 0.071   (b) 0.827   (c) 0.173
26. (a) 0.140   (b) 0.042; unusual
    (c) 0.407

17. **Glass Manufacturer** A glass manufacturer finds that 1 in every 500 glass items produced is warped. Find the probability that (a) the first warped glass item is the tenth item produced, (b) the first warped glass item is the first, second, or third item produced, and (c) none of the first 10 glass items produced are defective.

18. **Winning a Prize** A cereal maker places a game piece in each of its cereal boxes. The probability of winning a prize in the game is 1 in 4. Find the probability that you (a) win your first prize with your fourth purchase, (b) win your first prize with your first, second, or third purchase, and (c) do not win a prize with your first four purchases.

19. **Hurricanes** The mean number of hurricanes to strike the U.S. mainland per year from 1851 through 2015 was about 1.7. Find the probability that the number of hurricanes striking the U.S. mainland in any given year from 1851 through 2015 is (a) exactly one, (b) at most one, and (c) more than one. *(Source: National Oceanic & Atmospheric Administration)*

20. **Living Donor Transplants** The mean number of organ transplants from living donors performed per day in the United States in 2016 was about 16. Find the probability that the number of organ transplants from living donors performed on any given day is (a) exactly 12, (b) at least eight, and (c) no more than 10. *(Source: United Network for Organ Sharing)*

21. **Fracking** Fifty-one percent of U.S. adults oppose hydraulic fracturing (fracking) as a means of increasing the production of natural gas and oil in the United States. You randomly select eight U.S. adults. Find the probability that the number of U.S. adults who oppose fracking as a means of increasing the production of natural gas and oil in the United States is (a) exactly four, (b) less than five, and (c) at least three. *(Source: Gallup)*

22. **Teen Instagram Use** Sixty-three percent of U.S. teenagers say that they use Instagram daily. You randomly select seven U.S. teenagers. Find the probability that the number of U.S. teenagers who say that they use Instagram daily is (a) exactly two, (b) more than three, and (c) between one and four, inclusive. *(Source: eMarketer)*

23. **Paying for College Education** Sixty-eight percent of parents of children ages 8–14 say they are willing to get a second or part-time job to pay for their children's college eduction. You randomly select five parents. Find the probability that the number of parents who say they are willing to get a second or part-time job to pay for their children's college eduction is (a) exactly three, (b) less than four, and (c) at least three. *(Source: T. Rowe Price Group, Inc.)*

24. **Cheating** Sixty-eight percent of undergraduate students admit to cheating on tests or in written work. You randomly select six undergraduate students. Find the probability that the number of undergraduate students who admit to cheating on tests or in written work is (a) exactly four, (b) more than two, and (c) at most five. *(Source: The Atlantic)*

25. **Precipitation** In Akron, Ohio, the mean number of days in April with 0.01 inch or more of precipitation is 14. Find the probability that the number of days in April with 0.01 inch or more of precipitation in Akron is (a) exactly 17 days, (b) at most 17 days, and (c) more than 17 days. *(Source: National Climatic Data Center)*

26. **Oil Tankers** The mean number of oil tankers at a port city is eight per day. Find the probability that the number of oil tankers on any given day is (a) exactly eight, (b) at most three, and (c) more than eight.

Extending Concepts

27. Comparing Binomial and Poisson Distributions An automobile manufacturer finds that 1 in every 2500 automobiles produced has a specific manufacturing defect. (a) Use a binomial distribution to find the probability of finding 4 cars with the defect in a random sample of 6000 cars. (b) The Poisson distribution can be used to approximate the binomial distribution for large values of n and small values of p. Repeat part (a) using the Poisson distribution and compare the results.

28. Hypergeometric Distribution Binomial experiments require that any sampling be done with replacement because each trial must be independent of the others. The **hypergeometric distribution** also has two outcomes: success and failure. The sampling, however, is done without replacement. For a population of N items having k successes and $N - k$ failures, the probability of selecting a sample of size n that has x successes and $n - x$ failures is given by

$$P(x) = \frac{(_kC_x)(_{N-k}C_{n-x})}{_NC_n}.$$

In a shipment of 15 microchips, 2 are defective and 13 are not defective. A sample of three microchips is chosen at random. Use the above formula to find the probability that (a) all three microchips are not defective, (b) one microchip is defective and two are not defective, and (c) two microchips are defective and one is not defective.

Geometric Distribution: Mean and Variance *In Exercises 29 and 30, use the fact that the mean of a geometric distribution is $\mu = 1/p$ and the variance is $\sigma^2 = q/p^2$.*

29. Daily Lottery A daily number lottery chooses three balls numbered 0 to 9. The probability of winning the lottery is 1/1000. Let x be the number of times you play the lottery before winning the first time. (a) Find the mean, variance, and standard deviation. (b) How many times would you expect to have to play the lottery before winning? (c) The price to play is $1 and winners are paid $500. Would you expect to make or lose money playing this lottery? Explain.

30. Paycheck Errors A company assumes that 0.5% of the paychecks for a year were calculated incorrectly. The company has 200 employees and examines the payroll records from one month. (a) Find the mean, variance, and standard deviation. (b) How many employee payroll records would you expect to examine before finding one with an error?

Poisson Distribution: Variance *In Exercises 31 and 32, use the fact that the variance of the Poisson distribution is $\sigma^2 = \mu$.*

31. Golf In a recent year, the mean number of strokes per hole for golfer Steven Bowditch was about 4.1. (a) Find the variance and standard deviation. Interpret the results. (b) Find the probability that he would play an 18-hole round and have more than 72 strokes. *(Source: PGATour.com)*

32. Bankruptcies The mean number of bankruptcies filed per hour by businesses in the United States in 2016 was about 2.8. (a) Find the variance and the standard deviation. Interpret the results. (b) Find the probability that at most five businesses will file bankruptcy in any given hour. *(Source: Administrative Office of the U.S. Courts)*

Uses

There are countless occurrences of Poisson probability distributions in business, sociology, computer science, and many other fields.

For instance, suppose you work for the fire department in the city of Erie, Pennsylvania. You have to make sure the department has enough personnel and vehicles on hand to respond to fires, medical emergencies, and other situations where they provide aid. The fire department's records show that they respond to an average of 15 incidents per day, but one day the department responds to 19 incidents. Is this an unusual event? If so, they may need to update their guidelines so that they are prepared to respond to more incidents.

Knowing the characteristics of the Poisson distribution will help you answer this type of question. By the time you have completed this course, you will be able make educated decisions about the reasonableness of the fire department's guidelines.

Abuses

A common misuse of the Poisson distribution is to think that the "most likely" outcome is the outcome that will occur most of the time. For instance, suppose you are planning a typical day of responding to emergencies for the fire department. The most likely number of incidents the department will need to respond to is 15. Although this is the most likely outcome, the probability that it will occur is only about 0.102. There is about a 0.183 chance the department will need to respond to 16 or 17 incidents, and about a 0.251 chance of 18 or more incidents. So, it would be a mistake to simply plan for 15 incidents every day, thinking that days with less incidents and days with more incidents will balance out over time.

Citizens' safety and even lives can depend on the fire department, so it is important to be ready for any likely scenario. The lowest number of incidents that is unlikely ($P < 0.05$) is 20, with a probability of about 0.0418, so the fire department should be prepared to respond at least 19 incidents per day.

EXERCISES

In Exercises 1–3, assume the fire department guidelines are correct and that they respond to an average of 15 emergency incidents per day. Use the graph of the Poisson distribution and technology to answer the questions. Explain your reasoning.

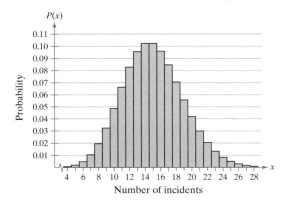

Number of incidents

1. On a random day, what is more likely, 15 emergency incidents or at least 20 incidents?

2. On a random day, what is more likely, 14 to 16 emergency incidents or less than 14 incidents?

3. On the 4th of July, the fire department responds to 21 incidents. Is there reason to believe the guidelines should be adjusted for this holiday?

Chapter Summary

What Did You Learn?	Example(s)	Review Exercises

Section 4.1

▶ How to distinguish between discrete random variables and continuous random variables — Example(s): 1 — Review Exercises: 1, 2

▶ How to construct and graph a discrete probability distribution — Example(s): 2 — Review Exercises: 3, 4

▶ How to determine whether a distribution is a probability distribution — Example(s): 3, 4 — Review Exercises: 5, 6

▶ How to find the mean, variance, and standard deviation of a discrete probability distribution — Example(s): 5, 6 — Review Exercises: 7, 8

$$\mu = \Sigma x P(x) \qquad \text{Mean of a discrete random variable}$$
$$\sigma^2 = \Sigma (x - \mu)^2 P(x) \qquad \text{Variance of a discrete random variable}$$
$$\sigma = \sqrt{\sigma^2} = \sqrt{\Sigma (x - \mu)^2 P(x)} \qquad \text{Standard deviation of a discrete random variable}$$

▶ How to find the expected value of a discrete probability distribution — Example(s): 7 — Review Exercises: 9, 10

$$E(x) = \mu = \Sigma x P(x) \qquad \text{Expected value}$$

Section 4.2

▶ How to determine whether a probability experiment is a binomial experiment — Example(s): 1 — Review Exercises: 11, 12

▶ How to find binomial probabilities using the binomial probability formula, a binomial probability table, and technology — Example(s): 2, 4–6 — Review Exercises: 13–16, 23, 26

$$P(x) = {}_nC_x p^x q^{n-x} = \frac{n!}{(n-x)!x!} p^x q^{n-x} \qquad \text{Binomial probability formula}$$

▶ How to construct and graph a binomial distribution — Example(s): 3, 7 — Review Exercises: 17, 18

▶ How to find the mean, variance, and standard deviation of a binomial probability distribution — Example(s): 8 — Review Exercises: 19, 20

$$\mu = np \qquad \text{Mean of a binomial distribution}$$
$$\sigma^2 = npq \qquad \text{Variance of a binomial distribution}$$
$$\sigma = \sqrt{npq} \qquad \text{Standard deviation of a binomial distribution}$$

Section 4.3

▶ How to find probabilities using the geometric distribution — Example(s): 1 — Review Exercises: 21, 24

$$P(x) = pq^{x-1} \qquad \text{Probability that the first success will occur on trial number } x$$

▶ How to find probabilities using the Poisson distribution — Example(s): 2, 3 — Review Exercises: 22, 25

$$P(x) = \frac{\mu^x e^{-\mu}}{x!} \qquad \text{Probability of exactly } x \text{ occurrences in an interval}$$

Review Exercises

1. Discrete; The number of pumps in use at a gas station is a random variable that is countable.

2. Continuous; The weight of a truck at a weigh station is a random variable that cannot be counted.

3. See Odd Answers, page A58.

4. (a)

x	P(x)
4	0.016
5	0.095
6	0.206
7	0.365
8	0.222
9	0.063
10	0.032

(b)

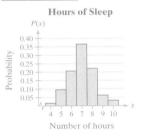

Hours of Sleep

Approximately symmetric

5. Yes

6. No, $P(5) > 1$ and $\Sigma P(x) \neq 1$.

7. (a) $\mu \approx 2.8$; $\sigma^2 \approx 1.7$; $\sigma \approx 1.3$

(b) The mean is 2.8, so the average number of cell phones per household is about 3. The standard deviation is 1.3, so most of the households differ from the mean by no more than about 1 cell phone.

8. (a) $\mu \approx 31.8$; $\sigma^2 \approx 266.2$; $\sigma \approx 16.3$

(b) The mean is 31.8, so the average length of an advertising block is 31.8 seconds. The standard deviation is 16.3, so most of the commercials differ from the mean by no more than about 16 seconds.

Section 4.1

In Exercises 1 and 2, determine whether the random variable x is discrete or continuous. Explain.

1. Let *x* represent the number of pumps in use at a gas station.

2. Let *x* represent the weight of a truck at a weigh station.

In Exercises 3 and 4, (a) construct a probability distribution, and (b) graph the probability distribution using a histogram and describe its shape.

3. The number of hits per game played by a Major League Baseball player

Hits	0	1	2	3	4	5
Games	29	62	33	12	3	1

4. The number of hours students in a college class slept the previous night

Hours	4	5	6	7	8	9	10
Students	1	6	13	23	14	4	2

In Exercises 5 and 6, determine whether the distribution is a probability distribution. If it is not a probability distribution, explain why.

5. The random variable *x* represents the number of tickets a police officer writes out each shift.

x	0	1	2	3	4	5
P(x)	0.09	0.23	0.29	0.16	0.21	0.02

6. The random variable *x* represents the number of classes in which a student is enrolled in a given semester at a university.

x	1	2	3	4	5	6	7	8
P(x)	$\frac{1}{80}$	$\frac{2}{75}$	$\frac{1}{10}$	$\frac{12}{25}$	$\frac{27}{20}$	$\frac{1}{5}$	$\frac{2}{25}$	$\frac{1}{120}$

In Exercises 7 and 8, (a) find the mean, variance, and standard deviation of the probability distribution, and (b) interpret the results.

7. The number of cell phones per household in a small town

Cell phones	0	1	2	3	4	5	6
Probability	0.020	0.140	0.272	0.292	0.168	0.076	0.032

8. A television station sells advertising in 15-, 30-, 60-, 90-, and 120-second blocks. The distribution of sales for one 24-hour day is given.

Length (in seconds)	15	30	60	90	120
Probability	0.134	0.786	0.053	0.006	0.021

Prize	Probability
$100,000	$\dfrac{1}{100,000}$
$100	$\dfrac{1}{100}$
$50	$\dfrac{1}{50}$

TABLE FOR EXERCISE 10

9. −$3.13

10. −$2

11. Binomial experiment
Success: a green candy is selected
$n = 12$, $p = 0.16$, $q = 0.84$,
$x = 0, 1, 2, 3, 4, 5, 6, 7, 8, 9, 10, 11, 12$

12. Not a binomial experiment because the experiment is not repeated for a fixed number of trials.

13. (a) 0.191   (b) 0.891   (c) 0.700

14. (a) 0.072   (b) 0.977   (c) 0.905

15. (a) 0.067   (b) 0.984   (c) 0.917

16. (a) 0.211   (b) 0.927   (c) 0.717

17. (a)

x	$P(x)$
0	0.0008
1	0.0126
2	0.0798
3	0.2529
4	0.4003
5	0.2536

(b)

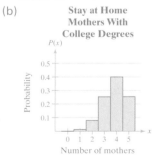

Stay at Home Mothers With College Degrees

Skewed left

(c) The values 0 and 1 are unusual because their probabilities are less than 0.05.

18. See Selected Answers, page A97.

In Exercises 9 and 10, find the expected net gain to the player for one play of the game.

9. It costs $25 to bet on a horse race. The horse has a $\frac{1}{8}$ chance of winning and a $\frac{1}{4}$ chance of placing 2nd or 3rd. You win $125 if the horse wins and receive your money back if the horse places 2nd or 3rd.

10. A scratch-off lottery ticket costs $5. The table at the left shows the probability of winning various prizes on the ticket.

Section 4.2

In Exercises 11 and 12, determine whether the experiment is a binomial experiment. If it is, identify a success, specify the values of n, p, and q, and list the possible values of the random variable x. If it is not a binomial experiment, explain why.

11. Bags of milk chocolate M&M's contain 16% green candies. One candy is selected from each of 12 bags. The random variable represents the number of green candies selected. *(Source: Mars, Inc.)*

12. A fair coin is tossed repeatedly until 15 heads are obtained. The random variable x counts the number of tosses.

In Exercises 13–16, find the indicated binomial probabilities. If convenient, use technology or Table 2 in Appendix B.

13. Fifty-three percent of U.S. adults want to lose weight. You randomly select eight U.S. adults. Find the probability that the number of U.S. adults who want to lose weight is (a) exactly three, (b) at least three, and (c) more than three. *(Source: Gallup)*

14. Thirty-nine percent of U.S. adults have a gun in their home. You randomly select 12 U.S. adults. Find the probability that the number of U.S. adults who have a gun in their home is (a) exactly two, (b) at least two, and (c) more than two. *(Source: Gallup)*

15. Eighty-eight percent of U.S. civilian full-time employees have access to medical care benefits. You randomly select nine civilian full-time employees. Find the probability that the number of civilian full-time employees who have access to medical care benefits is (a) exactly six, (b) at least six, and (c) more than six. *(Source: U.S. Bureau of Labor Statistics)*

16. Sixty-two percent of U.S. adults get news on social media sites. You randomly select five U.S. adults. Find the probability that the number of U.S. adults who get news on social media sites is (a) exactly two, (b) at least two, and (c) more than two. *(Source: Pew Research Center)*

In Exercises 17 and 18, (a) construct a binomial distribution, (b) graph the binomial distribution using a histogram and describe its shape, and (c) identify any values of the random variable x that you would consider unusual. Explain your reasoning.

17. Seventy-six percent of stay-at-home mothers have a college degree or higher. You randomly select five stay-at-home mothers and ask them whether they have a college degree or higher. The random variable represents the number of stay-at-home mothers who have a college degree or higher. *(Source: Hulafrog)*

18. Eighty-eight percent of U.S. adults use the Internet. You randomly select six U.S. adults and ask them whether they use the Internet. The random variable represents the number of U.S. adults who use the Internet. *(Source: Pew Research Center)*

19. $\mu \approx 1.0$; $\sigma^2 \approx 0.9$; $\sigma \approx 1.0$;
On average, 1 out of every
8 drivers is uninsured. The
standard deviation is 1.0, so
most samples of 8 drivers
would differ from the mean by
at most 1 driver.

20. $\mu = 2.8$; $\sigma^2 \approx 1.2$; $\sigma \approx 1.1$;
On average, 2.8 out of every
5 college student-athletes
receive athletics scholarships.
The standard deviation is 1.1,
so most samples of 5 college
student-athletes would differ
from the mean by at most
1.1 college student-athletes.

21. (a) 0.148   (b) 0.006; unusual
(c) 0.820

22. (a) 0.763   (b) 0.997
(c) 0.031; unusual

23. (a) 0.154   (b) 0.217
(c) 0.011; unusual

24. (a) 0.474   (b) 0.249
(c) 0.854   (d) 0.146

25. (a) 0.085   (b) 0.410   (c) 0.430

26. (a) 0.298   (b) 0.912   (c) 0.088

In Exercises 19 and 20, find the mean, variance, and standard deviation of the binomial distribution for the given random variable. Interpret the results.

19. About 13% of U.S. drivers are uninsured. You randomly select eight U.S. drivers and ask them whether they are uninsured. The random variable represents the number of U.S. drivers who are uninsured. *(Source: Insurance Research Council)*

20. Fifty-six percent of college student-athletes receive athletics scholarships. You randomly select five college student-athletes and ask whether they receive athletics scholarships. The random variable represents the number of college student-athletes who receive athletics scholarships. *(National Collegiate Athletic Association)*

Section 4.3

In Exercises 21–26, find the indicated probabilities using the geometric distribution, the Poisson distribution, or the binomial distribution. Then determine whether the events are unusual. If convenient, use a table or technology to find the probabilities.

21. Eighty-two percent of people using electronic cigarettes (vapers) are ex-smokers of conventional cigarettes. You randomly select 10 vapers. Find the probability that the first vaper who is an ex-smoker of conventional cigarettes is (a) the second person selected, (b) the fourth or fifth person selected, and (c) not one of the second through seventh persons selected. *(Source: ChurnMag)*

22. During a 77-year period, tornadoes killed about 0.27 people per day in the United States. Assume this rate holds true today and is constant throughout the year. Find the probability that the number of people in the United States killed by a tornado tomorrow is (a) exactly zero, (b) at most two, and (c) more than one. *(Source: National Weather Service)*

23. Thirty-six percent of Americans think there is still a need for the practice of changing their clocks for Daylight Savings Time. You randomly select seven Americans. Find the probability that the number of Americans who say there is still a need for changing their clocks for Daylight Savings Time is (a) exactly four, (b) less than two, and (c) at least six. *(Source: Rasmussen Reports)*

24. In a recent season, hockey player Evgeni Malkin scored 27 goals in 57 games he played. Assume that his goal production stayed at that level for the next season. Find the probability that he would get his first goal (a) in the first game of the season, (b) in the second game of the season, (c) within the first three games of the season, and (d) not within the first three games of the season. *(Source: National Hockey League)*

25. During a 10-year period, sharks killed an average of 6.1 people each year worldwide. Find the probability that the number of people killed by sharks next year is (a) exactly three, (b) more than six, and (c) at most five. *(Source: International Shark Attack File)*

26. Eighty-two percent of U.S. adults think that healthy children should be required to be vaccinated to attend school. You randomly select 10 U.S. adults. Find the probability that the number of U.S. adults who think that healthy children should be required to be vaccinated to attend school is (a) exactly eight, (b) more than six, and (c) at most six. *(Source: Pew Research Center)*

Chapter Quiz

1. (a) Discrete; The number of lightning strikes that occur in Wyoming during the month of June is a random variable that is countable.

   (b) Continuous; The fuel (in gallons) used by a jet during takeoff is a random variable that has an infinite number of possible outcomes and cannot be counted.

   (c) Discrete; The number of die rolls required for an individual to roll a five is a random variable that is countable.

2. (a) and (b) See Odd Answers, page A59.

   (c) $\mu \approx 1.3$; $\sigma^2 \approx 1.4$; $\sigma \approx 1.2$; The mean is 1.3, so the average number of wireless devices per household is 1.3. The standard deviation is 1.2, so most households will differ from the mean by no more than 1.2 wireless devices.

   (d) 0.058

3. (a) 0.269   (b) 0.811   (c) 0.061

4. (a) and (b) See Odd Answers, page A59.

   (c) $\mu \approx 5.2$; $\sigma^2 \approx 0.7$; $\sigma \approx 0.8$; On average, 5.2 out of every 6 patients have a successful surgery. The standard deviation is 0.8, so most samples of 6 surgeries would differ from the mean by at most 0.8 surgery.

5. (a) 0.175   (b) 0.440   (c) 0.007

6. (a) 0.048   (b) 0.355   (c) 0.085

7. Event (a) is unusual because its probability is less than 0.05.

Take this quiz as you would take a quiz in class. After you are done, check your work against the answers given in the back of the book.

1. Determine whether the random variable x is discrete or continuous. Explain your reasoning.

   (a) Let x represent the number of lightning strikes that occur in Wyoming during the month of June.

   (b) Let x represent the amount of fuel (in gallons) used by a jet during takeoff.

   (c) Let x represent the total number of die rolls required for an individual to roll a five.

2. The table lists the number of wireless devices per household in a small town in the United States.

Wireless devices	0	1	2	3	4	5
Number of households	277	471	243	105	46	22

   (a) Construct a probability distribution.

   (b) Graph the probability distribution using a histogram and describe its shape.

   (c) Find the mean, variance, and standard deviation of the probability distribution and interpret the results.

   (d) Find the probability of randomly selecting a household that has at least four wireless devices.

3. Thirty-six percent of U.S. adults have postponed medical checkups or procedures to save money. You randomly select nine U.S. adults. Find the probability that the number of U.S. adults who have postponed medical checkups or procedures to save money is (a) exactly three, (b) at most four, and (c) more than five. *(Source: Rasmussen Reports)*

4. The five-year success rate of kidney transplant surgery from living donors is 86%. The surgery is performed on six patients. *(Source: Mayo Clinic)*

   (a) Construct a binomial distribution.

   (b) Graph the binomial distribution using a histogram and describe its shape.

   (c) Find the mean, variance, and standard deviation of the binomial distribution and interpret the results.

5. An online magazine finds that the mean number of typographical errors per page is five. Find the probability that the number of typographical errors found on any given page is (a) exactly five, (b) less than five, and (c) exactly zero.

6. Basketball player Dwight Howard makes a free throw shot about 56% of the time. Find the probability that (a) the first free throw shot he makes is the fourth shot, (b) the first free throw shot he makes is the second or third shot, and (c) he does not make his first three shots. *(Source: ESPN)*

7. Which event(s) in Exercise 6 can be considered unusual? Explain your reasoning.

4 Chapter Test

Take this test as you would take a test in class.

In Exercises 1–3, find the indicated probabilities using the geometric distribution, the Poisson distribution, or the binomial distribution. Then determine whether the events are unusual. If convenient, use a table or technology to find the probabilities.

1. One out of every 119 tax returns that a tax auditor examines requires an audit. Find the probability that (a) the first return requiring an audit is the 25th return the tax auditor examines, (b) the first return requiring an audit is the first or second return the tax auditor examines, and (c) none of the first five returns the tax auditor examines require an audit. *(Source: Kiplinger)*

2. About 60% of U.S. full-time college students drank alcohol within a one-month period. You randomly select six U.S. full-time college students. Find the probability that the number of U.S. full-time college students who drank alcohol within a one-month period is (a) exactly two, (b) at least three, and (c) less than four. *(Source: National Center for Biotechnology Information)*

3. The mean increase in the U.S. population is about four people per minute. Find the probability that the increase in the U.S. population in any given minute is (a) exactly six people, (b) more than eight people, and (c) at most four people. *(Source: U.S. Census Bureau)*

4. Determine whether the distribution is a probability distribution. If it is not a probability distribution, explain why.

(a)

x	0	5	10	15	20
$P(x)$	0.03	0.09	0.19	0.32	0.37

(b)

x	1	2	3	4	5	6
$P(x)$	$\frac{1}{20}$	$\frac{1}{10}$	$\frac{2}{5}$	$\frac{3}{10}$	$\frac{1}{5}$	$\frac{1}{25}$

5. The table shows the ages of students in a freshman orientation course.

Age	17	18	19	20	21	22
Students	2	13	4	3	2	1

(a) Construct a probability distribution.

(b) Graph the probability distribution using a histogram and describe its shape.

(c) Find the mean, variance, and standard deviation of the probability distribution and interpret the results.

(d) Find the probability that a randomly selected student is less than 20 years old.

6. Seventy-seven percent of U.S. college students pay their bills on time. You randomly select five U.S. college students and ask them whether they pay their bills on time. The random variable represents the number of U.S. college students who pay their bills on time. *(Source: Sallie Mae)*

(a) Construct a probability distribution.

(b) Graph the probability distribution using a histogram and describe its shape.

(c) Find the mean, variance, and standard deviation of the probability distribution and interpret the results.

---

1. (a) 0.007; unusual
   (b) 0.017; unusual   (c) 0.959
2. (a) 0.138   (b) 0.821   (c) 0.456
3. (a) 0.104   (b) 0.021; unusual
   (c) 0.629
4. (a) Yes   (b) No, $\Sigma P(x) \neq 1$.
5. (a)

x	$P(x)$
17	0.08
18	0.52
19	0.16
20	0.12
21	0.08
22	0.04

(b)

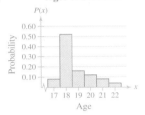

Ages of Students

Skewed right

(c) $\mu \approx 18.7$; $\sigma^2 \approx 1.6$; $\sigma \approx 1.2$; The mean is 18.7, so the average student in the course is about 19 years old. The standard deviation is 1.2, so most of the ages differ from the mean by no more than about 1 year.

(d) 0.76

6. (a) and (b) See Selected Answers, page A97.

(c) $\mu \approx 3.9$; $\sigma^2 \approx 0.9$; $\sigma \approx 0.9$; On average, 3.9 out of every 5 U.S. college students pay their bills on time. The standard deviation is 0.9, so most samples of 5 U.S. college students would differ from the mean by at most 0.9 U.S. college student.

The Centers for Disease Control and Prevention (CDC) is required by law to publish a report on assisted reproductive technology (ART). ART includes all fertility treatments in which both the egg and the sperm are used. These procedures generally involve removing eggs from a woman's ovaries, combining them with sperm in the laboratory, and returning them to the woman's body or giving them to another woman.

You are helping to prepare the CDC report and select at random 10 ART cycles for a special review. None of the cycles resulted in a clinical pregnancy. Your manager feels it is impossible to select at random 10 ART cycles that do not result in a clinical pregnancy. Use the pie chart at the right and your knowledge of statistics to determine whether your manager is correct.

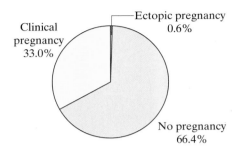

Results of ART Cycles Using Fresh Nondonor Eggs or Embryos

(Source: Centers for Disease Control and Prevention)

EXERCISES

1. *How Would You Do It?*

(a) How would you determine whether your manager is correct, that it is impossible to select at random 10 ART cycles that do not result in a clinical pregnancy?

(b) What probability distribution do you think best describes the situation? Do you think the distribution of the number of clinical pregnancies is discrete or continuous? Explain your reasoning.

2. *Answering the Question*

Write an explanation that answers the question, "Is it possible to select at random 10 ART cycles that do not result in a clinical pregnancy?" Include in your explanation the appropriate probability distribution and your calculation of the probability of no clinical pregnancies in 10 ART cycles.

3. *Suspicious Samples?*

A lab worker tells you that the samples below were selected at random. Using the graph at the right, which of the samples would you consider suspicious? Would you believe that the samples were selected at random? Explain your reasoning.

(a) A sample of 10 ART cycles among women of age 40, eight of which resulted in clinical pregnancies

(b) A sample of 10 ART cycles among women of age 41, none of which resulted in clinical pregnancies

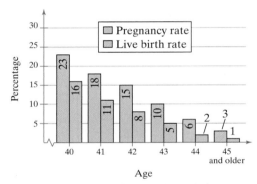

Pregnancy and Live Birth Rates for ART Cycles Among Women of Age 40 and Older

(Source: Centers for Disease Control and Prevention)

Using Poisson Distributions as Queuing Models

Queuing means waiting in line to be served. There are many examples of queuing in everyday life: waiting at a traffic light, waiting in line at a grocery checkout counter, waiting for an elevator, holding for a telephone call, and so on.

Poisson distributions are used to model and predict the number of people (calls, computer programs, vehicles) arriving at the line. In the exercises below, you are asked to use Poisson distributions to analyze the queues at a grocery store checkout counter.

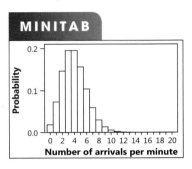

MINITAB

EXERCISES

In Exercises 1–7, consider a grocery store that can process a total of four customers at its checkout counters each minute.

1. The mean number of customers who arrive at the checkout counters each minute is 4. Create a Poisson distribution with $\mu = 4$ for $x = 0$ to 20. Compare your results with the histogram shown at the upper right.

2. Minitab was used to generate 20 random numbers with a Poisson distribution for $\mu = 4$. Let the random number represent the number of arrivals at the checkout counter each minute for 20 minutes.

   3 3 3 3 5 5 6 7 3 6
   3 5 6 3 4 6 2 2 4 1

   During each of the first four minutes, only three customers arrived. These customers could all be processed, so there were no customers waiting after four minutes.

   (a) How many customers were waiting after 5 minutes? 6 minutes? 7 minutes? 8 minutes?

   (b) Create a table that shows the number of customers waiting at the end of 1 through 20 minutes.

3. Generate a list of 20 random numbers with a Poisson distribution for $\mu = 4$. Create a table that shows the number of customers waiting at the end of 1 through 20 minutes.

4. The mean increases to 5 arrivals per minute, but the store can still process only four per minute. Generate a list of 20 random numbers with a Poisson distribution for $\mu = 5$. Then create a table that shows the number of customers waiting at the end of 20 minutes.

5. The mean number of arrivals per minute is 5. What is the probability that 10 customers will arrive during the first minute?

6. The mean number of arrivals per minute is 4. Find the probability that

   (a) three, four, or five customers will arrive during the third minute.

   (b) more than four customers will arrive during the first minute.

   (c) more than four customers will arrive during each of the first four minutes.

7. The mean number of arrivals per minute is 4. Find the probability that

   (a) no customers are waiting in line after one minute.

   (b) one customer is waiting in line after one minute.

   (c) one customer is waiting in line after one minute and no customers are waiting in line after the second minute.

   (d) no customers are waiting in line after two minutes.

Extended solutions are given in the technology manuals that accompany this text. Technical instruction is provided for Minitab, Excel, and the TI-84 Plus.

CHAPTER 5

Normal Probability Distributions

The average dairy cow in the United States produced 22,770 pounds of milk in 2016, more than twice the average from 50 years ago.

In Chapters 1 through 4, you learned how to collect and describe data, find the probability of an event, and analyze discrete probability distributions. You also learned that when a sample is used to make inferences about a population, it is critical that the sample not be biased. For instance, how would you organize a study to determine which breed of dairy cow is the most profitable?

When the U.S. Department of Agriculture performs this study, it uses random sampling and then records the measures of various milk production and physical traits such as pounds produced, fat percentage, protein percentage, productive life, somatic cell count, and calving ability. The studies have repeatedly shown Holstein cows to be the most profitable breed of dairy cow. Other top breeds are Jersey, Brown Swiss, and Ayrshire cows.

In Chapter 5, you will learn how to recognize normal (bell-shaped) distributions and how to use their properties in real-life applications. Suppose that you are a farmer planning to buy 20 Holstein cows and 10 Jersey cows from a breeder. You want to know the probabilities that the groups of cows will produce certain average daily amounts of milk. You will learn how to calculate this type of probability using a *sampling distribution of sample means* and the *Central Limit Theorem* in Section 5.4. The graphs below show the distributions of sample means of milk produced daily by the two breeds of cows.

The table shows the information given to you by the breeder. Assume that the amounts of milk produced are normally distributed.

Amount of milk produced per day (in pounds)		
Breed	**Mean**	**Standard deviation**
Holstein	69.3	11.7
Jersey	49.7	10.1

You can use this information to make calculations about average amounts of milk produced daily by the cows. For instance, the probability that the 20 Holstein cows will produce an average of at least 65 pounds of milk per day is about 94.95%, and the probability that the 10 Jersey cows will produce an average of between 50 and 60 pounds of milk per day is about 46.35%.

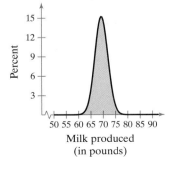

Average Daily Milk Production by Holstein Cows

Percent / Milk produced (in pounds)

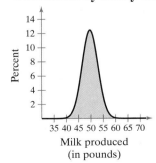

Average Daily Milk Production by Jersey Cows

Percent / Milk produced (in pounds)

5.1 Introduction to Normal Distributions and the Standard Normal Distribution

What You Should Learn

▸ How to interpret graphs of normal probability distributions

▸ How to find areas under the standard normal curve

Note to Instructor

Draw several different continuous probability curves. Then point out that the normal (or Gaussian) curve is graphed using the formula shown at the bottom of the page. Have students discuss measures in nature that are normally distributed. Mention that often grades in a statistics class are not normally distributed.

Study Tip

A normal curve with mean μ and standard deviation σ can be graphed using the normal probability density function

$$y = \frac{1}{\sigma\sqrt{2\pi}}\, e^{-(x-\mu)^2/(2\sigma^2)}.$$

(This formula will not be used in the text.) Because $e \approx 2.718$ and $\pi \approx 3.14$, a normal curve depends completely on μ and σ.

■ Properties of a Normal Distribution   ■ The Standard Normal Distribution

Properties of a Normal Distribution

In Section 4.1, you distinguished between discrete and continuous random variables, and learned that a continuous random variable has an infinite number of possible values that can be represented by an interval on a number line. Its probability distribution is called a **continuous probability distribution.** In this chapter, you will study the most important continuous probability distribution in statistics—the **normal distribution.** Normal distributions can be used to model many sets of measurements in nature, industry, and business. For instance, the systolic blood pressures of humans, the lifetimes of smartphones, and housing costs are all normally distributed random variables.

DEFINITION

A **normal distribution** is a continuous probability distribution for a random variable x. The graph of a normal distribution is called the **normal curve.** A normal distribution has these properties.

1. The mean, median, and mode are equal.
2. The normal curve is bell-shaped and is symmetric about the mean.
3. The total area under the normal curve is equal to 1.
4. The normal curve approaches, but never touches, the x-axis as it extends farther and farther away from the mean.
5. Between $\mu - \sigma$ and $\mu + \sigma$ (in the center of the curve), the graph curves downward. The graph curves upward to the left of $\mu - \sigma$ and to the right of $\mu + \sigma$. The points at which the curve changes from curving upward to curving downward are called **inflection points.**

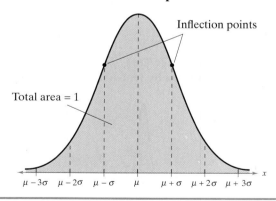

You have learned that a discrete probability distribution can be graphed with a histogram. For a continuous probability distribution, you can use a **probability density function (pdf).** A probability density function has two requirements: (1) the total area under the curve is equal to 1, and (2) the function can never be negative.

A normal distribution can have any mean and any positive standard deviation. These two parameters, μ and σ, determine the shape of the normal curve. The mean gives the location of the line of symmetry, and the standard deviation describes how much the data are spread out.

For instance, in the figures below, curves A and B have the same mean, and curves B and C have the same standard deviation. The total area under each curve is 1. Also, in each graph, one of the inflection points occurs one standard deviation to the left of the mean, and the other occurs one standard deviation to the right of the mean.

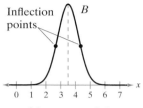

Mean: $\mu = 3.5$
Standard deviation:
$\sigma = 1.5$

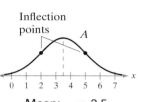

Mean: $\mu = 3.5$
Standard deviation:
$\sigma = 0.7$

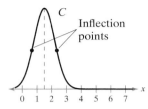

Mean: $\mu = 1.5$
Standard deviation:
$\sigma = 0.7$

Picturing the World

According to the National Center for Health Statistics, the number of births in the United States in a recent year was 3,978,497. The weights of the newborns can be approximated by a normal distribution, as shown in the figure. (Adapted from National Center for Health Statistics)

Weights of Newborns

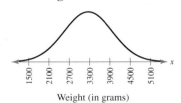

Weight (in grams)

What is the mean weight of the newborns? Estimate the standard deviation of this normal distribution.

The mean is about 3300 grams and the standard deviation is about 600 grams.

EXAMPLE 1

Understanding Mean and Standard Deviation

1. Which normal curve has a greater mean?

2. Which normal curve has a greater standard deviation?

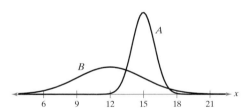

SOLUTION

1. The line of symmetry of curve A occurs at $x = 15$. The line of symmetry of curve B occurs at $x = 12$. So, curve A has a greater mean.

2. Curve B is more spread out than curve A. So, curve B has a greater standard deviation.

TRY IT YOURSELF 1

1. Which normal curve has the greatest mean?

2. Which normal curve has the greatest standard deviation?

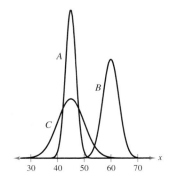

Answer: Page A35

EXAMPLE 2

Interpreting Graphs of Normal Distributions

The scaled test scores for the New York State Grade 4 Common Core Mathematics Test are normally distributed. The normal curve shown below represents this distribution. What is the mean test score? Estimate the standard deviation of this normal distribution. *(Adapted from New York State Education Department)*

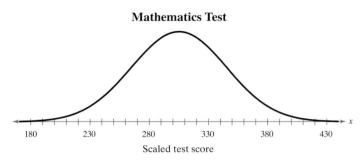

Mathematics Test

Scaled test score

SOLUTION

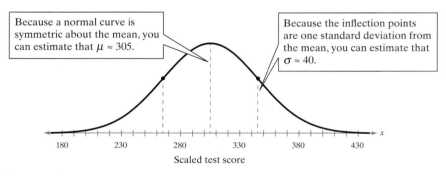

Because a normal curve is symmetric about the mean, you can estimate that $\mu \approx 305$.

Because the inflection points are one standard deviation from the mean, you can estimate that $\sigma \approx 40$.

Scaled test score

The scaled test scores for the New York State Grade 4 Common Core Mathematics Test are normally distributed with a mean of about 305 and a standard deviation of about 40.

Interpretation Using the Empirical Rule (see Section 2.4), you know that about 68% of the scores are between 265 and 345, about 95% of the scores are between 225 and 385, and about 99.7% of the scores are between 185 and 425.

TRY IT YOURSELF 2

The scaled test scores for the New York State Grade 4 Common Core English Language Arts Test are normally distributed. The normal curve shown below represents this distribution. What is the mean test score? Estimate the standard deviation of this normal distribution. *(Adapted from New York State Education Department)*

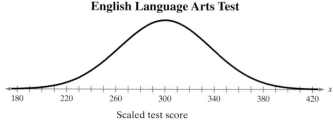

English Language Arts Test

Scaled test score

Answer: Page A35

Tech Tip

You can use technology to graph a normal curve. For instance, you can use a TI-84 Plus to graph the normal curve in Example 2.

```
Plot1 Plot2 Plot3
\Y1⌐normalpdf(X,
305,40)
\Y2=
\Y3=
\Y4=
\Y5=
\Y6=
```

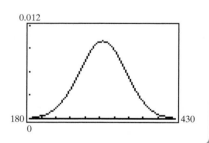

The Standard Normal Distribution

Study Tip

Because every normal distribution can be transformed to the standard normal distribution, you can use *z*-scores and the standard normal curve to find areas (and therefore probabilities) under any normal curve.

There are infinitely many normal distributions, each with its own mean and standard deviation. The normal distribution with a mean of 0 and a standard deviation of 1 is called the **standard normal distribution.** The horizontal scale of the graph of the standard normal distribution corresponds to *z*-scores. In Section 2.5, you learned that a *z*-score is a measure of position that indicates the number of standard deviations a value lies from the mean. Recall that you can transform an *x*-value to a *z*-score using the formula

$$z = \frac{\text{Value} - \text{Mean}}{\text{Standard deviation}}$$

$$= \frac{x - \mu}{\sigma}.$$ Round to the nearest hundredth.

Note to Instructor

Mention that the formula for a normal probability density function on page 234 is greatly simplified when $\mu = 0$ and $\sigma = 1$.

$$y = \frac{e^{-x^2/2}}{\sqrt{2\pi}}$$

DEFINITION

The **standard normal distribution** is a normal distribution with a mean of 0 and a standard deviation of 1. The total area under its normal curve is 1.

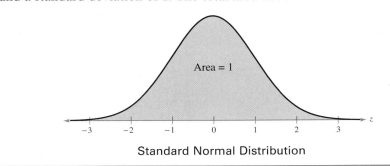

Standard Normal Distribution

Study Tip

It is important that you know the difference between *x* and *z*. The random variable *x* is sometimes called a raw score and represents values in a nonstandard normal distribution, whereas *z* represents values in the standard normal distribution.

When each data value of a normally distributed random variable *x* is transformed into a *z*-score, the result will be the standard normal distribution. After this transformation takes place, the area that falls in the interval under the nonstandard normal curve is the *same* as that under the standard normal curve within the corresponding *z*-boundaries.

In Section 2.4, you learned to use the Empirical Rule to approximate areas under a normal curve when the values of the random variable *x* corresponded to $-3, -2, -1, 0, 1, 2,$ or 3 standard deviations from the mean. Now, you will learn to calculate areas corresponding to other *x*-values. After you use the formula above to transform an *x*-value to a *z*-score, you can use the Standard Normal Table (Table 4 in Appendix B). The table lists the cumulative area under the standard normal curve to the left of *z* for *z*-scores from -3.49 to 3.49. As you examine the table, notice the following.

Properties of the Standard Normal Distribution

1. The cumulative area is close to 0 for *z*-scores close to $z = -3.49$.
2. The cumulative area increases as the *z*-scores increase.
3. The cumulative area for $z = 0$ is 0.5000.
4. The cumulative area is close to 1 for *z*-scores close to $z = 3.49$.

In addition to using the table, you can use technology to find the cumulative area that corresponds to a *z*-score. For instance, the next example shows how to use the Standard Normal Table and a TI-84 Plus to find the cumulative area that corresponds to a *z*-score.

Note to Instructor

If you prefer that your students use a 0-to-*z* table, refer them to Appendix A, where an alternative presentation of this material is given.

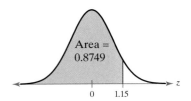

Area = 0.8749

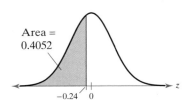

```
normalcdf(-10000
,1.15,0,1)
     .8749280114
```

EXAMPLE 3

Using the Standard Normal Table

1. Find the cumulative area that corresponds to a *z*-score of 1.15.

2. Find the cumulative area that corresponds to a *z*-score of −0.24.

SOLUTION

1. Find the area that corresponds to $z = 1.15$ by finding 1.1 in the left column and then moving across the row to the column under 0.05. The number in that row and column is 0.8749. So, the area to the left of $z = 1.15$ is 0.8749, as shown in the figure at the left.

z	.00	.01	.02	.03	.04	.05	.06
0.0	.5000	.5040	.5080	.5120	.5160	.5199	.5239
0.1	.5398	.5438	.5478	.5517	.5557	.5596	.5636
0.2	.5793	.5832	.5871	.5910	.5948	.5987	.6026
0.9	.8159	.8186	.8212	.8238	.8264	.8289	.8315
1.0	.8413	.8438	.8461	.8485	.8508	.8531	.8554
1.1	.8643	.8665	.8686	.8708	.8729	.8749	.8770
1.2	.8849	.8869	.8888	.8907	.8925	.8944	.8962
1.3	.9032	.9049	.9066	.9082	.9099	.9115	.9131
1.4	.9192	.9207	.9222	.9236	.9251	.9265	.9279

You can use technology to find the cumulative area that corresponds to $z = 1.15$, as shown at the left. Note that to specify the lower bound, use −10,000.

2. Find the area that corresponds to $z = -0.24$ by finding −0.2 in the left column and then moving across the row to the column under 0.04. The number in that row and column is 0.4052. So, the area to the left of $z = -0.24$ is 0.4052, as shown in the figure at the left.

Area = 0.4052

```
normalcdf(-10000
,-.24,0,1)
      .405165175
```

z	.09	.08	.07	.06	.05	.04	.03
−3.4	.0002	.0003	.0003	.0003	.0003	.0003	.0003
−3.3	.0003	.0004	.0004	.0004	.0004	.0004	.0004
−3.2	.0005	.0005	.0005	.0006	.0006	.0006	.0006
−0.5	.2776	.2810	.2843	.2877	.2912	.2946	.2981
−0.4	.3121	.3156	.3192	.3228	.3264	.3300	.3336
−0.3	.3483	.3520	.3557	.3594	.3632	.3669	.3707
−0.2	.3859	.3897	.3936	.3974	.4013	.4052	.4090
−0.1	.4247	.4286	.4325	.4364	.4404	.4443	.4483
−0.0	.4641	.4681	.4721	.4761	.4801	.4840	.4880

You can use technology to find the cumulative area that corresponds to $z = -0.24$, as shown at the left. Note that to specify the lower bound, use −10,000.

TRY IT YOURSELF 3

1. Find the cumulative area that corresponds to a *z*-score of −2.19.

2. Find the cumulative area that corresponds to a *z*-score of 2.17.

Answer: Page A35

When the *z*-score is not in the table, use the entry closest to it. For a *z*-score that is exactly midway between two *z*-scores, use the area midway between the corresponding areas.

You can use the following guidelines to find various types of areas under the standard normal curve.

Note to Instructor

Students find these three options easy to work with. If you have previously used a 0-to-z table, you will appreciate that students never need be confused about whether to add 0.5, subtract it from 0.5, or use the table entry to find a required probability.

GUIDELINES

Finding Areas Under the Standard Normal Curve

1. Sketch the standard normal curve and shade the appropriate area under the curve.

2. Find the area by following the directions for each case shown.

   a. To find the area to the *left* of z, find the area that corresponds to z in the Standard Normal Table.

   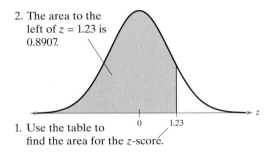

   2. The area to the left of z = 1.23 is 0.8907.

   1. Use the table to find the area for the z-score.

   b. To find the area to the *right* of z, use the Standard Normal Table to find the area that corresponds to z. Then subtract the area from 1.

   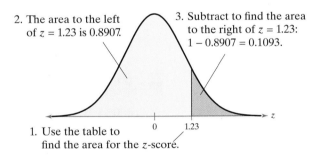

   2. The area to the left of z = 1.23 is 0.8907.

   3. Subtract to find the area to the right of z = 1.23: 1 − 0.8907 = 0.1093.

   1. Use the table to find the area for the z-score.

   c. To find the area *between* two z-scores, find the area corresponding to each z-score in the Standard Normal Table. Then subtract the smaller area from the larger area.

   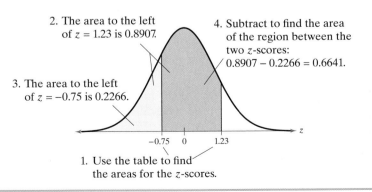

   2. The area to the left of z = 1.23 is 0.8907.

   4. Subtract to find the area of the region between the two z-scores: 0.8907 − 0.2266 = 0.6641.

   3. The area to the left of z = −0.75 is 0.2266.

   1. Use the table to find the areas for the z-scores.

Tech Tip

You can use technology to find the area under the standard normal curve. For instance, you can use the *ShadeNorm* feature on a TI-84 Plus to graph the area under the standard normal curve between z = −0.75 and z = 1.23, as shown below. The area between the two z-scores is shown below the graph. (Note that when you use technology, your answers may differ slightly from those found using the Standard Normal Table.)

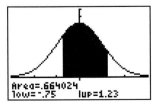

EXAMPLE 4

Finding Area Under the Standard Normal Curve

Find the area under the standard normal curve to the left of $z = -0.99$.

SOLUTION

The area under the standard normal curve to the left of $z = -0.99$ is shown.

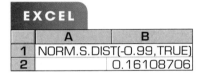

From the Standard Normal Table, this area is equal to

0.1611.   Area to the left of $z = -0.99$

You can use technology to find the area to the left of $z = -0.99$, as shown below.

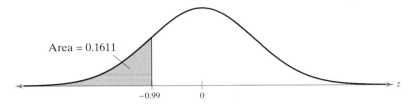

EXCEL

	A	B
1	NORM.S.DIST(-0.99,TRUE)	
2		0.16108706

TRY IT YOURSELF 4

Find the area under the standard normal curve to the left of $z = 2.13$.

Answer: Page A35

EXAMPLE 5

Finding Area Under the Standard Normal Curve

Find the area under the standard normal curve to the right of $z = 1.06$.

SOLUTION

The area under the standard normal curve to the right of $z = 1.06$ is shown.

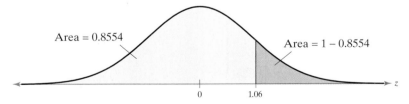

From the Standard Normal Table, the area to the left of $z = 1.06$ is 0.8554. Because the total area under the curve is 1, the area to the right of $z = 1.06$ is

Area $= 1 - 0.8554 = 0.1446$.

You can use technology to find the area to the right of $z = 1.06$, as shown at the left.

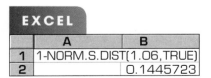

EXCEL

	A	B
1	1-NORM.S.DIST(1.06,TRUE)	
2		0.1445723

TRY IT YOURSELF 5

Find the area under the standard normal curve to the right of $z = -2.16$.

Answer: Page A35

EXAMPLE 6

Finding Area Under the Standard Normal Curve

Find the area under the standard normal curve between $z = -1.5$ and $z = 1.25$.

SOLUTION

The area under the standard normal curve between $z = -1.5$ and $z = 1.25$ is shown.

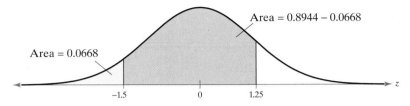

From the Standard Normal Table, the area to the left of $z = 1.25$ is 0.8944 and the area to the left of $z = -1.5$ is 0.0668. So, the area between $z = -1.5$ and $z = 1.25$ is

$$\text{Area} = 0.8944 - 0.0668 = 0.8276.$$

Note that when you use technology, your answers may differ slightly from those found using the Standard Normal Table. For instance, when finding the area on a TI-84 Plus, you get the result shown at the right.

Interpretation So, 82.76% of the area under the curve falls between $z = -1.5$ and $z = 1.25$.

TI-84 PLUS

```
normalcdf(-1.5,1
.25,0,1)
         .8275429323
```

TRY IT YOURSELF 6

Find the area under the standard normal curve between $z = -2.165$ and $z = -1.35$.

Answer: Page A35

Because the normal distribution is a continuous probability distribution, the area under the standard normal curve to the left of a z-score gives the probability that z is less than that z-score. For instance, in Example 4, the area to the left of $z = -0.99$ is 0.1611. So,

$$P(z < -0.99) = 0.1611$$

which is read as "the probability that z is less than -0.99 is 0.1611." The table shows the probabilities for Examples 5 and 6. (You will learn more about finding probabilities in the next section.)

	Area	**Probability**
Example 5	To the right of $z = 1.06$: 0.1446	$P(z > 1.06) = 0.1446$
Example 6	Between $z = -1.5$ and $z = 1.25$: 0.8276	$P(-1.5 < z < 1.25) = 0.8276$

Recall from Section 2.4 that values lying more than two standard deviations from the mean are considered unusual. Values lying more than three standard deviations from the mean are considered *very* unusual. So, a z-score greater than 2 or less than -2 is unusual. A z-score greater than 3 or less than -3 is *very* unusual.

5.1 EXERCISES

For Extra Help: **MyLab Statistics**

Answers column (left):

1. Answers will vary.

2. Neither. In a normal distribution, the mean and median are equal.

3. 1

4. Points at which the curve changes from curving upward to curving downward; $\mu - \sigma$ and $\mu + \sigma$

5. Answers will vary.

   Similarities: The two curves will have the same line of symmetry.

   Differences: The curve with the larger standard deviation will be more spread out than the curve with the smaller standard deviation.

6. Answers will vary.

   Similarities: The two curves will have the same shape because they have equal standard deviations.

   Differences: The two curves will have different lines of symmetry.

7. $\mu = 0$, $\sigma = 1$

8. Transform each data value x into a z-score by subtracting the mean from x and dividing the result by the standard deviation. In symbols, $z = \dfrac{x - \mu}{\sigma}$.

9. "The" standard normal distribution is used to describe one specific normal distribution ($\mu = 0$, $\sigma = 1$). "A" normal distribution is used to describe a normal distribution with any mean and standard deviation.

10. (c) is true because a z-score equal to 0 indicates that the corresponding x-value is equal to the mean.

11. No, the graph is skewed left.

12. Yes, the graph fulfills the properties of the normal distribution.

    $\mu \approx 18.5$, $\sigma \approx 2$

13. No, the graph crosses the x-axis.

14. No, the graph is not symmetric.

15. Yes, the graph fulfills the properties of the normal distribution.

    $\mu \approx 11.5$, $\sigma \approx 1.5$

16. No, the graph is skewed right.

Building Basic Skills and Vocabulary

1. Find three real-life examples of a continuous variable. Which do you think may be normally distributed? Why?

2. In a normal distribution, which is greater, the mean or the median? Explain.

3. What is the total area under the normal curve?

4. What do the inflection points on a normal distribution represent? Where do they occur?

5. Draw two normal curves that have the same mean but different standard deviations. Describe the similarities and differences.

6. Draw two normal curves that have different means but the same standard deviation. Describe the similarities and differences.

7. What is the mean of the standard normal distribution? What is the standard deviation of the standard normal distribution?

8. Describe how you can transform a nonstandard normal distribution to the standard normal distribution.

9. **Getting at the Concept** Why is it correct to say "a" normal distribution and "the" standard normal distribution?

10. **Getting at the Concept** A z-score is 0. Which of these statements must be true? Explain your reasoning.
    (a) The mean is 0.
    (b) The corresponding x-value is 0.
    (c) The corresponding x-value is equal to the mean.

Graphical Analysis *In Exercises 11–16, determine whether the graph could represent a variable with a normal distribution. Explain your reasoning. If the graph appears to represent a normal distribution, estimate the mean and standard deviation.*

11.
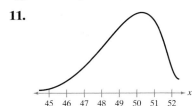

12.
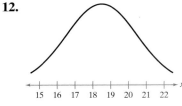

13.

14.
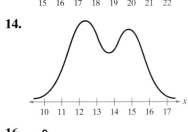

15.
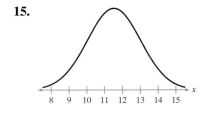

16.

17. 0.9032
18. 0.0668
19. 0.0228
20. 0.9893
21. 0.6429
22. 0.4878
23. 0.5675
24. 0.0008
25. 0.0050
26. 0.9139
27. 0.7422
28. 0.0006
29. 0.6387
30. 0.0688
31. 0.4979
32. 0.4370
33. 0.8788
34. 0.9802
35. 0.2006 (*Tech:* 0.2005)
36. 0.0885
37. (a)

Life Spans of Tires

It is reasonable to assume that the life spans are normally distributed because the histogram is symmetric and bell-shaped.

(b) 37,234.7, 6259.2

(c) The sample mean of 37,234.7 hours is less than the claimed mean, so, on average, the tires in the sample lasted for a shorter time. The sample standard deviation of 6259.2 is greater than the claimed standard deviation, so the tires in the sample had a greater variation in life span than the manufacturer's claim.

Using and Interpreting Concepts

Finding Area *In Exercises 17–22, find the area of the shaded region under the standard normal curve. If convenient, use technology to find the area.*

17.
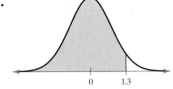

18.
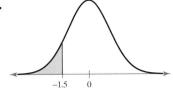

19.
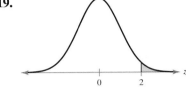

20.
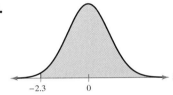

21.
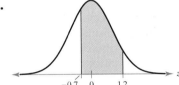

22.
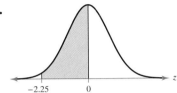

Finding Area *In Exercises 23–36, find the indicated area under the standard normal curve. If convenient, use technology to find the area.*

23. To the left of $z = 0.17$

24. To the left of $z = -3.16$

25. To the left of $z = -2.575$

26. To the left of $z = 1.365$

27. To the right of $z = -0.65$

28. To the right of $z = 3.25$

29. To the right of $z = -0.355$

30. To the right of $z = 1.485$

31. Between $z = 0$ and $z = 2.86$

32. Between $z = -1.53$ and $z = 0$

33. Between $z = -1.55$ and $z = 1.55$

34. Between $z = -2.33$ and $z = 2.33$

35. To the left of $z = -1.28$ and to the right of $z = 1.28$

36. To the left of $z = -1.44$ and to the right of $z = 2.21$

37. **Manufacturer Claims** You work for a consumer watchdog publication and are testing the advertising claims of a tire manufacturer. The manufacturer claims that the life spans of the tires are normally distributed, with a mean of 40,000 miles and a standard deviation of 4000 miles. You test 16 tires and record the life spans shown below.

48,778	41,046	29,083	36,394	32,302	42,787	41,972	37,229
25,314	31,920	38,030	38,445	30,750	38,886	36,770	46,049

(a) Draw a frequency histogram to display these data. Use five classes. Do the life spans appear to be normally distributed? Explain.

(b) Find the mean and standard deviation of your sample.

(c) Compare the mean and standard deviation of your sample with those in the manufacturer's claim. Discuss the differences.

38. (a)

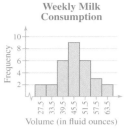

Weekly Milk Consumption

It is reasonable to assume that the weekly milk consumptions are normally distributed because the histogram is nearly symmetric and bell-shaped.

(b) 45.9, 9.5

(c) The mean of your sample is 2.8 fluid ounces less than that of the previous study, so the average milk consumption from the sample is less than in the previous study. The standard deviation is about 0.9 fluid ounce greater than that of the previous study, so the milk consumptions are slightly more spread out than in the previous study.

39. (a) $x = 162 \rightarrow z \approx 1.37$

$x = 168 \rightarrow z \approx 2.06$

$x = 155 \rightarrow z \approx 0.57$

$x = 138 \rightarrow z \approx -1.37$

(b) $x = 168$ is unusual because its corresponding z-score (2.06) lies more than 2 standard deviations from the mean.

40. (a) $x = 168 \rightarrow z \approx 2.44$

$x = 161 \rightarrow z \approx 1.05$

$x = 144 \rightarrow z \approx -2.31$

$x = 155 \rightarrow z \approx -0.14$

(b) $x = 168$ and $x = 144$ are unusual because their corresponding z-scores (2.44 and −2.31, respectively) lie more than 2 standard deviations from the mean.

41. 0.9750

42. 0.2660

43. 0.9832

44. 0.1003

45. 0.6826 (*Tech:* 0.6827)

46. 0.4535

38. Milk Consumption You are performing a study about weekly per capita milk consumption. A previous study found weekly per capita milk consumption to be normally distributed, with a mean of 48.7 fluid ounces and a standard deviation of 8.6 fluid ounces. You randomly sample 30 people and record the weekly milk consumptions shown below.

40 45 54 41 43 31 47 30 33 37 48 57 52 45 38
65 25 39 53 51 58 52 40 46 44 48 61 47 49 57

(a) Draw a frequency histogram to display these data. Use seven classes. Do the consumptions appear to be normally distributed? Explain.

(b) Find the mean and standard deviation of your sample.

(c) Compare the mean and standard deviation of your sample with those of the previous study. Discuss the differences.

Computing and Interpreting z-Scores *In Exercises 39 and 40, (a) find the z-score that corresponds to each value and (b) determine whether any of the values are unusual.*

39. GRE Scores The test scores for the verbal reasoning and the quantitative reasoning sections of the Graduate Record Examination (GRE) are normally distributed. In a recent year, the mean test score was 150 and the standard deviation was 8.75. The test scores of four students selected at random are 162, 168, 155, and 138. (*Source: Educational Testing Service*)

40. LSAT Scores The test scores for the Law School Admission Test (LSAT) are normally distributed. In a recent year, the mean test score was 155.69 and the standard deviation was 5.05. The test scores of four students selected at random are 168, 161, 144, and 155. (*Source: Law School Admission Council*)

Finding Probability *In Exercises 41–46, find the probability of z occurring in the shaded region of the standard normal distribution. If convenient, use technology to find the probability.*

41.

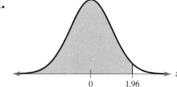

42.

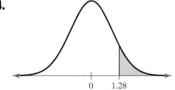

43.

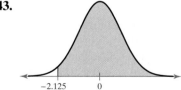

44.

45.

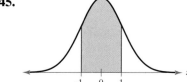

46.

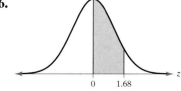

47. 0.8770

48. 0.4286

49. 0.0148

50. 0.9678

51. 0.3133

52. 0.2002

53. 0.9250 (*Tech:* 0.9249)

54. 0.8764

55. 0.0098 (*Tech:* 0.0099)

56. 0.2047

57.

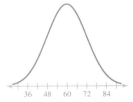

The normal distribution curve is centered at its mean (60) and has 2 points of inflection (48 and 72) representing $\mu \pm \sigma$.

58.

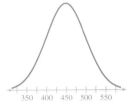

The normal distribution curve is centered at its mean (450) and has 2 points of inflection (400 and 500) representing $\mu \pm \sigma$.

59. (1) The area under the curve is

$(b - a)\left(\dfrac{1}{b - a}\right) = \dfrac{b - a}{b - a} = 1.$

(Because $a < b$, you do not have to worry about division by 0.)

(2) All of the values of the probability density function are positive because $\dfrac{1}{b - a}$ is positive when $a < b$.

60. (a) 0.25 (b) 0.333

(c) 0.5 (d) 0.25

Finding Probability *In Exercises 47–56, find the indicated probability using the standard normal distribution. If convenient, use technology to find the probability.*

47. $P(z < 1.16)$

48. $P(z < -0.18)$

49. $P(z > 2.175)$

50. $P(z > -1.85)$

51. $P(-0.89 < z < 0)$

52. $P(0 < z < 0.525)$

53. $P(-1.78 < z < 1.78)$

54. $P(-1.54 < z < 1.54)$

55. $P(z < -2.58 \text{ or } z > 2.58)$

56. $P(z < -1.22 \text{ or } z > 1.32)$

Extending Concepts

57. Writing Draw a normal curve with a mean of 60 and a standard deviation of 12. Describe how you constructed the curve and discuss its features.

58. Writing Draw a normal curve with a mean of 450 and a standard deviation of 50. Describe how you constructed the curve and discuss its features.

Uniform Distribution *A **uniform distribution** is a continuous probability distribution for a random variable x between two values a and b ($a < b$), where $a \leq x \leq b$ and all of the values of x are equally likely to occur. The graph of a uniform distribution is shown below.*

The probability density function of a uniform distribution is

$$y = \dfrac{1}{b - a}$$

on the interval from $x = a$ to $x = b$. For any value of x less than a or greater than b, $y = 0$. In Exercises 59 and 60, use this information.

59. Show that the probability density function of a uniform distribution satisfies the two conditions for a probability density function.

60. For two values c and d, where $a \leq c < d \leq b$, the probability that x lies between c and d is equal to the area under the curve between c and d, as shown below.

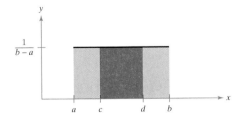

So, the area of the red region equals the probability that x lies between c and d. For a uniform distribution from $a = 1$ to $b = 25$, find the probability that

(a) x lies between 2 and 8.

(b) x lies between 4 and 12.

(c) x lies between 5 and 17.

(d) x lies between 8 and 14.

5.2 Normal Distributions: Finding Probabilities

What You Should Learn

▶ How to find probabilities for normally distributed variables using a table and using technology

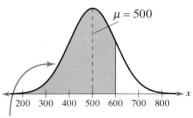

$\mu = 500$

200 300 400 500 600 700 800 x

Same area

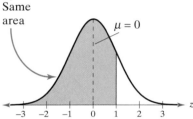

$\mu = 0$

−3   −2   −1   0   1   2   3 z

Study Tip

To learn how to determine whether a random sample is taken from a normal distribution, see Appendix C.

Probability and Normal Distributions

Probability and Normal Distributions

When a random variable x is normally distributed, you can find the probability that x will lie in an interval by calculating the area under the normal curve for the interval. To find the area under any normal curve, first convert the upper and lower bounds of the interval to z-scores. Then use the standard normal distribution to find the area. For instance, consider a normal curve with $\mu = 500$ and $\sigma = 100$, as shown at the upper left. The value of x one standard deviation above the mean is $\mu + \sigma = 500 + 100 = 600$. Now consider the standard normal curve shown at the lower left. The value of z one standard deviation above the mean is $\mu + \sigma = 0 + 1 = 1$. Because a z-score of 1 corresponds to an x-value of 600, and areas are not changed with a transformation to a standard normal curve, the shaded areas in the figures at the left are equal.

EXAMPLE 1

Finding Probabilities for Normal Distributions

A national study found that college students with jobs worked an average of 22 hours per week. The standard deviation is 9 hours. A college student with a job is selected at random. Find the probability that the student works for less than 4 hours per week. Assume that the lengths of time college students work are normally distributed and are represented by the variable x. *(Adapted from Sallie Mae/Ipsos Public Affairs)*

SOLUTION

The figure shows a normal curve with $\mu = 22$, $\sigma = 9$, and the shaded area for x less than 4. The z-score that corresponds to 4 hours is

$$z = \frac{x - \mu}{\sigma} = \frac{4 - 22}{9} = -2.$$

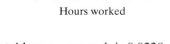

$\mu = 22$

0   8   16   28   36   44 x

Hours worked

The Standard Normal Table shows that

$P(z < -2) = 0.0228.$

The probability that the student works for less that 4 hours per week is 0.0228.

Interpretation So, 2.28% of college students with jobs worked for less than 4 hours per week. Because 2.28% is less than 5%, this is an unusual event.

TRY IT YOURSELF 1

The average speed of vehicles traveling on a stretch of highway is 67 miles per hour with a standard deviation of 3.5 miles per hour. A vehicle is selected at random. What is the probability that it is violating the speed limit of 70 miles per hour? Assume the speeds are normally distributed and are represented by the variable x.

Answer: Page A35

In Example 1, because $P(z < -2) = P(x < 4)$, another way to write the probability is $P(x < 4) = 0.0228$.

EXAMPLE 2

Finding Probabilities for Normal Distributions

A survey indicates that for each trip to a supermarket, a shopper spends an average of 43 minutes with a standard deviation of 12 minutes in the store. The lengths of time spent in the store are normally distributed and are represented by the variable x. A shopper enters the store. (a) Find the probability that the shopper will be in the store for each interval of time listed below. (b) When 200 shoppers enter the store, how many shoppers would you expect to be in the store for each interval of time listed below? *(Adapted from Time Use Institute)*

1. Between 22 and 52 minutes

2. More than 37 minutes

SOLUTION

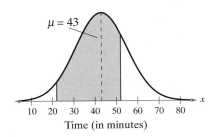

1. (a) The figure at the left shows a normal curve with $\mu = 43$ minutes, $\sigma = 12$ minutes, and the shaded area for x between 22 and 52 minutes. The z-scores that correspond to 22 minutes and to 52 minutes are

$$z_1 = \frac{22 - 43}{12} = -1.75 \quad \text{and} \quad z_2 = \frac{52 - 43}{12} = 0.75.$$

So, the probability that a shopper will be in the store between 22 and 52 minutes is

$$
\begin{aligned}
P(22 < x < 52) &= P(-1.75 < z < 0.75) \\
&= P(z < 0.75) - P(z < -1.75) \\
&= 0.7734 - 0.0401 \\
&= 0.7333.
\end{aligned}
$$

(b) *Interpretation* When 200 shoppers enter the store, you would expect about $200(0.7333) = 146.66 \approx 147$ shoppers to be in the store between 22 and 52 minutes.

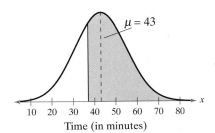

2. (a) The figure at the left shows a normal curve with $\mu = 43$ minutes, $\sigma = 12$ minutes, and the shaded area for x greater than 37. The z-score that corresponds to 37 minutes is

$$z = \frac{37 - 43}{12} = -0.5.$$

So, the probability that a shopper will be in the store more than 37 minutes is

$$
\begin{aligned}
P(x > 37) &= P(z > -0.5) \\
&= 1 - P(z < -0.5) \\
&= 1 - 0.3085 \\
&= 0.6915.
\end{aligned}
$$

(b) *Interpretation* When 200 shoppers enter the store, you would expect about $200(0.6915) = 138.3 \approx 138$ shoppers to be in the store more than 37 minutes.

TRY IT YOURSELF 2

What is the probability that the shopper in Example 2 will be in the supermarket between 31 and 58 minutes? When 200 shoppers enter the store, how many shoppers would you expect to be in the store between 31 and 58 minutes?

Answer: Page A35

Another way to find normal probabilities is to use technology. You can find normal probabilities using Minitab, Excel, StatCrunch, and the TI-84 Plus.

Picturing the World

In baseball, a batting average is the number of hits divided by the number of at bats. The batting averages of all Major League Baseball players in a recent year can be approximated by a normal distribution, as shown in the figure. The mean of the batting averages is 0.255 and the standard deviation is 0.010. (Source: Major League Baseball)

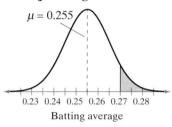

Major League Baseball

$\mu = 0.255$

0.23 0.24 0.25 0.26 0.27 0.28
Batting average

What percent of the players have a batting average of 0.270 or greater? Out of 40 players on a roster, how many would you expect to have a batting average of 0.270 or greater?

6.68%

Out of 40 players on a roster, you would expect $40(0.0668) \approx 2.67$, or about 3, players to have an average of 0.270 or greater.

EXAMPLE 3

Using Technology to Find Normal Probabilities

Triglycerides are a type of fat in the bloodstream. The mean triglyceride level for U.S. adults (ages 20 and older) is 97 milligrams per deciliter. Assume the triglyceride levels of U.S. adults who are at least 20 years old are normally distributed, with a standard deviation of 25 milligrams per deciliter. You randomly select a U.S. adult who is at least 20 years old. What is the probability that the person's triglyceride level is less than 100? (Triglyceride levels less than 150 milligrams per deciliter are considered normal.) Use technology to find the probability. *(Adapted from JAMA Cardiology)*

SOLUTION

Minitab, Excel, StatCrunch, and the TI-84 Plus each have features that allow you to find normal probabilities without first converting to standard z-scores. Note that to use these features, you must specify the mean and standard deviation of the population, as well as any x-values that determine the interval. You are given that $\mu = 97$ and $\sigma = 25$, and you want to find the probability that the person's triglyceride level is less than 100, or $P(x < 100)$.

MINITAB

Cumulative Distribution Function

Normal with mean = 97 and standard deviation = 25

x	P(X ≤ x)
100	0.547758

EXCEL

	A	B	C
1	NORM.DIST(100,97,25,TRUE)		
2			0.547758426

TI-84 PLUS

normalcdf(-10000, 100, 97, 25)

.547758471

STATCRUNCH

Normal Distribution

Mean: 97   Std. Dev.: 25
P(x ≤ 100) = 0.54775843

From the displays, you can see that $P(x < 100) \approx 0.548$.

Interpretation The probability that the person's triglyceride level is less than 100 is about 0.548, or 54.8%.

TRY IT YOURSELF 3

A U.S. adult who is at least 20 years old is selected at random. What is the probability that the person's triglyceride level is between 100 and 150? Use technology to find the probability.

Answer: Page A35

5.2 EXERCISES

1. 0.4207

2. 0.9032

3. 0.3446

4. 0.8289

5. 0.1787 (*Tech:* 0.1788)

6. 0.3557 (*Tech:* 0.3558)

7. (a) 0.2611 (*Tech:* 0.2623)

   (b) 0.3453 (*Tech:* 0.3452)

   (c) 0.1190 (*Tech:* 0.1186)

   (d) No unusual events because all of the probabilities are greater than 0.05.

8. (a) 0.00395 (*Tech:* 0.00396)

   (b) 0.6983 (*Tech:* 0.6979)

   (c) 0.0150 (*Tech:* 0.0149)

   The events in parts (a) and (c) are unusual because their probabilities are less than 0.05.

9. (a) 0.1492 (*Tech:* 0.1497)

   (b) 0.4262 (*Tech:* 0.4269)

   (c) 0.0188 (*Tech:* 0.0190)

   (d) The event in part (c) is unusual because its probability is less than 0.05.

10. (a) 0.0516 (*Tech:* 0.0512)

   (b) 0.6825 (*Tech:* 0.6824)

   (c) 0.0446

   The event in part (c) is unusual because its probability is less than 0.05.

11. (a) 0.0062

   (b) 0.7492 (*Tech:* 0.7499)

   (c) 0.0004

12. (a) 0.2743

   (b) 0.4452

   (c) 0.0228

Building Basic Skills and Vocabulary

Computing Probabilities for Normal Distributions *In Exercises 1–6, the random variable x is normally distributed with mean $\mu = 174$ and standard deviation $\sigma = 20$. Find the indicated probability.*

1. $P(x < 170)$

2. $P(x < 200)$

3. $P(x > 182)$

4. $P(x > 155)$

5. $P(160 < x < 170)$

6. $P(172 < x < 192)$

Using and Interpreting Concepts

Finding Probabilities for Normal Distributions *In Exercises 7–12, find the indicated probabilities. If convenient, use technology to find the probabilities.*

7. World Happiness In a recent study on world happiness, participants were asked to evaluate their current lives on a scale from 0 to 10, where 0 represents the worst possible life and 10 represents the best possible life. The responses were normally distributed, with a mean of 5.4 and a standard deviation of 2.2. Find the probability that a randomly selected study participant's response was (a) less than 4, (b) between 4 and 6, and (c) more than 8. Identify any unusual events in parts (a)–(c). Explain your reasoning. *(Source: The Earth Institute, Columbia University)*

8. Heights of Women In a survey of U.S. women, the heights in the 20- to 29-year age group were normally distributed, with a mean of 64.2 inches and a standard deviation of 2.9 inches. Find the probability that a randomly selected study participant has a height that is (a) less than 56.5 inches, (b) between 61 and 67 inches, and (c) more than 70.5 inches. Identify any unusual events in parts (a)–(c). Explain your reasoning. *(Adapted from National Center for Health Statistics)*

9. MCAT Scores In a recent year, the MCAT total scores were normally distributed, with a mean of 500 and a standard deviation of 10.6. Find the probability that a randomly selected medical student who took the MCAT has a total score that is (a) less than 489, (b) between 495 and 507, and (c) more than 522. Identify any unusual events in parts (a)–(c). Explain your reasoning. *(Source: Association of American Medical Colleges)*

10. MCAT Scores In a recent year, the MCAT scores for the critical analysis and reasoning skills portion of the test were normally distributed, with a mean of 124.9 and a standard deviation of 3.0. Find the probability that a randomly selected medical student who took the MCAT has a critical analysis and reasoning skills score that is (a) less than 120, (b) between 122 and 128, and (c) more than 130. Identify any unusual events in parts (a)–(c). Explain your reasoning. *(Source: Association of American Medical Colleges)*

11. Utility Bills The monthly utility bills in a city are normally distributed, with a mean of $100 and a standard deviation of $12. Find the probability that a randomly selected utility bill is (a) less than $70, (b) between $90 and $120, and (c) more than $140.

12. Health Club Schedule The amounts of time per workout an athlete uses a stairclimber are normally distributed, with a mean of 20 minutes and a standard deviation of 5 minutes. Find the probability that a randomly selected athlete uses a stairclimber for (a) less than 17 minutes, (b) between 20 and 28 minutes, and (c) more than 30 minutes.

13. 0.2918 (*Tech:* 0.2914)

14. 0.0839 (*Tech:* 0.0846)

15. 0.0324 (*Tech:* 0.0325)

16. 0.5865

17. (a) 86.86% (*Tech:* 86.96%)

   (b) 464 scores (*Tech:* 465 scores)

18. (a) 37.45% (*Tech:* 37.39%)

   (b) 726 scores (*Tech:* 729 scores)

19. (a) 98.93%

   (b) 75.75% (*Tech:* 75.76%)

   (c) 6 mothers

20. (a) 93.32%

   (b) 36.12%

   (c) 187 adult males

Graphical Analysis   *In Exercises 13–16, a member is selected at random from the population represented by the graph. Find the probability that the member selected at random is from the shaded region of the graph. Assume the variable x is normally distributed.*

13.   **SAT Total Scores**

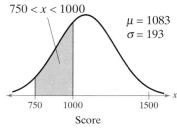

(*Source: The College Board*)

14.   **ACT Composite Scores**

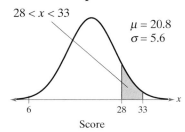

(*Source: ACT, Inc.*)

15.   **Pregnancy Length in a Population of New Mothers**

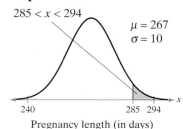

16.   **Red Blood Cell Count in a Population of Adult Males**

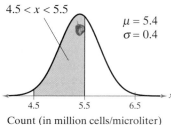

Using Normal Distributions   *In Exercises 17–20, answer the questions about the specified normal distribution.*

17. SAT Total Scores   Use the normal distribution in Exercise 13.

(a) What percent of the SAT total scores are less than 1300?

(b) Out of 1000 randomly selected SAT total scores, about how many would you expect to be greater than 1100?

18. ACT Composite Scores   Use the normal distribution in Exercise 14.

(a) What percent of the ACT composite scores are less than 19?

(b) Out of 1500 randomly selected ACT composite scores, about how many would you expect to be greater than 21?

19. Pregnancy Length   Use the normal distribution in Exercise 15.

(a) What percent of the new mothers had a pregnancy length of less than 290 days?

(b) What percent of the new mothers had a pregnancy length of between 260 and 300 days?

(c) Out of 250 randomly selected new mothers, about how many would you expect to have had a pregnancy length of greater than 287 days?

20. Red Blood Cell Count   Use the normal distribution in Exercise 16.

(a) What percent of the adult males have a red blood cell count less than 6 million cells per microliter?

(b) What percent of the adult males have a red blood cell count between 4.7 and 5.3 million cells per microliter?

(c) Out of 200 randomly selected adult males, about how many would you expect to have a red blood cell count greater than 4.8 million cells per microliter?

Extending Concepts

21. Out of control, because there is a point more than three standard deviations beyond the mean.

22. Out of control, because two out of three consecutive points lie more than two standard deviations from the mean.

23. Out of control, because there are nine consecutive points below the mean, and two out of three consecutive points lie more than two standard deviations from the mean.

24. In control, because none of the three warning signals detected a change.

Control Charts *Statistical process control (SPC) is the use of statistics to monitor and improve the quality of a process, such as manufacturing an engine part. In SPC, information about a process is gathered and used to determine whether a process is meeting all of the specified requirements. One tool used in SPC is a* **control chart.** *When individual measurements of a variable x are normally distributed, a control chart can be used to detect processes that are possibly out of statistical control. Three warning signals that a control chart uses to detect a process that may be out of control are listed below.*

(1) A point lies beyond three standard deviations of the mean.

(2) There are nine consecutive points that fall on one side of the mean.

(3) At least two of three consecutive points lie more than two standard deviations from the mean.

In Exercises 21–24, a control chart is shown. Each chart has horizontal lines drawn at the mean μ, at $\mu \pm 2\sigma$, and at $\mu \pm 3\sigma$. Determine whether the process shown is in control or out of control. Explain.

21. A gear has been designed to have a diameter of 3 inches. The standard deviation of the process is 0.2 inch.

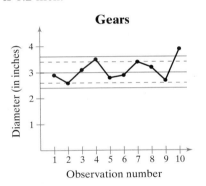

22. A nail has been designed to have a length of 4 inches. The standard deviation of the process is 0.12 inch.

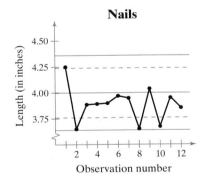

23. A liquid-dispensing machine has been designed to fill bottles with 1 liter of liquid. The standard deviation of the process is 0.1 liter.

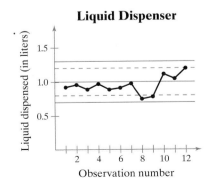

24. An engine part has been designed to have a diameter of 55 millimeters. The standard deviation of the process is 0.001 millimeter.

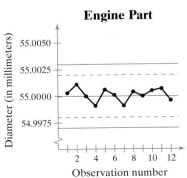

<table>
<tr><td></td></tr>
</table>

5.3 Normal Distributions: Finding Values

What You Should Learn

▶ How to find a z-score given the area under the normal curve

▶ How to transform a z-score to an x-value

▶ How to find a specific data value of a normal distribution given the probability

Finding z-Scores ∎ Transforming a z-Score to an x-Value ∎ Finding a Specific Data Value for a Given Probability

Finding z-Scores

In Section 5.2, you were given a normally distributed random variable x and you found the probability that x would lie in an interval by calculating the area under the normal curve for the interval.

But what if you are given a probability and want to find a value? For instance, a university might want to know the lowest test score a student can have on an entrance exam and still be in the top 10%, or a medical researcher might want to know the cutoff values for selecting the middle 90% of patients by age. In this section, you will learn how to find a value given an area under a normal curve (or a probability), as shown in the next example.

Note to Instructor

Have students note that as the z-scores increase, the cumulative areas increase. The CDF is one-to-one and as such has an inverse function. Discuss how this inverse function (INVCDF) can be used when a cumulative area (percentile) is known and the z-score must be found.

EXAMPLE 1

Finding a z-Score Given an Area

1. Find the z-score that corresponds to a cumulative area of 0.3632.

2. Find the z-score that has 10.75% of the distribution's area to its right.

SOLUTION

1. Find the z-score that corresponds to an area of 0.3632 by locating 0.3632 in the Standard Normal Table. The values at the beginning of the corresponding row and at the top of the corresponding column give the z-score. For this area, the row value is −0.3 and the column value is 0.05. So, the z-score is −0.35, as shown in the figure at the left.

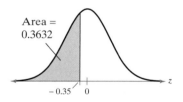

Area = 0.3632

z	.09	.08	.07	.06	.05	.04	.03
−3.4	.0002	.0003	.0003	.0003	.0003	.0003	.0003
−0.5	.2776	.2810	.2843	.2877	.2912	.2946	.2981
−0.4	.3121	.3156	.3192	.3228	.3264	.3300	.3336
−0.3	.3483	.3520	.3557	.3594	.3632	.3669	.3707
−0.2	.3859	.3897	.3936	.3974	.4013	.4052	.4090

2. Because the area to the right is 0.1075, the cumulative area is $1 - 0.1075 = 0.8925$. Find the z-score that corresponds to an area of 0.8925 by locating 0.8925 in the Standard Normal Table. For this area, the row value is 1.2 and the column value is 0.04. So, the z-score is 1.24, as shown in the figure at the left.

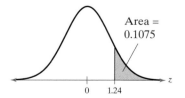

Area = 0.1075

z	.00	.01	.02	.03	.04	.05	.06
0.0	.5000	.5040	.5080	.5120	.5160	.5199	.5239
1.0	.8413	.8438	.8461	.8485	.8508	.8531	.8554
1.1	.8643	.8665	.8686	.8708	.8729	.8749	.8770
1.2	.8849	.8869	.8888	.8907	.8925	.8944	.8962
1.3	.9032	.9049	.9066	.9082	.9099	.9115	.9131

Tech Tip

You can use technology to find the z-scores that correspond to cumulative areas. For instance, you can use a TI-84 Plus to find the z-scores in Example 1, as shown below.

```
invNorm(.3632,0,
1)
        -.3499183227
invNorm(.8925,0,
1)
        1.239933478
```

TRY IT YOURSELF 1

1. Find the z-score that has 96.16% of the distribution's area to its right.

2. Find the positive z-score for which 95% of the distribution's area lies between $-z$ and z.

Answer: Page A35

In Example 1, the given areas correspond to entries in the Standard Normal Table. In most cases, the area will not be an entry in the table. In these cases, use the entry closest to it (or use technology, as shown at the left and in Example 2). When the area is halfway between two area entries, use the z-score halfway between the corresponding z-scores.

In Section 2.5, you learned that percentiles divide a data set into 100 equal parts. To find a z-score that corresponds to a percentile, you can use the Standard Normal Table. Recall that if a value x represents the 83rd percentile P_{83}, then 83% of the data values are below x and 17% of the data values are above x.

EXAMPLE 2

Finding a z-Score Given a Percentile

Find the z-score that corresponds to each percentile.

1. P_5 **2.** P_{50} **3.** P_{90}

SOLUTION

1. To find the z-score that corresponds to P_5, find the z-score that corresponds to an area of 0.05 (see upper figure) by locating 0.05 in the Standard Normal Table. The areas closest to 0.05 in the table are 0.0495 ($z = -1.65$) and 0.0505 ($z = -1.64$). Because 0.05 is halfway between the two areas in the table, use the z-score that is halfway between -1.64 and -1.65. So, the z-score that corresponds to an area of 0.05 is -1.645.

2. To find the z-score that corresponds to P_{50}, find the z-score that corresponds to an area of 0.5 (see middle figure) by locating 0.5 in the Standard Normal Table. The area closest to 0.5 in the table is 0.5000, so the z-score that corresponds to an area of 0.5 is 0.

3. To find the z-score that corresponds to P_{90}, find the z-score that corresponds to an area of 0.9 (see lower figure) by locating 0.9 in the Standard Normal Table. The area closest to 0.9 in the table is 0.8997, so the z-score that corresponds to an area of 0.9 is about 1.28.

You can use technology to find the z-score that corresponds to each percentile, as shown below. Remember that when you use technology, your answers may differ slightly from those found using the Standard Normal Table.

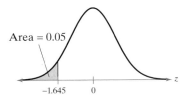

Area = 0.05

−1.645   0

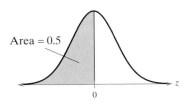

Area = 0.5

0

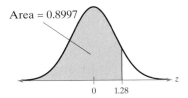

Area = 0.8997

0   1.28

EXCEL

	A	B
1	**1.**NORM.INV(0.05,0,1)	
2		−1.644853627
3	**2.**NORM.INV(0.50,0,1)	
4		0
5	**3.**NORM.INV(0.90,0,1)	
6		1.281551566

TRY IT YOURSELF 2

Find the z-score that corresponds to each percentile.

1. P_{10} **2.** P_{20} **3.** P_{99}

Answer: Page A35

Transforming a z-Score to an x-Value

Recall that to transform an x-value to a z-score, you can use the formula

$$z = \frac{x - \mu}{\sigma}.$$

This formula gives z in terms of x. When you solve this formula for x, you get a new formula that gives x in terms of z.

$z = \dfrac{x - \mu}{\sigma}$ Formula for z in terms of x

$z\sigma = x - \mu$ Multiply each side by σ.

$\mu + z\sigma = x$ Add μ to each side.

$x = \mu + z\sigma$ Interchange sides.

Transforming a z-Score to an x-Value

To transform a standard z-score to an x-value in a given population, use the formula

$$x = \mu + z\sigma.$$

EXAMPLE 3

Finding an x-Value Corresponding to a z-Score

A veterinarian records the weights of cats treated at a clinic. The weights are normally distributed, with a mean of 9 pounds and a standard deviation of 2 pounds. Find the weight x corresponding to each z-score. Interpret the results.

1. $z = 1.96$ **2.** $z = -0.44$ **3.** $z = 0$

SOLUTION

The x-value that corresponds to each standard z-score is calculated using the formula $x = \mu + z\sigma$. Note that $\mu = 9$ and $\sigma = 2$.

1. For $z = 1.96$, the corresponding weight x is

$$x = 9 + 1.96(2) = 12.92 \text{ pounds.}$$

2. For $z = -0.44$, the corresponding weight x is

$$x = 9 + (-0.44)(2) = 8.12 \text{ pounds.}$$

3. For $z = 0$, the corresponding weight x is

$$x = 9 + 0(2) = 9 \text{ pounds.}$$

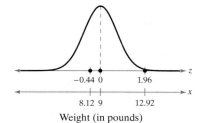

Weight (in pounds)

Interpretation   From the figure at the left, you can see that 12.92 pounds is to the right of the mean, 8.12 pounds is to the left of the mean, and 9 pounds is equal to the mean.

TRY IT YOURSELF 3

A veterinarian records the weights of dogs treated at a clinic. The weights are normally distributed, with a mean of 52 pounds and a standard deviation of 15 pounds. Find the weight x corresponding to each z-score. Interpret the results.

1. $z = -2.33$ **2.** $z = 3$ **3.** $z = 0.58$

Answer: Page A35

Picturing the World

Many investors choose mutual funds as a way to invest in the stock market. The mean annual rate of return for large growth mutual funds during a recent five-year period was about 12.1% with a standard deviation of 1.8%. *(Adapted from Morningstar)*

Annual Rate of Return for Large Growth Mutual Funds

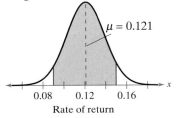

Rate of return

Between what two values does the middle 90% of the data lie?

The middle 90% of the data lies between 0.09 and 0.15, or 9% and 15%.

TI-84 PLUS

```
invNorm(.9,50,10
)
        62.81551567
```

Finding a Specific Data Value for a Given Probability

You can also use the normal distribution to find a specific data value (*x*-value) for a given probability, as shown in Examples 4 and 5.

EXAMPLE 4

Finding a Specific Data Value

Scores for the California Peace Officer Standards and Training test are normally distributed, with a mean of 50 and a standard deviation of 10. An agency will only hire applicants with scores in the top 10%. What is the lowest score an applicant can earn and still be eligible to be hired by the agency? *(Source: State of California)*

SOLUTION

Exam scores in the top 10% correspond to the shaded region shown.

Scores for the California Peace Officer Standards and Training Test

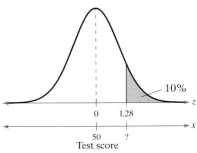

A test score in the top 10% is any score above the 90th percentile. To find the score that represents the 90th percentile, you must first find the *z*-score that corresponds to a cumulative area of 0.9. In the Standard Normal Table, the area closest to 0.9 is 0.8997. So, the *z*-score that corresponds to an area of 0.9 is $z = 1.28$. To find the *x*-value, note that $\mu = 50$ and $\sigma = 10$, and use the formula $x = \mu + z\sigma$, as shown.

$$x = \mu + z\sigma$$
$$= 50 + 1.28(10)$$
$$= 62.8$$

You can check this answer using technology. For instance, you can use a TI-84 Plus to find the *x*-value, as shown at the left.

Interpretation The lowest score an applicant can earn and still be eligible to be hired by the agency is about 63.

TRY IT YOURSELF 4

A researcher tests the braking distances of several cars. The braking distance from 60 miles per hour to a complete stop on dry pavement is measured in feet. The braking distances of a sample of cars are normally distributed, with a mean of 129 feet and a standard deviation of 5.18 feet. What is the longest braking distance one of these cars could have and still be in the bottom 1%? *(Adapted from Consumer Reports)*

Answer: Page A35

EXAMPLE 5

Finding a Specific Data Value

In a randomly selected sample of women ages 20–34, the mean total cholesterol level is 179 milligrams per deciliter with a standard deviation of 38.9 milligrams per deciliter. Assume the total cholesterol levels are normally distributed. Find the highest total cholesterol level a woman in this 20–34 age group can have and still be in the bottom 1%. *(Adapted from National Center for Health Statistics)*

SOLUTION

Total cholesterol levels in the lowest 1% correspond to the shaded region shown.

Total Cholesterol Levels in Women Ages 20–34

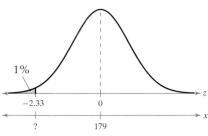

Total cholesterol level (in mg/dL)

A total cholesterol level in the lowest 1% is any level below the 1st percentile. To find the level that represents the 1st percentile, you must first find the z-score that corresponds to a cumulative area of 0.01. In the Standard Normal Table, the area closest to 0.01 is 0.0099. So, the z-score that corresponds to an area of 0.01 is $z = -2.33$. To find the x-value, note that $\mu = 179$ and $\sigma = 38.9$, and use the formula $x = \mu + z\sigma$, as shown.

$$x = \mu + z\sigma$$
$$= 179 + (-2.33)(38.9)$$
$$\approx 88.36$$

You can check this answer using technology. For instance, you can use Excel to find the x-value, as shown below.

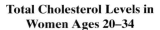

	A	B	C
1	NORM.INV(0.01,179,38.9)		
2			88.5050677

Interpretation The value that separates the lowest 1% of total cholesterol levels for women in the 20–34 age group from the highest 99% is about 88 milligrams per deciliter.

TRY IT YOURSELF 5

The lengths of time employees have worked at a corporation are normally distributed, with a mean of 11.2 years and a standard deviation of 2.1 years. In a company cutback, the lowest 10% in seniority are laid off. What is the maximum length of time an employee could have worked and still be laid off?

Answer: Page A35

5.3 EXERCISES

For Extra Help: **MyLab Statistics**

1. −0.81
2. −0.16
3. 0.45
4. 0.84
5. −1.645
6. 1.04
7. 1.555
8. −2.605
9. −1.04
10. −0.52
11. 1.175
12. 0.44
13. −0.67
14. −0.25
15. 1.34
16. 0.67
17. −0.38
18. 0.25
19. 1.99
20. −0.58
21. −1.96, 1.96
22. −1.645, 1.645
23. −1.18
24. 0.79
25. −0.35
26. 0.81
27. −2.00
28. −1.00
29. 1.28
30. 0.15

Building Basic Skills and Vocabulary

Finding a z-Score *In Exercises 1–16, use the Standard Normal Table or technology to find the z-score that corresponds to the cumulative area or percentile.*

1. 0.2090   2. 0.4364   3. 0.6736   4. 0.7995
5. 0.05   6. 0.85   7. 0.94   8. 0.0046
9. P_{15} 10. P_{30} 11. P_{88} 12. P_{67}
13. P_{25} 14. P_{40} 15. P_{91} 16. P_{75}

Graphical Analysis *In Exercises 17–22, find the indicated z-score(s) shown in the graph.*

17.
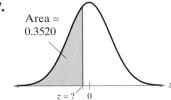
Area = 0.3520

18.
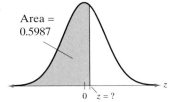
Area = 0.5987

19.

Area = 0.0233

20.
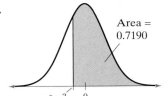
Area = 0.7190

21.
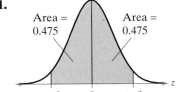
Area = 0.475   Area = 0.475

22.
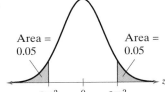
Area = 0.05   Area = 0.05

Finding a z-Score Given an Area *In Exercises 23–30, find the indicated z-score.*

23. Find the z-score that has 11.9% of the distribution's area to its left.

24. Find the z-score that has 78.5% of the distribution's area to its left.

25. Find the z-score that has 63.7% of the distribution's area to its right.

26. Find the z-score that has 20.9% of the distribution's area to its right.

27. Find the z-score that has 2.275% of the distribution's area to its left.

28. Find the z-score that has 84.1345% of the distribution's area to its right.

29. Find the positive z-score for which 80% of the distribution's area lies between −z and z.

30. Find the positive z-score for which 12% of the distribution's area lies between −z and z.

31. (a) 68.97 inches
    (b) 63.71 inches
       (*Tech:* 63.69 inches)
    (c) 62.26 inches
       (*Tech:* 62.24 inches)

32. (a) 7.99 (*Tech:* 7.98)
    (b) 6.02 (*Tech:* 6.01)
    (c) 3.93 (*Tech:* 3.92)

33. (a) 1315.99 kilowatt-hours
       (*Tech:* 1316.08 kilowatt-hours)
    (b) 1719.67 kilowatt-hours
       (*Tech:* 1719.58 kilowatt-hours)
    (c) 2671.34 kilowatt-hours
       (*Tech:* 2671.04 kilowatt-hours)

34. (a) 4.03 mega gallons
    (b) 0.07 mega gallon
    (c) 3.56 mega gallons

35. (a) 3.66
    (b) 3.24 and 3.48

Using and Interpreting Concepts

Finding Specified Data Values   *In Exercises 31–38, answer the questions about the specified normal distribution.*

31. **Heights of Women**   In a survey of women in the United States (ages 20–29), the mean height was 64.2 inches with a standard deviation of 2.9 inches. *(Adapted from National Center for Health Statistics)*
    (a) What height represents the 95th percentile?
    (b) What height represents the 43rd percentile?
    (c) What height represents the first quartile?

32. **World Happiness**   In a recent study on world happiness, participants were asked to evaluate their current lives on a scale from 0 to 10, where 0 represents the worst possible life and 10 represents the best possible life. The mean response was 5.4 with a standard deviation of 2.2. *(Source: The Earth Institute, Columbia University)*
    (a) What response represents the 88th percentile?
    (b) What response represents the 61st percentile?
    (c) What response represents the first quartile?

33. **Energy Consumption**   The per capita energy consumption level (in kilowatt-hours) in Venezuela for a recent year can be approximated by a normal distribution, as shown in the figure. *(Source: Latin American Journal of Economics)*
    (a) What consumption level represents the 5th percentile?
    (b) What consumption level represents the 17th percentile?
    (c) What consumption level represents the third quartile?

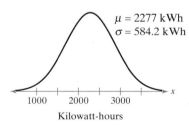

Per Capita Energy Consumption Level in Venezuela

$\mu = 2277$ kWh
$\sigma = 584.2$ kWh

Kilowatt-hours

FIGURE FOR EXERCISE 33

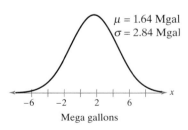

Water Footprint in the U.S.

$\mu = 1.64$ Mgal
$\sigma = 2.84$ Mgal

Mega gallons

FIGURE FOR EXERCISE 34

34. **Water Footprint**   A *water footprint* is a measure of the appropriation of fresh water. The per capita water footprint (in mega gallons) in the U.S. for a recent year can be approximated by a normal distribution, as shown in the figure. *(Source: Water Resources Research)*
    (a) What water footprint represents the 80th percentile?
    (b) What water footprint represents the 29th percentile?
    (c) What water footprint represents the third quartile?

35. **Undergraduate Grade Point Average**   The undergraduate grade point averages (UGPA) of students taking the Law School Admission Test in a recent year can be approximated by a normal distribution, as shown in the figure. *(Source: Law School Admission Council)*
    (a) What is the minimum UGPA that would still place a student in the top 5% of UGPAs?
    (b) Between what two values does the middle 50% of the UGPAs lie?

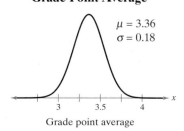

Undergraduate Grade Point Average

$\mu = 3.36$
$\sigma = 0.18$

Grade point average

FIGURE FOR EXERCISE 35

GRE Analytical Writing Scores

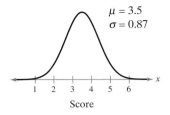

FIGURE FOR EXERCISE 36

36. (a) 2.39
    (b) 2.39 and 4.61
37. (a) 5.67 millions of cells per microliter
    (b) 4.98 millions of cells per microliter (*Tech:* 4.99 millions of cells per microliter)
38. (a) 279.8 days
    (b) 250.6 days
39. 32.61 ounces
40. Tires that wear out by 26,800 miles (*Tech:* 26,796 miles) will be replaced free of charge.
41. 7.93 ounces
42. (a) $A = 83.52$ (*Tech:* 83.53)
    (b) $B = 76.68$ (*Tech:* 76.72)
    (c) $C = 67.32$ (*Tech:* 67.28)
    (d) $D = 60.48$ (*Tech:* 60.47)

36. GRE Scores The test scores for the analytical writing section of the Graduate Record Examination (GRE) can be approximated by a normal distribution, as shown in the figure. *(Source: Educational Testing Service)*

(a) What is the maximum score that can be in the bottom 10% of scores?

(b) Between what two values does the middle 80% of the scores lie?

37. Red Blood Cell Count The red blood cell counts (in millions of cells per microliter) for a population of adult males can be approximated by a normal distribution, with a mean of 5.4 million cells per microliter and a standard deviation of 0.4 million cells per microliter.

(a) What is the minimum red blood cell count that can be in the top 25% of counts?

(b) What is the maximum red blood cell count that can be in the bottom 15% of counts?

38. Pregnancy Length The pregnancy length (in days) for a population of new mothers can be approximated by a normal distribution, with a mean of 267 days and a standard deviation of 10 days.

(a) What is the minimum pregnancy length that can be in the top 10% of pregnancy lengths?

(b) What is the maximum pregnancy length that can be in the bottom 5% of pregnancy lengths?

39. Bags of Baby Carrots The weights of bags of baby carrots are normally distributed, with a mean of 32 ounces and a standard deviation of 0.36 ounce. Bags in the upper 4.5% are too heavy and must be repackaged. What is the most a bag of baby carrots can weigh and not need to be repackaged?

40. Writing a Guarantee You sell a brand of automobile tire that has a life expectancy that is normally distributed, with a mean life of 30,000 miles and a standard deviation of 2500 miles. You want to give a guarantee for free replacement of tires that do not wear well. You are willing to replace approximately 10% of the tires. How should you word your guarantee?

Extending Concepts

41. Vending Machine A vending machine dispenses coffee into an eight-ounce cup. The amounts of coffee dispensed into the cup are normally distributed, with a standard deviation of 0.03 ounce. You can allow the cup to overflow 1% of the time. What amount should you set as the mean amount of coffee to be dispensed?

42. History Grades In a large section of a history class, the points for the final exam are normally distributed, with a mean of 72 and a standard deviation of 9. Grades are assigned according to the rule below.

- The top 10% receive an A.
- The next 20% receive a B.
- The middle 40% receive a C.
- The next 20% receive a D.
- The bottom 10% receive an F.

Find the lowest score on the final exam that would qualify a student for (a) an A, (b) a B, (c) a C, and (d) a D.

Final Exam Grades

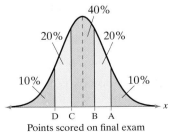

Points scored on final exam

FIGURE FOR EXERCISE 42

CASE STUDY — Birth Weights in America

The National Center for Health Statistics (NCHS) keeps records of many health-related aspects of people, including the birth weights of all babies born in the United States.

   The birth weight of a baby is related to its gestation period (the time between conception and birth). For a given gestation period, the birth weights can be approximated by a normal distribution. The means and standard deviations of the birth weights for various gestation periods are shown in the table below.

   One of the many goals of the NCHS is to reduce the percentage of babies born with low birth weights. The figure below shows the percents of preterm births and low birth weights from 2007 to 2015.

Gestation period	Mean birth weight	Standard deviation
Under 28 weeks	1.60 lb	0.76 lb
28 to 31 weeks	3.20 lb	1.02 lb
32 to 33 weeks	4.31 lb	0.97 lb
34 to 36 weeks	5.74 lb	1.13 lb
37 to 38 weeks	6.92 lb	1.07 lb
39 to 40 weeks	7.60 lb	0.99 lb
41 weeks	8.00 lb	0.99 lb
42 weeks and over	8.16 lb	1.12 lb

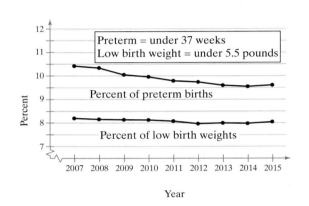

EXERCISES

1. The distributions of birth weights for three gestation periods are shown. Match each curve with a gestation period. Explain your reasoning.

   (a)
   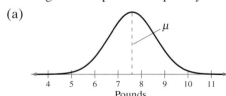

   (b)
   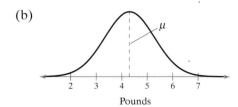

   (c)
   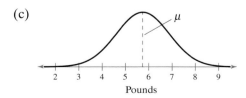

2. What percent of the babies born within each gestation period have a low birth weight (under 5.5 pounds)?
   (a) 28 to 31 weeks   (b) 32 to 33 weeks
   (c) 39 to 40 weeks   (d) 42 weeks and over

3. Describe the weights of the top 10% of the babies born within each gestation period.
   (a) Under 28 weeks   (b) 34 to 36 weeks
   (c) 41 weeks   (d) 42 weeks and over

4. For each gestation period, what is the probability that a baby will weigh between 6 and 9 pounds at birth?
   (a) Under 28 weeks   (b) 32 to 33 weeks
   (c) 37 to 38 weeks   (d) 41 weeks

5. A birth weight of less than 3.25 pounds is classified by the NCHS as a "very low birth weight." What is the probability that a baby has a very low birth weight for each gestation period?
   (a) Under 28 weeks   (b) 28 to 31 weeks
   (c) 32 to 33 weeks   (d) 39 to 40 weeks

5.4 Sampling Distributions and the Central Limit Theorem

What You Should Learn

▸ How to find sampling distributions and verify their properties

▸ How to interpret the Central Limit Theorem

▸ How to apply the Central Limit Theorem to find the probability of a sample mean

Sampling Distributions ▪ The Central Limit Theorem ▪ Probability and the Central Limit Theorem

Sampling Distributions

In previous sections, you studied the relationship between the mean of a population and values of a random variable. In this section, you will study the relationship between a population mean and the means of random samples taken from the population.

DEFINITION

A **sampling distribution** is the probability distribution of a sample statistic that is formed when random samples of size n are repeatedly taken from a population. If the sample statistic is the sample mean, then the distribution is the **sampling distribution of sample means.** Every sample statistic has a sampling distribution.

Study Tip

Sample means can vary from one another and can also vary from the population mean. This type of variation is to be expected and is called *sampling error.* You will learn more about this topic in Section 6.1.

Consider the Venn diagram below. The rectangle represents a large population, and each circle represents a random sample of size n. Because the sample entries can differ, the sample means can also differ. The mean of Random Sample 1 is $\overline{x_1}$; the mean of Random Sample 2 is $\overline{x_2}$; and so on. The sampling distribution of the sample means for samples of size n for this population consists of $\overline{x_1}$, $\overline{x_2}$, $\overline{x_3}$, and so on. If the samples are drawn with replacement, then an infinite number of samples can be drawn from the population.

Population with Mean μ and Standard Deviation σ

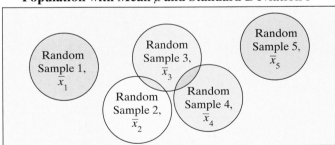

Note to Instructor

A good exercise that can be used in conjunction with the Venn diagram is to have each student randomly select a place in the random number table and write down the next five digits horizontally. Students can verify that the population of digits $\{0, 1, 2, \ldots, 9\}$ is uniform and has a mean of 4.5 and standard deviation of 2.87. Have each student calculate the mean of his or her sample and write that result on the board. Students can easily see that the sample means vary but are not dispersed as much as the population (range 0 to 9) is. Construct a histogram of the sample means; find the mean of these means and the standard deviation of the means. (With a TI-84 Plus, this takes little time even if only one student does the calculations.) Because the population standard deviation is known for this simulation, the results will be approximately normal.

Properties of Sampling Distributions of Sample Means

1. The mean of the sample means $\mu_{\overline{x}}$ is equal to the population mean μ.
$$\mu_{\overline{x}} = \mu$$

2. The standard deviation of the sample means $\sigma_{\overline{x}}$ is equal to the population standard deviation σ divided by the square root of the sample size n.
$$\sigma_{\overline{x}} = \frac{\sigma}{\sqrt{n}}$$

The standard deviation of the sampling distribution of the sample means is called the **standard error of the mean.**

**Probability Histogram
of Population of x**

Number of grocery shopping trips

**Probability Distribution
of Sample Means**

$\bar{x}$	f	Probability
1	1	1/16
2	2	2/16
3	3	3/16
4	4	4/16
5	3	3/16
6	2	2/16
7	1	1/16

**Probability Histogram of
Sampling Distribution of $\bar{x}$**

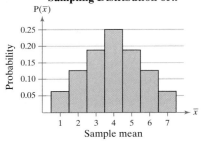

Sample mean

5.4 To explore this topic further, see **Activity 5.4** on page 274.

Study Tip

Review Section 4.1 to find the mean and standard deviation of a probability distribution.

EXAMPLE 1

A Sampling Distribution of Sample Means

The number of times four people go grocery shopping in a month is given by the population values $\{1, 3, 5, 7\}$. A probability histogram for the data is shown at the left. You randomly choose two of the four people, with replacement. List all possible samples of size $n = 2$ and calculate the mean of each. These means form the sampling distribution of the sample means. Find the mean, variance, and standard deviation of the sample means. Compare your results with the mean $\mu = 4$, variance $\sigma^2 = 5$, and standard deviation $\sigma = \sqrt{5} \approx 2.2$ of the population.

SOLUTION

List all 16 samples of size 2 from the population and the mean of each sample.

Sample	Sample mean, $\bar{x}$	Sample	Sample mean, $\bar{x}$
1, 1	1	5, 1	3
1, 3	2	5, 3	4
1, 5	3	5, 5	5
1, 7	4	5, 7	6
3, 1	2	7, 1	4
3, 3	3	7, 3	5
3, 5	4	7, 5	6
3, 7	5	7, 7	7

After constructing a probability distribution of the sample means, you can graph the sampling distribution using a probability histogram as shown at the left. Notice that the shape of the histogram is bell-shaped and symmetric, similar to a normal curve. The mean, variance, and standard deviation of the 16 sample means are

$\mu_{\bar{x}} = 4$ Mean of the sample means

$(\sigma_{\bar{x}})^2 = \dfrac{5}{2} = 2.5$ Variance of the sample means

and

$\sigma_{\bar{x}} = \sqrt{\dfrac{5}{2}} = \sqrt{2.5} \approx 1.6.$ Standard deviation of the sample means

These results satisfy the properties of sampling distributions because

$\mu_{\bar{x}} = \mu = 4$

and

$\sigma_{\bar{x}} = \dfrac{\sigma}{\sqrt{n}} = \dfrac{\sqrt{5}}{\sqrt{2}} \approx 1.6.$

TRY IT YOURSELF 1

List all possible samples of size $n = 3$, with replacement, from the population $\{1, 3, 5\}$. Calculate the mean of each sample. Find the mean, variance, and standard deviation of the sample means. Compare your results with the mean $\mu = 3$, variance $\sigma^2 = 8/3$, and standard deviation $\sigma = \sqrt{8/3} \approx 1.6$ of the population.

Answer: Page A35

The Central Limit Theorem

The Central Limit Theorem forms the foundation for the inferential branch of statistics. This theorem describes the relationship between the sampling distribution of sample means and the population that the samples are taken from. The Central Limit Theorem is an important tool that provides the information you will need to use sample statistics to make inferences about a population mean.

The Central Limit Theorem

1. If random samples of size n, where $n \geq 30$, are drawn from any population with a mean μ and a standard deviation σ, then the sampling distribution of sample means approximates a normal distribution. The greater the sample size, the better the approximation. (See figures for "Any Population Distribution" below.)

2. If random samples of size n are drawn from a population that is normally distributed, then the sampling distribution of sample means is normally distributed for *any* sample size n. (See figures for "Normal Population Distribution" below.)

In either case, the sampling distribution of sample means has a mean equal to the population mean.

$$\mu_{\bar{x}} = \mu \qquad \text{Mean of the sample means}$$

The sampling distribution of sample means has a variance equal to $1/n$ times the variance of the population and a standard deviation equal to the population standard deviation divided by the square root of n.

$$\sigma_{\bar{x}}^2 = \frac{\sigma^2}{n} \qquad \text{Variance of the sample means}$$

$$\sigma_{\bar{x}} = \frac{\sigma}{\sqrt{n}} \qquad \text{Standard deviation of the sample means}$$

Recall that the standard deviation of the sampling distribution of the sample means, $\sigma_{\bar{x}}$, is also called the standard error of the mean.

Study Tip

The distribution of sample means has the same mean as the population. But its standard deviation is less than the standard deviation of the population.

This tells you that the distribution of sample means has the same center as the population, but it is not as spread out.

Moreover, the distribution of sample means becomes less and less spread out (tighter concentration about the mean) as the sample size n increases.

1. Any Population Distribution

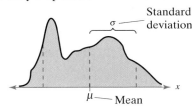

Distribution of Sample Means, $n \geq 30$

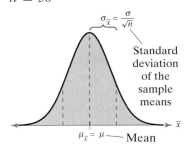

2. Normal Population Distribution

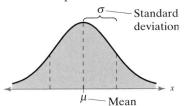

Distribution of Sample Means (any n)

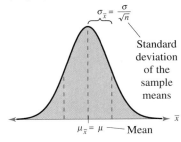

EXAMPLE 2

Interpreting the Central Limit Theorem

A study analyzed the sleep habits of college students. The study found that the mean sleep time was 6.8 hours, with a standard deviation of 1.4 hours. Random samples of 100 sleep times are drawn from this population, and the mean of each sample is determined. Find the mean and standard deviation of the sampling distribution of sample means. Then sketch a graph of the sampling distribution. *(Adapted from The Journal of American College Health)*

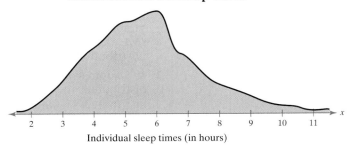

Distribution for All Sleep Times

Individual sleep times (in hours)

SOLUTION

The mean of the sampling distribution is equal to the population mean, and the standard deviation of the sample means is equal to the population standard deviation divided by $\sqrt{n}$. So,

$$\mu_{\bar{x}} = \mu = 6.8 \qquad \text{Mean of the sample means}$$

and

$$\sigma_{\bar{x}} = \frac{\sigma}{\sqrt{n}}$$
$$= \frac{1.4}{\sqrt{100}}$$
$$= 0.14. \qquad \text{Standard deviation of the sample means}$$

Interpretation From the Central Limit Theorem, because the sample size is greater than 30, the sampling distribution can be approximated by a normal distribution with a mean of 6.8 hours and a standard deviation of 0.14 hour, as shown in the figure.

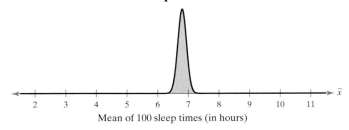

Distribution of Sample Means with *n* = 100

Mean of 100 sleep times (in hours)

TRY IT YOURSELF 2

Random samples of size 64 are drawn from the population in Example 2. Find the mean and standard deviation of the sampling distribution of sample means. Then sketch a graph of the sampling distribution and compare it with the sampling distribution in Example 2.

Answer: Page A35

Picturing the World

In a recent year, there were about 4.2 million parents in the United States who received child support payments. The histogram shows the distribution of children per custodial parent. The mean number of children was 1.7 and the standard deviation was 0.8. (Adapted from U.S. Census Bureau)

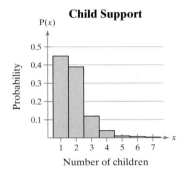

Child Support

You randomly select 35 parents who receive child support and ask how many children in their custody are receiving child support payments. What is the probability that the mean of the sample is between 1.5 and 1.9 children?

The probability that the mean of the sample is between 1.5 children and 1.9 children is 0.861.

EXAMPLE 3

Interpreting the Central Limit Theorem

Assume the training heart rates of all 20-year-old athletes are normally distributed, with a mean of 135 beats per minute and a standard deviation of 18 beats per minute, as shown in the figure. Random samples of size 4 are drawn from this population, and the mean of each sample is determined. Find the mean and standard deviation of the sampling distribution of sample means. Then sketch a graph of the sampling distribution.

Distribution of Population Training Heart Rates

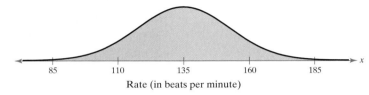

Rate (in beats per minute)

SOLUTION

$$\mu_{\bar{x}} = \mu = 135 \text{ beats per minute} \qquad \text{Mean of the sample means}$$

and

$$\sigma_{\bar{x}} = \frac{\sigma}{\sqrt{n}}$$
$$= \frac{18}{\sqrt{4}}$$
$$= 9 \text{ beats per minute} \qquad \text{Standard deviation of the sample means}$$

Interpretation From the Central Limit Theorem, because the population is normally distributed, the sampling distribution of the sample means is also normally distributed, as shown in the figure.

Distribution of Sample Means with *n* = 4

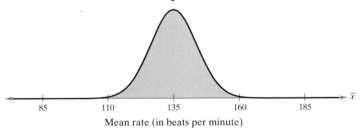

Mean rate (in beats per minute)

TRY IT YOURSELF 3

The diameters of fully grown white oak trees are normally distributed, with a mean of 3.5 feet and a standard deviation of 0.2 foot, as shown in the figure. Random samples of size 16 are drawn from this population, and the mean of each sample is determined. Find the mean and standard deviation of the sampling distribution of sample means. Then sketch a graph of the sampling distribution.

Distribution of Population Diameters

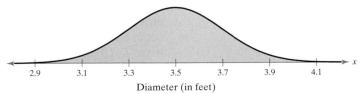

Diameter (in feet)

Answer: Page A35

Probability and the Central Limit Theorem

In Section 5.2, you learned how to find the probability that a random variable x will lie in a given interval of population values. In a similar manner, you can find the probability that a sample mean $\bar{x}$ will lie in a given interval of the $\bar{x}$ sampling distribution. To transform $\bar{x}$ to a z-score, you can use the formula

$$z = \frac{\text{Value} - \text{Mean}}{\text{Standard error}} = \frac{\bar{x} - \mu_{\bar{x}}}{\sigma_{\bar{x}}} = \frac{\bar{x} - \mu}{\sigma/\sqrt{n}}.$$

EXAMPLE 4

Finding Probabilities for Sampling Distributions

The figure at the right shows the mean distances traveled by drivers each day. You randomly select 50 drivers ages 16 to 19. What is the probability that the mean distance traveled each day is between 19.4 and 22.5 miles? Assume $\sigma = 6.5$ miles.

Miles to go
The average miles driven each day, by age group:

16-19   **20.7 miles**
20-29   **31.0**
30-49   **37.0**
50-64   **30.4**
65-74   **30.4**

Source: American Automobile Association

SOLUTION

The sample size is greater than 30, so you can use the Central Limit Theorem to conclude that the distribution of sample means is approximately normal, with a mean and a standard deviation of

$$\mu_{\bar{x}} = \mu = 20.7 \text{ miles} \quad \text{and} \quad \sigma_{\bar{x}} = \frac{\sigma}{\sqrt{n}} = \frac{6.5}{\sqrt{50}} \approx 0.9 \text{ mile.}$$

The graph of this distribution is shown at the left with a shaded area between 19.4 and 22.5 miles. The z-scores that correspond to sample means of 19.4 and 22.5 miles are found as shown.

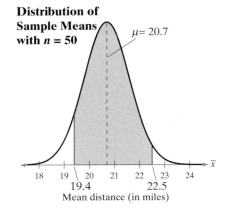

Distribution of Sample Means with $n = 50$

$\mu = 20.7$

18   19 | 20   21   22 | 23   24 $\bar{x}$
        19.4   22.5
Mean distance (in miles)

$$z_1 = \frac{19.4 - 20.7}{6.5/\sqrt{50}} \approx -1.41 \qquad \text{Convert 19.4 to } z\text{-score}$$

$$z_2 = \frac{22.5 - 20.7}{6.5/\sqrt{50}} \approx 1.96 \qquad \text{Convert 22.5 to } z\text{-score}$$

So, the probability that the mean distance driven each day by the sample of 50 people is between 19.4 and 22.5 miles is

$$\begin{aligned} P(19.4 < \bar{x} < 22.5) &= P(-1.41 < z < 1.96) \\ &= P(z < 1.96) - P(z < -1.41) \\ &= 0.9750 - 0.0793 \\ &= 0.8957. \end{aligned}$$

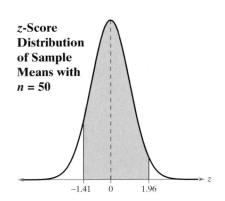

z-Score Distribution of Sample Means with $n = 50$

−1.41   0   1.96 z

Interpretation   Of all samples of 50 drivers ages 16 to 19, about 90% will drive a mean distance each day between 19.4 and 22.5 miles, as shown in the graph at the left. This implies that, assuming the value of $\mu = 20.7$ is correct, about 10% of such sample means will lie outside the given interval.

TRY IT YOURSELF 4

You randomly select 100 drivers ages 16 to 19 from Example 4. What is the probability that the mean distance traveled each day is between 19.4 and 22.5 miles? Use $\mu = 20.7$ miles and $\sigma = 6.5$ miles.

Answer: Page A35

Study Tip

Before you find probabilities for intervals of the sample mean $\bar{x}$, use the Central Limit Theorem to determine the mean and the standard deviation of the sampling distribution of the sample means. That is, calculate $\mu_{\bar{x}}$ and $\sigma_{\bar{x}}$.

EXAMPLE 5

Finding Probabilities for Sampling Distributions

The mean room and board expense per year at four-year colleges is $10,453. You randomly select 9 four-year colleges. What is the probability that the mean room and board is less than $10,750? Assume that the room and board expenses are normally distributed with a standard deviation of $1650. *(Adapted from National Center for Education Statistics)*

SOLUTION

Because the population is normally distributed, you can use the Central Limit Theorem to conclude that the distribution of sample means is normally distributed, with a mean and a standard deviation of

$$\mu_{\bar{x}} = \mu = \$10,453 \quad \text{and} \quad \sigma_{\bar{x}} = \frac{\sigma}{\sqrt{n}} = \frac{\$1650}{\sqrt{9}} = \$550.$$

The graph of this distribution is shown at the left. The area to the left of $10,750 is shaded. The z-score that corresponds to $10,750 is

$$z = \frac{10{,}750 - 10{,}453}{1650/\sqrt{9}} = \frac{297}{550} = 0.54.$$

So, the probability that the mean room and board expense is less than $10,750 is

$$P(\bar{x} < 10{,}750) = P(z < 0.54)$$
$$= 0.7054.$$

You can check this answer using technology. For instance, you can use a TI-84 Plus to find the x-value, as shown below.

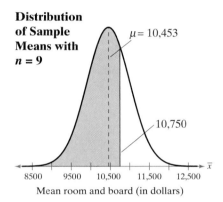

Distribution of Sample Means with n = 9

Mean room and board (in dollars)

```
TI-84 PLUS
normalcdf(-10000
,10750,10453,550
)
        .7054015111
```

Interpretation   So, about 71% of such samples with $n = 9$ will have a mean less than $10,750 and about 29% of these sample means will be greater than $10,750.

TRY IT YOURSELF 5

The average sales price of a single-family house in the United States is $235,500. You randomly select 12 single-family houses. What is the probability that the mean sales price is more than $225,000? Assume that the sales prices are normally distributed with a standard deviation of $50,000. *(Adapted from National Association of Realtors)*

Answer: Page A35

The Central Limit Theorem can also be used to investigate unusual events. An unusual event is one that occurs with a probability of less than 5%.

Study Tip

To find probabilities for individual members of a population with a normally distributed random variable x, use the formula

$$z = \frac{x - \mu}{\sigma}.$$

To find probabilities for the mean $\bar{x}$ of a sample of size n, use the formula

$$z = \frac{\bar{x} - \mu_{\bar{x}}}{\sigma_{\bar{x}}}.$$

EXAMPLE 6

Finding Probabilities for x and x̄

Some college students use credit cards to pay for school-related expenses. For this population, the amount paid is normally distributed, with a mean of $1615 and a standard deviation of $550. *(Adapted from Sallie Mae/Ipsos Public Affairs)*

1. What is the probability that a randomly selected college student, who uses a credit card to pay for school-related expenses, paid less than $1400?

2. You randomly select 25 college students who use credit cards to pay for school-related expenses. What is the probability that their mean amount paid is less than $1400?

3. Compare the probabilities from parts 1 and 2.

SOLUTION

1. In this case, you are asked to find the probability associated with a certain value of the random variable x. The z-score that corresponds to $x = \$1400$ is

$$z = \frac{x - \mu}{\sigma} = \frac{1400 - 1615}{550} = \frac{-215}{550} \approx -0.39.$$

So, the probability that the student paid less than $1400 is

$$P(x < 1400) = P(z < -0.39) = 0.3483.$$

You can check this answer using technology. For instance, you can use Excel to find the probability, as shown at the left. (The answer differs slightly due to rounding.)

2. Here, you are asked to find the probability associated with a sample mean $\bar{x}$. The z-score that corresponds to $\bar{x} = \$1400$ is

$$z = \frac{\bar{x} - \mu_{\bar{x}}}{\sigma_{\bar{x}}} = \frac{\bar{x} - \mu}{\sigma / \sqrt{n}} = \frac{1400 - 1615}{550 / \sqrt{25}} = \frac{-215}{110} \approx -1.95.$$

So, the probability that the mean credit card balance of the 25 card holders is less than $1400 is

$$P(\bar{x} < 1400) = P(z < -1.95) = 0.0256.$$

You can check this answer using technology. For instance, you can use Excel to find the probability, as shown at the left. (The answer differs slightly due to rounding.)

3. ***Interpretation***   Although there is about a 35% chance that a college student who uses a credit card to pay for school-related expenses will pay less than $1400, there is only about a 3% chance that the mean amount a sample of 25 college students will pay is less than $1400. Because there is only a 3% chance that the mean amount a sample of 25 college students will pay is less than $1400, this is an unusual event.

TRY IT YOURSELF 6

A consumer price analyst claims that prices for liquid crystal display (LCD) computer monitors are normally distributed, with a mean of $190 and a standard deviation of $48. What is the probability that a randomly selected LCD computer monitor costs less than $200? You randomly select 10 LCD computer monitors. What is the probability that their mean cost is less than $200? Compare these two probabilities. *Answer: Page A35*

5.4 EXERCISES

For Extra Help: **MyLab Statistics**

1. 150, 3.536

2. 45, 1.5

3. 790, 3.036

4. 1275, 0.190

5. False. As the size of a sample increases, the mean of the distribution of sample means does not change.

6. False. As the size of a sample increases, the standard deviation of the distribution of sample means decreases.

7. False. A sampling distribution is normal when either $n \geq 30$ or the population is normal.

8. True

9. (c), because $\mu_{\bar{x}} = 16.5$, $\sigma_{\bar{x}} = 1.19$, and the graph approximates a normal curve.

Building Basic Skills and Vocabulary

In Exercises 1–4, a population has a mean μ and a standard deviation σ. Find the mean and standard deviation of the sampling distribution of sample means with sample size n.

1. $\mu = 150, \sigma = 25, n = 50$

2. $\mu = 45, \sigma = 15, n = 100$

3. $\mu = 790, \sigma = 48, n = 250$

4. $\mu = 1275, \sigma = 6, n = 1000$

True or False? *In Exercises 5–8, determine whether the statement is true or false. If it is false, rewrite it as a true statement.*

5. As the sample size increases, the mean of the distribution of sample means increases.

6. As the sample size increases, the standard deviation of the distribution of sample means increases.

7. A sampling distribution is normal only when the population is normal.

8. If the sample size is at least 30, then you can use *z*-scores to determine the probability that a sample mean falls in a given interval of the sampling distribution.

Graphical Analysis *In Exercises 9 and 10, the graph of a population distribution is shown with its mean and standard deviation. Random samples of size 100 are drawn from the population. Determine which of the figures labeled (a)–(c) would most closely resemble the sampling distribution of sample means. Explain your reasoning.*

9. The waiting time (in seconds) to turn left at an intersection

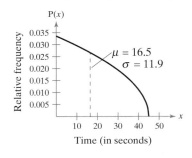

(a)

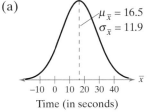

Time (in seconds)

(b)

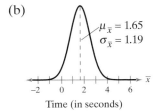

Time (in seconds)

(c)

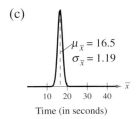

Time (in seconds)

10. (b), because $\mu_{\bar{x}} = 5.8$, $\sigma_{\bar{x}} = 0.23$, and the graph approximates a normal curve.

11. (a) $\mu = 53.2$, $\sigma \approx 19.9$

(b)

Sample	Mean
19, 19	19
19, 48	33.5
19, 56	37.5
19, 64	41.5
19, 79	49
48, 19	33.5
48, 48	48
48, 56	52
48, 64	56
48, 79	63.5
56, 19	37.5
56, 48	52
56, 56	56
56, 64	60
56, 79	67.5
64, 19	41.5
64, 48	56
64, 56	60
64, 64	64
64, 79	71.5
79, 19	49
79, 48	63.5
79, 56	67.5
79, 64	71.5
79, 79	79

(c) $\mu_{\bar{x}} = 53.2$, $\sigma_{\bar{x}} \approx 14.1$

The means are equal, but the standard deviation of the sampling distribution is smaller.

12. See Selected Answers, page A97.

13. See Odd Answers, page A61.

14. See Selected Answers, page A97.

15. 0.9726; not unusual

16. 0.0082; unusual

17. 0.0351 (*Tech:* 0.0349); unusual

18. 0.5; not unusual

10. The annual snowfall (in feet) for a central New York state county

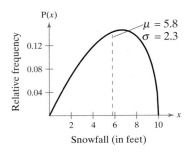

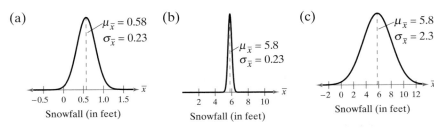

(a) Snowfall (in feet)   (b) Snowfall (in feet)   (c) Snowfall (in feet)

A Sampling Distribution of Sample Means *In Exercises 11–14, a population and sample size are given.*

(a) Find the mean and standard deviation of the population.

(b) List all samples (with replacement) of the given size from the population and find the mean of each.

(c) Find the mean and standard deviation of the sampling distribution of sample means and compare them with the mean and standard deviation of the population.

11. The load-bearing capacities (in thousands of pounds) of five transmission line insulators are 64, 48, 19, 79, and 56. Use a sample size of 2.

12. The diameters (in inches) of four machine parts are 1.000, 1.004, 1.001, and 1.003. Use a sample size of 2.

13. The melting points (in degrees Celsius) of three industrial lubricants are 350, 399, and 418. Use a sample size of 3.

14. The lifetimes (in hours) of four diamond-tipped cutting tools are 70, 85, 81, and 67. Use a sample size of 3.

Finding Probabilities *In Exercises 15–18, the population mean and standard deviation are given. Find the indicated probability and determine whether the given sample mean would be considered unusual.*

15. For a random sample of $n = 64$, find the probability of a sample mean being less than 24.3 when $\mu = 24$ and $\sigma = 1.25$.

16. For a random sample of $n = 100$, find the probability of a sample mean being greater than 24.3 when $\mu = 24$ and $\sigma = 1.25$.

17. For a random sample of $n = 45$, find the probability of a sample mean being greater than 551 when $\mu = 550$ and $\sigma = 3.7$.

18. For a random sample of $n = 36$, find the probability of a sample mean being less than 12,750 or greater than 12,753 when $\mu = 12,750$ and $\sigma = 1.7$.

19. $\mu_{\bar{x}} = 495$, $\sigma_{\bar{x}} \approx 26.83$

425 475 525 575
Mean score

20. $\mu_{\bar{x}} = 493$, $\sigma_{\bar{x}} = 19$

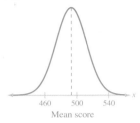

460 500 540
Mean score

21. $\mu_{\bar{x}} = 23$, $\sigma_{\bar{x}} = 0.26$

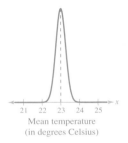

21 22 23 24 25
Mean temperature
(in degrees Celsius)

22. $\mu_{\bar{x}} = 87$, $\sigma_{\bar{x}} \approx 2.65$

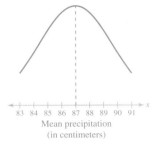

83 84 85 86 87 88 89 90 91
Mean precipitation
(in centimeters)

23. $\mu_{\bar{x}} = 1.64$, $\sigma_{\bar{x}} \approx 0.83$

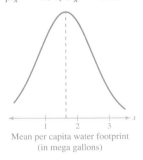

1 2 3
Mean per capita water footprint
(in mega gallons)

24. See Selected Answers, page A97.

25. See Odd Answers, page A61.

26. See Selected Answers, page A97.

27. See Odd Answers, page A61.

28. See Selected Answers, page A97.

Using and Interpreting Concepts

Interpreting the Central Limit Theorem *In Exercises 19–26, find the mean and standard deviation of the indicated sampling distribution of sample means. Then sketch a graph of the sampling distribution.*

19. SAT Critical Reading Scores: Males The scores for males on the critical reading portion of the SAT in 2016 are normally distributed, with a mean of 495 and a standard deviation of 120. Random samples of size 20 are drawn from this population, and the mean of each sample is determined. *(Source: The College Board)*

20. SAT Critical Reading Scores: Females The scores for females on the critical reading portion of the SAT in 2016 are normally distributed, with a mean of 493 and a standard deviation of 114. Random samples of size 36 are drawn from this population, and the mean of each sample is determined. *(Source: The College Board)*

21. Temperature The monthly growing season temperatures across villages in Tanzania are normally distributed, with a mean of 23°C and a standard deviation of 1.3°C. Random samples of size 25 are drawn from this population, and the mean of each sample is determined. *(Source: Agricultural and Applied Economics Association)*

22. Precipitation The monthly growing season precipitation across villages in Tanzania is normally distributed, with a mean of 87 centimeters and a standard deviation of 14.5 centimeters. Random samples of size 30 are drawn from this population, and the mean of each sample is determined. *(Source: Agricultural and Applied Economics Association)*

23. Water Footprint A *water footprint* is a measure of the appropriation of fresh water. The per capita water footprint in the United States for a recent year is approximately normally distributed, with a mean of 1.64 mega gallons and a standard deviation of 2.89 mega gallons. Random samples of size 12 are drawn from this population, and the mean of each sample is determined. *(Source: Water Resources Research)*

24. Water Use in Hospitals The amounts of cold water for patient consumption in hospitals in Spain are normally distributed, with a mean of 196 cubic meters per bed and a standard deviation of 70 cubic meters per bed. Random samples of size 15 are drawn from this population, and the mean of each sample is determined. *(Source: Journal of Healthcare Engineering)*

25. Salaries The annual salary for senior-level chemical engineers is normally distributed, with a mean of about $132,000 and a standard deviation of about $18,000. Random samples of 35 are drawn from this population, and the mean of each sample is determined. *(Adapted from Salary.com)*

26. Salaries The annual salary for clinical pharmacists is normally distributed, with a mean of about $111,000 and a standard deviation of about $13,000. Random samples of 48 are drawn from this population, and the mean of each sample is determined. *(Adapted from Salary.com)*

27. Repeat Exercise 19 for samples of size 40 and 60. What happens to the mean and the standard deviation of the distribution of sample means as the sample size increases?

28. Repeat Exercise 20 for samples of size 72 and 108. What happens to the mean and the standard deviation of the distribution of sample means as the sample size increases?

29. 0.4623 (*Tech:* 0.4645); About 46% of samples of 32 years will have a mean gain between 200 and 500.

30. 0.1497 (*Tech:* 0.1490); About 15% of samples of 38 years will have a mean return between 9.1% and 10.3%.

31. 0.0708 (*Tech:* 0.0702); About 7% of samples of 30 Chinese cities will have a mean childhood asthma rate greater than 2.6%.

32. 0.4840 (*Tech:* 0.4833); About 48% of samples of 44 countries will have a mean per capita carbon dioxide emission greater than 28 metric tons.

33. It is more likely to select a sample of 10 cities with a mean childhood asthma prevalence less than 3.2% because the sample of 10 has a higher probability.

34. It is more likely to select a sample of 15 countries with mean carbon dioxide emissions less than 30 metric tons because the sample of 15 has a higher probability.

35. Yes, it is very unlikely that you would have randomly sampled 40 cans with a mean equal to 127.9 ounces because it is more than 3 standard deviations from the mean of the sample means.

36. Yes, it is very unlikely that you would have randomly sampled 40 containers with a mean equal to 64.05 ounces because it is more than 2 standard deviations from the mean of the sample means.

37. (a) 0.3085   (b) 0.0008

38. (a) 0.3372   (b) 0.0179

Finding Probabilities for Sampling Distributions   *In Exercises 29–32, find the indicated probability and interpret the results.*

29. **Dow Jones Industrial Average**   From 1975 through 2016, the mean gain of the Dow Jones Industrial Average was 456. A random sample of 32 years is selected from this population. What is the probability that the mean gain for the sample was between 200 and 500? Assume $\sigma = 1215$.

30. **Standard & Poor's 500**   From 1871 through 2016, the mean return of the Standard & Poor's 500 was 10.72%. A random sample of 38 years is selected from this population. What is the probability that the mean return for the sample was between 9.1% and 10.3%? Assume $\sigma = 18.60\%$.

31. **Childhood Asthma Prevalence**   The mean percent of childhood asthma prevalence of 43 Chinese cities is 2.25%. A random sample of 30 Chinese cities is selected. What is the probability that the mean childhood asthma prevalence for the sample is greater than 2.6%? Assume $\sigma = 1.30\%$. *(Source: BioMed Research International)*

32. **Carbon Dioxide Emissions**   The mean per capita carbon dioxide emissions in 58 industrialized countries over a 22-year period is 25.5 metric tons. A random sample of 44 countries is selected. What is the probability that the mean carbon dioxide emissions for the sample is greater than 28 metric tons? Assume $\sigma = 395.1$ metric tons. *(Source: Energy Reports)*

33. **Which Is More Likely?**   Assume that the childhood asthma prevalences in Exercise 31 are normally distributed. Are you more likely to randomly select 1 city with childhood asthma prevalence less than 3.2% or to randomly select a sample of 10 cities with a mean childhood asthma prevalence less than 3.2%? Explain.

34. **Which Is More Likely?**   Assume that the carbon dioxide emissions in Exercise 32 are normally distributed. Are you more likely to randomly select 1 country with carbon dioxide emissions less than 30 metric tons or to randomly select a sample of 15 countries with mean carbon dioxide emissions less than 30 metric tons? Explain.

35. **Paint Cans**   A machine is set to fill paint cans with a mean of 128 ounces and a standard deviation of 0.2 ounce. A random sample of 40 cans has a mean of 127.9 ounces. The machine needs to be reset when the mean of a random sample is unusual. Does the machine need to be reset? Explain.

36. **Milk Containers**   A machine is set to fill milk containers with a mean of 64 ounces and a standard deviation of 0.11 ounce. A random sample of 40 containers has a mean of 64.05 ounces. The machine needs to be reset when the mean of a random sample is unusual. Does the machine need to be reset? Explain.

37. **Lumber Cutter**   The lengths of lumber a machine cuts are normally distributed, with a mean of 96 inches and a standard deviation of 0.5 inch.
    (a) What is the probability that a randomly selected board cut by the machine has a length greater than 96.25 inches?
    (b) You randomly select 40 boards. What is the probability that their mean length is greater than 96.25 inches?

38. **Ice Cream**   The weights of ice cream cartons are normally distributed with a mean weight of 10 ounces and a standard deviation of 0.5 ounce.
    (a) What is the probability that a randomly selected carton has a weight greater than 10.21 ounces?
    (b) You randomly select 25 cartons. What is the probability that their mean weight is greater than 10.21 ounces?

Extending Concepts

Finite Correction Factor *The formula for the standard deviation of the sampling distribution of sample means*

$$\sigma_{\bar{x}} = \frac{\sigma}{\sqrt{n}}$$

*given in the Central Limit Theorem is based on an assumption that the population has infinitely many members. This is the case whenever sampling is done with replacement (each member is put back after it is selected), because the sampling process could be continued indefinitely. The formula is also valid when the sample size is small in comparison with the population. When sampling is done without replacement and the sample size n is more than 5% of the finite population of size N ($n/N > 0.05$), however, there is a finite number of possible samples. A **finite correction factor**,*

$$\sqrt{\frac{N - n}{N - 1}}$$

should be used to adjust the standard deviation. The sampling distribution of the sample means will be normal with a mean equal to the population mean, and the standard deviation will be

$$\sigma_{\bar{x}} = \frac{\sigma}{\sqrt{n}}\sqrt{\frac{N - n}{N - 1}}.$$

In Exercises 39 and 40, determine whether the finite correction factor should be used. If so, use it in your calculations when you find the probability.

39. **Parking Infractions**   In a sample of 1000 fines issued by the city of Toronto for parking infractions, the mean fine was $47.12 and the standard deviation was $48.24. A random sample of size 55 is selected from this population. What is the probability that the mean fine is less than $50? *(Adapted from City of Toronto)*

40. **Old Faithful**   In a sample of 100 eruptions of the Old Faithful geyser at Yellowstone National Park, the mean interval between eruptions was 101.56 minutes and the standard deviation was 42.69 minutes. A random sample of size 30 is selected from this population. What is the probability that the mean interval between eruptions is between 95 minutes and 110 minutes? *(Adapted from Geyser Times)*

Sampling Distribution of Sample Proportions   *For a random sample of size n, the **sample proportion** is the number of individuals in the sample with a specified characteristic divided by the sample size. The **sampling distribution of sample proportions** is the distribution formed when sample proportions of size n are repeatedly taken from a population where the probability of an individual with a specified characteristic is p. The sampling distribution of sample proportions has a mean equal to the population proportion p and a standard deviation equal to $\sqrt{pq/n}$. In Exercises 41 and 42, assume the sampling distribution of sample proportions is a normal distribution.*

41. **Construction**   About 63% of the residents in a town are in favor of building a new high school. One hundred five residents are randomly selected. What is the probability that the sample proportion in favor of building a new school is less than 55%? Interpret your result.

42. **Conservation**   About 74% of the residents in a town say that they are making an effort to conserve water or electricity. One hundred ten residents are randomly selected. What is the probability that the sample proportion making an effort to conserve water or electricity is greater than 80%? Interpret your result.

---

39. Yes, the finite correction factor should be used; 0.6772 (*Tech:* 0.6755)

40. Yes, the finite correction factor should be used; 0.7428 (*Tech:* 0.7427)

41. 0.0446 (*Tech:* 0.0448); The probability that less than 55% of a sample of 105 residents are in favor of building a new high school is about 4.5%. Because the probability is less than 0.05, this is an unusual event.

42. 0.0764 (*Tech:* 0.0757); The probability that more than 80% of a sample of 110 residents are making an effort to conserve water or electricity is about 7.6%.

APPLET

You can find the interactive applet for this activity within **MyLab Statistics** or at *www.pearsonhighered.com/ mathstatsresources.*

The *sampling distributions* applet allows you to investigate sampling distributions by repeatedly taking random samples from a population. The top plot displays the distribution of a population. Several options are available for the population distribution (Uniform, Bell-shaped, Skewed, Binary, and Custom). When SAMPLE is clicked, N random samples of size n will be repeatedly selected from the population. The sample statistics specified in the bottom two plots will be updated for each sample. When N is set to 1 and n is less than or equal to 50, the display will show, in an animated fashion, the points selected from the population dropping into the second plot and the corresponding summary statistic values dropping into the third and fourth plots. Click RESET to stop an animation and clear existing results. Summary statistics for each plot are shown in the panel at the left of the plot.

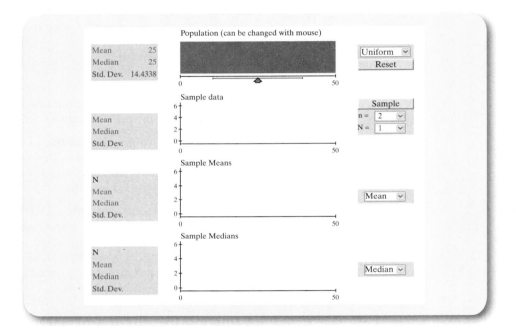

EXPLORE

Step 1   Specify a distribution.
Step 2   Specify values of n and N.
Step 3   Specify what to display in the bottom two graphs.
Step 4   Click SAMPLE to generate the sampling distributions.

DRAW CONCLUSIONS

APPLET

1. Run the simulation using $n = 30$ and $N = 10$ for a uniform, a bell-shaped, and a skewed distribution. What is the mean of the sampling distribution of the sample means for each distribution? For each distribution, is this what you would expect?

2. Run the simulation using $n = 50$ and $N = 10$ for a bell-shaped distribution. What is the standard deviation of the sampling distribution of the sample means? According to the formula, what should the standard deviation of the sampling distribution of the sample means be? Is this what you would expect?

5.5 | Normal Approximations to Binomial Distributions

What You Should Learn

▶ How to determine when a normal distribution can approximate a binomial distribution

▶ How to find the continuity correction

▶ How to use a normal distribution to approximate binomial probabilities

Approximating a Binomial Distribution ▪ Continuity Correction ▪ Approximating Binomial Probabilities

Approximating a Binomial Distribution

In Section 4.2, you learned how to find binomial probabilities. For instance, consider a surgical procedure that has an 85% chance of success. When a doctor performs this surgery on 10 patients, you can use the binomial formula to find the probability of exactly two successful surgeries.

But what if the doctor performs the surgical procedure on 150 patients and you want to find the probability of *fewer than 100* successful surgeries? To do this using the techniques described in Section 4.2, you would have to use the binomial formula 100 times and find the sum of the resulting probabilities. This approach is not practical, of course. A better approach is to use a normal distribution to approximate the binomial distribution.

Normal Approximation to a Binomial Distribution

If $np \geq 5$ and $nq \geq 5$, then the binomial random variable x is approximately normally distributed, with mean

$$\mu = np$$

and standard deviation

$$\sigma = \sqrt{npq}$$

where n is the number of independent trials, p is the probability of success in a single trial, and q is the probability of failure in a single trial.

Study Tip

Here are some properties of binomial experiments (see Section 4.2).

- n independent trials
- Two possible outcomes: success or failure
- Probability of success is p; probability of failure is $q = 1 - p$
- p is the same for each trial

To see why a normal approximation is valid, look at the binomial distributions for $p = 0.25$, $q = 1 - 0.25 = 0.75$, and $n = 4$, $n = 10$, $n = 25$, and $n = 50$ shown below. Notice that as n increases, the shape of the binomial distribution becomes more similar to a normal distribution.

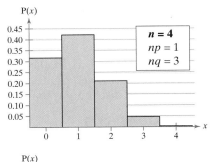

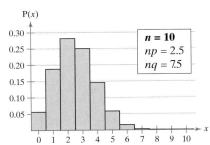

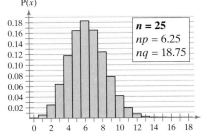

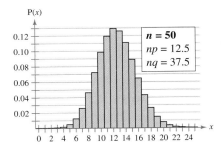

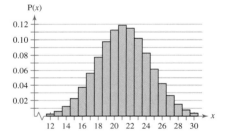

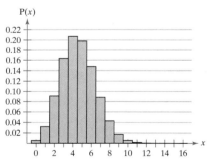

EXAMPLE 1

Approximating a Binomial Distribution

Two binomial experiments are listed. Determine whether you can use a normal distribution to approximate the distribution of x, the number of people who reply yes. If you can, find the mean and standard deviation. If you cannot, explain why.

1. In a survey of 8- to 18-year-old heavy media users in the United States, 47% said they get fair or poor grades (C and below). You randomly select forty-five 8- to 18-year-old heavy media users in the United States and ask them whether they get fair or poor grades. *(Source: Kaiser Family Foundation)*

2. In a survey of 8- to 18-year-old light media users in the United States, 23% said they get fair or poor grades (C and below). You randomly select twenty 8- to 18-year-old light media users in the United States and ask them whether they get fair or poor grades. *(Source: Kaiser Family Foundation)*

SOLUTION

1. In this binomial experiment, $n = 45$, $p = 0.47$, and $q = 0.53$. So,

$$np = 45(0.47) = 21.15$$

and

$$nq = 45(0.53) = 23.85.$$

Because np and nq are greater than 5, you can use a normal distribution with

$$\mu = np = 21.15$$

and

$$\sigma = \sqrt{npq} = \sqrt{45(0.47)(0.53)} \approx 3.35$$

to approximate the distribution of x. In the figure at the left, notice that the binomial distribution is approximately bell-shaped, which supports the conclusion that you can use a normal distribution to approximate the distribution of x.

2. In this binomial experiment, $n = 20$, $p = 0.23$, and $q = 0.77$. So,

$$np = 20(0.23) = 4.6$$

and

$$nq = 20(0.77) = 15.4.$$

Because $np < 5$, you cannot use a normal distribution to approximate the distribution of x. In the figure at the left, notice that the binomial distribution is skewed right, which supports the conclusion that you cannot use a normal distribution to approximate the distribution of x.

TRY IT YOURSELF 1

A binomial experiment is listed. Determine whether you can use a normal distribution to approximate the distribution of x, the number of people who reply yes. If you can, find the mean and standard deviation. If you cannot, explain why.

In a survey of adults in the United States, 29% said they have seen a person using a mobile device walk in front of a moving vehicle without looking. You randomly select 100 adults in the United States and ask them whether they have seen a person using a mobile device walk in front of a moving vehicle without looking. *(Source: Consumer Reports)*

Answer: Page A36

Continuity Correction

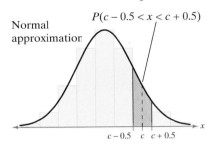

Exact binomial probability

$P(x = c)$

Normal approximation

$P(c - 0.5 < x < c + 0.5)$

$c - 0.5 \quad c \quad c + 0.5$

A binomial distribution is discrete and can be represented by a probability histogram. To calculate *exact* binomial probabilities, you can use the binomial formula for each value of x and add the results. Geometrically, this corresponds to adding the areas of bars in the probability histogram (see top figure at the left). Remember that each bar has a width of one unit and x is the midpoint of the interval.

When you use a *continuous* normal distribution to approximate a binomial probability, you need to move 0.5 unit to the left and right of the midpoint to include all possible x-values in the interval (see bottom figure at the left). When you do this, you are making a **continuity correction.**

EXAMPLE 2

Using a Continuity Correction

Use a continuity correction to convert each binomial probability to a normal distribution probability.

1. The probability of getting between 270 and 310 successes, inclusive

2. The probability of getting at least 158 successes

3. The probability of getting fewer than 63 successes

SOLUTION

1. The discrete midpoint values are 270, 271, . . ., 310. The corresponding interval for the continuous normal distribution is $269.5 < x < 310.5$ and the normal distribution probability is $P(269.5 < x < 310.5)$.

2. The discrete midpoint values are 158, 159, 160, ... The corresponding interval for the continuous normal distribution is $x > 157.5$ and the normal distribution probability is $P(x > 157.5)$.

3. The discrete midpoint values are . . ., 60, 61, 62. The corresponding interval for the continuous normal distribution is $x < 62.5$ and the normal distribution probability is $P(x < 62.5)$.

TRY IT YOURSELF 2

Use a continuity correction to convert each binomial probability to a normal distribution probability.

1. The probability of getting between 57 and 83 successes, inclusive

2. The probability of getting at most 54 successes

Answer: Page A36

Study Tip

In a discrete distribution, there is a difference between $P(x \geq c)$ and $P(x > c)$. This is true because the probability that x is exactly c is not 0. In a continuous distribution, however, there is no difference between $P(x \geq c)$ and $P(x > c)$ because the probability that x is exactly c is 0.

Shown below are several cases of binomial probabilities involving the number c and how to convert each to a normal distribution probability.

Binomial	Normal	Notes
Exactly c	$P(c - 0.5 < x < c + 0.5)$	Includes c
At most c	$P(x < c + 0.5)$	Includes c
Fewer than c	$P(x < c - 0.5)$	Does not include c
At least c	$P(x > c - 0.5)$	Includes c
More than c	$P(x > c + 0.5)$	Does not include c

Picturing the World

Have You Ever Hidden Purchases from Your Spouse?

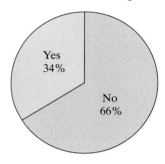

Yes 34%

No 66%

Assume that this survey is a true indication of the proportion of the population who say they have hidden purchases from their spouses. You sample 50 adults with spouses at random. What is the probability that between 20 and 25, inclusive, would say they have hidden purchases from their spouses?

For a sample of 50 adults, the probability that between 20 and 25, inclusive, would say they have hidden purchases from their spouses is about 0.2211 (*Tech:* 0.2221).

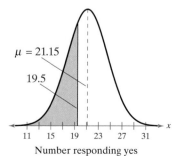

$\mu = 21.15$

19.5

11   15   19   23   27   31

Number responding yes

Approximating Binomial Probabilities

GUIDELINES

Using a Normal Distribution to Approximate Binomial Probabilities

In Words	In Symbols
1. Verify that a binomial distribution applies.	Specify n, p, and q.
2. Determine whether you can use a normal distribution to approximate x, the binomial variable.	Is $np \geq 5$? Is $nq \geq 5$?
3. Find the mean μ and standard deviation σ for the distribution.	$\mu = np$ $\sigma = \sqrt{npq}$
4. Apply the appropriate continuity correction. Shade the corresponding area under the normal curve.	Add 0.5 to (or subtract 0.5 from) the binomial probability.
5. Find the corresponding z-score(s).	$z = \dfrac{x - \mu}{\sigma}$
6. Find the probability.	Use the Standard Normal Table.

EXAMPLE 3

Approximating a Binomial Probability

In a survey of 8- to 18-year-old heavy media users in the United States, 47% said they get fair or poor grades (C and below). You randomly select forty-five 8- to 18-year-old heavy media users in the United States and ask them whether they get fair or poor grades. What is the probability that fewer than 20 of them respond yes? *(Source: Kaiser Family Foundation)*

SOLUTION

From Example 1, you know that you can use a normal distribution with

$$\mu = 21.15 \quad \text{and} \quad \sigma \approx 3.35$$

to approximate the binomial distribution. To use a normal distribution, note that the probability is "fewer than 20." So, apply the continuity correction by subtracting 0.5 from 20 and write the probability as

$$P(x < 20 - 0.5) = P(x < 19.5).$$

The figure at the left shows a normal curve with $\mu = 21.15$, $\sigma \approx 3.35$, and the shaded area to the left of 19.5. The z-score that corresponds to $x = 19.5$ is

$$z = \frac{x - \mu}{\sigma}$$

$$z \approx \frac{19.5 - 21.15}{3.35}$$

$$\approx -0.49.$$

Using the Standard Normal Table,

$$P(z < -0.49) = 0.3121.$$

Interpretation The probability that fewer than twenty 8- to 18-year-olds respond yes is approximately 0.3121, or about 31.21%.

TRY IT YOURSELF 3

In a survey of adults in the United States, 29% said they have seen a person using a mobile device walk in front of a moving vehicle without looking. You randomly select 100 adults in the United States and ask them whether they have seen a person using a mobile device walk in front of a moving vehicle without looking. What is the probability that more than 30 respond yes? *(Source: Consumer Reports)*

Answer: Page A36

EXAMPLE 4

Approximating a Binomial Probability

A study on aggressive driving found that 47% of drivers say they have yelled at another driver. You randomly select 200 drivers in the United States and ask them whether they have yelled at another driver. What is the probability that at least 100 drivers will say yes, they have yelled at another driver? *(Source: American Automobile Association)*

SOLUTION

Because $np = 200(0.47) = 94$ and $nq = 200(0.53) = 106$, the binomial variable x is approximately normally distributed, with

$$\mu = np = 94 \quad \text{and} \quad \sigma = \sqrt{npq} = \sqrt{200(0.47)(0.53)} \approx 7.06.$$

Using the continuity correction, you can rewrite the discrete probability $P(x \geq 100)$ as the continuous probability $P(x > 99.5)$. The figure shows a normal curve with $\mu = 94$, $\sigma = 7.06$, and the shaded area to the right of 99.5.

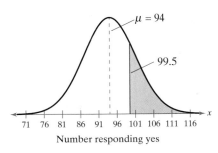

Number responding yes

The z-score that corresponds to 99.5 is

$$z = \frac{99.5 - 94}{\sqrt{200(0.47)(0.53)}} \approx 0.78.$$

So, the probability that at least 100 drivers will say "yes" is approximately

$$
\begin{aligned}
P(x > 99.5) &= P(z > 0.78) \\
&= 1 - P(z < 0.78) \\
&= 1 - 0.7823 \\
&= 0.2177.
\end{aligned}
$$

Interpretation The probability that at least 100 drivers will say "yes" is approximately 0.2177, or about 21.8%.

TRY IT YOURSELF 4

In Example 4, what is the probability that at most 80 drivers will say yes, they have yelled at another driver?

Answer: Page A36

Tech Tip

Recall that you can use technology to find a normal probability. For instance, in Example 4, you can use a TI-84 Plus to find the probability once the mean, standard deviation, and continuity correction are calculated. (Use 10,000 for the upper bound.)

```
normalcdf(99.5,1
0000,94,7.06)
        .2179789416
```

EXAMPLE 5

Approximating a Binomial Probability

A study of National Football League (NFL) retirees, ages 50 and older, found that 62.4% have arthritis. You randomly select 75 NFL retirees who are at least 50 years old and ask them whether they have arthritis. What is the probability that exactly 48 will say yes? *(Source: University of Michigan, Institute for Social Research)*

SOLUTION

Because $np = 75(0.624) = 46.8$ and $nq = 75(0.376) = 28.2$, the binomial variable x is approximately normally distributed, with

$$\mu = np = 46.8 \quad \text{and} \quad \sigma = \sqrt{npq} = \sqrt{75(0.624)(0.376)} \approx 4.19.$$

Using the continuity correction, you can rewrite the discrete probability $P(x = 48)$ as the continuous probability $P(47.5 < x < 48.5)$. The figure shows a normal curve with $\mu = 46.8$, $\sigma \approx 4.19$, and the shaded area under the curve between 47.5 and 48.5.

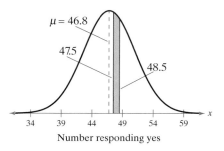

Number responding yes

The z-score that corresponds to 47.5 is

$$z_1 = \frac{47.5 - 46.8}{\sqrt{75(0.624)(0.376)}} \approx 0.17$$

and the z-score that corresponds to 48.5 is

$$z_2 = \frac{48.5 - 46.8}{\sqrt{75(0.624)(0.376)}} \approx 0.41.$$

So, the probability that exactly 48 NFL retirees will say they have arthritis is

$$
\begin{aligned}
P(47.5 < x < 48.5) &= P(0.17 < z < 0.41) \\
&= P(z < 0.41) - P(z < 0.17) \\
&= 0.6591 - 0.5675 \\
&= 0.0916.
\end{aligned}
$$

Interpretation The probability that exactly 48 NFL retirees will say they have arthritis is approximately 0.0916, or about 9.2%.

TRY IT YOURSELF 5

The study in Example 5 found that 32.0% of all men in the United States ages 50 and older have arthritis. You randomly select 75 men in the United States who are at least 50 years old and ask them whether they have arthritis. What is the probability that exactly 15 will say yes? *(Source: University of Michigan, Institute for Social Research)*

Answer: Page A36

Tech Tip

The approximation in Example 5 is almost the same as the probability found using the binomial probability feature of a technology tool. For instance, compare the result in Example 5 with the one found on a TI-84 Plus shown below.

```
binompdf(75,.624
,48)
        .0917597587
```

5.5 EXERCISES

1. Cannot use normal distribution
2. Cannot use normal distribution
3. Cannot use normal distribution
4. Can use normal distribution
5. a
6. d
7. c
8. b
9. The probability of getting fewer than 25 successes; $P(x < 24.5)$
10. The probability of getting at least 110 successes; $P(x > 109.5)$
11. The probability of getting exactly 33 successes; $P(32.5 < x < 33.5)$
12. The probability of getting more than 65 successes; $P(x > 65.5)$
13. The probability of getting at most 150 successes; $P(x < 150.5)$
14. The probability of getting between 55 and 60 successes; $P(55.5 < x < 59.5)$
15. Binomial: $P(5 \le x \le 7) \approx 0.549$
    Normal: $P(4.5 < x < 7.5) = 0.5463$ (*Tech:* 0.5466)
    The results are about the same.
16. Binomial: $P(3 \le x \le 5) \approx 0.3679$
    Normal: $P(2.5 < x < 5.5) = 0.3642$ (*Tech:* 0.3648)
    The results are about the same.
17. Can use normal distribution; $\mu = 9.3$, $\sigma \approx 2.533$
18. Cannot use normal distribution because $nq < 5$.

Building Basic Skills and Vocabulary

In Exercises 1–4, the sample size n, probability of success p, and probability of failure q are given for a binomial experiment. Determine whether you can use a normal distribution to approximate the distribution of x.

1. $n = 24, p = 0.85, q = 0.15$
2. $n = 15, p = 0.70, q = 0.30$
3. $n = 18, p = 0.90, q = 0.10$
4. $n = 20, p = 0.65, q = 0.35$

In Exercises 5–8, match the binomial probability statement with its corresponding normal distribution probability statement (a)–(d) after a continuity correction.

5. $P(x > 109)$
6. $P(x \ge 109)$
7. $P(x \le 109)$
8. $P(x < 109)$

(a) $P(x > 109.5)$
(b) $P(x < 108.5)$
(c) $P(x < 109.5)$
(d) $P(x > 108.5)$

In Exercises 9–14, write the binomial probability in words. Then, use a continuity correction to convert the binomial probability to a normal distribution probability.

9. $P(x < 25)$
10. $P(x \ge 110)$
11. $P(x = 33)$
12. $P(x > 65)$
13. $P(x \le 150)$
14. $P(55 < x < 60)$

Graphical Analysis *In Exercises 15 and 16, write the binomial probability and the normal probability for the shaded region of the graph. Find the value of each probability and compare the results.*

15.
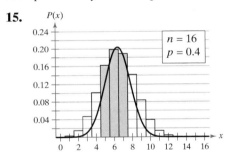

16.
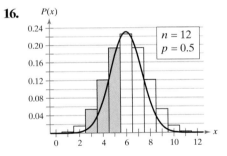

Using and Interpreting Concepts

Approximating a Binomial Distribution *In Exercises 17 and 18, a binomial experiment is given. Determine whether you can use a normal distribution to approximate the binomial distribution. If you can, find the mean and standard deviation. If you cannot, explain why.*

17. **Alcohol-Impaired Driving** In a recent year, alcohol-impaired driving was the cause of 31% of motor vehicle fatalities. You randomly select 30 motor vehicle fatalities and determine whether alcohol-impaired driving was the cause. *(Source: WalletHub)*

18. **Cell Phone and Internet Privileges** Sixty-five percent of parents of teenagers have taken their teenager's cell phone or Internet privileges away as a punishment. You randomly select 10 parents of teenagers and ask them whether they have taken their teenager's cell phone or Internet privileges away as a punishment. *(Source: Pew Research Center)*

19. Can use normal distribution

(a) 0.0793

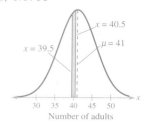

(b) 0.6198

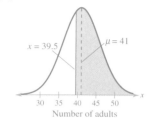

(c) 0.3802

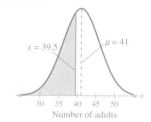

No unusual events because all of the probabilities are greater than 0.05.

20. Can use normal distribution

(a) 0.0174 (*Tech:* 0.0176)

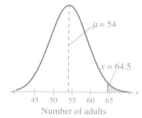

(b) and (c) See Selected Answers, page A98.

The event in part (a) is unusual because its probability is less than 0.05.

21. See Odd Answers, page A62.

22. See Selected Answers, page A98.

23. See Odd Answers, page A62.

24. See Selected Answers, page A98.

25. See Odd Answers, page A62.

26. See Selected Answers, page A98.

Approximating Binomial Probabilities *In Exercises 19–26, determine whether you can use a normal distribution to approximate the binomial distribution. If you can, use the normal distribution to approximate the indicated probabilities and sketch their graphs. If you cannot, explain why and use a binomial distribution to find the indicated probabilities. Identify any unusual events. Explain.*

19. Fraudulent Credit Card Charges A survey of U.S. adults found that 41% have encountered fraudulent charges on their credit cards. You randomly select 100 U.S. adults. Find the probability that the number who have encountered fraudulent charges on their credit cards is (a) exactly 40, (b) at least 40, and (c) fewer than 40. *(Source: Pew Research Center)*

20. Public Wi-Fi A survey of U.S. adults found that 54% of those who use the Internet access Wi-Fi networks in public places. You randomly select 100 U.S. adults who use the Internet. Find the probability that the number who access Wi-Fi networks in public places is (a) at least 65, (b) fewer than 50, and (c) more than 60. *(Source: Pew Research Center)*

21. Screen Lock A survey of U.S. adults found that 28% of those who own smartphones do not use a screen lock or other security features to access their phone. You randomly select 150 U.S. adults who own smartphones. Find the probability that the number who do not use a screen lock or other security features to access their phone is (a) at most 40, (b) more than 50, and (c) between 20 and 30, inclusive. *(Source: Pew Research Center)*

22. Online Account Passwords A survey of U.S. adults found that 39% of those who have online accounts use the same or very similar passwords for many of their accounts. You randomly select 500 U.S. adults who have online accounts. Find the probability that the number who use the same or very similar passwords for many of their accounts is (a) exactly 175, (b) no more than 225, and (c) at most 200. *(Source: Pew Research Center)*

23. Favorite Sport A survey of U.S. adults found that 33% name professional football as their favorite sport. You randomly select 14 U.S. adults and ask them to name their favorite sport. Find the probability that the number who name professional football as their favorite sport is (a) exactly 8, (b) at most 4, and (c) fewer than 6. *(Source: The Harris Poll)*

24. Favorite Beverage A survey of U.S. adults ages 21 and older who are regular drinkers found that 38% name beer as their favorite beverage. You randomly select 25 U.S. adults ages 21 and older who are regular drinkers and ask them to name their favorite beverage. Find the probability that the number who name beer as their favorite beverage is (a) no fewer than 10, (b) at least 13, and (c) between 6 and 8, inclusive. *(Source: The Harris Poll)*

25. College Graduates Fifty-one percent of U.S. college graduates consider themselves underemployed. You randomly select 250 U.S. college graduates and ask them whether they consider themselves underemployed. Find the probability that the number who consider themselves underemployed is (a) no more than 125, (b) no fewer than 135, and (c) between 100 and 125, inclusive. *(Source: Accenture)*

26. College Graduates Fourteen percent of U.S. college graduates want to work for a large company. You randomly select 1500 U.S. college graduates and ask them whether they want to work for a large company. Find the probability that the number who want to work for a large company is (a) exactly 175, (b) no more than 225, and (c) at most 200. *(Source: Accenture)*

27. (a) 0.0885 (*Tech:* 0.0878)

(b) 0.1660 (*Tech:* 0.1658)

(c) 0.7324 (*Tech:* 0.7322)

28. (a) 0.0084

(b) 0.4801 (*Tech:* 0.4792)

(c) 0.9686 (*Tech:* 0.9683)

29. Highly unlikely. Answers will vary.

30. Probable. Answers will vary.

31. 0.1020

32. 0.1736 (*Tech:* 0.1727)

27. **Minimum Wage** About 3.3% of hourly paid U.S. workers earn the prevailing minimum wage or less. A grocery chain offers discount rates to companies that have at least 30 employees who earn the prevailing minimum wage or less. Find the probability that each company will get the discount. *(Source: U.S. Bureau of Labor Statistics)*

(a) Company A has 700 employees.

(b) Company B has 750 employees.

(c) Company C has 1000 employees.

28. **Education** A survey of U.S. adults found that 8% believe the biggest problem in schools today is poor teaching. You randomly select a sample of U.S. adults. Find the probability that more than 100 U.S. adults believe the biggest problem in schools today is poor teaching. *(Source: Rasmussen Reports)*

(a) You select 1000 U.S. adults.

(b) You select 1250 U.S. adults.

(c) You select 1500 U.S. adults.

Extending Concepts

Getting Physical *The figure shows the results of a survey of U.S. adults ages 33 to 51 who were asked whether they participated in a sport. Seventy percent of U.S. adults ages 33 to 51 said they regularly participated in at least one sport, and they gave their favorite sport. Use this information in Exercises 29 and 30.*

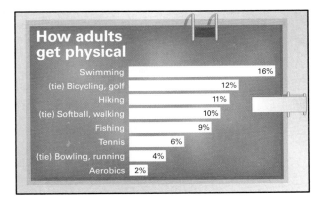

29. You randomly select 250 U.S. adults ages 33 to 51 and ask them whether they regularly participate in at least one sport. You find that 60% say no. How likely is this result? Do you think this sample is a good one? Explain your reasoning.

30. You randomly select 300 U.S. adults ages 33 to 51 and ask them whether they regularly participate in at least one sport. Of the 200 who say yes, 9% say they participate in hiking. How likely is this result? Do you think this sample is a good one? Explain your reasoning.

Testing a Drug *A drug manufacturer claims that a drug cures a rare skin disease 75% of the time. The claim is checked by testing the drug on 100 patients. If at least 70 patients are cured, then this claim will be accepted. Use this information in Exercises 31 and 32.*

31. Find the probability that the claim will be rejected, assuming that the manufacturer's claim is true.

32. Find the probability that the claim will be accepted, assuming that the actual probability that the drug cures the skin disease is 65%.

Uses

Normal distributions can be used to describe many real-life situations and are widely used in the fields of science, business, and psychology. They are the most important probability distributions in statistics and can be used to approximate other distributions, such as discrete binomial distributions.

The most incredible application of the normal distribution lies in the Central Limit Theorem. This theorem states that no matter what type of distribution a population may have, as long as the size of each random sample is at least 30, the distribution of sample means will be approximately normal. When a population is normal, the distribution of sample means is normal for any random sample of size n.

The normal distribution is essential to sampling theory. Sampling theory forms the basis of statistical inference, which you will study in the next chapter.

Abuses

Consider a population that is normally distributed, with a mean of 100 and standard deviation of 15. It would not be unusual for an individual value taken from this population to be 115 or more. In fact, this will happen almost 16% of the time. It *would* be, however, highly unusual to take a random sample of 100 values from that population and obtain a sample mean of 115 or more. Because the population is normally distributed, the mean of the sampling distribution of sample means will be 100, and the standard deviation will be 1.5. A sample mean of 115 lies 10 standard deviations above the mean. This would be an extremely unusual event. When an event this unusual occurs, it is a good idea to question the original parameters or the assumption that the population is normally distributed.

Although normal distributions are common in many populations, you should not try to make *nonnormal* statistics fit a normal distribution. The statistics used for normal distributions are often inappropriate when the distribution is nonnormal. For instance, some economists argue that financial risk managers' reliance on normal distributions to model stock market behavior is a mistake because the normal distributions do not accurately predict unusual events like market crashes.

EXERCISES

1. *Is It Unusual?* A population is normally distributed, with a mean of 100 and a standard deviation of 15. Determine whether either event is unusual. Explain your reasoning.

   a. The mean of a sample of 3 is 112 or more.

   b. The mean of a sample of 75 is 105 or more.

2. *Find the Error* The mean age of students at a high school is 16.5, with a standard deviation of 0.7. You use the Standard Normal Table to determine that the probability of selecting one student at random whose age is more than 17.5 years is about 8%. What is the error in this problem?

3. Give an example of a distribution that might be nonnormal.

5 Chapter Summary

What Did You Learn?	Example(s)	Review Exercises
Section 5.1		
▶ How to interpret graphs of normal probability distributions	1, 2	1–4
▶ How to find areas under the standard normal curve	3–6	5–26
Section 5.2		
▶ How to find probabilities for normally distributed variables using a table and using technology	1–3	27–36
Section 5.3		
▶ How to find a z-score given the area under the normal curve	1, 2	37–44
▶ How to transform a z-score to an x-value	3	45, 46
$x = \mu + z\sigma$		
▶ How to find a specific data value of a normal distribution given the probability	4, 5	47–50
Section 5.4		
▶ How to find sampling distributions and verify their properties	1	51, 52
▶ How to interpret the Central Limit Theorem	2, 3	53, 54
$\mu_{\bar{x}} = \mu$ Mean of the sample means		
$\sigma_{\bar{x}} = \dfrac{\sigma}{\sqrt{n}}$ Standard deviation of the sample means		
▶ How to apply the Central Limit Theorem to find the probability of a sample mean	4–6	55–60
Section 5.5		
▶ How to determine when a normal distribution can approximate a binomial distribution	1	61, 62
$\mu = np$ Mean		
$\sigma = \sqrt{npq}$ Standard deviation		
▶ How to find the continuity correction	2	63–68
▶ How to use a normal distribution to approximate binomial probabilities	3–5	69, 70

5 Review Exercises

1. $\mu = 15$, $\sigma = 3$

2. $\mu = 56$, $\sigma = 5$

3. Curve *B* has the greatest mean because its line of symmetry occurs the farthest to the right.

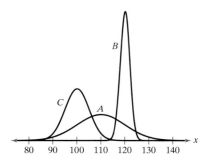

FIGURE FOR EXERCISES 3 AND 4

4. Curve *A* has the greatest standard deviation because it is the most spread out.

5. 0.6772

6. 0.2025

7. 0.6293

8. 0.0256

9. 0.7157

10. 0.0006

11. 0.00235 (*Tech:* 0.00236)

12. 0.4940

13. 0.4495

14. 0.7902 (*Tech:* 0.7903)

15. 0.4365 (*Tech:* 0.4364)

16. 0.9926

17. 0.1336

18. 0.7392

19. $x = 17 \rightarrow z \approx -0.66$
$x = 29 \rightarrow z \approx 1.18$
$x = 8 \rightarrow z \approx -2.05$
$x = 23 \rightarrow z \approx 0.26$

20. $x = 8$ is unusual because its corresponding *z*-score (-2.05) lies more than 2 standard deviations from the mean.

Section 5.1

In Exercises 1 and 2, use the normal curve to estimate the mean and standard deviation.

1.

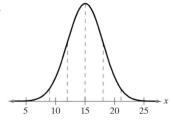

2.

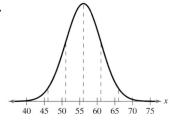

In Exercises 3 and 4, use the normal curves shown at the left.

3. Which normal curve has the greatest mean? Explain your reasoning.

4. Which normal curve has the greatest standard deviation? Explain your reasoning.

In Exercises 5 and 6, find the area of the indicated region under the standard normal curve. If convenient, use technology to find the area.

5.

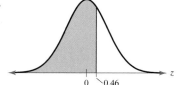

6.

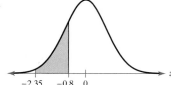

In Exercises 7–18, find the indicated area under the standard normal curve. If convenient, use technology to find the area.

7. To the left of $z = 0.33$

8. To the left of $z = -1.95$

9. To the right of $z = -0.57$

10. To the right of $z = 3.22$

11. To the left of $z = -2.825$

12. To the right of $z = 0.015$

13. Between $z = -1.64$ and $z = 0$

14. Between $z = -1.55$ and $z = 1.04$

15. Between $z = 0.05$ and $z = 1.71$

16. Between $z = -2.68$ and $z = 2.68$

17. To the left of $z = -1.5$ and to the right of $z = 1.5$

18. To the left of $z = 0.64$ and to the right of $z = 3.415$

The scores for the reading portion of the ACT test are normally distributed. In a recent year, the mean test score was 21.3 and the standard deviation was 6.5. The test scores of four students selected at random are 17, 29, 8, and 23. Use this information in Exercises 19 and 20. (*Source: ACT, Inc.*)

19. Find the *z*-score that corresponds to each value.

20. Determine whether any of the values are unusual.

21. 0.8997

22. 0.7704

23. 0.9236 (*Tech:* 0.9237)

24. 0.3364

25. 0.0124

26. 0.5465

27. 0.8944

28. 0.0088

29. 0.2266

30. 0.6179

31. 0.2684 (*Tech:* 0.2685)

32. 0.4400 (*Tech:* 0.4401)

33. (a) 0.4168 (*Tech:* 0.4173)

(b) 0.1425 (*Tech:* 0.1407)

(c) 0.3974 (*Tech:* 0.3971)

34. (a) 0.8435 (*Tech:* 0.8438)

(b) 0.1632 (*Tech:* 0.1643)

(c) 0.2843 (*Tech:* 0.2839)

35. No unusual events because all of the probabilities are greater than 0.05.

36. No unusual events because all of the probabilities are greater than 0.05.

37. −0.07

38. −1.28

39. 2.455 (*Tech:* 2.457)

40. −2.05

41. 1.04

42. −0.10

43. 0.51

44. 1.88

In Exercises 21–26, find the indicated probability using the standard normal distribution. If convenient, use technology to find the probability.

21. $P(z < 1.28)$

22. $P(z > -0.74)$

23. $P(-2.15 < z < 1.55)$

24. $P(0.42 < z < 3.15)$

25. $P(z < -2.50 \text{ or } z > 2.50)$

26. $P(z < 0 \text{ or } z > 1.68)$

Section 5.2

In Exercises 27–32, the random variable x is normally distributed with mean $\mu = 74$ and standard deviation $\sigma = 8$. Find the indicated probability.

27. $P(x < 84)$

28. $P(x < 55)$

29. $P(x > 80)$

30. $P(x > 71.6)$

31. $P(60 < x < 70)$

32. $P(72 < x < 82)$

In Exercises 33 and 34, find the indicated probabilities. If convenient, use technology to find the probabilities.

33. Yearly amounts of black carbon emissions from cars in India are normally distributed, with a mean of 14.7 gigagrams per year and a standard deviation of 11.5 gigagrams per year. Find the probability that the amount of black carbon emissions from cars in India for a randomly selected year are

(a) less than 12.3 gigagrams per year.

(b) between 15.4 and 19.6 gigagrams per year.

(c) greater than 17.7 gigagrams per year. *(Adapted from Atmospheric Chemistry and Physics)*

34. The daily surface concentration of carbonyl sulfide on the Indian Ocean is normally distributed, with a mean of 9.1 picomoles per liter and a standard deviation of 3.5 picomoles per liter. Find the probability that on a randomly selected day, the surface concentration of carbonyl sulfide on the Indian Ocean is

(a) between 5.1 and 15.7 picomoles per liter.

(b) between 10.5 and 12.3 picomoles per liter.

(c) more than 11.1 picomoles per liter. *(Source: Atmospheric Chemistry and Physics)*

35. Determine whether any of the events in Exercise 33 are unusual. Explain your reasoning.

36. Determine whether any of the events in Exercise 34 are unusual. Explain your reasoning.

Section 5.3

In Exercises 37–42, use the Standard Normal Table or technology to find the z-score that corresponds to the cumulative area or percentile.

37. 0.4721

38. 0.1

39. 0.993

40. P_2

41. P_{85}

42. P_{46}

43. Find the z-score that has 30.5% of the distribution's area to its right.

44. Find the positive z-score for which 94% of the distribution's area lies between −z and z.

Braking Distance of a Sedan

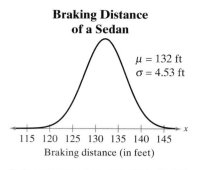

$\mu = 132$ ft
$\sigma = 4.53$ ft

Braking distance (in feet)

FIGURE FOR EXERCISES 45–50

On a dry surface, the braking distances (in feet), from 60 miles per hour to a complete stop, of a sedan can be approximated by a normal distribution, as shown in the figure at the left. Use this information in Exercises 45–50.

45. Find the braking distance of a sedan that corresponds to $z = -2.75$.

46. Find the braking distance of a sedan that corresponds to $z = 1.6$.

47. What braking distance of a sedan represents the 90th percentile?

48. What braking distance of a sedan represents the first quartile?

49. What is the shortest braking distance of a sedan that can be in the top 15% of braking distances?

50. What is the longest braking distance of a sedan that can be in the bottom 20% of braking distances?

Section 5.4

In Exercises 51 and 52, a population and sample size are given. (a) Find the mean and standard deviation of the population. (b) List all samples (with replacement) of the given size from the population and find the mean of each. (c) Find the mean and standard deviation of the sampling distribution of sample means and compare them with the mean and standard deviation of the population.

51. The goals scored in a season by the four starting defenders on a soccer team are 1, 2, 0, and 3. Use a sample size of 2.

52. The minutes of overtime reported by each of the three executives at a corporation are 90, 120, and 210. Use a sample size of 3.

In Exercises 53 and 54, find the mean and standard deviation of the indicated sampling distribution of sample means. Then sketch a graph of the sampling distribution.

53. The per capita electric power consumption level in a recent year in Ecuador is normally distributed, with a mean of 471.5 kilowatt-hours and a standard deviation of 187.9 kilowatt-hours. Random samples of size 35 are drawn from this population, and the mean of each sample is determined. *(Source: Latin America Journal of Economics)*

54. The test scores for the Law School Admission Test (LSAT) in a recent year are normally distributed, with a mean of 155.69 and a standard deviation of 5.05. Random samples of size 40 are drawn from this population, and the mean of each sample is determined. *(Source: Law School Admission Council)*

In Exercises 55–60, find the indicated probabilities and interpret the results.

55. Refer to Exercise 33. A random sample of 2 years is selected. Find the probability that the mean amount of black carbon emissions for the sample is (a) less than 12.3 gigagrams per year, (b) between 15.4 and 19.6 gigagrams per year, and (c) greater than 17.7 gigagrams per year. Compare your answers with those in Exercise 33.

56. Refer to Exercise 34. A random sample of six days is selected. Find the probability that the mean surface concentration of carbonyl sulfide for the sample is (a) between 5.1 and 15.7 picomoles per liter, (b) between 10.5 and 12.3 picomoles per liter, and (c) more than 11.1 picomoles per liter. Compare your answers with those in Exercise 34.

45. 119.54 feet   **46.** 139.25 feet

47. 137.80 feet (*Tech:* 137.81 feet)

48. 128.94 feet

49. 136.71 feet (*Tech:* 136.70 feet)

50. 128.19 feet

51. (a) $\mu = 1.5$, $\sigma \approx 1.118$

(b) See Odd Answers, page A63.

(c) $\mu_{\bar{x}} = 1.5$, $\sigma_{\bar{x}} \approx 0.791$

The means are equal, but the standard deviation of the sampling distribution is smaller.

52. (a) $\mu = 140$, $\sigma \approx 50.990$

(b) See Selected Answers, page A99.

(c) $\mu_{\bar{x}} = 140$, $\sigma_{\bar{x}} \approx 29.439$

The means are equal, but the standard deviation of the sampling distribution is smaller.

53. $\mu_{\bar{x}} = 471.5$, $\sigma_{\bar{x}} \approx 31.761$

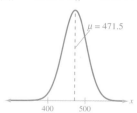

$\mu = 471.5$

Mean electric power consumption (in kilowatt-hours)

54. See Selected Answers, page A99.

55. (a) 0.3840 (*Tech:* 0.3839)

(b) 0.1898 (*Tech:* 0.1923)

(c) 0.3557 (*Tech:* 0.3561)

The probabilities in parts (a) and (c) are smaller, and the probability in part (b) is larger.

56. (a) 0.9974   (b) 0.1510

(c) 0.0808

The probability in part (a) is larger, and the probabilities in parts (b) and (c) are smaller.

57. (a) 0.8051 (*Tech:* 0.8043)

(b) 0.8577 (*Tech:* 0.8580)

(c) 0.3993 (*Tech:* 0.3994)

58. (a) 0.9452 (*Tech:* 0.9453)

(b) 0.1423 (*Tech:* 0.1429)

(c) 0.5596 (*Tech:* 0.5603)

59. (a) 0.2709 (*Tech:* 0.2710)

(b) 0.1112 (*Tech:* 0.1113)

60. (a) 0.3632 (*Tech:* 0.3618)

(b) 0.0384 (*Tech:* 0.0385)

61. Can use normal distribution; $\mu = 15$, $\sigma \approx 1.936$

62. Cannot use normal distribution because $np < 5$.

63. The probability of getting at least 25 successes; $P(x > 24.5)$

64. The probability of getting at most 36 successes; $P(x < 36.5)$

65. The probability of getting exactly 45 successes; $P(44.5 < x < 45.5)$

66. The probability of getting more than 14 successes; $P(x > 14.5)$

67. The probability of getting less than 60 successes; $P(x < 59.5)$

68. The probability of getting between 54 and 64 successes; $P(54.5 < x < 63.5)$

69. Can use normal distribution

(a) 0.0384 (*Tech:* 0.0385)

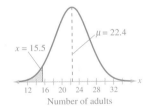

(b) 0.0798 (*Tech:* 0.0818)

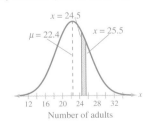

(c) See Odd Answers, page A63.

The events in parts (a) and (c) are unusual because their probabilities are less than 0.05.

70. See Selected Answers, page A99.

57. The mean ACT composite score in a recent year is 20.8. A random sample of 36 ACT composite scores is selected. What is the probability that the mean score for the sample is (a) less than 21.6, (b) more than 19.8, and (c) between 20.5 and 21.5? Assume $\sigma = 5.6$. *(Source: ACT, Inc)*

58. The mean MCAT total score in a recent year is 500. A random sample of 32 MCAT total scores is selected. What is the probability that the mean score for the sample is (a) less than 503, (b) more than 502, and (c) between 498 and 501? Assume $\sigma = 10.6$. *(Source: Association of American Medical Colleges)*

59. The mean annual salary for intermediate level life insurance underwriters is about $61,000. A random sample of 45 intermediate level life insurance underwriters is selected. What is the probability that the mean annual salary of the sample is (a) less than $60,000 and (b) more than $63,000? Assume $\sigma = \$11,000$. *(Adapted from Salary.com)*

60. The mean annual salary for magnetic resonance imaging (MRI) technologists is about $72,000. A random sample of 50 MRI technologists is selected. What is the probability that the mean annual salary of the sample is (a) less than $71,500 and (b) more than $74,500? Assume $\sigma = \$10,000$. *(Adapted from Salary.com)*

Section 5.5

In Exercises 61 and 62, a binomial experiment is given. Determine whether you can use a normal distribution to approximate the binomial distribution. If you can, find the mean and standard deviation. If you cannot, explain why.

61. A survey of U.S. adults found that 75% support labeling legislation for genetically modified organisms (GMOs). You randomly select 20 U.S. adults and ask them whether they support labeling legislation for genetically modified organisms (GMOs). *(Source: The Harris Poll)*

62. A survey of U.S. likely voters found that 11% think Congress is doing a good or excellent job. You randomly select 45 U.S. likely voters and ask them whether they think Congress is doing a good or excellent job. *(Source: Rasmussen Reports)*

In Exercises 63–68, write the binomial probability in words. Then, use a continuity correction to convert the binomial probability to a normal distribution probability.

63. $P(x \geq 25)$ **64.** $P(x \leq 36)$ **65.** $P(x = 45)$

66. $P(x > 14)$ **67.** $P(x < 60)$ **68.** $P(54 < x < 64)$

In Exercises 69 and 70, determine whether you can use a normal distribution to approximate the binomial distribution. If you can, use the normal distribution to approximate the indicated probabilities and sketch their graphs. If you cannot, explain why and use a binomial distribution to find the indicated probabilities.

69. A survey of U.S. adults found that 32% have an online account with their healthcare provider. You randomly select 70 U.S. adults and ask them whether they have an online account with their healthcare provider. Find the probability that the number who have an online account with their healthcare provider is (a) at most 15, (b) exactly 25, and (c) greater than 30. Identify any unusual events. Explain. *(Source: Pew Research Center)*

70. Sixty-five percent of U.S. college graduates are employed in their field of study. You randomly select 20 U.S. college graduates and ask them whether they are employed in their field of study. Find the probability that the number who are employed in their field of study is (a) exactly 15, (b) less than 10, and (c) between 20 and 35. Identify any unusual events. Explain. *(Source: Accenture)*

5 Chapter Quiz

Answers (left column)

1. (a) 0.9535
   (b) 0.9871
   (c) 0.3616
   (d) 0.7703 (*Tech:* 0.7702)

2. (a) 0.0233 (*Tech:* 0.0231)
   (b) 0.9929 (*Tech:* 0.9928)
   (c) 0.9198 (*Tech:* 0.9199)
   (d) 0.3607 (*Tech:* 0.3610)

3. 0.0475 (*Tech:* 0.0478); Yes, the event is unusual because its probability is less than 0.05.

4. 0.2586 (*Tech:* 0.2611); No, the event is not unusual because its probability is greater than 0.05.

5. 21.19%

6. 503 people (*Tech:* 505 people)

7. 125

8. 80

9. 0.0049; About 0.5% of samples of 60 people will have a mean IQ score greater than 105. This is a very unusual event.

10. More likely to select one person with an IQ score greater than 105 because the standard error of the mean is less than the standard deviation.

11. Can use normal distribution; $\mu = 40$, $\sigma \approx 5.797$

12. (a) 0.5359 (*Tech:* 0.5344)
    (b) 0.7823 (*Tech:* 0.7812)
    (c) 0.0277 (*Tech:* 0.0266)

    The event in part (c) is unusual because its probability is less than 0.05.

Take this quiz as you would take a quiz in class. After you are done, check your work against the answers given in the back of the book.

1. Find each probability using the standard normal distribution.
   (a) $P(z > -1.68)$
   (b) $P(z < 2.23)$
   (c) $P(-0.47 < z < 0.47)$
   (d) $P(z < -1.992 \text{ or } z > -0.665)$

2. The random variable x is normally distributed with the given parameters. Find each probability.
   (a) $\mu = 9.2, \sigma \approx 1.62, P(x < 5.97)$
   (b) $\mu = 87, \sigma \approx 19, P(x > 40.5)$
   (c) $\mu = 5.5, \sigma \approx 0.08, P(5.36 < x < 5.64)$
   (d) $\mu = 18.5, \sigma \approx 4.25, P(19.6 < x < 26.1)$

In a standardized IQ test, scores are normally distributed, with a mean score of 100 and a standardized deviation of 15. Use this information in Exercises 3–10. (Adapted from 123test)

3. Find the probability that a randomly selected person has an IQ score higher than 125. Is this an unusual event? Explain.

4. Find the probability that a randomly selected person has an IQ score between 95 and 105. Is this an unusual event? Explain.

5. What percent of the IQ scores are greater than 112?

6. Out of 2000 randomly selected people, about how many would you expect to have IQ scores less than 90?

7. What is the lowest score that would still place a person in the top 5% of the scores?

8. What is the highest score that would still place a person in the bottom 10% of the scores?

9. A random sample of 60 people is selected from this population. What is the probability that the mean IQ score of the sample is greater than 105? Interpret the result.

10. Are you more likely to randomly select one person with an IQ score greater than 105 or are you more likely to randomly select a sample of 15 people with a mean IQ score greater than 105? Explain.

In a survey of U.S. adults, 16% say they have had someone take over their email accounts without their permission. You randomly select 250 U.S. adults and ask them whether they have had someone take over their email accounts without their permission. Use this information in Exercises 11 and 12. (Source: Pew Research Center)

11. Determine whether you can use a normal distribution to approximate the binomial distribution. If you can, find the mean and standard deviation. If you cannot, explain why.

12. Find the probability that the number of U.S. adults who say they have had someone take over their email accounts without their permission is (a) at most 40, (b) less than 45, and (c) exactly 48. Identify any unusual events. Explain.

5 Chapter Test

Answers (left column)

1. (a) $\mu_{\bar{x}} = 83$, $\sigma_{\bar{x}} \approx 1.683$
   (b) 0.1170 (*Tech:* 0.1173)
   (c) 0.2401 (*Tech:* 0.2389)
2. (a) 0.3974 (*Tech:* 0.3962)
   (b) 0.0347
   (c) 0.2380 (*Tech:* 0.2363)
3. 27.044 (*Tech:* 27.045)
4. 15.112 (*Tech:* 15.113)
5. Can use normal distribution
   (a) 0.0954 (*Tech:* 0.0956)

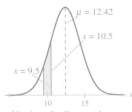

   (b) and (c) See Selected Answers, page A99.

   The event in part (b) is unusual because its probability is less than 0.05.
6. Cannot use normal distribution because $nq < 5$.
   (a) 0.1766
   (b) 0.5871
   (c) 0.2363

   No unusual events because all of the probabilities are greater than 0.05.
7. 0.3409 (*Tech:* 0.3416); No, the event is not unusual because its probability is greater than 0.05.
8. 124 residents
9. Between $41,938.08 and $46,061.92 (*Tech:* $41,938.03 and $46,061.97)
10. Yes; By the Central Limit Theorem, when a population is normally distributed, the sampling distribution of sample means is normally distributed for any sample size.

Test

Take this test as you would take a test in class.

1. The mean per capita daily water consumption in a village in Bangladesh is about 83 liters per person and the standard deviation is about 11.9 liters per person. Random samples of size 50 are drawn from this population and the mean of each sample is determined. *(Source: Journal of Education and Social Sciences)*
   (a) Find the mean and standard deviation of the sampling distribution of sample means.
   (b) What is the probability that the mean per capita daily water consumption for a given sample is more than 85 liters per person?
   (c) What is the probability that the mean per capita daily water consumption for a given sample is between 80 and 82 liters per person?

In Exercises 2–4, the random variable x is normally distributed with mean $\mu = 18$ and standard deviation $\sigma = 7.6$.

2. Find each probability.
   (a) $P(x > 20)$ (b) $P(0 < x < 5)$ (c) $P(x < 9 \text{ or } x > 27)$
3. Find the value of x that has 88.3% of the distribution's area to its left.
4. Find the value of x that has 64.8% of the distribution's area to its right.

In Exercises 5 and 6, determine whether you can use a normal distribution to approximate the binomial distribution. If you can, use the normal distribution to approximate the indicated probabilities and sketch their graphs. If you cannot, explain why and use a binomial distribution to find the indicated probabilities.

5. Sixty-nine percent of U.S. college graduates expect to stay at their first employer for three or more years. You randomly select 18 U.S. college graduates and ask them whether they expect to stay at their first employer for three or more years. Find the probability that the number who expect to stay at their first employer for three or more years is (a) exactly 10, (b) less than 7, and (c) at least 15. Identify any unusual events. Explain. *(Source: Accenture)*
6. A survey of U.S. adults found that 86% of those who use the Internet keep track of their online passwords in their heads. You randomly select 30 U.S. adults who use the Internet. Find the probability that the number who keep track of their online passwords in their heads is (a) exactly 25, (b) more than 25, and (c) less than 25. Identify any unusual events. Explain. *(Source: Pew Research Center)*

The per capita disposable income for residents of a U.S. city in a recent year is normally distributed, with a mean of about $44,000 and a standard deviation of about $2450. Use this information in Exercises 7–10.

7. Find the probability that the disposable income of a resident is more than $45,000. Is this an unusual event? Explain.
8. Out of 800 residents, about how many would you expect to have a disposable income of between $40,000 and $42,000?
9. Between what two values does the middle 60% of disposable incomes lie?
10. Random samples of size 8 are drawn from the population and the mean of each sample is determined. Is the sampling distribution of sample means normally distributed? Explain.

You work for a pharmaceuticals company as a statistical process analyst. Your job is to analyze processes and make sure they are in statistical control. In one process, a machine is supposed to add 9.8 milligrams of a compound to a mixture in a vial. (Assume this process can be approximated by a normal distribution with a standard deviation of 0.05.) The acceptable range of amounts of the compound added is 9.65 milligrams to 9.95 milligrams, inclusive.

Because of an error with the release valve, the setting on the machine "shifts" from 9.8 milligrams. To check that the machine is adding the correct amount of the compound into the vials, you select at random three samples of five vials and find the mean amount of the compound added for each sample. A coworker asks why you take 3 samples of size 5 and find the mean instead of randomly choosing and measuring the amounts in 15 vials individually to check the machine's settings. (*Note:* Both samples are chosen without replacement.)

EXERCISES

1. Sampling Individuals

Assume the machine shifts and the distribution of the amount of the compound added now has a mean of 9.96 milligrams and a standard deviation of 0.05 milligram. You select one vial and determine how much of the compound was added.

(a) What is the probability that you select a vial that is within the acceptable range (in other words, you do not detect that the machine has shifted)? (See figure.)

(b) You randomly select 15 vials. What is the probability that you select at least one vial that is within the acceptable range?

2. Sampling Groups of Five

Assume the machine shifts and is filling the vials with a mean amount of 9.96 milligrams and a standard deviation of 0.05 milligram. You select five vials and find the mean amount of compound added.

(a) What is the probability that you select a sample of five vials that has a mean that is within the acceptable range? (See figure.)

(b) You randomly select three samples of five vials. What is the probability that you select at least one sample of five vials that has a mean that is within the acceptable range?

(c) Which is more sensitive to a shift of parameters—an individual random selection or a randomly selected sample mean?

3. Writing an Explanation

Write a paragraph to your coworker explaining why you take 3 samples of size 5 and find the mean of each sample instead of randomly choosing and measuring the amounts in 15 vials individually to check the machine's setting.

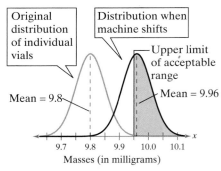

FIGURE FOR EXERCISE 1

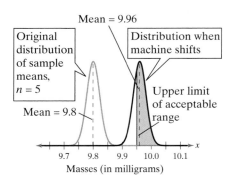

FIGURE FOR EXERCISE 2

Age Distribution in California

One of the jobs of the U.S. Census Bureau is to keep track of the age distribution in the country and in each of the states. The estimated age distribution in California is 2016 is shown in the table and the histogram.

Age Distribution in California

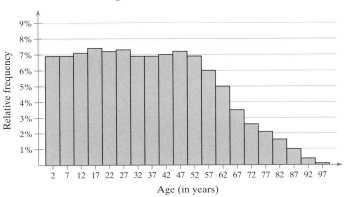

Age (in years)

Class	Class midpoint	Relative frequency
0–4	2	6.9%
5–9	7	6.9%
10–14	12	7.1%
15–19	17	7.4%
20–24	22	7.2%
25–29	27	7.3%
30–34	32	6.9%
35–39	37	6.9%
40–44	42	7.0%
45–49	47	7.2%
50–54	52	6.9%
55–59	57	6.0%
60–64	62	5.0%
65–69	67	3.5%
70–74	72	2.6%
75–79	77	2.1%
80–84	82	1.6%
85–89	87	1.0%
90–94	92	0.4%
95–99	97	0.1%

EXERCISES

 The means of 36 randomly selected samples generated by technology with n = 40 are shown below.

> 28.14, 31.56, 36.86, 32.37, 36.12, 39.53,
> 36.19, 39.02, 35.62, 36.30, 34.38, 32.98,
> 36.41, 30.24, 34.19, 44.72, 38.84, 42.87,
> 38.90, 34.71, 34.13, 38.25, 38.04, 34.07,
> 39.74, 40.91, 42.63, 35.29, 35.91, 34.36,
> 36.51, 36.47, 32.88, 37.33, 31.27, 35.80

1. Use technology and the age distribution to estimate the mean age in California.

2. Use technology to find the mean of the set of 36 sample means. How does it compare with the mean age in California found in Exercise 1? Does this agree with the result predicted by the Central Limit Theorem?

3. Are the ages of people in California normally distributed? Explain your reasoning.

4. Sketch a relative frequency histogram for the 36 sample means. Use nine classes. Is the histogram approximately bell-shaped and symmetric? Does this agree with the result predicted by the Central Limit Theorem?

5. Use technology and the age distribution to find the standard deviation of the ages of people in California.

6. Use technology to find the standard deviation of the set of 36 sample means. How does it compare with the standard deviation of the ages found in Exercise 5? Does this agree with the result predicted by the Central Limit Theorem?

Extended solutions are given in the technology manuals that accompany this text. Technical instruction is provided for Minitab, Excel, and the TI-84 Plus.

CHAPTERS 3-5
CUMULATIVE REVIEW

Section 5.5

1. A survey of adults in the United States found that 61% ate at a restaurant at least once in the past week. You randomly select 30 adults and ask them whether they ate at a restaurant at least once in the past week. *(Source: Gallup)*

   (a) Verify that a normal distribution can be used to approximate the binomial distribution.

   (b) Find the probability that at most 14 adults ate at a restaurant at least once in the past week.

   (c) Is it unusual for exactly 14 out of 30 adults to have eaten in a restaurant at least once in the past week? Explain your reasoning.

Section 4.1

In Exercises 2 and 3, find the (a) mean, (b) variance, (c) standard deviation, and (d) expected value of the probability distribution. Interpret the results.

2. The table shows the distribution of household sizes in the United States for a recent year. *(Source: U.S. Census Bureau)*

x	1	2	3	4	5	6	7
$P(x)$	0.281	0.340	0.154	0.129	0.060	0.022	0.013

3. The table shows the distribution of personal fouls per game for Garrett Temple in a recent NBA season. *(Source: National Basketball Association)*

x	0	1	2	3	4	5	6
$P(x)$	0.113	0.188	0.188	0.288	0.150	0.038	0.038

Section 4.1

4. Use the probability distribution in Exercise 3 to find the probability of randomly selecting a game in which Garrett Temple had (a) fewer than four personal fouls, (b) at least three personal fouls, and (c) between two and four personal fouls, inclusive.

Section 3.4

5. From a pool of 16 candidates, 9 men and 7 women, the offices of president, vice president, secretary, and treasurer will be filled. (a) In how many different ways can the offices be filled? (b) What is the probability that all four of the offices are filled by women?

Section 5.1

In Exercises 6–11, find the indicated area under the standard normal curve. If convenient, use technology to find the area.

6. To the left of $z = 0.72$

7. To the left of $z = -3.08$

8. To the right of $z = -0.84$

9. Between $z = 0$ and $z = 2.95$

10. Between $z = -1.22$ and $z = -0.26$

11. To the left of $z = 0.12$ or to the right of $z = 1.72$

Section 4.2

12. Twenty-eight percent of U.S. adults think that climate scientists understand the causes of climate change very well. You randomly select 25 U.S. adults. Find the probability that the number of U.S. adults who think that climate scientists understand the causes of climate change very well is (a) exactly three, (b) between 8 and 11, inclusive, and (c) less than two. (d) Are any of these events unusual? Explain your reasoning. *(Source: Pew Research Center)*

13. An auto parts seller finds that 1 in every 200 parts sold is defective. Use the geometric distribution to find the probability that (a) the first defective part is the fifth part sold, (b) the first defective part is the first, second, or third part sold, and (c) none of the first 20 parts sold are defective.

14. The table shows the results of a survey in which 3,405,100 public and 489,900 private school teachers were asked about their full-time teaching experience. *(Adapted from National Center for Education Statistics)*

	Public	Private	Total
Less than 3 years	304,650	90,675	395,325
3 to 9 years	1,127,205	145,545	1,272,750
10 to 20 years	1,232,140	128,805	1,360,945
More than 20 years	721,005	99,510	820,515
Total	3,385,000	464,535	3,849,535

(a) Find the probability that a randomly selected private school teacher has 10 to 20 years of full-time teaching experience.

(b) Find the probability that a randomly selected teacher is at a public school, given that the teacher has 3 to 9 years of full-time experience.

(c) Are the events "being a public school teacher" and "having more than 20 years of full-time teaching experience" independent? Explain.

(d) Find the probability that a randomly selected teacher has 3 to 9 years of full-time teaching experience or is at a private school.

15. The initial pressures for bicycle tires when first filled are normally distributed, with a mean of 70 pounds per square inch (psi) and a standard deviation of 1.2 psi.

(a) Random samples of size 40 are drawn from this population, and the mean of each sample is determined. Find the mean and standard deviation of the sampling distribution of sample means. Then sketch a graph of the sampling distribution.

(b) A random sample of 15 tires is drawn from this population. What is the probability that the mean tire pressure of the sample is less than 69 psi?

16. The life spans of car batteries are normally distributed, with a mean of 44 months and a standard deviation of 5 months.

(a) Find the probability that the life span of a randomly selected battery is less than 36 months.

(b) Find the probability that the life span of a randomly selected battery is between 42 and 60 months.

(c) What is the shortest life expectancy a car battery can have and still be in the top 5% of life expectancies?

17. A florist has 12 different flowers from which floral arrangements can be made. A centerpiece is made using four different flowers. (a) How many different centerpieces can be made? (b) What is the probability that the four flowers in the centerpiece are roses, daisies, hydrangeas, and lilies?

18. Seventy percent of U.S. adults anticipate major cyberattacks on public infrastructure in the next five years. You randomly select 10 U.S. adults. (a) Construct a binomial distribution for the random variable x, the number of U.S. adults who anticipate major cyberattacks on public infrastructure in the next five years (b) Graph the binomial distribution using a histogram and describe its shape. (c) Identify any values of the random variable x that you would consider unusual. Explain. *(Source: Pew Research Center)*

Confidence Intervals

David Wechsler was one of the most influential psychologists of the 20th century. He is known for developing intelligence tests, such as the Wechsler Adult Intelligence Scale and the Wechsler Intelligence Scale for Children.

In Chapters 1 through 5, you studied descriptive statistics (how to collect and describe data) and probability (how to find probabilities and analyze discrete and continuous probability distributions). For instance, psychologists use descriptive statistics to analyze the data collected during experiments and tests.

One of the most commonly administered psychological tests is the Wechsler Adult Intelligence Scale. It is an intelligence quotient (IQ) test that is standardized to have a normal distribution with a mean of 100 and a standard deviation of 15.

In this chapter, you will begin your study of inferential statistics—the second major branch of statistics. For instance, a chess club wants to estimate the mean IQ of its members. The mean of a random sample of members is 115. Because this estimate consists of a single number represented by a point on a number line, it is called a *point estimate*. The problem with using a point estimate is that it is rarely equal to the exact parameter (mean, standard deviation, or proportion) of the population.

In this chapter, you will learn how to make a more meaningful estimate by specifying an interval of values on a number line, together with a statement of how confident you are that your interval contains the population parameter. Suppose the club wants to be 90% confident of its estimate for the mean IQ of its members. Here is an overview of how to construct an interval estimate.

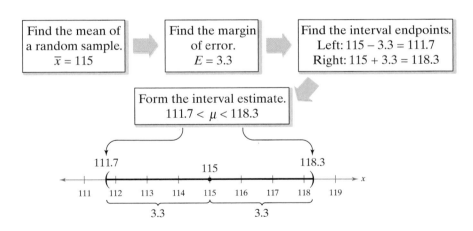

So, the club can be 90% confident that the mean IQ of its members is between 111.7 and 118.3.

6.1 Confidence Intervals for the Mean (σ Known)

What You Should Learn

▶ How to find a point estimate and a margin of error

▶ How to construct and interpret confidence intervals for a population mean when σ is known

▶ How to determine the minimum sample size required when estimating a population mean

Estimating Population Parameters ■ Confidence Intervals for a Population Mean ■ Sample Size

Estimating Population Parameters

In this chapter, you will learn an important technique of statistical inference—to use sample statistics to estimate the value of an unknown population parameter. In this section and the next, you will learn how to use sample statistics to make an estimate of the population parameter μ when the population standard deviation σ is known (this section) or when σ is unknown (Section 6.2). To make such an inference, begin by finding a **point estimate.**

DEFINITION

A **point estimate** is a single value estimate for a population parameter. The most unbiased point estimate of the population mean μ is the sample mean $\bar{x}$.

The validity of an estimation method is increased when you use a sample statistic that is unbiased and has low variability. A statistic is unbiased if it does not overestimate or underestimate the population parameter. In Chapter 5, you learned that the mean of all possible sample means of the same size equals the population mean. As a result, $\bar{x}$ is an **unbiased estimator** of μ. When the standard error $\sigma / \sqrt{n}$ of a sample mean is decreased by increasing n, it becomes less variable.

EXAMPLE 1

Finding a Point Estimate

A researcher is collecting data about a college athletic conference and its student-athletes. A random sample of 40 student-athletes is selected and their numbers of hours spent on required athletic activities for one week are recorded (see table at left). Find a point estimate for the population mean μ, the mean number of hours spent on required athletic activities by all student-athletes in the conference. *(Adapted from Penn Schoen Berland)*

Number of hours							
19	25	15	21	22	20	20	22
22	21	21	23	22	16	21	18
25	23	23	21	22	24	18	19
23	20	19	19	24	25	17	21
21	25	23	18	22	20	21	21

SOLUTION

The sample mean of the data is

$$\bar{x} = \frac{\Sigma x}{n} = \frac{842}{40} \approx 21.1.$$

So, the point estimate for the mean number of hours spent on required athletic activities by all student-athletes in the conference is about 21.1 hours.

TRY IT YOURSELF 1

In Example 1, the researcher selects a second random sample of 30 student-athletes and records their numbers of hours spent on required athletic activities (see table at left). Use this sample to find another point estimate of the population mean μ. *(Adapted from Penn Schoen Berland)*

Number of hours					
21	17	21	18	23	25
22	21	23	22	19	20
20	23	20	22	21	19
20	21	23	16	19	20
21	20	22	19	24	24

Answer: Page A36

In Example 1, the probability that the population mean is exactly 21.1 is virtually zero. So, instead of estimating μ to be exactly 21.1 using a point estimate, you can estimate that μ lies in an interval. This is called making an **interval estimate.**

DEFINITION

An **interval estimate** is an interval, or range of values, used to estimate a population parameter.

Although you can assume that the point estimate in Example 1 is not equal to the actual population mean, it is probably close to it. To form an interval estimate, use the point estimate as the center of the interval, and then add and subtract a margin of error. For instance, if the margin of error is 0.6, then an interval estimate would be given by

$$21.1 \pm 0.6 \quad \text{or} \quad 20.5 < \mu < 21.7.$$

The point estimate and interval estimate are shown in the figure.

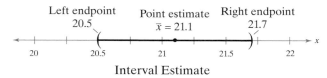

Left endpoint Point estimate Right endpoint
20.5 $\bar{x} = 21.1$ 21.7

Interval Estimate

Before finding a margin of error for an interval estimate, you should first determine how confident you need to be that your interval estimate contains the population mean μ.

DEFINITION

The **level of confidence c** is the probability that the interval estimate contains the population parameter, assuming that the estimation process is repeated a large number of times.

You know from the Central Limit Theorem that when $n \geq 30$, the sampling distribution of sample means approximates a normal distribution. The level of confidence c is the area under the standard normal curve between the *critical values*, $-z_c$ and z_c. **Critical values** are values that separate sample statistics that are probable from sample statistics that are improbable, or unusual. You can see from the figure shown below that c is the percent of the area under the normal curve between $-z_c$ and z_c. The area remaining is $1 - c$, so the area in one tail is

$$\tfrac{1}{2}(1 - c). \qquad \text{Area in one tail}$$

For instance, if $c = 90\%$, then 5% of the area lies to the left of $-z_c = -1.645$ and 5% lies to the right of $z_c = 1.645$, as shown in the table.

Study Tip

In this text, you will usually use 90%, 95%, and 99% levels of confidence. Here are the z-scores that correspond to these levels of confidence.

Level of Confidence	z_c
90%	1.645
95%	1.96
99%	2.575

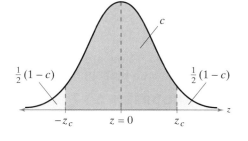

$\tfrac{1}{2}(1 - c)$ $\qquad\qquad$ $\tfrac{1}{2}(1 - c)$

$-z_c$ $\quad$ $z = 0$ $\quad$ z_c

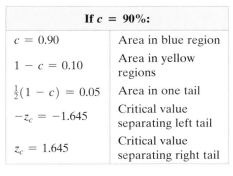

	If $c = 90\%$:
$c = 0.90$	Area in blue region
$1 - c = 0.10$	Area in yellow regions
$\tfrac{1}{2}(1 - c) = 0.05$	Area in one tail
$-z_c = -1.645$	Critical value separating left tail
$z_c = 1.645$	Critical value separating right tail

Picturing the World

A survey of a random sample of 1000 smartphone owners found that the mean daily time spent communicating on a smartphone was 131.4 minutes. From previous studies, it is assumed that the population standard deviation is 21.2 minutes. Communicating on a smartphone includes text, email, social media, and phone calls. (Adapted from International Data Corporation)

Daily Time Spent on Smartphone

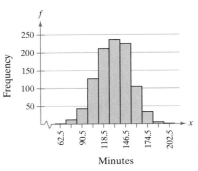

For a 95% confidence interval, what would be the margin of error for the population mean daily time spent communicating on a smartphone?

$E \approx 1.3$ minutes

The difference between the point estimate and the actual parameter value is called the **sampling error**. When μ is estimated, the sampling error is the difference $\bar{x} - \mu$. In most cases, of course, μ is unknown, and $\bar{x}$ varies from sample to sample. However, you can calculate a maximum value for the error when you know the level of confidence and the sampling distribution.

DEFINITION

Given a level of confidence c, the **margin of error E** (sometimes also called the maximum error of estimate or error tolerance) is the greatest possible distance between the point estimate and the value of the parameter it is estimating. For a population mean μ where σ is known, the margin of error is

$$E = z_c \sigma_{\bar{x}} = z_c \frac{\sigma}{\sqrt{n}} \qquad \text{Margin of error for } \mu \ (\sigma \text{ known})$$

when these conditions are met.

1. The sample is random.
2. At least one of the following is true: The population is normally distributed or $n \geq 30$. (Recall from the Central Limit Theorem that when $n \geq 30$, the sampling distribution of sample means approximates a normal distribution.)

EXAMPLE 2

Finding the Margin of Error

Use the data in Example 1 and a 95% confidence level to find the margin of error for the mean number of hours spent on required athletic activities by all student-athletes in the conference. Assume the population standard deviation is 2.3 hours.

SOLUTION

Because σ is known ($\sigma = 2.3$), the sample is random (see Example 1), and $n = 40 \geq 30$, use the formula for E given above. The z-score that corresponds to a 95% confidence level is 1.96. This implies that 95% of the area under the standard normal curve falls within 1.96 standard deviations of the mean, as shown in the figure below. (You can approximate the distribution of the sample means with a normal curve by the Central Limit Theorem because $n = 40 \geq 30$.)

Using the values $z_c = 1.96$, $\sigma = 2.3$, and $n = 40$,

$$E = z_c \frac{\sigma}{\sqrt{n}}$$
$$= 1.96 \cdot \frac{2.3}{\sqrt{40}}$$
$$\approx 0.7.$$

Interpretation   You are 95% confident that the margin of error for the population mean is about 0.7 hour.

TRY IT YOURSELF 2

Use the data in Try It Yourself 1 and a 95% confidence level to find the margin of error for the mean number of hours spent on required athletic activities by all student-athletes in the conference. Assume the population standard deviation is 2.3 hours.

Answer: Page A36

Confidence Intervals for a Population Mean

Using a point estimate and a margin of error, you can construct an interval estimate of a population parameter such as μ. This interval estimate is called a **confidence interval.**

Study Tip

When you construct a confidence interval for a population mean, the general *round-off rule* is to round off to the same number of decimal places as the sample mean.

DEFINITION

A *c*-**confidence interval for a population mean** μ is

$$\bar{x} - E < \mu < \bar{x} + E.$$

The probability that the confidence interval contains μ is c, assuming that the estimation process is repeated a large number of times.

GUIDELINES

Constructing a Confidence Interval for a Population Mean (σ Known)

In Words	In Symbols
1. Verify that σ is known, the sample is random, and either the population is normally distributed or $n \geq 30$.	
2. Find the sample statistics n and $\bar{x}$.	$\bar{x} = \dfrac{\Sigma x}{n}$
3. Find the critical value z_c that corresponds to the given level of confidence.	Use Table 4 in Appendix B.
4. Find the margin of error E.	$E = z_c \dfrac{\sigma}{\sqrt{n}}$
5. Find the left and right endpoints and form the confidence interval.	Left endpoint: $\bar{x} - E$ Right endpoint: $\bar{x} + E$ Interval: $\bar{x} - E < \mu < \bar{x} + E$

EXAMPLE 3

See Minitab steps on page 344.

Constructing a Confidence Interval

Use the data in Example 1 to construct a 95% confidence interval for the mean number of hours spent on required athletic activities by all student-athletes in the conference.

Study Tip

Other ways to represent a confidence interval are $(\bar{x} - E, \bar{x} + E)$ and $\bar{x} \pm E$. For instance, in Example 3, you could write the confidence interval as $(20.4, 21.8)$ or 21.1 ± 0.7.

SOLUTION

In Examples 1 and 2, you found that $\bar{x} \approx 21.1$ and $E \approx 0.7$. The confidence interval is constructed as shown.

Left Endpoint	Right Endpoint
$\bar{x} - E \approx 21.1 - 0.7$	$\bar{x} + E \approx 21.1 + 0.7$
$= 20.4$	$= 21.8$

$$20.4 < \mu < 21.8$$

```
            20.4          21.1       21.8
       +----+--(--+--+----●----+--+--)--+--+---→ x
      20.0      20.5      21.0     21.5     22
```

Interpretation With 95% confidence, you can say that the population mean number of hours spent on required athletic activities is between 20.4 and 21.8 hours.

Study Tip

The width of a confidence interval is $2E$. Examine the formula for E to see why a larger sample size tends to give you a narrower confidence interval for the same level of confidence.

TRY IT YOURSELF 3

Use the data in Try It Yourself 1 to construct a 95% confidence interval for the mean number of hours spent on required athletic activities by all student-athletes in the conference. Compare your result with the interval found in Example 3. *Answer: Page A36*

EXAMPLE 4

Constructing a Confidence Interval Using Technology

Use the data in Example 1 and technology to construct a 99% confidence interval for the mean number of hours spent on required athletic activities by all student-athletes in the conference.

SOLUTION

Minitab and StatCrunch each have features that allow you to construct a confidence interval. You can construct a confidence interval by entering the original data or by using the descriptive statistics. The original data was used to construct the confidence intervals shown below. From the displays, a 99% confidence interval for μ is $(20.1, 22.0)$. Note that this interval is rounded to the same number of decimals places as the sample mean.

Tech Tip

Here are instructions for constructing a confidence interval on a TI-84 Plus. First, either enter the original data into a list or enter the descriptive statistics.

STAT

Choose the TESTS menu.

7: ZInterval...

Select the *Data* input option when you use the original data. Select the *Stats* input option when you use the descriptive statistics. In each case, enter the appropriate values, then select *Calculate*. Your results may differ slightly depending on the method you use. For Example 4, the original data were entered.

```
ZInterval
(20.113,21.987)
x̄=21.05
Sx=2.438473671
n=40
```

MINITAB

One-Sample Z: Hours

The assumed standard deviation = 2.3

Variable	N	Mean	StDev	SE Mean	99% CI
Hours	40	21.050	2.438	0.364	(20.113, 21.987)

STATCRUNCH

One sample Z confidence interval:

μ : Mean of variable
Standard deviation = 2.3

99% confidence interval results:

Variable	n	Sample Mean	Std. Err.	L. Limit	U. Limit
Hours	40	21.05	0.36366193	20.113269	21.986731

Interpretation With 99% confidence, you can say that the population mean number of hours spent on required athletic activities is between 20.1 and 22.0 hours.

TRY IT YOURSELF 4

Use the data in Example 1 and technology to construct 75%, 85%, and 90% confidence intervals for the mean number of hours spent on required athletic activities by all student-athletes in the conference. How does the width of the confidence interval change as the level of confidence increases?

Answer: Page A36

In Examples 3 and 4, and Try It Yourself 4, the same sample data were used to construct confidence intervals with different levels of confidence. Notice that as the level of confidence increases, the width of the confidence interval also increases. In other words, when the same sample data are used, *the greater the level of confidence, the wider the interval.*

For a normally distributed population with σ known, you may use the normal sampling distribution for any sample size (even when $n < 30$), as shown in Example 5.

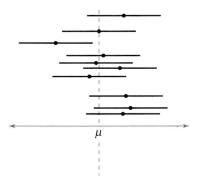

Tech Tip

Here are instructions for constructing a confidence interval in Excel. First, click *Formulas* at the top of the screen and click *Insert Function* in the *Function Library* group. Select the category *Statistical* and select the *Confidence.Norm* function. In the dialog box, enter the values of alpha, the standard deviation, and the sample size (see below). Then click OK. The value returned is the margin of error, which is used to construct the confidence interval.

	A	B
1	CONFIDENCE.NORM(0.1,1.5,20)	
2		0.551700678

Alpha is the *level of significance,* which will be explained in Chapter 7. When using Excel in Chapter 6, you can think of alpha as the complement of the level of confidence. So, for a 90% confidence interval, alpha is equal to $1 - 0.90 = 0.10$.

EXAMPLE 5

See TI-84 Plus steps on page 345.

Constructing a Confidence Interval

A college admissions director wishes to estimate the mean age of all students currently enrolled. In a random sample of 20 students, the mean age is found to be 22.9 years. From past studies, the standard deviation is known to be 1.5 years, and the population is normally distributed. Construct a 90% confidence interval for the population mean age.

SOLUTION

Because σ is known, the sample is random, and the population is normally distributed, use the formula for E given in this section. Using $n = 20$, $\bar{x} = 22.9$, $\sigma = 1.5$, and $z_c = 1.645$, the margin of error at the 90% confidence level is

$$E = z_c \frac{\sigma}{\sqrt{n}}$$

$$= 1.645 \cdot \frac{1.5}{\sqrt{20}}$$

$$\approx 0.6.$$

The 90% confidence interval can be written as $\bar{x} \pm E \approx 22.9 \pm 0.6$ or as shown below.

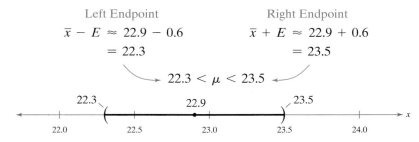

Left Endpoint
$\bar{x} - E \approx 22.9 - 0.6$
$= 22.3$

Right Endpoint
$\bar{x} + E \approx 22.9 + 0.6$
$= 23.5$

$22.3 < \mu < 23.5$

Interpretation With 90% confidence, you can say that the mean age of all the students is between 22.3 and 23.5 years.

TRY IT YOURSELF 5

Construct a 90% confidence interval for the population mean age for the college students in Example 5 with the sample size increased to 30 students. Compare your answer with Example 5. *Answer: Page A36*

After constructing a confidence interval, it is important that you interpret the results correctly. Consider the 90% confidence interval constructed in Example 5. Because μ is a fixed value predetermined by the population, it is either in the interval or not. It is *not* correct to say, "There is a 90% probability that the actual mean will be in the interval (22.3, 23.5)." This statement is wrong because it suggests that the value of μ can vary, which is not true. The correct way to interpret this confidence interval is to say, "With 90% confidence, the mean is in the interval (22.3, 23.5)." This means that when a large number of samples is collected and a confidence interval is created for each sample, approximately 90% of these intervals will contain μ, as shown in the figure at the left. This correct interpretation refers to the success rate of the process being used.

The horizontal segments represent 90% confidence intervals for different samples of the same size. In the long run, 9 of every 10 such intervals will contain μ.

Sample Size

For the same sample statistics, as the level of confidence increases, the confidence interval widens. As the confidence interval widens, the precision of the estimate decreases. One way to improve the precision of an estimate without decreasing the level of confidence is to increase the sample size. But how large a sample size is needed to guarantee a certain level of confidence for a given margin of error? By using the formula for the margin of error

$$E = z_c \frac{\sigma}{\sqrt{n}}$$

a formula can be derived (see Exercise 59) to find the minimum sample size n, as shown in the next definition.

> ### Finding a Minimum Sample Size to Estimate μ
>
> Given a c-confidence level and a margin of error E, the minimum sample size n needed to estimate the population mean μ is
>
> $$n = \left(\frac{z_c \sigma}{E} \right)^2.$$
>
> If n is not a whole number, then round n up to the next whole number (see Example 6). Also, when σ is unknown, you can estimate it using s, provided you have a preliminary random sample with at least 30 members.

EXAMPLE 6

Determining a Minimum Sample Size

The researcher in Example 1 wants to estimate the mean number of hours spent on required athletic activities by all student-athletes in the conference. How many student-athletes must be included in the sample to be 95% confident that the sample mean is within 0.5 hour of the population mean?

SOLUTION

Using $c = 0.95$, $z_c = 1.96$, $\sigma = 2.3$ (from Example 2), and $E = 0.5$, you can solve for the minimum sample size n.

$$n = \left(\frac{z_c \sigma}{E} \right)^2$$

$$= \left(\frac{1.96 \cdot 2.3}{0.5} \right)^2$$

$$\approx 81.29.$$

Because n is not a whole number, round up to 82. So, the researcher needs at least 82 student-athletes in the sample.

Interpretation The researcher already has 40 student-athletes, so the sample needs 42 more members. Note that 82 is the *minimum* number of student-athletes to include in the sample. The researcher could include more, if desired.

TRY IT YOURSELF 6

In Example 6, how many student-athletes must the researcher include in the sample to be 95% confident that the sample mean is within 0.75 hour of the population mean? Compare your answer with Example 6.

Answer: Page A36

6.1 EXERCISES

For Extra Help: MyLab Statistics

Building Basic Skills and Vocabulary

1. When estimating a population mean, are you more likely to be correct when you use a point estimate or an interval estimate? Explain your reasoning.

2. Which statistic is the best unbiased estimator for μ?

(a) s (b) $\bar{x}$ (c) the median   (d) the mode

3. For the same sample statistics, which level of confidence would produce the widest confidence interval? Explain your reasoning.

(a) 90%   (b) 95%   (c) 98%   (d) 99%

4. You construct a 95% confidence interval for a population mean using a random sample. The confidence interval is $24.9 < \mu < 31.5$. Is the probability that μ is in this interval 0.95? Explain.

In Exercises 5–8, find the critical value z_c necessary to construct a confidence interval at the level of confidence c.

5. $c = 0.80$ **6.** $c = 0.85$ **7.** $c = 0.75$ **8.** $c = 0.97$

Graphical Analysis   *In Exercises 9–12, use the values on the number line to find the sampling error.*

9. $\bar{x} = 3.8$ $\mu = 4.27$

3.4 3.6 3.8 4.0 4.2 4.4 4.6

10. $\mu = 8.76$ $\bar{x} = 9.5$

8.6 8.8 9.0 9.2 9.4 9.6 9.8

11. $\mu = 24.67$ $\bar{x} = 26.43$

24   25   26   27

12. $\bar{x} = 46.56$ $\mu = 48.12$

46   47   48   49

In Exercises 13–16, find the margin of error for the values of c, σ, and n.

13. $c = 0.95, \sigma = 5.2, n = 30$ **14.** $c = 0.90, \sigma = 2.9, n = 50$

15. $c = 0.80, \sigma = 1.3, n = 75$ **16.** $c = 0.975, \sigma = 4.6, n = 100$

Matching   *In Exercises 17–20, match the level of confidence c with the appropriate confidence interval. Assume each confidence interval is constructed for the same sample statistics.*

17. $c = 0.88$ **18.** $c = 0.90$ **19.** $c = 0.95$ **20.** $c = 0.98$

(a)   54.9   57.2   59.5

54 55 56 57 58 59 60

(b)   55.2   57.2   59.2

54 55 56 57 58 59 60

(c)   55.6   57.2   58.8

54 55 56 57 58 59 60

(d)   55.5   57.2   58.9

54 55 56 57 58 59 60

In Exercises 21–24, construct the indicated confidence interval for the population mean μ.

21. $c = 0.90, \bar{x} = 12.3, \sigma = 1.5, n = 50$

22. $c = 0.95, \bar{x} = 31.39, \sigma = 0.80, n = 82$

23. $c = 0.99, \bar{x} = 10.50, \sigma = 2.14, n = 45$

24. $c = 0.80, \bar{x} = 20.6, \sigma = 4.7, n = 100$

1. You are more likely to be correct using an interval estimate because it is unlikely that a point estimate will exactly equal the population mean.

2. b

3. d; As the level of confidence increases, z_c increases, causing wider intervals.

4. No; The 95% confidence interval means that with 95% confidence, you can say that the population mean is in this interval. If a large number of samples is collected and a confidence interval created for each, approximately 95% of these intervals will contain the population mean.

5. 1.28

6. 1.44

7. 1.15

8. 2.17

9. −0.47

10. 0.74

11. 1.76

12. −1.56

13. 1.861

14. 0.675

15. 0.192

16. 1.030

17. c

18. d

19. b

20. a

21. (12.0, 12.6)

22. (31.22, 31.56)

23. (9.7, 11.3)

24. (20.0, 21.2)

25. $E = 1.4$, $\bar{x} = 13.4$

26. $E = 4.27$, $\bar{x} = 25.88$

27. $E = 0.17$, $\bar{x} = 1.88$

28. $E = 0.016$, $\bar{x} = 3.16$

29. 126   **30.** 25

31. 7   **32.** 139

33. $E = 1.95$. $\bar{x} = 28.15$

34. $E = 18.45$, $\bar{x} = 262.52$

35. (1320.4, 1416.6); (1311.2, 1425.8)

With 90% confidence, you can say that the population mean price is between $1320.40 and $1416.60. With 95% confidence, you can say that the population mean price is between $1311.20 and $1425.80. The 95% CI is wider.

36. (113.34, 118.98); (112.81, 119.51)

With 90% confidence, you can say that the population mean price is between $113.34 and $118.98. With 95% confidence, you can say that the population mean price is between $112.81 and $119.51. The 95% CI is wider.

37. (81.14, 87.12); (80.56, 87.70)

With 90% confidence, you can say that the population mean temperature is between 81.14°F and 87.12°F. With 95% confidence, you can say that the population mean temperature is between 80.56°F and 87.70°F. The 95% CI is wider.

38. (62, 138); (54, 146)

With 90% confidence, you can say that the population mean tornadoes per month is between 62 and 138 tornadoes. With 95% confidence, you can say that the population mean tornadoes per month is between 54 and 146 tornadoes. The 95% CI is wider.

39. No; The margin of error is large ($E = 48.1$).

40. Yes; The margin of error is small ($E = 2.82$).

41. No; The right endpoint of the 95% CI is 87.70.

42. Yes; The midpoint of both confidence intervals is 100. Therefore, it is equally likely for the population mean to be less than 100 as it is likely to be greater than 100.

In Exercises 25–28, use the confidence interval to find the margin of error and the sample mean.

25. (12.0, 14.8)

26. (21.61, 30.15)

27. (1.71, 2.05)

28. (3.144, 3.176)

In Exercises 29–32, determine the minimum sample size n needed to estimate μ for the values of c, σ, and E.

29. $c = 0.90$, $\sigma = 6.8$, $E = 1$

30. $c = 0.95$, $\sigma = 2.5$, $E = 1$

31. $c = 0.80$, $\sigma = 4.1$, $E = 2$

32. $c = 0.98$, $\sigma = 10.1$, $E = 2$

Using and Interpreting Concepts

Finding the Margin of Error   *In Exercises 33 and 34, use the confidence interval to find the estimated margin of error. Then find the sample mean.*

33. Commute Times   A government agency reports a confidence interval of (26.2, 30.1) when estimating the mean commute time (in minutes) for the population of workers in a city.

34. Book Prices   A store manager reports a confidence interval of (244.07, 280.97) when estimating the mean price (in dollars) for the population of textbooks.

Constructing Confidence Intervals   *In Exercises 35–38, you are given the sample mean and the population standard deviation. Use this information to construct 90% and 95% confidence intervals for the population mean. Interpret the results and compare the widths of the confidence intervals.*

35. Gold Prices   From a random sample of 48 business days from January 4, 2010, through February 24, 2017, U.S. gold prices had a mean of $1368.48. Assume the population standard deviation is $202.60. *(Source: Federal Reserve Bank of St. Louis)*

36. Stock Prices   From a random sample of 36 business days from February 24, 2016, through February 24, 2017, the mean closing price of Apple stock was $116.16. Assume the population standard deviation is $10.27. *(Source: Nasdaq)*

37. Maximum Daily Temperature   From a random sample of 64 dates, the mean record high daily temperature in the Chicago, Illinois, area has a mean of 84.13°F. Assume the population standard deviation is 14.56°F. *(Source: NOAA)*

38. Tornadoes per Month   From a random sample of 24 months from January 2006 through December 2016, the mean number of tornadoes per month in the United States was about 100. Assume the population standard deviation is 114. *(Source: NOAA)*

39. In Exercise 35, does it seem possible that the population mean could equal the sample mean? Explain.

40. In Exercise 36, does it seem possible that the population mean could be within 1% of the sample mean? Explain.

41. In Exercise 37, does it seem possible that the population mean could be greater than 90°F? Explain.

42. In Exercise 38, does it seem possible that the population mean could be less than 100? Explain.

43. (a) An increase in the level of confidence will widen the confidence interval and the less certain you can be about a point estimate.

(b) An increase in the sample size will narrow the confidence interval because it decreases the standard error.

(c) An increase in the population standard deviation will widen the confidence interval because small standard deviations produce more precise intervals, which are smaller.

44. Answers will vary.

45. (20.9, 24.8); (19.8, 25.9)

With 90% confidence, you can say that the population mean length of time is between 20.9 and 24.8 minutes. With 99% confidence, you can say that the population mean length of time is between 19.8 and 25.9 minutes. The 99% CI is wider.

46. (20.01, 24.19); (18.83, 25.37)

With 90% confidence, you can say that the population mean concentration is between 20.01 and 24.19 grams per liter. With 99% confidence, you can say that the population mean concentration is between 18.83 and 25.37 grams per liter. The 99% CI is wider.

47. 89

48. 4

49. (a) 66 servings

(b) No; Yes; The 95% CI is (28.252, 29.748). If the population mean is within 3% of the sample mean, then it falls outside the CI. If the population mean is within 0.3% of the sample mean, then it falls within the CI.

43. When all other quantities remain the same, how does the indicated change affect the width of a confidence interval? Explain.

(a) Increase in the level of confidence

(b) Increase in the sample size

(c) Increase in the population standard deviation

44. Describe how you would construct a 90% confidence interval to estimate the population mean age for students at your school.

Constructing Confidence Intervals *In Exercises 45 and 46, use the information to construct 90% and 99% confidence intervals for the population mean. Interpret the results and compare the widths of the confidence intervals.*

45. DVR and Other Time-Shifted Viewing A group of researchers estimates the mean length of time (in minutes) the average U.S. adult spends watching television using digital video recorders (DVRs) and other forms of time-shifted television each day. To do so, the group takes a random sample of 30 U.S. adults and obtains the times (in minutes) below.

29	12	23	24	33	24	28	31	18	27	27	32	17	13	17
12	21	32	26	16	28	28	21	24	29	13	20	13	21	27

From past studies, the research council assumes that σ is 6.5 minutes. *(Adapted from the Nielsen Company)*

46. Sodium Chloride Concentrations The sodium chloride concentrations (in grams per liter) for 36 randomly selected seawater samples are listed. Assume that σ is 7.61 grams per liter.

30.63	33.47	26.76	15.23	13.21	10.57
16.57	27.32	27.06	15.07	28.98	34.66
10.22	22.43	17.33	28.40	35.70	14.09
11.77	33.60	27.09	26.78	22.39	30.35
11.83	13.05	22.22	13.45	18.86	24.92
32.86	31.10	18.84	10.86	15.69	22.35

47. Determining a Minimum Sample Size Determine the minimum sample size required when you want to be 95% confident that the sample mean is within one unit of the population mean and $\sigma = 4.8$. Assume the population is normally distributed.

48. Determining a Minimum Sample Size Determine the minimum sample size required when you want to be 99% confident that the sample mean is within two units of the population mean and $\sigma = 1.4$. Assume the population is normally distributed.

49. Cholesterol Contents of Cheese A cheese processing company wants to estimate the mean cholesterol content of all one-ounce servings of a type of cheese. The estimate must be within 0.75 milligram of the population mean.

(a) Determine the minimum sample size required to construct a 95% confidence interval for the population mean. Assume the population standard deviation is 3.10 milligrams.

(b) The sample mean is 29 milligrams. Using the minimum sample size with a 95% level of confidence, does it seem possible that the population mean could be within 3% of the sample mean? within 0.3% of the sample mean? Explain.

50. (a) 4 students

(b) No; No; The 90% CI is (18.684, 21.316). If the population mean is within 7% or 8% of the sample mean, then it falls outside the CI.

Error tolerance = 0.5 oz

Volume = 1 gal (128 oz)

FIGURE FOR EXERCISE 51

Error tolerance = 0.25 fl oz

Volume = 1/2 gal (64 fl oz)

FIGURE FOR EXERCISE 52

51. (a) 7 cans

(b) Yes; The 90% CI is (127.3, 128.2) and 128 ounces falls within that interval.

52. (a) 62 bottles

(b) Yes; The 95% CI is (63.762, 64.238) and there are amounts greater than 63.85 fluid ounces that fall within that interval.

53. (a) 74 balls

(b) Yes; The 99% CI is (27.360, 27.640) and there are amounts less than 27.6 inches that fall within that interval.

54. (a) 27 balls

(b) Yes; The 99% CI is (8.2558, 8.3442) and 8.258 inches falls within that interval.

50. Ages of College Students An admissions director wants to estimate the mean age of all students enrolled at a college. The estimate must be within 1.5 years of the population mean. Assume the population of ages is normally distributed.

(a) Determine the minimum sample size required to construct a 90% confidence interval for the population mean. Assume the population standard deviation is 1.6 years.

(b) The sample mean is 20 years of age. Using the minimum sample size with a 90% level of confidence, does it seem possible that the population mean could be within 7% of the sample mean? within 8% of the sample mean? Explain.

51. Paint Can Volumes A paint manufacturer uses a machine to fill gallon cans with paint (see figure). The manufacturer wants to estimate the mean volume of paint the machine is putting in the cans within 0.5 ounce. Assume the population of volumes is normally distributed.

(a) Determine the minimum sample size required to construct a 90% confidence interval for the population mean. Assume the population standard deviation is 0.75 ounce.

(b) The sample mean is 127.75 ounces. With a sample size of 8, a 90% level of confidence, and a population standard deviation of 0.75 ounce, does it seem possible that the population mean could be exactly 128 ounces? Explain.

52. Juice Dispensing Machine A beverage company uses a machine to fill half-gallon bottles with fruit juice (see figure). The company wants to estimate the mean volume of water the machine is putting in the bottles within 0.25 fluid ounce.

(a) Determine the minimum sample size required to construct a 95% confidence interval for the population mean. Assume the population standard deviation is 1 fluid ounce.

(b) The sample mean is exactly 64 fluid ounces. With a sample size of 68, a 95% level of confidence, and a population standard deviation of 1 fluid ounce, does it seem possible that the population mean could be greater than 63.85 fluid ounces? Explain.

53. Soccer Balls A soccer ball manufacturer wants to estimate the mean circumference of soccer balls within 0.15 inch.

(a) Determine the minimum sample size required to construct a 99% confidence interval for the population mean. Assume the population standard deviation is 0.5 inch.

(b) The sample mean is 27.5 inches. With a sample size of 84, a 99% level of confidence, and a population standard deviation of 0.5 inch, does it seem possible that the population mean could be less than 27.6 inches? Explain.

54. Tennis Balls A tennis ball manufacturer wants to estimate the mean circumference of tennis balls within 0.05 inch. Assume the population of circumferences is normally distributed.

(a) Determine the minimum sample size required to construct a 99% confidence interval for the population mean. Assume the population standard deviation is 0.10 inch.

(b) The sample mean is 8.3 inches. With a sample size of 34, a 99% level of confidence, and a population standard deviation of 0.10 inch, does it seem possible that the population mean could be exactly 8.258 inches? Explain.

55. *Sample answer:* A 99% CI may not be practical to use in all situations. It may produce a CI so wide that it has no practical application.

56. (a) An increase in the level of confidence will increase the minimum sample size required because the more data you have, the narrower the interval.

(b) An increase in the error tolerance will decrease the minimum sample size required because the less data you have, the less accurate your results will be.

(c) An increase in the population standard deviation will increase the minimum sample size required because the data is more spread out.

57. (a) 0.707   (b) 0.949
(c) 0.962   (d) 0.975
(e) 0.711   (f) 0.937
(g) 0.964   (h) 0.979

The finite population correction factor approaches 1 as the sample size decreases and the population size remains the same.

The finite population correction factor approaches 1 as the population size increases and the sample size remains the same.

58. (a) (6.2, 11.0)
(b) (10.3, 11.5)
(c) (40.2, 40.4)
(d) (54.7, 58.7)

59. *Sample answer:*

$E = \dfrac{z_c \sigma}{\sqrt{n}}$ Write the original equation.

$E\sqrt{n} = z_c \sigma$ Multiply each side by $\sqrt{n}$.

$\sqrt{n} = \dfrac{z_c \sigma}{E}$ Divide each side by E.

$n = \left(\dfrac{z_c \sigma}{E}\right)^2$ Square each side.

55. When estimating the population mean, why not construct a 99% confidence interval every time?

56. When all other quantities remain the same, how does the indicated change affect the minimum sample size requirement? Explain.

(a) Increase in the level of confidence

(b) Increase in the error tolerance

(c) Increase in the population standard deviation

Extending Concepts

Finite Population Correction Factor *In Exercises 57 and 58, use the information below.*

In this section, you studied the construction of a confidence interval to estimate a population mean. In each case, the underlying assumption was that the sample size n was small in comparison to the population size N. When $n \geq 0.05N$, however, the formula that determines the standard error of the mean $\sigma_{\bar{x}}$ needs to be adjusted, as shown below.

$$\sigma_{\bar{x}} = \frac{\sigma}{\sqrt{n}}\sqrt{\frac{N-n}{N-1}}$$

Recall from the Section 5.4 exercises that the expression $\sqrt{(N-n)/(N-1)}$ is called a ***finite population correction factor.*** The margin of error is

$$E = z_c\frac{\sigma}{\sqrt{n}}\sqrt{\frac{N-n}{N-1}}.$$

57. Determine the finite population correction factor for each value of N and n.

(a) $N = 1000$ and $n = 500$ (b) $N = 1000$ and $n = 100$
(c) $N = 1000$ and $n = 75$ (d) $N = 1000$ and $n = 50$
(e) $N = 100$ and $n = 50$ (f) $N = 400$ and $n = 50$
(g) $N = 700$ and $n = 50$ (h) $N = 1200$ and $n = 50$

What happens to the finite population correction factor as the sample size n decreases but the population size N remains the same? as the population size N increases but the sample size n remains the same?

58. Use the finite population correction factor to construct each confidence interval for the population mean.

(a) $c = 0.99, \bar{x} = 8.6, \sigma = 4.9, N = 200, n = 25$
(b) $c = 0.90, \bar{x} = 10.9, \sigma = 2.8, N = 500, n = 50$
(c) $c = 0.95, \bar{x} = 40.3, \sigma = 0.5, N = 300, n = 68$
(d) $c = 0.80, \bar{x} = 56.7, \sigma = 9.8, N = 400, n = 36$

59. Sample Size The equation for determining the sample size

$$n = \left(\frac{z_c \sigma}{E}\right)^2$$

can be obtained by solving the equation for the margin of error

$$E = \frac{z_c \sigma}{\sqrt{n}}$$

for n. Show that this is true and justify each step.

6.2 Confidence Intervals for the Mean (σ Unknown)

What You Should Learn

▶ How to interpret the *t*-distribution and use a *t*-distribution table

▶ How to construct and interpret confidence intervals for a population mean when σ is not known

The *t*-Distribution ■ Confidence Intervals and *t*-Distributions

The *t*-Distribution

In many real-life situations, the population standard deviation is unknown. So, how can you construct a confidence interval for a population mean when σ is *not* known? For a simple random sample that is drawn from a population that is normally distributed or has a sample size of 30 or more, you can use the sample standard deviation *s* to estimate the population standard deviation σ. However, when using *s*, the sampling distribution of $\bar{x}$ does not follow a normal distribution. In this case, the sampling distribution of $\bar{x}$ follows a *t*-distribution.

> ### DEFINITION
>
> If the distribution of a random variable *x* is approximately normal, then
> $$t = \frac{\bar{x} - \mu}{s / \sqrt{n}}$$
> follows a *t*-distribution. Critical values of *t* are denoted by t_c. Here are several properties of the *t*-distribution.
>
> 1. The mean, median, and mode of the *t*-distribution are equal to 0.
> 2. The *t*-distribution is bell-shaped and symmetric about the mean.
> 3. The total area under the *t*-distribution curve is equal to 1.
> 4. The tails in the *t*-distribution are "thicker" than those in the standard normal distribution.
> 5. The standard deviation of the *t*-distribution varies with the sample size, but it is greater than 1.
> 6. The *t*-distribution is a family of curves, each determined by a parameter called the *degrees of freedom*. The **degrees of freedom** (sometimes abbreviated as d.f.) are the number of free choices left after a sample statistic such as $\bar{x}$ is calculated. When you use a *t*-distribution to estimate a population mean, the degrees of freedom are equal to one less than the sample size.
> $$\text{d.f.} = n - 1 \qquad \text{Degrees of freedom}$$
> 7. As the degrees of freedom increase, the *t*-distribution approaches the standard normal distribution, as shown in the figure. For 30 or more degrees of freedom, the *t*-distribution is close to the standard normal distribution.

Study Tip

Here is an example that illustrates the concept of degrees of freedom.

The number of chairs in a classroom equals the number of students: 25 chairs and 25 students. Each of the first 24 students to enter the classroom has a choice on which chair he or she will sit. There is no freedom of choice, however, for the 25th student who enters the room.

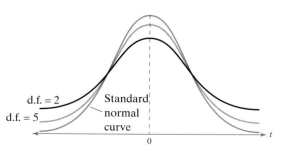

Table 5 in Appendix B lists critical values of t for selected confidence intervals and degrees of freedom.

EXAMPLE 1

Finding Critical Values of *t*

Find the critical value t_c for a 95% confidence level when the sample size is 15.

SOLUTION

Because $n = 15$, the degrees of freedom are d.f. $= n - 1 = 15 - 1 = 14$. A portion of Table 5 is shown. Using d.f. $= 14$ and $c = 0.95$, you can find the critical value t_c, as shown by the highlighted areas in the table.

Study Tip

Critical values in the *t*-distribution table for a specific confidence interval can be found in the column headed by *c* in the appropriate d.f. row. (The symbol α will be explained in Chapter 7.)

	Level of confidence, *c*	0.80	0.90	0.95	0.98	0.99
	One tail, α	0.10	0.05	0.025	0.01	0.005
d.f.	Two tails, α	0.20	0.10	0.05	0.02	0.01
1		3.078	6.314	12.706	31.821	63.657
2		1.886	2.920	4.303	6.965	9.925
3		1.638	2.353	3.182	4.541	5.841
12		1.356	1.782	2.179	2.681	3.055
13		1.350	1.771	2.160	2.650	3.012
14		1.345	1.761	2.145	2.624	2.977
15		1.341	1.753	2.131	2.602	2.947
16		1.337	1.746	2.120	2.583	2.921

From the table, you can see that $t_c = 2.145$. The figure shows the *t*-distribution for 14 degrees of freedom, $c = 0.95$, and $t_c = 2.145$.

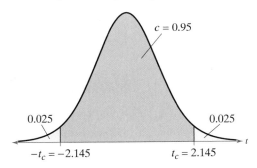

TI-84 PLUS

```
invT(.975,14)
       2.144786681
```

You can use technology to find t_c. To use a TI-84 Plus, you need to know the area under the curve to the left of t_c, which is

$$0.95 + 0.025 = 0.975.$$ Area to the left of t_c

From the TI-84 Plus display at the left, $t_c \approx 2.145$.

Interpretation So, for a *t*-distribution curve with 14 degrees of freedom, 95% of the area under the curve lies between $t = \pm 2.145$.

TRY IT YOURSELF 1

Find the critical value t_c for a 90% confidence level when the sample size is 22.

Answer: Page A36

When the number of degrees of freedom you need is not in the table, use the closest number in the table that is *less than* the value you need (or use technology, as shown in Example 1). For instance, for d.f. $= 57$, use 50 degrees of freedom. This conservative approach will yield a larger confidence interval with a slightly higher level of confidence c.

Confidence Intervals and *t*-Distributions

Constructing a confidence interval for μ when σ is *not* known using the *t*-distribution is similar to constructing a confidence interval for μ when σ is known using the standard normal distribution—both use a point estimate $\bar{x}$ and a margin of error E. When σ is not known, the margin of error E is calculated using the sample standard deviation s and the critical value t_c. So, the formula for E is

$$E = t_c \frac{s}{\sqrt{n}}.$$ Margin of error for μ (σ unknown)

Before using this formula, verify that the sample is random, and either the population is normally distributed or $n \geq 30$.

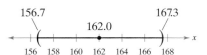

Study Tip

Remember that you can calculate the sample standard deviation s using the formula

$$s = \sqrt{\frac{\Sigma(x - \bar{x})^2}{n - 1}}$$

or the alternate formula

$$s = \sqrt{\frac{\Sigma x^2 - (\Sigma x)^2/n}{n - 1}}.$$

However, the most convenient way to find the sample standard deviation is to use technology.

GUIDELINES

Constructing a Confidence Interval for a Population Mean (σ Unknown)

In Words	In Symbols
1. Verify that σ is not known, the sample is random, and either the population is normally distributed or $n \geq 30$.	
2. Find the sample statistics n, $\bar{x}$, and s.	$\bar{x} = \dfrac{\Sigma x}{n}, s = \sqrt{\dfrac{\Sigma(x - \bar{x})^2}{n - 1}}$
3. Identify the degrees of freedom, the level of confidence c, and the critical value t_c.	d.f. $= n - 1$ Use Table 5 in Appendix B.
4. Find the margin of error E.	$E = t_c \dfrac{s}{\sqrt{n}}$
5. Find the left and right endpoints and form the confidence interval.	Left endpoint: $\bar{x} - E$ Right endpoint: $\bar{x} + E$ Interval: $\bar{x} - E < \mu < \bar{x} + E$

EXAMPLE 2

See Minitab steps on page 344.

Constructing a Confidence Interval

You randomly select 16 coffee shops and measure the temperature of the coffee sold at each. The sample mean temperature is 162.0°F with a sample standard deviation of 10.0°F. Construct a 95% confidence interval for the population mean temperature of coffee sold. Assume the temperatures are approximately normally distributed.

SOLUTION   Because σ is unknown, the sample is random, and the temperatures are approximately normally distributed, use the *t*-distribution. Using $n = 16$, $\bar{x} = 162.0$, $s = 10.0$, $c = 0.95$, and d.f. $= 15$, you can use Table 5 to find that $t_c = 2.131$. The margin of error at the 95% confidence level is

$$E = t_c \frac{s}{\sqrt{n}} = 2.131 \cdot \frac{10.0}{\sqrt{16}} \approx 5.3.$$

The confidence interval is shown below and in the figure at the left.

156.7 ⟍
   162.0
            ⟋ 167.3

←─┼─(─┼──┼──●──┼──┼──)─┼─→ *x*
156 158 160 162 164 166 168

	Left Endpoint	Right Endpoint
	$\bar{x} - E \approx 162 - 5.3 = 156.7$	$\bar{x} + E \approx 162 + 5.3 = 167.3$

$$156.7 < \mu < 167.3$$

Interpretation   With 95% confidence, you can say that the population mean temperature of coffee sold is between 156.7°F and 167.3°F.

TRY IT YOURSELF 2

Construct 90% and 99% confidence intervals for the population mean temperature of coffee sold in Example 2.

Answer: Page A36

6.2 To explore this topic further, see **Activity 6.2** on page 318.

EXAMPLE 3

See TI-84 Plus steps on page 345.

Constructing a Confidence Interval

You randomly select 36 cars of the same model that were sold at a car dealership and determine the number of days each car sat on the dealership's lot before it was sold. The sample mean is 9.75 days, with a sample standard deviation of 2.39 days. Construct a 99% confidence interval for the population mean number of days the car model sits on the dealership's lot.

SOLUTION

Because σ is unknown, the sample is random, and $n = 36 \geq 30$, use the t-distribution. Using $n = 36$, $\bar{x} = 9.75$, $s = 2.39$, $c = 0.99$, and d.f. = 35, you can use Table 5 to find that $t_c = 2.724$. The margin of error at the 99% confidence level is

$$E = t_c \frac{s}{\sqrt{n}} = 2.724 \cdot \frac{2.39}{\sqrt{36}} \approx 1.09.$$

The confidence interval is constructed as shown.

Left Endpoint
$\bar{x} - E \approx 9.75 - 1.09$
$= 8.66$

Right Endpoint
$\bar{x} + E \approx 9.75 + 1.09$
$= 10.84$

$8.66 < \mu < 10.84$

You can check this answer using technology, as shown below. (When using technology, your answers may differ slightly from those found using Table 5.)

STATCRUNCH

One sample T confidence interval:
μ : Mean of population

99% confidence interval results:

Mean	Sample Mean	Std. Err.	DF	L. Limit	U. Limit
μ	9.75	0.39833333	35	8.6650174	10.834983

Interpretation With 99% confidence, you can say that the population mean number of days the car model sits on the dealership's lot is between 8.66 and 10.84.

TRY IT YOURSELF 3

Construct 90% and 95% confidence intervals for the population mean number of days the car model sits on the dealership's lot in Example 3. Compare the widths of the confidence intervals.

Answer: Page A36

HISTORICAL REFERENCE

William S. Gosset (1876–1937)

Developed the *t*-distribution while employed by the Guinness Brewing Company in Dublin, Ireland. Gosset published his findings using the pseudonym Student. The *t*-distribution is sometimes referred to as Student's *t*-distribution. (See page 35 for others who were important in the history of statistics.)

Picturing the World

Two footballs, one filled with air and the other filled with helium, were kicked on a windless day at Ohio State University. The footballs were alternated with each kick. After 10 practice kicks, each football was kicked 29 more times. The distances (in yards) are listed. (Source: The Columbus Dispatch)

Air Filled

1	9
2	0 0 2 2 2
2	5 5 5 5 6 6
2	7 7 7 8 8 8 8 8 9 9 9
3	1 1 1 2
3	3 4 Key: 1|9 = 19

Helium Filled

1	1 2
1	4
1	
2	2
2	3 4 6 6 6
2	7 8 8 8 9 9 9 9
3	0 0 0 0 1 1 2 2
3	3 4 5
3	9 Key: 1|1 = 11

Assume that the distances are normally distributed for each football. Apply the flowchart at the right to each sample. Construct a 95% confidence interval for the population mean distance each football traveled. Do the confidence intervals overlap? What does this result tell you?

Air: (25.3, 28.3), Helium: (25.1, 29.9); Yes; Because the confidence intervals overlap, the means may be the same.

The flowchart describes when to use the standard normal distribution and when to use the *t*-distribution to construct a confidence interval for a population mean.

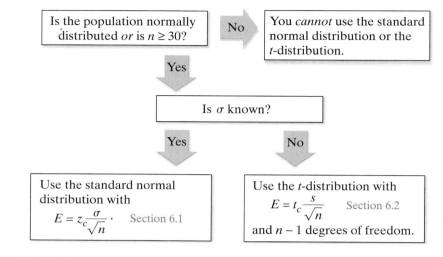

Notice in the flowchart that when both $n < 30$ and the population is *not* normally distributed, you *cannot* use the standard normal distribution or the *t*-distribution.

EXAMPLE 4

Choosing the Standard Normal Distribution or the *t*-Distribution

You randomly select 25 newly constructed houses. The sample mean construction cost is $181,000 and the population standard deviation is $28,000. Assuming construction costs are normally distributed, should you use the standard normal distribution, the *t*-distribution, or neither to construct a 95% confidence interval for the population mean construction cost? Explain your reasoning.

SOLUTION

Is the population normally distributed or is n ≥ 30?
Yes, the population is normally distributed. Note that even though

$$n = 25 < 30$$

you can still use either the standard normal distribution or the *t*-distribution because the population is normally distributed.

Is σ known?
Yes.

Decision:
Use the standard normal distribution.

TRY IT YOURSELF 4

You randomly select 18 adult male athletes and measure the resting heart rate of each. The sample mean heart rate is 64 beats per minute, with a sample standard deviation of 2.5 beats per minute. Assuming the heart rates are normally distributed, should you use the standard normal distribution, the *t*-distribution, or neither to construct a 90% confidence interval for the population mean heart rate? Explain your reasoning.

Answer: Page A36

6.2 EXERCISES

Building Basic Skills and Vocabulary

Finding Critical Values of *t*   *In Exercises 1–4, find the critical value t_c for the level of confidence c and sample size n.*

1. $c = 0.90, n = 10$

2. $c = 0.95, n = 12$

3. $c = 0.99, n = 16$

4. $c = 0.98, n = 40$

In Exercises 5–8, find the margin of error for the values of c, s, and n.

5. $c = 0.95, s = 5, n = 16$

6. $c = 0.99, s = 3, n = 6$

7. $c = 0.90, s = 2.4, n = 35$

8. $c = 0.98, s = 4.7, n = 9$

In Exercises 9–12, construct the indicated confidence interval for the population mean μ using the t-distribution. Assume the population is normally distributed.

9. $c = 0.90, \bar{x} = 12.5, s = 2.0, n = 6$

10. $c = 0.95, \bar{x} = 13.4, s = 0.85, n = 8$

11. $c = 0.98, \bar{x} = 4.3, s = 0.34, n = 14$

12. $c = 0.99, \bar{x} = 24.7, s = 4.6, n = 50$

In Exercises 13–16, use the confidence interval to find the margin of error and the sample mean.

13. $(14.7, 22.1)$

14. $(6.17, 8.53)$

15. $(64.6, 83.6)$

16. $(16.2, 29.8)$

Using and Interpreting Concepts

Constructing a Confidence Interval   *In Exercises 17–20, you are given the sample mean and the sample standard deviation. Assume the population is normally distributed and use the t-distribution to find the margin of error and construct a 95% confidence interval for the population mean. Interpret the results.*

17. Commute Time   In a random sample of eight people, the mean commute time to work was 35.5 minutes and the standard deviation was 7.2 minutes.

18. Driving Distance   In a random sample of five people, the mean driving distance to work was 22.2 miles and the standard deviation was 5.8 miles.

19. Cell Phone Prices   In a random sample of eight cell phones, the mean full retail price was $526.50 and the standard deviation was $184.00.

20. Mobile Device Repair Costs   In a random sample of 12 mobile devices, the mean repair cost was $90.42 and the standard deviation was $33.61.

21. You research commute times to work and find that the population standard deviation is 9.3 minutes. Repeat Exercise 17 using the standard normal distribution with the appropriate calculations for a standard deviation that is known. Compare the results.

Answers (margin column)

1. 1.833
2. 2.201
3. 2.947
4. 2.426
5. 2.664
6. 4.938
7. 0.686
8. 4.537
9. (10.9, 14.1)
10. (12.7, 14.1)
11. (4.1, 4.5)
12. (23.0, 26.4)
13. $E = 3.7, \bar{x} = 18.4$
14. $E = 1.18, \bar{x} = 7.35$
15. $E = 9.5, \bar{x} = 74.1$
16. $E = 6.8, \bar{x} = 23$
17. 6.0; (29.5, 41.5); With 95% confidence, you can say that the population mean commute time is between 29.5 and 41.5 minutes.
18. 7.2; (15.0, 29.4); With 95% confidence, you can say that the population mean driving distance is between 15.0 and 29.4 miles.
19. 153.83; (372.67, 680.33); With 95% confidence, you can say that the population mean cell phone price is between $372.67 and $680.33.
20. 21.35; (69.07, 111.77); With 95% confidence, you can say that the population mean repair cost is between $69.07 and $111.77.
21. 6.4; (29.1, 41.9); With 95% confidence, you can say that the population mean commute time is between 29.1 and 41.9 minutes. This confidence interval is slightly wider than the one found in Exercise 17.

22. 4.6; (17.6, 26.8); With 95% confidence, you can say that the population mean driving distance is between 17.6 and 26.8 miles. This confidence interval is narrower than the one found in Exercise 18.

23. Yes

24. Yes

25. (a) 1185
    (b) 168.1
    (c) (1034.3, 1335.7)

26. (a) 2.51
    (b) 0.87
    (c) (1.81, 3.21)

27. (a) 7.49
    (b) 1.64
    (c) (6.28, 8.70)

28. (a) 12.19
    (b) 1.75
    (c) (10.99, 13.39)

29. No

30. No

31. (a) 68,757.94
    (b) 15,834.18
    (c) (61,892.21, 75,623.67)

32. (a) 61,327.45
    (b) 17,690.14
    (c) (54,542.23, 68,112.67)

22. You research driving distances to work and find that the population standard deviation is 5.2 miles. Repeat Exercise 18 using the standard normal distribution with the appropriate calculations for a standard deviation that is known. Compare the results.

23. You research prices of cell phones and find that the population mean is $431.61. In Exercise 19, does the t-value fall between $-t_{0.95}$ and $t_{0.95}$?

24. You research repair costs of mobile devices and find that the population mean is $89.56. In Exercise 20, does the t-value fall between $-t_{0.95}$ and $t_{0.95}$?

Constructing a Confidence Interval *In Exercises 25–28, use the data set to (a) find the sample mean, (b) find the sample standard deviation, and (c) construct a 99% confidence interval for the population mean. Assume the population is normally distributed.*

25. **SAT Scores** The SAT scores of 12 randomly selected high school seniors

| 1130 | 1290 | 1010 | 1320 | 950 | 1250 |
| 1340 | 1100 | 1260 | 1180 | 1470 | 920 |

26. **Grade Point Averages** The grade point averages of 14 randomly selected college students

2.3 3.3 2.6 1.8 3.1 4.0 0.7 2.3 2.0 3.1 3.4 1.3 2.6 2.6

27. **College Football** The weekly time (in hours) spent weight lifting for 16 randomly selected college football players

| 7.4 | 5.8 | 7.3 | 7.0 | 8.9 | 9.4 | 8.3 | 9.3 |
| 6.9 | 7.5 | 9.0 | 5.8 | 5.5 | 8.6 | 9.3 | 3.8 |

28. **Homework** The weekly time spent (in hours) on homework for 18 randomly selected high school students

| 12.0 | 11.3 | 13.5 | 11.7 | 12.0 | 13.0 | 15.5 | 10.8 | 12.5 |
| 12.3 | 14.0 | 9.5 | 8.8 | 10.0 | 12.8 | 15.0 | 11.8 | 13.0 |

29. In Exercise 25, the population mean SAT score is 1020. Does the t-value fall between $-t_{0.99}$ and $t_{0.99}$? *(Source: The College Board)*

30. In Exercise 28, the population mean weekly time spent on homework by students is 7.8 hours. Does the t-value fall between $-t_{0.99}$ and $t_{0.99}$?

Constructing a Confidence Interval *In Exercises 31 and 32, use the data set to (a) find the sample mean, (b) find the sample standard deviation, and (c) construct a 98% confidence interval for the population mean.*

31. **Earnings** The annual earnings (in dollars) of 32 randomly selected magnetic resonance imaging technologists *(Adapted from Salary.com)*

90,198	65,357	62,108	78,201	80,882	74,759	46,382	75,196
56,412	74,610	65,460	68,066	53,610	92,391	57,579	99,477
42,363	47,315	67,769	50,110	92,437	58,188	76,515	65,997
42,713	77,745	57,928	94,066	61,748	91,868	62,445	70,359

32. **Earnings** The annual earnings (in dollars) of 40 randomly selected intermediate level life insurance underwriters *(Adapted from Salary.com)*

31,530	79,749	53,851	92,386	45,186	64,852	72,372	61,498
92,630	62,000	65,504	62,355	44,100	73,498	59,171	40,991
70,008	37,570	47,075	37,682	84,368	40,257	74,415	63,601
39,446	59,610	45,225	40,161	91,934	51,608	81,686	76,180
71,477	42,612	56,210	84,443	34,487	70,809	81,521	69,040

33. Yes

34. Yes

35. Use a *t*-distribution because σ is unknown and $n \geq 30$.

(26.0, 29.4); With 95% confidence, you can say that the population mean BMI is between 26.0 and 29.4.

36. Use a *t*-distribution because σ is unknown and the interest rates are normally distributed.

(3.99, 4.73); With 95% confidence, you can say that the population mean interest rate is between 3.99% and 4.73%.

37. Neither distribution can be used because $n < 30$ and the mileages are not normally distributed.

38. Use the standard normal distribution because σ is known and the yards per carry are normally distributed.

(3.19, 5.15); With 95% confidence, you can say that the population mean yards per carry is between 3.19 and 5.15 yards.

39. No; Half the sample mean is 2.18%, which falls outside the confidence interval.

40. Yes; For the population mean to be within 10% of the sample mean, its value would be in the interval (3.753, 4.587). This interval falls within the confidence interval.

41. No; They are not making good tennis balls because the *t*-value for the sample is $t = 10$, which is not between $-t_{0.99} = -2.797$ and $t_{0.99} = 2.797$.

42. Yes; They are making good light bulbs because the *t*-value for the sample is $t = 2.4$, which is between $-t_{0.99} = -2.947$ and $t_{0.99} = 2.947$.

33. In Exercise 31, the population mean salary is $72,000. Does the *t*-value fall between $-t_{0.98}$ and $t_{0.98}$? *(Source: Salary.com)*

34. In Exercise 32, the population mean salary is $61,000. Does the *t*-value fall between $-t_{0.98}$ and $t_{0.98}$? *(Source: Salary.com)*

Choosing a Distribution *In Exercises 35–38, use the standard normal distribution or the t-distribution to construct a 95% confidence interval for the population mean. Justify your decision. If neither distribution can be used, explain why. Interpret the results.*

35. Body Mass Index In a random sample of 50 people, the mean body mass index (BMI) was 27.7 and the standard deviation was 6.12.

36. Mortgages In a random sample of 18 months from June 2008 through September 2016, the mean interest rate for 30-year fixed rate conventional home mortgages was 4.36% and the standard deviation was 0.75%. Assume the interest rates are normally distributed. *(Source: Board of Governors of the Federal Reserve System)*

37. Gas Mileage The gas mileages (in miles per gallon) of 28 randomly selected sports cars are listed. Assume the mileages are not normally distributed.

21 30 19 20 21 24 18 24 27 20 22 30 25 26
22 17 21 24 22 20 24 21 20 18 20 21 20 27

38. Yards Per Carry In a recent season, the population standard deviation of the yards per carry for all running backs was 1.80. The yards per carry of 13 randomly selected running backs are listed. Assume the yards per carry are normally distributed. *(Source: National Football League)*

4.4 3.6 3.9 4.1 4.3 5.9 5.7 5.2 3.5 1.8 7.5 2.3 2.0

39. In Exercise 36, does it seem possible that the population mean could equal half the sample mean? Explain.

40. In Exercise 38, does it seem possible that the population mean could be within 10% of the sample mean? Explain.

Extending Concepts

41. Tennis Ball Manufacturing A company manufactures tennis balls. When its tennis balls are dropped onto a concrete surface from a height of 100 inches, the company wants the mean height the balls bounce upward to be 55.5 inches. This average is maintained by periodically testing random samples of 25 tennis balls. If the *t*-value falls between $-t_{0.99}$ and $t_{0.99}$, then the company will be satisfied that it is manufacturing acceptable tennis balls. For a random sample, the mean bounce height of the sample is 56.0 inches and the standard deviation is 0.25 inch. Assume the bounce heights are approximately normally distributed. Is the company making acceptable tennis balls? Explain.

42. Light Bulb Manufacturing A company manufactures light bulbs. The company wants the bulbs to have a mean life span of 1000 hours. This average is maintained by periodically testing random samples of 16 light bulbs. If the *t*-value falls between $-t_{0.99}$ and $t_{0.99}$, then the company will be satisfied that it is manufacturing acceptable light bulbs. For a random sample, the mean life span of the sample is 1015 hours and the standard deviation is 25 hours. Assume the life spans are approximately normally distributed. Is the company making acceptable light bulbs? Explain.

APPLET

You can find the interactive applet for this activity within **MyLab Statistics** or at *www.pearsonhighered.com/ mathstatsresources.*

The *confidence intervals for a mean (the impact of not knowing the standard deviation)* applet allows you to visually investigate confidence intervals for a population mean. You can specify the sample size n, the shape of the distribution (Normal or Right-skewed), the population mean (Mean), and the true population standard deviation (Std. Dev.). When you click SIMULATE, 100 separate samples of size n will be selected from a population with these population parameters. For each of the 100 samples, a 95% Z confidence interval (known standard deviation) and a 95% T confidence interval (unknown standard deviation) are displayed in the plot at the right. The 95% Z confidence interval is displayed in green and the 95% T confidence interval is displayed in blue. When an interval does not contain the population mean, it is displayed in red. Additional simulations can be carried out by clicking SIMULATE multiple times. The cumulative number of times that each type of interval contains the population mean is also shown. Press CLEAR to clear existing results and start a new simulation.

EXPLORE

Step 1 Specify a value for n.
Step 2 Specify a distribution.
Step 3 Specify a value for the mean.
Step 4 Specify a value for the standard deviation.
Step 5 Click SIMULATE to generate the confidence intervals.

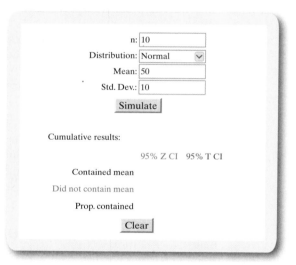

DRAW CONCLUSIONS

APPLET

1. Set $n = 30$, Mean = 25, Std. Dev. = 5, and the distribution to Normal. Run the simulation so that at least 1000 confidence intervals are generated. Compare the proportion of the 95% Z confidence intervals and 95% T confidence intervals that contain the population mean. Is this what you would expect? Explain.

2. In a random sample of 24 high school students, the mean number of hours of sleep per night during the school week was 7.26 hours and the standard deviation was 1.19 hours. Assume the sleep times are normally distributed. Run the simulation for $n = 10$ so that at least 500 confidence intervals are generated. What proportion of the 95% Z confidence intervals and 95% T confidence intervals contain the population mean? Should you use a Z confidence interval or a T confidence interval for the mean number of hours of sleep? Explain.

Marathon Training

A marathon is a foot race with a distance of 26.22 miles. It was one of the original events of the modern Olympics, where it was a men's only event. The women's marathon became an Olympic event in 1984. The Olympic record for the men's marathon was set during the 2008 Olympics by Samuel Kamau Wanjiru of Kenya, with a time of 2 hours, 6 minutes, 32 seconds. The Olympic record for the women's marathon was set during the 2012 Olympics by Tiki Gelana of Ethiopa, with a time of 2 hours, 23 minutes, 7 seconds.

Training for a marathon typically lasts at least 6 months. The training is gradual, with increases in distance about every 2 weeks. About 1 to 3 weeks before the race, the distance run is decreased slightly. The stem-and-leaf plots below show the marathon training times (in minutes) for a random sample of 30 male runners and 30 female runners.

**Training Times (in minutes)
of Male Runners**

```
15 | 5  8  9  9  9              Key: 15|5 = 155
16 | 0  0  0  0  1  2  3  4  4  5  8  9
17 | 0  1  1  3  5  6  6  7  7  9
18 | 0  1  5
```

**Training Times (in minutes)
of Female Runners**

```
17 | 8  9  9                    Key: 17|8 = 178
18 | 0  0  0  0  1  2  3  4  6  6  7  9
19 | 0  0  0  1  3  4  5  5  6  6
20 | 0  0  1  2  3
```

EXERCISES

1. Use the sample to find a point estimate for the mean training time of the
   (a) male runners.
   (b) female runners.

2. Find the sample standard deviation of the training times for the
   (a) male runners.
   (b) female runners.

3. Use the sample to construct a 95% confidence interval for the population mean training time of the
   (a) male runners.
   (b) female runners.

4. Interpret the results of Exercise 3.

5. Use the sample to construct a 95% confidence interval for the population mean training time of all runners. How do your results differ from those in Exercise 3? Explain.

6. A trainer wants to estimate the population mean running times for both male and female runners within 2 minutes. Determine the minimum sample size required to construct a 99% confidence interval for the population mean training time of
   (a) male runners. Assume the population standard deviation is 8.9 minutes.
   (b) female runners. Assume the population standard deviation is 8.4 minutes.

What You Should Learn

▶ How to find a point estimate for a population proportion

▶ How to construct and interpret confidence intervals for a population proportion

▶ How to determine the minimum sample size required when estimating a population proportion

Point Estimate for a Population Proportion ■ Confidence Intervals for a Population Proportion ■ Finding a Minimum Sample Size

Point Estimate for a Population Proportion

Recall from Section 4.2 that the probability of success in a single trial of a binomial experiment is p. This probability is a **population proportion.** In this section, you will learn how to estimate a population proportion p using a confidence interval. As with confidence intervals for μ, you will start with a point estimate.

DEFINITION

The **point estimate for p,** the population proportion of successes, is given by the proportion of successes in a sample and is denoted by

$$\hat{p} = \frac{x}{n} \qquad \text{Sample proportion}$$

where x is the number of successes in the sample and n is the sample size. The point estimate for the population proportion of failures is $\hat{q} = 1 - \hat{p}$. The symbols $\hat{p}$ and $\hat{q}$ are read as "p hat" and "q hat."

EXAMPLE 1

Finding a Point Estimate for p

In a survey of 1550 U.S. adults, 1054 said that they use the social media website Facebook. Find a point estimate for the population proportion of U.S. adults who use Facebook. *(Adapted from Pew Research Center)*

SOLUTION

The number of successes is the number of adults who use Facebook, so $x = 1054$. The sample size is $n = 1550$. So, the sample proportion is

$$\hat{p} = \frac{x}{n} \qquad \text{Formula for sample proportion}$$

$$= \frac{1054}{1550} \qquad \text{Substitute 1054 for } x \text{ and 1550 for } n.$$

$$= 0.68 \qquad \text{Divide.}$$

$$= 68\%. \qquad \text{Write as a percent.}$$

So, the point estimate for the population proportion of U.S. adults who use Facebook is 0.68 or 68%.

Study Tip

In Sections 6.1 and 6.2, estimates were made for quantitative data. In this section, sample proportions are used to make estimates for qualitative data.

TRY IT YOURSELF 1

A poll surveyed 4780 U.S. adults about how often they shop online. The results are shown in the table. Find a point estimate for the population proportion of U.S. adults who shop online at least once a week. *(Adapted from Pew Research Center)*

How often do you shop online?	Number responding yes
At least once a week	717
A few times a month	1338
Less often	1769
Never	956

Answer: Page A36

Confidence Intervals for a Population Proportion

Constructing a confidence interval for a population proportion p is similar to constructing a confidence interval for a population mean. You start with a point estimate and calculate a margin of error.

DEFINITION

A *c*-confidence interval for a population proportion *p* is

$$\hat{p} - E < p < \hat{p} + E$$

where

$$E = z_c \sqrt{\frac{\hat{p}\hat{q}}{n}}. \qquad \text{Margin of error for } p$$

The probability that the confidence interval contains p is c, assuming that the estimation process is repeated a large number of times.

In Section 5.5, you learned that a binomial distribution can be approximated by a normal distribution when $np \geq 5$ and $nq \geq 5$. When $n\hat{p} \geq 5$ and $n\hat{q} \geq 5$, the sampling distribution of $\hat{p}$ is approximately normal with a mean of

$$\mu_{\hat{p}} = p \qquad \text{Mean of the sample proportions}$$

and a standard error of

$$\sigma_{\hat{p}} = \sqrt{\frac{pq}{n}}. \qquad \text{Standard error of the sample proportions}$$

$$\left(\text{Notice } \sigma_{\hat{p}} = \frac{\sigma}{n} = \frac{\sqrt{npq}}{n} = \frac{\sqrt{npq}}{\sqrt{n^2}} = \sqrt{\frac{npq}{n^2}} = \sqrt{\frac{pq}{n}}. \right)$$

GUIDELINES

Constructing a Confidence Interval for a Population Proportion

In Words	In Symbols
1. Identify the sample statistics n and x.	
2. Find the point estimate $\hat{p}$.	$\hat{p} = \dfrac{x}{n}$
3. Verify that the sampling distribution of $\hat{p}$ can be approximated by a normal distribution.	$n\hat{p} \geq 5, n\hat{q} \geq 5$
4. Find the critical value z_c that corresponds to the given level of confidence c.	Use Table 4 in Appendix B.
5. Find the margin of error E.	$E = z_c \sqrt{\dfrac{\hat{p}\hat{q}}{n}}$
6. Find the left and right endpoints and form the confidence interval.	Left endpoint: $\hat{p} - E$ Right endpoint: $\hat{p} + E$ Interval: $\hat{p} - E < p < \hat{p} + E$

In Step 4 above, note that the critical value z_c is found the same way it was found in Section 6.1, by either using Table 4 in Appendix B or using technology.

Picturing the World

A poll surveyed 1519 U.S. adults about global climate change. Of those surveyed, 936 said that they expect to make major changes in their lives to address problems from climate change in the next 50 years. (Adapted from Pew Research Center)

In the Next 50 Years, Do You Think You Will Make Major Changes to Your Way of Life in Order to Address Problems from Global Climate Change?

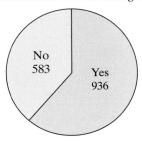

No
583

Yes
936

Find a 90% confidence interval for the population proportion of people who expect to make major changes in their lives to address problems from climate change in the next 50 years.

(0.595, 0.637)

[*Tech:* (0.596, 0.637)]

Tech Tip

Here are instructions for constructing a confidence interval for a population proportion on a TI-84 Plus.

STAT

Choose the TESTS menu.

   A: 1–PropZInt . . .

Enter the values of x, n, and the level of confidence c (C-Level). Then select *Calculate*.

Note to Instructor

Point out that the value of *E* is calculated by multiplying the *z*-score (the number of standard deviations from the mean) by the standard error of $\hat{p}$. The *z*-scores are found the same way they were found in Section 6.1.

Study Tip

Notice in Example 2 that the confidence interval for the population proportion *p* is rounded to three decimal places. This *round-off rule* will be used throughout the text.

> **EXAMPLE 2**

> Minitab and TI-84 Plus steps are shown on pages 344 and 345.

Constructing a Confidence Interval for *p*

Use the data in Example 1 to construct a 95% confidence interval for the population proportion of U.S. adults who use Facebook.

SOLUTION

From Example 1, $\hat{p} = 0.68$. So, the point estimate for the population proportion of failures is

$$\hat{q} = 1 - 0.68 = 0.32.$$

Using $n = 1550$, you can verify that the sampling distribution of $\hat{p}$ can be approximated by a normal distribution.

$$n\hat{p} = (1550)(0.68) = 1054 > 5$$

and

$$n\hat{q} = (1550)(0.32) = 496 > 5$$

Using $z_c = 1.96$, the margin of error is

$$E = z_c \sqrt{\frac{\hat{p}\hat{q}}{n}} = 1.96 \sqrt{\frac{(0.68)(0.32)}{1550}} \approx 0.023.$$

Next, find the left and right endpoints and form the 95% confidence interval.

Left Endpoint   Right Endpoint
$\hat{p} - E \approx 0.68 - 0.023$ $\hat{p} + E \approx 0.68 + 0.023$
$= 0.657$ $= 0.703$

$$0.657 < p < 0.703$$

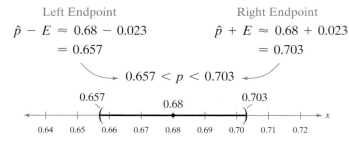

You can check this answer using technology, as shown below. (When using technology, your answers may differ slightly from those found using Table 4.)

> **STATCRUNCH**

> **95% confidence interval results:**

> | Proportion | Count | Total | Sample Prop. | Std. Err. | L. Limit | U. Limit |
> |---|---|---|---|---|---|---|
> | p | 1054 | 1550 | 0.68 | 0.011849 | 0.65678 | 0.70322 |

Interpretation With 95% confidence, you can say that the population proportion of U.S. adults who use Facebook is between 65.7% and 70.3%.

TRY IT YOURSELF 2

Use the data in Try It Yourself 1 to construct a 90% confidence interval for the population proportion of U.S. adults who shop online at least once a week.

Answer: Page A36

The confidence level of 95% used in Example 2 is typical of opinion polls. The result, however, is usually not stated as a confidence interval. Instead, the result of Example 2 would be stated as shown.

> *A survey found that 68% of U.S. adults use Facebook.*
> *The margin of error for the survey is ±2.3%.*

EXAMPLE 3

Constructing a Confidence Interval for *p*

The figure below is from a survey of 800 U.S. adults ages 18 to 29. Construct a 99% confidence interval for the population proportion of 18- to 29-year-olds who get their news on television. *(Adapted from Pew Research Center)*

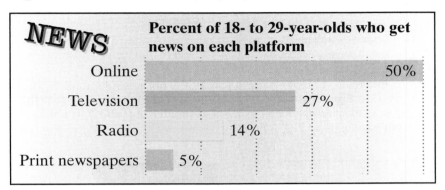

NEWS **Percent of 18- to 29-year-olds who get news on each platform**

Online 50%
Television 27%
Radio 14%
Print newspapers 5%

SOLUTION

From the figure, $\hat{p} = 0.27$. So, $\hat{q} = 1 - 0.27 = 0.73$. Using $n = 800$, note that

$$n\hat{p} = (800)(0.27) = 216 > 5$$

and

$$n\hat{q} = (800)(0.73) = 584 > 5.$$

So, the sampling distribution of $\hat{p}$ is approximately normal. Using $z_c = 2.575$, the margin of error is

$$E = z_c\sqrt{\frac{\hat{p}\hat{q}}{n}}$$

$$\approx 2.575\sqrt{\frac{(0.27)(0.73)}{800}}$$ Use Table 4 in Appendix B to estimate that z_c is halfway between 2.57 and 2.58.

$$\approx 0.040.$$

Next, find the left and right endpoints and form the 99% confidence interval.

Left Endpoint
$\hat{p} - E \approx 0.27 - 0.040$
$= 0.230$

Right Endpoint
$\hat{p} + E \approx 0.27 + 0.040$
$= 0.310$

$$0.230 < p < 0.310$$

0.230 0.27 0.310

0.21 0.22 0.23 0.24 0.25 0.26 0.27 0.28 0.29 0.30 0.31 0.32

You can check this answer using technology, as shown at the left.

Interpretation With 99% confidence, you can say that the population proportion of 18- to 29-year-olds who get their news on television is between 23.0% and 31.0%.

TRY IT YOURSELF 3

Use the data in Example 3 to construct a 99% confidence interval for the population proportion of 18- to 29-year-olds who get their news online.

Answer: Page A36

6.3 To explore this topic further, see **Activity 6.3** on page 329.

TI-84 PLUS

```
1-PropZInt
(.22957,.31043)
p̂=.27
n=800
```

Finding a Minimum Sample Size

One way to increase the precision of a confidence interval without decreasing the level of confidence is to increase the sample size.

Finding a Minimum Sample Size to Estimate p

Given a c-confidence level and a margin of error E, the minimum sample size n needed to estimate the population proportion p is

$$n = \hat{p}\hat{q}\left(\frac{z_c}{E}\right)^2.$$

If n is not a whole number, then round n up to the next whole number (see Example 4). Also, note that this formula assumes that you have preliminary estimates of $\hat{p}$ and $\hat{q}$. If not, use $\hat{p} = 0.5$ and $\hat{q} = 0.5$.

EXAMPLE 4

Determining a Minimum Sample Size

You are running a political campaign and wish to estimate, with 95% confidence, the population proportion of registered voters who will vote for your candidate. Your estimate must be accurate within 3% of the population proportion. Find the minimum sample size needed when (1) no preliminary estimate is available and (2) a preliminary estimate gives $\hat{p} = 0.31$. Compare your results.

SOLUTION

1. Because you do not have a preliminary estimate of $\hat{p}$, use $\hat{p} = 0.5$ and $\hat{q} = 0.5$. Using $z_c = 1.96$ and $E = 0.03$, you can solve for n.

$$n = \hat{p}\hat{q}\left(\frac{z_c}{E}\right)^2 = (0.5)(0.5)\left(\frac{1.96}{0.03}\right)^2 \approx 1067.11$$

Because n is not a whole number, round up to the next whole number, 1068.

2. You have a preliminary estimate of $\hat{p} = 0.31$. So, $\hat{q} = 0.69$. Using $z_c = 1.96$ and $E = 0.03$, you can solve for n.

$$n = \hat{p}\hat{q}\left(\frac{z_c}{E}\right)^2 = (0.31)(0.69)\left(\frac{1.96}{0.03}\right)^2 \approx 913.02$$

Because n is not a whole number, round up to the next whole number, 914.

Interpretation With no preliminary estimate, the minimum sample size should be at least 1068 registered voters. With a preliminary estimate of $\hat{p} = 0.31$, the sample size should be at least 914 registered voters. So, you will need a larger sample size when no preliminary estimate is available.

TRY IT YOURSELF 4

A researcher is estimating the population proportion of people in the United States who delayed seeking medical care during the last 12 months due to costs. The estimate must be accurate within 2% of the population proportion with 90% confidence. Find the minimum sample size needed when (1) no preliminary estimate is available and (2) a previous survey found that 6.3% of people in the United States delayed seeking medical care during the last 12 months due to costs. *(Source: NCHS, National Health Interview Survey)*

Answer: Page A36

6.3 EXERCISES

Building Basic Skills and Vocabulary

True or False? *In Exercises 1 and 2, determine whether the statement is true or false. If it is false, rewrite it as a true statement.*

1. To estimate the value of p, the population proportion of successes, use the point estimate x.

2. The point estimate for the population proportion of failures is $1 - \hat{p}$.

Finding $\hat{p}$ and $\hat{q}$ *In Exercises 3–6, let p be the population proportion for the situation. Find point estimates of p and q.*

3. **Tax Fraud** In a survey of 1040 U.S. adults, 62 have had someone impersonate them to try to claim tax refunds. *(Adapted from Pew Research Center)*

4. **Investigating Crimes** In a survey of 1040 U.S. adults, 478 believe the government should be able to access encrypted communications when investigating crimes. *(Adapted from Pew Research Center)*

5. **Mainstream Media** In a survey of 2016 U.S. adults, 1310 think mainstream media is more interested in making money than in telling the truth. *(Adapted from Ipsos Public Affairs)*

6. **Terrorism** In a survey of 2016 U.S. adults, 665 believe America should stop terrorism at all costs. *(Adapted from Ipsos Public Affairs)*

In Exercises 7–10, use the confidence interval to find the margin of error and the sample proportion.

7. $(0.905, 0.933)$

8. $(0.245, 0.475)$

9. $(0.512, 0.596)$

10. $(0.087, 0.263)$

Using and Interpreting Concepts

Constructing Confidence Intervals *In Exercises 11 and 12, construct 90% and 95% confidence intervals for the population proportion. Interpret the results and compare the widths of the confidence intervals.*

11. **New Year's Resolutions** In a survey of 2241 U.S. adults in a recent year, 1322 say they have made a New Year's resolution. *(Adapted from The Harris Poll)*

12. **New Year's Resolutions** In a survey of 2241 U.S. adults in a recent year, 650 made a New Year's resolution to eat healthier. *(Adapted from The Harris Poll)*

Constructing Confidence Intervals *In Exercises 13 and 14, construct a 99% confidence interval for the population proportion. Interpret the results.*

13. **Police Body Cameras** In a survey of 1000 U.S. adults, 700 think police officers should be required to wear body cameras while on duty. *(Adapted from Rasmussen Reports)*

14. **Teacher Body Cameras** In a survey of 600 United Kingdom teachers, 226 say they would wear a body camera in school. *(Adapted from Times Education Supplement)*

1. False. To estimate the value of p, the population proportion of successes, use the point estimate $\hat{p} = x/n$.
2. True
3. 0.060, 0.940
4. 0.460, 0.540
5. 0.650, 0.350
6. 0.330, 0.670
7. $E = 0.014$, $\hat{p} = 0.919$
8. $E = 0.115$, $\hat{p} = 0.360$
9. $E = 0.042$, $\hat{p} = 0.554$
10. $E = 0.088$, $\hat{p} = 0.175$
11. $(0.573, 0.607)$; $(0.570, 0.610)$
    With 90% confidence, you can say that the population proportion of U.S. adults who say they have made a New Year's resolution is between 57.3% and 60.7%. With 95% confidence, you can say it is between 57.0% and 61.0%. The 95% confidence interval is slightly wider.
12. $(0.274, 0.306)$; $(0.271, 0.309)$
    With 90% confidence, you can say that the population proportion of U.S. adults who say they made a New Year's resolution to eat healthier is between 27.4% and 30.6%. With 95% confidence, you can say it is between 27.1% and 30.9%. The 95% confidence interval is slightly wider.
13. $(0.663, 0.737)$
    With 99% confidence, you can say that the population proportion of U.S. adults who say they think police officers should be required to wear body cameras while on duty is between 66.3% and 73.7%.
14. $(0.326, 0.428)$
    With 99% confidence, you can say that the population proportion of United Kingdom teachers who say they would wear a body camera in school is between 32.6% and 42.8%.

15. LGBT Identification In a survey of 1,626,773 U.S. adults, 49,311 personally identify as lesbian, gay, bisexual, or transgender. Construct a 95% confidence interval for the population proportion of U.S. adults who personally identify as lesbian, gay, bisexual, or transgender. *(Source: Gallup)*

16. Transgender Bathroom Policy In a survey of 1000 U.S. adults, 490 oppose allowing transgender students to use the bathrooms of the opposite biological sex. Construct a 90% confidence interval for the population proportion of U.S. adults who oppose allowing transgender students to use the bathrooms of the opposite biological sex. *(Adapted from Rasmussen Reports)*

17. Congress You wish to estimate, with 95% confidence, the population proportion of U.S. adults who think Congress is doing a good or excellent job. Your estimate must be accurate within 4% of the population proportion.

  (a) No preliminary estimate is available. Find the minimum sample size needed.

  (b) Find the minimum sample size needed, using a prior survey that found that 25% of U.S. adults think Congress is doing a good or excellent job. *(Source: Rasmussen Reports)*

  (c) Compare the results from parts (a) and (b).

18. Genetically Modified Organisms You wish to estimate, with 99% confidence, the population proportion of U.S. adults who support labeling legislation for genetically modified organisms (GMOs). Your estimate must be accurate within 2% of the population proportion.

  (a) No preliminary estimate is available. Find the minimum sample size needed.

  (b) Find the minimum sample size needed, using a prior survey that found that 75% of U.S. adults support labeling legislation for GMOs. *(Source: The Harris Poll)*

  (c) Compare the results from parts (a) and (b).

19. Fast Food You wish to estimate, with 90% confidence, the population proportion of U.S. adults who eat fast food four to six times per week. Your estimate must be accurate within 3% of the population proportion.

  (a) No preliminary estimate is available. Find the minimum sample size needed.

  (b) Find the minimum sample size needed, using a prior study that found that 11% of U.S. adults eat fast food four to six times per week. *(Source: Statista)*

  (c) Compare the results from parts (a) and (b).

20. Alcohol-Impaired Driving You wish to estimate, with 95% confidence, the population proportion of motor vehicle fatalities that were caused by alcohol-impaired driving. Your estimate must be accurate within 5% of the population proportion.

  (a) No preliminary estimate is available. Find the minimum sample size needed.

  (b) Find the minimum sample size needed, using a prior study that found that 31% of motor vehicle fatalities were caused by alcohol-impaired driving. *(Source: WalletHub)*

  (c) Compare the results from parts (a) and (b).

21. Yes; It falls within both confidence intervals.

22. Yes; For the population proportion to be within 1% of the point estimate, its value would be in the interval (0.373, 0.381). This interval falls within the confidence interval.

23. No; The minimum sample size needed is 451 adults.

24. Yes; The minimum sample size needed is 329 adults.

25. United States: (0.282, 0.358)

 Canada: (0.177, 0.243)

 France: (0.215, 0.285)

 Japan: (0.459, 0.541)

 Australia: (0.103, 0.157)

26. Yes; It is possible that the population proportion for the United States is the same as the population proportion for France, and/or the population proportion for France is the same as the population proportion for Canada, because the confidence intervals overlap.

27. (a) Expect to stay at first employer for 3 or more years: (0.670, 0.710)

 Completed an apprenticeship or internship: (0.660, 0.700)

 Employed in field of study: (0.629, 0.671)

 Feel underemployed: (0.488, 0.532)

 Prefer to work for a large company: (0.125, 0.155)

 (b) Expect to stay at first employer for 3 or more years: (0.663, 0.717)

 Completed an apprenticeship or internship: (0.653, 0.707)

 Employed in field of study: (0.623, 0.677)

 Feel underemployed: (0.481, 0.539)

 Prefer to work for a large company: (0.120, 0.160)

28. Yes; The confidence intervals in parts (a) and (b) for "expect to stay at first employer for 3 or more years," "completed an apprenticeship or internship," and "employed in field of study" overlap.

21. In Exercise 11, does it seem possible that the population proportion could equal 0.59? Explain.

22. In Exercise 14, does it seem possible that the population proportion could be within 1% of the point estimate? Explain.

23. In Exercise 17(b), would a sample size of 200 be acceptable? Explain.

24. In Exercise 20(b), would a sample size of 600 be acceptable? Explain.

Constructing Confidence Intervals *In Exercises 25 and 26, use the figure, which shows the results of a survey in which 1003 adults from the United States, 1020 adults from Canada, 999 adults from France, 1000 adults from Japan, and 1000 adults from Australia were asked whether national identity is strongly tied to birthplace.* *(Source: Pew Research Center)*

National Identity and Birthplace
People from different countries who believe national identity is strongly tied to birthplace

United States	32%
Canada	21%
France	25%
Japan	50%
Australia	13%

25. **National Identity** Construct a 99% confidence interval for the population proportion of adults who say national identity is strongly tied to birthplace for each country listed.

26. In Exercise 25, does it seem possible that any of the population proportions could be equal? Explain.

Constructing Confidence Intervals *In Exercises 27 and 28, use the figure, which shows the results of a survey in which 2000 U.S. college graduates from the year 2016 were asked questions about employment.* *(Source: Accenture)*

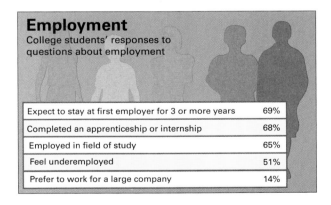

Employment
College students' responses to questions about employment

Expect to stay at first employer for 3 or more years	69%
Completed an apprenticeship or internship	68%
Employed in field of study	65%
Feel underemployed	51%
Prefer to work for a large company	14%

27. **Employment** Construct (a) a 95% confidence interval and (b) a 99% confidence interval for the population proportion of college students who gave each response.

28. In Exercise 27, does it seem possible that any of the population proportions could be equal? Explain.

29. (0.666, 0.734) is approximately a 98.1% CI.

30. (0.761, 0.819) is approximately a 99.4% CI.

31. (0.68, 0.74) is approximately a 96.3% CI.

32. (0.33, 0.41) is approximately a 99.2% CI.

33. (0.45, 0.49) is approximately a 98.3% CI.

(0.51, 0.55) is approximately a 98.3% CI.

34. (0.65, 0.71) is approximately a 96.3% CI.

(0.39, 0.45) is approximately a 95.1% CI.

35. If $n\hat{p} < 5$ or $n\hat{q} < 5$, the sampling distribution of $\hat{p}$ may not be normally distributed, so z_c cannot be used to calculate the confidence interval.

36. *Sample answer:*

$$E = z_c\sqrt{\frac{\hat{p}\hat{q}}{n}} \quad \text{Write the original equation.}$$

$$\frac{E}{z_c} = \sqrt{\frac{\hat{p}\hat{q}}{n}} \quad \text{Divide each side by } z_c.$$

$$\left(\frac{E}{z_c}\right)^2 = \frac{\hat{p}\hat{q}}{n} \quad \text{Square each side.}$$

$$\hat{p}\hat{q}\left(\frac{z_c}{E}\right)^2 = n \quad \text{Solve for } n.$$

37. See Odd Answers, page A66.

$\hat{p} = 0.5$ gives the maximum value of $\hat{p}\hat{q}$.

Extending Concepts

Translating Statements *In Exercises 29–34, translate the statement into a confidence interval. Approximate the level of confidence.*

29. In a survey of 1003 U.S. adults, 70% said being able to speak English is at the core of national identity. The survey's margin of error is ±3.4%. *(Source: Pew Research Center)*

30. In a survey of 1503 U.S. adults, 79% say people have the right to nonviolent protest. The survey's margin of error is ±2.9%. *(Source: Pew Research Center)*

31. In a survey of 1000 U.S. adults, 71% think teaching is one of the most important jobs in our country today. The survey's margin of error is ±3%. *(Source: Rasmussen Reports)*

32. In a survey of 1035 U.S. adults, 37% say the U.S. spends too little on defense. The survey's margin of error is ±4%. *(Source: Gallup)*

33. In a survey of 3539 U.S. adults, 47% believe the economy is getting better. Three weeks prior to this survey, 53% believed the economy was getting better. The survey's margin of error is ±2%. *(Source: Gallup)*

34. In a survey of 1052 parents of children ages 8–14, 68% say they are willing to get a second or part-time job to pay for their children's college education, and 42% say they lose sleep worrying about college costs. The survey's margin of error is ±3%. *(Source: T. Rowe Price Group, Inc.)*

35. Why Check It? Why is it necessary to check that $n\hat{p} \geq 5$ and $n\hat{q} \geq 5$?

36. Sample Size The equation for determining the sample size

$$n = \hat{p}\hat{q}\left(\frac{z_c}{E}\right)^2$$

can be obtained by solving the equation for the margin of error

$$E = z_c\sqrt{\frac{\hat{p}\hat{q}}{n}}$$

for n. Show that this is true and justify each step.

37. Maximum Value of $\hat{p}\hat{q}$ Complete the tables for different values of $\hat{p}$ and $\hat{q} = 1 - \hat{p}$. From the tables, which value of $\hat{p}$ appears to give the maximum value of the product $\hat{p}\hat{q}$?

$\hat{p}$	$\hat{q} = 1 - \hat{p}$	$\hat{p}\hat{q}$
0.0	1.0	0.00
0.1	0.9	0.09
0.2	0.8	
0.3		
0.4		
0.5		
0.6		
0.7		
0.8		
0.9		
1.0		

$\hat{p}$	$\hat{q} = 1 - \hat{p}$	$\hat{p}\hat{q}$
0.45		
0.46		
0.47		
0.48		
0.49		
0.50		
0.51		
0.52		
0.53		
0.54		
0.55		

APPLET

You can find the interactive applet for this activity within MyLab Statistics or at *www.pearsonhighered.com/ mathstatsresources.*

The *confidence intervals for a proportion* applet allows you to visually investigate confidence intervals for a population proportion. You can specify the sample size n and the population proportion p. When you click SIMULATE, 100 separate samples of size n will be selected from a population with a proportion of successes equal to p. For each of the 100 samples, a 95% confidence interval (in green) and a 99% confidence interval (in blue) are displayed in the plot at the right. Each of these intervals is computed using the standard normal approximation. When an interval does not contain the population proportion, it is displayed in red. Note that the 99% confidence interval is always wider than the 95% confidence interval. Additional simulations can be carried out by clicking SIMULATE multiple times. The cumulative number of times that each type of interval contains the population proportion is also shown. Press CLEAR to clear existing results and start a new simulation.

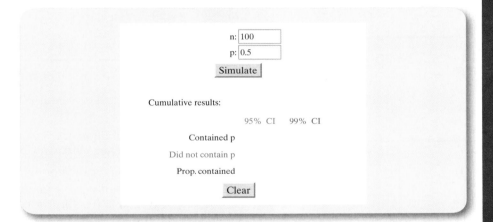

EXPLORE

Step 1 Specify a value for n.
Step 2 Specify a value for p.
Step 3 Click SIMULATE to generate the confidence intervals.

DRAW CONCLUSIONS

APPLET

1. Run the simulation for $p = 0.6$ and $n = 10, 20, 40,$ and 100. Clear the results after each trial. What proportion of the confidence intervals for each confidence level contains the population proportion? What happens to the proportion of confidence intervals that contains the population proportion for each confidence level as the sample size increases?

2. Run the simulation for $p = 0.4$ and $n = 100$ so that at least 1000 confidence intervals are generated. Compare the proportion of confidence intervals that contains the population proportion for each confidence level. Is this what you would expect? Explain.

6.4 Confidence Intervals for Variance and Standard Deviation

The Chi-Square Distribution ■ Confidence Intervals for σ^2 and σ

The Chi-Square Distribution

In manufacturing, it is necessary to control the amount that a process varies. For instance, an automobile part manufacturer must produce thousands of parts to be used in the manufacturing process. It is important that the parts vary little or not at all. How can you measure, and consequently control, the amount of variation in the parts? You can start with a point estimate.

> **DEFINITION**
>
> The **point estimate for σ^2** is s^2 and the **point estimate for σ** is s. The most unbiased estimate for σ^2 is s^2.

You can use a **chi-square distribution** to construct a confidence interval for the variance and standard deviation.

Study Tip

The Greek letter χ is pronounced "*ki*," which rhymes with the more familiar Greek letter π.

> **DEFINITION**
>
> If a random variable x has a normal distribution, then the distribution of
>
> $$\chi^2 = \frac{(n-1)s^2}{\sigma^2}$$
>
> forms a **chi-square distribution** for samples of any size $n > 1$. Here are several properties of the chi-square distribution.
>
> 1. All values of χ^2 are greater than or equal to 0.
> 2. The chi-square distribution is a family of curves, each determined by the degrees of freedom. To form a confidence interval for σ^2, use the chi-square distribution with degrees of freedom equal to one less than the sample size.
>
>   d.f. $= n - 1$ Degrees of freedom
>
> 3. The total area under each chi-square distribution curve is equal to 1.
> 4. The chi-square distribution is positively skewed and therefore the distribution is not symmetric.
> 5. The chi-square distribution is different for each number of degrees of freedom, as shown in the figure. As the degrees of freedom increase, the chi-square distribution approaches a normal distribution.

Note to Instructor

If you are short of time, this section can be omitted. Or, if you prefer, it can be covered with Chapter 10 when additional chi-square applications are presented.

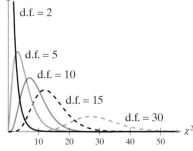

Chi-Square Distribution for Different Degrees of Freedom

Study Tip

For chi-square critical values with a c-confidence level, the values shown below, χ_L^2 and χ_R^2 are what you look up in Table 6 in Appendix B.

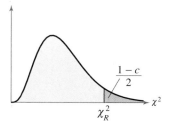

Area to the right of χ_R^2

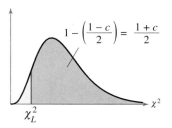

Area to the right of χ_L^2

The result is that you can conclude that the area between the left and right critical values is c.

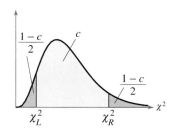

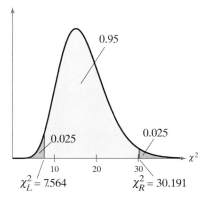

There are two critical values for each level of confidence. The value χ_R^2 represents the right-tail critical value and χ_L^2 represents the left-tail critical value. Table 6 in Appendix B lists critical values of χ^2 for various degrees of freedom and areas. Each area listed in the top row of the table represents the region under the chi-square curve to the *right* of the critical value.

EXAMPLE 1

Finding Critical Values for χ^2

Find the critical values χ_R^2 and χ_L^2 for a 95% confidence interval when the sample size is 18.

SOLUTION

Because the sample size is 18,

$$\text{d.f.} = n - 1 = 18 - 1 = 17. \qquad \textit{Degrees of freedom}$$

The area to the right of χ_R^2 is

$$\text{Area to the right of } \chi_R^2 = \frac{1 - c}{2} = \frac{1 - 0.95}{2} = 0.025$$

and the area to the right of χ_L^2 is

$$\text{Area to the right of } \chi_L^2 = \frac{1 + c}{2} = \frac{1 + 0.95}{2} = 0.975.$$

A portion of Table 6 is shown. Using d.f. = 17 and the areas 0.975 and 0.025, you can find the critical values, as shown by the highlighted areas in the table. (Note that the top row in the table lists areas to the right of the critical value. The entries in the table are critical values.)

Degrees of freedom	0.995	0.99	0.975	0.95	0.90	0.10	0.05	0.025
1	—	—	0.001	0.004	0.016	2.706	3.841	5.024
2	0.010	0.020	0.051	0.103	0.211	4.605	5.991	7.378
3	0.072	0.115	0.216	0.352	0.584	6.251	7.815	9.348
15	4.601	5.229	6.262	7.261	8.547	22.307	24.996	27.488
16	5.142	5.812	6.908	7.962	9.312	23.542	26.296	28.845
17	5.697	6.408	7.564	8.672	10.085	24.769	27.587	30.191
18	6.265	7.015	8.231	9.390	10.865	25.989	28.869	31.526
19	6.844	7.633	8.907	10.117	11.651	27.204	30.144	32.852
20	7.434	8.260	9.591	10.851	12.443	28.412	31.410	34.170

From the table, you can see that the critical values are

$$\chi_R^2 = 30.191 \quad \text{and} \quad \chi_L^2 = 7.564.$$

Interpretation So, for a chi-square distribution curve with 17 degrees of freedom, 95% of the area under the curve lies between 7.564 and 30.191, as shown in the figure at the left.

TRY IT YOURSELF 1

Find the critical values χ_R^2 and χ_L^2 for a 90% confidence interval when the sample size is 30.

Answer: Page A36

Picturing the World

The Florida panther is one of the most endangered mammals on Earth. In the southeastern United States, the only breeding population (with an estimated population of about 100 to 180 panthers) can be found on the southern tip of Florida. Most of the panthers live in (1) the Big Cypress National Preserve, (2) Everglades National Park, and (3) the Florida Panther National Wildlife Refuge, as shown on the map. In a study of 7 female panthers, it was found that the mean litter size was 2.14 kittens, with a standard deviation of 0.69.
(Source: Florida Fish and Wildlife Conservation Commission)

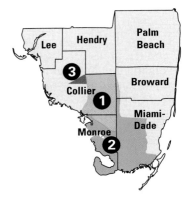

Construct a 90% confidence interval for the standard deviation of the litter size for female Florida panthers. Assume the litter sizes are normally distributed.

(0.48, 1.32)

Confidence Intervals for σ^2 and σ

You can use the critical values χ_R^2 and χ_L^2 to construct confidence intervals for a population variance and standard deviation. The best point estimate for the variance is s^2 and the best point estimate for the standard deviation is s. Because the chi-square distribution is not symmetric, the confidence interval for σ^2 *cannot* be written as $s^2 \pm E$. You must do separate calculations for the endpoints of the confidence interval, as shown in the next definition.

DEFINITION

The c-confidence intervals for the population variance and standard deviation are shown.

Confidence Interval for σ^2:

$$\frac{(n-1)s^2}{\chi_R^2} < \sigma^2 < \frac{(n-1)s^2}{\chi_L^2}$$

Confidence Interval for σ:

$$\sqrt{\frac{(n-1)s^2}{\chi_R^2}} < \sigma < \sqrt{\frac{(n-1)s^2}{\chi_L^2}}$$

The probability that the confidence intervals contain σ^2 or σ is c, assuming that the estimation process is repeated a large number of times.

GUIDELINES

Constructing a Confidence Interval for a Variance and Standard Deviation

In Words	In Symbols
1. Verify that the population has a normal distribution.	
2. Identify the sample statistic n and the degrees of freedom.	d.f. $= n - 1$
3. Find the point estimate s^2.	$s^2 = \dfrac{\Sigma(x - \bar{x})^2}{n - 1}$
4. Find the critical values χ_R^2 and χ_L^2 that correspond to the given level of confidence c and the degrees of freedom.	Use Table 6 in Appendix B.

5. Find the left and right endpoints and form the confidence interval for the population variance.

Left Endpoint   Right Endpoint

$$\frac{(n-1)s^2}{\chi_R^2} < \sigma^2 < \frac{(n-1)s^2}{\chi_L^2}$$

6. Find the confidence interval for the population standard deviation by taking the square root of each endpoint.

Left Endpoint   Right Endpoint

$$\sqrt{\frac{(n-1)s^2}{\chi_R^2}} < \sigma < \sqrt{\frac{(n-1)s^2}{\chi_L^2}}$$

EXAMPLE 2

Constructing Confidence Intervals

You randomly select and weigh 30 samples of an allergy medicine. The sample standard deviation is 1.20 milligrams. Assuming the weights are normally distributed, construct 99% confidence intervals for the population variance and standard deviation.

SOLUTION

The area to the right of χ_R^2 is

$$\text{Area to the right of } \chi_R^2 = \frac{1 - c}{2} = \frac{1 - 0.99}{2} = 0.005$$

and the area to the right of χ_L^2 is

$$\text{Area to the right of } \chi_L^2 = \frac{1 + c}{2} = \frac{1 + 0.99}{2} = 0.995.$$

Using the values $n = 30$, d.f. $= 29$, and $c = 0.99$, the critical values χ_R^2 and χ_L^2 are

$$\chi_R^2 = 52.336 \quad \text{and} \quad \chi_L^2 = 13.121.$$

Using these critical values and $s = 1.20$, the confidence interval for σ^2 is

Left Endpoint

$$\frac{(n - 1)s^2}{\chi_R^2} = \frac{(30 - 1)(1.20)^2}{52.336}$$

$$\approx 0.80$$

Right Endpoint

$$\frac{(n - 1)s^2}{\chi_L^2} = \frac{(30 - 1)(1.20)^2}{13.121}$$

$$\approx 3.18$$

$$0.80 < \sigma^2 < 3.18.$$

The confidence interval for σ is

Left Endpoint

$$\sqrt{\frac{(30 - 1)(1.20)^2}{52.336}} < \sigma < \sqrt{\frac{(30 - 1)(1.20)^2}{13.121}}$$

Right Endpoint

$$0.89 < \sigma < 1.78.$$

You can check your answer using technology, as shown below using Minitab.

MINITAB

Test and CI for One Variance
99% Confidence Intervals

Method	CI for StDev	CI for Variance
Chi-Square	(0.89, 1.78)	(0.80, 3.18)

Interpretation With 99% confidence, you can say that the population variance is between 0.80 and 3.18, and the population standard deviation is between 0.89 and 1.78 milligrams.

TRY IT YOURSELF 2

Construct the 90% and 95% confidence intervals for the population variance and standard deviation of the medicine weights. *Answer: Page A36*

Note in Example 2 that the confidence interval for the population standard deviation *cannot* be written as $s \pm E$ because the confidence interval does not have s as its center. (The same is true for the population variance.)

Note to Instructor

Point out that the left endpoint requires using χ_R^2 and the right endpoint requires using χ_L^2. This is true because $\chi_R^2 > \chi_L^2$ and dividing the same numerator by a larger value will produce a smaller quotient.

Study Tip

When you construct a confidence interval for a population variance or standard deviation, the general *round-off rule* is to round off to the same number of decimal places as the sample variance or standard deviation.

6.4 EXERCISES

Answers (left column)

1. Yes
2. It approaches the shape of a normal curve.
3. $\chi_R^2 = 14.067, \chi_L^2 = 2.167$
4. $\chi_R^2 = 31.319, \chi_L^2 = 4.075$
5. $\chi_R^2 = 32.852, \chi_L^2 = 8.907$
6. $\chi_R^2 = 44.314, \chi_L^2 = 11.524$
7. $\chi_R^2 = 52.336, \chi_L^2 = 13.121$
8. $\chi_R^2 = 63.167, \chi_L^2 = 37.689$
9. (a) (7.33, 20.89) (b) (2.71, 4.57)
10. (a) (0.21, 5.68) (b) (0.46, 2.38)
11. (a) (755, 2401) (b) (27, 49)
12. (a) (48,571.8, 139,577.0)
    (b) (220.4, 373.6)
13. (a) (0.0426, 0.1699)
    (b) (0.2063, 0.4122)

With 95% confidence, you can say that the population variance is between 0.0426 and 0.1699, and the population standard deviation is between 0.2063 and 0.4122 inch.

14. (a) (0.0006, 0.0022)
    (b) (0.0247, 0.0469)

With 90% confidence, you can say that the population variance is between 0.0006 and 0.0022, and the population standard deviation is between 0.0247 and 0.0469 fluid ounce.

15. (a) (181.50, 976.54)
    (b) (13.47, 31.25)

With 99% confidence, you can say that the population variance is between 181.50 and 976.54, and the population standard deviation is between 13.47 and 31.25 thousand dollars.

16. See Selected Answers, page A99.
17. See Odd Answers, page A66.

Final exam scores					
61	73	59	99	83	60
68	69	97	43	61	87
55	40	67	48	87	64
55	90	59	71	65	59

TABLE FOR EXERCISE 16

Building Basic Skills and Vocabulary

1. Does a population have to be normally distributed in order to use the chi-square distribution?

2. What happens to the shape of the chi-square distribution as the degrees of freedom increase?

Finding Critical Values for χ^2 *In Exercises 3–8, find the critical values χ_R^2 and χ_L^2 for the level of confidence c and sample size n.*

3. $c = 0.90, n = 8$
4. $c = 0.99, n = 15$
5. $c = 0.95, n = 20$
6. $c = 0.98, n = 26$
7. $c = 0.99, n = 30$
8. $c = 0.80, n = 51$

In Exercises 9–12, construct the indicated confidence intervals for (a) the population variance σ^2 and (b) the population standard deviation σ. Assume the sample is from a normally distributed population.

9. $c = 0.95, s^2 = 11.56, n = 30$
10. $c = 0.99, s^2 = 0.64, n = 7$
11. $c = 0.90, s = 35, n = 18$
12. $c = 0.98, s = 278.1, n = 41$

Using and Interpreting Concepts

Constructing Confidence Intervals *In Exercises 13–24, assume the sample is from a normally distributed population and construct the indicated confidence intervals for (a) the population variance σ^2 and (b) the population standard deviation σ. Interpret the results.*

13. **Bolts** The diameters (in inches) of 18 randomly selected bolts produced by a machine are listed. Use a 95% level of confidence.

    4.477 4.425 4.034 4.317 4.003 3.760
    3.818 3.749 4.240 3.941 4.131 4.545
    3.958 3.741 3.859 3.816 4.448 4.206

14. **Cough Syrup** The volumes (in fluid ounces) of the contents of 15 randomly selected bottles of cough syrup are listed. Use a 90% level of confidence.

    4.211 4.246 4.269 4.241 4.260 4.293 4.189 4.248
    4.220 4.239 4.253 4.209 4.300 4.256 4.290

15. **Earnings** The annual earnings (in thousands of dollars) of 21 randomly selected clinical pharmacists are listed. Use a 99% level of confidence. *(Adapted from Salary.com)*

    91.8 90.6 101.5 119.2 110.5 117.0 138.6
    112.1 136.6 123.6 111.4 80.5 105.7 99.9
    138.3 113.6 81.4 89.4 94.8 146.6 106.6

16. **Final Exam Scores** The final exam scores of 24 randomly selected students in a statistics class are shown in the table at the left. Use a 95% level of confidence.

17. **Space Shuttle Flights** The durations (in days) of 14 randomly selected space shuttle flights have a sample standard deviation of 3.54 days. Use a 99% level of confidence. *(Source: NASA)*

18. (a) (54.68, 179.69)
(b) (7.39, 13.40)
With 80% confidence, you can say that the population variance is between 54.68 and 179.69, and the population standard deviation is between 7.39 and 13.40 touchdowns.

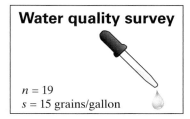

FIGURE FOR EXERCISE 19

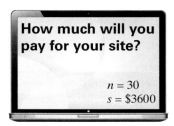

FIGURE FOR EXERCISE 20

19. (a) (128, 492)   (b) (11, 22)
With 95% confidence, you can say that the population variance is between 128 and 492, and the population standard deviation is between 11 and 22 grains per gallon.

20. (a) (8,831,450, 21,224,305)
(b) (2972, 4607)
With 90% confidence, you can say that the population variance is between 8,831,450 and 21,224,305, and the population standard deviation is between $2972 and $4607.

21. See Odd Answers, page A66.

22. See Selected Answers, page A99.

23. See Odd Answers, page A66.

24. See Selected Answers, page A99.

25. Yes, because all of the values in the confidence interval are less than 0.5.

26. No, because 0.025 is contained in the confidence interval.

27. No, because 0.25 is contained in the confidence interval.

28. No, because all of the values in the confidence interval are greater than 2.5.

29. See Odd Answers, page A66.

18. College Football The numbers of touchdowns scored by 11 randomly selected NCAA Division I Subdivision teams in a recent season have a sample standard deviation of 9.35. Use an 80% level of confidence. *(Source: National Collegiate Athletic Association)*

19. Water Quality As part of a water quality survey, you test the water hardness in several randomly selected streams. The results are shown in the figure at the left. Use a 95% level of confidence.

20. Website Costs As part of a survey, you ask a random sample of business owners how much they would be willing to pay for a website for their company. The results are shown in the figure at the left. Use a 90% level of confidence.

21. Car Batteries The reserve capacities (in hours) of 18 randomly selected automotive batteries have a sample standard deviation of 0.25 hour. Use an 80% level of confidence.

22. Maximum Daily Temperature The record high daily temperatures (in degrees Fahrenheit) of a random sample of 64 days of the year in Grand Junction, Colorado, have a sample standard deviation of 16.8°F. Use a 98% level of confidence. *(Source: NOAA)*

23. Waiting Times The waiting times (in minutes) of a random sample of 22 people at a bank have a sample standard deviation of 3.6 minutes. Use a 98% level of confidence.

24. Motorcycles The prices of a random sample of 20 new motorcycles have a sample standard deviation of $3900. Use a 90% level of confidence.

Extending Concepts

25. Bolt Diameters You are analyzing the sample of bolts in Exercise 13. The population standard deviation of the bolts' diameters should be less than 0.5 inch. Does the confidence interval you constructed for σ suggest that the variation in the bolts' diameters is at an acceptable level? Explain your reasoning.

26. Cough Syrup Bottle Contents You are analyzing the sample of cough syrup bottles in Exercise 14. The population standard deviation of the volumes of the bottles' contents should be less than 0.025 fluid ounce. Does the confidence interval you constructed for σ suggest that the variation in the volumes of the bottles' contents is at an acceptable level? Explain your reasoning.

27. Battery Reserve Capacities You are analyzing the sample of car batteries in Exercise 21. The population standard deviation of the batteries' reserve capacities should be less than 0.25 hour. Does the confidence interval you constructed for σ suggest that the variation in the batteries' reserve capacities is at an acceptable level? Explain your reasoning.

28. Waiting Times You are analyzing the sample of waiting times in Exercise 23. The population standard deviation of the waiting times should be less than 2.5 minutes. Does the confidence interval you constructed for σ suggest that the variation in the waiting times is at an acceptable level? Explain your reasoning.

29. In your own words, explain how finding a confidence interval for a population variance is different from finding a confidence interval for a population mean or proportion.

Statistics in the Real World

Uses

By now, you know that complete information about population parameters is often not available. The techniques of this chapter can be used to make interval estimates of these parameters so that you can make informed decisions.

From what you learned in this chapter, you know that point estimates (sample statistics) of population parameters are usually close but rarely equal to the actual values of the parameters they are estimating. Remembering this can help you make good decisions in your career and in everyday life. For instance, the results of a survey tell you that 52% of registered voters plan to vote in favor of the rezoning of a portion of a town from residential to commercial use. You know that this is only a point estimate of the actual proportion that will vote in favor of rezoning. If the margin of error is 3%, then the interval estimate is $0.49 < p < 0.55$ and it is possible that the item will not receive a majority vote.

Abuses

Registered voters

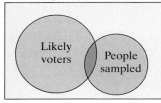

Unrepresentative Samples There are many ways that surveys can result in incorrect predictions. When you read the results of a survey, remember to question the sample size, the sampling technique, and the questions asked. For instance, you want to know the proportion of people who will vote in favor of rezoning. From the diagram at the left, you can see that even when your sample is large enough, it may not consist of people who are likely to vote.

Biased Survey Questions In surveys, it is also important to analyze the wording of the questions. For instance, the question about rezoning might be presented as: "Knowing that rezoning will result in more businesses contributing to school taxes, would you support the rezoning?"

Misinterpreted Polls Some political pundits and voters vowed never to trust polls again after they failed to predict Donald Trump's win over Hillary Clinton in the 2016 U.S. presidential election. However, nationwide polls the week of the election were only off by about 1%—the polls showed Clinton ahead by about 3% and she ended up ahead in votes by about 2%.

Many state polls were inaccurate, most of them in the same direction, with Trump receiving up to 10% more of the vote than expected in some states. This was enough to give him the majority of electoral votes and the presidency. Analysts are still debating the reasons so many state polls were unrepresentative of the people who actually voted.

EXERCISES

1. ***Unrepresentative Samples*** Find an example of a survey that is reported in a newspaper, in a magazine, or on a website. Describe different ways that the sample could have been unrepresentative of the population.

2. ***Biased Survey Questions*** Find an example of a survey that is reported in a newspaper, in a magazine, or on a website. Describe different ways that the survey questions could have been biased.

3. ***Misinterpreted Polls*** Determine whether each state election poll below was misleading. Assume the margin of error is 4% for each poll.
   (a) Michigan poll leader: Clinton by 3.4%; Election winner: Trump by 0.3%
   (b) Wisconsin poll leader: Clinton by 6.5%; Election winner: Trump by 0.7%

6 Chapter Summary

What Did You Learn?	Example(s)	Review Exercises
Section 6.1		
▷ How to find a point estimate and a margin of error	1, 2	1, 2
$E = z_c \dfrac{\sigma}{\sqrt{n}}$ Margin of error		
▷ How to construct and interpret confidence intervals for a population mean when σ is known	3–5	3–6
$\bar{x} - E < \mu < \bar{x} + E$		
▷ How to determine the minimum sample size required when estimating a population mean	6	7, 8
Section 6.2		
▷ How to interpret the t-distribution and use a t-distribution table	1	9–12
$t = \dfrac{\bar{x} - \mu}{s/\sqrt{n}}, \quad \text{d.f.} = n - 1$		
▷ How to construct and interpret confidence intervals for a population mean when σ is not known	2–4	13–18
$\bar{x} - E < \mu < \bar{x} + E, \quad E = t_c \dfrac{s}{\sqrt{n}}$		
Section 6.3		
▷ How to find a point estimate for a population proportion	1	19–24
$\hat{p} = \dfrac{x}{n}$		
▷ How to construct and interpret confidence intervals for a population proportion	2, 3	19–24
$\hat{p} - E < p < \hat{p} + E, \quad E = z_c \sqrt{\dfrac{\hat{p}\hat{q}}{n}}$		
▷ How to determine the minimum sample size required when estimating a population proportion	4	25, 26
Section 6.4		
▷ How to interpret the chi-square distribution and use a chi-square distribution table	1	27–30
$\chi^2 = \dfrac{(n-1)s^2}{\sigma^2}, \quad \text{d.f.} = n - 1$		
▷ How to construct and interpret confidence intervals for a population variance and standard deviation	2	31, 32
$\dfrac{(n-1)s^2}{\chi_R^2} < \sigma^2 < \dfrac{(n-1)s^2}{\chi_L^2}, \quad \sqrt{\dfrac{(n-1)s^2}{\chi_R^2}} < \sigma < \sqrt{\dfrac{(n-1)s^2}{\chi_L^2}}$		

6 Review Exercises

Waking times (in minutes past 5:00 A.M.)

135	145	95	140	135	95	110
50	90	165	110	125	80	125
130	110	25	75	65	100	60
125	115	135	95	90	140	40
75	50	130	85	100	160	135
45	135	115	75	130		

TABLE FOR EXERCISE 1

Driving distances to work (in miles)

12	9	7	2	8	7
3	27	21	10	13	7
2	30	7	6	13	6
4	1	10	3	13	6
2	9	2	12	16	18

TABLE FOR EXERCISE 2

1. (a) 103.5   (b) 11.7
2. (a) 9.5   (b) 2.9 .
3. (a) (91.8, 115.2); With 90% confidence, you can say that the population mean waking time is between 91.8 and 115.2 minutes past 5:00 a.m.

   (b) Yes; If the population mean is within 10% of the sample mean, then it falls inside the confidence interval.

4. (a) (6.6, 12.4); With 95% confidence, you can say that the population mean driving distance is between 6.6 and 12.4 miles.

   (b) No; 12.5 is outside the confidence interval.

5. $E = 1.675, \bar{x} = 22.425$
6. $E = 0.067, \bar{x} = 7.495$
7. 78 people   8. 107 people
9. 1.383   10. 2.069
11. 2.624   12. 2.756
13. (a) 11.2   (b) (60.9, 83.3)
14. (a) 0.5   (b) (3.0, 4.0)
15. (a) 0.7   (b) (6.1, 7.5)
16. (a) 10.6   (b) (14.6, 35.8)
17. See Odd Answers, page A66.
18. Yes

Section 6.1

 1. The waking times (in minutes past 5:00 A.M.) of 40 people who start work at 8:00 A.M. are shown in the table at the left. Assume the population standard deviation is 45 minutes. Find (a) the point estimate of the population mean μ and (b) the margin of error for a 90% confidence interval.

 2. The driving distances (in miles) to work of 30 people are shown in the table at the left. Assume the population standard deviation is 8 miles. Find (a) the point estimate of the population mean μ and (b) the margin of error for a 95% confidence interval.

3. (a) Construct a 90% confidence interval for the population mean in Exercise 1. Interpret the results. (b) Does it seem possible that the population mean could be within 10% of the sample mean? Explain.

4. (a) Construct a 95% confidence interval for the population mean in Exercise 2. Interpret the results. (b) Does it seem possible that the population mean could be greater than 12.5 miles? Explain.

In Exercises 5 and 6, use the confidence interval to find the margin of error and the sample mean.

5. (20.75, 24.10)   **6.** (7.428, 7.562)

7. Determine the minimum sample size required to be 95% confident that the sample mean waking time is within 10 minutes of the population mean waking time. Use the population standard deviation from Exercise 1.

8. Determine the minimum sample size required to be 99% confident that the sample mean driving distance to work is within 2 miles of the population mean driving distance to work. Use the population standard deviation from Exercise 2.

Section 6.2

In Exercises 9–12, find the critical value t_c for the level of confidence c and sample size n.

9. $c = 0.80, n = 10$ **10.** $c = 0.95, n = 24$
11. $c = 0.98, n = 15$ **12.** $c = 0.99, n = 30$

In Exercises 13–16, (a) find the margin of error for the values of c, s, and n, and (b) construct the confidence interval for μ using the t-distribution. Assume the population is normally distributed.

13. $c = 0.90, s = 25.6, n = 16, \bar{x} = 72.1$
14. $c = 0.95, s = 1.1, n = 25, \bar{x} = 3.5$
15. $c = 0.98, s = 0.9, n = 12, \bar{x} = 6.8$
16. $c = 0.99, s = 16.5, n = 20, \bar{x} = 25.2$

17. In a random sample of 36 top-rated roller coasters, the average height is 165 feet and the standard deviation is 67 feet. Construct a 90% confidence interval for μ. Interpret the results. *(Source: POP World Media, LLC)*

18. You research the heights of top-rated roller coasters and find that the population mean is 160 feet. In Exercise 17, does the t-value fall between $-t_{0.95}$ and $t_{0.95}$?

19. See Odd Answers, page A66.

20. See Selected Answers, page A99.

21. See Odd Answers, page A67.

22. (a) 0.600, 0.400

    (b) (0.583, 0.617); (0.580, 0.620)

    (c) With 90% confidence, you can say that the population proportion of U.S. adults who say an occupation as an athlete is prestigious is between 58.3% and 61.7%. With 95% confidence, you can say it is between 58.0% and 62.0%. The 95% confidence interval is slightly wider.

23. No; It falls outside both confidence intervals.

24. Yes; If the population proportion is within 1% of the point estimate, then it falls inside both confidence intervals.

25. (a) 385 adults   (b) 335 adults

    (c) Having an estimate of the population proportion reduces the minimum sample size needed.

26. Yes, because the minimum sample size needed is 335 adults.

27. $\chi_R^2 = 23.337$, $\chi_L^2 = 4.404$

28. $\chi_R^2 = 42.980$, $\chi_L^2 = 10.856$

29. $\chi_R^2 = 24.996$, $\chi_L^2 = 7.261$

30. $\chi_R^2 = 23.589$, $\chi_L^2 = 1.735$

31. (a) (185.1, 980.8)

    (b) (13.6, 31.3)

    With 95% confidence, you can say that the population variance is between 185.1 and 980.8, and the population standard deviation is between 13.6 and 31.3 knots.

32. (a) (4.92, 16.09)

    (b) (2.22, 4.01)

    With 98% confidence, you can say that the population variance is between 4.92 and 16.09, and the population standard deviation is between 2.22 and 4.01 seconds.

Section 6.3

In Exercises 19–22, let p be the population proportion for the situation. (a) Find point estimates of p and q, (b) construct 90% and 95% confidence intervals for p, and (c) interpret the results of part (b) and compare the widths of the confidence intervals.

19. In a survey of 1035 U.S. adults, 745 say they want the U.S. to play a leading or major role in global affairs. *(Adapted from Gallup)*

20. In a survey of 1003 U.S. adults, 451 believe that for a person to be considered truly American, it is very important that he or she share American customs and traditions. *(Adapted from Pew Research Center)*

21. In a survey of 2202 U.S. adults, 1167 think antibiotics are effective against viral infections. *(Adapted from The Harris Poll)*

22. In a survey of 2223 U.S. adults, 1334 say an occupation as an athlete is prestigious. *(Adapted from The Harris Poll)*

23. In Exercise 19, does it seem possible that the population proportion could equal 0.75? Explain.

24. In Exercise 22, does it seem possible that the population proportion could be within 1% of the point estimate? Explain.

25. You wish to estimate, with 95% confidence, the population proportion of U.S. adults who have taken or planned to take a winter vacation in a recent year. Your estimate must be accurate within 5% of the population proportion.

    (a) No preliminary estimate is available. Find the minimum sample size needed.

    (b) Find the minimum sample size needed, using a prior study that found that 32% of U.S. adults have taken or planned to take a winter vacation in a recent year. *(Source: Rasmussen Reports)*

    (c) Compare the results from parts (a) and (b).

26. In Exercise 25(b), would a sample size of 369 be acceptable? Explain.

Section 6.4

In Exercises 27–30, find the critical values χ_R^2 and χ_L^2 for the level of confidence c and sample size n.

27. $c = 0.95$, $n = 13$ **28.** $c = 0.98$, $n = 25$

29. $c = 0.90$, $n = 16$ **30.** $c = 0.99$, $n = 10$

In Exercises 31 and 32, assume the sample is from a normally distributed population and construct the indicated confidence intervals for (a) the population variance σ^2 and (b) the population standard deviation σ. Interpret the results.

31. The maximum wind speeds (in knots) of 13 randomly selected hurricanes that have hit the U.S. mainland are listed. Use a 95% level of confidence. *(Source: National Oceanic & Atmospheric Administration)*

    70   85   70   75   100   100   110   105   130   75   85   75   70

32. The acceleration times (in seconds) from 0 to 60 miles per hour for 33 randomly selected sedans are listed. Use a 98% level of confidence. *(Source: Zero to 60 Times)*

    6.5   5.0   5.2   3.3   6.6   6.3   5.1   5.3   5.4   9.5   7.5

    4.5   5.8   8.6   6.9   8.1   6.0   6.7   7.9   8.8   7.1   7.9

    7.2   18.4   9.1   6.8   12.5   4.2   7.1   9.9   9.5   2.8   4.9

6 | Chapter Quiz

Women's Open Division winning times (in hours)				
3.36	3.45	3.50	3.14	2.79
2.79	2.75	2.59	2.45	2.38
2.57	2.42	2.41	2.42	2.41
2.42	2.45	2.44	2.39	2.44
2.40	2.35	2.41	2.39	2.49
2.42	2.54	2.44	2.44	2.49

TABLE FOR EXERCISE 1

1. (a) 2.598   (b) 0.123
   (c) (2.475, 2.721); With 95% confidence, you can say that the population mean winning time is between 2.475 and 2.721 hours.
   (d) No; It falls outside the confidence interval.

2. 42 champions

3. (a) $\bar{x} = 6.61$, $s \approx 3.38$
   (b) (4.65, 8.57); With 90% confidence, you can say that the population mean amount of time is between 4.65 and 8.57 minutes.
   (c) (4.79, 8.43); With 90% confidence, you can say that the population mean amount of time is between 4.79 and 8.43 minutes. This confidence interval is narrower than the one in part (b).

4. (109,990, 156,662); With 95% confidence, you can say that the population mean annual earnings is between $109,990 and $156,662.

5. Yes

6. See Odd Answers, page A67.

7. (a) (5.41, 38.08)
   (b) (2.32, 6.17); With 95% confidence, you can say that the population standard deviation is between 2.32 and 6.17 minutes.

Take this quiz as you would take a quiz in class. After you are done, check your work against the answers given in the back of the book.

 1. The winning times (in hours) for a sample of 30 randomly selected Boston Marathon Women's Open Division champions are shown in the table at the left. *(Source: Boston Athletic Association)*

   (a) Find the point estimate of the population mean.
   (b) Find the margin of error for a 95% confidence level.
   (c) Construct a 95% confidence interval for the population mean. Interpret the results.
   (d) Does it seem possible that the population mean could be greater than 2.75 hours? Explain.

2. You wish to estimate the mean winning time for Boston Marathon Women's Open Division champions. The estimate must be within 0.13 hour of the population mean. Determine the minimum sample size required to construct a 99% confidence interval for the population mean. Use the population standard deviation from Exercise 1.

3. The data set represents the amounts of time (in minutes) spent checking email for a random sample of employees at a company.

   7.5   2.0   12.1   8.8   9.4   7.3   1.9   2.8   7.0   7.3

   (a) Find the sample mean and the sample standard deviation.
   (b) Construct a 90% confidence interval for the population mean. Interpret the results. Assume the times are normally distributed.
   (c) Repeat part (b), assuming $\sigma = 3.5$ minutes. Compare the results.

4. In a random sample of 12 senior-level chemical engineers, the mean annual earnings was $133,326 and the standard deviation was $36,729. Assume the annual earnings are normally distributed and construct a 95% confidence interval for the population mean annual earnings for senior-level chemical engineers. Interpret the results. *(Adapted from Salary.com)*

5. You research the salaries of senior-level chemical engineers and find that the population mean is $131,935. In Exercise 4, does the *t*-value fall between $-t_{0.95}$ and $t_{0.95}$?

6. In a survey of 1018 U.S. adults, 753 say that the energy situation in the United States is very or fairly serious. *(Adapted from Gallup)*

   (a) Find the point estimate for the population proportion.
   (b) Construct a 90% confidence interval for the population proportion. Interpret the results.
   (c) Does it seem possible that the population proportion could be between 90% and 95% of the point estimate? Explain.
   (d) Find the minimum sample size needed to estimate the population proportion at the 99% confidence level in order to ensure that the estimate is accurate within 4% of the population proportion.

7. Refer to the data set in Exercise 3. Assume the population of times spent checking email is normally distributed. Construct a 95% confidence interval for (a) the population variance and (b) the population standard deviation. Interpret the results.

Chapter Test

Take this test as you would take a test in class.

1. In a survey of 2096 U.S. adults, 1740 think football teams of all levels should require players who suffer a head injury to take a set amount of time off from playing to recover. *(Adapted from The Harris Poll)*

 (a) Find the point estimate for the population proportion.

 (b) Construct a 95% confidence interval for the population proportion. Interpret the results.

 (c) Does it seem possible that the population proportion could be within 99% of the point estimate? Explain.

 (d) Find the minimum sample size needed to estimate the population proportion at the 99% confidence level in order to ensure that the estimate is accurate within 3% of the population proportion.

2. The data set represents the weights (in pounds) of 10 randomly selected black bears from northeast Pennsylvania. Assume the weights are normally distributed. *(Source: Pennsylvania Game Commission)*

   170 225 183 137 287 191 268 185 211 284

 (a) Find the sample mean and the sample standard deviation.

 (b) Construct a 95% confidence interval for the population mean. Interpret the results.

 (c) Construct a 99% confidence interval for the population standard deviation. Interpret the results.

3. The data set represents the scores of 12 randomly selected students on the SAT Physics Subject Test. Assume the population test scores are normally distributed and the population standard deviation is 104. *(Adapted from The College Board)*

   590 450 490 680 380 500 570 620 640 530 780 720

 (a) Find the point estimate of the population mean.

 (b) Construct a 90% confidence interval for the population mean. Interpret the results.

 (c) Does it seem possible that the population mean could equal 667? Explain.

 (d) Determine the minimum sample size required to be 95% confident that the sample mean test score is within 10 points of the population mean test score.

4. Use the standard normal distribution or the *t*-distribution to construct the indicated confidence interval for the population mean of each data set. Justify your decision. If neither distribution can be used, explain why. Interpret the results.

 (a) In a random sample of 40 patients, the mean waiting time at a dentist's office was 20 minutes and the standard deviation was 7.5 minutes. Construct a 95% confidence interval for the population mean.

 (b) In a random sample of 15 cereal boxes, the mean weight was 11.89 ounces. Assume the weights of the cereal boxes are normally distributed and the population standard deviation is 0.05 ounce. Construct a 90% confidence interval for the population mean.

---

1. (a) 0.830

 (b) (0.814, 0.846); With 95% confidence, you can say that the population proportion of U.S. adults who think football teams of all levels should require players who suffer a head injury to take a set amount of time off from playing to recover is between 81.4% and 84.6%.

 (c) Yes; 99% of the point estimate lies within the confidence interval.

 (d) 1040 adults

2. (a) $\bar{x} = 214.1$, $s \approx 51.1$

 (b) (177.6, 250.7); With 95% confidence, you can say that the population mean weight is between 177.6 and 250.7 pounds.

 (c) (31.6, 116.4); With 99% confidence, you can say that the population standard deviation is between 31.6 and 116.4 pounds.

3. (a) 579.2

 (b) (529.8, 628.6); With 90% confidence, you can say that the population mean test score is between 529.8 and 628.6.

 (c) No; It falls outside the confidence interval.

 (d) 416 students

4. (a) Use a *t*-distribution because σ is unknown and $n \geq 30$.

   (17.6, 22.4); With 95% confidence, you can say that the population mean waiting time is between 17.6 and 22.4 minutes.

 (b) Use the standard normal distribution because σ is known and the weights are normally distributed.

   (11.87, 11.91); With 90% confidence, you can say that the population mean weight is between 11.87 and 11.91 ounces.

The Safe Drinking Water Act, which was passed in 1974, allows the Environmental Protection Agency (EPA) to regulate the levels of contaminants in drinking water. The EPA requires that water utilities give their customers water quality reports annually. These reports include the results of daily water quality monitoring, which is performed to determine whether drinking water is safe for consumption.

A water department tests for contaminants at water treatment plants and at customers' taps. These contaminants include microorganisms, organic chemicals, and inorganic chemicals, such as cyanide. Cyanide's presence in drinking water is the result of discharges from steel, plastics, and fertilizer factories. For drinking water, the maximum contaminant level of cyanide is 0.2 part per million.

As part of your job for your city's water department, you are preparing a report that includes an analysis of the results shown in the figure at the right. The figure shows the point estimates for the population mean concentration and the 95% confidence intervals for μ for cyanide over a three-year period. The data are based on random water samples taken by the city's three water treatment plants.

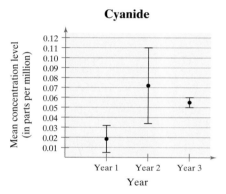

EXERCISES

1. Interpreting the Results

Use the figure to determine whether there has been a change in the mean concentration level of cyanide for each time period. Explain your reasoning.

(a) From Year 1 to Year 2   (b) From Year 2 to Year 3

(c) From Year 1 to Year 3

2. What Can You Conclude?

Using the results of Exercise 1, what can you conclude about the concentrations of cyanide in the drinking water?

3. What Do You Think?

The confidence interval for Year 2 is much larger than the other years. What do you think may have caused this larger confidence level?

4. How Can You Improve the Report?

What can the water department do to decrease the size of the confidence intervals, regardless of the amount of variance in cyanide levels?

5. How Do You Think They Did It?

How do you think the water department constructed the 95% confidence intervals for the population mean concentration of cyanide in the water? Include answers to the questions below in your explanation.

(a) What sampling distribution do you think they used? Why?

(b) Do you think they used the population standard deviation in calculating the margin of error? Why or why not? If not, what could they have used?

United States Foreign Policy Polls

THE GALLUP ORGANIZATION

www.gallup.com

Since 1935, the Gallup Organization has conducted public opinion polls in the United States and around the world. The table shows the results of four polls of randomly selected U.S. adults from 2015 through 2017. The remaining percentages not shown in the results are adults who were not sure.

Question	Results	Number Polled
Do you think the U.S. made a mistake sending troops to Iraq?	Yes: 51% No: 46%	1527
In the Middle East situation, are your sympathies more with the Israelis or the Palestinians?	Israelis: 62% Palestinians: 19%	1035
Do you have a favorable or unfavorable opinion of Russian president Vladimir Putin?	Favorable: 22% Unfavorable: 72%	1035
Should the NATO alliance be maintained or is it not necessary anymore?	Should be maintained: 80% Not necessary: 16%	485

EXERCISES

1. Use technology to find a 95% confidence interval for the population proportion of adults who

    (a) think sending troops to Iraq was a mistake.

    (b) sympathize more with the Israelis than the Palestinians.

    (c) have a favorable opinion of Vladimir Putin.

    (d) think the NATO alliance is not necessary.

    (e) do not sympathize with either the Israelis or the Palestinians more than the other.

2. Find the minimum sample size needed to estimate, with 95% confidence, the population proportion of adults who have a favorable opinion of Vladimir Putin. Your estimate must be accurate within 2% of the population proportion.

3. Use technology to simulate a poll. Assume that the actual population proportion of adults who think the U.S. made a mistake sending troops to Iraq is 54%. Run the simulation several times using $n = 1527$.

    (a) What was the least value you obtained for $\hat{p}$?

    (b) What was the greatest value you obtained for $\hat{p}$?

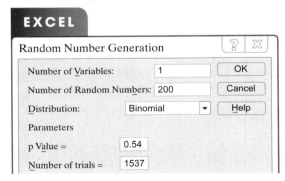

4. Is it probable that the population proportion of adults who think the U.S. made a mistake sending troops to Iraq is 54%? Explain your reasoning.

Extended solutions are given in the technology manuals that accompany this text. Technical instruction is provided for Minitab, Excel, and the TI-84 Plus.

Using Technology to Construct Confidence Intervals

Here are some Minitab and TI-84 Plus printouts for some examples in this chapter. Answers may be slightly different because of rounding.

See Example 3, page 301.

Display Descriptive Statistics...
Store Descriptive Statistics...
Graphical Summary...

1-Sample Z...
1-Sample t...
2-Sample t...
Paired t...

1 Proportion...
2 Proportions...

19	25	15	21	22	20	20	22	22	21
21	23	22	16	21	18	25	23	23	21
22	24	18	19	23	20	19	19	24	25
17	21	21	25	23	18	22	20	21	21

MINITAB

One-Sample Z: Hours

The assumed standard deviation = 2.3

Variable	N	Mean	StDev	SE Mean	95% CI
Hours	40	21.050	2.438	0.364	(20.337, 21.763)

See Example 2, page 312.

Display Descriptive Statistics...
Store Descriptive Statistics...
Graphical Summary...

1-Sample Z...
1-Sample t...
2-Sample t...
Paired t...

1 Proportion...
2 Proportions...

MINITAB

One-Sample T

N	Mean	StDev	SE Mean	95% CI
16	162.00	10.00	2.50	(156.67, 167.33)

See Example 2, page 322.

Display Descriptive Statistics...
Store Descriptive Statistics...
Graphical Summary...

1-Sample Z...
1-Sample t...
2-Sample t...
Paired t...

1 Proportion...
2 Proportions...

MINITAB

Test and CI for One Proportion

Sample	X	N	Sample p	95% CI
1	1054	1550	0.680000	(0.656130, 0.703186)

See Example 5, page 303.

TI-84 PLUS

EDIT CALC **TESTS**
1: Z–Test...
2: T–Test...
3: 2–SampZTest...
4: 2–SampTTest...
5: 1–PropZTest...
6: 2–PropZTest...
7↓ ZInterval...

TI-84 PLUS

ZInterval
Inpt:Data **Stats**
σ:1.5
x̄:22.9
n:20
C–Level:.9
Calculate

TI-84 PLUS

ZInterval
(22.348, 23.452)
x̄=22.9
n=20

See Example 3, page 313.

TI-84 PLUS

EDIT CALC **TESTS**
2↑ T–Test...
3: 2–SampZTest...
4: 2–SampTTest...
5: 1–PropZTest...
6: 2–PropZTest...
7: ZInterval...
8↓ TInterval...

TI-84 PLUS

TInterval
Inpt:Data **Stats**
x̄:9.75
Sx:2.39
n:36
C–Level:.99
Calculate

TI-84 PLUS

TInterval
(8.665, 10.835)
x̄=9.75
Sx=2.39
n=36

See Example 2, page 322.

TI-84 PLUS

EDIT CALC **TESTS**
5↑ 1–PropZTest...
6: 2–PropZTest...
7: ZInterval...
8: TInterval...
9: 2–SampZInt...
0: 2–SampTInt...
A↓ 1–PropZInt...

TI-84 PLUS

1-PropZInt
x:1054
n:1550
C–Level:.95
Calculate

TI-84 PLUS

1-PropZInt
(.65678, .70322)
p̂=.68
n=1550

CHAPTER 7

Hypothesis Testing with One Sample

The Entertainment Software Rating Board (ESRB) assigns ratings to video games to indicate the appropriate ages for players. These ratings include EC (early childhood), E (everyone), E10+ (everyone 10+), T (teen), M (mature), and AO (adults only).

In Chapter 6, you began your study of inferential statistics. There, you learned how to form a confidence interval to estimate a population parameter, such as the proportion of people in the United States who agree with a certain statement. For instance, in a nationwide poll conducted by Pew Research Center, 2001 U.S. adults were asked whether they agreed or disagreed with the statement, "People who play violent video games are more likely to be violent themselves." Out of those surveyed, 800 adults agreed with the statement.

You have learned how to use these results to state with 95% confidence that the population proportion of U.S. adults who agree that people who play violent video games are more likely to be violent themselves is between 37.9% and 42.1%.

Where You're Going

In this chapter, you will continue your study of inferential statistics. But now, instead of making an estimate about a population parameter, you will learn how to test a claim about a parameter.

For instance, suppose that you work for Pew Research Center and are asked to test a claim that the proportion of U.S. adults who agree that people who play violent video games are more likely to be violent themselves is $p = 0.35$. To test the claim, you take a random sample of $n = 2001$ U.S. adults and find that 800 of them think that people who play violent video games are more likely to be violent themselves. Your sample statistic is $\hat{p} \approx 0.400$.

Is your sample statistic different enough from the claim ($p = 0.35$) to decide that the claim is false? The answer lies in the sampling distribution of sample proportions taken from a population in which $p = 0.35$. The figure below shows that your sample statistic is more than 4 standard errors from the claimed value. If the claim is true, then the probability of the sample statistic being 4 standard errors or more from the claimed value is extremely small. Something is wrong! If your sample was truly random, then you can conclude that the actual proportion of the adult population is not 0.35. In other words, you tested the original claim (hypothesis), and you decided to reject it.

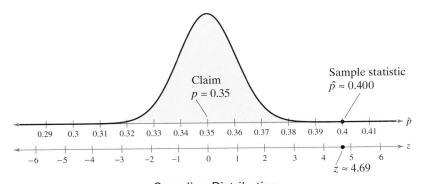

Sampling Distribution

What You Should Learn

▶ A practical introduction to hypothesis tests

▶ How to state a null hypothesis and an alternative hypothesis

▶ How to identify type I and type II errors and interpret the level of significance

▶ How to know whether to use a one-tailed or two-tailed statistical test and find a *P*-value

▶ How to make and interpret a decision based on the results of a statistical test

▶ How to write a claim for a hypothesis test

Study Tip

As you study this chapter, do not get confused regarding concepts of certainty and importance. For instance, even if you were very certain that the mean gas mileage of a type of hybrid vehicle is not 50 miles per gallon, the actual mean mileage might be very close to this value and the difference might not be important.

Hypothesis Tests ▪ Stating a Hypothesis ▪ Types of Errors and Level of Significance ▪ Statistical Tests and *P*-Values ▪ Making a Decision and Interpreting the Decision ▪ Strategies for Hypothesis Testing

Hypothesis Tests

Throughout the remainder of this text, you will study an important technique in inferential statistics called hypothesis testing. A **hypothesis test** is a process that uses sample statistics to test a claim about the value of a population parameter. Researchers in fields such as medicine, psychology, and business rely on hypothesis testing to make informed decisions about new medicines, treatments, and marketing strategies.

For instance, consider a manufacturer that advertises its new hybrid car has a mean gas mileage of 50 miles per gallon. If you suspect that the mean mileage is not 50 miles per gallon, how could you show that the advertisement is false?

Obviously, you cannot test *all* the vehicles, but you can still make a reasonable decision about the mean gas mileage by taking a random sample from the population of vehicles and measuring the mileage of each. If the sample mean differs enough from the advertisement's mean, you can decide that the advertisement is wrong.

For instance, to test that the mean gas mileage of all hybrid vehicles of this type is $\mu = 50$ miles per gallon, you take a random sample of $n = 30$ vehicles and measure the mileage of each. You obtain a sample mean of $\bar{x} = 47$ miles per gallon with a sample standard deviation of $s = 5.5$ miles per gallon. Does this indicate that the manufacturer's advertisement is false?

To decide, you do something unusual—*you assume the advertisement is correct!* That is, you assume that $\mu = 50$. Then, you examine the sampling distribution of sample means (with $n = 30$) taken from a population in which $\mu = 50$ and $\sigma = 5.5$. From the Central Limit Theorem, you know this sampling distribution is normal with a mean of 50 and standard error of

$$\frac{5.5}{\sqrt{30}} \approx 1.$$

In the figure below, notice that the sample mean of $\bar{x} = 47$ miles per gallon is highly unlikely—it is about 3 standard errors ($z \approx -2.99$) from the claimed mean! Using the techniques you studied in Chapter 5, you can determine that if the advertisement is true, then the probability of obtaining a sample mean of 47 or less is about 0.001. This is an unusual event! Your assumption that the company's advertisement is correct has led you to an improbable result. So, either you had a very unusual sample, or the advertisement is probably false. The logical conclusion is that the advertisement is probably false.

Sampling Distribution of $\bar{x}$

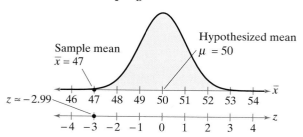

Study Tip

The term *null hypothesis* was introduced by Ronald Fisher (see page 35). If the statement in the null hypothesis is not true, then the alternative hypothesis must be true.

Picturing the World

A study was done on the effect of a wearable fitness device combined with a low-calorie diet on weight loss. The study used a random sample of 237 adults. At the end of the study, the adults had a mean weight loss of 3.5 kilograms. So, it is claimed that the mean weight loss is 3.5 kilograms for all adults who use a wearable fitness device combined with a low-calorie diet. (Adapted from The Journal of the American Medical Association)

Determine a null hypothesis and alternative hypothesis for this claim.

$H_0: \mu = 3.5, H_a: \mu \neq 3.5$

Stating a Hypothesis

A statement about a population parameter is called a **statistical hypothesis.** To test a population parameter, you should carefully state a pair of hypotheses—one that represents the claim and the other, its complement. When one of these hypotheses is false, the other must be true. Either hypothesis—the **null hypothesis** or the **alternative hypothesis**—may represent the original claim.

DEFINITION

1. A **null hypothesis** H_0 is a statistical hypothesis that contains a statement of equality, such as $\leq$, $=$, or $\geq$.
2. The **alternative hypothesis** H_a is the complement of the null hypothesis. It is a statement that must be true if H_0 is false and it contains a statement of strict inequality, such as $>$, $\neq$, or $<$.

The symbol H_0 is read as "H sub-zero" or "H naught" and H_a is read as "H sub-a."

To write the null and alternative hypotheses, translate the claim made about the population parameter from a verbal statement to a mathematical statement. Then, write its complement. For instance, if the claim value is k and the population parameter is μ, then some possible pairs of null and alternative hypotheses are

$$\begin{cases} H_0: \mu \leq k \\ H_a: \mu > k \end{cases} \quad \begin{cases} H_0: \mu \geq k \\ H_a: \mu < k \end{cases} \quad \text{and} \quad \begin{cases} H_0: \mu = k \\ H_a: \mu \neq k \end{cases}.$$

Regardless of which of the three pairs of hypotheses you use, you always assume $\mu = k$ and examine the sampling distribution on the basis of this assumption. Within this sampling distribution, you will determine whether or not a sample statistic is unusual.

The table shows the relationship between possible verbal statements about the parameter μ and the corresponding null and alternative hypotheses. Similar statements can be made to test other population parameters, such as p, σ, or σ^2.

Verbal Statement H_0 The mean is . . .	Mathematical Statements	Verbal Statement H_a The mean is . . .
. . . greater than or equal to k. . . . at least k. . . . not less than k. . . . not shorter than k.	$\begin{cases} H_0: \mu \geq k \\ H_a: \mu < k \end{cases}$	. . . less than k. . . . below k. . . . fewer than k. . . . shorter than k.
. . . less than or equal to k. . . . at most k. . . . not more than k. . . . not longer than k.	$\begin{cases} H_0: \mu \leq k \\ H_a: \mu > k \end{cases}$	. . . greater than k. . . . above k. . . . more than k. . . . longer than k.
. . . equal to k. . . . k. . . . exactly k. . . . the same as k. . . . not changed from k.	$\begin{cases} H_0: \mu = k \\ H_a: \mu \neq k \end{cases}$	. . . not equal to k. . . . different from k. . . . not k. . . . different from k. . . . changed from k.

Note to Instructor

Begin with a hypothesis statement and ask students to state its logical complement. Some students will have difficulty with the fact that the complement of $\mu \neq k$ is $\mu = k$. Discuss the role of a double negative in English. The important point is that if you conclude that H_0 is false, then you are also concluding that H_a is true.

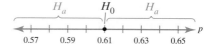

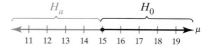

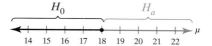

EXAMPLE 1

Stating the Null and Alternative Hypotheses

Write each claim as a mathematical statement. State the null and alternative hypotheses, and identify which represents the claim.

1. A school publicizes that the proportion of its students who are involved in at least one extracurricular activity is 61%.

2. A car dealership announces that the mean time for an oil change is less than 15 minutes.

3. A company advertises that the mean life of its furnaces is more than 18 years.

SOLUTION

1. The claim "the proportion . . . is 61%" can be written as $p = 0.61$. Its complement is $p \neq 0.61$, as shown in the figure at the left. Because $p = 0.61$ contains the statement of equality, it becomes the null hypothesis. In this case, the null hypothesis represents the claim. You can write the null and alternative hypotheses as shown.

H_0: $p = 0.61$ (Claim)

H_a: $p \neq 0.61$

2. The claim "the mean . . . is less than 15 minutes" can be written as $\mu < 15$. Its complement is $\mu \geq 15$, as shown in the figure at the left. Because $\mu \geq 15$ contains the statement of equality, it becomes the null hypothesis. In this case, the alternative hypothesis represents the claim. You can write the null and alternative hypotheses as shown.

H_0: $\mu \geq 15$ minutes

H_a: $\mu < 15$ minutes (Claim)

3. The claim "the mean . . . is more than 18 years" can be written as $\mu > 18$. Its complement is $\mu \leq 18$, as shown in the figure at the left. Because $\mu \leq 18$ contains the statement of equality, it becomes the null hypothesis. In this case, the alternative hypothesis represents the claim. You can write the null and alternative hypotheses as shown.

H_0: $\mu \leq 18$ years

H_a: $\mu > 18$ years (Claim)

In the three figures at the left, notice that each point on the number line is in either H_0 or H_a, but no point is in both.

TRY IT YOURSELF 1

Write each claim as a mathematical statement. State the null and alternative hypotheses, and identify which represents the claim.

1. A consumer analyst reports that the mean life of a certain type of automobile battery is not 74 months.
2. An electronics manufacturer publishes that the variance of the life of its home theater systems is less than or equal to 2.7.
3. A realtor publicizes that the proportion of homeowners who feel their house is too small for their family is more than 24%.

Answer: Page A36

In Example 1, notice that the claim is represented by either the null hypothesis *or* the alternative hypothesis.

Types of Errors and Level of Significance

No matter which hypothesis represents the claim, you always begin a hypothesis test by assuming that the equality condition in the null hypothesis is true. So, when you perform a hypothesis test, you make one of two decisions:

1. reject the null hypothesis

or

2. fail to reject the null hypothesis.

Because your decision is based on a sample rather than the entire population, there is always the possibility you will make the wrong decision.

For instance, you claim that a coin is not fair. To test your claim, you toss the coin 100 times and get 49 heads and 51 tails. You would probably agree that you do not have enough evidence to support your claim. Even so, it is possible that the coin is actually not fair and you had an unusual sample.

But then you toss the coin 100 times and get 21 heads and 79 tails. It would be a rare occurrence to get only 21 heads out of 100 tosses with a fair coin. So, you probably have enough evidence to support your claim that the coin is not fair. However, you cannot be 100% sure. It is possible that the coin is fair and you had an unusual sample.

Letting p represent the proportion of heads, the claim that "the coin is not fair" can be written as the mathematical statement $p \neq 0.5$. Its complement, "the coin is fair," is written as $p = 0.5$, as shown in the figure.

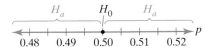

So, the null hypothesis is

$$H_0\colon p = 0.5$$

and the alternative hypothesis is

$$H_a\colon p \neq 0.5. \text{ (Claim)}$$

Remember, the only way to be absolutely certain of whether H_0 is true or false is to test the entire population. Because your decision—to reject H_0 or to fail to reject H_0—is based on a sample, you must accept the fact that your decision might be incorrect. You might reject a null hypothesis when it is actually true. Or, you might fail to reject a null hypothesis when it is actually false. These types of errors are summarized in the next definition.

DEFINITION

A **type I error** occurs if the null hypothesis is rejected when it is true.

A **type II error** occurs if the null hypothesis is not rejected when it is false.

The table shows the four possible outcomes of a hypothesis test.

Decision	Truth of H_0	
	H_0 is true.	H_0 is false.
Do not reject H_0.	Correct decision	Type II error
Reject H_0.	Type I error	Correct decision

Hypothesis testing is sometimes compared to the legal system used in the United States. Under this system, these steps are used.

1. A carefully worded accusation is written.
2. The defendant is assumed innocent (H_0) until proven guilty. The burden of proof lies with the prosecution. If the evidence is not strong enough, then there is no conviction. A "not guilty" verdict does not prove that a defendant is innocent.
3. The evidence needs to be conclusive beyond a reasonable doubt. The system assumes that more harm is done by convicting the innocent (type I error) than by not convicting the guilty (type II error).

The table at the left shows the four possible outcomes.

	Truth about defendant	
Verdict	**Innocent**	**Guilty**
Not guilty	Justice	Type II error
Guilty	Type I error	Justice

EXAMPLE 2

Identifying Type I and Type II Errors

The USDA limit for salmonella contamination for ground beef is 7.5%. A meat inspector reports that the ground beef produced by a company exceeds the USDA limit. You perform a hypothesis test to determine whether the meat inspector's claim is true. When will a type I or type II error occur? Which error is more serious? *(Source: U.S. Department of Agriculture)*

SOLUTION

Let p represent the proportion of the ground beef that is contaminated. The meat inspector's claim is "more than 7.5% is contaminated." You can write the null hypothesis as

$H_0: p \leq 0.075$ The proportion is less than or equal to 0.075.

and the alternative hypothesis is

$H_a: p > 0.075$. (Claim)   The proportion is greater than 0.075.

You can visualize the null and alternative hypotheses using a number line, as shown below.

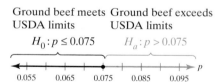

A type I error will occur when the actual proportion of contaminated ground beef is less than or equal to 0.075, but you reject H_0. A type II error will occur when the actual proportion of contaminated ground beef is greater than 0.075, but you do not reject H_0. With a type I error, you might create a health scare and hurt the sales of ground beef producers who were actually meeting the USDA limits. With a type II error, you could be allowing ground beef that exceeded the USDA contamination limit to be sold to consumers. A type II error is more serious because it could result in sickness or even death.

TRY IT YOURSELF 2

A company specializing in parachute assembly states that its main parachute failure rate is not more than 1%. You perform a hypothesis test to determine whether the company's claim is false. When will a type I or type II error occur? Which error is more serious?

Answer: Page A36

You will reject the null hypothesis when the sample statistic from the sampling distribution is unusual. You have already identified unusual events to be those that occur with a probability of 0.05 or less. When statistical tests are used, an unusual event is sometimes required to have a probability of 0.10 or less, 0.05 or less, or 0.01 or less. Because there is variation from sample to sample, there is always a possibility that you will reject a null hypothesis when it is actually true. In other words, although the null hypothesis is true, your sample statistic is determined to be an unusual event in the sampling distribution. You can decrease the probability of this happening by lowering the **level of significance.**

Study Tip

When you decrease α (the maximum allowable probability of making a type I error), you are likely to be increasing β. The value $1 - \beta$ is called the *power of the test.* It represents the probability of rejecting the null hypothesis when it is false. The value of the power is difficult (and sometimes impossible) to find in most cases.

Note to Instructor

You can use an example of "false positive" and "false negative" results for a medical test (for example, cancer) to discuss type I and type II errors. You might also want to point out that the computation of β is beyond the scope of this text.

DEFINITION

In a hypothesis test, the **level of significance** is your maximum allowable probability of making a type I error. It is denoted by α, the lowercase Greek letter alpha.

The probability of a type II error is denoted by β, the lowercase Greek letter beta.

By setting the level of significance at a small value, you are saying that you want the probability of rejecting a true null hypothesis to be small. Three commonly used levels of significance are

$$\alpha = 0.10, \qquad \alpha = 0.05, \qquad \text{and} \qquad \alpha = 0.01.$$

Statistical Tests and *P*-Values

After stating the null and alternative hypotheses and specifying the level of significance, the next step in a hypothesis test is to obtain a random sample from the population and calculate the sample statistic (such as $\bar{x}$, $\hat{p}$, or s^2) corresponding to the parameter in the null hypothesis (such as μ, p, or σ^2). This sample statistic is called the **test statistic.** With the assumption that the null hypothesis is true, the test statistic is then converted to a **standardized test statistic,** such as z, t, or χ^2. The standardized test statistic is used in making the decision about the null hypothesis.

In this chapter, you will learn about several one-sample statistical tests. The table shows the relationships between population parameters and their corresponding test statistics and standardized test statistics.

Population parameter	Test statistic	Standardized test statistic
μ	$\bar{x}$	z (Section 7.2, σ known), t (Section 7.3, σ unknown)
p	$\hat{p}$	z (Section 7.4)
σ^2	s^2	χ^2 (Section 7.5)

One way to decide whether to reject the null hypothesis is to determine whether the probability of obtaining the standardized test statistic (or one that is more extreme) is less than the level of significance.

DEFINITION

If the null hypothesis is true, then a ***P*-value** (or **probability value**) of a hypothesis test is the probability of obtaining a sample statistic with a value as extreme or more extreme than the one determined from the sample data.

The *P*-value of a hypothesis test depends on the nature of the test. There are three types of hypothesis tests—**left-tailed, right-tailed,** and **two-tailed.** The type of test depends on the location of the region of the sampling distribution that favors a rejection of H_0. This region is indicated by the alternative hypothesis.

DEFINITION

1. If the alternative hypothesis H_a contains the less-than inequality symbol $(<)$, then the hypothesis test is a **left-tailed test.**

H_0: $\mu \geq k$
H_a: $\mu < k$

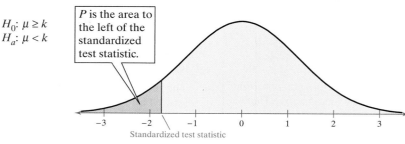

P is the area to the left of the standardized test statistic.

Standardized test statistic

Left-Tailed Test

2. If the alternative hypothesis H_a contains the greater-than inequality symbol $(>)$, then the hypothesis test is a **right-tailed test.**

H_0: $\mu \leq k$
H_a: $\mu > k$

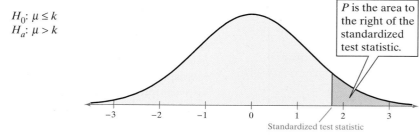

P is the area to the right of the standardized test statistic.

Standardized test statistic

Right-Tailed Test

3. If the alternative hypothesis H_a contains the not-equal-to symbol $(\neq)$, then the hypothesis test is a **two-tailed test.** In a two-tailed test, each tail has an area of $\frac{1}{2}P$.

H_0: $\mu = k$
H_a: $\mu \neq k$

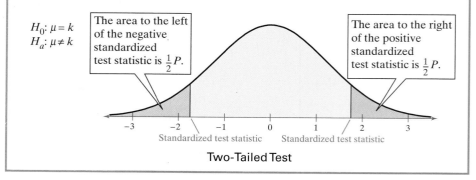

The area to the left of the negative standardized test statistic is $\frac{1}{2}P$.

The area to the right of the positive standardized test statistic is $\frac{1}{2}P$.

Standardized test statistic   Standardized test statistic

Two-Tailed Test

Study Tip

The third type of test is called a two-tailed test because evidence that would support the alternative hypothesis could lie in either tail of the sampling distribution.

The smaller the *P*-value of the test, the more evidence there is to reject the null hypothesis. A very small *P*-value indicates an unusual event. Remember, however, that even a very low *P*-value does not constitute proof that the null hypothesis is false, only that it is probably false.

EXAMPLE 3

Identifying the Nature of a Hypothesis Test

For each claim, state H_0 and H_a in words and in symbols. Then determine whether the hypothesis test is a left-tailed test, right-tailed test, or two-tailed test. Sketch a normal sampling distribution and shade the area for the *P*-value.

1. A school publicizes that the proportion of its students who are involved in at least one extracurricular activity is 61%.
2. A car dealership announces that the mean time for an oil change is less than 15 minutes.
3. A company advertises that the mean life of its furnaces is more than 18 years.

SOLUTION

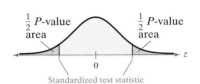

Standardized test statistic

In Symbols	*In Words*
1. $H_0: p = 0.61$	The proportion of students who are involved in at least one extracurricular activity is 61%.
$H_a: p \neq 0.61$	The proportion of students who are involved in at least one extracurricular activity is not 61%.

Because H_a contains the $\neq$ symbol, the test is a two-tailed hypothesis test. The figure at the left shows the normal sampling distribution with a shaded area for the *P*-value.

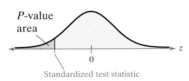

Standardized test statistic

In Symbols	*In Words*
2. $H_0: \mu \geq 15$ min	The mean time for an oil change is greater than or equal to 15 minutes.
$H_a: \mu < 15$ min	The mean time for an oil change is less than 15 minutes.

Because H_a contains the $<$ symbol, the test is a left-tailed hypothesis test. The figure at the left shows the normal sampling distribution with a shaded area for the *P*-value.

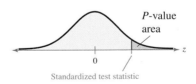

Standardized test statistic

In Symbols	*In Words*
3. $H_0: \mu \leq 18$ yr	The mean life of the furnaces is less than or equal to 18 years.
$H_a: \mu > 18$ yr	The mean life of the furnaces is more than 18 years.

Because H_a contains the $>$ symbol, the test is a right-tailed hypothesis test. The figure at the left shows the normal sampling distribution with a shaded area for the *P*-value.

TRY IT YOURSELF 3

For each claim, state H_0 and H_a in words and in symbols. Then determine whether the hypothesis test is a left-tailed test, right-tailed test, or two-tailed test. Sketch a normal sampling distribution and shade the area for the *P*-value.

1. A consumer analyst reports that the mean life of a certain type of automobile battery is not 74 months.
2. An electronics manufacturer publishes that the variance of the life of its home theater systems is less than or equal to 2.7.
3. A realtor publicizes that the proportion of homeowners who feel their house is too small for their family is more than 24%.

Answer: Page A36

Making a Decision and Interpreting the Decision

To conclude a hypothesis test, you make a decision and interpret that decision. For any hypothesis test, there are two possible outcomes: (1) reject the null hypothesis or (2) fail to reject the null hypothesis. To decide to reject H_0 or fail to reject H_0, you can use the following **decision rule.**

Study Tip

In this chapter, you will learn that there are two types of decision rules for deciding whether to reject H_0 or fail to reject H_0. The decision rule described on this page is based on *P*-values. The second type of decision rule is based on rejection regions. When the standardized test statistic falls in the rejection region, the observed probability (*P*-value) of a type I error is less than α. You will learn more about rejection regions in the next section.

Decision Rule Based on *P*-Value

To use a *P*-value to make a decision in a hypothesis test, compare the *P*-value with α.

1. If $P \leq \alpha$, then reject H_0.
2. If $P > \alpha$, then fail to reject H_0.

Failing to reject the null hypothesis does not mean that you have accepted the null hypothesis as true. It simply means that there is not enough evidence to reject the null hypothesis. To support a claim, state it so that it becomes the alternative hypothesis. To reject a claim, state it so that it becomes the null hypothesis. The table will help you interpret your decision.

	Claim	
Decision	**Claim is H_0.**	**Claim is H_a.**
Reject H_0.	There is enough evidence to reject the claim.	There is enough evidence to support the claim.
Fail to reject H_0.	There is not enough evidence to reject the claim.	There is not enough evidence to support the claim.

EXAMPLE 4

Interpreting a Decision

You perform a hypothesis test for each claim. How should you interpret your decision if you reject H_0? If you fail to reject H_0?

1. H_0 (Claim): A school publicizes that the proportion of its students who are involved in at least one extracurricular activity is 61%.
2. H_a (Claim): A car dealership announces that the mean time for an oil change is less than 15 minutes.

SOLUTION

1. The claim is represented by H_0. If you reject H_0, then you should conclude "there is enough evidence to reject the school's claim that the proportion of students who are involved in at least one extracurricular activity is 61%." If you fail to reject H_0, then you should conclude "there is not enough evidence to reject the school's claim that the proportion of students who are involved in at least one extracurricular activity is 61%."
2. The claim is represented by H_a, so the null hypothesis is "the mean time for an oil change is greater than or equal to 15 minutes." If you reject H_0, then you should conclude "there is enough evidence to support the dealership's claim that the mean time for an oil change is less than 15 minutes." If you fail to reject H_0, then you should conclude "there is not enough evidence to support the dealership's claim that the mean time for an oil change is less than 15 minutes."

TRY IT YOURSELF 4

You perform a hypothesis test for each claim. How should you interpret your decision if you reject H_0? If you fail to reject H_0?

1. A consumer analyst reports that the mean life of a certain type of automobile battery is not 74 months.

2. H_a (Claim): A realtor publicizes that the proportion of homeowners who feel their house is too small for their family is more than 24%.

Answer: Page A36

The general steps for a hypothesis test using *P*-values are summarized below. Note that when performing a hypothesis test, you should always state the null and alternative hypotheses before collecting data. You should not collect the data first and then create a hypothesis based on something unusual in the data.

Steps for Hypothesis Testing

1. State the claim mathematically and verbally. Identify the null and alternative hypotheses.

    H_0: **?** H_a: **?**

2. Specify the level of significance.

    $\alpha =$ **?**

3. Determine the standardized sampling distribution and sketch its graph.

    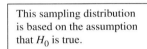

    This sampling distribution is based on the assumption that H_0 is true.

4. Calculate the test statistic and its corresponding standardized test statistic. Add it to your sketch.

    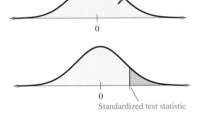

    Standardized test statistic

5. Find the *P*-value.
6. Use this decision rule.

    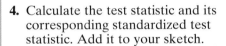

    | Is the *P*-value less than or equal to the level of significance? | No → | Fail to reject H_0. |

    Yes ↓

    Reject H_0.

7. Write a statement to interpret the decision in the context of the original claim.

In Step 4 above, the figure shows a right-tailed test. However, the same basic steps also apply to left-tailed and two-tailed tests.

Strategies for Hypothesis Testing

In a courtroom, the strategy used by an attorney depends on whether the attorney is representing the defense or the prosecution. In a similar way, the strategy that you will use in hypothesis testing should depend on whether you are trying to support or reject a claim. Remember that you cannot use a hypothesis test to support your claim when your claim is the null hypothesis. So, as a researcher, to perform a hypothesis test where the possible outcome will support a claim, word the claim so it is the alternative hypothesis. To perform a hypothesis test where the possible outcome will reject a claim, word it so the claim is the null hypothesis.

EXAMPLE 5

Writing the Hypotheses

A medical research team is investigating the benefits of a new surgical treatment. One of the claims is that the mean recovery time for patients after the new treatment is less than 96 hours.

1. How would you write the null and alternative hypotheses when you are on the research team and want to support the claim? How should you interpret a decision that rejects the null hypothesis?

2. How would you write the null and alternative hypotheses when you are on an opposing team and want to reject the claim? How should you interpret a decision that rejects the null hypothesis?

SOLUTION

1. To answer the question, first think about the context of the claim. Because you want to support this claim, make the alternative hypothesis state that the mean recovery time for patients is less than 96 hours. So, H_a: $\mu < 96$ hours. Its complement, H_0: $\mu \geq 96$ hours, would be the null hypothesis. If you reject H_0, then you will support the claim that the mean recovery time is less than 96 hours.

$$H_0: \mu \geq 96 \quad \text{and} \quad H_a: \mu < 96 \text{ (Claim)}$$

2. First think about the context of the claim. As an opposing researcher, you do not want the recovery time to be less than 96 hours. Because you want to reject this claim, make it the null hypothesis. So, H_0: $\mu \leq 96$ hours. Its complement, H_a: $\mu > 96$ hours, would be the alternative hypothesis. If you reject H_0, then you will reject the claim that the mean recovery time is less than or equal to 96 hours.

$$H_0: \mu \leq 96 \text{ (Claim)} \quad \text{and} \quad H_a: \mu > 96$$

TRY IT YOURSELF 5

1. You represent a chemical company that is being sued for paint damage to automobiles. You want to support the claim that the mean repair cost per automobile is less than $650. How would you write the null and alternative hypotheses? How should you interpret a decision that rejects the null hypothesis?

2. You are on a research team that is investigating the mean temperature of adult humans. The commonly accepted claim is that the mean temperature is about 98.6°F. You want to show that this claim is false. How would you write the null and alternative hypotheses? How should you interpret a decision that rejects the null hypothesis?

Answer: Page A36

7.1 EXERCISES

For Extra Help: MyLab Statistics

Building Basic Skills and Vocabulary

1. What are the two types of hypotheses used in a hypothesis test? How are they related?

2. Describe the two types of errors possible in a hypothesis test decision.

3. What are the two decisions that you can make from performing a hypothesis test?

4. Does failing to reject the null hypothesis mean that the null hypothesis is true? Explain.

True or False? *In Exercises 5–10, determine whether the statement is true or false. If it is false, rewrite it as a true statement.*

5. In a hypothesis test, you assume the alternative hypothesis is true.

6. A statistical hypothesis is a statement about a sample.

7. If you decide to reject the null hypothesis, then you can support the alternative hypothesis.

8. The level of significance is the maximum probability you allow for rejecting a null hypothesis when it is actually true.

9. A large P-value in a test will favor rejection of the null hypothesis.

10. To support a claim, state it so that it becomes the null hypothesis.

Stating Hypotheses *In Exercises 11–16, the statement represents a claim. Write its complement and state which is H_0 and which is H_a.*

11. $\mu \leq 645$
12. $\mu < 128$

13. $\sigma \neq 5$
14. $\sigma^2 \geq 1.2$

15. $p < 0.45$
16. $p = 0.21$

Graphical Analysis *In Exercises 17–20, match the alternative hypothesis with its graph. Then state the null hypothesis and sketch its graph.*

17. $H_a: \mu > 3$

18. $H_a: \mu < 3$

19. $H_a: \mu \neq 3$

20. $H_a: \mu > 2$

(a)
(b)
(c)
(d)

Identifying a Test *In Exercises 21–24, determine whether the hypothesis test is left-tailed, right-tailed, or two-tailed.*

21. $H_0: \mu \leq 8.0$
    $H_a: \mu > 8.0$

22. $H_0: \sigma \geq 5.2$
    $H_a: \sigma < 5.2$

23. $H_0: \sigma^2 = 142$
    $H_a: \sigma^2 \neq 142$

24. $H_0: p = 0.25$
    $H_a: p \neq 0.25$

Answers (margin column)

1. The two types of hypotheses used in a hypothesis test are the null hypothesis and the alternative hypothesis.

   The alternative hypothesis is the complement of the null hypothesis.

2. A type I error occurs if the null hypothesis is rejected when it is true.

   A type II error occurs if the null hypothesis is not rejected when it is false.

3. You can reject the null hypothesis, or you can fail to reject the null hypothesis.

4. No; Failing to reject the null hypothesis means that there is not enough evidence to reject it.

5. False. In a hypothesis test, you assume the null hypothesis is true.

6. False. A statistical hypothesis is a statement about a population.

7. True   8. True

9. False. A small P-value in a test will favor rejection of the null hypothesis.

10. False. To support a claim, state it so that it becomes the alternative hypothesis.

11. $H_0: \mu \leq 645$ (claim); $H_a: \mu > 645$

12. $H_0: \mu \geq 128$; $H_a: \mu < 128$ (claim)

13. $H_0: \sigma = 5$; $H_a: \sigma \neq 5$ (claim)

14. $H_0: \sigma^2 \geq 1.2$ (claim); $H_a: \sigma^2 < 1.2$

15. $H_0: p \geq 0.45$; $H_a: p < 0.45$ (claim)

16. $H_0: p = 0.21$ (claim); $H_a: p \neq 0.21$

17. c; $H_0: \mu \leq 3$

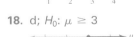

18. d; $H_0: \mu \geq 3$

19. b; $H_0: \mu = 3$

20. a; $H_0: \mu \leq 2$

21. Right-tailed   22. Left-tailed

23. Two-tailed   24. Two-tailed

25. $\mu > 8$
H_0: $\mu \leq 8$; H_a: $\mu > 8$ (claim)

26. $\sigma < 3$
H_0: $\sigma \geq 3$; H_a: $\sigma < 3$ (claim)

27. $\sigma \leq 320$
H_0: $\sigma \leq 320$ (claim); H_a: $\sigma > 320$

28. $\mu \geq 20{,}000$
H_0: $\mu \geq 20{,}000$ (claim);
H_a: $\mu < 20{,}000$

29. $p = 0.73$
H_0: $p = 0.73$ (claim);
H_a: $p \neq 0.73$

30. $p = 0.52$
H_0: $p = 0.52$ (claim);
H_a: $p \neq 0.52$

31. A type I error will occur when the actual proportion of new customers who return to buy their next textbook is at least 0.60, but you reject H_0: $p \geq 0.60$.

A type II error will occur when the actual proportion of new customers who return to buy their next textbook is less than 0.60, but you fail to reject H_0: $p \geq 0.60$.

32. A type I error will occur when the actual noontime mean traffic flow rate is 35 cars per minute, but you reject H_0: $\mu = 35$.

A type II error will occur when the actual noontime mean traffic flow rate is not 35 cars per minute, but you fail to reject H_0: $\mu = 35$.

33. A type I error will occur when the actual standard deviation of the length of time to play a game is less than or equal to 12 minutes, but you reject H_0: $\sigma \leq 12$.

A type II error will occur when the actual standard deviation of the length of time to play a game is greater than 12 minutes, but you fail to reject H_0: $\sigma \leq 12$.

34. See Selected Answers, page A99.

35. See Odd Answers, page A68.

36. See Selected Answers, page A99.

37. See Odd Answers, page A68.

38. See Selected Answers, page A100.

39. See Odd Answers, page A68.

40. See Selected Answers, page A100.

Using and Interpreting Concepts

Stating the Null and Alternative Hypotheses *In Exercises 25–30, write the claim as a mathematical statement. State the null and alternative hypotheses, and identify which represents the claim.*

25. Tablets A tablet manufacturer claims that the mean life of the battery for a certain model of tablet is more than 8 hours.

26. Shipping Errors As stated by a company's shipping department, the number of shipping errors per million shipments has a standard deviation that is less than 3.

27. Base Price of an ATV The standard deviation of the base price of an all-terrain vehicle is no more than $320.

28. Attendance An amusement park claims that the mean daily attendance at the park is at least 20,000 people.

29. Paying for College According to a recent survey, 73% of college students did not use student loans to pay for college. *(Source: Sallie Mae)*

30. Paying for College According to a recent survey, 52% of college students used their own income or savings to pay for college. *(Source: Sallie Mae)*

Identifying Type I and Type II Errors *In Exercises 31–36, describe type I and type II errors for a hypothesis test of the indicated claim.*

31. Repeat Customers A used textbook selling website claims that at least 60% of its new customers will return to buy their next textbook.

32. Flow Rate An urban planner claims that the noontime mean traffic flow rate on a busy downtown college campus street is 35 cars per minute.

33. Chess A local chess club claims that the length of time to play a game has a standard deviation of more than 12 minutes.

34. Video Game Systems A researcher claims that the percentage of adults in the United States who own a video game system is not 26%.

35. Security A campus security department publicizes that at most 25% of applicants become campus security officers.

36. Phone Repairs A cellphone repair shop advertises that the mean cost of repairing a phone screen is less than $75.

Identifying the Nature of a Hypothesis Test *In Exercises 37–42, state H_0 and H_a in words and in symbols. Then determine whether the hypothesis test is left-tailed, right-tailed, or two-tailed. Explain your reasoning. Sketch a normal sampling distribution and shade the area for the P-value.*

37. Security Alarms A security expert claims that at least 14% of all homeowners have a home security alarm.

38. Clocks A manufacturer of grandfather clocks claims that the mean time its clocks lose is no more than 0.02 second per day.

39. Golf A golf analyst claims that the standard deviation of the 18-hole scores for a golfer is less than 2.1 strokes.

40. Lung Cancer A report claims that lung cancer accounts for 25% of all cancer diagnoses. *(Source: American Cancer Society)*

41. See Odd Answers, page A68.
42. See Selected Answers, page A100.
43. See Odd Answers, page A68.
44. See Selected Answers, page A100.
45. See Odd Answers, page A68.
46. Null hypothesis
    (a) There is enough evidence to reject the automotive manufacturer's claim that the standard deviation for the gas mileage of its vehicles is 3.9 miles per gallon.
    (b) There is not enough evidence to reject the automotive manufacturer's claim that the standard deviation for the gas mileage of its vehicles is 3.9 miles per gallon.
47. Null hypothesis
    (a) There is enough evidence to reject the report's claim that at least 65% of individuals convicted of terrorism or terrorism-related offenses in the United States are foreign born.
    (b) There is not enough evidence to reject the report's claim that at least 65% of individuals convicted of terrorism or terrorism-related offenses in the United States are foreign born.
48. Null hypothesis
    (a) There is enough evidence to reject the organization's claim that none of its employees are paid minimum wage.
    (b) There is not enough evidence to reject the organization's claim that none of its employees are paid minimum wage.
49. H_0: $\mu \geq 60$; H_a: $\mu < 60$
50. H_0: $\mu = 16$; H_a: $\mu \neq 16$
51. (a) H_0: $\mu \geq 5$; H_a: $\mu < 5$
    (b) H_0: $\mu \leq 5$; H_a: $\mu > 5$
52. (a) H_0: $\mu \leq 28$; H_a: $\mu > 28$
    (b) H_0: $\mu \geq 28$; H_a: $\mu < 28$

41. High School Graduation Rate A high school claims that its mean graduation rate is more than 97%.

42. Survey A polling organization reports that the number of responses to a survey mailed to 100,000 U.S. residents is not 100,000.

Interpreting a Decision *In Exercises 43–48, determine whether the claim represents the null hypothesis or the alternative hypothesis. If a hypothesis test is performed, how should you interpret a decision that*

(a) rejects the null hypothesis?

(b) fails to reject the null hypothesis?

43. Swans A scientist claims that the mean incubation period for swan eggs is less than 40 days.

44. Affording Basic Necessities A report claims that more than 40% of households in a New York county struggle to afford basic necessities. *(Source: Niagara Frontier Publications)*

45. Lawn Mowers A researcher claims that the standard deviation of the life of a brand of lawn mower is at most 2.8 years.

46. Gas Mileage An automotive manufacturer claims that the standard deviation for the gas mileage of one of the vehicles it manufactures is 3.9 miles per gallon.

47. Terrorism Convictions A report claims that at least 65% of individuals convicted of terrorism or terrorism-related offenses in the United States are foreign born. *(Source: Hannity.com)*

48. Minimum Wage A marketing organization claims that none of its employees are paid minimum wage.

49. Writing Hypotheses: Medicine A medical research team is investigating the mean cost of a 30-day supply of a heart medication. A pharmaceutical company thinks that the mean cost is less than $60. You want to support this claim. How would you write the null and alternative hypotheses?

50. Writing Hypotheses: Transportation Network Company A transportation network company claims that the mean travel time between two destinations is about 16 minutes. You work for one of the company's competitors and want to reject this claim. How would you write the null and alternative hypotheses?

51. Writing Hypotheses: Backpack Manufacturer A backpack manufacturer claims that the mean life of its competitor's backpacks is less than 5 years. You are asked to perform a hypothesis test to test this claim. How would you write the null and alternative hypotheses when

(a) you represent the manufacturer and want to support the claim?

(b) you represent the competitor and want to reject the claim?

52. Writing Hypotheses: Internet Provider An Internet provider is trying to gain advertising deals and claims that the mean time a customer spends online per day is greater than 28 minutes. You are asked to test this claim. How would you write the null and alternative hypotheses when

(a) you represent the Internet provider and want to support the claim?

(b) you represent a competing advertiser and want to reject the claim?

53. If you decrease α, then you are decreasing the probability that you will reject H_0. Therefore, you are increasing the probability of failing to reject H_0. This could increase β, the probability of failing to reject H_0 when H_0 is false.

54. If $\alpha = 0$, then the null hypothesis cannot be rejected and the hypothesis test is useless.

55. Yes; If the P-value is less than $\alpha = 0.05$, then it is also less than $\alpha = 0.10$.

56. Not necessarily; A P-value less than $\alpha = 0.10$ may or may not also be less than $\alpha = 0.05$.

57. (a) Fail to reject H_0 because the confidence interval includes values greater than 70.

(b) Reject H_0 because the confidence interval is located entirely to the left of 70.

(c) Fail to reject H_0 because the confidence interval includes values greater than 70.

58. (a) Fail to reject H_0 because the confidence interval includes values less than 54.

(b) Fail to reject H_0 because the confidence interval includes values less than 54.

(c) Reject H_0 because the confidence interval is located entirely to the right of 54.

59. (a) Reject H_0 because the confidence interval is located entirely to the right of 0.20.

(b) Fail to reject H_0 because the confidence interval includes values less than 0.20.

(c) Fail to reject H_0 because the confidence interval includes values less than 0.20.

60. (a) Fail to reject H_0 because the confidence interval includes values greater than 0.73.

(b) Reject H_0 because the confidence interval is located entirely to the left of 0.73.

(c) Fail to reject H_0 because the confidence interval includes values greater than 0.73.

Extending Concepts

53. Getting at the Concept Why can decreasing the probability of a type I error cause an increase in the probability of a type II error?

54. Getting at the Concept Explain why a level of significance of $\alpha = 0$ is not used.

55. Writing A null hypothesis is rejected with a level of significance of 0.05. Is it also rejected at a level of significance of 0.10? Explain.

56. Writing A null hypothesis is rejected with a level of significance of 0.10. Is it also rejected at a level of significance of 0.05? Explain.

Graphical Analysis *In Exercises 57–60, you are given a null hypothesis and three confidence intervals that represent three samplings. Determine whether each confidence interval indicates that you should reject H_0. Explain your reasoning.*

57.

$H_0: \mu \geq 70$

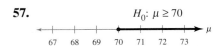

(a) $67 < \mu < 71$

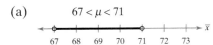

(b) $67 < \mu < 69$

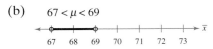

(c) $69.5 < \mu < 72.5$

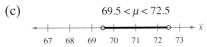

58.

$H_0: \mu \leq 54$

(a) $53.5 < \mu < 56.5$

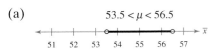

(b) $51.5 < \mu < 54.5$

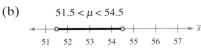

(c) $54.5 < \mu < 55.5$

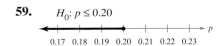

59.

$H_0: p \leq 0.20$

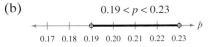

(a) $0.21 < p < 0.23$

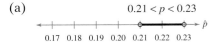

(b) $0.19 < p < 0.23$

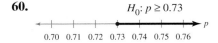

(c) $0.175 < p < 0.205$

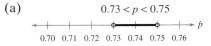

60.

$H_0: p \geq 0.73$

(a) $0.73 < p < 0.75$

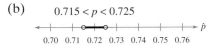

(b) $0.715 < p < 0.725$

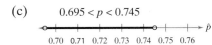

(c) $0.695 < p < 0.745$

7.2 Hypothesis Testing for the Mean (σ Known)

What You Should Learn

▶ How to find and interpret P-values

▶ How to use P-values for a z-test for a mean μ when σ is known

▶ How to find critical values and rejection regions in the standard normal distribution

▶ How to use rejection regions for a z-test for a mean μ when σ is known

Using P-Values to Make Decisions ■ Using P-Values for a z-Test ■ Rejection Regions and Critical Values ■ Using Rejection Regions for a z-Test

Using P-Values to Make Decisions

In Chapter 5, you learned that when the sample size is at least 30, the sampling distribution for $\bar{x}$ (the sample mean) is normal. In Section 7.1, you learned that a way to reach a conclusion in a hypothesis test is to use a P-value for the sample statistic, such as $\bar{x}$. Recall that when you assume the null hypothesis is true, a P-value (or probability value) of a hypothesis test is the probability of obtaining a sample statistic with a value as extreme or more extreme than the one determined from the sample data. The decision rule for a hypothesis test based on a P-value is shown below.

Decision Rule Based on P-Value

To use a P-value to make a decision in a hypothesis test, compare the P-value with α.

1. If $P \leq \alpha$, then reject H_0.
2. If $P > \alpha$, then fail to reject H_0.

Note to Instructor

If a P-value is less than 0.01, then the null hypothesis will be rejected at the common levels of $\alpha = 0.01$, $\alpha = 0.05$, and $\alpha = 0.10$. If the P-value is greater than 0.10, then you would fail to reject H_0 for these common levels. Make sure students know that the same conclusion will be reached regardless of whether they use the critical value method or the P-value method.

EXAMPLE 1

Interpreting a P-Value

The P-value for a hypothesis test is $P = 0.0237$. What is your decision when the level of significance is (1) $\alpha = 0.05$ and (2) $\alpha = 0.01$?

SOLUTION

1. Because $0.0237 < 0.05$, you reject the null hypothesis.
2. Because $0.0237 > 0.01$, you fail to reject the null hypothesis.

TRY IT YOURSELF 1

The P-value for a hypothesis test is $P = 0.0745$. What is your decision when the level of significance is (1) $\alpha = 0.05$ and (2) $\alpha = 0.10$?   *Answer: Page A37*

The lower the P-value, the more evidence there is in favor of rejecting H_0. The P-value gives you the lowest level of significance for which the sample statistic allows you to reject the null hypothesis. In Example 1, you would reject H_0 at any level of significance greater than or equal to 0.0237.

Finding the P-Value for a Hypothesis Test

After determining the hypothesis test's standardized test statistic and the standardized test statistic's corresponding area, do one of the following to find the P-value.

a. For a left-tailed test, $P = $ (Area in left tail).
b. For a right-tailed test, $P = $ (Area in right tail).
c. For a two-tailed test, $P = 2$(Area in tail of standardized test statistic).

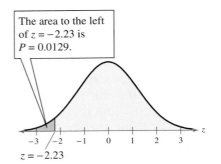

The area to the left of $z = -2.23$ is $P = 0.0129$.

$z = -2.23$

Left-Tailed Test

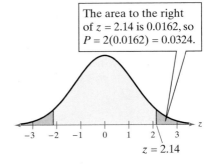

The area to the right of $z = 2.14$ is 0.0162, so $P = 2(0.0162) = 0.0324$.

$z = 2.14$

Two-Tailed Test

EXAMPLE 2

Finding a *P*-Value for a Left-Tailed Test

Find the *P*-value for a left-tailed hypothesis test with a standardized test statistic of $z = -2.23$. Decide whether to reject H_0 when the level of significance is $\alpha = 0.01$.

SOLUTION

The figure at the left shows the standard normal curve with a shaded area to the left of $z = -2.23$. For a left-tailed test,

$P = (\text{Area in left tail})$.

Using Table 4 in Appendix B, the area corresponding to $z = -2.23$ is 0.0129, which is the area in the left tail. So, the *P*-value for a left-tailed hypothesis test with a standardized test statistic of $z = -2.23$ is $P = 0.0129$. You can check your answer using technology, as shown below.

	A	B
EXCEL		
1	NORM.DIST(-2.23,0,1,TRUE)	
2		0.012873721

Interpretation   Because the *P*-value of 0.0129 is greater than 0.01, you fail to reject H_0.

TRY IT YOURSELF 2

Find the *P*-value for a left-tailed hypothesis test with a standardized test statistic of $z = -1.71$. Decide whether to reject H_0 when the level of significance is $\alpha = 0.05$.

Answer: Page A37

EXAMPLE 3

Finding a *P*-Value for a Two-Tailed Test

Find the *P*-value for a two-tailed hypothesis test with a standardized test statistic of $z = 2.14$. Decide whether to reject H_0 when the level of significance is $\alpha = 0.05$.

SOLUTION

The figure at the left shows the standard normal curve with shaded areas to the left of $z = -2.14$ and to the right of $z = 2.14$. For a two-tailed test,

$P = 2(\text{Area in tail of standardized test statistic})$.

Using Table 4, the area corresponding to $z = 2.14$ is 0.9838. The area in the right tail is $1 - 0.9838 = 0.0162$. So, the *P*-value for a two-tailed hypothesis test with a standardized test statistic of $z = 2.14$ is

$P = 2(0.0162) = 0.0324$.

Interpretation   Because the *P*-value of 0.0324 is less than 0.05, you reject H_0.

TRY IT YOURSELF 3

Find the *P*-value for a two-tailed hypothesis test with a standardized test statistic of $z = 1.64$. Decide whether to reject H_0 when the level of significance is $\alpha = 0.10$.

Answer: Page A37

Using *P*-Values for a *z*-Test

You will now learn how to perform a hypothesis test for a mean μ assuming the standard deviation σ is known. When σ is known, you can use a *z*-test for the mean. To use the *z*-test, you need to find the standardized value for the test statistic $\bar{x}$. The standardized test statistic takes the form of

$$z = \frac{(\text{Sample mean}) - (\text{Hypothesized mean})}{\text{Standard error}}.$$

z-Test for a Mean μ

The **z-test for a mean μ** is a statistical test for a population mean. The **test statistic** is the sample mean $\bar{x}$. The **standardized test statistic** is

$$z = \frac{\bar{x} - \mu}{\sigma / \sqrt{n}} \qquad \text{Standardized test statistic for } \mu \ (\sigma \text{ known})$$

when these conditions are met.

1. The sample is random.
2. At least one of the following is true: The population is normally distributed or $n \geq 30$.

Recall that $\sigma / \sqrt{n}$ is the standard error of the mean, $\sigma_{\bar{x}}$.

Note to Instructor

We use the same format for all hypothesis testing throughout the text. Using the same format makes it easier for students to understand the logic of the test. Emphasize that the sampling distribution and, consequently, the logic of the test are based on the assumption that the equality condition of the null hypothesis is true.

GUIDELINES

Using *P*-Values for a *z*-Test for a Mean μ (σ Known)

In Words	In Symbols
1. Verify that σ is known, the sample is random, and either the population is normally distributed or $n \geq 30$.	
2. State the claim mathematically and verbally. Identify the null and alternative hypotheses.	State H_0 and H_a.
3. Specify the level of significance.	Identify α.
4. Find the standardized test statistic.	$z = \dfrac{\bar{x} - \mu}{\sigma / \sqrt{n}}$
5. Find the area that corresponds to z.	Use Table 4 in Appendix B.

6. Find the *P*-value.
   a. For a left-tailed test, $P = (\text{Area in left tail})$.
   b. For a right-tailed test, $P = (\text{Area in right tail})$.
   c. For a two-tailed test, $P = 2(\text{Area in tail of standardized test statistic})$.

7. Make a decision to reject or fail to reject the null hypothesis.	If $P \leq \alpha$, then reject H_0. Otherwise, fail to reject H_0.
8. Interpret the decision in the context of the original claim.	

With all hypothesis tests, it is helpful to sketch the sampling distribution. Your sketch should include the standardized test statistic.

EXAMPLE 4

Hypothesis Testing Using a *P*-Value

In auto racing, a pit stop is where a racing vehicle stops for new tires, fuel, repairs, and other mechanical adjustments. The efficiency of a pit crew that makes these adjustments can affect the outcome of a race. A pit crew claims that its mean pit stop time (for 4 new tires and fuel) is less than 13 seconds. A random sample of 32 pit stop times has a sample mean of 12.9 seconds. Assume the population standard deviation is 0.19 second. Is there enough evidence to support the claim at $\alpha = 0.01$? Use a *P*-value.

SOLUTION

Because σ is known ($\sigma = 0.19$), the sample is random, and $n = 32 \geq 30$, you can use the *z*-test. The claim is "the mean pit stop time is less than 13 seconds." So, the null and alternative hypotheses are

$$H_0: \mu \geq 13 \text{ seconds} \quad \text{and} \quad H_a: \mu < 13 \text{ seconds.} \quad \text{(Claim)}$$

The level of significance is $\alpha = 0.01$. The standardized test statistic is

$$z = \frac{\bar{x} - \mu}{\sigma / \sqrt{n}} \qquad \text{Because } \sigma \text{ is known and } n \geq 30, \text{ use the z-test.}$$

$$= \frac{12.9 - 13}{0.19 / \sqrt{32}} \qquad \text{Assume } \mu = 13.$$

$$\approx -2.98. \qquad \text{Round to two decimal places.}$$

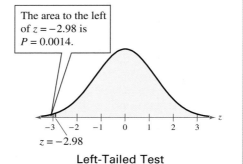

The area to the left of $z = -2.98$ is $P = 0.0014$.

$z = -2.98$

Left-Tailed Test

Using Table 4 in Appendix B, the area corresponding to $z = -2.98$ is 0.0014. Because this test is a left-tailed test, the *P*-value is equal to the area to the left of $z = -2.98$, as shown in the figure at the left. So, $P = 0.0014$. Because the *P*-value is less than $\alpha = 0.01$, you reject the null hypothesis. You can check your answer using technology, as shown below. Note that the *P*-value differs slightly from the one you found due to rounding.

STATCRUNCH

One sample Z hypothesis test:
μ : Mean of population
$H_0 : \mu = 13$
$H_A : \mu < 13$
Standard deviation = 0.19

Hypothesis test results:

Mean	n	Sample Mean	Std. Err.	Z-Stat	P-value
μ	32	12.9	0.033587572	−2.9772917	0.0015

Interpretation There is enough evidence at the 1% level of significance to support the claim that the mean pit stop time is less than 13 seconds.

TRY IT YOURSELF 4

Homeowners claim that the mean speed of automobiles traveling on their street is greater than the speed limit of 35 miles per hour. A random sample of 100 automobiles has a mean speed of 36 miles per hour. Assume the population standard deviation is 4 miles per hour. Is there enough evidence to support the claim at $\alpha = 0.05$? Use a *P*-value.

Answer: Page A37

EXAMPLE 5

See Minitab steps on page 414.

Hypothesis Testing Using a *P*-Value

According to a study of U.S. homes that use heating equipment, the mean indoor temperature at night during winter is 68.3°F. You think this information is incorrect. You randomly select 25 U.S. homes that use heating equipment in the winter and find that the mean indoor temperature at night is 67.2°F. From past studies, the population standard deviation is known to be 3.5°F and the population is normally distributed. Is there enough evidence to support your claim at $\alpha = 0.05$? Use a *P*-value. *(Adapted from U.S. Energy Information Administration)*

SOLUTION

Because σ is known ($\sigma = 3.5°F$), the sample is random, and the population is normally distributed, you can use the z-test. The claim is "the mean is different from 68.3°F." So, the null and alternative hypotheses are

$$H_0: \mu = 68.3°F \quad \text{and} \quad H_a: \mu \neq 68.3°F. \quad \text{(Claim)}$$

The level of significance is $\alpha = 0.05$. The standardized test statistic is

$$z = \frac{\bar{x} - \mu}{\sigma / \sqrt{n}}$$

Because σ is known and the population is normally distributed, use the z-test.

$$= \frac{67.2 - 68.3}{3.5 / \sqrt{25}}$$

Assume $\mu = 68.3°F$.

$$\approx -1.57.$$

Round to two decimal places.

In Table 4, the area corresponding to $z = -1.57$ is 0.0582. Because the test is a two-tailed test, the *P*-value is equal to twice the area to the left of $z = -1.57$, as shown in the figure.

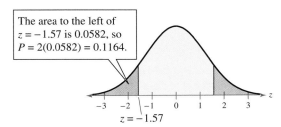

The area to the left of $z = -1.57$ is 0.0582, so $P = 2(0.0582) = 0.1164$.

$z = -1.57$

Two-Tailed Test

So, the *P*-value is $P = 2(0.0582) = 0.1164$. Because the *P*-value is greater than $\alpha = 0.05$, you fail to reject the null hypothesis.

Interpretation   There is not enough evidence at the 5% level of significance to support the claim that the mean indoor temperature at night during winter is different from 68.3°F for U.S. homes that use heating equipment.

TRY IT YOURSELF 5

According to a study of employed U.S. adults ages 18 and over, the mean number of workdays missed due to illness or injury in the past 12 months is 3.5 days. You randomly select 25 employed U.S. adults ages 18 and over and find that the mean number of workdays missed is 4 days. Assume the population standard deviation is 1.5 days and the population is normally distributed. Is there enough evidence to doubt the study's claim at $\alpha = 0.01$? Use a *P*-value. *(Adapted from U.S. National Center for Health Statistics)*

Answer: Page A37

Tech Tip

Using a TI-84 Plus, you can either enter the original data into a list to find a *P*-value or enter the descriptive statistics.

STAT

Choose the TESTS menu.

1: Z-Test...

Select the *Data* input option when you use the original data. Select the *Stats* input option when you use the descriptive statistics. In each case, enter the appropriate values including the corresponding type of hypothesis test indicated by the alternative hypothesis. Then select *Calculate*.

EXAMPLE 6

Using Technology to Find a *P*-Value

Use the TI-84 Plus displays to make a decision to reject or fail to reject the null hypothesis at a level of significance of $\alpha = 0.05$.

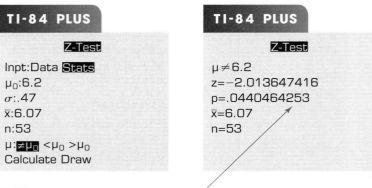

SOLUTION

The *P*-value for this test is 0.0440464253. Because the *P*-value is less than $\alpha = 0.05$, you reject the null hypothesis.

TRY IT YOURSELF 6

Repeat Example 6 using a level of significance of $\alpha = 0.01$.

Answer: Page A37

Rejection Regions and Critical Values

Another method to decide whether to reject the null hypothesis is to determine whether the standardized test statistic falls within a range of values called the **rejection region** of the sampling distribution.

DEFINITION

A **rejection region** (or **critical region**) of the sampling distribution is the range of values for which the null hypothesis is not probable. If a standardized test statistic falls in this region, then the null hypothesis is rejected. A **critical value** z_0 separates the rejection region from the nonrejection region.

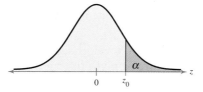

Left-Tailed Test

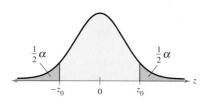

Right-Tailed Test

GUIDELINES

Finding Critical Values in the Standard Normal Distribution

1. Specify the level of significance α.
2. Determine whether the test is left-tailed, right-tailed, or two-tailed.
3. Find the critical value(s) z_0. When the hypothesis test is
   a. *left-tailed,* find the *z*-score that corresponds to an area of α.
   b. *right-tailed,* find the *z*-score that corresponds to an area of $1 - \alpha$.
   c. *two-tailed,* find the *z*-scores that correspond to $\frac{1}{2}\alpha$ and $1 - \frac{1}{2}\alpha$.
4. Sketch the standard normal distribution. Draw a vertical line at each critical value and shade the rejection region(s). (See the figures at the left.)

Note that a standardized test statistic that falls in a rejection region is considered an unusual event.

Two-Tailed Test

When you cannot find the exact area in Table 4, use the area that is closest. For an area that is exactly midway between two areas in the table, use the z-score midway between the corresponding z-scores.

EXAMPLE 7

Finding a Critical Value for a Left-Tailed Test

Find the critical value and rejection region for a left-tailed test with $\alpha = 0.01$.

SOLUTION

The figure shows the standard normal curve with a shaded area of 0.01 in the left tail. In Table 4, the z-score that is closest to an area of 0.01 is -2.33. So, the critical value is

$$z_0 = -2.33.$$

The rejection region is to the left of this critical value. You can check your answer using technology, as shown below.

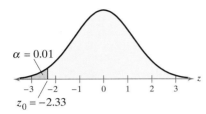

$\alpha = 0.01$

$z_0 = -2.33$

1% Level of Significance

EXCEL

	A	B
1	NORM.S.INV(0.01)	
2		−2.32634787

TRY IT YOURSELF 7

Find the critical value and rejection region for a left-tailed test with $\alpha = 0.10$.

Answer: Page A37

Because normal distributions are symmetric, in a two-tailed test the critical values are opposites, as shown in the next example.

EXAMPLE 8

Finding Critical Values for a Two-Tailed Test

Find the critical values and rejection regions for a two-tailed test with $\alpha = 0.05$.

SOLUTION

The figure shows the standard normal curve with shaded areas of $\frac{1}{2}\alpha = 0.025$ in each tail. The area to the left of $-z_0$ is $\frac{1}{2}\alpha = 0.025$, and the area to the left of z_0 is $1 - \frac{1}{2}\alpha = 0.975$. In Table 4, the z-scores that correspond to the areas 0.025 and 0.975 are -1.96 and 1.96, respectively. So, the critical values are

$$-z_0 = -1.96 \quad \text{and} \quad z_0 = 1.96.$$

The rejection regions are to the left of -1.96 and to the right of 1.96.

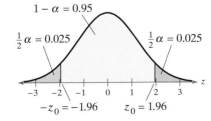

$1 - \alpha = 0.95$

$\frac{1}{2}\alpha = 0.025$ $\frac{1}{2}\alpha = 0.025$

$-z_0 = -1.96$ $z_0 = 1.96$

5% Level of Significance

TRY IT YOURSELF 8

Find the critical values and rejection regions for a two-tailed test with $\alpha = 0.08$.

Answer: Page A37

Study Tip

The table lists the critical values for commonly used levels of significance.

Alpha	Tail	z
0.10	Left	−1.28
	Right	1.28
	Two	± 1.645
0.05	Left	−1.645
	Right	1.645
	Two	± 1.96
0.01	Left	−2.33
	Right	2.33
	Two	± 2.575

Using Rejection Regions for a z-Test

To conclude a hypothesis test using rejection region(s), you make a decision and interpret the decision according to the next rule.

Decision Rule Based on Rejection Region

To use a rejection region to conduct a hypothesis test, calculate the standardized test statistic z. If the standardized test statistic

1. is in the rejection region, then reject H_0.

2. is *not* in the rejection region, then fail to reject H_0.

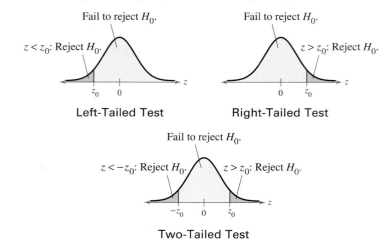

Left-Tailed Test   Right-Tailed Test

Two-Tailed Test

Remember, failing to reject the null hypothesis does not mean that you have accepted the null hypothesis as true. It simply means that there is not enough evidence to reject the null hypothesis.

GUIDELINES

Using Rejection Regions for a z-Test for a Mean μ (σ Known)

In Words	In Symbols
1. Verify that σ is known, the sample is random, and either the population is normally distributed or $n \geq 30$.	
2. State the claim mathematically and verbally. Identify the null and alternative hypotheses.	State H_0 and H_a.
3. Specify the level of significance.	Identify α.
4. Determine the critical value(s).	Use Table 4 in Appendix B.
5. Determine the rejection region(s).	
6. Find the standardized test statistic and sketch the sampling distribution.	$z = \dfrac{\bar{x} - \mu}{\sigma / \sqrt{n}}$
7. Make a decision to reject or fail to reject the null hypothesis.	If z is in the rejection region, then reject H_0. Otherwise, fail to reject H_0.
8. Interpret the decision in the context of the original claim.	

See TI-84 Plus
steps on page 415.

Picturing the World

Each year, the Environmental Protection Agency (EPA) publishes reports of gas mileage for all makes and models of passenger vehicles. In a recent year, the small station wagon with an automatic transmission that posted the best mileage had a mean mileage of 52 miles per gallon (city) and 49 miles per gallon (highway). An auto manufacturer claims its station wagons exceed 49 miles per gallon on the highway. To support its claim, it tests 36 vehicles on highway driving and obtains a sample mean of 51.2 miles per gallon. Assume the population standard deviation is 4.8 miles per gallon. (Source: U.S. Department of Energy)

Is the evidence strong enough to support the claim that the station wagon's highway miles per gallon exceeds the EPA estimate? Use a z-test with $\alpha = 0.01$.

There is enough evidence at the 1% level of significance to conclude that the mean mileage of the station wagon is greater than 49 miles per gallon on the highway.

EXAMPLE 9

Hypothesis Testing Using a Rejection Region

Employees at a construction and mining company claim that the mean salary of the company's mechanical engineers is less than that of one of its competitors, which is $88,200. A random sample of 20 of the company's mechanical engineers has a mean salary of $85,900. Assume the population standard deviation is $9500 and the population is normally distributed. At $\alpha = 0.05$, test the employees' claim.

SOLUTION

Because σ is known ($\sigma = \$9500$), the sample is random, and the population is normally distributed, you can use the z-test. The claim is "the mean salary is less than $88,200." So, the null and alternative hypotheses can be written as

$$H_0: \mu \geq \$88{,}200 \quad \text{and} \quad H_a: \mu < \$88{,}200. \text{ (Claim)}$$

Because the test is a left-tailed test and the level of significance is $\alpha = 0.05$, the critical value is $z_0 = -1.645$ and the rejection region is $z < -1.645$. The standardized test statistic is

$$z = \frac{\bar{x} - \mu}{\sigma / \sqrt{n}}$$ Because σ is known and the population is normally distributed, use the z-test.

$$= \frac{85{,}900 - 88{,}200}{9500 / \sqrt{20}}$$ Assume $\mu = \$88{,}200$.

$$\approx -1.08.$$ Round to two decimal places.

The figure shows the location of the rejection region and the standardized test statistic z. Because z is not in the rejection region, you fail to reject the null hypothesis.

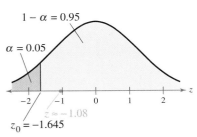

5% Level of Significance

Interpretation There is not enough evidence at the 5% level of significance to support the employees' claim that the mean salary is less than $88,200.

Be sure you understand the decision made in this example. Even though your sample has a mean of $85,900, you cannot (at a 5% level of significance) support the claim that the mean of all the mechanical engineers' salaries is less than $88,200. The difference between your test statistic ($\bar{x} = \$85{,}900$) and the hypothesized mean ($\mu = \$88{,}200$) is probably due to sampling error.

TRY IT YOURSELF 9

The CEO of the company in Example 9 claims that the mean workday of the company's mechanical engineers is less than 8.5 hours. A random sample of 25 of the company's mechanical engineers has a mean workday of 8.2 hours. Assume the population standard deviation is 0.5 hour and the population is normally distributed. At $\alpha = 0.01$, test the CEO's claim.

Answer: Page A37

EXAMPLE 10

Hypothesis Testing Using Rejection Regions

A researcher claims that the mean annual cost of raising a child (age 2 and under) by married-couple families in the U.S. is $14,050. In a random sample of married-couple families in the U.S., the mean annual cost of raising a child (age 2 and under) is $13,795. The sample consists of 500 children. Assume the population standard deviation is $2875. At $\alpha = 0.10$, is there enough evidence to reject the claim? *(Adapted from U.S. Department of Agriculture Center for Nutrition Policy and Promotion)*

SOLUTION

Because σ is known ($\sigma = \$2875$), the sample is random, and $n = 500 \geq 30$, you can use the z-test. The claim is "the mean annual cost is $14,050." So, the null and alternative hypotheses are

$$H_0: \mu = \$14,050 \text{ (Claim)} \quad \text{and} \quad H_a: \mu \neq \$14,050.$$

Because the test is a two-tailed test and the level of significance is $\alpha = 0.10$, the critical values are $-z_0 = -1.645$ and $z_0 = 1.645$. The rejection regions are $z < -1.645$ and $z > 1.645$. The standardized test statistic is

$$z = \frac{\bar{x} - \mu}{\sigma / \sqrt{n}} \qquad \text{Because } \sigma \text{ is known and } n \geq 30, \text{ use the } z\text{-test.}$$

$$= \frac{13,795 - 14,050}{2875 / \sqrt{500}} \qquad \text{Assume } \mu = \$14,050.$$

$$\approx -1.98. \qquad \text{Round to two decimal places.}$$

The figure shows the location of the rejection regions and the standardized test statistic z. Because z is in the rejection region, you reject the null hypothesis.

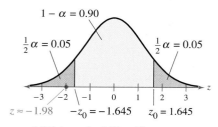

10% Level of Significance

You can check your answer using technology, as shown below.

MINITAB

One-Sample Z

Test of μ = 14050 vs ≠ 14050
The assumed standard deviation = 2875

N	Mean	SE Mean	90% CI	Z	P
500	13795	129	(13584, 14006)	-1.98	0.047

Interpretation There is enough evidence at the 10% level of significance to reject the claim that the mean annual cost of raising a child (age 2 and under) by married-couple families in the U.S. is $14,050.

TRY IT YOURSELF 10

In Example 10, at $\alpha = 0.01$, is there enough evidence to reject the claim?

Answer: Page A37

7.2 EXERCISES

Building Basic Skills and Vocabulary

1. Explain the difference between the z-test for μ using a P-value and the z-test for μ using rejection region(s).

2. In hypothesis testing, does using the critical value method or the P-value method affect your conclusion? Explain.

Interpreting a P-Value   *In Exercises 3–8, the P-value for a hypothesis test is shown. Use the P-value to decide whether to reject H_0 when the level of significance is (a) $\alpha = 0.01$, (b) $\alpha = 0.05$, and (c) $\alpha = 0.10$.*

3. $P = 0.0461$ **4.** $P = 0.0691$

5. $P = 0.1271$ **6.** $P = 0.0107$

7. $P = 0.0838$ **8.** $P = 0.0062$

Finding a P-Value   *In Exercises 9–14, find the P-value for the hypothesis test with the standardized test statistic z. Decide whether to reject H_0 for the level of significance α.*

9. Left-tailed test
$z = -1.32$
$\alpha = 0.10$

10. Left-tailed test
$z = -1.55$
$\alpha = 0.05$

11. Right-tailed test
$z = 2.46$
$\alpha = 0.01$

12. Right-tailed test
$z = 1.23$
$\alpha = 0.10$

13. Two-tailed test
$z = -1.68$
$\alpha = 0.05$

14. Two-tailed test
$z = 1.95$
$\alpha = 0.08$

Graphical Analysis   *In Exercises 15 and 16, match each P-value with the graph that displays its area without performing any calculations. Explain your reasoning.*

15. $P = 0.0089$ and $P = 0.3050$

(a) 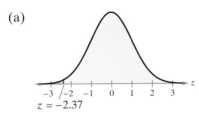   (b)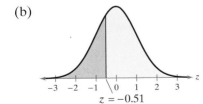

16. $P = 0.0688$ and $P = 0.2802$

(a) 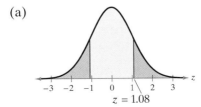   (b)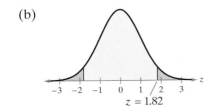

---

1. The z-test using a P-value compares the P-value with the level of significance α. In the z-test using rejection region(s), the test statistic is compared with critical values.

2. No; Both involve comparing the test statistic's probability with the level of significance. The P-value method converts the standardized test statistic to a probability (P-value) and compares this with the level of significance, whereas the critical value method converts the level of significance to a z-score and compares this with the standardized test statistic.

3. (a) Fail to reject H_0.
(b) Reject H_0.   (c) Reject H_0.

4. (a) Fail to reject H_0.
(b) Fail to reject H_0.
(c) Reject H_0.

5. (a) Fail to reject H_0.
(b) Fail to reject H_0.
(c) Fail to reject H_0.

6. (a) Fail to reject H_0.
(b) Reject H_0.   (c) Reject H_0.

7. (a) Fail to reject H_0.
(b) Fail to reject H_0.
(c) Reject H_0.

8. (a) Reject H_0.   (b) Reject H_0.
(c) Reject H_0.

9. $P = 0.0934$; Reject H_0.

10. $P = 0.0606$; Fail to reject H_0.

11. $P = 0.0069$; Reject H_0.

12. $P = 0.1093$; Fail to reject H_0.

13. $P = 0.0930$; Fail to reject H_0.

14. $P = 0.0512$; Reject H_0.

15. (a) $P = 0.0089$
(b) $P = 0.3050$
The larger P-value corresponds to the larger area.

16. (a) $P = 0.2802$
(b) $P = 0.0688$
The larger P-value corresponds to the larger area.

17. Fail to reject H_0.

18. Reject H_0.

19. Critical value: $z_0 = -1.88$
Rejection region: $z < -1.88$

20. Critical value: $z_0 = -1.34$
Rejection region: $z < -1.34$

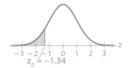

21. Critical value: $z_0 = 1.645$
Rejection region: $z > 1.645$

22. See Selected Answers, page A100.

23. See Odd Answers, page A69.

24. See Selected Answers, page A100.

25. (a) Fail to reject H_0 because $z < 1.285$.

(b) Fail to reject H_0 because $z < 1.285$.

(c) Fail to reject H_0 because $z < 1.285$.

(d) Reject H_0 because $z > 1.285$.

26. (a) Reject H_0 because $z > 1.96$.

(b) Fail to reject H_0 because $-1.96 < z < 1.96$.

(c) Fail to reject H_0 because $-1.96 < z < 1.96$.

(d) Reject H_0 because $z < -1.96$.

27. Reject H_0. There is enough evidence at the 5% level of significance to reject the claim.

28. Fail to reject H_0. There is not enough evidence at the 7% level of significance to support the claim.

29. Fail to reject H_0. There is not enough evidence at the 3% level of significance to support the claim.

30. Reject H_0. There is enough evidence at the 1% level of significance to reject the claim.

In Exercises 17 and 18, use the TI-84 Plus displays to make a decision to reject or fail to reject the null hypothesis at the level of significance.

17. $\alpha = 0.05$

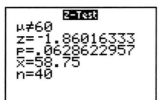

18. $\alpha = 0.01$

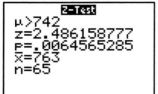

Finding Critical Values and Rejection Regions *In Exercises 19–24, find the critical value(s) and rejection region(s) for the type of z-test with level of significance α. Include a graph with your answer.*

19. Left-tailed test, $\alpha = 0.03$

20. Left-tailed test, $\alpha = 0.09$

21. Right-tailed test, $\alpha = 0.05$

22. Right-tailed test, $\alpha = 0.08$

23. Two-tailed test, $\alpha = 0.02$

24. Two-tailed test, $\alpha = 0.12$

Graphical Analysis *In Exercises 25 and 26, state whether each standardized test statistic z allows you to reject the null hypothesis. Explain your reasoning.*

25. (a) $z = -1.301$

(b) $z = 1.203$

(c) $z = 1.280$

(d) $z = 1.286$

26. (a) $z = 1.98$

(b) $z = -1.89$

(c) $z = 1.65$

(d) $z = -1.99$

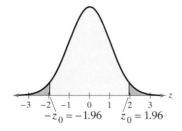

In Exercises 27–30, test the claim about the population mean μ at the level of significance α. Assume the population is normally distributed.

27. Claim: $\mu = 40$; $\alpha = 0.05$; $\sigma = 1.97$
Sample statistics: $\bar{x} = 39.2$, $n = 25$

28. Claim: $\mu \geq 1475$; $\alpha = 0.07$; $\sigma = 29$
Sample statistics: $\bar{x} = 1468$, $n = 26$

29. Claim: $\mu \neq 5880$; $\alpha = 0.03$; $\sigma = 413$
Sample statistics: $\bar{x} = 5771$, $n = 67$

30. Claim: $\mu \leq 22{,}500$; $\alpha = 0.01$; $\sigma = 1200$
Sample statistics: $\bar{x} = 23{,}500$, $n = 45$

31. (a) The claim is "the mean total score for the school's applicants is more than 499."

$H_0: \mu \leq 499$
$H_a: \mu > 499$ (claim)

(b) 2.83   (c) 0.0023

(d) Reject H_0.

(e) There is enough evidence at the 1% level of significance to support the report's claim that the mean total score for the school's applicants is more than 499.

32. (a) The claim is "the average activating temperature is at least 135°F."

$H_0: \mu \geq 135$ (claim)
$H_a: \mu < 135$

(b) −3.43   (c) 0.0003

(d) Reject H_0.

(e) There is enough evidence at the 10% level of significance to reject the manufacturer's claim that the average activating temperature is at least 135°F.

33. (a) The claim is "the mean winning times for Boston Marathon women's open division champions is at least 2.68 hours."

$H_0: \mu \geq 2.68$ (claim)
$H_a: \mu < 2.68$

(b) −1.37   (c) 0.0853

(d) Fail to reject H_0.

(e) There is not enough evidence at the 5% level of significance to reject the statistician's claim that the mean winning times for Boston Marathon women's open division champions is at least 2.68 hours.

34. See Selected Answers, page A100.

35. See Odd Answers, page A69.

36. See Selected Answers, page A100.

Using and Interpreting Concepts

Hypothesis Testing Using a *P*-Value   *In Exercises 31–36,*

(a) identify the claim and state H_0 and H_a.

(b) find the standardized test statistic z.

(c) find the corresponding P-value.

(d) decide whether to reject or fail to reject the null hypothesis.

(e) interpret the decision in the context of the original claim.

31. MCAT Scores   A random sample of 100 medical school applicants at a university has a mean total score of 502 on the MCAT. According to a report, the mean total score for the school's applicants is more than 499. Assume the population standard deviation is 10.6. At $\alpha = 0.01$, is there enough evidence to support the report's claim? *(Source: Association of American Medical Colleges)*

32. Sprinkler Systems   A manufacturer of sprinkler systems designed for fire protection claims that the average activating temperature is at least 135°F. To test this claim, you randomly select a sample of 32 systems and find the mean activation temperature to be 133°F. Assume the population standard deviation is 3.3°F. At $\alpha = 0.10$, do you have enough evidence to reject the manufacturer's claim?

33. Boston Marathon   A sports statistician claims that the mean winning times for Boston Marathon women's open division champions is at least 2.68 hours. The mean winning time of a sample of 30 randomly selected Boston Marathon women's open division champions is 2.60 hours. Assume the population standard deviation is 0.32 hour. At $\alpha = 0.05$, can you reject the claim? *(Source: Boston Athletic Association)*

34. Acceleration Times   A consumer group claims that the mean acceleration time from 0 to 60 miles per hour for a sedan is 6.3 seconds. A random sample of 33 sedans has a mean acceleration time from 0 to 60 miles per hour of 7.2 seconds. Assume the population standard deviation is 2.5 seconds. At $\alpha = 0.05$, can you reject the claim? *(Source: Zero to 60 Times)*

 35. Roller Coasters   The heights (in feet) of 36 randomly selected top-rated roller coasters are listed. Assume the population standard deviation is 71.6 feet. At $\alpha = 0.05$, is there enough evidence to reject the claim that the mean height of top-rated roller coasters is 160 feet? *(Source: POP World Media, LLC)*

325	188	306	107	208	167	105	78	140
232	230	170	170	205	305	135	200	200
100	223	135	195	80	90	120	210	82
161	245	88	70	116	121	146	149	124

36. Salaries   An analyst claims that the mean annual salary for intermediate level architects in Wichita, Kansas, is more than the national mean, $52,000. The annual salaries (in dollars) for a random sample of 21 intermediate level architects in Wichita are listed. Assume the population is normally distributed and the population standard deviation is $8000. At $\alpha = 0.09$, is there enough evidence to support the analyst's claim? *(Adapted from Salary.com)*

47,066	58,955	59,774	56,016	52,487	41,258	43,806
44,291	44,063	44,365	40,120	49,853	50,233	43,827
56,085	48,967	57,983	60,295	57,776	46,500	47,658

37. (a) The claim is "the mean
caffeine content per
12-ounce bottle of a
population of caffeinated soft
drinks is 37.7 milligrams."

H_0: $\mu = 37.7$ (claim)
H_a: $\mu \neq 37.7$

(b) $-z_0 = -2.575$, $z_0 = 2.575$
Rejection regions:
$z < -2.575$, $z > 2.575$

(c) -0.72 (d) Fail to reject H_0.

(e) There is not enough
evidence at the 1% level
of significance to reject
the consumer research
organization's claim that
the mean caffeine content
per 12-ounce bottle of a
population of caffeinated
soft drinks is 37.7 milligrams.

38. (a) The claim is "the mean high
school graduation rate per
state in the United States is
80%."

H_0: $\mu = 80$ (claim)
H_a: $\mu \neq 80$

(b) $-z_0 = -1.96$, $-z_0 = 1.96$
Rejection regions: $z < -1.96$,
$z > 1.96$

(c) 2.15 (d) Reject H_0.

(e) There is enough evidence at
the 5% level of significance
to reject the education
researcher's claim that the
mean high school graduation
rate per state in the United
States is 80%.

39. See Odd Answers, page A69.

40. See Selected Answers, page A100.

41. See Odd Answers, page A69.

Carbon dioxide emissions (in megatons)					
340	76	46	44	75	1617
34	43	23	0.5	0.3	6
0.3	0.7	11	0.1	0.2	7.6
0.6	0.6	26	9.9	2.3	8.2
3.4	0.1	472	4.2	4.2	0
113	21	7.2	5	0.1	16
0.2	45	5.1	175	0	4.1

TABLE FOR EXERCISE 42

42. See Selected Answers, page A100.

43. See Odd Answers, page A69.

44. See Selected Answers, page A100.

Hypothesis Testing Using Rejection Region(s) *In Exercises 37–42,
(a) identify the claim and state H_0 and H_a, (b) find the critical value(s) and identify
the rejection region(s), (c) find the standardized test statistic z, (d) decide whether
to reject or fail to reject the null hypothesis, and (e) interpret the decision in the
context of the original claim.*

37. Caffeine Content   A consumer research organization states that the mean
caffeine content per 12-ounce bottle of a population of caffeinated soft
drinks is 37.7 milligrams. You want to test this claim. During your tests,
you find that a random sample of thirty-six 12-ounce bottles of caffeinated
soft drinks has a mean caffeine content of 36.4 milligrams. Assume the
population standard deviation is 10.8 milligrams. At $\alpha = 0.01$, can you reject
the research organization's claim? *(Source: National Soft Drink Association)*

38. High School Graduation Rate   An education researcher claims that the
mean high school graduation rate per state in the United States is 80%. You
want to test this claim. You find that a random sample of 30 states has a
mean high school graduation rate of 82%. Assume the population standard
deviation is 5.1%. At $\alpha = 0.05$, do you have enough evidence to support the
researcher's claim? *(Source: U.S. Department of Education)*

39. Fast Food   A fast food restaurant estimates that the mean sodium content
in one of its breakfast sandwiches is no more than 920 milligrams. A
random sample of 44 breakfast sandwiches has a mean sodium content of
925 milligrams. Assume the population standard deviation is 18 milligrams.
At $\alpha = 0.10$, do you have enough evidence to reject the restaurant's claim?

40. Light Bulbs   A light bulb manufacturer guarantees that the mean life
of a certain type of light bulb is at least 750 hours. A random sample of
25 light bulbs has a mean life of 745 hours. Assume the population is
normally distributed and the population standard deviation is 60 hours. At
$\alpha = 0.02$, do you have enough evidence to reject the manufacturer's claim?

41. Fluorescent Lamps   A fluorescent lamp manufacturer guarantees that
the mean life of a fluorescent lamp is at least 10,000 hours. You want to
test this guarantee. To do so, you record the lives of a random sample
of 32 fluorescent lamps. The results (in hours) are listed. Assume the
population standard deviation is 1850 hours. At $\alpha = 0.11$, do you have
enough evidence to reject the manufacturer's claim?

8,800	9,155	13,001	10,250	10,002	11,413	8,234	10,402
10,016	8,015	6,110	11,005	11,555	9,254	6,991	12,006
10,420	8,302	8,151	10,980	10,186	10,003	8,814	11,445
6,277	8,632	7,265	10,584	9,397	11,987	7,556	10,380

42. Carbon Dioxide Emissions   A scientist estimates that the mean
carbon dioxide emissions per country in a recent year are greater than
150 megatons. You want to test this estimate. To do so, you determine
the carbon dioxide emissions for 42 randomly selected countries for that
year. The results (in megatons) are shown in the table at the left. Assume
the population standard deviation is 816 megatons. At $\alpha = 0.06$, can
you support the scientist's estimate? *(Source: Global Carbon Project)*

Extending Concepts

43. Writing   When $P > \alpha$, does the standardized test statistic lie inside or
outside of the rejection region(s)? Explain your reasoning.

44. Writing   In a right-tailed test where $P < \alpha$, does the standardized test
statistic lie to the left or the right of the critical value? Explain your reasoning.

7.3 Hypothesis Testing for the Mean (σ Unknown)

What You Should Learn

▶ How to find critical values in a t-distribution

▶ How to use the t-test to test a mean μ when σ is not known

▶ How to use technology to find P-values and use them with a t-test to test a mean μ when σ is not known

Critical Values in a *t*-Distribution ■ The *t*-Test for a Mean μ ■ Using P-Values with *t*-Tests

Critical Values in a *t*-Distribution

In Section 7.2, you learned how to perform a hypothesis test for a population mean when the population standard deviation is known. In many real-life situations, the population standard deviation in *not* known. When either the population has a normal distribution or the sample size is at least 30, you can still test the population mean μ. To do so, you can use the *t*-distribution with $n - 1$ degrees of freedom.

> **GUIDELINES**
>
> **Finding Critical Values in a *t*-Distribution**
> 1. Specify the level of significance α.
> 2. Identify the degrees of freedom, d.f. $= n - 1$.
> 3. Find the critical value(s) using Table 5 in Appendix B in the row with $n - 1$ degrees of freedom. When the hypothesis test is
>   a. *left-tailed*, use the "One Tail, α" column with a negative sign.
>   b. *right-tailed*, use the "One Tail, α" column with a positive sign.
>   c. *two-tailed*, use the "Two Tails, α" column with a negative and a positive sign.
>
> See the figures below.

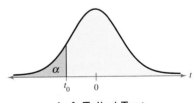

Left-Tailed Test

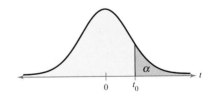

Right-Tailed Test

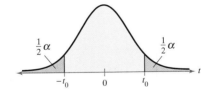

Two-Tailed Test

> **EXAMPLE 1**
>
> ### Finding a Critical Value for a Left-Tailed Test
>
> Find the critical value t_0 for a left-tailed test with $\alpha = 0.05$ and $n = 21$.
>
> **SOLUTION**
>
> The degrees of freedom are
>
> $$\text{d.f.} = n - 1 = 21 - 1 = 20.$$
>
> To find the critical value, use Table 5 in Appendix B with d.f. $= 20$ and $\alpha = 0.05$ in the "One Tail, α" column. Because the test is left-tailed, the critical value is negative. So, $t_0 = -1.725$, as shown in the figure at the left.
>
> **TRY IT YOURSELF 1**
>
> Find the critical value t_0 for a left-tailed test with $\alpha = 0.01$ and $n = 14$.
>
> *Answer: Page A37*

$\alpha = 0.05$

$t_0 = -1.725$

5% Level of Significance

EXAMPLE 2

Finding a Critical Value for a Right-Tailed Test

Find the critical value t_0 for a right-tailed test with $\alpha = 0.01$ and $n = 17$.

SOLUTION

The degrees of freedom are

$$\text{d.f.} = n - 1$$
$$= 17 - 1$$
$$= 16.$$

To find the critical value, use Table 5 with d.f. $= 16$ and $\alpha = 0.01$ in the "One Tail, α" column. Because the test is right-tailed, the critical value is positive. So,

$$t_0 = 2.583$$

as shown in the figure.

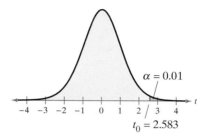

1% Level of Significance

TRY IT YOURSELF 2

Find the critical value t_0 for a right-tailed test with $\alpha = 0.10$ and $n = 9$.

Answer: Page A37

Because *t*-distributions are symmetric, in a two-tailed test the critical values are opposites, as shown in the next example.

EXAMPLE 3

Finding Critical Values for a Two-Tailed Test

Find the critical values $-t_0$ and t_0 for a two-tailed test with $\alpha = 0.10$ and $n = 26$.

SOLUTION

The degrees of freedom are

$$\text{d.f.} = n - 1$$
$$= 26 - 1$$
$$= 25.$$

To find the critical values, use Table 5 with d.f. $= 25$ and $\alpha = 0.10$ in the "Two Tails, α" column. Because the test is two-tailed, one critical value is negative and one is positive. So,

$$-t_0 = -1.708 \qquad \text{and} \qquad t_0 = 1.708$$

as shown in the figure at the left. You can check your answer using technology, as shown below.

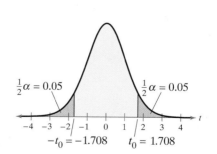

10% Level of Significance

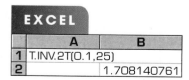

	A	B
1	T.INV.2T(0.1,25)	
2		1.708140761

TRY IT YOURSELF 3

Find the critical values $-t_0$ and t_0 for a two-tailed test with $\alpha = 0.05$ and $n = 16$.

Answer: Page A37

The *t*-Test for a Mean μ

To test a claim about a mean μ when σ is *not* known, you can use a *t*-sampling distribution. The standardized test statistic takes the form of

$$t = \frac{(\text{Sample mean}) - (\text{Hypothesized mean})}{\text{Standard error}}.$$

Because σ is not known, the standardized test statistic is calculated using the sample standard deviation s, as shown in the next definition.

t-Test for a Mean μ

The ***t*-test for a mean μ** is a statistical test for a population mean. The **test statistic** is the sample mean $\bar{x}$. The **standardized test statistic** is

$$t = \frac{\bar{x} - \mu}{s/\sqrt{n}}$$ Standardized test statistic for μ (σ unknown)

when these conditions are met.

1. The sample is random.
2. At least one of the following is true: The population is normally distributed or $n \geq 30$.

The degrees of freedom are d.f. $= n - 1$.

GUIDELINES

Using the *t*-Test for a Mean μ (σ Unknown)

In Words	In Symbols
1. Verify that σ is not known, the sample is random, and either the population is normally distributed or $n \geq 30$.	
2. State the claim mathematically and verbally. Identify the null and alternative hypotheses.	State H_0 and H_a.
3. Specify the level of significance.	Identify α.
4. Identify the degrees of freedom.	d.f. $= n - 1$
5. Determine the critical value(s).	Use Table 5 in Appendix B.
6. Determine the rejection region(s).	
7. Find the standardized test statistic and sketch the sampling distribution.	$t = \dfrac{\bar{x} - \mu}{s/\sqrt{n}}$
8. Make a decision to reject or fail to reject the null hypothesis.	If t is in the rejection region, then reject H_0. Otherwise, fail to reject H_0.
9. Interpret the decision in the context of the original claim.	

In Step 8 of the guidelines, the decision rule uses rejection regions. You can also test a claim using *P*-values, as shown on page 382. Also, when the number of degrees of freedom you need is not in Table 5, use the closest number in the table that is less than the value you need (or use technology). For instance, for d.f. $= 57$, use 50 degrees of freedom.

EXAMPLE 4

See Minitab steps on page 414.

Hypothesis Testing Using a Rejection Region

A used car dealer says that the mean price of used cars sold in the last 12 months is at least $21,000. You suspect this claim is incorrect and find that a random sample of 14 used cars sold in the last 12 months has a mean price of $19,189 and a standard deviation of $2950. Is there enough evidence to reject the dealer's claim at $\alpha = 0.05$? Assume the population is normally distributed. *(Adapted from Edmunds.com)*

SOLUTION

Because σ is unknown, the sample is random, and the population is normally distributed, you can use the *t*-test. The claim is "the mean price is at least $21,000." So, the null and alternative hypotheses are

$$H_0: \mu \geq \$21{,}000 \ \text{(Claim)}$$

and

$$H_a: \mu < \$21{,}000.$$

The test is a left-tailed test, the level of significance is $\alpha = 0.05$, and the degrees of freedom are

$$\text{d.f.} = 14 - 1 = 13.$$

So, using Table 5, the critical value is $t_0 = -1.771$. The rejection region is $t < -1.771$. The standardized test statistic is

$$t = \frac{\bar{x} - \mu}{s / \sqrt{n}}$$

Because σ is unknown and the population is normally distributed, use the *t*-test.

$$= \frac{19{,}189 - 21{,}000}{2950 / \sqrt{14}}$$

Assume $\mu = 21{,}000$.

$$\approx -2.297.$$

Round to three decimal places.

7.3 To explore this topic further, see **Activity 7.3** on page 386.

The figure shows the location of the rejection region and the standardized test statistic *t*. Because *t* is in the rejection region, you reject the null hypothesis.

Interpretation   There is enough evidence at the 5% level of significance to reject the claim that the mean price of used cars sold in the last 12 months is at least $21,000.

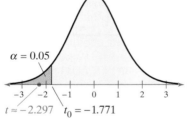

$\alpha = 0.05$

$t \approx -2.297$ $t_0 = -1.771$

5% Level of Significance

TRY IT YOURSELF 4

An industry analyst says that the mean age of a used car sold in the last 12 months is less than 4.1 years. A random sample of 25 used cars sold in the last 12 months has a mean age of 3.7 years and a standard deviation of 1.3 years. Is there enough evidence to support the analyst's claim at $\alpha = 0.10$? Assume the population is normally distributed. *(Adapted from Edmunds.com)*

Answer: Page A37

Remember that when you make a decision, the possibility of a type I or a type II error exists. For instance, in Example 4, a type I error is possible when you reject H_0, because $\mu \geq \$21{,}000$ may be true.

EXAMPLE 5

See TI-84 Plus
steps on page 415.

Hypothesis Testing Using Rejection Regions

An industrial company claims that the mean pH level of the water in a nearby river is 6.8. You randomly select 39 water samples and measure the pH of each. The sample mean and standard deviation are 6.7 and 0.35, respectively. Is there enough evidence to reject the company's claim at $\alpha = 0.05$?

SOLUTION

Because σ is unknown, the sample is random, and $n = 39 \geq 30$, you can use the *t*-test. The claim is "the mean pH level is 6.8." So, the null and alternative hypotheses are

$$H_0: \mu = 6.8 \quad \text{(Claim)} \qquad \text{and} \qquad H_a: \mu \neq 6.8.$$

The test is a two-tailed test, the level of significance is $\alpha = 0.05$, and the degrees of freedom are d.f. $= 39 - 1 = 38$. So, using Table 5, the critical values are $-t_0 = -2.024$ and $t_0 = 2.024$. The rejection regions are $t < -2.024$ and $t > 2.024$. The standardized test statistic is

$$t = \frac{\bar{x} - \mu}{s / \sqrt{n}} \qquad \text{Because } \sigma \text{ is unknown and } n \geq 30, \text{ use the } t\text{-test.}$$

$$= \frac{6.7 - 6.8}{0.35 / \sqrt{39}} \qquad \text{Assume } \mu = 6.8.$$

$$\approx -1.784. \qquad \text{Round to three decimal places.}$$

The figure shows the location of the rejection regions and the standardized test statistic t. Because t is not in the rejection region, you fail to reject the null hypothesis. You can confirm this decision using technology, as shown below. Note that the standardized statistic t differs from the one found using Table 5 due to rounding.

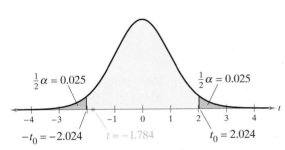

$\frac{1}{2}\alpha = 0.025$ $\frac{1}{2}\alpha = 0.025$

$-t_0 = -2.024$ $t \approx -1.784$ $t_0 = 2.024$

5% Level of Significance

MINITAB

One-Sample T

Test of $\mu = 6.8$ vs $\neq 6.8$

N	Mean	StDev	SE Mean	95% CI	T	P
39	6.7000	0.3500	0.0560	(6.5865, 6.8135)	-1.78	0.082

Interpretation There is not enough evidence at the 5% level of significance to reject the claim that the mean pH level is 6.8.

TRY IT YOURSELF 5

The company in Example 5 claims that the mean conductivity of the river is 1890 milligrams per liter. The conductivity of a water sample is a measure of the total dissolved solids in the sample. You randomly select 39 water samples and measure the conductivity of each. The sample mean and standard deviation are 2350 milligrams per liter and 900 milligrams per liter, respectively. Is there enough evidence to reject the company's claim at $\alpha = 0.01$?

Answer: Page A37

Using *P*-Values With *t*-Tests

You can also use *P*-values for a *t*-test for a mean μ. For instance, consider finding a *P*-value given $t = 1.98$, 15 degrees of freedom, and a right-tailed test. Using Table 5 in Appendix B, you can determine that *P* falls between

$$\alpha = 0.025 \quad \text{and} \quad \alpha = 0.05$$

but you cannot determine an exact value for *P*. In such cases, you can use technology to perform a hypothesis test and find exact *P*-values.

EXAMPLE 6

Using *P*-Values with a *t*-Test

A department of motor vehicles office claims that the mean wait time is less than 14 minutes. A random sample of 10 people has a mean wait time of 13 minutes with a standard deviation of 3.5 minutes. At $\alpha = 0.10$, test the office's claim. Assume the population is normally distributed.

SOLUTION

Because σ is unknown, the sample is random, and the population is normally distributed, you can use the *t*-test. The claim is "the mean wait time is less than 14 minutes." So, the null and alternative hypotheses are

H_0: $\mu \geq 14$ minutes

and

H_a: $\mu < 14$ minutes.   (Claim)

The TI-84 Plus display at the far left shows how to set up the hypothesis test. The two displays on the right show the possible results, depending on whether you select *Calculate* or *Draw*.

TI-84 PLUS

T-Test
Inpt:Data **Stats**
μ_0:14
$\bar{x}$:13
Sx:3.5
n:10
μ:$\neq\mu_0$ **<μ_0** >μ_0
Calculate Draw

TI-84 PLUS

T-Test
μ<14
t=−.9035079029
p=.1948994027
$\bar{x}$=13
Sx=3.5
n=10

TI-84 PLUS

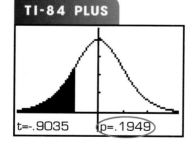

From the displays, you can see that

$P \approx 0.1949.$

Because the *P*-value is greater than $\alpha = 0.10$, you fail to reject the null hypothesis.

Interpretation There is not enough evidence at the 10% level of significance to support the office's claim that the mean wait time is less than 14 minutes.

TRY IT YOURSELF 6

Another department of motor vehicles office claims that the mean wait time is at most 18 minutes. A random sample of 12 people has a mean wait time of 15 minutes with a standard deviation of 2.2 minutes. At $\alpha = 0.05$, test the office's claim. Assume the population is normally distributed.

Answer: Page A37

7.3 EXERCISES

For Extra Help: **MyLab Statistics**

Building Basic Skills and Vocabulary

1. See Odd Answers, page A69.

2. See Selected Answers, page A100.

3. Critical value: $t_0 = -1.328$
   Rejection region: $t < -1.328$

4. Critical value: $t_0 = -2.441$
   Rejection region: $t < -2.441$

5. Critical value: $t_0 = 1.717$
   Rejection region: $t > 1.717$

6. Critical value: $t_0 = 2.457$
   Rejection region: $t > 2.457$

7. Critical values: $-t_0 = -2.056$,
   $t_0 = 2.056$
   Rejection regions: $t < -2.056$,
   $t > 2.056$

8. Critical values: $-t_0 = -1.687$,
   $t_0 = 1.687$
   Rejection regions: $t < -1.687$,
   $t > 1.687$

9. (a) Fail to reject H_0 because
       $t > -2.086$.
   (b) Fail to reject H_0 because
       $t > -2.086$.
   (c) Reject H_0 because
       $t < -2.086$.

10. (a) Fail to reject H_0 because
        $t < 1.402$.
    (b) Reject H_0 because $t > 1.402$.
    (c) Fail to reject H_0 because
        $t < 1.402$.

11. (a) Reject H_0 because
        $t < -1.725$.
    (b) Fail to reject H_0 because
        $-1.725 < t < 1.725$.
    (c) Reject H_0 because $t > 1.725$.

12. (a) Reject H_0 because
        $t < -1.071$.
    (b) Fail to reject H_0 because
        $-1.071 < t < 1.071$.
    (c) Reject H_0 because $t > 1.071$.

13. Fail to reject H_0. There is not
    enough evidence at the 1%
    level of significance to reject the
    claim.

14. See Selected Answers, page A100.

15. See Odd Answers, page A69.

16. See Selected Answers, page A100.

17. See Odd Answers, page A69.

18. See Selected Answers, page A100.

1. Explain how to find critical values for a t-distribution.

2. Explain how to use a t-test to test a hypothesized mean μ when σ is unknown. What assumptions are necessary?

In Exercises 3–8, find the critical value(s) and rejection region(s) for the type of t-test with level of significance α and sample size n.

3. Left-tailed test, $\alpha = 0.10, n = 20$

4. Left-tailed test, $\alpha = 0.01, n = 35$

5. Right-tailed test, $\alpha = 0.05, n = 23$

6. Right-tailed test, $\alpha = 0.01, n = 31$

7. Two-tailed test, $\alpha = 0.05, n = 27$

8. Two-tailed test, $\alpha = 0.10, n = 38$

Graphical Analysis *In Exercises 9–12, state whether each standardized test statistic t allows you to reject the null hypothesis. Explain.*

9. (a) $t = 2.091$
    (b) $t = 0$
    (c) $t = -2.096$

10. (a) $t = 1.4$
     (b) $t = 1.42$
     (c) $t = -1.402$

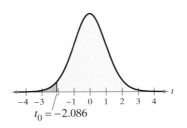

11. (a) $t = -1.755$
     (b) $t = -1.585$
     (c) $t = 1.745$

12. (a) $t = -1.1$
     (b) $t = 1.01$
     (c) $t = 1.7$

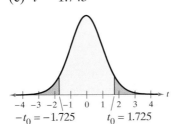

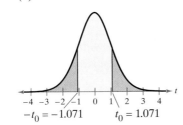

In Exercises 13–18, test the claim about the population mean μ at the level of significance α. Assume the population is normally distributed.

13. Claim: $\mu = 15$; $\alpha = 0.01$. Sample statistics: $\bar{x} = 13.9, s = 3.23, n = 36$

14. Claim: $\mu > 25$; $\alpha = 0.05$. Sample statistics: $\bar{x} = 26.2, s = 2.32, n = 17$

15. Claim: $\mu \geq 8000$; $\alpha = 0.01$. Sample statistics: $\bar{x} = 7700, s = 450, n = 25$

16. Claim: $\mu \leq 1600$; $\alpha = 0.02$. Sample statistics: $\bar{x} = 1550, s = 165, n = 46$

17. Claim: $\mu < 4915$; $\alpha = 0.02$. Sample statistics: $\bar{x} = 5017, s = 5613, n = 51$

18. Claim: $\mu \neq 52{,}200$; $\alpha = 0.05$. Sample statistics: $\bar{x} = 53{,}220, s = 2700, n = 34$

Using and Interpreting Concepts

Hypothesis Testing Using Rejection Regions *In Exercises 19–26, (a) identify the claim and state H_0 and H_a, (b) find the critical value(s) and identify the rejection region(s), (c) find the standardized test statistic t, (d) decide whether to reject or fail to reject the null hypothesis, and (e) interpret the decision in the context of the original claim. Assume the population is normally distributed.*

19. Used Car Cost A used car dealer says that the mean price of a three-year-old sport utility vehicle (in good condition) is $20,000. You suspect this claim is incorrect and find that a random sample of 22 similar vehicles has a mean price of $20,640 and a standard deviation of $1990. Is there enough evidence to reject the claim at $\alpha = 0.05$?

20. DMV Wait Times A state Department of Transportation claims that the mean wait time for various services at its different locations is at most 6 minutes. A random sample of 34 services at different locations has a mean wait time of 10.3 minutes and a standard deviation of 8.0 minutes. Is there enough evidence to reject the claim at $\alpha = 0.01$?

21. Credit Card Debt A credit reporting agency claims that the mean credit card debt by state is greater than $5500 per person. You want to test this claim. You find that a random sample of 30 states has a mean credit card debt of $5594 per person and a standard deviation of $597 per person. At $\alpha = 0.05$, can you support the claim? *(Adapted from TransUnion)*

22. Battery Life A company claims that the mean battery life of their MP3 player is at least 30 hours. You suspect this claim is incorrect and find that a random sample of 18 MP3 players has a mean battery life of 28.5 hours and a standard deviation of 1.7 hours. Is there enough evidence to reject the claim at $\alpha = 0.01$?

23. Carbon Monoxide Levels As part of your work for an environmental awareness group, you want to test a claim that the mean amount of carbon monoxide in the air in U.S. cities is less than 2.34 parts per million. You find that the mean amount of carbon monoxide in the air for a random sample of 64 U.S. cities is 2.37 parts per million and the standard deviation is 2.11 parts per million. At $\alpha = 0.10$, can you support the claim? *(Adapted from U.S. Environmental Protection Agency)*

24. Lead Levels As part of your work for an environmental awareness group, you want to test a claim that the mean amount of lead in the air in U.S. cities is less than 0.036 microgram per cubic meter. You find that the mean amount of lead in the air for a random sample of 56 U.S. cities is 0.039 microgram per cubic meter and the standard deviation is 0.069 microgram per cubic meter. At $\alpha = 0.01$, can you support the claim? *(Adapted from U.S. Environmental Protection Agency)*

25. Annual Salary An employment information service claims the mean annual salary for senior level product engineers is $98,000. The annual salaries (in dollars) for a random sample of 16 senior level product engineers are shown in the table at the left. At $\alpha = 0.05$, test the claim that the mean salary is $98,000. *(Adapted from Salary.com)*

26. Annual Salary An employment information service claims the mean annual salary for home care physical therapists is more than $80,000. The annual salaries (in dollars) for a random sample of 12 home care physical therapists are shown in the table at the left. At $\alpha = 0.10$, is there enough evidence to support the claim that the mean salary is more than $80,000? *(Adapted from Salary.com)*

---

19. (a) The claim is "the mean price of a three-year-old sport utility vehicle (in good condition) is $20,000."

H_0: $\mu = 20,000$ (claim)
H_a: $\mu \neq 20,000$

(b) $-t_0 = -2.080$, $t_0 = 2.080$
Rejection regions:
$t < -2.080$, $t > 2.080$

(c) 1.51   **(d)** Fail to reject H_0.

(e) There is not enough evidence at the 5% level of significance to reject the claim that the mean price of a three-year-old sport utility vehicle (in good condition) is $20,000.

20. (a) The claim is "the mean wait time for various services at different locations is at most 6 minutes."

H_0: $\mu \leq 6$ (claim); H_a: $\mu > 6$

(b) $t_0 = 2.445$
Rejection region: $t > 2.445$

(c) 3.13   **(d)** Reject H_0.

(e) There is enough evidence at the 1% level of significance to reject the state Department of Transportation's claim that the mean wait time for various services at different locations is at most 6 minutes.

21. See Odd Answers, page A70.

22. See Selected Answers, page A100.

23. See Odd Answers, page A70.

24. See Selected Answers, page A101.

25. See Odd Answers, page A70.

26. See Selected Answers, page A101.

Annual salaries

100,651	82,505	102,450	91,091
96,309	74,193	76,184	82,088
93,551	77,012	104,020	85,063
112,717	80,970	103,982	110,316

TABLE FOR EXERCISE 25

Annual salaries

89,245	86,013	83,151	69,771
87,834	67,964	76,523	90,268
90,440	93,538	76,999	68,257

TABLE FOR EXERCISE 26

Using a *P*-Value with a *t*-Test *In Exercises 27–30, (a) identify the claim and state H_0 and H_a, (b) use technology to find the P-value, (c) decide whether to reject or fail to reject the null hypothesis, and (d) interpret the decision in the context of the original claim. Assume the population is normally distributed.*

27. Quarter Mile Times A consumer group claims that the mean minimum time it takes for a sedan to travel a quarter mile is greater than 14.7 seconds. A random sample of 22 sedans has a mean minimum time to travel a quarter mile of 15.4 seconds and a standard deviation of 2.10 seconds. At $\alpha = 0.10$, do you have enough evidence to support the consumer group's claim? *(Adapted from Zero to 60 Times)*

28. Dive Duration An oceanographer claims that the mean dive duration of a North Atlantic right whale is 11.5 minutes. A random sample of 34 dive durations has a mean of 12.2 minutes and a standard deviation of 2.2 minutes. Is there enough evidence to reject the claim at $\alpha = 0.10$? *(Source: Marine Ecology Progress Series)*

29. Class Size You receive a brochure from a large university. The brochure indicates that the mean class size for full-time faculty is fewer than 32 students. You want to test this claim. You randomly select 18 classes taught by full-time faculty and determine the class size of each. The results are shown in the table at the left. At $\alpha = 0.05$, can you support the university's claim?

Class sizes					
35	28	29	33	32	40
26	25	29	28	30	36
33	29	27	30	28	25

TABLE FOR EXERCISE 29

30. Faculty Classroom Hours The dean of a university estimates that the mean number of classroom hours per week for full-time faculty is 11.0. As a member of the student council, you want to test this claim. A random sample of the number of classroom hours for eight full-time faculty for one week is shown in the table at the left. At $\alpha = 0.01$, can you reject the dean's claim?

Classroom hours			
11.8	8.6	12.6	7.9
6.4	10.4	13.6	9.1

TABLE FOR EXERCISE 30

Extending Concepts

Deciding on a Distribution *In Exercises 31 and 32, decide whether you should use the standard normal sampling distribution or a t-sampling distribution to perform the hypothesis test. Justify your decision. Then use the distribution to test the claim. Write a short paragraph about the results of the test and what you can conclude about the claim.*

31. Gas Mileage A car company claims that the mean gas mileage for its luxury sedan is at least 23 miles per gallon. You believe the claim is incorrect and find that a random sample of 5 cars has a mean gas mileage of 22 miles per gallon and a standard deviation of 4 miles per gallon. At $\alpha = 0.05$, test the company's claim. Assume the population is normally distributed.

32. Tuition and Fees An education publication claims that the mean in-state tuition and fees at public four-year institutions by state is more than $9000 per year. A random sample of 30 states has a mean in-state tuition and fees at public four-year institutions of $9231 per year. Assume the population standard deviation is $2380. At $\alpha = 0.01$, test the publication's claim. *(Adapted from The College Board)*

33. Writing You are testing a claim and incorrectly use the standard normal sampling distribution instead of the *t*-sampling distribution. Does this make it more or less likely to reject the null hypothesis? Is this result the same no matter whether the test is left-tailed, right-tailed, or two-tailed? Explain your reasoning.

27. (a) The claim is "the mean minimum time it takes for a sedan to travel a quarter mile is greater than 14.7 seconds."
H_0: $\mu \le 14.7$
H_a: $\mu > 14.7$ (claim)
(b) 0.0664 (c) Reject H_0.
(d) There is enough evidence at the 10% level of significance to support the consumer group's claim that the mean minimum time it takes for a sedan to travel a quarter mile is greater than 14.7 seconds.

28. (a) The claim is "the mean dive duration of a North Atlantic right whale is 11.5 minutes."
H_0: $\mu = 11.5$ (claim)
H_a: $\mu \ne 11.5$
(b) 0.0725 (c) Reject H_0.
(d) There is enough evidence at the 10% level of significance to reject the oceanographer's claim that the mean dive duration of a North Atlantic right whale is 11.5 minutes.

29. See Odd Answers, page A70.
30. See Selected Answers, page A101.
31. See Odd Answers, page A70.
32. See Selected Answers, page A101.
33. See Odd Answers, page A70.

APPLET

You can find the interactive applet for this activity within MyLab Statistics or at *www.pearsonhighered.com/ mathstatsresources.*

The *hypothesis tests for a mean* applet allows you to visually investigate hypothesis tests for a mean. You can specify the sample size n, the shape of the distribution (Normal or Right skewed), the true population mean (Mean), the true population standard deviation (Std. Dev.), the null value for the mean (Null mean), and the alternative for the test (Alternative). When you click SIMULATE, 100 separate samples of size n will be selected from a population with these population parameters. For each of the 100 samples, a hypothesis test based on the T statistic is performed, and the results from each test are displayed in the plots at the right. The test statistic for each test is shown in the top plot and the P-value is shown in the bottom plot. The green and blue lines represent the cutoffs for rejecting the null hypothesis with the 0.05 and 0.01 level tests, respectively. Additional simulations can be carried out by clicking SIMULATE multiple times. The cumulative number of times that each test rejects the null hypothesis is also shown. Press CLEAR to clear existing results and start a new simulation.

EXPLORE

Step 1 Specify a value for n.
Step 2 Specify a distribution.
Step 3 Specify a value for the mean.
Step 4 Specify a value for the standard deviation.
Step 5 Specify a value for the null mean.
Step 6 Specify an alternative hypothesis.
Step 7 Click SIMULATE to generate the hypothesis tests.

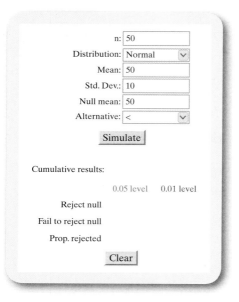

DRAW CONCLUSIONS

APPLET

1. Set $n = 15$, Mean = 40, Std. Dev. = 5, and the distribution to "Normal." Test the claim that the mean is equal to 40. What are the null and alternative hypotheses? Run the simulation so that at least 1000 hypothesis tests are run. Compare the proportion of null hypothesis rejections for the 0.05 level and the 0.01 level. Is this what you would expect? Explain.

2. Suppose a null hypothesis is rejected at the 0.01 level. Will it be rejected at the 0.05 level? Explain. Suppose a null hypothesis is rejected at the 0.05 level. Will it be rejected at the 0.01 level? Explain.

3. Set $n = 25$, Mean = 25, Std. Dev. = 3, and the distribution to "Normal." Test the claim that the mean is at least 27. What are the null and alternative hypotheses? Run the simulation so that at least 1000 hypothesis tests are run. Compare the proportion of null hypothesis rejections for the 0.05 level and the 0.01 level. Is this what you would expect? Explain.

Human Body Temperature: What's Normal?

In an article in the *Journal of Statistics Education* (vol. 4, no. 2), Allen Shoemaker describes a study that was reported in the Journal of the American Medical Association (JAMA).* It is generally accepted that the mean body temperature of an adult human is 98.6°F. In his article, Shoemaker uses the data from the JAMA article to test this hypothesis. Here is a summary of his test.

Claim: The body temperature of adults is 98.6°F.

$$H_0: \mu = 98.6°F \text{ (Claim)} \qquad H_a: \mu \neq 98.6°F$$

Sample Size: $n = 130$

Population: Adult human temperatures (Fahrenheit)

Distribution: Approximately normal

Test Statistics: $\bar{x} \approx 98.25, s \approx 0.73$

* Data for the JAMA article were collected from healthy men and women, ages 18 to 40, at the University of Maryland Center for Vaccine Development, Baltimore.

Men's Temperatures (in degrees Fahrenheit)

```
 96 | 3
 96 | 7 9
 97 | 0 1 1 1 2 3 4 4 4 4
 97 | 5 5 6 6 6 7 8 8 8 8 9 9
 98 | 0 0 0 0 0 0 1 1 2 2 2 2 3 3 4 4 4 4
 98 | 5 5 6 6 6 6 6 6 7 7 8 8 8 9
 99 | 0 0 0 1 2 3 4
 99 | 5
100 |
100 |          Key: 96|3 = 96.3
```

Women's Temperatures (in degrees Fahrenheit)

```
 96 | 4
 96 | 7 8
 97 | 2 2 4
 97 | 6 7 7 8 8 8 9 9 9
 98 | 0 0 0 0 0 1 2 2 2 2 2 2 3 3 3 4 4 4 4 4
 98 | 5 6 6 6 6 7 7 7 7 7 7 8 8 8 8 8 8 8 9
 99 | 0 0 1 1 2 2 3 4
 99 | 9
100 | 0
100 | 8          Key: 96|4 = 96.4
```

EXERCISES

1. Complete the hypothesis test for all adults (men and women) by performing the following steps. Use a level of significance of $\alpha = 0.05$.

   (a) Sketch the sampling distribution.

   (b) Determine the critical values and add them to your sketch.

   (c) Determine the rejection regions and shade them in your sketch.

   (d) Find the standardized test statistic. Plot and label it in your sketch.

   (e) Make a decision to reject or fail to reject the null hypothesis.

   (f) Interpret the decision in the context of the original claim.

2. If you lower the level of significance to $\alpha = 0.01$, does your decision change? Explain your reasoning.

3. Test the hypothesis that the mean temperature of men is 98.6°F. What can you conclude at a level of significance of $\alpha = 0.01$?

4. Test the hypothesis that the mean temperature of women is 98.6°F. What can you conclude at a level of significance of $\alpha = 0.01$?

5. Use the sample of 130 temperatures to form a 99% confidence interval for the mean body temperature of adult humans.

6. The conventional "normal" body temperature was established by Carl Wunderlich over 100 years ago. What were possible sources of error in Wunderlich's sampling procedure?

7.4 Hypothesis Testing for Proportions

What You Should Learn

▶ How to use the z-test to test a population proportion p

Hypothesis Test for Proportions

Hypothesis Test for Proportions

In Sections 7.2 and 7.3, you learned how to perform a hypothesis test for a population mean μ. In this section, you will learn how to test a population proportion p.

Hypothesis tests for proportions can be used when politicians want to know the proportion of their constituents who favor a certain bill or when quality assurance engineers test the proportion of parts that are defective.

If $np \geq 5$ and $nq \geq 5$ for a binomial distribution, then the sampling distribution for $\hat{p}$ is approximately normal with a mean of $\mu_{\hat{p}} = p$ and a standard error of

$$\sigma_{\hat{p}} = \sqrt{pq/n}.$$

z-Test for a Proportion p

The **z-test for a proportion p** is a statistical test for a population proportion. The z-test can be used when a binomial distribution is given such that $np \geq 5$ and $nq \geq 5$. The **test statistic** is the sample proportion $\hat{p}$ and the **standardized test statistic** is

$$z = \frac{\hat{p} - \mu_{\hat{p}}}{\sigma_{\hat{p}}} = \frac{\hat{p} - p}{\sqrt{pq/n}}. \qquad \text{Standardized test statistic for } p$$

GUIDELINES

Using a z-Test for a Proportion p

In Words	In Symbols
1. Verify that the sampling distribution of $\hat{p}$ can be approximated by a normal distribution.	$np \geq 5, nq \geq 5$
2. State the claim mathematically and verbally. Identify the null and alternative hypotheses.	State H_0 and H_a.
3. Specify the level of significance.	Identify α.
4. Determine the critical value(s).	Use Table 4 in Appendix B.
5. Determine the rejection region(s).	
6. Find the standardized test statistic and sketch the sampling distribution.	$z = \dfrac{\hat{p} - p}{\sqrt{pq/n}}.$
7. Make a decision to reject or fail to reject the null hypothesis.	If z is in the rejection region, then reject H_0. Otherwise, fail to reject H_0.
8. Interpret the decision in the context of the original claim.	

Study Tip

A hypothesis test for a proportion p can also be performed using P-values. Use the guidelines on page 365 for using P-values for a z-test for a mean μ, but in Step 4 find the standardized test statistic by using the formula

$$z = \frac{\hat{p} - p}{\sqrt{pq/n}}.$$

The other steps in the test are the same.

In Step 7 of the guidelines, the decision rule uses rejection regions. You can also test a claim using P-values, as shown in the Study Tip at the left.

7.4 To explore this topic further, see **Activity 7.4** on page 393.

EXAMPLE 1

See TI-84 Plus steps on page 415.

Hypothesis Test for a Proportion

A researcher claims that less than 45% of U.S. adults use passwords that are less secure because complicated ones are too hard to remember. In a random sample of 100 adults, 41% say they use passwords that are less secure because complicated ones are too hard to remember. At $\alpha = 0.01$, is there enough evidence to support the researcher's claim? *(Adapted from Pew Research Center)*

SOLUTION

The products $np = 100(0.45) = 45$ and $nq = 100(0.55) = 55$ are both greater than 5. So, you can use a z-test. The claim is "less than 45% of U.S. adults use passwords that are less secure because complicated ones are too hard to remember." So, the null and alternative hypotheses are

$$H_0: p \geq 0.45 \quad \text{and} \quad H_a: p < 0.45. \text{ (Claim)}$$

Because the test is a left-tailed test and the level of significance is $\alpha = 0.01$, the critical value is $z_0 = -2.33$ and the rejection region is $z < -2.33$. The standardized test statistic is

$$z = \frac{\hat{p} - p}{\sqrt{pq/n}} \qquad \text{Because } np \geq 5 \text{ and } n \geq 5, \text{ you can use the } z\text{-test.}$$

$$= \frac{0.41 - 0.45}{\sqrt{(0.45)(0.55)/100}} \qquad \text{Assume } p = 0.45.$$

$$\approx -0.80. \qquad \text{Round to two decimal places.}$$

The figure shows the location of the rejection region and the standardized test statistic z. Because z is not in the rejection region, you fail to reject the null hypothesis.

Interpretation There is not enough evidence at the 1% level of significance to support the claim that less than 45% of U.S. adults use passwords that are less secure because complicated ones are too hard to remember.

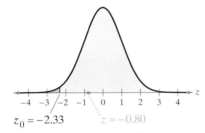

$z_0 = -2.33$ $z \approx -0.80$

1% Level of Significance

Study Tip

Remember that when you fail to reject H_0, a type II error is possible. For instance, in Example 1 the null hypothesis, $p \geq 0.45$, may be false.

TRY IT YOURSELF 1

A researcher claims that more than 90% of U.S. adults have access to a smartphone. In a random sample of 150 adults, 87% say they have access to a smartphone. At $\alpha = 0.01$, is there enough evidence to support the researcher's claim? *(Adapted from Nielsen Mobile Insights)*

Answer: Page A37

To use a P-value to perform the hypothesis test in Example 1, you can use technology, as shown at the right, or you can use Table 4. Using Table 4, the area corresponding to $z = -0.80$ is 0.2119. Because this is a left-tailed test, the P-value is equal to the area to the left of $z = -0.80$. So, $P = 0.2119$. (This value differs from the one found using technology due to rounding.) Because the P-value is greater than $\alpha = 0.01$, you fail to reject the null hypothesis. Note that this is the same result obtained in Example 1.

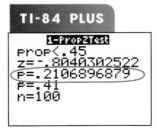

```
TI-84 PLUS

1-PropZTest
prop<.45
z=-.8040302522
p=.2106896879
p̂=.41
n=100
```

Picturing the World

According to a survey, at least 35% of smartphone owners say the first thing they access on their phones each day is texts or instant messages. To test this claim, you randomly select 300 smartphone owners. In the sample, you find that 93 of them say the first thing they access on their phones each day is texts or instant messages. (Adapted from Deloitte's 2016 Global Mobile Consumer Survey: U.S. edition)

At $\alpha = 0.05$, is there enough evidence to reject the claim?

No, there is not enough evidence at the 5% level of significance to reject the claim that at least 35% of smartphone owners say the first thing they access on their phones each day is texts or instant messages.

Recall from Section 6.3 that when the sample proportion is not given, you can find it using the formula

$$\hat{p} = \frac{x}{n} \qquad \text{Sample proportion}$$

where x is the number of successes in the sample and n is the sample size.

EXAMPLE 2

See Minitab steps on page 414.

Hypothesis Test for a Proportion

A researcher claims that 51% of U.S. adults believe, incorrectly, that antibiotics are effective against viruses. In a random sample of 2202 adults, 1161 say antibiotics are effective against viruses. At $\alpha = 0.10$, is there enough evidence to support the researcher's claim? *(Source: HealthDay/Harris Poll)*

SOLUTION

The products $np = 2202(0.51) \approx 1123$ and $nq = 2202(0.49) \approx 1079$ are both greater than 5. So, you can use a z-test. The claim is "51% of U.S. adults believe, incorrectly, that antibiotics are effective against viruses." So, the null and alternative hypotheses are

$$H_0: p = 0.51 \text{ (Claim)} \qquad \text{and} \qquad H_a: p \neq 0.51.$$

Because the test is a two-tailed test and the level of significance is $\alpha = 0.10$, the critical values are $-z_0 = -1.645$ and $z_0 = 1.645$. The rejection regions are $z < -1.645$ and $z > 1.645$. Because the number of successes is $x = 1161$ and $n = 2202$, the sample proportion is

$$\hat{p} = \frac{x}{n} = \frac{1161}{2202} \approx 0.527.$$

The standardized test statistic is

$$
\begin{aligned}
z &= \frac{\hat{p} - p}{\sqrt{pq/n}} && \text{Because } np \geq 5 \text{ and } nq \geq 5, \\
&&& \text{you can use the } z\text{-test.} \\
&= \frac{0.527 - 0.51}{\sqrt{(0.51)(0.49)/2202}} && \text{Assume } p = 0.51. \\
&\approx 1.60. && \text{Round to two decimal places.}
\end{aligned}
$$

The figure shows the location of the rejection regions and the standardized test statistic z. Because z is not in the rejection region, you fail to reject the null hypothesis.

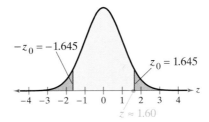

10% Level of Significance

Interpretation There is not enough evidence at the 10% level of significance to reject the claim that 51% of U.S. adults believe, incorrectly, that antibiotics are effective against viruses.

TRY IT YOURSELF 2

A researcher claims that 67% of U.S. adults believe that doctors prescribing antibiotics for viral infections for which antibiotics are not effective is a significant cause of drug-resistant superbugs. (Superbugs are bacterial infections that are resistant to many or all antibiotics.) In a random sample of 1768 adults, 1150 say they believe that doctors prescribing antibiotics for viral infections for which antibiotics are not effective is a significant cause of drug-resistant superbugs. At $\alpha = 0.10$, is there enough evidence to support the researcher's claim? *(Source: HealthDay/Harris Poll)*   *Answer: Page A37*

7.4 EXERCISES

Answers column (left):

1. If $np \geq 5$ and $nq \geq 5$, then the normal distribution can be used.

2. Verify that $np \geq 5$ and $nq \geq 5$. State H_0 and H_a. Specify the level of significance α. Determine the critical value(s) and rejection region(s). Find the standardized test statistic. Make a decision and interpret it in the context of the original claim.

3. Cannot use normal distribution.

4. Can use normal distribution.

   Reject H_0. There is enough evidence at the 8% level of significance to reject the claim.

5. Can use normal distribution.

   Fail to reject H_0. There is not enough evidence at the 5% level of significance to support the claim.

6. Can use normal distribution.

   Fail to reject H_0. There is not enough evidence at the 4% level of significance to support the claim.

7. (a) The claim is "less than 80% of U.S. adults think that healthy children should be required to be vaccinated."

   H_0: $p \geq 0.80$
   H_a: $p < 0.80$ (claim)

   (b) $z_0 = -1.645$
   Rejection region: $z < -1.645$

   (c) 0.707   (d) Fail to reject H_0.

   (e) There is not enough evidence at the 5% level of significance to support the medical researcher's claim that less than 80% of U.S. adults think that healthy children should be required to be vaccinated.

8. See Selected Answers, page A101.

9. See Odd Answers, page A70.

10. See Selected Answers, page A101.

11. See Odd Answers, page A70.

12. See Selected Answers, page A101.

Building Basic Skills and Vocabulary

1. Explain how to determine whether a normal distribution can be used to approximate a binomial distribution.

2. Explain how to test a population proportion p.

In Exercises 3–6, determine whether a normal sampling distribution can be used. If it can be used, test the claim.

3. Claim: $p < 0.12$; $\alpha = 0.01$. Sample statistics: $\hat{p} = 0.10$, $n = 40$

4. Claim: $p \geq 0.48$; $\alpha = 0.08$. Sample statistics: $\hat{p} = 0.40$, $n = 90$

5. Claim: $p \neq 0.15$; $\alpha = 0.05$. Sample statistics: $\hat{p} = 0.12$, $n = 500$

6. Claim: $p > 0.70$; $\alpha = 0.04$. Sample statistics: $\hat{p} = 0.64$, $n = 225$

Using and Interpreting Concepts

Hypothesis Testing Using Rejection Regions *In Exercises 7–12, (a) identify the claim and state H_0 and H_a, (b) find the critical value(s) and identify the rejection region(s), (c) find the standardized test statistic z, (d) decide whether to reject or fail to reject the null hypothesis, and (e) interpret the decision in the context of the original claim.*

7. **Vaccination Requirement** A medical researcher says that less than 80% of U.S. adults think that healthy children should be required to be vaccinated. In a random sample of 200 U.S. adults, 82% think that healthy children should be required to be vaccinated. At $\alpha = 0.05$, is there enough evidence to support the researcher's claim? *(Adapted from Pew Research Center)*

8. **Internal Revenue Service Audits** A research center claims that at least 27% of U.S. adults think that the IRS will audit their taxes. In a random sample of 1000 U.S. adults in a recent year, 23% say they are concerned that the IRS will audit their taxes. At $\alpha = 0.01$, is there enough evidence to reject the center's claim? *(Source: Rasmussen Reports)*

9. **Student Employment** An eduction researcher claims that at most 3% of working college students are employed as teachers or teaching assistants. In a random sample of 200 working college students, 4% are employed as teachers or teaching assistants. At $\alpha = 0.01$, is there enough evidence to reject the researcher's claim? *(Adapted from Sallie Mae)*

10. **Working Students** An education researcher claims that 57% of college students work year-round. In a random sample of 300 college students, 171 say they work year-round. At $\alpha = 0.10$, is there enough evidence to support the researcher's claim? *(Adapted from Sallie Mae)*

11. **Zika Virus** A researcher claims that 85% percent of Americans think they are unlikely to contract the Zika virus. In a random sample of 250 Americans, 225 think they are unlikely to contract the Zika virus. At $\alpha = 0.05$, is there enough evidence to reject the researcher's claim? *(Adapted from Gallup)*

12. **Changing Jobs** A research center claims that more than 29% of U.S. employees have changed jobs in the past three years. In a random sample of 180 U.S. employees, 63 have changed jobs in the past three years. At $\alpha = 0.10$, is there enough evidence to support the center's claim? *(Adapted from Gallup)*

13. (a) The claim is "27% of U.S. adults would travel into space on a commercial flight if they could afford it."

H_0: $p = 0.27$ (claim)
H_a: $p \neq 0.27$

(b) 0.03   (c) Reject H_0.

(d) There is enough evidence at the 5% level of significance to reject the research center's claim that 27% of U.S. adults would travel into space on a commercial flight if they could afford it.

14. See Selected Answers, page A101.

15. See Odd Answers, page A71.

16. See Selected Answers, page A101.

17. Fail to reject H_0. There is not enough evidence at the 5% level of significance to reject the claim that at least 63% of adults make an effort to live in ways that help protect the environment some of the time.

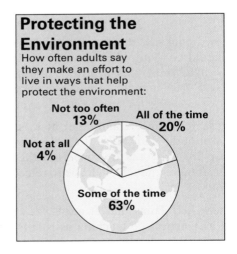

Protecting the Environment

How often adults say they make an effort to live in ways that help protect the environment:

Not too often 13%
All of the time 20%
Not at all 4%
Some of the time 63%

FIGURE FOR EXERCISES 17 AND 18

18. Answers will vary.

19. See Odd Answers, page A71.

20. See Selected Answers, page A101.

Hypothesis Testing Using a *P*-Value   *In Exercises 13–16, (a) identify the claim and state H_0 and H_a, (b) use technology to find the P-value, (c) decide whether to reject or fail to reject the null hypothesis, and (d) interpret the decision in the context of the original claim.*

13. Space Travel   A research center claims that 27% of U.S. adults would travel into space on a commercial flight if they could afford it. In a random sample of 1000 U.S. adults, 30% say that they would travel into space on a commercial flight if they could afford it. At $\alpha = 0.05$, is there enough evidence to reject the research center's claim? *(Source: Rasmussen Reports)*

14. Purchasing Food Online   A research center claims that at most 18% of U.S. adults' online food purchases are for snacks. In a random sample of 1995 U.S. adults, 20% say their online food purchases are for snacks. At $\alpha = 0.10$, is there enough evidence to support the center's claim? *(Source: The Harris Poll)*

15. Pet Ownership   A humane society claims that less than 67% of U.S. households own a pet. In a random sample of 600 U.S. households, 390 say they own a pet. At $\alpha = 0.10$, is there enough evidence to support the society's claim? *(Adapted from The Humane Society of the United States)*

16. Stray dogs   A humane society claims that 5% of U.S. households have taken in a stray dog. In a random sample of 200 U.S. households, 12 say they have taken in a stray dog. At $\alpha = 0.05$, is there enough evidence to reject the society's claim? *(Adapted from The Humane Society of the United States)*

Protecting the Environment   *In Exercises 17 and 18, use the figure at the left, which suggests what adults think about protecting the environment. (Source: Pew Research Center)*

17. Are People Concerned About Protecting the Environment?   You interview a random sample of 100 adults. The results of the survey show that 59% of the adults said they live in ways that help protect the environment some of the time. At $\alpha = 0.05$, can you reject the claim that at least 63% of adults make an effort to live in ways that help protect the environment some of the time?

18. What Are People's Attitudes About Protecting the Environment?   Use your conclusion from Exercise 17 to write a paragraph on people's attitudes about protecting the environment.

Extending Concepts

Alternative Formula   *In Exercises 19 and 20, use the information below. When you know the number of successes x, the sample size n, and the population proportion p, it can be easier to use the formula*

$$z = \frac{x - np}{\sqrt{npq}}$$

to find the standardized test statistic when using a z-test for a population proportion p.

19. Rework Exercise 7 using the alternative formula and verify that the results are the same.

20. The alternative formula is derived from the formula

$$z = \frac{\hat{p} - p}{\sqrt{pq/n}} = \frac{(x/n) - p}{\sqrt{pq/n}}.$$

Use this formula to derive the alternative formula. Justify each step.

7.4 ACTIVITY
Hypothesis Tests for a Proportion

APPLET

You can find the interactive applet for this activity within **MyLab Statistics** or at *www.pearsonhighered.com/ mathstatsresources.*

The *hypothesis tests for a proportion* applet allows you to visually investigate hypothesis tests for a population proportion. You can specify the sample size n, the population proportion (True p), the null value for the proportion (Null p), and the alternative for the test (Alternative). When you click SIMULATE, 100 separate samples of size n will be selected from a population with a proportion of successes equal to True p. For each of the 100 samples, a hypothesis test based on the Z statistic is performed, and the results from each test are displayed in plots at the right. The standardized test statistic for each test is shown in the top plot and the P-value is shown in the bottom plot. The green and blue lines represent the cutoffs for rejecting the null hypothesis with the 0.05 and 0.01 level tests, respectively. Additional simulations can be carried out by clicking SIMULATE multiple times. The cumulative number of times that each test rejects the null hypothesis is also shown. Press CLEAR to clear existing results and start a new simulation.

EXPLORE

Step 1 Specify a value for n.
Step 2 Specify a value for True p.
Step 3 Specify a value for Null p.
Step 4 Specify an alternative hypothesis.
Step 5 Click SIMULATE to generate the hypothesis tests.

DRAW CONCLUSIONS

APPLET

1. Set $n = 25$ and True $p = 0.35$. Test the claim that the proportion is equal to 35%. What are the null and alternative hypotheses? Run the simulation so that at least 1000 tests are run. Compare the proportion of null hypothesis rejections for the 0.05 and 0.01 levels. Is this what you would expect? Explain.

2. Set $n = 50$ and True $p = 0.6$. Test the claim that the proportion is at least 40%. What are the null and alternative hypotheses? Run the simulation so that at least 1000 tests are run. Compare the proportion of null hypothesis rejections for the 0.05 and 0.01 levels. Perform a hypothesis test for each level. Use the results of the hypothesis tests to explain the results of the simulation.

7.5 Hypothesis Testing for Variance and Standard Deviation

What You Should Learn

▶ How to find critical values for a chi-square test

▶ How to use the chi-square test to test a variance σ^2 or a standard deviation σ

Critical Values for a Chi-Square Test ■ The Chi-Square Test

Critical Values for a Chi-Square Test

In real life, it is important to produce consistent, predictable results. For instance, consider a company that manufactures golf balls. The manufacturer must produce millions of golf balls, each having the same size and the same weight. There is a very low tolerance for variation. For a normally distributed population, you can test the variance and standard deviation of the process using the chi-square distribution with $n - 1$ degrees of freedom. Before learning how to do the test, you must know how to find the critical values, as shown in the guidelines.

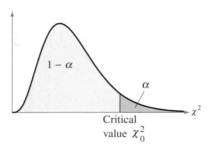

Right-Tailed Test

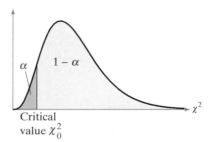

Left-Tailed Test

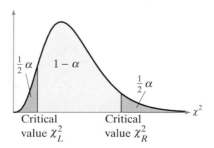

Two-Tailed Test

GUIDELINES

Finding Critical Values for a Chi-Square Test

1. Specify the level of significance α.
2. Identify the degrees of freedom, d.f. $= n - 1$.
3. The critical values for the chi-square distribution are found in Table 6 in Appendix B. To find the critical value(s) for a
   a. *right-tailed test,* use the value that corresponds to d.f. and α.
   b. *left-tailed test,* use the value that corresponds to d.f. and $1 - \alpha$.
   c. *two-tailed test,* use the values that correspond to d.f. and $\frac{1}{2}\alpha$, and d.f. and $1 - \frac{1}{2}\alpha$.

See the figures at the left.

EXAMPLE 1

Finding a Critical Value for a Right-Tailed Test

Find the critical value χ_0^2 for a right-tailed test when $n = 26$ and $\alpha = 0.10$.

SOLUTION

The degrees of freedom are d.f. $= n - 1 = 26 - 1 = 25$. The figure below shows a chi-square distribution with 25 degrees of freedom and a shaded area of $\alpha = 0.10$ in the right tail. Using Table 6 in Appendix B with d.f. $= 25$ and $\alpha = 0.10$, the critical value is $\chi_0^2 = 34.382$.

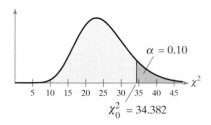

$$\chi_0^2 = 34.382$$

TRY IT YOURSELF 1

Find the critical value χ_0^2 for a right-tailed test when $n = 18$ and $\alpha = 0.01$.

Answer: Page A37

Note to Instructor

This section can be omitted or covered later (with Chapter 10) without loss of continuity.

EXAMPLE 2

Finding a Critical Value for a Left-Tailed Test

Find the critical value χ_0^2 for a left-tailed test when $n = 11$ and $\alpha = 0.01$.

SOLUTION

The degrees of freedom are

$$\text{d.f.} = n - 1 = 11 - 1 = 10.$$

The figure at the left shows a chi-square distribution with 10 degrees of freedom and a shaded area of $\alpha = 0.01$ in the left tail. The area to the right of the critical value is

$$1 - \alpha = 1 - 0.01 = 0.99.$$

Using Table 6 with d.f. $= 10$ and the area 0.99, the critical value is $\chi_0^2 = 2.558$. You can check your answer using technology, as shown below.

MINITAB

Inverse Cumulative Distribution Function

Chi-Square with 10 DF

P (X ≤ x)   x
      0.01   2.55821

TRY IT YOURSELF 2

Find the critical value χ_0^2 for a left-tailed test when $n = 30$ and $\alpha = 0.05$.

Answer: Page A37

Note that because chi-square distributions are not symmetric (like normal or t-distributions), in a two-tailed test the two critical values are not opposites. Each critical value must be calculated separately, as shown in the next example.

EXAMPLE 3

Finding Critical Values for a Two-Tailed Test

Find the critical values χ_L^2 and χ_R^2 for a two-tailed test when $n = 9$ and $\alpha = 0.05$.

SOLUTION

The degrees of freedom are

$$\text{d.f.} = n - 1 = 9 - 1 = 8.$$

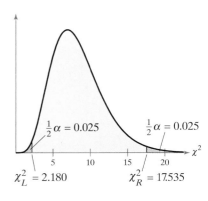

The figure shows a chi-square distribution with 8 degrees of freedom and a shaded area of $\frac{1}{2}\alpha = 0.025$ in each tail. The area to the right of χ_R^2 is $\frac{1}{2}\alpha = 0.025$, and the area to the right of χ_L^2 is $1 - \frac{1}{2}\alpha = 0.975$. Using Table 6 with d.f. $= 8$ and the areas 0.025 and 0.975, the critical values are $\chi_R^2 = 17.535$ and $\chi_L^2 = 2.180$. You can check you answer using technology, as shown at the left.

TRY IT YOURSELF 3

Find the critical values χ_L^2 and χ_R^2 for a two-tailed test when $n = 51$ and $\alpha = 0.01$.

Answer: Page A37

EXCEL

	A	B
1	CHISQ.INV(0.025,8)	
2		2.179730747
3	CHISQ.INV.RT(0.025,8)	
4		17.53454614

$\chi_0^2 = 2.558$

The Chi-Square Test

To test a variance σ^2 or a standard deviation σ of a population that is normally distributed, you can use the chi-square test. The chi-square test for a variance or standard deviation is not as robust as the tests for the population mean μ or the population proportion p. So, it is essential in performing a chi-square test for a variance or standard deviation that the population be normally distributed. The results can be misleading when the population is not normal.

Chi-Square Test for a Variance σ^2 or Standard Deviation σ

The **chi-square test for a variance σ^2 or standard deviation σ** is a statistical test for a population variance or standard deviation. The chi-square test can only be used when the population is normal. The **test statistic** is s^2 and the **standardized test statistic**

$$\chi^2 = \frac{(n-1)s^2}{\sigma^2}$$ Standardized test statistic for σ^2 or σ

follows a chi-square distribution with degrees of freedom

d.f. $= n - 1$.

Note to Instructor

Review the properties of chi-square distributions. Tell students that this family of distributions will be used in later chapters, but the degrees of freedom for those tests are not necessarily $n - 1$.

In Step 8 of the guidelines below, the decision rule uses rejection regions. You can also test a claim using P-values (see Exercises 31-34).

GUIDELINES

Using the Chi-Square Test for a Variance σ^2 or a Standard Deviation σ

In Words	In Symbols
1. Verify that the sample is random and the population is normally distributed.	
2. State the claim mathematically and verbally. Identify the null and alternative hypotheses.	State H_0 and H_a.
3. Specify the level of significance.	Identify α.
4. Identify the degrees of freedom.	d.f. $= n - 1$
5. Determine the critical value(s).	Use Table 6 in Appendix B.
6. Determine the rejection region(s).	
7. Find the standardized test statistic and sketch the sampling distribution.	$\chi^2 = \frac{(n-1)s^2}{\sigma^2}$
8. Make a decision to reject or fail to reject the null hypothesis.	If χ^2 is in the rejection region, then reject H_0. Otherwise, fail to reject H_0.
9. Interpret the decision in the context of the original claim.	

For Step 5 of the guidelines, in addition to using Table 6 in Appendix B, you can use technology to find the critical value(s). Also, some technology tools allow you to perform a hypothesis test for a variance (or a standard deviation) using only the descriptive statistics.

EXAMPLE 4

Using a Hypothesis Test for the Population Variance

A dairy processing company claims that the variance of the amount of fat in the whole milk processed by the company is no more than 0.25. You suspect this is wrong and find that a random sample of 41 milk containers has a variance of 0.27. At $\alpha = 0.05$, is there enough evidence to reject the company's claim? Assume the population is normally distributed.

SOLUTION

Because the sample is random and the population is normally distributed, you can use the chi-square test. The claim is "the variance is no more than 0.25." So, the null and alternative hypotheses are

$$H_0: \sigma^2 \leq 0.25 \text{ (Claim)} \quad \text{and} \quad H_a: \sigma^2 > 0.25.$$

The test is a right-tailed test, the level of significance is $\alpha = 0.05$, and the degrees of freedom are d.f. $= 41 - 1 = 40$. So, using Table 6, the critical value is

$$\chi_0^2 = 55.758.$$

The rejection region is $\chi^2 > 55.758$. The standardized test statistic is

$$
\begin{aligned}
\chi^2 &= \frac{(n-1)s^2}{\sigma^2} &&\text{Use the chi-square test.}\\
&= \frac{(41-1)(0.27)}{0.25} &&\text{Assume } \sigma^2 = 0.25.\\
&= 43.2.
\end{aligned}
$$

The figure at the left shows the location of the rejection region and the standardized test statistic χ^2. Because χ^2 is not in the rejection region, you fail to reject the null hypothesis. You can check your answer using technology, as shown below. Note that the test statistic, 43.2, is the same as what you found above.

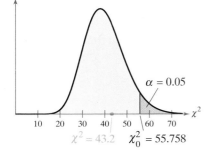

$\alpha = 0.05$

$\chi^2 = 43.2 \qquad \chi_0^2 = 55.758$

STATCRUNCH

One sample variance hypothesis test:
σ^2 : Variance of population
$H_0 : \sigma^2 = 0.25$
$H_A : \sigma^2 > 0.25$

Hypothesis test results:

Variance	Sample Var.	DF	Chi-square Stat	P-value
σ^2	0.27	40	43.2	0.3362

Interpretation   There is not enough evidence at the 5% level of significance to reject the company's claim that the variance of the amount of fat in the whole milk is no more than 0.25.

TRY IT YOURSELF 4

A bottling company claims that the variance of the amount of sports drink in a 12-ounce bottle is no more than 0.40. A random sample of 31 bottles has a variance of 0.75. At $\alpha = 0.01$, is there enough evidence to reject the company's claim? Assume the population is normally distributed.

Answer: Page A37

EXAMPLE 5

Using a Hypothesis Test for the Standard Deviation

A company claims that the standard deviation of the lengths of time it takes an incoming telephone call to be transferred to the correct office is less than 1.4 minutes. A random sample of 25 incoming telephone calls has a standard deviation of 1.1 minutes. At $\alpha = 0.10$, is there enough evidence to support the company's claim? Assume the population is normally distributed.

SOLUTION

Because the sample is random and the population is normally distributed, you can use the chi-square test. The claim is "the standard deviation is less than 1.4 minutes." So, the null and alternative hypotheses are

$$H_0: \sigma \geq 1.4 \text{ minutes} \quad \text{and} \quad H_a: \sigma < 1.4 \text{ minutes.} \text{ (Claim)}$$

The test is a left-tailed test, the level of significance is $\alpha = 0.10$, and the degrees of freedom are

$$\text{d.f.} = 25 - 1 = 24.$$

So, using Table 6, the critical value is

$$\chi_0^2 = 15.659.$$

The rejection region is $\chi^2 < 15.659$. The standardized test statistic is

$$
\begin{aligned}
\chi^2 &= \frac{(n-1)s^2}{\sigma^2} && \text{Use the chi-square test.} \\
&= \frac{(25-1)(1.1)^2}{(1.4)^2} && \text{Assume } \sigma = 1.4. \\
&\approx 14.816. && \text{Round to three decimal places.}
\end{aligned}
$$

Study Tip

Although you are testing a standard deviation in Example 5, the standardized test statistic χ^2 requires variance. Remember to square the standard deviation to calculate the variance.

The figure below shows the location of the rejection region and the standardized test statistic χ^2. Because χ^2 is in the rejection region, you reject the null hypothesis.

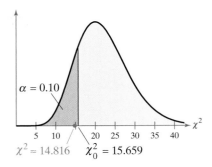

Interpretation There is enough evidence at the 10% level of significance to support the claim that the standard deviation of the lengths of time it takes an incoming telephone call to be transferred to the correct office is less than 1.4 minutes.

TRY IT YOURSELF 5

A police chief claims that the standard deviation of the lengths of response times is less than 3.7 minutes. A random sample of 9 response times has a standard deviation of 3.0 minutes. At $\alpha = 0.05$, is there enough evidence to support the police chief's claim? Assume the population is normally distributed.

Answer: Page A37

Picturing the World

A community center claims that the chlorine level in its pool has a standard deviation of 0.46 parts per million (ppm). A sampling of the pool's chlorine levels at 25 random times during a month yields a standard deviation of 0.61 ppm. (Adapted from American Pool Supply)

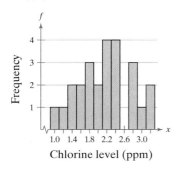

Chlorine level (ppm)

At 0.05, is there enough evidence to reject the claim?

Yes, there is enough evidence at the 5% level of significance to reject the claim that the chlorine level in the pool has a standard deviation of 0.46 parts per million.

EXAMPLE 6

Using a Hypothesis Test for the Population Variance

A sporting goods manufacturer claims that the variance of the strengths of a certain fishing line is 15.9. A random sample of 15 fishing line spools has a variance of 21.8. At $\alpha = 0.05$, is there enough evidence to reject the manufacturer's claim? Assume the population is normally distributed.

SOLUTION

Because the sample is random and the population is normally distributed, you can use the chi-square test. The claim is "the variance is 15.9." So, the null and alternative hypotheses are

$$H_0: \sigma^2 = 15.9 \text{ (Claim)} \quad \text{and} \quad H_a: \sigma^2 \neq 15.9.$$

The test is a two-tailed test, the level of significance is $\alpha = 0.05$, and the degrees of freedom are

$$\text{d.f.} = 15 - 1$$
$$= 14.$$

Using Table 6, the critical values are $\chi^2_L = 5.629$ and $\chi^2_R = 26.119$. The rejection regions are

$$\chi^2 < 5.629 \quad \text{and} \quad \chi^2 > 26.119.$$

The standardized test statistic is

$$\chi^2 = \frac{(n-1)s^2}{\sigma^2} \qquad \text{Use the chi-square test.}$$
$$= \frac{(15-1)(21.8)}{(15.9)} \qquad \text{Assume } \sigma^2 = 15.9.$$
$$\approx 19.195. \qquad \text{Round to three decimal places.}$$

The figure below shows the location of the rejection regions and the standardized test statistic χ^2. Because χ^2 is not in the rejection regions, you fail to reject the null hypothesis.

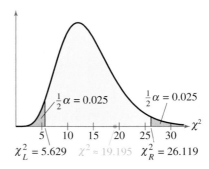

Interpretation There is not enough evidence at the 5% level of significance to reject the claim that the variance of the strengths of the fishing line is 15.9.

TRY IT YOURSELF 6

A company that offers dieting products and weight loss services claims that the variance of the weight losses of their users is 25.5. A random sample of 13 users has a variance of 10.8. At $\alpha = 0.10$, is there enough evidence to reject the company's claim? Assume the population is normally distributed.

Answer: Page A37

7.5 EXERCISES

For Extra Help: **MyLab Statistics**

1. See Odd Answers, page A71.
2. See Selected Answers, page A101.
3. See Odd Answers, page A71.
4. See Selected Answers, page A101.
5. Critical value: $\chi_0^2 = 38.885$
   Rejection region: $\chi^2 > 38.885$
6. Critical value: $\chi_0^2 = 14.684$
   Rejection region: $\chi^2 > 14.684$
7. Critical value: $\chi_0^2 = 0.872$
   Rejection region: $\chi^2 < 0.872$
8. Critical value: $\chi_0^2 = 13.091$
   Rejection region: $\chi^2 < 13.091$
9. Critical values: $\chi_L^2 = 60.391$,
   $\chi_R^2 = 101.879$
   Rejection regions: $\chi^2 < 60.391$,
   $\chi^2 > 101.879$
10. Critical values: $\chi_L^2 = 35.534$,
    $\chi_R^2 = 91.952$
    Rejection regions: $\chi^2 < 35.534$,
    $\chi^2 > 91.952$
11. Critical value: $\chi_0^2 = 49.588$
    Rejection region: $\chi^2 > 49.588$
12. Critical values: $\chi_L^2 = 16.791$,
    $\chi_R^2 = 46.979$
    Rejection regions: $\chi^2 < 16.791$,
    $\chi^2 > 46.979$
13. (a) Fail to reject H_0 because
        $\chi^2 < 6.251$.
    (b) Fail to reject H_0 because
        $\chi^2 < 6.251$.
    (c) Fail to reject H_0 because
        $\chi^2 < 6.251$.
    (d) Reject H_0 because $\chi^2 > 6.251$.
14. (a) Fail to reject H_0 because
        $8.547 < \chi^2 < 22.307$.
    (b) Reject H_0 because
        $\chi^2 > 22.307$.
    (c) Reject H_0 because $\chi^2 < 8.547$.
    (d) Fail to reject H_0 because
        $8.547 < \chi^2 < 22.307$.
15. See Odd Answers, page A71.
16. See Selected Answers, page A101.
17. See Odd Answers, page A71.
18. See Selected Answers, page A101.
19. See Odd Answers, page A71.
20. See Selected Answers, page A101.
21. See Odd Answers, page A71.
22. Reject H_0. There is enough
    evidence at the 10% level of
    significance to reject the claim.

Building Basic Skills and Vocabulary

1. Explain how to find critical values in a chi-square distribution.

2. Can a critical value for the chi-square test be negative? Explain.

3. How do the requirements for a chi-square test for a variance or standard deviation differ from a z-test or a t-test for a mean?

4. Explain how to test a population variance or a population standard deviation.

In Exercises 5–12, find the critical value(s) and rejection region(s) for the type of chi-square test with sample size n and level of significance α.

5. Right-tailed test,
   $n = 27, \alpha = 0.05$

6. Right-tailed test,
   $n = 10, \alpha = 0.10$

7. Left-tailed test,
   $n = 7, \alpha = 0.01$

8. Left-tailed test,
   $n = 24, \alpha = 0.05$

9. Two-tailed test,
   $n = 81, \alpha = 0.10$

10. Two-tailed test,
    $n = 61, \alpha = 0.01$

11. Right-tailed test,
    $n = 30, \alpha = 0.01$

12. Two-tailed test,
    $n = 31, \alpha = 0.05$

Graphical Analysis *In Exercises 13 and 14, state whether each standardized test statistic χ^2 allows you to reject the null hypothesis. Explain.*

13. (a) $\chi^2 = 2.091$
    (b) $\chi^2 = 0$
    (c) $\chi^2 = 1.086$
    (d) $\chi^2 = 6.3471$

14. (a) $\chi^2 = 22.302$
    (b) $\chi^2 = 23.309$
    (c) $\chi^2 = 8.457$
    (d) $\chi^2 = 8.577$

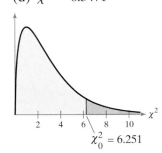

$\chi_0^2 = 6.251$

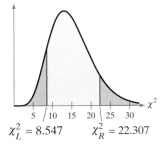

$\chi_L^2 = 8.547$ $\quad \chi_R^2 = 22.307$

In Exercises 15–22, test the claim about the population variance σ^2 or standard deviation σ at the level of significance α. Assume the population is normally distributed.

15. Claim: $\sigma^2 = 0.52$; $\alpha = 0.05$. Sample statistics: $s^2 = 0.508, n = 18$

16. Claim: $\sigma^2 \geq 8.5$; $\alpha = 0.05$. Sample statistics: $s^2 = 7.45, n = 23$

17. Claim: $\sigma^2 \leq 17.6$; $\alpha = 0.01$. Sample statistics: $s^2 = 28.33, n = 41$

18. Claim: $\sigma^2 > 19$; $\alpha = 0.1$. Sample statistics: $s^2 = 28, n = 17$

19. Claim: $\sigma^2 \neq 32.8$; $\alpha = 0.1$. Sample statistics: $s^2 = 40.9, n = 101$

20. Claim: $\sigma^2 = 63$; $\alpha = 0.01$. Sample statistics: $s^2 = 58, n = 29$

21. Claim: $\sigma < 40$; $\alpha = 0.01$. Sample statistics: $s = 40.8, n = 12$

22. Claim: $\sigma = 24.9$; $\alpha = 0.10$. Sample statistics: $s = 29.1, n = 51$

23. See Odd Answers, page A71.
24. See Selected Answers, page A102.
25. See Odd Answers, page A71.
26. See Selected Answers, page A102.
27. See Odd Answers, page A71.
28. See Selected Answers, page A102.
29. See Odd Answers, page A71.
30. See Selected Answers, page A102.
31. P-value $= 0.4524$
Fail to reject H_0.
32. P-value $= 0.4014$
Fail to reject H_0.
33. P-value $= 0.0033$
Reject H_0.
34. P-value $= 0.0060$
Reject H_0.

Annual salaries

47,262	67,363	81,246
65,876	59,649	78,268
88,549	52,130	73,955
91,288	54,476	86,787
66,923	48,337	70,172

TABLE FOR EXERCISE 29

Annual salaries

59,922	99,493	98,221
90,143	65,106	78,975
74,644	107,817	85,492
87,179	90,505	71,090

TABLE FOR EXERCISE 30

TI-84 PLUS

X²cdf(0,43.2,40)
           .6637768667

Using and Interpreting Concepts

Hypothesis Testing Using Rejection Regions *In Exercises 23–30, (a) identify the claim and state H_0 and H_a, (b) find the critical value(s) and identify the rejection region(s), (c) find the standardized test statistic χ^2, (d) decide whether to reject or fail to reject the null hypothesis, and (e) interpret the decision in the context of the original claim. Assume the population is normally distributed.*

23. **Tires** A tire manufacturer claims that the variance of the diameters in a tire model is 8.6. A random sample of 10 tires has a variance of 4.3. At $\alpha = 0.01$, is there enough evidence to reject the claim?

24. **Gas Mileage** An auto manufacturer claims that the variance of the gas mileages in a model of hybrid vehicle is 0.16. A random sample of 30 vehicles has a variance of 0.26. At $\alpha = 0.05$, is there enough evidence to reject the claim? *(Adapted from Green Hybrid)*

25. **Mathematics Assessment Tests** A school administrator claims that the standard deviation for grade 12 students on a mathematics assessment test is less than 35 points. A random sample of 28 grade 12 test scores has a standard deviation of 34 points. At $\alpha = 0.10$, is there enough evidence to support the claim? *(Adapted from National Center for Educational Statistics)*

26. **Vocabulary Assessment Tests** A school administrator claims that the standard deviation for grade 12 students on a vocabulary assessment test is greater than 45 points. A random sample of 25 grade 12 test scores has a standard deviation of 46 points. At $\alpha = 0.01$, is there enough evidence to support the claim? *(Adapted from National Center for Educational Statistics)*

27. **Waiting Times** A hospital claims that the standard deviation of the waiting times for patients in its emergency department is no more than 0.5 minute. A random sample of 25 waiting times has a standard deviation of 0.7 minute. At $\alpha = 0.10$, is there enough evidence to reject the claim?

28. **Hotel Room Rates** A travel analyst claims that the standard deviation of the room rates for two adults at three-star hotels in Denver is at least $68. A random sample of 18 three-star hotels has a standard deviation of $40. At $\alpha = 0.01$, is there enough evidence to reject the claim? *(Adapted from Expedia)*

29. **Salaries** The annual salaries (in dollars) of 15 randomly chosen senior level graphic design specialists are shown in the table at the left. At $\alpha = 0.05$, is there enough evidence to support the claim that the standard deviation of the annual salaries is different from $10,300? *(Adapted from Salary.com)*

30. **Salaries** The annual salaries (in dollars) of 12 randomly chosen nursing supervisors are shown in the table at the left. At $\alpha = 0.10$, is there enough evidence to reject the claim that the standard deviation of the annual salaries is $16,500? *(Adapted from Salary.com)*

Extending Concepts

***P*-Values** *You can calculate the P-value for a chi-square test using technology. After calculating the standardized test statistic, use the cumulative distribution function (CDF) to calculate the area under the curve. From Example 4 on page 397, $\chi^2 = 43.2$. Using a TI-84 Plus (choose 8 from the DISTR menu), enter 0 for the lower bound, 43.2 for the upper bound, and 40 for the degrees of freedom, as shown at the left. Because it is a right-tailed test, the P-value is approximately $1 - 0.6638 = 0.3362$. Because $P > \alpha = 0.05$, fail to reject H_0. In Exercises 31–34, use the P-value method to perform the hypothesis test for the indicated exercise.*

31. Exercise 25   32. Exercise 26   33. Exercise 27   34. Exercise 28

7   A Summary of Hypothesis Testing

With hypothesis testing, perhaps more than any other area of statistics, it can be difficult to see the forest for all the trees. To help you see the forest—the overall picture—a summary of what you studied in this chapter is provided.

Writing the Hypotheses

■ You are given a claim about a population parameter μ, p, σ^2, or σ.

■ Rewrite the claim and its complement using $\leq, \geq, =$ and $>, <, \neq$.
  $\underbrace{\qquad}_{H_0}$ $\underbrace{\qquad}_{H_a}$

■ Identify the claim. Is it H_0 or H_a?

Specifying a Level of Significance

■ Specify α, the maximum acceptable probability of rejecting a valid H_0 (a type I error).

Specifying the Sample Size

■ Specify your sample size n.

Study Tip

Large sample sizes will usually increase the cost and effort of testing a hypothesis, but they also tend to make your decision more reliable.

Choosing the Test   ▲ Normally distributed population   ● Any population

■ **Mean:** H_0 describes a hypothesized population mean μ.
  ▲ Use a **z-test** when σ is known and the population is normal.
  ● Use a **z-test** for any population when σ is known and $n \geq 30$.
  ▲ Use a **t-test** when σ is not known and the population is normal.
  ● Use a **t-test** for any population when σ is not known and $n \geq 30$.

■ **Proportion:** H_0 describes a hypothesized population proportion p.
  ● Use a **z-test** for any population when $np \geq 5$ and $nq \geq 5$.

■ **Variance or Standard Deviation:** H_0 describes a hypothesized population variance σ^2 or standard deviation σ.
  ▲ Use a **chi-square test** when the population is normal.

Sketching the Sampling Distribution

■ Use H_a to decide whether the test is left-tailed, right-tailed, or two-tailed.

Finding the Standardized Test Statistic

■ Take a random sample of size n from the population.

■ Compute the test statistic $\bar{x}$, $\hat{p}$, or s^2.

■ Find the standardized test statistic z, t, or χ^2.

Making a Decision

Option 1. Decision based on rejection region

■ Use α to find the critical value(s) z_0, t_0, or χ_0^2 and rejection region(s).

■ **Decision Rule:**

  Reject H_0 when the standardized test statistic is in the rejection region.
  Fail to reject H_0 when the standardized test statistic is not in the rejection region.

Option 2. Decision based on P-value

■ Use the standardized test statistic or technology to find the P-value.

■ **Decision Rule:**

  Reject H_0 when $P \leq \alpha$.
  Fail to reject H_0 when $P > \alpha$.

z-Test for a Hypothesized Mean μ (σ Known)   *(Section 7.2)*

Test statistic: $\bar{x}$
Critical value: z_0 (Use Table 4.)
Sampling distribution of sample means is a normal distribution.

Standardized test statistic: z

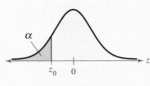

$$z = \frac{\bar{x} - \mu}{\sigma / \sqrt{n}}$$

Sample mean ⟶
Hypothesized mean
Population standard deviation ⟶
⟵ Sample size

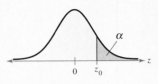

Left-Tailed   Two-Tailed   Right-Tailed

z-Test for a Hypothesized Proportion p *(Section 7.4)*

Test statistic: $\hat{p}$
Critical value: z_0 (Use Table 4.)
Sampling distribution of sample proportions is a normal distribution.

Standardized test statistic: z

$$z = \frac{\hat{p} - p}{\sqrt{pq/n}}$$

Sample proportion ⟶
Hypothesized proportion
$q = 1 - p$ ⟶
⟵ Sample size

Study Tip

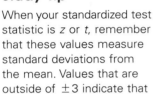

When your standardized test statistic is z or t, remember that these values measure standard deviations from the mean. Values that are outside of ± 3 indicate that H_0 is very unlikely. Values that are outside of ± 5 indicate that H_0 is almost impossible.

t-Test for a Hypothesized Mean μ (σ Unknown)   *(Section 7.3)*

Test statistic: $\bar{x}$
Critical value: t_0 (Use Table 5.)
Sampling distribution of sample means is approximated by a t-distribution with d.f. $= n - 1$.

Standardized test statistic: t

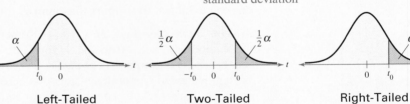

$$t = \frac{\bar{x} - \mu}{s / \sqrt{n}}$$

Sample mean ⟶
Hypothesized mean
Sample standard deviation ⟶
⟵ Sample size

Left-Tailed   Two-Tailed   Right-Tailed

Chi-Square Test for a Hypothesized Variance σ^2 or Standard Deviation σ *(Section 7.5)*

Test statistic: s^2
Critical value: χ_0^2 (Use Table 6.)
Sampling distribution is approximated by a chi-square distribution with d.f. $= n - 1$.

Standardized test statistic: χ^2

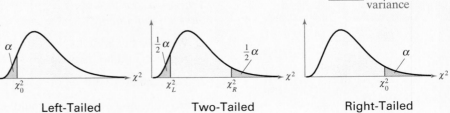

$$\chi^2 = \frac{(n - 1)s^2}{\sigma^2}$$

Sample size ⟶
Sample variance ⟶
Hypothesized variance ⟶

Left-Tailed   Two-Tailed   Right-Tailed

USES AND ABUSES

Statistics in the Real World

Uses

Hypothesis testing is important in many different fields because it gives a scientific procedure for assessing the validity of a claim about a population. Some of the concepts in hypothesis testing are intuitive, but some are not. For instance, the *American Journal of Clinical Nutrition* suggests that eating dark chocolate can help prevent heart disease. A random sample of healthy volunteers were assigned to eat 3.5 ounces of dark chocolate each day for 15 days. After 15 days, the mean systolic blood pressure of the volunteers was 6.4 millimeters of mercury lower. A hypothesis test could show whether this drop in systolic blood pressure is significant or simply due to sampling error.

Careful inferences must be made concerning the results. The study only examined the effects of dark chocolate, so the inference of health benefits cannot be extended to all types of chocolate. You also would not infer that you should eat large quantities of chocolate because the benefits must be weighed against known risks, such as weight gain and acid reflux.

Abuses

Not Using a Random Sample The entire theory of hypothesis testing is based on the fact that the sample is randomly selected. If the sample is not random, then you cannot use it to infer anything about a population parameter.

Attempting to Prove the Null Hypothesis When the *P*-value for a hypothesis test is greater than the level of significance, you have not proven the null hypothesis is true—only that there is not enough evidence to reject it. For instance, with a *P*-value higher than the level of significance, a researcher could not prove that there is no benefit to eating dark chocolate—only that there is not enough evidence to support the claim that there is a benefit.

Making Type I or Type II Errors Remember that a type I error is rejecting a null hypothesis that is true and a type II error is failing to reject a null hypothesis that is false. You can decrease the probability of a type I error by lowering the level of significance α. Generally, when you decrease the probability of making a type I error, you increase the probability β of making a type II error. Which error is more serious? It depends on the situation. In a criminal trial, a type I error is considered worse, as explained on page 352. If you are testing a person for a disease and they are assumed to be disease-free (H_0), then a type II error is more serious because you would fail to detect the disease even though the person has it. You can decrease the chance of making both types of errors by increasing the sample size.

Do You Consider the Amount of Federal Income Tax You Pay as Too High, About Right, or Too Low?

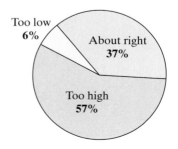

EXERCISES

In Exercises 1–3, assume that you work for the Internal Revenue Service. You are asked to write a report about the claim that 57% of U.S. adults think the amount of federal income tax they pay is too high. (Source: Gallup)

1. What is the null hypothesis in this situation? Describe how your report could be incorrect by trying to prove the null hypothesis.

2. Describe how your report could make a type I error.

3. Describe how your report could make a type II error.

7 | Chapter Summary

What Did You Learn?	Example(s)	Review Exercises
Section 7.1		
▶ How to state a null hypothesis and an alternative hypothesis	1	1–6
▶ How to identify type I and type II errors	2	7–10
▶ How to know whether to use a one-tailed or a two-tailed statistical test and find a P-value	3	7–10
▶ How to interpret a decision based on the results of a statistical test	4	7–10
▶ How to write a claim for a hypothesis test	5	7–10
Section 7.2		
▶ How to find and interpret P-values	1–3	11, 12
▶ How to use P-values for a z-test for a mean μ when σ is known	4–6	25, 26
▶ How to find critical values and rejection regions in the standard normal distribution	7, 8	13–16
▶ How to use rejection regions for a z-test for a mean μ when σ is known	9, 10	17–24, 27, 28
Section 7.3		
▶ How to find critical values in a t-distribution	1–3	29–34
▶ How to use the t-test to test a mean μ when σ is not known	4, 5	35–42
▶ How to use technology to find P-values and use them with a t-test to test a mean μ when σ is not known	6	43, 44
Section 7.4		
▶ How to use the z-test to test a population proportion p	1, 2	45–50
Section 7.5		
▶ How to find critical values for a chi-square test	1–3	51–54
▶ How to use the chi-square test to test a variance σ^2 or a standard deviation σ	4–6	55–62

Review Exercises

1. $H_0: \mu \leq 375$ (claim); $H_a: \mu > 375$

2. $H_0: \mu = 82$ (claim); $H_a: \mu \neq 82$

3. $H_0: p \geq 0.205$
$H_a: p < 0.205$ (claim)

4. $H_0: \mu = 150{,}020$
$H_a: \mu \neq 150{,}020$ (claim)

5. $H_0: \sigma \leq 1.9$; $H_a: \sigma > 1.9$ (claim)

6. $H_0: p \geq 0.64$ (claim); $H_a: p < 0.64$

7. See Odd Answers, page A72.

8. See Selected Answers, page A102.

9. See Odd Answers, page A72.

10. See Selected Answers, page A102.

11. 0.1736; Fail to reject H_0.

12. 0.0102; Reject H_0.

13. See Odd Answers, page A72.

14. See Selected Answers, page A102.

15. See Odd Answers, page A72.

16. See Selected Answers, page A102.

17. Fail to reject H_0 because $-1.645 < z < 1.645$.

18. Reject H_0 because $z > 1.645$.

19. Fail to reject H_0 because $-1.645 < z < 1.645$.

20. Reject H_0 because $z < -1.645$.

21. Fail to reject H_0. There is not enough evidence at the 5% level of significance to reject the claim.

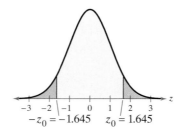

FIGURE FOR EXERCISES 17–20

22. Reject H_0. There is enough evidence at the 3% level of significance to support the claim.

23. Fail to reject H_0. There is not enough evidence at the 1% level of significance to support the claim.

24. See Selected Answers, page A102.

Section 7.1

In Exercises 1–6, the statement represents a claim. Write its complement and state which is H_0 and which is H_a.

1. $\mu \leq 375$ **2.** $\mu = 82$ **3.** $p < 0.205$

4. $\mu \neq 150{,}020$ **5.** $\sigma > 1.9$ **6.** $p \geq 0.64$

In Exercises 7–10, (a) state the null and alternative hypotheses and identify which represents the claim, (b) describe type I and type II errors for a hypothesis test of the claim, (c) explain whether the hypothesis test is left-tailed, right-tailed, or two-tailed, (d) explain how you should interpret a decision that rejects the null hypothesis, and (e) explain how you should interpret a decision that fails to reject the null hypothesis.

7. A polling organization reports that the proportion of U.S. adults who have volunteered their time or donated money to help clean up the environment is 65%. *(Source: Rasmussen Reports)*

8. An agricultural cooperative guarantees that the mean shelf life of a type of dried fruit is at least 400 days.

9. A nonprofit consumer organization says that the standard deviation of the fuel economies of its top-rated vehicles for a recent year is no more than 9.5 miles per gallon. *(Adapted from Consumer Reports)*

10. An energy bar maker claims that the mean number of grams of carbohydrates in one bar is less than 25.

Section 7.2

In Exercises 11 and 12, find the P-value for the hypothesis test with the standardized test statistic z. Decide whether to reject H_0 for the level of significance α.

11. Left-tailed test, $z = -0.94$, $\alpha = 0.05$

12. Two-tailed test, $z = 2.57$, $\alpha = 0.10$

In Exercises 13–16, find the critical value(s) and rejection region(s) for the type of z-test with level of significance α. Include a graph with your answer.

13. Left-tailed test, $\alpha = 0.02$ **14.** Two-tailed test, $\alpha = 0.005$

15. Right-tailed test, $\alpha = 0.025$ **16.** Two-tailed test, $\alpha = 0.03$

In Exercises 17–20, state whether the standardized test statistic z allows you to reject the null hypothesis. Explain your reasoning.

17. $z = 1.631$ **18.** $z = 1.723$ **19.** $z = -1.464$ **20.** $z = -1.655$

In Exercises 21–24, test the claim about the population mean μ at the level of significance α. Assume the population is normally distributed.

21. Claim: $\mu \leq 45$; $\alpha = 0.05$; $\sigma = 6.7$. Sample statistics: $\bar{x} = 47.2$, $n = 22$

22. Claim: $\mu \neq 8.45$; $\alpha = 0.03$; $\sigma = 1.75$. Sample statistics: $\bar{x} = 7.88$, $n = 60$

23. Claim: $\mu < 5.500$; $\alpha = 0.01$; $\sigma = 0.011$. Sample statistics: $\bar{x} = 5.497$, $n = 36$

24. Claim: $\mu = 7450$; $\alpha = 0.10$; $\sigma = 243$. Sample statistics: $\bar{x} = 7495$, $n = 27$

25. (a) The claim is "the mean annual production of cotton is 3.5 million bales per country."

H_0: $\mu = 3.5$ (claim)
H_a: $\mu \neq 3.5$

(b) -2.06 (c) 0.0394

(d) Reject H_0.

(e) There is enough evidence at the 5% level of significance to reject the researcher's claim that the mean annual production of cotton is 3.5 million bales per country.

26. See Selected Answers, page A102.

27. See Odd Answers, page A72.

28. See Selected Answers, page A102.

29. Critical values: $-t_0 = -2.093$, $t_0 = 2.093$
Rejection regions: $t < -2.093$, $t > 2.093$

30. Critical value: $t_0 = 2.449$
Rejection region: $t > 2.449$

31. Critical value: $t_0 = 2.098$
Rejection region: $t > 2.098$

32. Critical value: $t_0 = -1.678$
Rejection region: $t < -1.678$

33. Critical value: $t_0 = -2.977$
Rejection region: $t < -2.977$

34. Critical values: $-t_0 = -2.718$, $t_0 = 2.718$
Rejection regions: $t < -2.718$, $t > 2.718$

35. Reject H_0. There is enough evidence at the 0.5% level of significance to support the claim.

36. Fail to reject H_0. There is not enough evidence at the 10% level of significance to reject the claim.

37. Reject H_0. There is enough evidence at the 1% level of significance to reject the claim.

38. Fail to reject H_0. There is not enough evidence at the 2.5% level of significance to support the claim.

39. Fail to reject H_0. There is not enough evidence at the 10% level of significance to reject the claim.

40. Reject H_0. There is enough evidence at the 5% level of significance to support the claim.

In Exercises 25 and 26, (a) identify the claim and state H_0 and H_a, (b) find the standardized test statistic z, (c) find the corresponding P-value, (d) decide whether to reject or fail to reject the null hypothesis, and (e) interpret the decision in the context of the original claim.

25. Cotton Production A researcher claims that the mean annual production of cotton is 3.5 million bales per country. A random sample of 44 countries has a mean annual production of 2.1 million bales. Assume the population standard deviation is 4.5 million bales. At $\alpha = 0.05$, can you reject the claim? *(Source: U.S. Department of Agriculture)*

26. Cotton Consumption A researcher claims that the mean annual consumption of cotton is greater than 1.1 million bales per country. A random sample of 67 countries has a mean annual consumption of 1.0 million bales. Assume the population standard deviation is 4.3 million bales. At $\alpha = 0.01$, can you support the claim? *(Source: U.S. Department of Agriculture)*

In Exercises 27 and 28, (a) identify the claim and state H_0 and H_a, (b) find the critical value(s) and identify the rejection region(s), (c) find the standardized test statistic z, (d) decide whether to reject or fail to reject the null hypothesis, and (e) interpret the decision in the context of the original claim.

27. An environmental researcher claims that the mean amount of sulfur dioxide in the air in U.S. cities is 1.15 parts per billion. In a random sample of 134 U.S. cities, the mean amount of sulfur dioxide in the air is 0.93 parts per billion. Assume the population standard deviation is 2.62 parts per billion. At $\alpha = 0.01$, is there enough evidence to reject the claim? *(Source: U.S. Environmental Protection Agency)*

28. A travel analyst claims that the mean price of a round trip flight from New York City to Los Angeles is less than $507. In a random sample of 55 round trip flights from New York City to Los Angeles, the mean price is $502. Assume the population standard deviation is $111. At $\alpha = 0.05$, is there enough evidence to support the travel analyst's claim? *(Adapted from Expedia)*

Section 7.3

In Exercises 29–34, find the critical value(s) and rejection region(s) for the type of t-test with level of significance α and sample size n.

29. Two-tailed test, $\alpha = 0.05$, $n = 20$

30. Right-tailed test, $\alpha = 0.01$, $n = 33$

31. Right-tailed test, $\alpha = 0.02$, $n = 63$

32. Left-tailed test, $\alpha = 0.05$, $n = 48$

33. Left-tailed test, $\alpha = 0.005$, $n = 15$

34. Two-tailed test, $\alpha = 0.02$, $n = 12$

In Exercises 35–40, test the claim about the population mean μ at the level of significance α. Assume the population is normally distributed.

35. Claim: $\mu > 12{,}700$; $\alpha = 0.005$.
Sample statistics: $\bar{x} = 12{,}855$, $s = 248$, $n = 21$

36. Claim: $\mu \geq 0$; $\alpha = 0.10$. Sample statistics: $\bar{x} = -0.45$, $s = 2.38$, $n = 31$

37. Claim: $\mu \leq 51$; $\alpha = 0.01$. Sample statistics: $\bar{x} = 52$, $s = 2.5$, $n = 40$

38. Claim: $\mu < 850$; $\alpha = 0.025$. Sample statistics: $\bar{x} = 875$, $s = 25$, $n = 14$

39. Claim: $\mu = 195$; $\alpha = 0.10$. Sample statistics: $\bar{x} = 190$, $s = 36$, $n = 101$

40. Claim: $\mu \neq 3{,}330{,}000$; $\alpha = 0.05$.
Sample statistics: $\bar{x} = 3{,}293{,}995$, $s = 12{,}801$, $n = 35$

41. (a) The claim is "the mean monthly cost of joining a health club is $25."

H_0: $\mu = 25$ (claim)
H_a: $\mu \neq 25$

(b) $-t_0 = -1.740$, $t_0 = 1.740$
Rejection regions:
$t < -1.740$, $t > 1.740$

(c) 1.64   (d) Fail to reject H_0.

(e) There is not enough evidence at the 10% level of significance to reject the advertisement's claim that the mean monthly cost of joining a health club is $25.

42. (a) The claim is "the mean cost of a yoga session is no more than $14."

H_0: $\mu \leq 14$ (claim)
H_a: $\mu > 14$

(b) $t_0 = 2.040$
Rejection region: $t > 2.040$

(c) 3.46   (d) Reject H_0.

(e) There is enough evidence at the 2.5% level of significance to reject the fitness magazine's claim that the mean cost of a yoga session is no more than $14.

43. (a) The claim is "the mean score for grade 12 students on a science achievement test is more than 145."

H_0: $\mu \leq 145$
H_a: $\mu > 145$ (claim)

(b) 0.0824   (c) Reject H_0.

(d) There is enough evidence at the 10% level of significance to support the education publication's claim that the mean score for grade 12 students on a science achievement test is more than 145.

44. See Selected Answers, page A102.

45. See Odd Answers, page A72.

46. Can use normal distribution. Reject H_0. There is enough evidence at the 3% level of significance to reject the claim.

47. Can use normal distribution. Reject H_0. There is enough evidence at the 1% level of significance to support the claim.

48. Cannot use normal distribution.

In Exercises 41 and 42, (a) identify the claim and state H_0 and H_a, (b) find the critical value(s) and identify the rejection region(s), (c) find the standardized test statistic t, (d) decide whether to reject or fail to reject the null hypothesis, and (e) interpret the decision in the context of the original claim. Assume the population is normally distributed.

41. A fitness magazine advertises that the mean monthly cost of joining a health club is $25. You want to test this claim. You find that a random sample of 18 clubs has a mean monthly cost of $26.25 and a standard deviation of $3.23. At $\alpha = 0.10$, do you have enough evidence to reject the advertisement's claim?

42. A fitness magazine claims that the mean cost of a yoga session is no more than $14. You want to test this claim. You find that a random sample of 32 yoga sessions has a mean cost of $15.59 and a standard deviation of $2.60. At $\alpha = 0.025$, do you have enough evidence to reject the magazine's claim?

In Exercises 43 and 44, (a) identify the claim and state H_0 and H_a, (b) use technology to find the P-value, (c) decide whether to reject or fail to reject the null hypothesis, and (d) interpret the decision in the context of the original claim. Assume the population is normally distributed.

 43. An education publication claims that the mean score for grade 12 students on a science achievement test is more than 145. You want to test this claim. You randomly select 36 grade 12 test scores. The results are listed below. At $\alpha = 0.1$, can you support the publication's claim? *(Adapted from National Center for Education Statistics)*

188	80	175	195	201	143	119	81	118	119	165	222
109	134	200	110	199	181	79	135	124	205	90	120
216	167	198	183	173	187	143	166	147	219	206	97

44. An education researcher claims that the overall average score of 15-year-old students on an international mathematics literacy test is 494. You want to test this claim. You randomly select the average scores of 33 countries. The results are listed below. At $\alpha = 0.05$, do you have enough evidence to reject the researcher's claim? *(Source: National Center for Education Statistics)*

561	554	536	531	523	518	515	511	506	500	499
493	490	489	485	482	482	479	477	466	453	448
439	432	423	421	413	407	394	388	386	376	368

Section 7.4

In Exercises 45–48, determine whether a normal sampling distribution can be used to approximate the binomial distribution. If it can, test the claim.

45. Claim: $p = 0.15$; $\alpha = 0.05$
Sample statistics: $\hat{p} = 0.09$, $n = 40$

46. Claim: $p = 0.65$; $\alpha = 0.03$
Sample statistics: $\hat{p} = 0.76$, $n = 116$

47. Claim: $p < 0.70$; $\alpha = 0.01$
Sample statistics: $\hat{p} = 0.50$, $n = 68$

48. Claim: $p \geq 0.04$; $\alpha = 0.10$
Sample statistics: $\hat{p} = 0.03$, $n = 30$

49. (a) The claim is "over 40% of U.S. adults say they are less likely to travel to Europe in the next six months for fear of terrorist attacks."

H_0: $p \le 0.40$
H_a: $p > 0.40$ (claim)

(b) $z_0 = 2.33$
Rejection region: $z > 2.33$

(c) 1.29   (d) Fail to reject H_0.

(e) There is not enough evidence at the 1% level of significance to support the polling agency's claim that over 40% of U.S. adults say they are less likely to travel to Europe in the next six months for fear of terrorist attacks.

50. See Selected Answers, page A103.

51. Critical value: $\chi_0^2 = 30.144$
Rejection region: $\chi^2 > 30.144$

52. Critical values: $\chi_L^2 = 3.565$,
$\chi_R^2 = 29.819$
Rejection regions: $\chi^2 < 3.565$,
$\chi^2 > 29.819$

53. Critical values: $\chi_L^2 = 26.509$,
$\chi_R^2 = 55.758$
Rejection regions: $\chi^2 < 26.509$,
$\chi^2 > 55.758$

54. Critical value: $\chi_0^2 = 1.145$
Rejection region: $\chi^2 < 1.145$

55. Reject H_0. There is enough evidence at the 10% level of significance to support the claim.

56. Fail to reject H_0. There is not enough evidence at the 2.5% level of significance to reject the claim.

57. Fail to reject H_0. There is not enough evidence at the 5% level of significance to reject the claim.

58. Fail to reject H_0. There is not enough evidence at the 1% level of significance to support the claim.

59. See Odd Answers, page A73.

60. See Selected Answers, page A103.

61. You can reject H_0 at the 1% level of significance because $\chi^2 = 172.8 > 46.963$.

62. You can reject H_0 at the 5% level of significance because $\chi^2 = 43.94 > 41.923$.

In Exercises 49 and 50, (a) identify the claim and state H_0 and H_a, (b) find the critical value(s) and identify the rejection region(s), (c) find the standardized test statistic z, (d) decide whether to reject or fail to reject the null hypothesis, and (e) interpret the decision in the context of the original claim.

49. A polling agency reports that over 40% of U.S. adults say they are less likely to travel to Europe in the next six months for fear of terrorist attacks. In a random sample of 1000 U.S. adults, 42% said they are less likely to travel to Europe in the next six months for fear of terrorist attacks. At $\alpha = 0.01$, is there enough evidence to support the agency's claim? *(Adapted from Rasmussen Reports)*

50. A labor researcher claims that 6% of U.S. employees say it is likely they will be laid off in the next year. In a random sample of 547 U.S. employees, 44 said it is likely they will be laid off in the next year. At $\alpha = 0.05$, is there enough evidence to reject the researcher's claim? *(Adapted from Gallup)*

Section 7.5

In Exercises 51–54, find the critical value(s) and rejection region(s) for the type of chi-square test with sample size n and level of significance α.

51. Right-tailed test, $n = 20$, $\alpha = 0.05$

52. Two-tailed test, $n = 14$, $\alpha = 0.01$

53. Two-tailed test, $n = 41$, $\alpha = 0.10$

54. Left-tailed test, $n = 6$, $\alpha = 0.05$

In Exercises 55–58, test the claim about the population variance σ^2 or standard deviation σ at the level of significance α. Assume the population is normally distributed.

55. Claim: $\sigma^2 > 2$; $\alpha = 0.10$. Sample statistics: $s^2 = 2.95$, $n = 18$

56. Claim: $\sigma^2 \le 60$; $\alpha = 0.025$. Sample statistics: $s^2 = 72.7$, $n = 15$

57. Claim: $\sigma = 1.25$; $\alpha = 0.05$. Sample statistics: $s = 1.03$, $n = 6$

58. Claim: $\sigma \ne 0.035$; $\alpha = 0.01$. Sample statistics: $s = 0.026$, $n = 16$

In Exercises 59 and 60, (a) identify the claim and state H_0 and H_a, (b) find the critical value(s) and identify the rejection region(s), (c) find the standardized test statistic χ^2, (d) decide whether to reject or fail to reject the null hypothesis, and (e) interpret the decision in the context of the original claim. Assume the population is normally distributed.

59. A bolt manufacturer makes a type of bolt to be used in airtight containers. The manufacturer claims that the variance of the bolt widths is at most 0.01. A random sample of 28 bolts has a variance of 0.064. At $\alpha = 0.005$, is there enough evidence to reject the claim?

60. A restaurant claims that the standard deviation of the lengths of serving times is 3 minutes. A random sample of 27 serving times has a standard deviation of 3.9 minutes. At $\alpha = 0.01$, is there enough evidence to reject the claim?

61. In Exercise 59, is there enough evidence to reject the claim at the $\alpha = 0.01$ level? Explain.

62. In Exercise 60, is there enough evidence to reject the claim at the $\alpha = 0.05$ level? Explain.

Chapter Quiz

1. (a) The claim is "the mean hat size for a male is at least 7.25."

H_0: $\mu \geq 7.25$ (claim)
H_a: $\mu < 7.25$

(b) Left-tailed because the alternative hypothesis contains $<$; z-test because σ is known and the population is normally distributed.

(c) *Sample answer:* $z_0 = -2.33$; Rejection region: $z < -2.33$; -1.28

(d) Fail to reject H_0.

(e) There is not enough evidence at the 1% level of significance to reject the company's claim that the mean hat size for a male is at least 7.25.

2. (a) The claim is "the mean daily base price for renting a full-size or less expensive vehicle in Vancouver, Washington, is more than $36."

H_0: $\mu \leq 36$
H_a: $\mu > 36$ (claim)

(b) Right-tailed because the alternative hypothesis contains $>$; z-test because σ is known and $n \geq 30$.

(c) *Sample answer:* $z_0 = 1.28$; Rejection region: $z > 1.28$; 1.997

(d) Reject H_0.

(e) There is enough evidence at the 10% level of significance to support the travel analyst's claim that the mean daily base price for renting a full-size or less expensive vehicle in Vancouver, Washington, is more than $36.

3. See Odd Answers, page A73.

4. See Odd Answers, page A73.

5. See Odd Answers, page A73.

6. See Odd Answers, page A73.

Take this quiz as you would take a quiz in class. After you are done, check your work against the answers given in the back of the book.

For each exercise, perform the steps below.

(a) Identify the claim and state H_0 and H_a.

(b) Determine whether the hypothesis test is left-tailed, right-tailed, or two-tailed, and whether to use a z-test, a t-test, or a chi-square test. Explain your reasoning.

(c) Choose one of the options.

Option 1: Find the critical value(s), identify the rejection region(s), and find the appropriate standardized test statistic.

Option 2: Find the appropriate standardized test statistic and the P-value.

(d) Decide whether to reject or fail to reject the null hypothesis.

(e) Interpret the decision in the context of the original claim.

1. A hat company claims that the mean hat size for a male is at least 7.25. A random sample of 12 hat sizes has a mean of 7.15. At $\alpha = 0.01$, can you reject the company's claim? Assume the population is normally distributed and the population standard deviation is 0.27.

2. A travel analyst claims the mean daily base price for renting a full-size or less expensive vehicle in Vancouver, Washington, is more than $36. You want to test this claim. In a random sample of 40 full-size or less expensive vehicles available to rent in Vancouver, Washington, the mean daily base price is $42. Assume the population standard deviation is $19. At $\alpha = 0.10$, do you have enough evidence to support the analyst's claim? *(Adapted from Expedia)*

3. A government agency reports that the mean amount of earnings for full-time workers ages 18 to 24 with a bachelor's degree in a recent year is $47,254. In a random sample of 15 full-time workers ages 18 to 24 with a bachelor's degree, the mean amount of earnings is $50,781 and the standard deviation is $5290. At $\alpha = 0.05$, is there enough evidence to support the claim? Assume the population is normally distributed. *(Adapted from U.S. Census Bureau)*

4. A weight loss program claims that program participants have a mean weight loss of at least 10.5 pounds after 1 month. The weight losses after 1 month (in pounds) of a random sample of 40 program participants are listed below. At $\alpha = 0.01$, is there enough evidence to reject the program's claim?

4.7	6.0	7.2	8.3	9.2	10.1	14.0	11.7	12.8	10.8
11.0	7.2	8.0	4.7	11.8	10.7	6.1	8.8	7.7	8.5
9.5	10.2	5.6	6.9	7.9	8.6	10.5	9.6	5.7	9.6
12.6	12.9	6.8	12.0	5.1	14.0	9.7	10.8	9.1	12.9

5. A nonprofit consumer organization says that less than 18% of the vehicles the organization rated in a recent year have an overall score of 78 or more. In a random sample of 90 vehicles the organization rated in a recent year, 20% have an overall score of 78 or more. At $\alpha = 0.05$, can you support the organization's claim? *(Adapted from Consumer Reports)*

6. In Exercise 5, the nonprofit consumer organization says that the standard deviation of the vehicle rating scores is 11.90. A random sample of 90 vehicle rating scores has a standard deviation of 11.96. At $\alpha = 0.10$, is there enough evidence to reject the organization's claim? Assume the population is normally distributed. *(Adapted from Consumer Reports)*

7 Chapter Test

1.
(a) The claim is "more than 30% of adults have purchased a meal kit in a recent year."
H_0: $p \le 0.30$
H_a: $p > 0.30$ (claim)

(b) Right-tailed because the alternative hypothesis contains $>$; z-test because $np \ge 5$ and $nq \ge 5$.

(c) *Sample answer:* $z_0 = 1.28$; Rejection region: $z > 1.28$; -0.65

(d) Fail to reject H_0.

(e) There is not enough evidence at the 10% level of significance to support the retail grocery chain's claim that more than 30% of adults have purchased a meal kit in a recent year.

2.
(a) The claim is "the mean of the room rates for two adults at three-star hotels in Salt Lake City is $134."
H_0: $\mu = 134$ (claim)
H_a: $\mu \ne 134$

(b) Two-tailed because the alternative hypothesis contains $\ne$; z-test because σ is known and $n \ge 30$.

(c) *Sample answer:* $-z_0 = -1.645$, $z_0 = 1.645$; Rejection regions: $z < -1.645$, $z > 1.645$; 1.82

(d) Reject H_0.

(e) There is enough evidence at the 10% level of significance to reject the travel analyst's claim that the mean of the room rates for two adults at three-star hotels in Salt Lake City is $134.

3. See Selected Answers, page A103.
4. See Selected Answers, page A103.
5. See Selected Answers, page A103.
6. See Selected Answers, page A103.
7. See Selected Answers, page A103.

Take this test as you would take a test in class.

For each exercise, perform the steps below.

(a) *Identify the claim and state H_0 and H_a.*

(b) *Determine whether the hypothesis test is left-tailed, right-tailed, or two-tailed, and whether to use a z-test, a t-test, or a chi-square test. Explain your reasoning.*

(c) *Choose one of the options.*

Option 1: Find the critical value(s), identify the rejection region(s), and find the appropriate standardized test statistic.

Option 2: Find the appropriate standardized test statistic and the P-value.

(d) *Decide whether to reject or fail to reject the null hypothesis.*

(e) *Interpret the decision in the context of the original claim.*

1. A retail grocery chain owner claims that more than 30% of adults have purchased a meal kit in a recent year. In a random sample of 36 adults, 25% have purchased a meal kit in a recent year. At $\alpha = 0.10$, is there enough evidence to support the owner's claim? *(Adapted from Harris Interactive)*

2. A travel analyst claims that the mean of the room rates for two adults at three-star hotels in Salt Lake City is $134. In a random sample of 37 three-star hotels in Salt Lake City, the mean room rate for two adults is $143. Assume the population standard deviation is $30. At $\alpha = 0.10$, is there enough evidence to reject the analyst's claim? *(Adapted from Expedia)*

3. A travel analyst says that the mean price of a meal for a family of 4 in a resort restaurant is at most $100. A random sample of 33 meal prices for families of 4 has a mean of $110 and a standard deviation of $19. At $\alpha = 0.01$, is there enough evidence to reject the analyst's claim?

4. A research center claims that more than 80% of U.S. adults think that mothers should have paid maternity leave. In a random sample of 50 U.S. adults, 82% think that mothers should have paid maternity leave. At $\alpha = 0.05$, is there enough evidence to support the center's claim? *(Adapted from Pew Research Center)*

5. A nutrition bar manufacturer claims that the standard deviation of the number of grams of carbohydrates in a bar is 1.11 grams. A random sample of 26 bars has a standard deviation of 1.19 grams. At $\alpha = 0.05$, is there enough evidence to reject the manufacturer's claim? Assume the population is normally distributed.

6. A nonprofit consumer organization says that the mean price of the vehicles the organization rated in a recent year is at least $41,000. In a random sample of 150 vehicles the organization rated in a recent year, the mean price is $40,600 and the standard deviation is $17,300. At $\alpha = 0.01$, is there enough evidence to reject the organization's claim? *(Adapted from Consumer Reports)*

7. A researcher claims that the mean age of the residents of a small town is more than 38 years. The ages (in years) of a random sample of 30 residents are listed below. At $\alpha = 0.10$, is there enough evidence to support the researcher's claim? Assume the population standard deviation is 9 years.

41	44	40	30	29	46	42	53	21	29	43	46	39	35	33
42	35	43	35	24	21	29	24	25	85	56	82	87	72	31

The charts show results of studies on four-year colleges in the United States. You want to portray your college in a positive light for an advertising campaign designed to attract high school students. You decide to use hypothesis tests to show that your college is better than the average in certain aspects.

EXERCISES

1. *What Would You Test?*

What claims could you test if you wanted to convince a student to come to your college? Suppose the student you are trying to convince is mainly concerned with (a) affordability, (b) having a good experience, and (c) graduating and starting a career. List one claim for each case. State the null and alternative hypotheses for each claim.

2. *Choosing a Random Sample*

Classmates suggest conducting the following sampling techniques to test various claims. Determine whether the sample will be random. If not, suggest an alternative.

(a) Survey all the students you have class with and ask about the average time they spend daily on different activities.

(b) Randomly select former students from a list of recent graduates and ask whether they are employed.

(c) Randomly select students from a directory, ask how much debt money they borrowed to pay for college this year, and multiply by four.

3. *Supporting a Claim*

You want your test to support a positive claim about your college, not just fail to reject one. Should you state your claim so that the null hypothesis contains the claim or the alternate hypothesis contains the claim? Explain.

4. *Testing a Claim*

You want to claim that students at your college graduate with an average debt of less than $25,000. A random sample of 40 recent graduates has a mean amount borrowed of $23,475 and a standard deviation of $8000. At $\alpha = 0.05$, is there enough evidence to support your claim?

5. *Testing a Claim*

You want to claim that your college has a freshmen retention rate of at least 80%. You take a random sample of 60 of last year's freshmen and find that 54 of them still attend your college. At $\alpha = 0.05$, is there enough evidence to reject your claim?

6. *Conclusion*

Test one of the claims you listed in Exercise 1 and interpret the results. Discuss any limits of your sampling process.

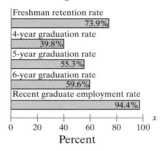

College Success

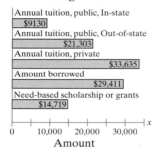

College Cost

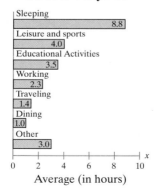

Student Daily Life

The Case of the Vanishing Women

53% ➡ 29% ➡ 9% ➡ 0%

From 1966 to 1968, Dr. Benjamin Spock and others were tried for conspiracy to violate the Selective Service Act by encouraging resistance to the Vietnam War. By a series of three selections, no women ended up being on the jury. In 1969, Hans Zeisel wrote an article in *The University of Chicago Law Review* using statistics and hypothesis testing to argue that the jury selection was biased against Dr. Spock. Dr. Spock was a well-known pediatrician and author of books about raising children. Millions of mothers had read his books and followed his advice. Zeisel argued that, by keeping women off the jury, the court prejudiced the verdict.

The jury selection process for Dr. Spock's trial is shown at the right.

Stage 1. The clerk of the Federal District Court selected 350 people "at random" from the Boston City Directory. The directory contained several hundred names, 53% of whom were women. However, only 102 of the 350 people selected were women.

Stage 2. The trial judge, Judge Ford, selected 100 people "at random" from the 350 people. This group was called a venire and it contained only nine women.

Stage 3. The court clerk assigned numbers to the members of the venire and, one by one, they were interrogated by the attorneys for the prosecution and defense until 12 members of the jury were chosen. At this stage, only one potential female juror was questioned, and she was eliminated by the prosecutor under his quota of peremptory challenges (for which he did not have to give a reason).

EXERCISES

1. The Minitab display below shows a hypothesis test for a claim that the proportion of women in the city directory is $p = 0.53$. In the test, $n = 350$ and $\hat{p} \approx 0.291$. Should you reject the claim? What is the level of significance? Explain.

2. In Exercise 1, you rejected the claim that $p = 0.53$. But this claim was true. What type of error is this?

3. When you reject a true claim with a level of significance that is virtually zero, what can you infer about the randomness of your sampling process?

4. Describe a hypothesis test for Judge Ford's "random" selection of the venire. Use a claim of

$$p = \frac{102}{350} \approx 0.291.$$

   (a) Write the null and alternative hypotheses.

   (b) Use technology to perform the test.

   (c) Make a decision.

   (d) Interpret the decision in the context of the original claim. Could Judge Ford's selection of 100 venire members have been random?

MINITAB

Test and CI for One Proportion

Test of p = 0.53 vs p ≠ 0.53

Sample	X	N	Sample p	99 % CI	Z-Value	P-Value
1	102	350	0.291429	(0.228862, 0.353995)	−8.94	0.000

Using the normal approximation.

Extended solutions are given in the technology manuals that accompany this text.
Technical instruction is provided for Minitab, Excel, and the TI-84 Plus.

7 Using Technology to Perform Hypothesis Tests

Here are some Minitab and TI-84 Plus printouts for some of the examples in this chapter.

See Example 5, page 367.

Display Descriptive Statistics...
Store Descriptive Statistics...
Graphical Summary...

1-Sample Z...
1-Sample t...
2-Sample t...
Paired t...

1 Proportion...
2 Proportions...

MINITAB

One-Sample Z

Test of $\mu = 68.3$ vs $\neq 68.3$
The assumed standard deviation = 3.5

N	Mean	SE Mean	95% CI	Z	P
25	67.200	0.700	(65.828, 68.572)	−1.57	0.116

See Example 4, page 380.

Display Descriptive Statistics...
Store Descriptive Statistics...
Graphical Summary...

1-Sample Z...
1-Sample t...
2-Sample t...
Paired t...

1 Proportion...
2 Proportions...

MINITAB

One-Sample T

Test of $\mu = 21000$ vs < 21000

N	Mean	StDev	SE Mean	95% Upper Bound	T	P
14	19189	2950	788	20585	−2.30	0.019

See Example 2, page 390.

Display Descriptive Statistics...
Store Descriptive Statistics...
Graphical Summary...

1-Sample Z...
1-Sample t...
2-Sample t...
Paired t...

1 Proportion...
2 Proportions...

MINITAB

Test and CI for One Proportion

Test of $p = 0.51$ vs $p \neq 0.51$

Sample	X	N	Sample p	90% CI	Z-Value	P-Value
1	1161	2202	0.527248	(0.509748, 0.544748)	1.62	0.105

Using the normal approximation.

See Example 9, page 371.

TI-84 PLUS

EDIT CALC **TESTS**
1: Z–Test...
2: T–Test...
3: 2–SampZTest...
4: 2–SampTTest...
5: 1–PropZTest...
6: 2–PropZTest...
7↓ ZInterval...

⬇

TI-84 PLUS

Z-Test
Inpt:Data **Stats**
μ_0:88200
σ:9500
$\bar{x}$:85900
n:20
μ:$\neq\mu_0$ **<μ_0** >μ_0
Calculate Draw

⬇

TI-84 PLUS

Z-Test
μ<88200
z=−1.082727652
p=.1394646984
$\bar{x}$=85900
n=20

⬇

TI-84 PLUS

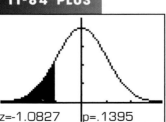

z=−1.0827 |p=.1395

See Example 5, page 381.

TI-84 PLUS

EDIT CALC **TESTS**
1: Z–Test...
2: T–Test...
3: 2–SampZTest...
4: 2–SampTTest...
5: 1–PropZTest...
6: 2–PropZTest...
7↓ ZInterval...

⬇

TI-84 PLUS

T-Test
Inpt:Data **Stats**
μ_0:6.8
$\bar{x}$:6.7
Sx:.35
n:39
μ:$\neq\mu_0$ <μ_0 >μ_0
Calculate Draw

⬇

TI-84 PLUS

T-Test
$\mu\neq$6.8
t=−1.784285142
p=.0823638462
$\bar{x}$=6.7
Sx=.35
n=39

⬇

TI-84 PLUS

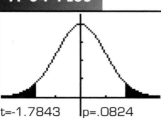

t=−1.7843 |p=.0824

See Example 1, page 389.

TI-84 PLUS

EDIT CALC **TESTS**
1: Z–Test...
2: T–Test...
3: 2–SampZTest...
4: 2–SampTTest...
5: 1–PropZTest...
6: 2–PropZTest...
7↓ ZInterval...

⬇

TI-84 PLUS

1-PropZTest
p_0:.45
x:41
n:100
prop$\neq p_0$ **<p_0** >p_0
Calculate Draw

⬇

TI-84 PLUS

1-PropZTest
prop<.45
z=−.8040302522
p=.2106896879
$\hat{p}$=.41
n=100

⬇

TI-84 PLUS

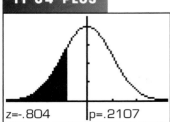

z=−.804 |p=.2107

CHAPTER 8

Hypothesis Testing with Two Samples

According to a study published in the *Journal of General Internal Medicine,* 50% of yoga users are college-educated while only 23% of non-yoga users are college-educated.

In Chapter 6, you were introduced to inferential statistics and you learned how to form a confidence interval to estimate a population parameter. Then, in Chapter 7, you learned how to test a claim about a population parameter, basing your decision on sample statistics and their sampling distributions.

Using data from the National Health Interview Survey, a study was conducted to analyze the characteristics of yoga users and non-yoga users. The study was published in the *Journal of General Internal Medicine.* Some of the results are shown below for a random sample of yoga users.

Yoga Users ($n = 1593$)

Characteristic	Frequency	Proportion
40 to 49 years old	367	0.2304
Income of $20,000 to $34,999	239	0.1500
Non-smoking	1323	0.8305

In this chapter, you will continue your study of inferential statistics and hypothesis testing. Now, however, instead of testing a hypothesis about a single population, you will learn how to test a hypothesis that compares two populations.

For instance, in the yoga study, a random sample of non-yoga users was also surveyed. Here are the study's findings for this second group.

Non-Yoga Users ($n = 29,948$)

Characteristic	Frequency	Proportion
40 to 49 years old	6,290	0.2100
Income of $20,000 to $34,999	5,990	0.2000
Non-smoking	23,360	0.7800

From these two samples, can you conclude that there is a difference in the proportion of 40- to 49-year-olds, people with an income of $20,000 to $34,999, or non-smokers between yoga users and non-yoga users? Or, might the differences in the proportions be due to chance?

In this chapter, you will learn to answer these questions by testing the hypothesis that the two proportions are equal. For instance, for non-smokers, you can conclude that the proportion of yoga users is different from the proportion of non-yoga users.

8.1 Testing the Difference Between Means (Independent Samples, σ_1 and σ_2 Known)

What You Should Learn

▶ How to determine whether two samples are independent or dependent

▶ An introduction to two-sample hypothesis testing for the difference between two population parameters

▶ How to perform a two-sample z-test for the difference between two means μ_1 and μ_2 using independent samples with σ_1 and σ_2 known

Independent and Dependent Samples ■ An Overview of Two-Sample Hypothesis Testing ■ Two-Sample z-Test for the Difference Between Means

Independent and Dependent Samples

In Chapter 7, you studied methods for testing a claim about the value of a population parameter. In this chapter, you will learn how to test a claim comparing parameters from two populations. Before learning how to test the difference between two parameters, you need to understand the distinction between **independent samples** and **dependent samples.**

DEFINITION

Two samples are **independent** when the sample selected from one population is not related to the sample selected from the second population (see top figure at the left). Two samples are **dependent** when each member of one sample corresponds to a member of the other sample (see bottom figure at the left). Dependent samples are also called **paired samples** or **matched samples.**

Independent Samples

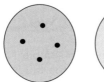

Sample 1   Sample 2

Dependent Samples

Sample 1   Sample 2

EXAMPLE 1

Independent and Dependent Samples

Classify each pair of samples as independent or dependent.

1. Sample 1: Triglyceride levels of 70 patients
   Sample 2: Triglyceride levels of the same 70 patients after using a triglyceride-lowering drug for 6 months.

2. Sample 1: Scores for 38 adult males on a psychological screening test for attention-deficit/hyperactivity disorder
   Sample 2: Scores for 50 adult females on a psychological screening test for attention-deficit/hyperactivity disorder

SOLUTION

1. These samples are dependent. Because the triglyceride levels of the same patients are taken, the samples are related. The samples can be paired with respect to each patient.

2. These samples are independent. It is not possible to form a pairing between the members of samples, the sample sizes are different, and the data represent scores for different individuals.

TRY IT YOURSELF 1

Classify each pair of samples as independent or dependent.

1. Sample 1: Systolic blood pressures of 30 adult females
   Sample 2: Systolic blood pressures of 30 adult males
2. Sample 1: Midterm exam scores of 14 chemistry students
   Sample 2: Final exam scores of the same 14 chemistry students

Answer: Page A37

Study Tip

Dependent samples often involve before and after results for the same person or object (such as a person's weight before starting a diet and after 6 weeks), or results of individuals matched for specific characteristics (such as identical twins).

An Overview of Two-Sample Hypothesis Testing

In this section and the next, you will learn how to test a claim comparing the means of two different populations using independent samples.

For instance, an advertiser is developing a marketing plan and wants to determine whether there is a difference in the amounts of time adults ages 18 to 34 and adults ages 35 to 49 spend on social media each day. The only way to conclude with certainty that there is a difference is to take a census of all adults in both age groups, calculate their mean daily times spent on social media, and find the difference. Of course, it is not practical to take such a census. However, it is possible to determine with some degree of certainty whether such a difference exists.

To determine whether a difference exists, the advertiser begins by assuming that there is no difference in the mean times of the two populations. That is,

$$\mu_1 - \mu_2 = 0. \qquad \text{Assume there is no difference.}$$

Then, by taking a random sample from each population, a two-sample hypothesis test is performed using the test statistic

$$\bar{x}_1 - \bar{x}_2 = 0. \qquad \text{Test statistic}$$

The advertiser obtains the results shown in the next two figures.

Study Tip

In the figures at the right, the members in the two samples, adults ages 18 to 34 and adults ages 35 to 49, are not matched or paired, so the samples are independent.

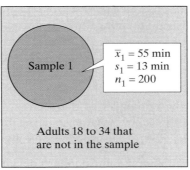

Adults 18 to 34

Sample 1

$\bar{x}_1 = 55$ min
$s_1 = 13$ min
$n_1 = 200$

Adults 18 to 34 that are not in the sample

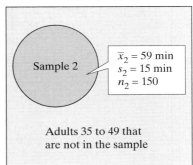

Adults 35 to 49

Sample 2

$\bar{x}_2 = 59$ min
$s_2 = 15$ min
$n_2 = 150$

Adults 35 to 49 that are not in the sample

The figure below shows the sampling distribution of $\bar{x}_1 - \bar{x}_2$ for many similar samples taken from two populations for which $\mu_1 - \mu_2 = 0$. The figure also shows the test statistic and the standardized test statistic.

Note to Instructor

In the figure at the right, the standardized test statistic is from a two-sample *t*-test. Students will study this test in the next section.

Sampling Distribution

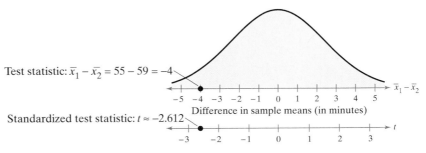

Test statistic: $\bar{x}_1 - \bar{x}_2 = 55 - 59 = -4$

Difference in sample means (in minutes)

Standardized test statistic: $t \approx -2.612$

From the figure, you can see that it is quite unlikely to obtain sample means that differ by 4 minutes assuming the actual difference is 0. The difference of the sample means would be more than 2.5 standard errors from the hypothesized difference of 0! Performing a two-sample hypothesis test using a level of significance of $\alpha = 0.10$, the advertiser can conclude that there is a difference in the amounts of time adults ages 18 to 34 and adults ages 35 to 49 spend on social media each day.

It is important to remember that when you perform a two-sample hypothesis test using independent samples, you are testing a claim concerning the difference between the parameters in two populations, not the values of the parameters themselves.

DEFINITION

For a two-sample hypothesis test with independent samples,

1. the **null hypothesis** H_0 is a statistical hypothesis that usually states there is no difference between the parameters of two populations. The null hypothesis always contains the symbol $\leq$, $=$, or $\geq$.
2. the **alternative hypothesis** H_a is a statistical hypothesis that is true when H_0 is false. The alternative hypothesis contains the symbol $>$, $\neq$, or $<$.

To write the null and alternative hypotheses for a two-sample hypothesis test with independent samples, translate the claim made about the population parameters from a verbal statement to a mathematical statement. Then, write its complementary statement. For instance, for a claim about two population parameters μ_1 and μ_2, some possible pairs of null and alternative hypotheses are

$$\begin{cases} H_0: \mu_1 = \mu_2 \\ H_a: \mu_1 \neq \mu_2 \end{cases}, \quad \begin{cases} H_0: \mu_1 \leq \mu_2 \\ H_a: \mu_1 > \mu_2 \end{cases}, \quad \text{and} \quad \begin{cases} H_0: \mu_1 \geq \mu_2 \\ H_a: \mu_1 < \mu_2 \end{cases}.$$

Regardless of which hypotheses you use, you always assume there is no difference between the population means ($\mu_1 = \mu_2$).

Two-Sample z-Test for the Difference Between Means

In the remainder of this section, you will learn how to perform a z-test for the difference between two population means μ_1 and μ_2 when the samples are *independent*. These conditions are necessary to perform such a test.

1. The population standard deviations are known.
2. The samples are randomly selected.
3. The samples are independent.
4. The populations are normally distributed *or* each sample size is at least 30.

When these conditions are met, the **sampling distribution for $\bar{x}_1 - \bar{x}_2$, the difference of the sample means,** is a normal distribution with mean and standard error as shown in the table below and the figure at the left.

Study Tip

You can also write the null and alternative hypotheses as shown below.

$$\begin{cases} H_0: \mu_1 - \mu_2 = 0 \\ H_a: \mu_1 - \mu_2 \neq 0 \end{cases}$$

$$\begin{cases} H_0: \mu_1 - \mu_2 \leq 0 \\ H_a: \mu_1 - \mu_2 > 0 \end{cases}$$

$$\begin{cases} H_0: \mu_1 - \mu_2 \geq 0 \\ H_a: \mu_1 - \mu_2 < 0 \end{cases}$$

Sampling Distribution for $\bar{x}_1 - \bar{x}_2$

In Words	In Symbols
The mean of the difference of the sample means is the assumed difference between the two population means. When no difference is assumed, the mean is 0.	$\text{Mean} = \mu_{\bar{x}_1 - \bar{x}_2}$ $= \mu_{\bar{x}_1} - \mu_{\bar{x}_2}$ $= \mu_1 - \mu_2$
The variance of the sampling distribution is the sum of the variances of the individual sampling distributions for $\bar{x}_1$ and $\bar{x}_2$. The standard error is the square root of the sum of the variances.	$\text{Standard error} = \sigma_{\bar{x}_1 - \bar{x}_2}$ $= \sqrt{\sigma_{\bar{x}_1}^2 + \sigma_{\bar{x}_2}^2}$ $= \sqrt{\dfrac{\sigma_1^2}{n_1} + \dfrac{\sigma_2^2}{n_2}}$

Note to Instructor

Point out the similarities between two-sample hypothesis tests and the one-sample tests presented in Chapter 7. Also, point out that P-values can be used in this chapter in much the same way as they were used in Chapter 7.

When the conditions on the preceding page are met and the sampling distribution for $\bar{x}_1 - \bar{x}_2$ is a normal distribution, you can use the z-test to test the difference between two population means μ_1 and μ_2. The standardized test statistic takes the form of

$$z = \frac{(\text{Observed difference}) - (\text{Hypothesized difference})}{\text{Standard error}}.$$

As you read the definition and guidelines for a two-sample z-test, note that if the null hypothesis states $\mu_1 = \mu_2$, $\mu_1 \leq \mu_2$, or $\mu_1 \geq \mu_2$, then $\mu_1 = \mu_2$ is assumed and the expression $\mu_1 - \mu_2$ is equal to 0.

Two-Sample z-Test for the Difference Between Means

A **two-sample z-test** can be used to test the difference between two population means μ_1 and μ_2 when these conditions are met.

1. Both σ_1 and σ_2 are known.
2. The samples are random.
3. The samples are independent.
4. The populations are normally distributed *or* both $n_1 \geq 30$ and $n_2 \geq 30$.

The **test statistic** is $\bar{x}_1 - \bar{x}_2$. The **standardized test statistic** is

$$z = \frac{(\bar{x}_1 - \bar{x}_2) - (\mu_1 - \mu_2)}{\sigma_{\bar{x}_1 - \bar{x}_2}} \quad \text{where} \quad \sigma_{\bar{x}_1 - \bar{x}_2} = \sqrt{\frac{\sigma_1^2}{n_1} + \frac{\sigma_2^2}{n_2}}.$$

Picturing the World

There are about 110,799 public elementary and secondary school teachers in Georgia and about 107,385 in Ohio. In a survey, 200 public elementary and secondary school teachers in each state were asked to report their salary. The results are shown below. It is claimed that the mean salary in Ohio is greater than the mean salary in Georgia. (Source: National Education Association)

Georgia
$\bar{x}_1 = \$53,375$
$n_1 = 200$

Ohio
$\bar{x}_2 = \$56,150$
$n_2 = 200$

Determine a null hypothesis and alternative hypothesis for this claim.

H_0: $\mu_1 \geq \mu_2$
H_a: $\mu_1 < \mu_2$ (claim)

GUIDELINES

Using a Two-Sample z-Test for the Difference Between Means (Independent Samples, σ_1 and σ_2 Known)

In Words	In Symbols
1. Verify that σ_1 and σ_2 are known, the samples are random and independent, and either the populations are normally distributed *or* both $n_1 \geq 30$ and $n_2 \geq 30$.	
2. State the claim mathematically and verbally. Identify the null and alternative hypotheses.	State H_0 and H_a.
3. Specify the level of significance.	Identify α.
4. Determine the critical value(s).	Use Table 4 in Appendix B.
5. Determine the rejection region(s).	
6. Find the standardized test statistic and sketch the sampling distribution.	$z = \dfrac{(\bar{x}_1 - \bar{x}_2) - (\mu_1 - \mu_2)}{\sigma_{\bar{x}_1 - \bar{x}_2}}$
7. Make a decision to reject or fail to reject the null hypothesis.	If z is in the rejection region, then reject H_0. Otherwise, fail to reject H_0.
8. Interpret the decision in the context of the original claim.	

A hypothesis test for the difference between means can also be performed using P-values. Use the guidelines above, skipping Steps 4 and 5. After finding the standardized test statistic, use Table 4 in Appendix B to calculate the P-value. Then make a decision to reject or fail to reject the null hypothesis. If P is less than or equal to α, then reject H_0. Otherwise, fail to reject H_0.

Sample Statistics for Credit Card Debt

California	Florida
$\bar{x}_1 = \$3060$	$\bar{x}_2 = \$2910$
$n_1 = 250$	$n_2 = 250$

EXAMPLE 2

See TI-84 Plus steps on page 465.

A Two-Sample *z*-Test for the Difference Between Means

A credit card watchdog group claims that there is a difference in the mean credit card debts of people in California and Florida. The results of a random survey of 250 people from each state are shown at the left. The two samples are independent. Assume that $\sigma_1 = \$960$ for California and $\sigma_2 = \$845$ for Florida. Do the results support the group's claim? Use $\alpha = 0.05$. *(Adapted from Federal Reserve Bank of New York)*

SOLUTION

Note that σ_1 and σ_2 are known, the samples are random and independent, and both n_1 and n_2 are at least 30. So, you can use the *z*-test. The claim is "there is a difference in the mean credit card debts of people in California and Florida." So, the null and alternative hypotheses are

$$H_0: \mu_1 = \mu_2 \quad \text{and} \quad H_a: \mu_1 \neq \mu_2. \quad \text{(Claim)}$$

Because the test is a two-tailed test and the level of significance is $\alpha = 0.05$, the critical values are $-z_0 = -1.96$ and $z_0 = 1.96$. The rejection regions are $z < -1.96$ and $z > 1.96$. The standardized test statistic is

$$z = \frac{(\bar{x}_1 - \bar{x}_2) - (\mu_1 - \mu_2)}{\sqrt{\dfrac{\sigma_1^2}{n_1} + \dfrac{\sigma_2^2}{n_2}}} \qquad \text{Use the } z\text{-test.}$$

$$= \frac{(3060 - 2910) - 0}{\sqrt{\dfrac{960^2}{250} + \dfrac{845^2}{250}}} \qquad \text{Assume } \mu_1 = \mu_2, \text{ so } \mu_1 - \mu_2 = 0.$$

$$\approx 1.85. \qquad \text{Round to two decimal places.}$$

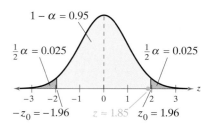

$1 - \alpha = 0.95$

$\frac{1}{2}\alpha = 0.025$ $\frac{1}{2}\alpha = 0.025$

$-z_0 = -1.96$ $z \approx 1.85$ $z_0 = 1.96$

The figure at the left shows the location of the rejection regions and the standardized test statistic *z*. Because *z* is not in the rejection region, you fail to reject the null hypothesis.

Interpretation   There is not enough evidence at the 5% level of significance to support the group's claim that there is a difference in the mean credit card debts of people in California and Florida.

TRY IT YOURSELF 2

A survey indicates that the mean annual wages for forensic science technicians working for local and state governments are $60,680 and $59,430, respectively. The survey includes a randomly selected sample of size 100 from each government branch. Assume that the population standard deviations are $6200 (local) and $5575 (state). The two samples are independent. At $\alpha = 0.10$, is there enough evidence to conclude that there is a difference in the mean annual wages? *(Adapted from U.S. Bureau of Labor Statistics)*

Answer: Page A37

In Example 2, you can also use a *P*-value to perform the hypothesis test. For instance, the test is a two-tailed test, so the *P*-value is equal to twice the area to the right of $z = 1.85$, or

$$P = 2(1 - 0.9678) = 2(0.0322) = 0.0644.$$

Because $0.0644 > 0.05$, you fail to reject H_0.

Sample Statistics for Daily Cost of Meals and Lodging for Two Adults

Texas	Virginia
$\bar{x}_1 = \$245$	$\bar{x}_2 = \$251$
$n_1 = 25$	$n_2 = 20$

EXAMPLE 3

Using Technology to Perform a Two-Sample z-Test

A travel agency claims that the average daily cost of meals and lodging for vacationing in Texas is less than the average daily cost in Virginia. The table at the left shows the results of a random survey of vacationers in each state. The two samples are independent. Assume that $\sigma_1 = \$20$ for Texas and $\sigma_2 = \$25$ for Virginia, and that both populations are normally distributed. At $\alpha = 0.01$, is there enough evidence to support the claim? *(Adapted from American Automobile Association)*

SOLUTION

Note that σ_1 and σ_2 are known, the samples are random and independent, and the populations are normally distributed. So, you can use the z-test. The claim is "the average daily cost of meals and lodging for vacationing in Texas is less than the average daily cost in Virginia." So, the null and alternative hypotheses are $H_0: \mu_1 \geq \mu_2$ and $H_a: \mu_1 < \mu_2$ (claim). The top two displays show how to set up the hypothesis test using a TI-84 Plus. The remaining displays show the results of selecting *Calculate* or *Draw*.

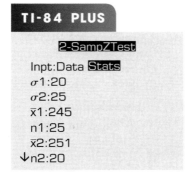

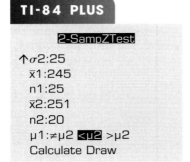

Tech Tip

Note that the TI-84 Plus displays $P \approx 0.1914$. Because $P > \alpha$, you fail to reject the null hypothesis.

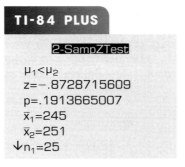

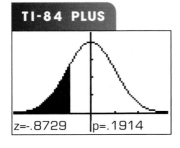

Because the test is a left-tailed test and $\alpha = 0.01$, the rejection region is $z < -2.33$. The standardized test statistic $z \approx -0.87$ is not in the rejection region, so you fail to reject the null hypothesis.

Interpretation There is not enough evidence at the 1% level of significance to support the travel agency's claim.

TRY IT YOURSELF 3

A travel agency claims that the average daily cost of meals and lodging for vacationing in Alaska is greater than the average daily cost in Colorado. The table at the left shows the results of a random survey of vacationers in each state. The two samples are independent. Assume that $\sigma_1 = \$25$ for Alaska and $\sigma_2 = \$20$ for Colorado, and that both populations are normally distributed. At $\alpha = 0.05$, is there enough evidence to support the claim? *(Adapted from American Automobile Association)* *Answer: Page A37*

Sample Statistics for Daily Cost of Meals and Lodging for Two Adults

Alaska	Colorado
$\bar{x}_1 = \$310$	$\bar{x}_2 = \$306$
$n_1 = 15$	$n_2 = 20$

8.1 EXERCISES

Building Basic Skills and Vocabulary

1. What is the difference between two samples that are dependent and two samples that are independent? Give an example of each.

2. Explain how to perform a two-sample z-test for the difference between two population means using independent samples with σ_1 and σ_2 known.

3. Describe another way you can perform a hypothesis test for the difference between the means of two populations using independent samples with σ_1 and σ_2 known that does not use rejection regions.

4. What conditions are necessary in order to use the z-test to test the difference between two population means?

Independent and Dependent Samples *In Exercises 5–8, classify the two samples as independent or dependent and justify your answer.*

5. Sample 1: The maximum bench press weights for 53 football players
Sample 2: The maximum bench press weights for the same 53 football players after completing a weight lifting program

6. Sample 1: The IQ scores of 60 females
Sample 2: The IQ scores of 60 males

7. Sample 1: The average speed of 23 powerboats using an old hull design
Sample 2: The average speed of 14 powerboats using a new hull design

8. Sample 1: The commute times of 10 workers when they use their own vehicles
Sample 2: The commute times of the same 10 workers when they use public transportation

In Exercises 9 and 10, use the TI-84 Plus display to make a decision to reject or fail to reject the null hypothesis at the level of significance. Make your decision using the standardized test statistic and using the P-value. Assume the sample sizes are equal.

9. $\alpha = 0.05$

```
    2-SampZTest
μ1≠μ2
z=2.956485408
p=.0031118068
x̄1=2500
x̄2=2425
↓n1=120
```

10. $\alpha = 0.01$

```
    2-SampZTest
μ1>μ2
z=1.941656065
p=.0260893059
x̄1=44
x̄2=42
↓n1=50
```

In Exercises 11–14, test the claim about the difference between two population means μ_1 and μ_2 at the level of significance α. Assume the samples are random and independent, and the populations are normally distributed.

11. Claim: $\mu_1 = \mu_2$; $\alpha = 0.1$
Population statistics: $\sigma_1 = 3.4$ and $\sigma_2 = 1.5$
Sample statistics: $\bar{x}_1 = 16$, $n_1 = 29$ and $\bar{x}_2 = 14$, $n_2 = 28$

Answers (margin)

1. Two samples are dependent when each member of one sample corresponds to a member of the other sample. Example: The weights of 22 people before starting an exercise program and the weights of the same 22 people 6 weeks after starting the exercise program.

Two samples are independent when the sample selected from one population is not related to the sample selected from the other population. Example: The weights of 25 cats and the weights of 20 dogs.

2. State the hypotheses and identify the claim. Specify the level of significance. Find the critical value(s) and identify the rejection region(s). Find the standardized test statistic. Make a decision and interpret it in the context of the claim.

3. Use P-values.

4. (1) The population standard deviations are known.
   (2) The samples are randomly selected.
   (3) The samples are independent.
   (4) The populations are normally distributed *or* each sample size is at least 30.

5. Dependent because the same football players were sampled.

6. Independent because different individuals were sampled.

7. Independent because different boats were sampled.

8. Dependent because the same workers were sampled.

9. Reject H_0.

10. Fail to reject H_0.

11. Reject H_0. There is enough evidence at the 1% level of significance to reject the claim.

12. Fail to reject H_0. There is not enough evidence at the 10% level of significance to support the claim.

13. Fail to reject H_0. There is not enough evidence at the 5% level of significance to support the claim.

14. Reject H_0. There is enough evidence at the 3% level of significance to reject the claim.

15. (a) The claim is "the mean braking distances are different for the two makes of automobiles."

    H_0: $\mu_1 = \mu_2$
    H_a: $\mu_1 \neq \mu_2$ (claim)

    (b) $-z_0 = -1.645$, $z_0 = 1.645$
    Rejection regions:
    $z < -1.645$, $z > 1.645$

    (c) 2.80

    (d) Reject H_0.

    (e) There is enough evidence at the 10% level of significance to support the safety engineer's claim that the mean braking distances are different for the two makes of automobiles.

16. (a) The claim is "the mean customer rating of online retailers is greater than the mean customer rating of walk-in retailers."

    H_0: $\mu_1 \leq \mu_2$
    H_a: $\mu_1 > \mu_2$ (claim)

    (b) $z_0 = 2.33$
    Rejection region: $z > 2.33$

    (c) 2.26

    (d) Fail to reject H_0.

    (e) There is not enough evidence at the 1% level of significance to support the researcher's claim that the mean customer rating of online retailers is greater than the mean customer rating of walk-in retailers.

17. See Odd Answers, page A74.

18. See Selected Answers, page A103.

12. Claim: $\mu_1 > \mu_2$; $\alpha = 0.10$
    Population statistics: $\sigma_1 = 40$ and $\sigma_2 = 15$
    Sample statistics: $\bar{x}_1 = 500$, $n_1 = 100$ and $\bar{x}_2 = 495$, $n_2 = 75$

13. Claim: $\mu_1 < \mu_2$; $\alpha = 0.05$
    Population statistics: $\sigma_1 = 75$ and $\sigma_2 = 105$
    Sample statistics: $\bar{x}_1 = 2435$, $n_1 = 35$ and $\bar{x}_2 = 2432$, $n_2 = 90$

14. Claim: $\mu_1 \leq \mu_2$; $\alpha = 0.03$
    Population statistics: $\sigma_1 = 136$ and $\sigma_2 = 215$
    Sample statistics: $\bar{x}_1 = 5004$, $n_1 = 144$ and $\bar{x}_2 = 4895$, $n_2 = 156$

Using and Interpreting Concepts

Testing the Difference Between Two Means *In Exercises 15–24, (a) identify the claim and state H_0, and H_a, (b) find the critical value(s) and identify the rejection region(s), (c) find the standardized test statistic z, (d) decide whether to reject or fail to reject the null hypothesis, and (e) interpret the decision in the context of the original claim. Assume the samples are random and independent, and the populations are normally distributed.*

15. **Braking Distances** To compare the dry braking distances from 60 to 0 miles per hour for two makes of automobiles, a safety engineer conducts braking tests for 23 models of Make A and 24 models of Make B. The mean braking distance for Make A is 137 feet. Assume the population standard deviation is 5.5 feet. The mean braking distance for Make B is 132 feet. Assume the population standard deviation is 6.7 feet. At $\alpha = 0.10$, can the engineer support the claim that the mean braking distances are different for the two makes of automobiles? *(Source: Consumer Reports)*

16. **Digital Gear Shopping** To compare customer satisfaction with holiday gift purchases of digital gear from online and walk-in retailers, a researcher randomly selects 30 customer ratings of online retailers and 31 customer ratings of walk-in retailers. The mean customer rating of online retailers is 90 out of 100. Assume the population standard deviation is 3.4. The mean customer rating of walk-in retailers is 88 out of 100. Assume the population standard deviation is 3.5. At $\alpha = 0.01$, can the researcher support the claim that the mean customer rating of online retailers is greater than the mean customer rating of walk-in retailers? *(Source: Consumer Reports)*

17. **Wind Energy** An energy company wants to choose between two regions in a state to install energy-producing wind turbines. A researcher claims that the wind speed in Region A is less than the wind speed in Region B. To test the regions, the average wind speed is calculated for 60 days in each region. The mean wind speed in Region A is 14.0 miles per hour. Assume the population standard deviation is 2.9 miles per hour. The mean wind speed in Region B is 15.1 miles per hour. Assume the population standard deviation is 3.3 miles per hour. At $\alpha = 0.05$, can the company support the researcher's claim?

18. **Repair Costs: Washing Machines** You want to buy a washing machine, and a salesperson tells you that the mean repair costs for Model A and Model B are equal. You research the repair costs. The mean repair cost of 24 Model A washing machines is $208. Assume the population standard deviation is $18. The mean repair cost of 26 Model B washing machines is $221. Assume the population standard deviation is $22. At $\alpha = 0.01$, can you reject the salesperson's claim?

19. (a) The claim is "ACT mathematics and science scores are equal."

$H_0: \mu_1 = \mu_2$ (claim)
$H_a: \mu_1 \neq \mu_2$

(b) $-z_0 = -2.575$, $z_0 = 2.575$
Rejection regions:
$z < -2.575$, $z > 2.575$

(c) -0.21

(d) Fail to reject H_0.

(e) There is not enough evidence at the 1% level of significance to reject the claim that ACT mathematics and science scores are equal.

20. (a) The claim is "ACT reading scores are higher than ACT English scores."

$H_0: \mu_1 \geq \mu_2$
$H_a: \mu_1 < \mu_2$ (claim)

(b) $z_0 = -1.28$
Rejection region: $z < -1.28$

(c) -1.47

(d) Reject H_0.

(e) There is enough evidence at the 10% level of significance to support the claim that ACT reading scores are higher than ACT English scores.

21. (a) The claim is "the mean home sales price in Casper, Wyoming, is the same as in Cheyenne, Wyoming."

$H_0: \mu_1 = \mu_2$ (claim)
$H_a: \mu_1 \neq \mu_2$

(b) $-z_0 = -2.575$, $z_0 = 2.575$
Rejection regions:
$z < -2.575$, $z > 2.575$

(c) 0.15

(d) Fail to reject H_0.

(e) There is not enough evidence at the 1% level of significance to reject the real estate agency's claim that the mean home sales price in Casper, Wyoming, is the same as in Cheyenne, Wyoming.

22. See Selected Answers, page A103.

23. See Odd Answers, page A74.

24. See Selected Answers, page A104.

19. **ACT Mathematics and Science Scores** The mean ACT mathematics score for 60 high school students is 20.6. Assume the population standard deviation is 5.4. The mean ACT science score for 75 high school students is 20.8. Assume the population standard deviation is 5.6. At $\alpha = 0.01$, can you reject the claim that ACT mathematics and science scores are equal? *(Source: ACT, Inc.)*

20. **ACT English and Reading Scores** The mean ACT English score for 120 high school students is 20.1. Assume the population standard deviation is 6.8. The mean ACT reading score for 150 high school students is 21.3. Assume the population standard deviation is 6.5. At $\alpha = 0.10$, can you support the claim that ACT reading scores are higher than ACT English scores? *(Source: ACT, Inc.)*

21. **Home Prices** A real estate agency says that the mean home sales price in Casper, Wyoming, is the same as in Cheyenne, Wyoming. The mean home sales price for 25 homes in Casper is $294,220. Assume the population standard deviation is $135,387. The mean home sales price for 25 homes in Cheyenne is $287,984. Assume the population standard deviation is $151,996. At $\alpha = 0.01$, is there enough evidence to reject the agency's claim? *(Adapted from RealtyTrac)*

22. **Home Prices** Refer to Exercise 21. Two more samples are taken, one from Casper and one from Cheyenne. For 50 homes in Casper, $\bar{x}_1 = \$231,581$. For 50 homes in Cheyenne, $\bar{x}_2 = \$315,706$. Use $\alpha = 0.01$. Do the new samples lead to a different conclusion? *(Adapted from RealtyTrac)*

23. **Precipitation** A climatologist claims that the precipitation in Seattle, Washington, was greater than in Birmingham, Alabama, in a recent year. The daily precipitation amounts (in inches) for 30 days in a recent year in Seattle are shown below. Assume the population standard deviation is 0.24 inch.

0.00	0.33	0.00	0.80	0.47	0.00	0.18	0.00	0.00	0.00
0.01	0.01	0.01	0.12	0.00	0.00	0.00	0.00	0.00	0.00
0.17	0.00	0.00	0.00	0.61	1.19	0.00	0.00	0.00	0.81

The daily precipitation amounts (in inches) for 30 days in a recent year in Birmingham are shown below. Assume the population standard deviation is 0.33 inch.

0.00	0.00	0.00	0.01	0.00	0.00	0.00	1.82	0.10	0.00
0.00	0.00	0.00	0.00	0.00	0.32	0.01	0.15	0.00	0.01
0.00	0.00	0.00	0.00	0.00	0.00	0.00	0.00	0.00	0.00

At $\alpha = 0.05$, can you support the climatologist's claim? *(Source: NOAA)*

24. **Temperature** A climatologist claims that the temperature in Seattle, Washington, was lower than in Birmingham, Alabama, in a recent year. The maximum daily temperatures (in degrees Fahrenheit) for 30 days in a recent year in Seattle are shown below. Assume the population standard deviation is 13.6°F.

51	49	47	48	49	50	75	61	87	72	62	72	84	75	68
73	72	92	64	72	66	59	59	61	57	57	45	46	46	37

The maximum daily temperatures (in degrees Fahrenheit) for 30 days in a recent year in Birmingham are shown below. Assume the population standard deviation is 15.4°F.

43	62	61	71	54	69	79	79	84	82	81	84	90	97	95
89	95	93	94	94	87	84	81	72	72	69	66	37	47	64

At $\alpha = 0.01$, can you support the climatologist's claim? *(Source: NOAA)*

25. They are equivalent through algebraic manipulation of the equation.

$\mu_1 = \mu_2 \Rightarrow \mu_1 - \mu_2 = 0$

26. They are equivalent through algebraic manipulation of the equation.

$\mu_1 \geq \mu_2 \Rightarrow \mu_1 - \mu_2 \geq 0$

Entry level software engineers in Raleigh, NC

$\bar{x}_1 = \$64,270$
$n_1 = 42$

• Raleigh

Entry level software engineers in Wichita, KS

$\bar{x}_2 = \$62,610$
$n_2 = 38$

Wichita •

FIGURE FOR EXERCISE 27

27. H_0: $\mu_1 - \mu_2 \leq 2000$
H_a: $\mu_1 - \mu_2 > 2000$ (claim)

Fail to reject H_0. There is not enough evidence at the 5% level of significance to support the claim that the difference between the mean annual salaries of entry-level software engineers in Raleigh, North Carolina, and Wichita, Kansas, is more than $2000.

28. H_0: $\mu_1 - \mu_2 = 10,000$ (claim);
H_a: $\mu_1 - \mu_2 \neq 10,000$

Reject H_0. There is enough evidence at the 1% level of significance to reject the claim that the difference between the mean annual salaries of entry-level architects in Denver, Colorado, and Los Angeles, California, is equal to $10,000.

29. $-\$3129 < \mu_1 - \mu_2 < \$6449$

30. $-\$8707 < \mu_1 - \mu_2 < \$247$

25. Getting at the Concept Explain why the null hypothesis H_0: $\mu_1 = \mu_2$ is equivalent to the null hypothesis H_0: $\mu_1 - \mu_2 = 0$.

26. Getting at the Concept Explain why the null hypothesis H_0: $\mu_1 \geq \mu_2$ is equivalent to the null hypothesis H_0: $\mu_1 - \mu_2 \geq 0$.

Extending Concepts

Testing a Difference Other Than Zero *Sometimes a researcher is interested in testing a difference in means other than zero. In Exercises 27 and 28, you will test the difference between two means using a null hypothesis of H_0: $\mu_1 - \mu_2 = k$, H_0: $\mu_1 - \mu_2 \geq k$, or H_0: $\mu_1 - \mu_2 \leq k$. The standardized test statistic is still*

$$z = \frac{(\bar{x}_1 - \bar{x}_2) - (\mu_1 - \mu_2)}{\sigma_{\bar{x}_1 - \bar{x}_2}} \quad where \quad \sigma_{\bar{x}_1 - \bar{x}_2} = \sqrt{\frac{\sigma_1^2}{n_1} + \frac{\sigma_2^2}{n_2}}.$$

27. Software Engineer Salaries Is the difference between the mean annual salaries of entry level software engineers in Raleigh, North Carolina, and Wichita, Kansas, more than $2000? To decide, you select a random sample of entry level software engineers from each city. The results of each survey are shown in the figure at the left. Assume the population standard deviations are $\sigma_1 = \$10,850$ and $\sigma_2 = \$10,970$. At $\alpha = 0.05$, what should you conclude? *(Adapted from Salary.com)*

28. Architect Salaries Is the difference between the mean annual salaries of entry level architects in Denver, Colorado, and Los Angeles, California, equal to $10,000? To decide, you select a random sample of entry level architects from each city. The results of each survey are shown in the figure. Assume the population standard deviations are $\sigma_1 = \$6520$ and $\sigma_2 = \$7130$. At $\alpha = 0.01$, what should you conclude? *(Adapted from Salary.com)*

Entry level architects in Denver, CO
$\bar{x}_1 = \$50,410$
$n_1 = 32$

• Denver

Entry level architects in Los Angeles, CA
$\bar{x}_2 = \$54,640$
$n_2 = 30$

Los Angeles •

Constructing Confidence Intervals for $\mu_1 - \mu_2$ *You can construct a confidence interval for the difference between two population means $\mu_1 - \mu_2$, as shown below, when both population standard deviations are known, and either both populations are normally distributed or both $n_1 \geq 30$ and $n_2 \geq 30$. Also, the samples must be randomly selected and independent.*

$$(\bar{x}_1 - \bar{x}_2) - z_c\sqrt{\frac{\sigma_1^2}{n_1} + \frac{\sigma_2^2}{n_2}} < \mu_1 - \mu_2 < (\bar{x}_1 - \bar{x}_2) + z_c\sqrt{\frac{\sigma_1^2}{n_1} + \frac{\sigma_2^2}{n_2}}$$

In Exercises 29 and 30, construct the indicated confidence interval for $\mu_1 - \mu_2$.

29. Software Engineer Salaries Construct a 95% confidence interval for the difference between the mean annual salaries of entry level software engineers in Raleigh, North Carolina, and Wichita, Kansas, using the data from Exercise 27.

30. Architect Salaries Construct a 99% confidence interval for the difference between the mean annual salaries of entry level architects in Denver, Colorado, and Los Angeles, California, using the data from Exercise 28.

8.2 Testing the Difference Between Means (Independent Samples, σ_1 and σ_2 Unknown)

What You Should Learn

▶ How to perform a two-sample *t*-test for the difference between two means μ_1 and μ_2 using independent samples with σ_1 and σ_2 unknown

The Two-Sample *t*-Test for the Difference Between Means

The Two-Sample *t*-Test for the Difference Between Means

In Section 8.1, you learned how to test the difference between means when both population standard deviations are known. In many real-life situations, both population standard deviations are *not* known. In this section, you will learn how to use a *t*-test to test the difference between two population means μ_1 and μ_2 using independent samples from each population when σ_1 and σ_2 are unknown. These conditions are necessary to perform such a test: (1) the population standard deviations are unknown, (2) the samples are randomly selected, (3) the samples are independent, and (4) the populations are normally distributed *or* each sample size is at least 30. When these conditions are met, the sampling distribution for the difference between the sample means $\bar{x}_1 - \bar{x}_2$ is approximated by a *t*-distribution with mean $\mu_1 - \mu_2$. So, you can use a two-sample *t*-test to test the difference between the population means μ_1 and μ_2. The standard error and the degrees of freedom of the sampling distribution depend on whether the population variances σ_1^2 and σ_2^2 are equal, as shown in the next definition.

Study Tip

To perform the two-sample *t*-test described at the right, you will need to know whether the variances of two populations are equal. In this chapter, each example and exercise will state whether the variances are equal. You will learn to test for differences between two population variances in Chapter 10.

> #### Two-Sample *t*-Test for the Difference Between Means
>
> A **two-sample *t*-test** is used to test the difference between two population means μ_1 and μ_2 when (1) σ_1 and σ_2 are unknown, (2) the samples are random, (3) the samples are independent, and (4) the populations are normally distributed *or* both $n_1 \geq 30$ and $n_2 \geq 30$. The **test statistic** is $\bar{x}_1 - \bar{x}_2$, and the **standardized test statistic** is
>
> $$ t = \frac{(\bar{x}_1 - \bar{x}_2) - (\mu_1 - \mu_2)}{s_{\bar{x}_1 - \bar{x}_2}}. $$
>
> ***Variances are equal:*** If the population variances are equal, then information from the two samples is combined to calculate a **pooled estimate of the standard deviation** $\hat{\sigma}$.
>
> $$ \hat{\sigma} = \sqrt{\frac{(n_1 - 1)s_1^2 + (n_2 - 1)s_2^2}{n_1 + n_2 - 2}} $$
>
> The standard error for the sampling distribution of $\bar{x}_1 - \bar{x}_2$ is
>
> $$ s_{\bar{x}_1 - \bar{x}_2} = \hat{\sigma} \cdot \sqrt{\frac{1}{n_1} + \frac{1}{n_2}} \qquad \text{Variances equal} $$
>
> and d.f. $= n_1 + n_2 - 2$.
>
> ***Variances are not equal:*** If the population variances are not equal, then the standard error is
>
> $$ s_{\bar{x}_1 - \bar{x}_2} = \sqrt{\frac{s_1^2}{n_1} + \frac{s_2^2}{n_2}} \qquad \text{Variances not equal} $$
>
> and d.f. $=$ smaller of $n_1 - 1$ and $n_2 - 1$.

Note to Instructor

If you choose, you can cover tests for equal variances (Section 10.3) before doing these *t*-tests. In case you cover Section 10.3 later or do not have time to cover it at all, students will be informed whether to assume equal variances in each example, Try It Yourself, and exercise in this section. In each instance, we have run an *F*-test at the level of significance stated in the problem and have reported the outcome of the test.

Picturing the World

A study published by the American Psychological Association in the journal *Neuropsychology* reported that children with musical training showed better verbal memory than children with no musical training. The study also showed that the longer the musical training, the better the verbal memory. Suppose you tried to duplicate the results as follows. A verbal memory test with a possible 100 points was administered to 90 children. Half had musical training, while the other half had no training and acted as the control group. The 45 children with training had an average score of 83.12 with a standard deviation of 5.7. The 45 students in the control group had an average score of 79.9 with a standard deviation of 6.2.

At $\alpha = 0.05$, is there enough evidence to support the claim that children with musical training have better verbal memory test scores than those without training? Assume the population variances are equal.

There is enough evidence at the 5% level of significance to support the claim that children with musical training have better verbal memory test scores than those without training.

The requirements for the z-test described in Section 8.1 and the t-test described in this section are shown in the flowchart below.

Two-Sample Tests for Independent Samples

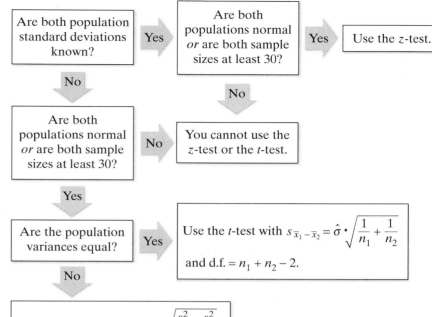

GUIDELINES

Using a Two-Sample t-Test for the Difference Between Means (Independent Samples, σ_1 and σ_2 Unknown)

In Words	In Symbols
1. Verify that σ_1 and σ_2 are unknown, the samples are random and independent, and either the populations are normally distributed *or* both $n_1 \geq 30$ and $n_2 \geq 30$.	
2. State the claim mathematically and verbally. Identify the null and alternative hypotheses.	State H_0 and H_a.
3. Specify the level of significance.	Identify α.
4. Determine the degrees of freedom.	d.f. $= n_1 + n_2 - 2$ or d.f. $=$ smaller of $n_1 - 1$ and $n_2 - 1$
5. Determine the critical value(s).	Use Table 5 in Appendix B.
6. Determine the rejection region(s).	
7. Find the standardized test statistic and sketch the sampling distribution.	$t = \dfrac{(\bar{x}_1 - \bar{x}_2) - (\mu_1 - \mu_2)}{s_{\bar{x}_1 - \bar{x}_2}}$
8. Make a decision to reject or fail to reject the null hypothesis.	If t is in the rejection region, then reject H_0. Otherwise, fail to reject H_0.
9. Interpret the decision in the context of the original claim.	

Sample Statistics for State Mathematics Test Scores

Teacher 1	Teacher 2
$\bar{x}_1 = 473$	$\bar{x}_2 = 459$
$s_1 = 39.7$	$s_2 = 24.5$
$n_1 = 8$	$n_2 = 18$

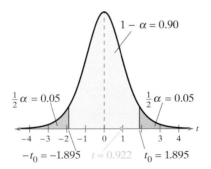

$1 - \alpha = 0.90$

$\frac{1}{2}\alpha = 0.05$ $\frac{1}{2}\alpha = 0.05$

$-t_0 = -1.895$ $t \approx 0.922$ $t_0 = 1.895$

Sample Statistics for Annual Earnings

High school diploma	Associate's degree
$\bar{x}_1 = \$36{,}875$	$\bar{x}_2 = \$44{,}900$
$s_1 = \$5475$	$s_2 = \$8580$
$n_1 = 25$	$n_2 = 16$

TI-84 PLUS

```
         2-SampTTest
μ1≠μ2
t=.9224141169
P=.37924039
df=9.458685946
x̄1=473
↓x̄2=459
```

> See Minitab steps on page 464.

EXAMPLE 1

A Two-Sample *t*-Test for the Difference Between Means

The results of a state mathematics test for random samples of students taught by two different teachers at the same school are shown at the left. Can you conclude that there is a difference in the mean mathematics test scores for the students of the two teachers? Use $\alpha = 0.10$. Assume the populations are normally distributed and the population variances are not equal.

SOLUTION

Note that σ_1 and σ_2 are unknown, the samples are random and independent, and the populations are normally distributed. So, you can use the *t*-test. The claim is "there is a difference in the mean mathematics test scores for the students of the two teachers." So, the null and alternative hypotheses are

$$H_0: \mu_1 = \mu_2 \quad \text{and} \quad H_a: \mu_1 \neq \mu_2. \quad \text{(Claim)}$$

Because the population variances are not equal and the smaller sample size is 8, use d.f. $= 8 - 1 = 7$. The test is a two-tailed test with d.f. $= 7$ and $\alpha = 0.10$, so the critical values are $-t_0 = -1.895$ and $t_0 = 1.895$. The rejection regions are $t < -1.895$ and $t > 1.895$. The standardized test statistic is

$$t = \frac{(\bar{x}_1 - \bar{x}_2) - (\mu_1 - \mu_2)}{\sqrt{\dfrac{s_1^2}{n_1} + \dfrac{s_2^2}{n_2}}} \qquad \text{Use the } t\text{-test (variances are } not \text{ equal).}$$

$$= \frac{(473 - 459) - 0}{\sqrt{\dfrac{(39.7)^2}{8} + \dfrac{(24.5)^2}{18}}} \qquad \text{Assume } \mu_1 = \mu_2, \text{ so } \mu_1 - \mu_2 = 0.$$

$$\approx 0.922. \qquad \text{Round to three decimal places.}$$

The figure at the left shows the location of the rejection regions and the standardized test statistic *t*. Because *t* is not in the rejection region, you fail to reject the null hypothesis.

Interpretation There is not enough evidence at the 10% level of significance to support the claim that the mean mathematics test scores for the students of the two teachers are different.

TRY IT YOURSELF 1

The annual earnings of 25 people with a high school diploma and 16 people with an associate's degree are shown at the left. Can you conclude that there is a difference in the mean annual earnings based on level of education? Use $\alpha = 0.05$. Assume the populations are normally distributed and the population variances are not equal. *(Adapted from U.S. Census Bureau)*

Answer: Page A37

You can also use technology and a *P*-value to perform a hypothesis test for the difference between means. For instance, in Example 1, you can enter the data in a TI-84 Plus, as shown at the left, and find $P \approx 0.379$. Because $P > \alpha$, you fail to reject the null hypothesis. Note that when using technology, the number of degrees of freedom for the *t*-test is often determined by the formula

$$\text{d.f.} = \frac{\left(s_1^2/n_1 + s_2^2/n_2\right)^2}{\left(s_1^2/n_1\right)^2/(n_1 - 1) + \left(s_2^2/n_2\right)^2/(n_2 - 1)}.$$

This formula will not be used in the text.

Sample Statistics for Sedan Driving Costs

Manufacturer	Competitor
$\bar{x}_1 = \$0.48/\text{mi}$	$\bar{x}_2 = \$0.51/\text{mi}$
$s_1 = \$0.05/\text{mi}$	$s_2 = \$0.07/\text{mi}$
$n_1 = 30$	$n_2 = 32$

EXAMPLE 2

See TI-84 Plus steps on page 465.

A Two-Sample *t*-Test for the Difference Between Means

A manufacturer claims that the mean driving cost per mile of its sedans is less than that of its leading competitor. You conduct a study using 30 randomly selected sedans from the manufacturer and 32 from the leading competitor. The results are shown at the left. At $\alpha = 0.05$, can you support the manufacturer's claim? Assume the population variances are equal. *(Adapted from American Automobile Association)*

SOLUTION

Note that σ_1 and σ_2 are unknown, the samples are random and independent, and both n_1 and n_2 are at least 30. So, you can use the *t*-test. The claim is "the mean driving cost per mile of the manufacturer's sedans is less than that of its leading competitor." So, the null and alternative hypotheses are

$$H_0: \mu_1 \geq \mu_2 \quad \text{and} \quad H_a: \mu_1 < \mu_2. \quad \text{(Claim)}$$

The population variances are equal, so d.f. $= n_1 + n_2 - 2 = 30 + 32 - 2 = 60$. Because the test is a left-tailed test with d.f. $= 60$ and $\alpha = 0.05$, the critical value is $t_0 = -1.671$. The rejection region is $t < -1.671$. To make the calculation of the standardized test statistic easier, first find the standard error.

$$s_{\bar{x}_1 - \bar{x}_2} = \sqrt{\frac{(n_1 - 1)s_1^2 + (n_2 - 1)s_2^2}{n_1 + n_2 - 2}} \cdot \sqrt{\frac{1}{n_1} + \frac{1}{n_2}}$$

$$= \sqrt{\frac{(30 - 1)(0.05)^2 + (32 - 1)(0.07)^2}{30 + 32 - 2}} \cdot \sqrt{\frac{1}{30} + \frac{1}{32}}$$

$$\approx 0.0155416$$

The standardized test statistic is

$$t = \frac{(\bar{x}_1 - \bar{x}_2) - (\mu_1 - \mu_2)}{s_{\bar{x}_1 - \bar{x}_2}} \qquad \text{Use the } t\text{-test (variances are equal).}$$

$$\approx \frac{(0.48 - 0.51) - 0}{0.0155416} \qquad \text{Assume } \mu_1 = \mu_2, \text{ so } \mu_1 - \mu_2 = 0.$$

$$\approx -1.930. \qquad \text{Round to three decimal places.}$$

The figure at the left shows the location of the rejection region and the standardized test statistic *t*. Because *t* is in the rejection region, you reject the null hypothesis.

Interpretation There is enough evidence at the 5% level of significance to support the manufacturer's claim that the mean driving cost per mile of its sedans is less than that of its competitor's.

> ### Tech Tip
> It is important to note that when using a TI-84 Plus for the two-sample *t*-test, select the *Pooled: Yes* input option when the variances are equal.

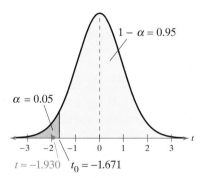

$1 - \alpha = 0.95$

$\alpha = 0.05$

$t \approx -1.930 \quad t_0 = -1.671$

TRY IT YOURSELF 2

A manufacturer claims that the mean driving cost per mile of its minivans is less than that of its leading competitor. You conduct a study using 34 randomly selected minivans from the manufacturer and 38 from the leading competitor. The results are shown at the right. At $\alpha = 0.10$, can you support the manufacturer's claim? Assume the population variances are equal. *(Adapted from American Automobile Association)*

Sample Statistics for Minivan Driving Costs

Manufacturer	Competitor
$\bar{x}_1 = \$0.52/\text{mi}$	$\bar{x}_2 = \$0.54/\text{mi}$
$s_1 = \$0.08/\text{mi}$	$s_2 = \$0.07/\text{mi}$
$n_1 = 34$	$n_2 = 38$

Answer: Page A37

8.2 EXERCISES

1. (1) The population standard deviations are unknown.

   (2) The samples are randomly selected.

   (3) The samples are independent.

   (4) The populations are normally distributed or each sample size is at least 30.

2. State hypotheses and identify the claim. Specify the level of significance. Determine the degrees of freedom. Find the critical value(s) and identify the rejection region(s). Find the standardized test statistic. Make a decision and interpret it in the context of the original claim.

3. (a) $-t_0 = -1.714$, $t_0 = 1.714$
   (b) $-t_0 = -1.812$, $t_0 = 1.812$

4. (a) $t_0 = 2.485$ (b) $t_0 = 2.718$

5. (a) $t_0 = -1.746$ (b) $t_0 = -1.943$

6. (a) $-t_0 = -2.708$, $t_0 = 2.708$
   (b) $-t_0 = -2.878$, $t_0 = 2.878$

7. (a) $t_0 = 1.729$ (b) $t_0 = 1.895$

8. (a) $t_0 = -1.296$ (b) $t_0 = -1.311$

9. Fail to reject H_0. There is not enough evidence at the 1% level of significance to reject the claim.

10. Reject H_0. There is enough evidence at the 10% level of significance to support the claim.

11. Reject H_0. There is enough evidence at the 5% level of significance to reject the claim.

12. Fail to reject H_0. There is not enough evidence at the 1% level of significance to support the claim.

13. See Odd Answers, page A74.

Sample Statistics for Annual Costs of Routine Veterinarian Visits

Dogs	Cats
$\bar{x}_1 = \$263$	$\bar{x}_2 = \$183$
$s_1 = \$30$	$s_2 = \$27$
$n_1 = 16$	$n_2 = 18$

TABLE FOR EXERCISE 13

Building Basic Skills and Vocabulary

1. What conditions are necessary in order to use the t-test to test the difference between two population means?

2. Explain how to perform a two-sample t-test for the difference between two population means.

In Exercises 3–8, use Table 5 in Appendix B to find the critical value(s) for the alternative hypothesis, level of significance α, and sample sizes n_1 and n_2. Assume that the samples are random and independent, the populations are normally distributed, and the population variances are (a) equal and (b) not equal.

3. $H_a: \mu_1 \neq \mu_2$, $\alpha = 0.10$, $n_1 = 11$, $n_2 = 14$

4. $H_a: \mu_1 > \mu_2$, $\alpha = 0.01$, $n_1 = 12$, $n_2 = 15$

5. $H_a: \mu_1 < \mu_2$, $\alpha = 0.05$, $n_1 = 7$, $n_2 = 11$

6. $H_a: \mu_1 \neq \mu_2$, $\alpha = 0.01$, $n_1 = 19$, $n_2 = 22$

7. $H_a: \mu_1 > \mu_2$, $\alpha = 0.05$, $n_1 = 13$, $n_2 = 8$

8. $H_a: \mu_1 < \mu_2$, $\alpha = 0.10$, $n_1 = 30$, $n_2 = 32$

In Exercises 9–12, test the claim about the difference between two population means μ_1 and μ_2 at the level of significance α. Assume the samples are random and independent, and the populations are normally distributed.

9. Claim: $\mu_1 = \mu_2$; $\alpha = 0.01$. Assume $\sigma_1^2 = \sigma_2^2$
   Sample statistics: $\bar{x}_1 = 33.7$, $s_1 = 3.5$, $n_1 = 12$ and $\bar{x}_2 = 35.5$, $s_2 = 2.2$, $n_2 = 17$

10. Claim: $\mu_1 < \mu_2$; $\alpha = 0.10$. Assume $\sigma_1^2 = \sigma_2^2$
    Sample statistics: $\bar{x}_1 = 0.345$, $s_1 = 0.305$, $n_1 = 11$ and $\bar{x}_2 = 0.515$, $s_2 = 0.215$, $n_2 = 9$

11. Claim: $\mu_1 \leq \mu_2$; $\alpha = 0.05$. Assume $\sigma_1^2 \neq \sigma_2^2$
    Sample statistics: $\bar{x}_1 = 2410$, $s_1 = 175$, $n_1 = 13$ and $\bar{x}_2 = 2305$, $s_2 = 52$, $n_2 = 10$

12. Claim: $\mu_1 > \mu_2$; $\alpha = 0.01$. Assume $\sigma_1^2 \neq \sigma_2^2$
    Sample statistics: $\bar{x}_1 = 52$, $s_1 = 4.8$, $n_1 = 32$ and $\bar{x}_2 = 50$, $s_2 = 1.2$, $n_2 = 40$

Using and Interpreting Concepts

Testing the Difference Between Two Means *In Exercises 13–22, (a) identify the claim and state H_0 and H_a, (b) find the critical value(s) and identify the rejection region(s), (c) find the standardized test statistic t, (d) decide whether to reject or fail to reject the null hypothesis, and (e) interpret the decision in the context of the original claim. Assume the samples are random and independent, and the populations are normally distributed.*

13. **Veterinarian Visits** A pet association claims that the mean annual costs of routine veterinarian visits for dogs and cats are the same. The results for samples of the two types of pets are shown at the left. At $\alpha = 0.10$, can you reject the pet association's claim? Assume the population variances are equal. *(Adapted from American Pet Products Association)*

Sample Statistics for Amount Spent by Customers

Burger Stop	Fry World
$\bar{x}_1 = \$5.46$	$\bar{x}_2 = \$5.12$
$s_1 = \$0.89$	$s_2 = \$0.79$
$n_1 = 22$	$n_2 = 30$

TABLE FOR EXERCISE 14

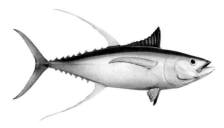

FIGURE FOR EXERCISE 16

14. (a) The claim is "the mean amount spent by a customer at Burger Stop is greater than the mean amount spent by a customer at Fry World."

H_0: $\mu_1 \le \mu_2$
H_a: $\mu_1 > \mu_2$ (claim)

(b) $t_0 = 1.676$
Rejection region: $t > 1.676$

(c) 1.45

(d) Fail to reject H_0.

(e) There is not enough evidence at the 5% level of significance to support the claim that the mean amount spent by a customer at Burger Stop is greater than the mean amount spent by a customer at Fry World.

15. See Odd Answers, page A75.

16. See Selected Answers, page A104.

17. See Odd Answers, page A75.

18. See Selected Answers, page A104.

19. See Odd Answers, page A75.

14. Transactions A magazine claims that the mean amount spent by a customer at Burger Stop is greater than the mean amount spent by a customer at Fry World. The results for samples of customer transactions for the two fast food restaurants are shown at the left. At $\alpha = 0.05$, can you support the magazine's claim? Assume the population variances are equal.

15. Blue Crabs A marine researcher claims that the stomachs of blue crabs from one location contain more fish than the stomachs of blue crabs from another location. The stomach contents of a sample of 25 blue crabs from Location A contain a mean of 320 milligrams of fish and a standard deviation of 60 milligrams. The stomach contents of a sample of 15 blue crabs from Location B contain a mean of 280 milligrams of fish and a standard deviation of 80 milligrams. At $\alpha = 0.01$, can you support the marine researcher's claim? Assume the population variances are equal.

16. Yellowfin Tuna A marine biologist claims that the mean fork length (see figure at the left) of yellowfin tuna is different in two zones in the eastern tropical Pacific Ocean. A sample of 26 yellowfin tuna collected in Zone A has a mean fork length of 76.2 centimeters and a standard deviation of 16.5 centimeters. A sample of 31 yellowfin tuna collected in Zone B has a mean fork length of 80.8 centimeters and a standard deviation of 23.4 centimeters. At $\alpha = 0.01$, can you support the marine biologist's claim? Assume the population variances are equal. *(Adapted from Fishery Bulletin)*

17. Annual Income A demographics researcher claims that the mean household income in a recent year is greater in Cuyahoga County, Ohio, than it is in Wayne County, Michigan. In Cuyahoga County, a sample of 19 residents has a mean household income of $45,600 and a standard deviation of $2800. In Wayne County, a sample of 15 residents has a mean household income of $41,500 and a standard deviation of $1310. At $\alpha = 0.05$, can you support the demographics researcher's claim? Assume the population variances are not equal. *(Adapted from U.S. Census Bureau)*

18. Annual Income A demographics researcher claims that the mean household income in a recent year is the same in Ada County, Idaho, and Cameron Parish, Louisiana. In Ada County, a sample of 18 residents has a mean household income of $58,300 and a standard deviation of $9000. In Cameron Parish, a sample of 20 residents has a mean household income of $56,600 and a standard deviation of $15,600. At $\alpha = 0.10$, can you reject the demographics researcher's claim? Assume the population variances are not equal. *(Adapted from U.S. Census Bureau)*

19. Tensile Strength The tensile strength of a metal is a measure of its ability to resist tearing when it is pulled lengthwise. An experimental method of treatment produced steel bars with the tensile strengths (in newtons per square millimeter) listed below.

Experimental Method:
391   383   333   378   368   401   339   376   366   348

The conventional method produced steel bars with the tensile strengths (in newtons per square millimeter) listed below.

Conventional Method:
362   382   368   398   381   391   400
410   396   411   385   385   395   371

At $\alpha = 0.01$, can you support the claim that the experimental method of treatment makes a difference in the tensile strength of steel bars? Assume the population variances are equal.

20. (a) The claim is "the experimental method produces steel with greater mean tensile strength."

    H_0: $\mu_1 \leq \mu_2$
    H_a: $\mu_1 > \mu_2$ (claim)

    (b) $t_0 = 1.350$
    Rejection region: $t > 1.350$

    (c) 3.543

    (d) Reject H_0.

    (e) There is enough evidence at the 10% level of significance to support the claim that the experimental method produces steel with greater mean tensile strength.

21. (a) The claim is "the new method of teaching reading produces higher reading test scores than the old method."

    H_0: $\mu_1 \geq \mu_2$
    H_a: $\mu_1 < \mu_2$ (claim)

    (b) $t_0 = -1.303$
    Rejection region: $t < -1.303$

    (c) -4.286

    (d) Reject H_0.

    (e) There is enough evidence at the 10% level of significance to support the claim that the new method of teaching reading produces higher reading test scores than the old method.

22. (a) The claim is "the mean science test score is lower for students taught using the traditional lab method than it is for students taught using the interactive simulation software."

    H_0: $\mu_1 \geq \mu_2$
    H_a: $\mu_1 < \mu_2$ (claim)

    (b) $t_0 = -2.426$
    Rejection region: $t < -2.426$

    (c) -1.722

    (d) Fail to reject H_0.

    (e) There is not enough evidence at the 5% level of significance to support the claim that the mean science test score is lower for students taught using the traditional lab method than it is for students taught using the interactive simulation software.

 20. Tensile Strength An engineer wants to compare the tensile strengths of steel bars that are produced using a conventional method and an experimental method. (The tensile strength of a metal is a measure of its ability to resist tearing when pulled lengthwise.) To do so, the engineer randomly selects steel bars that are manufactured using each method and records the tensile strengths (in newtons per square millimeter) listed below.

Experimental Method:
| 395 | 389 | 421 | 394 | 407 | 411 | 389 | 402 | 422 |
| 416 | 402 | 408 | 400 | 386 | 411 | 405 | 389 | 410 |

Conventional Method:
| 362 | 352 | 380 | 382 | 413 | 384 | 400 |
| 378 | 419 | 379 | 384 | 388 | 372 | 383 |

At $\alpha = 0.10$, can the engineer support the claim that the experimental method produces steel with a greater mean tensile strength? Assume the population variances are not equal.

 21. Teaching Methods A new method of teaching reading is being tested on third grade students. A group of third grade students is taught using the new curriculum. A control group of third grade students is taught using the old curriculum. The reading test scores for the two groups are shown in the back-to-back stem-and-leaf plot.

Old Curriculum		New Curriculum
9	3	
9 9	4	3
9 8 8 4 3 3 2 1	5	2 4
7 6 4 2 2 1 0 0	6	0 1 1 4 7 7 7 7 7 8 9 9
	7	0 1 1 2 3 3 4 9
	8	2 4

Key: $9|4|3 = 49$ for old curriculum and 43 for new curriculum

At $\alpha = 0.10$, is there enough evidence to support the claim that the new method of teaching reading produces higher reading test scores than the old method does? Assume the population variances are equal.

 22. Teaching Methods Two teaching methods and their effects on science test scores are being reviewed. A group of students is taught in traditional lab sessions. A second group of students is taught using interactive simulation software. The science test scores for the two groups are shown in the back-to-back stem-and-leaf plot.

Traditional Lab		Interactive Simulation Software
4	6	
9 9 8 8 7 6 6 3 2 1 0	7	0 4 5 5 7 7 8
9 8 5 1 1 1 0 0	8	0 0 3 4 7 8 8 9 9
2 0	9	1 3 9

Key: $0|9|1 = 90$ for traditional and 91 for interactive

At $\alpha = 0.01$, can you support the claim that the mean science test score is lower for students taught using the traditional lab method than it is for students taught using the interactive simulation software? Assume the population variances are equal.

Extending Concepts

Sample Statistics for Finishing Times of 10K Race Participants

Males	Females
$\bar{x}_1 = 0.73$ h	$\bar{x}_2 = 0.75$ h
$s_1 = 0.17$ h	$s_2 = 0.04$ h
$n_1 = 20$	$n_2 = 12$

TABLE FOR EXERCISE 23

Sample Statistics for Driving Distances

Golfer 1	Golfer 2
$\bar{x}_1 = 267$ yd	$\bar{x}_2 = 244$ yd
$s_1 = 6$ yd	$s_2 = 12$ yd
$n_1 = 9$	$n_2 = 5$

TABLE FOR EXERCISE 24

23. $-0.07 < \mu_1 - \mu_2 < 0.03$
24. $10.8 < \mu_1 - \mu_2 < 35.2$
25. $-20.8 < \mu_1 - \mu_2 < 24.8$
26. $-10.4 < \mu_1 - \mu_2 < 10.4$

Constructing Confidence Intervals for $\mu_1 - \mu_2$ *When the sampling distribution for $\bar{x}_1 - \bar{x}_2$ is approximated by a t-distribution and the population variances are not equal, you can construct a confidence interval for $\mu_1 - \mu_2$, as shown below.*

$$(\bar{x}_1 - \bar{x}_2) - t_c\sqrt{\frac{s_1^2}{n_1} + \frac{s_2^2}{n_2}} < \mu_1 - \mu_2 < (\bar{x}_1 - \bar{x}_2) + t_c\sqrt{\frac{s_1^2}{n_1} + \frac{s_2^2}{n_2}}$$

where d.f. is the smaller of $n_1 - 1$ and $n_2 - 1$

In Exercises 23 and 24, construct the indicated confidence interval for $\mu_1 - \mu_2$. Assume the populations are approximately normal with unequal variances.

23. **10K Race** To compare the mean finishing times of male and female participants in a 10K race, you randomly select several finishing times from both sexes. The results are shown at the left. Construct an 80% confidence interval for the difference in mean finishing times of male and female participants in the race. *(Adapted from Xact)*

24. **Golf** To compare the mean driving distances for two golfers, you randomly select several drives from each golfer. The results are shown at the left. Construct a 90% confidence interval for the difference in mean driving distances for the two golfers.

Constructing Confidence Intervals for $\mu_1 - \mu_2$ *When the sampling distribution for $\bar{x}_1 - \bar{x}_2$ is approximated by a t-distribution and the populations have equal variances, you can construct a confidence interval for $\mu_1 - \mu_2$, as shown below.*

$$(\bar{x}_1 - \bar{x}_2) - t_c\hat{\sigma} \cdot \sqrt{\frac{1}{n_1} + \frac{1}{n_2}} < \mu_1 - \mu_2 < (\bar{x}_1 - \bar{x}_2) + t_c\hat{\sigma} \cdot \sqrt{\frac{1}{n_1} + \frac{1}{n_2}}$$

where $\hat{\sigma} = \sqrt{\dfrac{(n_1 - 1)s_1^2 + (n_2 - 1)s_2^2}{n_1 + n_2 - 2}}$ *and d.f.* $= n_1 + n_2 - 2$

In Exercises 25 and 26, construct the indicated confidence interval for $\mu_1 - \mu_2$. Assume the populations are approximately normal with equal variances.

Sample Statistics for Number of Days Waiting for an Appointment with a Family Doctor

Miami	Seattle
$\bar{x}_1 = 28$ days	$\bar{x}_2 = 26$ days
$s_1 = 39.7$ days	$s_2 = 42.4$ days
$n_1 = 20$	$n_2 = 17$

TABLE FOR EXERCISE 25

25. **Family Doctor** To compare the mean number of days spent waiting to see a family doctor for two large cities, you randomly select several people in each city who have had an appointment with a family doctor. The results are shown at the left. Construct a 90% confidence interval for the difference in mean number of days spent waiting to see a family doctor for the two cities. *(Adapted from Merritt Hawkins)*

26. **10K Race** To compare the mean ages of male and female participants in a 10K race, you randomly select several ages from both sexes. The results are shown below. Construct a 95% confidence interval for the difference in mean ages of male and female participants in the race. *(Adapted from Xact)*

Sample Statistics for Ages of 10K Race Participants

Males	Females
$\bar{x}_1 = 41$ years	$\bar{x}_2 = 41$ years
$s_1 = 13.7$ years	$s_2 = 14.4$ years
$n_1 = 20$	$n_2 = 12$

In a study published in the *Journal of the American Medical Association,* three groups of 18- to 35-year-old participants overate for an 8-week period. The groups consumed different levels of protein in their diet. The low protein group's diet was 5% protein, the normal protein group's diet was 15% protein, and the high protein group's diet was 25% protein. The study found that the low protein group gained considerably less weight than the normal protein group or the high protein group.

You are a scientist working at a health research firm. The firm wants you to replicate the experiment. You conduct a similar experiment over an 8-week period. The results of the experiment are shown below.

	Low protein group	Normal protein group	High protein group
Weight gain (after 8 weeks)	$\bar{x}_1 = 6.8$ lb $s_1 = 1.7$ lb $n_1 = 12$	$\bar{x}_2 = 13.5$ lb $s_2 = 2.5$ lb $n_2 = 16$	$\bar{x}_3 = 14.2$ lb $s_3 = 2.1$ lb $n_3 = 15$

EXERCISES

In Exercises 1–3, perform a two-sample t-test to determine whether the mean weight gains of the two indicated studies are different. Assume the populations are normally distributed and the population variances are equal. For each exercise, write your conclusions as a sentence. Use $\alpha = 0.05$.

1. Test the weight gains of the low protein group against those in the normal protein group.

2. Test the weight gains of the low protein group against those in the high protein group.

3. Test the weight gains of the normal protein group against those in the high protein group.

4. In which comparisons in Exercises 1–3 did you find a difference in weight gains? Write a summary of your findings.

5. Construct a 95% confidence interval for $\mu_1 - \mu_2$, where μ_1 is the mean weight gain in the normal protein group and μ_2 is the mean weight gain in the high protein group. Assume the populations are normally distributed and the population variances are equal. (See Extending Concepts in Section 8.2 Exercises.)

8.3 Testing the Difference Between Means (Dependent Samples)

What You Should Learn

▷ How to perform a *t*-test to test the mean of the differences for a population of paired data

Study Tip

Recall from Section 8.1 that two samples are dependent when each member of one sample corresponds to a member of the other sample.

The *t*-Test for the Difference Between Means

The *t*-Test for the Difference Between Means

In Sections 8.1 and 8.2, you performed two-sample hypothesis tests with independent samples using the test statistic $\bar{x}_1 - \bar{x}_2$ (the difference between the means of the two samples). To perform a two-sample hypothesis test with dependent samples, you will use a different technique. You will first find the difference *d* for each data pair.

$$d = (\text{data entry in first sample}) - (\text{corresponding data entry in second sample})$$

The test statistic is the mean $\bar{d}$ of these differences

$$\bar{d} = \frac{\Sigma d}{n}.$$ Mean of the differences between paired data entries in the dependent samples

These conditions are necessary to conduct the test.

1. The samples are randomly selected.
2. The samples are dependent (paired).
3. The populations are normally distributed *or* the number *n* of pairs of data is at least 30.

When these conditions are met, the **sampling distribution for $\bar{d}$, the mean of the differences of the paired data entries in the dependent samples,** is approximated by a *t*-distribution with $n - 1$ degrees of freedom, where *n* is the number of data pairs.

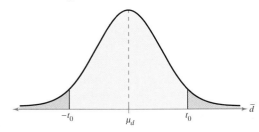

The symbols listed in the table are used for the *t*-test for μ_d. Although formulas are given for the mean and standard deviation of differences, you should use technology to calculate these statistics.

Study Tip

You can also calculate the standard deviation of the differences between paired data entries using the alternate formula

$$s_d = \sqrt{\frac{\Sigma d^2 - \frac{(\Sigma d)^2}{n}}{n - 1}}.$$

Symbol	Description
n	The number of pairs of data
d	The difference between entries in a data pair
μ_d	The hypothesized mean of the differences of paired data in the population
$\bar{d}$	The mean of the differences between the paired data entries in the dependent samples $$\bar{d} = \frac{\Sigma d}{n}$$
s_d	The standard deviation of the differences between the paired data entries in the dependent samples $$s_d = \sqrt{\frac{\Sigma (d - \bar{d})^2}{n - 1}}$$

Picturing the World

The manufacturer of an appetite suppressant claims that when its product is taken while following a low-fat diet with regular exercise for 4 months, the average weight loss is 20 pounds. To test this claim, you studied 12 randomly selected dieters taking an appetite suppressant for 4 months. The dieters followed a low-fat diet with regular exercise for all 4 months. The results are shown in the table. (Adapted from NetHealth, Inc.)

Weights (in pounds) of 12 Dieters

	Original weight	Weight after 4th month
1	185	168
2	194	177
3	213	196
4	198	180
5	244	229
6	162	144
7	211	197
8	273	252
9	178	161
10	192	178
11	181	161
12	209	193

At $\alpha = 0.10$, does your study provide enough evidence to reject the manufacturer's claim? Assume the weights are normally distributed.

There is enough evidence at the 10% level of significance to reject the manufacturer's claim that when its product is taken while following a low-fat diet with regular exercise for 4 months, the average weight loss is 20 pounds.

When you use a t-distribution to approximate the sampling distribution for $\bar{d}$, the mean of the differences between paired data entries, you can use a t-test to test a claim about the mean of the differences for a population of paired data.

t-Test for the Difference Between Means

A t-test can be used to test the difference of two population means when these conditions are met.

1. The samples are random.
2. The samples are dependent (paired).
3. The populations are normally distributed *or* $n \geq 30$.

The **test statistic** is

$$\bar{d} = \frac{\Sigma d}{n}$$

and the **standardized test statistic** is

$$t = \frac{\bar{d} - \mu_d}{s_d / \sqrt{n}}.$$

The degrees of freedom are

d.f. $= n - 1$.

GUIDELINES

Using the t-Test for the Difference Between Means (Dependent Samples)

In Words	In Symbols
1. Verify that the samples are random and dependent, and either the populations are normally distributed *or* $n \geq 30$.	
2. State the claim mathematically and verbally. Identify the null and alternative hypotheses.	State H_0 and H_a.
3. Specify the level of significance.	Identify α.
4. Identify the degrees of freedom.	d.f. $= n - 1$
5. Determine the critical value(s).	Use Table 5 in Appendix B.
6. Determine the rejection region(s).	
7. Calculate $\bar{d}$ and s_d.	$\bar{d} = \dfrac{\Sigma d}{n}$ $s_d = \sqrt{\dfrac{\Sigma (d - \bar{d})^2}{n - 1}}$
8. Find the standardized test statistic and sketch the sampling distribution.	$t = \dfrac{\bar{d} - \mu_d}{s_d / \sqrt{n}}$
9. Make a decision to reject or fail to reject the null hypothesis.	If t is in the rejection region, then reject H_0. Otherwise, fail to reject H_0.
10. Interpret the decision in the context of the original claim.	

Study Tip

To simplify the calculation of t, you can round the values of $\overline{d}$ and s_d to four decimal places, as shown in Examples 1 and 2.

Before	After	d	d^2
24	26	−2	4
22	25	−3	9
25	25	0	0
28	29	−1	1
35	33	2	4
32	34	−2	4
30	35	−5	25
27	30	−3	9
		$\Sigma = -14$	$\Sigma = 56$

Study Tip

You can also use a P-value to perform a hypothesis test for the difference between means. For instance, in Example 1, you can enter the data in Minitab (as shown on page 464) and find $P = 0.026$. Because $P < \alpha$, you reject the null hypothesis.

EXAMPLE 1

See Minitab steps on page 464.

The t-Test for the Difference Between Means

A shoe manufacturer claims that athletes can increase their vertical jump heights using the manufacturer's training shoes. The vertical jump heights of eight randomly selected athletes are measured. After the athletes have used the shoes for 8 months, their vertical jump heights are measured again. The vertical jump heights (in inches) for each athlete are shown in the table. At $\alpha = 0.10$, is there enough evidence to support the manufacturer's claim? Assume the vertical jump heights are normally distributed. *(Adapted from Coaches Sports Publishing)*

Athlete	1	2	3	4	5	6	7	8
Vertical jump height (before using shoes)	24	22	25	28	35	32	30	27
Vertical jump height (after using shoes)	26	25	25	29	33	34	35	30

SOLUTION

Because the samples are random and dependent, and the populations are normally distributed, you can use the t-test. The claim is that "athletes can increase their vertical jump heights." In other words, the manufacturer claims that an athlete's vertical jump height before using the shoes will be less than the athlete's vertical jump height after using the shoes. Each difference is given by

$$d = (\text{jump height before shoes}) - (\text{jump height after shoes}).$$

The null and alternative hypotheses are

$$H_0: \mu_d \geq 0 \quad \text{and} \quad H_a: \mu_d < 0. \quad \text{(Claim)}$$

Because the test is a left-tailed test, $\alpha = 0.10$, and d.f. $= 8 - 1 = 7$, the critical value is $t_0 = -1.415$. The rejection region is $t < -1.415$. Using the table at the left, you can calculate $\overline{d}$ and s_d as shown below. Notice that the alternate formula is used to calculate the standard deviation.

$$\overline{d} = \frac{\Sigma d}{n} = \frac{-14}{8} = -1.75$$

$$s_d = \sqrt{\frac{\Sigma d^2 - \left[\frac{(\Sigma d)^2}{n}\right]}{n-1}} = \sqrt{\frac{56 - \frac{(-14)^2}{8}}{8-1}} \approx 2.1213$$

The standardized test statistic is

$$t = \frac{\overline{d} - \mu_d}{s_d / \sqrt{n}}$$

$$\approx \frac{-1.75 - 0}{2.1213 / \sqrt{8}}$$

$$\approx -2.333.$$

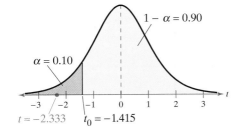

The figure shows the location of the rejection region and the standardized test statistic t. Because t is in the rejection region, you reject the null hypothesis.

Interpretation There is enough evidence at the 10% level of significance to support the shoe manufacturer's claim that athletes can increase their vertical jump heights using the manufacturer's training shoes.

TRY IT YOURSELF 1

A shoe manufacturer claims that athletes can decrease their times in the 40-yard dash using the manufacturer's training shoes. The 40-yard dash times of 12 randomly selected athletes are measured. After the athletes have used the shoes for 8 months, their 40-yard dash times are measured again. The times (in seconds) are listed in the table below. At $\alpha = 0.05$, is there enough evidence to support the manufacturer's claim? Assume the times are normally distributed. *(Adapted from Coaches Sports Publishing)*

Athlete	1	2	3	4	5	6	7	8	9	10	11	12
40-yard dash time (before using shoes)	4.85	4.90	5.08	4.72	4.62	4.54	5.25	5.18	4.81	4.57	4.63	4.77
40-yard dash time (after using shoes)	4.78	4.90	5.05	4.65	4.64	4.50	5.24	5.27	4.75	4.43	4.61	4.82

Answer: Page A37

Note in Example 1 that it is possible the vertical jump height improved because of other reasons. Many advertisements misuse statistical results by implying a cause-and-effect relationship that has not been substantiated by testing.

EXAMPLE 2

The *t*-Test for the Difference Between Means

The campaign staff for a state legislator wants to determine whether the legislator's performance rating (0–100) has changed from last year to this year. The table below shows the legislator's performance ratings from the same 16 randomly selected voters for last year and this year. At $\alpha = 0.01$, is there enough evidence to conclude that the legislator's performance rating has changed? Assume the performance ratings are normally distributed.

Tech Tip

One way to use technology to perform a hypothesis test for the difference between means is to enter the data in two columns and form a third column in which you calculate the difference for each pair. You can now perform a one-sample *t*-test on the difference column, as shown in Chapter 7.

Voter	1	2	3	4	5	6	7	8
Rating (last year)	60	54	78	84	91	25	50	65
Rating (this year)	56	48	70	60	85	40	40	55

Voter	9	10	11	12	13	14	15	16
Rating (last year)	68	81	75	45	62	79	58	63
Rating (this year)	80	75	78	50	50	85	53	60

SOLUTION

Because the samples are random and dependent, and the populations are normally distributed, you can use the *t*-test. If there is a change in the legislator's rating, then there will be a difference between last year's ratings and this year's ratings. Because the legislator wants to determine whether there is a difference, the null and alternative hypotheses are

$$H_0: \mu_d = 0 \qquad \text{and} \qquad H_a: \mu_d \neq 0. \quad \text{(Claim)}$$

Because the test is a two-tailed test, $\alpha = 0.01$, and d.f. $= 16 - 1 = 15$, the critical values are $-t_0 = -2.947$ and $t_0 = 2.947$. The rejection regions are $t < -2.947$ and $t > 2.947$.

Before	After	d	d^2
60	56	4	16
54	48	6	36
78	70	8	64
84	60	24	576
91	85	6	36
25	40	−15	225
50	40	10	100
65	55	10	100
68	80	−12	144
81	75	6	36
75	78	−3	9
45	50	−5	25
62	50	12	144
79	85	−6	36
58	53	5	25
63	60	3	9
		$\Sigma = 53$	$\Sigma = 1581$

Using the table at the left, you can calculate $\overline{d}$ and s_d as shown below.

$$\overline{d} = \frac{\Sigma d}{n} = \frac{53}{16} = 3.3125$$

$$s_d = \sqrt{\frac{\Sigma d^2 - \left[\frac{(\Sigma d)^2}{n}\right]}{n - 1}}$$

$$= \sqrt{\frac{1581 - \frac{53^2}{16}}{16 - 1}}$$

$$\approx 9.6797$$

The standardized test statistic is

$$t = \frac{\overline{d} - \mu_d}{s_d / \sqrt{n}} \qquad \text{Use the } t\text{-test.}$$

$$\approx \frac{3.3125 - 0}{9.6797 / \sqrt{16}} \qquad \text{Assume } \mu_d = 0.$$

$$\approx 1.369.$$

You can check this result using technology, as shown below using StatCrunch.

STATCRUNCH

Paired T hypothesis test:
$\mu_D = \mu_1 - \mu_2$: Mean of the difference between Last year and This year
$H_0 : \mu_D = 0$
$H_A : \mu_D \neq 0$

Hypothesis test results:

Difference	Mean	Std. Err.	DF	T-Stat	P-value
Last year−This year	3.3125	2.4199152	15	1.3688496	0.1912

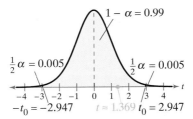

The figure at the left shows the location of the rejection region and the standardized test statistic t. Because t is not in the rejection region, you fail to reject the null hypothesis.

Interpretation   There is not enough evidence at the 1% level of significance to conclude that the legislator's performance rating has changed.

TRY IT YOURSELF 2

A medical researcher wants to determine whether a drug changes the body's temperature. Seven test subjects are randomly selected, and the body temperature (in degrees Fahrenheit) of each is measured. The subjects are then given the drug and, after 20 minutes, the body temperature of each is measured again. The results are listed below. At $\alpha = 0.05$, is there enough evidence to conclude that the drug changes the body's temperature? Assume the body temperatures are normally distributed.

Subject	1	2	3	4	5	6	7
Initial temperature	101.8	98.5	98.1	99.4	98.9	100.2	97.9
Second temperature	99.2	98.4	98.2	99.0	98.6	99.7	97.8

Answer: Page A37

8.3 EXERCISES

For Extra Help: **MyLab Statistics**

Building Basic Skills and Vocabulary

1. (1) The samples are randomly selected.

(2) The samples are dependent.

(3) The populations are normally distributed or the number n of pairs of data is at least 30.

2. The symbol $\overline{d}$ represents the mean of the differences between the paired data entries in the dependent samples.

The symbol s_d represents the standard deviation of the differences between the paired data entries in the dependent samples.

3. Fail to reject H_0. There is not enough evidence at the 5% level of significance to support the claim.

4. Fail to reject H_0. There is not enough evidence at the 1% level of significance to reject the claim.

5. Reject H_0. There is enough evidence at the 10% level of significance to reject the claim.

6. Reject H_0. There is enough evidence at the 5% level of significance to support the claim.

7. Reject H_0. There is enough evidence at the 1% level of significance to reject the claim.

8. Fail to reject H_0. There is not enough evidence at the 10% level of significance to support the claim.

9. See Odd Answers, page A75.

10. See Selected Answers, page A104.

1. What conditions are necessary in order to use the dependent samples t-test for the mean of the differences for a population of paired data?

2. Explain what the symbols $\overline{d}$ and s_d represent.

In Exercises 3–8, test the claim about the mean of the differences for a population of paired data at the level of significance α. Assume the samples are random and dependent, and the populations are normally distributed.

3. Claim: $\mu_d < 0$; $\alpha = 0.05$. Sample statistics: $\overline{d} = 1.5, s_d = 3.2, n = 14$

4. Claim: $\mu_d = 0$; $\alpha = 0.01$. Sample statistics: $\overline{d} = 3.2, s_d = 8.45, n = 8$

5. Claim: $\mu_d \leq 0$; $\alpha = 0.10$. Sample statistics: $\overline{d} = 6.5, s_d = 9.54, n = 16$

6. Claim: $\mu_d > 0$; $\alpha = 0.05$. Sample statistics: $\overline{d} = 0.55, s_d = 0.99, n = 28$

7. Claim: $\mu_d \geq 0$; $\alpha = 0.01$. Sample statistics: $\overline{d} = -2.3, s_d = 1.2, n = 15$

8. Claim: $\mu_d \neq 0$; $\alpha = 0.10$. Sample statistics: $\overline{d} = -1, s_d = 2.75, n = 20$

Using and Interpreting Concepts

Testing the Difference Between Two Means *In Exercises 9–20, (a) identify the claim and state H_0 and H_a, (b) find the critical value(s) and identify the rejection region(s), (c) calculate $\overline{d}$ and s_d, (d) find the standardized test statistic t, (e) decide whether to reject or fail to reject the null hypothesis, and (f) interpret the decision in the context of the original claim. Assume the samples are random and dependent, and the populations are normally distributed.*

9. Dow Jones Stocks A stock market analyst claims that seven of the stocks that make up the Dow Jones Industrial Average lost value from one hour to the next on one business day. The table shows the prices (in dollars per share) of the seven stocks at one time during the day and then an hour later. At $\alpha = 0.01$, is there enough evidence to support the analyst's claim? *(Source: MarketWatch)*

Stock	1	2	3	4	5	6	7
Price (first hour)	183.72	31.46	64.26	68.11	174.53	35.57	92.85
Price (second hour)	182.85	31.62	64.20	68.20	174.18	35.65	93.19

 10. SAT Scores An instructor for a SAT preparation course claims that the course will improve the test scores of students. The table shows the critical reading scores for 10 students the first two times they took the SAT. Before taking the SAT for the second time, the students took the instructor's course to try to improve their critical reading SAT scores. At $\alpha = 0.01$, is there enough evidence to support the instructor's claim?

Student	1	2	3	4	5	6	7	8	9	10
Score (first)	300	450	350	430	300	470	420	370	320	410
Score (second)	400	520	400	490	340	580	450	400	390	450

11. (a) The claim is "caffeine ingestion improves repeated freestyle sprints in trained male swimmers."

H_0: $\mu_d \leq 0$
H_a: $\mu_d > 0$ (claim)

(b) $t_0 = 3.365$
Rejection region: $t > 3.365$

(c) $\overline{d} \approx 0.533$; $s_d \approx 0.350$

(d) 3.730

(e) Reject H_0.

(f) There is enough evidence at the 1% level of significance to support the researcher's claim that caffeine ingestion improves repeated freestyle sprints in trained male swimmers.

12. (a) The claim is "a baseball clinic will help players raise their batting averages."

H_0: $\mu_d \geq 0$
H_a: $\mu_d < 0$ (claim)

(b) $t_0 = -1.771$
Rejection region: $t < -1.771$

(c) $\overline{d} \approx -0.002$; $s_d \approx 0.015$

(d) -0.499

(e) Fail to reject H_0.

(f) There is not enough evidence at the 5% level of significance to support the claim that the baseball clinic will help players raise their batting averages.

13. (a) The claim is "soft tissue therapy helps to reduce the numbers of days per week patients suffer from headaches."

H_0: $\mu_d \leq 0$
H_a: $\mu_d > 0$ (claim)

(b) $t_0 = 2.567$
Rejection region: $t > 2.567$

(c) $\overline{d} = 1.5$; $s_d \approx 1.249$

(d) 5.095

(e) Reject H_0.

(f) There is enough evidence at the 1% level of significance to support the physical therapist's claim that soft tissue therapy helps to reduce the numbers of days per week patients suffer from headaches.

11. Caffeine Ingestion A researcher claims that caffeine ingestion improves repeated freestyle sprints in trained male swimmers. The table shows the mean performance times (in seconds) for a group of trained male swimmers who complete six 75-meter maximal freestyle sprints after ingesting a placebo and after ingesting caffeine. At $\alpha = 0.01$, is there enough evidence to support the researcher's claim? *(Source: Journal of Sports Science & Medicine)*

Sprint number	1	2	3	4	5	6
Sprint time (with placebo)	40.2	40.3	40.7	41.0	40.7	40.6
Sprint time (with caffeine)	39.9	39.9	39.7	40.1	40.2	40.5

12. Batting Averages A coach claims that a baseball clinic will help players raise their batting averages. The table shows the batting averages of 14 players before participating in the clinic and two months after participating in the clinic. At $\alpha = 0.05$, is there enough evidence to support the coach's claim?

Player	1	2	3	4	5	6	7
Batting average (before clinic)	0.290	0.275	0.278	0.310	0.302	0.325	0.256
Batting average (after clinic)	0.295	0.320	0.280	0.300	0.298	0.330	0.260

Player	8	9	10	11	12	13	14
Batting average (before clinic)	0.350	0.380	0.316	0.270	0.300	0.330	0.340
Batting average (after clinic)	0.345	0.380	0.315	0.280	0.282	0.336	0.325

13. Headaches A physical therapist suggests that soft tissue massage therapy helps to reduce the numbers of days per week patients suffer from headaches. The table shows the numbers of days per week 18 patients suffered from headaches before and after 6 weeks of receiving massage therapy. At $\alpha = 0.01$, is there enough evidence to support the therapist's claim? *(Adapted from Annals of Musculoskeletal Medicine)*

Patient	1	2	3	4	5	6	7	8	9
Days (before)	4	5	6	5	5	6	5	4	4
Days (after)	5	3	3	4	3	4	3	5	2

Patient	10	11	12	13	14	15	16	17	18
Days (before)	5	5	3	4	4	5	5	5	5
Days (after)	4	1	3	2	2	3	4	4	3

14. (a) The claim is "the use of a specific type of therapeutic tape reduces pain in patients with chronic tennis elbow."

H_0: $\mu_d \le 0$
H_a: $\mu_d > 0$ (claim)

(b) $t_0 = 1.761$
Rejection region: $t > 1.761$

(c) $\bar{d} \approx 2.133$; $s_d \approx 1.885$

(d) 4.383

(e) Reject H_0.

(f) There is enough evidence at the 5% level of significance to support the physical therapist's claim that the use of a specific type of therapeutic tape reduces pain in patients with chronic tennis elbow.

15. (a) The claim is "student housing rates have increased from one academic year to the next."

H_0: $\mu_d \ge 0$
H_a: $\mu_d < 0$ (claim)

(b) $t_0 = -1.796$
Rejection region: $t < -1.796$

(c) $\bar{d} = -254.5$; $s_d \approx 291.767$

(d) -3.022

(e) Reject H_0.

(f) There is enough evidence at the 5% level of significance to support the college administrator's claim that student housing rates have increased from one academic year to the next.

16. (a) The claim is "stipends for PhD students have increased from one academic year to the next."

H_0: $\mu_d \ge 0$
H_a: $\mu_d < 0$ (claim)

(b) $t_0 = -1.943$
Rejection region: $t < -1.943$

(c) $\bar{d} \approx -664.4$; $s_d \approx 2472.373$

(d) -0.711

(e) Fail to reject H_0.

(f) There is not enough evidence at the 5% level of significance to support the education researcher's claim that stipends for PhD students have increased from one academic year to the next.

 14. Therapeutic Taping A physical therapist claims that the use of a specific type of therapeutic tape reduces pain in patients with chronic tennis elbow. The table shows the pain levels on a scale of 0 to 10, where 0 is no pain and 10 is the worst pain possible, for 15 patients with chronic tennis elbow when holding a 1 kilogram weight. At $\alpha = 0.05$, is there enough evidence to support the therapist's claim? *(Adapted from BioMed Central, Ltd.)*

Patient	1	2	3	4	5	6	7	8
Pain level (before taping)	7	7	4	4	8	5	3	9
Pain level (after taping)	6	5	1	1	4	0	0	5

Patient	9	10	11	12	13	14	15
Pain level (before taping)	3	1	1	3	5	4	2
Pain level (after taping)	0	0	3	2	3	1	3

 15. Student Housing A college administrator suggests that student housing rates have increased from one academic year to the next. The table shows the rates (in dollars per academic year) for 12 student housing arrangement options in two consecutive academic years. At $\alpha = 0.05$, is there enough evidence to support the college administrator's claim? *(Source: The University of Kansas)*

Option	1	2	3	4	5	6
Rate (first year)	4488	5738	5738	6064	6150	6150
Rate (second year)	4616	5910	5910	6246	6246	6246

Option	7	8	9	10	11	12
Rate (first year)	7288	9230	5738	5738	6064	7288
Rate (second year)	7518	9516	5910	5910	6246	8454

16. PhD Stipends An education researcher claims that stipends for PhD students have increased from one academic year to the next. The table shows the PhD stipends (in dollars per academic year) for seven different fields of study at various institutions in two consecutive academic years. At $\alpha = 0.05$, is there enough evidence to support the education researcher's claim? *(Source: PhDStipends.com)*

Field of Study	1	2	3	4	5	6	7
Stipend (first year)	30,600	29,658	15,200	26,233	11,900	33,000	33,590
Stipend (second year)	32,850	30,770	19,800	27,100	11,000	33,000	30,312

17. (a) The claim is "the product ratings have changed from last year to this year."

$H_0: \mu_d = 0$
$H_a: \mu_d \neq 0$ (claim)

(b) $-t_0 = -2.365$, $t_0 = 2.365$
Rejection regions:
$t < -2.365$, $t > 2.365$

(c) $\bar{d} = -1$; $s_d \approx 1.309$

(d) -2.161 (*Tech:* -2.160)

(e) Fail to reject H_0.

(f) There is not enough evidence at the 5% level of significance to support the claim that the product ratings have changed from last year to this year.

18. (a) The claim is "the pass completion percentages have changed."

$H_0: \mu_d = 0$
$H_a: \mu_d \neq 0$ (claim)

(b) $-t_0 = -1.833$, $t_0 = 1.833$
Rejection regions:
$t < -1.833$, $t > 1.833$

(c) $\bar{d} = -1.7$; $s_d \approx 4.020$

(d) -1.337

(e) Fail to reject H_0.

(f) There is not enough evidence at the 10% level of significance to support the claim that the pass completion percentages have changed.

19. (a) The claim is "eating a new cereal as part of a daily diet lowers total blood cholesterol levels."

$H_0: \mu_d \leq 0$
$H_a: \mu_d > 0$ (claim)

(b) $t_0 = 1.943$
Rejection region: $t > 1.943$

(c) $\bar{d} \approx 2.857$; $s_d \approx 4.451$

(d) 1.698

(e) Fail to reject H_0.

(f) There is not enough evidence at the 5% level of significance to support the claim that the new cereal lowers total blood cholesterol levels.

20. See Selected Answers, page A104.

---

17. Product Ratings A company claims that its consumer product ratings (0–10) have changed from last year to this year. The table shows the company's product ratings from the same eight consumers for last year and this year. At $\alpha = 0.05$, is there enough evidence to support the company's claim?

Consumer	1	2	3	4	5	6	7	8
Rating (last year)	5	7	2	3	9	10	8	7
Rating (this year)	5	9	4	6	9	9	9	8

18. Pass Completion Percentages The pass completion percentages of 10 college football quarterbacks for their freshman and sophomore seasons are shown in the table below. At $\alpha = 0.10$, is there enough evidence to support the claim that the pass completion percentages have changed? (*Source: Sports Reference, LLC*)

Player	1	2	3	4	5	6	7	8	9	10
Pass completion percentage (freshman)	67.9	61.5	56.8	60.0	63.6	50.0	57.0	63.1	54.7	58.5
Pass completion percentage (sophomore)	67.8	56.5	63.5	60.7	61.9	57.9	62.3	62.1	56.2	61.2

19. Cholesterol Levels A food manufacturer claims that eating its new cereal as part of a daily diet lowers total blood cholesterol levels. The table shows the total blood cholesterol levels (in milligrams per deciliter of blood) of seven patients before eating the cereal and after one year of eating the cereal as part of their diets. At $\alpha = 0.05$, is there enough evidence to support the food manufacturer's claim?

Patient	1	2	3	4	5	6	7
Total blood cholesterol level (before)	210	225	240	250	255	270	235
Total blood cholesterol level (after)	200	220	245	248	252	268	232

20. Obstacle Course On a television show, eight contestants try to lose the highest percentage of weight in order to win a cash prize. As part of the show, the contestants are timed as they run an obstacle course. The table shows the times (in seconds) of the contestants at the beginning of the season and at the end of the season. At $\alpha = 0.01$, is there enough evidence to support the claim that the contestants' times have changed?

Contestant	1	2	3	4	5	6	7	8
Time (beginning)	130.2	104.8	100.1	136.4	125.9	122.6	150.4	158.2
Time (end)	121.5	100.7	90.2	135.0	112.1	120.5	139.8	142.9

21. Yes; $P \approx 0.0058 < 0.05$, so you reject H_0.
22. No; $P \approx 0.2139 > 0.10$, so you reject H_0.
23. $-1.76 < \mu_d < -1.29$
24. $-0.83 < \mu_d < -0.05$

Extending Concepts

21. In Exercise 15, use technology to perform the hypothesis test with a *P*-value. Compare your result with the result obtained using rejection regions. Are they the same?

22. In Exercise 18, use technology to perform the hypothesis test with a *P*-value. Compare your result with the result obtained using rejection regions. Are they the same?

Constructing Confidence Intervals for μ_d *To construct a confidence interval for μ_d, use the inequality below.*

$$\bar{d} - t_c \frac{s_d}{\sqrt{n}} < \mu_d < \bar{d} + t_c \frac{s_d}{\sqrt{n}}$$

In Exercises 23 and 24, construct the indicated confidence interval for μ_d. Assume the populations are normally distributed.

23. Drug Testing A sleep disorder specialist wants to test the effectiveness of a new drug that is reported to increase the number of hours of sleep patients get during the night. To do so, the specialist randomly selects 16 patients and records the number of hours of sleep each gets with and without the new drug. The table shows the results of the two-night study. Construct a 90% confidence interval for μ_d.

Patient	1	2	3	4	5	6	7	8
Hours of sleep (without the drug)	1.8	2.0	3.4	3.5	3.7	3.8	3.9	3.9
Hours of sleep (using the drug)	3.0	3.6	4.0	4.4	4.5	5.2	5.5	5.7

Patient	9	10	11	12	13	14	15	16
Hours of sleep (without the drug)	4.0	4.9	5.1	5.2	5.0	4.5	4.2	4.7
Hours of sleep (using the drug)	6.2	6.3	6.6	7.8	7.2	6.5	5.6	5.9

24. Herbal Medicine Testing A sleep disorder specialist wants to test whether herbal medicine increases the number of hours of sleep patients get during the night. To do so, the specialist randomly selects 14 patients and records the number of hours of sleep each gets with and without the new drug. The table shows the results of the two-night study. Construct a 95% confidence interval for μ_d.

Patient	1	2	3	4	5	6	7
Hours of sleep (without medicine)	1.0	1.4	3.4	3.7	5.1	5.1	5.2
Hours of sleep (using medicine)	2.9	3.3	3.5	4.4	5.0	5.0	5.2

Patient	8	9	10	11	12	13	14
Hours of sleep (without medicine)	5.3	5.5	5.8	4.2	4.8	2.9	4.5
Hours of sleep (using medicine)	5.3	6.0	6.5	4.4	4.7	3.1	4.7

8.4   Testing the Difference Between Proportions

What You Should Learn

▶ How to perform a two-sample z-test for the difference between two population proportions p_1 and p_2

Study Tip

You can also write the null and alternative hypotheses as shown below.

$$\begin{cases} H_0\colon p_1 - p_2 = 0 \\ H_a\colon p_1 - p_2 \neq 0 \end{cases}$$

$$\begin{cases} H_0\colon p_1 - p_2 \leq 0 \\ H_a\colon p_1 - p_2 > 0 \end{cases}$$

$$\begin{cases} H_0\colon p_1 - p_2 \geq 0 \\ H_a\colon p_1 - p_2 < 0 \end{cases}$$

Study Tip

The symbols in the table below are used in the z-test for $p_1 - p_2$. See Sections 4.2 and 5.5 to review the binomial distribution.

Symbol	Description
p_1, p_2	Population proportions
x_1, x_2	Number of successes in each sample
n_1, n_2	Size of each sample
$\hat{p}_1, \hat{p}_2$	Sample proportions of successes
$\bar{p}$	Weighted estimate of p_1 and p_2
$\bar{q}$	Weighted estimate of q_1 and q_2, $\bar{q} = 1 - \bar{p}$

Two-Sample z-Test for the Difference Between Proportions

Two-Sample z-Test for the Difference Between Proportions

In this section, you will learn how to use a z-test to test the difference between two population proportions p_1 and p_2 using a sample proportion from each population. If a claim is about two population parameters p_1 and p_2, then some possible pairs of null and alternative hypotheses are

$$\begin{cases} H_0\colon p_1 = p_2, \\ H_a\colon p_1 \neq p_2, \end{cases} \quad \begin{cases} H_0\colon p_1 \leq p_2, \\ H_a\colon p_1 > p_2, \end{cases} \quad \text{and} \quad \begin{cases} H_0\colon p_1 \geq p_2 \\ H_a\colon p_1 < p_2. \end{cases}$$

Regardless of which hypotheses you use, you always assume there is no difference between the population proportions $(p_1 = p_2)$.

For instance, suppose you want to determine whether the proportion of female college students who earn a bachelor's degree in four years is different from the proportion of male college students who earn a bachelor's degree in four years. These conditions are necessary to use a z-test to test such a difference.

1. The samples are randomly selected.

2. The samples are independent.

3. The samples are large enough to use a normal sampling distribution. That is, $n_1 p_1 \geq 5$, $n_1 q_1 \geq 5$, $n_2 p_2 \geq 5$, and $n_2 q_2 \geq 5$.

When these conditions are met, the **sampling distribution for $\hat{p}_1 - \hat{p}_2$, the difference between the sample proportions,** is a normal distribution with mean

$$\mu_{\hat{p}_1 - \hat{p}_2} = p_1 - p_2$$

and standard error

$$\sigma_{\hat{p}_1 - \hat{p}_2} = \sqrt{\frac{p_1 q_1}{n_1} + \frac{p_2 q_2}{n_2}}.$$

Notice that you need to know the population proportions to calculate the standard error. Because a hypothesis test for $p_1 - p_2$ is based on the assumption that $p_1 = p_2$, you can calculate a weighted estimate of p_1 and p_2 using

$$\bar{p} = \frac{x_1 + x_2}{n_1 + n_2}$$

where $x_1 = n_1 \hat{p}_1$ and $x_2 = n_2 \hat{p}_2$. With the weighted estimate $\bar{p}$, the standard error of the sampling distribution for $\hat{p}_1 - \hat{p}_2$ is

$$\sigma_{\hat{p}_1 - \hat{p}_2} = \sqrt{\bar{p}\,\bar{q}\left(\frac{1}{n_1} + \frac{1}{n_2}\right)}$$

where $\bar{q} = 1 - \bar{p}$.

Also, you need to know the population proportions to verify that the samples are large enough to be approximated by the normal distribution. But when determining whether the z-test can be used for the difference between proportions for a binomial experiment, you should use $\bar{p}$ in place of p_1 and p_2 and use $\bar{q}$ in place of q_1 and q_2.

Picturing the World

A medical research team conducted a study to test whether a drug lowers the chance of getting diabetes. In the study, 2623 people took the drug and 2646 people took a placebo. The results are shown below. (Source: The New England Journal of Medicine)

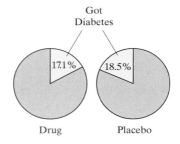

Got Diabetes

17.1%   18.5%

Drug   Placebo

At $\alpha = 0.05$, can you support the claim that the drug lowers the chance of getting diabetes?

There is not enough evidence at the 5% level of significance to support the claim that the drug lowers the chance of getting diabetes.

Study Tip

To simplify the calculation of z, you can round the values of $\bar{p}$, $\bar{q}$, $\hat{p}_1$, and $\hat{p}_2$ to four decimal places, as shown in Examples 1 and 2.

When the sampling distribution for $\hat{p}_1 - \hat{p}_2$ is normal, you can use a two-sample z-test to test the difference between two population proportions p_1 and p_2.

Two-Sample z-Test for the Difference Between Proportions

A two-sample z-test is used to test the difference between two population proportions p_1 and p_2 when these conditions are met.

1. The samples are random.
2. The samples are independent.
3. The quantities $n_1\bar{p}$, $n_1\bar{q}$, $n_2\bar{p}$, and $n_2\bar{q}$ are at least 5.

The **test statistic** is $\hat{p}_1 - \hat{p}_2$. The **standardized test statistic** is

$$z = \frac{(\hat{p}_1 - \hat{p}_2) - (p_1 - p_2)}{\sqrt{\bar{p}\,\bar{q}\left(\frac{1}{n_1} + \frac{1}{n_2}\right)}}$$

where $\bar{p} = \dfrac{x_1 + x_2}{n_1 + n_2}$ and $\bar{q} = 1 - \bar{p}$.

If the null hypothesis states $p_1 = p_2$, $p_1 \le p_2$, or $p_1 \ge p_2$, then $p_1 = p_2$ is assumed and the expression $p_1 - p_2$ is equal to 0 in the preceding test.

GUIDELINES

Using a Two-Sample z-Test for the Difference Between Proportions

In Words	In Symbols
1. Verify that the samples are random and independent.	
2. Find the weighted estimate of p_1 and p_2. Verify that $n_1\bar{p}$, $n_1\bar{q}$, $n_2\bar{p}$, and $n_2\bar{q}$ are at least 5.	$\bar{p} = \dfrac{x_1 + x_2}{n_1 + n_2}, \bar{q} = 1 - \bar{p}$
3. State the claim mathematically and verbally. Identify the null and alternative hypotheses.	State H_0 and H_a.
4. Specify the level of significance.	Identify α.
5. Determine the critical value(s).	Use Table 4 in Appendix B.
6. Determine the rejection region(s).	
7. Find the standardized test statistic and sketch the sampling distribution.	$z = \dfrac{(\hat{p}_1 - \hat{p}_2) - (p_1 - p_2)}{\sqrt{\bar{p}\,\bar{q}\left(\frac{1}{n_1} + \frac{1}{n_2}\right)}}$
8. Make a decision to reject or fail to reject the null hypothesis.	If z is in the rejection region, then reject H_0. Otherwise, fail to reject H_0.
9. Interpret the decision in the context of the original claim.	

A hypothesis test for the difference between proportions can also be performed using P-values. Use the guidelines above, skipping Steps 5 and 6. After finding the standardized test statistic, use Table 4 in Appendix B to calculate the P-value. Then make a decision to reject or fail to reject the null hypothesis. If P is less than or equal to α, then reject H_0. Otherwise, fail to reject H_0.

Study Tip

To find x_1 and x_2, use $x_1 = n_1\hat{p}_1$ and $x_2 = n_2\hat{p}_2$.

Sample Statistics for Vehicles

Passenger cars	Pickup trucks
$n_1 = 200$	$n_2 = 250$
$\hat{p}_1 = 0.910$	$\hat{p}_2 = 0.832$
$x_1 = 182$	$x_2 = 208$

> EXAMPLE 1

See TI-84 Plus steps on page 465.

A Two-Sample z-Test for the Difference Between Proportions

A study of 200 randomly selected occupants in passenger cars and 250 randomly selected occupants in pickup trucks shows that 91.0% of occupants in passenger cars and 83.2% of occupants in pickup trucks wear seat belts. At $\alpha = 0.10$, can you reject the claim that the proportion of occupants who wear seat belts is the same for passenger cars and pickup trucks? *(Adapted from National Highway Traffic Safety Administration)*

SOLUTION

The samples are random and independent. Also, the weighted estimate of p_1 and p_2 is

$$\bar{p} = \frac{x_1 + x_2}{n_1 + n_2} = \frac{182 + 208}{200 + 250} = \frac{390}{450} \approx 0.8667$$

and the value of $\bar{q}$ is

$$\bar{q} = 1 - \bar{p} \approx 1 - 0.8667 = 0.1333.$$

Because $n_1\bar{p} \approx 200(0.8667)$, $n_1\bar{q} \approx 200(0.1333)$, $n_2\bar{p} \approx 250(0.8667)$, and $n_2\bar{q} \approx 250(0.1333)$ are at least 5, you can use a two-sample z-test. The claim is "the proportion of occupants who wear seat belts is the same for passenger cars and pickup trucks." So, the null and alternative hypotheses are

$$H_0: p_1 = p_2 \quad \text{(Claim)} \quad \text{and} \quad H_a: p_1 \neq p_2.$$

Because the test is two-tailed and the level of significance is $\alpha = 0.10$, the critical values are $-z_0 = -1.645$ and $z_0 = 1.645$. The rejection regions are $z < -1.645$ and $z > 1.645$. The standardized test statistic is

$$z = \frac{(\hat{p}_1 - \hat{p}_2) - (p_1 - p_2)}{\sqrt{\bar{p}\,\bar{q}\left(\dfrac{1}{n_1} + \dfrac{1}{n_2}\right)}} \approx \frac{(0.910 - 0.832) - 0}{\sqrt{(0.8667)(0.1333)\left(\dfrac{1}{200} + \dfrac{1}{250}\right)}} \approx 2.42.$$

The figure below shows the location of the rejection regions and the standardized test statistic z. Because z is in the rejection region, you reject the null hypothesis.

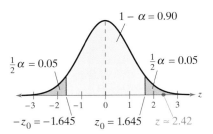

Interpretation There is enough evidence at the 10% level of significance to reject the claim that the proportion of occupants who wear seat belts is the same for passenger cars and pickup trucks.

TRY IT YOURSELF 1

Consider the results of the study discussed on page 417. At $\alpha = 0.05$, can you support the claim that there is a difference between the proportion of yoga users who are 40- to 49-year-olds and the proportion of non-yoga users who are 40- to 49-year-olds?

Answer: Page A37

Study Tip

To find $\hat{p}_1$ and $\hat{p}_2$ use

$$\hat{p}_1 = \frac{x_1}{n_1} \text{ and } \hat{p}_2 = \frac{x_2}{n_2}.$$

Sample Statistics for Cholesterol-Reducing Medication

Received medication	Received placebo
$n_1 = 4700$	$n_2 = 4300$
$x_1 = 301$	$x_2 = 357$
$\hat{p}_1 \approx 0.0640$	$\hat{p}_2 \approx 0.0830$

EXAMPLE 2

A Two-Sample z-Test for the Difference Between Proportions

A medical research team conducted a study to test the effect of a cholesterol-reducing medication. At the end of the study, the researchers found that of the 4700 randomly selected subjects who took the medication, 301 died of heart disease. Of the 4300 randomly selected subjects who took a placebo, 357 died of heart disease. At $\alpha = 0.01$, can you support the claim that the death rate due to heart disease is lower for those who took the medication than for those who took the placebo? *(Adapted from The New England Journal of Medicine)*

SOLUTION

The samples are random and independent. Also, the weighted estimate of p_1 and p_2 is

$$\bar{p} = \frac{x_1 + x_2}{n_1 + n_2} = \frac{301 + 357}{4700 + 4300} = \frac{658}{9000} \approx 0.0731$$

and the value of $\bar{q}$ is

$$\bar{q} = 1 - \bar{p} \approx 1 - 0.0731 = 0.9269.$$

Because $n_1\bar{p} = 4700(0.0731)$, $n_1\bar{q} = 4700(0.9269)$, $n_2\bar{p} = 4300(0.0731)$, and $n_2\bar{q} = 4300(0.9269)$ are at least 5, you can use a two-sample z-test. The claim is "the death rate due to heart disease is lower for those who took the medication than for those who took the placebo." So, the null and alternative hypotheses are

$$H_0: p_1 \geq p_2 \quad \text{and} \quad H_a: p_1 < p_2. \quad \text{(Claim)}$$

Because the test is left-tailed and the level of significance is $\alpha = 0.01$, the critical value is $z_0 = -2.33$. The rejection region is $z < -2.33$. The standardized test statistic is

$$z = \frac{(\hat{p}_1 - \hat{p}_2) - (p_1 - p_2)}{\sqrt{\bar{p}\,\bar{q}\left(\frac{1}{n_1} + \frac{1}{n_2}\right)}} \approx \frac{(0.0640 - 0.0830) - 0}{\sqrt{(0.0731)(0.9269)\left(\frac{1}{4700} + \frac{1}{4300}\right)}} \approx -3.46.$$

The figure below shows the location of the rejection region and the standardized test statistic z. Because z is in the rejection region, you reject the null hypothesis.

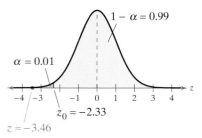

Interpretation There is enough evidence at the 1% level of significance to support the claim that the death rate due to heart disease is lower for those who took the medication than for those who took the placebo.

TRY IT YOURSELF 2

Consider the results of the study discussed on page 417. At $\alpha = 0.05$, can you support the claim that the proportion of yoga users with incomes of \$20,000 to \$34,999 is less than the proportion of non-yoga users with incomes of \$20,000 to \$34,999?

Answer: Page A37

8.4 EXERCISES

For Extra Help: MyLab Statistics

1. See Odd Answers, page A76.
2. See Selected Answers, page A104.
3. Can use normal sampling distribution; Fail to reject H_0. There is not enough evidence at the 1% level of significance to support the claim.
4. Can use normal sampling distribution; Reject H_0. There is enough evidence at the 5% level of significance to support the claim.
5. Can use normal sampling distribution; Reject H_0. There is enough evidence at the 10% level of significance to reject the claim.
6. Cannot use normal sampling distribution.
7. See Odd Answers, page A76.
8. See Selected Answers, page A104.
9. See Odd Answers, page A76.

Building Basic Skills and Vocabulary

1. What conditions are necessary in order to use the z-test to test the difference between two population proportions?

2. Explain how to perform a two-sample z-test for the difference between two population proportions.

In Exercises 3–6, determine whether a normal sampling distribution can be used. If it can be used, test the claim about the difference between two population proportions p_1 and p_2 at the level of significance α. Assume the samples are random and independent.

3. Claim: $p_1 \neq p_2$; $\alpha = 0.01$
   Sample statistics: $x_1 = 35$, $n_1 = 70$ and $x_2 = 36$, $n_2 = 60$

4. Claim: $p_1 < p_2$; $\alpha = 0.05$
   Sample statistics: $x_1 = 471$, $n_1 = 785$ and $x_2 = 372$, $n_2 = 465$

5. Claim: $p_1 = p_2$; $\alpha = 0.10$
   Sample statistics: $x_1 = 42$, $n_1 = 150$ and $x_2 = 76$, $n_2 = 200$

6. Claim: $p_1 > p_2$; $\alpha = 0.01$
   Sample statistics: $x_1 = 6$, $n_1 = 20$ and $x_2 = 4$, $n_2 = 30$

Using and Interpreting Concepts

Testing the Difference Between Two Proportions *In Exercises 7–12, (a) identify the claim and state H_0 and H_a, (b) find the critical value(s) and identify the rejection region(s), (c) find the standardized test statistic z, (d) decide whether to reject or fail to reject the null hypothesis, and (e) interpret the decision in the context of the original claim. Assume the samples are random and independent.*

7. **Multiple Sclerosis Drug** In a study to determine the effectiveness of using a drug to treat multiple sclerosis, 488 subjects were given the drug and 244 subjects were given a placebo. The numbers of subjects who had 12-week confirmed disability progression were tracked. The results are shown at the left. At $\alpha = 0.01$, can you support the claim that there is a difference in the proportion of subjects who had no 12-week confirmed disability progression? *(Adapted from The New England Journal of Medicine)*

8. **Cancer Drug** In a study, 760 men with recurrent prostate cancer underwent radiation with or without a type of hormone-based chemotherapy. For 24 months, 384 subjects were given the chemotherapy and 376 subjects were given a placebo. The numbers who survived and did not survive after 12 years were tracked. The results are shown at the left. At $\alpha = 0.10$, can you support the claim that the proportion of 12-year survivors is greater for subjects who were given the chemotherapy than for subjects who were given the placebo? *(Adapted from The New England Journal of Medicine)*

9. **Young Adults** In a survey of 1750 females ages 20 to 24 whose highest level of education is completing high school, 64.4% were employed. In a survey of 2000 males ages 20 to 24 whose highest level of education is completing high school, 73.2% were employed. At $\alpha = 0.01$, can you support the claim that there is a difference in the proportion of those employed between the two groups? *(Adapted from National Center for Education Statistics)*

How Many Subjects Had 12-Week Confirmed Disability Progression and How Many Did Not?

Disability progression

161   96

No disability progression 327 (Drug)   No disability progression 148 (Placebo)

Drug   Placebo

FIGURE FOR EXERCISE 7

How Many Subjects Survived After 12 Years and How Many Did Not?

Did not survive

91   108

Survived 293   Survived 268

Chemotherapy   Placebo

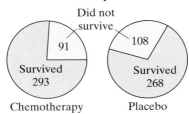

FIGURE FOR EXERCISE 8

10. See Selected Answers, page A104.

11. See Odd Answers, page A76.

12. See Selected Answers, page A104.

13. No, there is not enough evidence at the 5% level of significance to reject the claim that the proportion of newlywed Asians who have a spouse of a different race or ethnicity is the same as the proportion of newlywed Hispanics who have a spouse of a different race or ethnicity.

14. Yes, there is enough evidence at the 1% level of significance to support the claim that the proportion of newlywed blacks who have a spouse of a different race or ethnicity is less than the proportion of newlywed Asians who have a spouse of a different race or ethnicity.

15. Yes, there is enough evidence at the 1% level of significance to support the claim that the proportion of newlywed Asians who have a spouse of a different race or ethnicity is greater than the proportion of newlywed whites who have a spouse of a different race or ethnicity.

16. Yes, there is enough evidence at the 5% level of significance to support the claim that the proportion of newlywed Hispanics who have a spouse of a different race or ethnicity is different from the proportion of newlywed blacks who have a spouse of a different race or ethnicity.

17. Yes, there is enough evidence at the 1% level of significance to support the claim that the proportion of newlywed whites who have a spouse of a different race or ethnicity is less than the proportion of newlywed blacks who have a spouse of a different race or ethnicity.

18. Yes, there is enough evidence at the 5% level of significance to support the claim that the proportion of newlywed Hispanics who have a spouse of a different race or ethnicity is greater than the proportion of newlywed whites who have a spouse of a different race or ethnicity.

10. **Young Adults**   In a survey of 500 males ages 20 to 24, 15.8% were neither in school nor working. In a survey of 500 females ages 20 to 24, 17.8% were neither in school nor working. At $\alpha = 0.05$, can you support the claim that the proportion of males ages 20 to 24 who were neither in school nor working is less than the proportion of females ages 20 to 24 who were neither in school nor working? *(Adapted from National Center for Education Statistics)*

11. **Seat Belt Use**   In a survey of 1000 drivers from the West, 934 wear a seat belt. In a survey of 1000 drivers from the Northeast, 909 wear a seat belt. At $\alpha = 0.05$, can you support the claim that the proportion of drivers who wear seat belts is greater in the West than in the Northeast? *(Adapted from National Highway Traffic Safety Administration)*

12. **Seat Belt Use**   In a survey of 1000 drivers from the Midwest, 855 wear a seat belt. In a survey of 1000 drivers from the South, 909 wear a seat belt. At $\alpha = 0.10$, can you support the claim that the proportion of drivers who wear seat belts in the Midwest is less than the proportion of drivers who wear seat belts in the South? *(Adapted from National Highway Traffic Safety Administration)*

Intermarriages   *In Exercises 13–18, use the figure, which shows the percentages of newlyweds in the United States who have a spouse of a different race or ethnicity. The survey included random samples of 1000 Asian newlyweds, 1000 Hispanic newlyweds, 1000 black newlyweds, and 1000 white newlyweds. (Adapted from Pew Research Center)*

13. **Asians and Hispanics**   At $\alpha = 0.05$, can you reject the claim that the proportion of newlywed Asians who have a spouse of a different race or ethnicity is the same as the proportion of newlywed Hispanics who have a spouse of a different race or ethnicity?

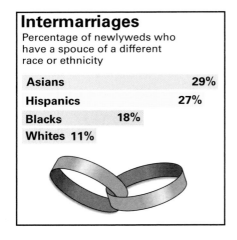

Intermarriages

Percentage of newlyweds who have a spouse of a different race or ethnicity

Asians	29%
Hispanics	27%
Blacks	18%
Whites	11%

14. **Blacks and Asians**   At $\alpha = 0.01$, can you support the claim that the proportion of newlywed blacks who have a spouse of a different race or ethnicity is less than the proportion of newlywed Asians who have a spouse of a different race or ethnicity?

15. **Asians and Whites**   At $\alpha = 0.01$, can you support the claim that the proportion of newlywed Asians who have a spouse of a different race or ethnicity is greater than the proportion of newlywed whites who have a spouse of a different race or ethnicity?

16. **Hispanics and Blacks**   At $\alpha = 0.05$, can you support the claim that the proportion of newlywed Hispanics who have a spouse of a different race or ethnicity is different from the proportion of newlywed blacks who have a spouse of a different race or ethnicity?

17. **Whites and Blacks**   At $\alpha = 0.01$, can you support the claim that the proportion of newlywed whites who have a spouse of a different race or ethnicity is less than the proportion of newlywed blacks who have a spouse of a different race or ethnicity?

18. **Hispanics and Whites**   At $\alpha = 0.05$, can you support the claim that the proportion of newlywed Hispanics who have a spouse of a different race or ethnicity is greater than the proportion of newlywed whites who have a spouse of a different race or ethnicity?

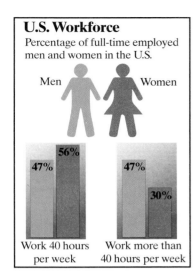

U.S. Workforce

Percentage of full-time employed men and women in the U.S.

Men   Women

56%
47%   47%
30%

Work 40 hours per week   Work more than 40 hours per week

FIGURE FOR EXERCISES 19–22

19. No, there is not enough evidence at the 1% level of significance to reject the claim that the proportion of men who work 40 hours per week is the same as the proportion of men who work more than 40 hours per week.

20. Yes, there is enough evidence at the 5% level of significance to support the claim that the proportion of women who work 40 hours per week is greater than the proportion of women who work more than 40 hours per week.

21. Yes, there is enough evidence at the 5% level of significance to support the claim that the U.S. workforce that works 40 hours per week is greater for women than for men.

22. Yes, there is enough evidence at the 10% level of significance to support the claim that the proportion of the U.S. workforce that works more than 40 hours per week is less for women than for men.

23. $-0.028 < p_1 - p_2 < -0.012$

24. $0.035 < p_1 - p_2 < 0.045$

25. $0.011 < p_1 - p_2 < 0.069$; Answers will vary.

26. $0.002 < p_1 - p_2 < 0.078$; Answers will vary.

U.S. Workforce   *In Exercises 19–22, use the figure shown at the left, which gives the percentages of full-time employed men and women in the United States who work 40 hours per week and who work more than 40 hours per week. Assume the survey included random samples of 300 men and 250 women.* (Adapted from Gallup)

19. **Men: Numbers of Hours Worked Per Week**   At $\alpha = 0.01$, can you reject the claim that the proportion of men who work 40 hours per week is the same as the proportion of men who work more than 40 hours per week?

20. **Women: Numbers of Hours Worked Per Week**   At $\alpha = 0.05$, can you support the claim that the proportion of women who work 40 hours per week is greater than the proportion of women who work more than 40 hours per week?

21. **Working 40 Hours Per Week: Men and Women**   At $\alpha = 0.05$, can you support the claim that the proportion of the U.S. workforce that works 40 hours per week is greater for women than for men?

22. **Working More Than 40 Hours Per Week: Men and Women**   At $\alpha = 0.10$, can you support the claim that the proportion of the U.S. workforce that works more than 40 hours per week is less for women than for men?

Extending Concepts

Constructing Confidence Intervals for $p_1 - p_2$ *You can construct a confidence interval for the difference between two population proportions $p_1 - p_2$ by using the inequality below.*

$$(\hat{p}_1 - \hat{p}_2) - z_c\sqrt{\frac{\hat{p}_1\hat{q}_1}{n_1} + \frac{\hat{p}_2\hat{q}_2}{n_2}} < p_1 - p_2 < (\hat{p}_1 - \hat{p}_2) + z_c\sqrt{\frac{\hat{p}_1\hat{q}_1}{n_1} + \frac{\hat{p}_2\hat{q}_2}{n_2}}$$

In Exercises 23–26, construct the indicated confidence interval for $p_1 - p_2$. Assume the samples are random and independent.

23. **Students Planning to Study Visual and Performing Arts**   In a survey of 10,000 students taking the SAT, 7% were planning to study visual and performing arts in college. In another survey of 8000 students taken 10 years before, 9% were planning to study visual and performing arts in college. Construct a 95% confidence interval for $p_1 - p_2$, where p_1 is the proportion from the recent survey and p_2 is the proportion from the survey taken 10 years ago. (Adapted from The College Board)

24. **Students Undecided on an Intended College Major**   In a survey of 10,000 students taking the SAT, 7% were undecided on an intended college major. In another survey of 8000 students taken 10 years before, 3% were undecided on an intended college major. Construct a 90% confidence interval for $p_1 - p_2$, where p_1 is the proportion from the recent survey and p_2 is the proportion from the survey taken 10 years ago. (Adapted from The College Board)

25. **Employment**   In Section 6.3, Exercises 27 and 28, let p_1 be the proportion of the population of U.S. college graduates who expect to stay at their first employer for 3 or more years and let p_2 be the proportion of the population of U.S. college graduates who are employed in their field of study. Construct a 95% confidence interval for $p_1 - p_2$. Compare your result with the result in Section 6.3, Exercise 27, part (a).

26. **Employment**   Repeat Exercise 25 but with a 99% confidence interval. Compare your result with the result in Section 6.3, Exercise 27, part (b).

Uses

Hypothesis Testing with Two Samples Hypothesis testing enables you to determine whether differences in samples indicate actual differences in populations or are merely due to sampling error. For instance, a study conducted on about 1400 American children in a variety of settings compared the behavior of the children who attended day care with the behavior of those who stayed home. Aggressive behavior such as stealing toys, pushing other children, and starting fights was measured in both groups. The study showed that children who attended day care for more than 30 hours per week were about three times more likely to be aggressive than those who stayed home. Although the aggressive behavior observed in the study was well within the normal range for healthy children, these statistics have been used to persuade parents to keep their children at home until they start school.

Abuses

Confounding Variables The U.S. study found that the results were the same regardless of quality of the day care center and income of the family. However, the overall quality of care experienced by most of the children studied could be the problem—a survey of American day care centers that measured aspects such as number and expertise of caregivers found that only 10 percent of American day care centers provided high-quality care.

A similar study of preschoolers and aggressive behavior in Norway, where day care centers are subject to strict standards and the ratio of adult caregivers to children is high, found that the link between day care attendance and aggressive behavior was minimal. Another Norwegian study included an additional variable, differences between siblings, and found no relationship between day care attendance and behavior problems. These additional variables that are often out of the researcher's control are known as *confounding variables*.

Study Funding A series of studies was conducted on various methods for reducing the number of cigarettes that smokers smoke. The study compared smokers who were simply told to smoke less and those who tried methods such as nicotine replacement therapy, electronic cigarettes, and using reduced tar, carbon, or nicotine cigarettes. Some methods were shown to be effective in reducing the number of cigarettes smoked.

Some of the studies were funded by the tobacco industry, which could profit from promoting strategies other than quitting as beneficial to smokers' health. When dealing with statistics, it is always good to know who is paying for a study, and whether the researchers are unbiased.

EXERCISES

1. ***Confounding Variables*** A pharmaceutical company has applied for approval to market a new arthritis medication. The research involved a test group that was given the medication and another test group that was given a placebo. Describe some possible confounding variables that could influence the results of the study.

2. Medical research often involves blind and double-blind testing. Explain what these two terms mean.

8 | Chapter Summary

What Did You Learn?	Example(s)	Review Exercises
Section 8.1		
▶ How to determine whether two samples are independent or dependent	1	1–4
▶ How to perform a two-sample z-test for the difference between two means μ_1 and μ_2 using independent samples with σ_1 and σ_2 known $z = \dfrac{(\bar{x}_1 - \bar{x}_2) - (\mu_1 - \mu_2)}{\sigma_{\bar{x}_1 - \bar{x}_2}}$	2, 3	5–10
Section 8.2		
▶ How to perform a two-sample t-test for the difference between two means μ_1 and μ_2 using independent samples with σ_1 and σ_2 unknown $t = \dfrac{(\bar{x}_1 - \bar{x}_2) - (\mu_1 - \mu_2)}{s_{\bar{x}_1 - \bar{x}_2}}$	1, 2	11–18
Section 8.3		
▶ How to perform a t-test to test the mean of the differences for a population of paired data $t = \dfrac{\bar{d} - \mu_d}{s_d / \sqrt{n}}$	1, 2	19–24
Section 8.4		
▶ How to perform a two-sample z-test for the difference between two population proportions p_1 and p_2 $z = \dfrac{(\hat{p}_1 - \hat{p}_2) - (p_1 - p_2)}{\sqrt{\bar{p}\,\bar{q}\left(\dfrac{1}{n_1} + \dfrac{1}{n_2}\right)}}$	1, 2	25–30

Two-Sample Hypothesis Testing for Population Means

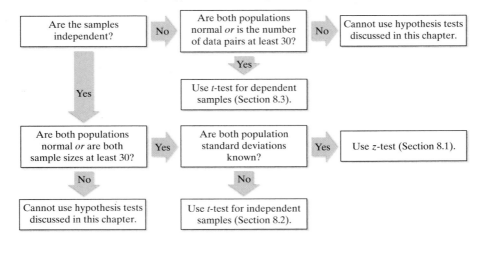

Review Exercises

Section 8.1

In Exercises 1–4, classify the two samples as independent or dependent and justify your answer.

1. Sample 1: The weights of 43 adults
   Sample 2: The weights of the same 43 adults after participating in a diet and exercise program

2. Sample 1: The weights of 39 dogs
   Sample 2: The weights of 39 cats

3. Sample 1: The fuel efficiencies of 20 sports utility vehicles
   Sample 2: The fuel efficiencies of 20 minivans

4. Sample 1: The fuel efficiencies of 12 cars
   Sample 2: The fuel efficiencies of the same 12 cars using an alternative fuel

In Exercises 5–8, test the claim about the difference between two population means μ_1 and μ_2 at the level of significance α. Assume the samples are random and independent, and the populations are normally distributed.

5. Claim: $\mu_1 \geq \mu_2$; $\alpha = 0.05$
   Population statistics: $\sigma_1 = 0.30$ and $\sigma_2 = 0.23$
   Sample statistics: $\bar{x}_1 = 1.28$, $n_1 = 96$ and $\bar{x}_2 = 1.34$, $n_2 = 85$

6. Claim: $\mu_1 = \mu_2$; $\alpha = 0.01$
   Population statistics: $\sigma_1 = 52$ and $\sigma_2 = 68$
   Sample statistics: $\bar{x}_1 = 5595$, $n_1 = 156$ and $\bar{x}_2 = 5575$, $n_2 = 216$

7. Claim: $\mu_1 < \mu_2$; $\alpha = 0.10$
   Population statistics: $\sigma_1 = 0.11$ and $\sigma_2 = 0.10$
   Sample statistics: $\bar{x}_1 = 0.28$, $n_1 = 41$ and $\bar{x}_2 = 0.33$, $n_2 = 34$

8. Claim: $\mu_1 \neq \mu_2$; $\alpha = 0.05$
   Population statistics: $\sigma_1 = 14$ and $\sigma_2 = 15$
   Sample statistics: $\bar{x}_1 = 87$, $n_1 = 410$ and $\bar{x}_2 = 85$, $n_2 = 340$

In Exercises 9 and 10, (a) identify the claim and state H_0 and H_a, (b) find the critical value(s) and identify the rejection region(s), (c) find the standardized test statistic z, (d) decide whether to reject or fail to reject the null hypothesis, and (e) interpret the decision in the context of the original claim. Assume the samples are random and independent, and the populations are normally distributed.

9. A researcher claims that the mean sodium content of sandwiches at Restaurant A is less than the mean sodium content of sandwiches at Restaurant B. The mean sodium content of 22 randomly selected sandwiches at Restaurant A is 670 milligrams. Assume the population standard deviation is 20 milligrams. The mean sodium content of 28 randomly selected sandwiches at Restaurant B is 690 milligrams. Assume the population standard deviation is 30 milligrams. At $\alpha = 0.05$, is there enough evidence to support the claim?

10. A career counselor claims that the mean annual salary of entry-level paralegals in Peoria, Illinois, and Gary, Indiana, is the same. The mean annual salary of 40 randomly selected entry-level paralegals in Peoria is $50,410. Assume the population standard deviation is $9320. The mean annual salary of 35 randomly selected entry-level paralegals in Gary is $47,350. Assume the population standard deviation is $9330. At $\alpha = 0.10$, is there enough evidence to reject the counselor's claim? *(Adapted from Salary.com)*

11. Reject H_0. There is enough evidence at the 5% level of significance to reject the claim.

12. Fail to reject H_0. There is not enough evidence at the 10% level of significance to support the claim.

13. Fail to reject H_0. There is not enough evidence at the 10% level of significance to reject the claim.

14. Reject H_0. There is enough evidence at the 1% level of significance to reject the claim.

15. Reject H_0. There is enough evidence at the 1% level of significance to support the claim.

16. Fail to reject H_0. There is not enough evidence at the 10% level of significance to support the claim.

17. (a) The claim is "the new method of teaching mathematics produces higher mathematics test scores than the old method does."

H_0: $\mu_1 \leq \mu_2$
H_a: $\mu_1 > \mu_2$ (claim)

(b) $t_0 = 1.667$
Rejection region: $t > 1.667$

(c) 2.313

(d) Reject H_0.

(e) There is enough evidence at the 5% level of significance to support the claim that the new method of teaching mathematics produces higher mathematics test scores than the old method does.

Section 8.2

In Exercises 11–16, test the claim about the difference between two population means μ_1 and μ_2 at the level of significance α. Assume the samples are random and independent, and the populations are normally distributed.

11. Claim: $\mu_1 = \mu_2$; $\alpha = 0.05$. Assume $\sigma_1^2 = \sigma_2^2$
Sample statistics: $\bar{x}_1 = 228$, $s_1 = 27$, $n_1 = 20$ and
$\bar{x}_2 = 207$, $s_2 = 25$, $n_2 = 13$

12. Claim: $\mu_1 < \mu_2$; $\alpha = 0.10$. Assume $\sigma_1^2 \neq \sigma_2^2$
Sample statistics: $\bar{x}_1 = 0.015$, $s_1 = 0.011$, $n_1 = 8$ and
$\bar{x}_2 = 0.019$, $s_2 = 0.004$, $n_2 = 6$

13. Claim: $\mu_1 \leq \mu_2$; $\alpha = 0.10$. Assume $\sigma_1^2 \neq \sigma_2^2$
Sample statistics: $\bar{x}_1 = 664.5$, $s_1 = 2.4$, $n_1 = 40$ and
$\bar{x}_2 = 665.5$, $s_2 = 4.1$, $n_2 = 40$

14. Claim: $\mu_1 \geq \mu_2$; $\alpha = 0.01$. Assume $\sigma_1^2 = \sigma_2^2$
Sample statistics: $\bar{x}_1 = 44.5$, $s_1 = 5.85$, $n_1 = 17$ and
$\bar{x}_2 = 49.1$, $s_2 = 5.25$, $n_2 = 18$

15. Claim: $\mu_1 \neq \mu_2$; $\alpha = 0.01$. Assume $\sigma_1^2 = \sigma_2^2$
Sample statistics: $\bar{x}_1 = 61$, $s_1 = 3.3$, $n_1 = 5$ and
$\bar{x}_2 = 55$, $s_2 = 1.2$, $n_2 = 7$

16. Claim: $\mu_1 > \mu_2$; $\alpha = 0.10$. Assume $\sigma_1^2 \neq \sigma_2^2$
Sample statistics: $\bar{x}_1 = 520$, $s_1 = 25$, $n_1 = 7$ and
$\bar{x}_2 = 500$, $s_2 = 55$, $n_2 = 6$

In Exercises 17 and 18, (a) identify the claim and state H_0 and H_a, (b) find the critical value(s) and identify the rejection region(s), (c) find the standardized test statistic t, (d) decide whether to reject or fail to reject the null hypothesis, and (e) interpret the decision in the context of the original claim. Assume the samples are random and independent, and the populations are normally distributed.

 17. A new method of teaching mathematics is being tested on sixth grade students. A group of sixth grade students is taught using the new curriculum. A control group of sixth grade students is taught using the old curriculum. The mathematics test scores for the two groups are shown in the back-to-back stem-and-leaf plot.

Old Curriculum		New Curriculum
4 5 8	0	
0 1 1 5 7	1	
1 6	2	2 4 5 7 7
0 1 2 8	3	4 7
0 2 6 9	4	2 5 6 7
1 3 4 9	5	1 5 7
0 7	6	2 3 5 6 6 7
3 3 3 4 4 6 8	7	0 0 2 5 5 6
1 9	8	2 3 6 6 9
4 4 4	9	0 1 4 6 8

Key: $6|2|2 = 26$ for old curriculum and 22 for new curriculum

At $\alpha = 0.05$, is there enough evidence to support the claim that the new method of teaching mathematics produces higher mathematics test scores than the old method does? Assume the population variances are equal.

18. (a) The claim is "there is no difference between the mean household incomes of two neighborhoods."

    H_0: $\mu_1 = \mu_2$ (claim)
    H_a: $\mu_1 \neq \mu_2$

    (b) $-t_0 = -2.845$, $t_0 = 2.845$
    Rejection regions:
    $t < -2.845$, $t > 2.845$

    (c) 1.383

    (d) Fail to reject H_0.

    (e) There is not enough evidence at the 1% level of significance to reject the agent's claim that there is no difference between the mean household incomes of the two neighborhoods.

19. Reject H_0. There is enough evidence at the 1% level of significance to reject the claim.

20. Fail to reject H_0. There is not enough evidence at the 10% level of significance to support the claim.

21. Reject H_0. There is enough evidence at the 10% level of significance to reject the claim.

22. Reject H_0. There is enough evidence at the 5% level of significance to support the claim.

23. (a) The claim is "the numbers of passing yards for college football quarterbacks change from their junior to their senior years."

    H_0: $\mu_d = 0$
    H_a: $\mu_d \neq 0$ (claim)

    (b) $-t_0 = -2.262$, $t_0 = 2.262$
    Rejection regions:
    $t < -2.262$, $t > 2.262$

    (c) $\bar{d} = -12.3$; $s_d \approx 553.0877$

    (d) -0.070

    (e) Fail to reject H_0.

    (f) There is not enough evidence at the 5% level of significance to support the sports statistician's claim that the numbers of passing yards for college football quarterbacks change from their junior to their senior years.

24. See Selected Answers, page A105.

18. A real estate agent claims that there is no difference between the mean household incomes of two neighborhoods. The mean income of 12 randomly selected households from the first neighborhood is \$52,750 with a standard deviation of \$2900. In the second neighborhood, 10 randomly selected households have a mean income of \$51,200 with a standard deviation of \$2225. At $\alpha = 0.01$, can you reject the real estate agent's claim? Assume the population variances are equal.

Section 8.3

In Exercises 19–22, test the claim about the mean of the differences for a population of paired data at the level of significance α. Assume the samples are random and dependent, and the populations are normally distributed.

19. Claim: $\mu_d = 0$; $\alpha = 0.01$. Sample statistics: $\bar{d} = 8.5$, $s_d = 10.7$, $n = 16$
20. Claim: $\mu_d < 0$; $\alpha = 0.10$. Sample statistics: $\bar{d} = 3.2$, $s_d = 5.68$, $n = 25$
21. Claim: $\mu_d \leq 0$; $\alpha = 0.10$. Sample statistics: $\bar{d} = 10.3$, $s_d = 18.19$, $n = 33$
22. Claim: $\mu_d \neq 0$; $\alpha = 0.05$. Sample statistics: $\bar{d} = 17.5$, $s_d = 4.05$, $n = 37$

In Exercises 23 and 24, (a) identify the claim and state H_0 and H_a, (b) find the critical value(s) and identify the rejection region(s), (c) calculate $\bar{d}$ and s_d, (d) find the standardized test statistic t, (e) decide whether to reject or fail to reject the null hypothesis, and (f) interpret the decision in the context of the original claim. Assume the samples are random and dependent, and the populations are normally distributed.

 23. A sports statistician claims that the numbers of passing yards for college football quarterbacks change from their junior to their senior years. The table shows the numbers of passing yards for 10 college football quarterbacks in their junior and senior years. At $\alpha = 0.05$, is there enough evidence to support the sports statistician's claim? *(Source: Sports Reference, LLC)*

Player	1	2	3	4	5
Passing yards (junior year)	2517	2291	3853	2827	2701
Passing yards (senior year)	2184	2946	3540	3557	2169

Player	6	7	8	9	10
Passing yards (junior year)	3145	4332	1001	2401	1984
Passing yards (senior year)	3328	3348	1464	2366	2273

24. A physical fitness instructor claims that a weight loss supplement will help users lose weight after two weeks. The table shows the weights (in pounds) of 9 adults before using the supplement and two weeks after using the supplement. At $\alpha = 0.10$, is there enough evidence to support the physical fitness instructor's claim?

User	1	2	3	4	5	6	7	8	9
Weight (before)	228	210	245	272	203	198	256	217	240
Weight (after)	225	208	242	270	205	196	250	220	240

25. Can use normal sampling distribution; Fail to reject H_0. There is not enough evidence at the 5% level of significance to reject the claim.

26. Can use normal sampling distribution; Reject H_0. There is enough evidence at the 1% level of significance to reject the claim.

27. Can use normal sampling distribution; Reject H_0. There is enough evidence at the 10% level of significance to support the claim.

28. Can use normal sampling distribution; Fail to reject H_0. There is not enough evidence at the 5% level of significance to support the claim.

29. (a) The claim is "the proportion of subjects who had at least 24 weeks of accrued remission is the same for the two groups."

H_0: $p_1 = p_2$ (claim)
H_a: $p_1 \neq p_2$

(b) $-z_0 = -1.96$, $z_0 = 1.96$
Rejection regions:
$z < -1.96$, $z > 1.96$

(c) 4.03   (d) Reject H_0.

(e) There is enough evidence at the 5% level of significance to reject the medical research team's claim that the proportion of subjects who had at least 24 weeks of accrued remission is the same for the two groups.

30. (a) The claim is "the proportion of motorcyclists who use helmets that are compliant with federal safety regulations increased from the first year to the second year."

H_0: $p_1 \geq p_2$
H_a: $p_1 < p_2$ (claim)

(b) $z_0 = -2.33$
Rejection region: $z < -2.33$

(c) -2.13 (d) Fail to reject H_0.

(e) There is not enough evidence at the 1% level of significance to support the traffic safety research team's claim that the proportion of motorcyclists who use helmets that are compliant with federal safety regulations increased from the first year to the second year.

Section 8.4

In Exercises 25–28, determine whether a normal sampling distribution can be used. If it can be used, test the claim about the difference between two population proportions p_1 and p_2 at the level of significance α. Assume the samples are random and independent.

25. Claim: $p_1 = p_2$; $\alpha = 0.05$
Sample statistics: $x_1 = 425$, $n_1 = 840$ and $x_2 = 410$, $n_2 = 760$

26. Claim: $p_1 \leq p_2$; $\alpha = 0.01$
Sample statistics: $x_1 = 36$, $n_1 = 100$ and $x_2 = 46$, $n_2 = 200$

27. Claim: $p_1 > p_2$; $\alpha = 0.10$
Sample statistics: $x_1 = 261$, $n_1 = 556$ and $x_2 = 207$, $n_2 = 483$

28. Claim: $p_1 < p_2$; $\alpha = 0.05$
Sample statistics: $x_1 = 86$, $n_1 = 900$ and $x_2 = 107$, $n_2 = 1200$

In Exercises 29 and 30, (a) identify the claim and state H_0 and H_a, (b) find the critical value(s) and identify the rejection region(s), (c) find the standardized test statistic z, (d) decide whether to reject or fail to reject the null hypothesis, and (e) interpret the decision in the context of the original claim. Assume the samples are random and independent.

29. A medical research team conducted a study to test the effect of a drug used to treat a type of inflammation. In the study, 68 subjects took the drug and 68 subjects took a placebo. The results are shown below. At $\alpha = 0.05$, can you reject the claim that the proportion of subjects who had at least 24 weeks of accrued remission is the same for the two groups? *(Source: The New England Journal of Medicine)*

Do You Have At Least 24 Weeks of Accrued Remission?

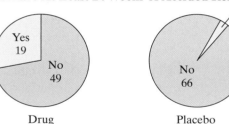

Drug   Placebo

30. A traffic safety research team conducted a survey over two years on the use of motorcycle helmets. In the survey, each year 1000 motorcyclists were asked whether they use helmets that are compliant with federal safety regulations. The results are shown below. At $\alpha = 0.01$, can you support the claim that the proportion of motorcyclists who use such helmets increased from the first year to the second year? *(Adapted from National Highway Traffic Safety Administration)*

Do You Use Helmets that Are Compliant with Federal Saftey Regulations?

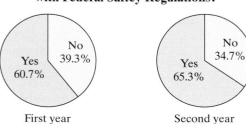

First year   Second year

8 Chapter Quiz

Take this quiz as you would take a quiz in class. After you are done, check your work against the answers given in the back of the book.

For each exercise, perform the steps below.

(a) *Identify the claim and state H_0 and H_a.*

(b) *Determine whether the hypothesis test is left-tailed, right-tailed, or two-tailed, and whether to use a z-test or a t-test. Explain your reasoning.*

(c) *Find the critical value(s) and identify the rejection region(s).*

(d) *Find the appropriate standardized test statistic.*

(e) *Decide whether to reject or fail to reject the null hypothesis.*

(f) *Interpret the decision in the context of the original claim.*

1. The mean score on a reading assessment test for 49 randomly selected male high school students was 279. Assume the population standard deviation is 41. The mean score on the same test for 50 randomly selected female high school students was 278. Assume the population standard deviation is 39. At $\alpha = 0.05$, can you support the claim that the mean score on the reading assessment test for male high school students is greater than the mean score for female high school students? *(Adapted from National Center for Education Statistics)*

2. A music teacher claims that the mean scores on a music assessment test for eighth grade boys and girls are equal. The mean score for 13 randomly selected boys is 142 with a standard deviation of 49, and the mean score for 15 randomly selected girls is 156 with a standard deviation of 42. At $\alpha = 0.1$, can you reject the teacher's claim? Assume the populations are normally distributed and the population variances are equal. *(Adapted from National Center for Education Statistics)*

 3. The table shows the credit scores for 12 randomly selected adults who are considered high-risk borrowers before and two years after they attend a personal finance seminar. At $\alpha = 0.01$, is there enough evidence to support the claim that the personal finance seminar helps adults increase their credit scores? Assume the populations are normally distributed.

Adult	1	2	3	4	5	6
Credit score (before seminar)	608	620	610	650	640	680
Credit score (after seminar)	646	692	715	669	725	786

Adult	7	8	9	10	11	12
Credit score (before seminar)	655	602	644	656	632	664
Credit score (after seminar)	700	650	660	650	680	702

4. In a random sample of 1020 U.S. adults in a recent year, 459 approve of the job the Supreme Court is doing. In another random sample of 1510 U.S. adults taken 3 years prior, 694 approve of the job the Supreme Court is doing. At $\alpha = 0.05$, can you support the claim that the proportion of U.S. adults who approve of the job the Supreme Court is doing is less than it was 3 years prior? *(Adapted from Gallup)*

1. (a) The claim is "the mean score on the reading assessment test for male high school students is greater than the mean score for female high school students."

H_0: $\mu_1 \leq \mu_2$
H_a: $\mu_1 > \mu_2$ (claim)

(b) Right-tailed because H_a contains $>$; z-test because σ_1 and σ_2 are known, the samples are random samples, the samples are independent, and $n_1 \geq 30$ and $n_2 \geq 30$.

(c) $z_0 = 1.645$
Rejection region: $z > 1.645$

(d) 0.12   (e) Fail to reject H_0.

(f) There is not enough evidence at the 5% level of significance to support the claim that the mean score on the reading assessment test for the male high school students was higher than for the female high school students.

2. (a) The claim is "the mean scores on a music assessment test for eighth grade boys and girls are equal."

H_0: $\mu_1 = \mu_2$ (claim)
H_a: $\mu_1 \neq \mu_2$

(b) Two-tailed because H_a contains $\neq$; t-test because σ_1 and σ_2 are unknown, the samples are random samples, the samples are independent, and the populations are normally distributed.

(c) $-t_0 = -1.706$, $t_0 = 1.706$
Rejection regions:
$t < -1.706$, $t > 1.706$

(d) -0.814 (e) Fail to reject H_0.

(f) There is not enough evidence at the 10% level of significance to reject the teacher's claim that the mean scores on the music assessment test are the same for eighth grade boys and girls.

3 See Odd Answers, page A77.

4. See Odd Answers, page A77.

8 Chapter Test

1. (a) The claim is "the proportion of students taking the SAT who are undecided on an intended college major has not changed."

H_0: $p_1 = p_2$ (claim)
H_a: $p_1 \neq p_2$

(b) Two-tailed because H_a contains $\neq$; z-test because you are testing proportions, the samples are random, the samples are independent, and the quantities $n_1 \overline{p}$, $n_2 \overline{p}$, $n_1 \overline{q}$, and $n_2 \overline{q}$ are at least 5.

(c) $-z_0 = -1.645$, $z_0 = 1.645$
Rejection regions:
$z < -1.645$, $z > 1.645$

(d) 11.88   (e) Reject H_0.

(f) There is enough evidence at the 10% level of significance to reject the claim that the proportion of students taking the SAT who are undecided on an intended college major has not changed.

2. (a) The claim is "the mean home sales price in Olathe, Kansas, is greater than in Rolla, Missouri."

H_0: $\mu_1 \leq \mu_2$
H_a: $\mu_1 > \mu_2$ (claim)

(b) Right-tailed because H_a contains $>$; z-test because σ_1 and σ_2 are known, the samples are random, the samples are independent, and $n_1 \geq 30$ and $n_2 \geq 30$.

(c) $z_0 = 1.645$
Rejection region: $z > 1.645$

(d) 2.40   (e) Reject H_0.

(f) There is enough evidence at the 5% level of significance to support the real estate agent's claim that the mean home sales price in Olathe, Kansas, is greater than in Rolla, Missouri.

3 See Selected Answers, page A105.

4. See Selected Answers, page A105.

Take this test as you would take a test in class.

For each exercise, perform the steps below.

(a) Identify the claim and state H_0 and H_a.

(b) Determine whether the hypothesis test is left-tailed, right-tailed, or two-tailed, and whether to use a z-test or a t-test. Explain your reasoning.

(c) Find the critical value(s) and identify the rejection region(s).

(d) Find the appropriate standardized test statistic.

(e) Decide whether to reject or fail to reject the null hypothesis.

(f) Interpret the decision in the context of the original claim.

1. In a survey of 5000 students taking the SAT, 350 were undecided on an intended college major. In another survey of 12,000 students taken 10 years before, 360 were undecided on an intended college major. At $\alpha = 0.10$, can you reject the claim that the proportion of students taking the SAT who are undecided on an intended college major has not changed? *(Adapted from The College Board)*

2. A real estate agency says that the mean home sales price in Olathe, Kansas, is greater than in Rolla, Missouri. The mean home sales price for 64 homes in Olathe is $356,889. Assume the population standard deviation is $537,407. The mean home sales price for 36 homes in Rolla is $189,389. Assume the population standard deviation is $113,555. At $\alpha = 0.05$, is there enough evidence to support the agency's claim? *(Adapted from RealtyTrac)*

3. A physical therapist suggests that soft tissue massage therapy helps to reduce the lengths of time patients suffer from headaches. The table shows the numbers of hours per day 18 patients suffered from headaches before and after 6 weeks of receiving treatment. At $\alpha = 0.05$, is there enough evidence to support the therapist's claim? Assume the populations are normally distributed. *(Adapted from Annals of Musculoskeletal Medicine)*

Patient	1	2	3	4	5	6	7	8	9
Hours (before)	5.2	5.1	4.9	1.6	6.1	2.3	4.6	5.2	3.1
Hours (after)	3.5	3.3	3.7	2.3	2.7	2.4	2.1	2.5	2.8

Patient	10	11	12	13	14	15	16	17	18
Hours (before)	4.4	4.2	5.4	3.3	5.2	3.7	2.6	2.7	2.6
Hours (after)	4.1	3.0	2.4	2.4	2.7	2.6	2.4	2.7	2.4

4. A demographics researcher claims that the mean household income in a recent year is different in Polk County, Iowa, than it is in Woodward County, Oklahoma. In Polk County, a sample of 13 residents has a mean household income of $61,300 and a standard deviation of $1770. In Woodward County, a sample of 15 residents has a mean household income of $59,800 and a standard deviation of $8350. At $\alpha = 0.01$, can you support the demographics researcher's claim? Assume the populations are normally distributed and the population variances are not equal. *(Adapted from U.S. Census Bureau)*

The U.S. Department of Health & Human Services (HHS) is a department of the U.S. federal government with the motto "Improving the health, safety, and well-being of America." The Centers for Medicare & Medicaid Services work within the HHS to help administer Medicare, Medicaid, and other health programs. They also gather information about health expenditure, program utilization, and other data. One area studied is the average amount of time that Medicare patients spend at short-stay hospitals.

You work for the Centers for Medicare & Medicaid Services. You want to test the claim that the mean length of stay for inpatients in 2015 is different than what it was in 2000 from a random sample of inpatient records. The results for several inpatients from 2000 and 2015 are shown in the histograms.

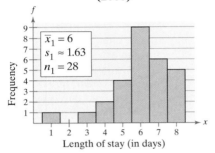

Inpatients Length of Stay (2000)

$\bar{x}_1 = 6$
$s_1 \approx 1.63$
$n_1 = 28$

Length of stay (in days)

EXERCISES

1. **How Could You Do It?**

   Explain how you could use each sampling technique to select the sample for the study.

   (a) stratified sample

   (b) cluster sample

   (c) systematic sample

   (d) simple random sample

2. **Choosing a Sampling Technique**

   (a) Which sampling technique in Exercise 1 would you choose to implement for the study? Why?

   (b) Identify possible flaws or biases in your study.

3. **Choosing a Test**

   To test the claim that there is a difference in the mean length of hospital stays, should you use a z-test or a t-test? Are the samples independent or dependent? Do you need to know what each population's distribution is? Do you need to know anything about the population variances?

4. **Testing a Mean**

   Test the claim that there is a difference in the mean length of hospital stays for inpatients. Assume the populations are normal and the population variances are equal. Use $\alpha = 0.05$. Interpret the test's decision. Does the decision support the claim?

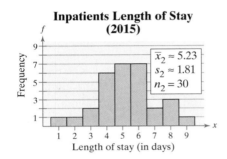

Inpatients Length of Stay (2015)

$\bar{x}_2 \approx 5.23$
$s_2 \approx 1.81$
$n_2 = 30$

Length of stay (in days)

TECHNOLOGY

MINITAB   EXCEL   TI-84 PLUS

Tails Over Heads

In the article "Tails over Heads" in the Washington Post (Oct. 13, 1996), journalist William Casey describes one of his hobbies—keeping track of every coin he finds on the street! From January 1, 1985 until the article was written, Casey found 11,902 coins.

As each coin is found, Casey records the time, date, location, value, mint location, and whether the coin is lying heads up or tails up. In the article, Casey notes that 6130 coins were found tails up and 5772 were found heads up. Of the 11,902 coins found, 43 were minted in San Francisco, 7133 were minted in Philadelphia, and 4726 were minted in Denver.

A simulation of Casey's experiment can be done in Minitab as shown below. A frequency histogram of one simulation's results is shown at the right.

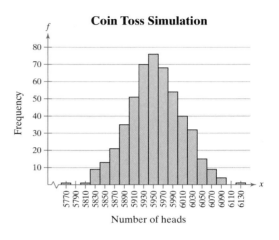

Coin Toss Simulation

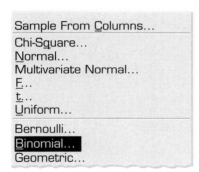

MINITAB

Number of rows of data to generate: 500

Store in column(s): C1

Number of trials: 11902

Event probability: .5

EXERCISES

1. Use technology to perform a one-sample z-test to test the hypothesis that the proportion of coins found lying heads up is 0.5. Use $\alpha = 0.01$. Use Casey's data as your sample and write your conclusion as a sentence.

2. Do Casey's data differ significantly from chance? If so, what might be the reason?

3. In the simulation shown above, what percent of the trials had heads less than or equal to the number of heads in Casey's data? Use technology to repeat the simulation. Are your results comparable?

In Exercises 4 and 5, use technology to perform a two-sample t-test to determine whether there is a difference in the mint dates and in the values of coins found on a street from 1985 through 1996 for the two mint locations. Write your conclusion as a sentence. Use $\alpha = 0.05$.

4. Mint dates of coins (years)
   Philadelphia: $\bar{x}_1 = 1984.8$ $s_1 = 8.6$
   Denver: $\bar{x}_2 = 1983.4$ $s_2 = 8.4$
   Assume population variances are equal.

5. Value of coins (dollars)
   Philadelphia: $\bar{x}_1 = \$0.034$ $s_1 = \$0.054$
   Denver: $\bar{x}_2 = \$0.033$ $s_2 = \$0.052$
   Assume population variances are not equal.

Extended solutions are given in the technology manuals that accompany this text. Technical instruction is provided for Minitab, Excel, and the TI-84 Plus.

8 Using Technology to Perform Two-Sample Hypothesis Tests

Here are some Minitab and TI-84 Plus printouts for several examples in this chapter.

See Example 1, page 430.

Display Descriptive Statistics...
Store Descriptive Statistics...
Graphical Summary...

1-Sample Z...
1-Sample t...
2-Sample t...
Paired t...

1 Proportion...
2 Proportions...

MINITAB

Two-Sample T-Test and CI

Sample	N	Mean	StDev	SE Mean
1	8	473.0	39.7	14
2	18	459.0	24.5	5.8

Difference = mu (1) − mu (2)
Estimate for difference: 14.0
90% CI for difference: (−13.8, 41.8)
T-Test of difference = 0 (vs not =): T-Value = 0.92 P-Value = 0.380 DF = 9

See Example 1, page 439.

Vertical Jump Heights, Before and After Using Shoes

Athlete	1	2	3	4	5	6	7	8
Vertical jump height (before using shoes)	24	22	25	28	35	32	30	27
Vertical jump height (after using shoes)	26	25	25	29	33	34	35	30

Display Descriptive Statistics...
Store Descriptive Statistics...
Graphical Summary...

1-Sample Z...
1-Sample t...
2-Sample t...
Paired t...

1 Proportion...
2 Proportions...

MINITAB

Paired T-Test and CI: Before, After

Paired T for Before − After

	N	Mean	StDev	SE Mean
Before	8	27.88	4.32	1.53
After	8	29.63	4.07	1.44
Difference	8	−1.750	2.121	0.750

90% upper bound for mean difference: −0.689
T-Test of mean difference = 0 (vs < 0): T-Value = −2.33 P-Value = 0.026

See Example 2, page 422.

TI-84 PLUS

EDIT CALC **TESTS**
1: Z–Test...
2: T–Test...
 2–SampZTest...
4: 2–SampTTest...
5: 1–PropZTest...
6: 2–PropZTest...
7↓ ZInterval...

TI-84 PLUS

2-SampZTest
Inpt:Data **Stats**
σ1:960
σ2:845
$\bar{x}$1:3060
n1:250
$\bar{x}$2:2910
↓n2:250

TI-84 PLUS

2-SampZTest
↑σ2:845
$\bar{x}$1:3060
n1:250
$\bar{x}$2:2910
n2:250
μ1:**≠μ2** <μ2 >μ2
Calculate Draw

TI-84 PLUS

2-SampZTest
$\mu_1 \neq \mu_2$
z=1.854468212
p=.0636720795
$\bar{x}_1$=3060
$\bar{x}_2$=2910
↓n_1=250

See Example 2, page 431.

TI-84 PLUS

EDIT CALC **TESTS**
1: Z–Test...
2: T–Test...
3: 2–SampZTest...
 2–SampTTest...
5: 1–PropZTest...
6: 2–PropZTest...
7↓ ZInterval...

TI-84 PLUS

2-SampTTest
Inpt:Data **Stats**
$\bar{x}$1:.48
Sx1:.05
n1:30
$\bar{x}$2:.51
Sx2:.07
↓n2:32

TI-84 PLUS

2-SampTTest
↑n1:30
$\bar{x}$2:.51
Sx2:.07
n2:32
μ1:≠μ2 **<μ2** >μ2
Pooled:No **Yes**
Calculate Draw

TI-84 PLUS

2-SampTTest
$\mu_1 < \mu_2$
t=−1.930301843
p=.0291499618
df=60
$\bar{x}_1$=.48
↓$\bar{x}_2$=.51

See Example 1, page 449.

TI-84 PLUS

EDIT CALC **TESTS**
1: Z–Test...
2: T–Test...
3: 2–SampZTest...
4: 2–SampTTest...
5: 1–PropZTest...
 2–PropZTest...
7↓ ZInterval...

TI-84 PLUS

2-PropZTest
x1:182
n1:200
x2:208
n2:250
p1:**≠p2** <p2 >p2
Calculate Draw

TI-84 PLUS

2-PropZTest
$p_1 \neq p_2$
z=2.418677324
p=.0155770453
$\hat{p}_1$=.91
$\hat{p}_2$=.832
↓$\hat{p}$=.8666666667

CHAPTERS 6-8
CUMULATIVE REVIEW

Sections 6.3 and 7.4

1. In a survey of 3015 U.S. adults, 80% say their household contains a desktop or laptop computer. *(Source: Pew Research Center)*

   (a) Construct a 95% confidence interval for the proportion of U.S. adults who say their household contains a desktop or laptop computer.

   (b) A researcher claims that more than 75% of U.S. adults say their household contains a desktop or laptop computer. At $\alpha = 0.05$, can you support the researcher's claim? Interpret the decision in the context of the original claim.

Section 8.3

2. **Gas Mileage**   The table shows the gas mileages (in miles per gallon) of eight cars with and without using a fuel additive. At $\alpha = 0.10$, is there enough evidence to conclude that the additive improved gas mileage? Assume the populations are normally distributed.

Car	1	2	3	4
Gas mileage (without fuel additive)	23.1	25.4	21.9	24.3
Gas mileage (with fuel additive)	23.6	27.7	23.6	26.8

Car	5	6	7	8
Gas mileage (without fuel additive)	19.9	21.2	25.9	24.8
Gas mileage (with fuel additive)	22.1	22.4	26.3	26.6

Sections 6.1 and 6.2

In Exercises 3–6, construct the indicated confidence interval for the population mean μ. Which distribution did you use to create the confidence interval?

3. $c = 0.95, \bar{x} = 26.97, \sigma = 3.4, n = 42$

4. $c = 0.95, \bar{x} = 3.46, s = 1.63, n = 16$

5. $c = 0.99, \bar{x} = 12.1, s = 2.64, n = 26$

6. $c = 0.90, \bar{x} = 8.21, \sigma = 0.62, n = 8$

Section 7.1

In Exercises 7–10, the statement represents a claim. Write its complement and state which is H_0 and which is H_a.

7. $\mu < 33$

8. $p \geq 0.19$

9. $\sigma = 0.63$

10. $\mu \neq 2.28$

Section 8.1

11. A pediatrician claims that the mean birth weight of a single-birth baby is greater than the mean birth weight of a baby that has a twin. The mean birth weight of a random sample of 85 single-birth babies is 3086 grams. Assume the population standard deviation is 563 grams. The mean birth weight of a random sample of 68 babies that have a twin is 2263 grams. Assume the population standard deviation is 624 grams. At $\alpha = 0.10$, can you support the pediatrician's claim? Interpret the decision in the context of the original claim.

12. The mean room rate for two adults for a random sample of 26 three-star hotels in Cincinnati has a sample standard deviation of $31. Assume the population is normally distributed. *(Adapted from Expedia)*

(a) Construct a 99% confidence interval for the population variance.

(b) Construct a 99% confidence interval for the population standard deviation.

(c) A travel analyst claims that the standard deviation of the mean room rate for two adults at three-star hotels in Cincinnati is at most $30. At $\alpha = 0.01$, can you reject the travel analyst's claim? Interpret the decision in the context of the original claim.

13. An education organization claims that the mean SAT scores for male athletes and male non-athletes at a college are different. A random sample of 26 male athletes at the college has a mean SAT score of 1189 and a standard deviation of 218. A random sample of 18 male non-athletes at the college has a mean SAT score of 1376 and a standard deviation of 186. At $\alpha = 0.05$, can you support the organization's claim? Interpret the decision in the context of the original claim. Assume the populations are normally distributed and the population variances are equal.

 14. The annual earnings (in dollars) for 30 randomly selected locksmiths are shown below. Assume the population is normally distributed. *(Adapted from Salary.com)*

44,044	42,206	38,262	57,022	66,462	50,211
64,804	67,191	55,101	64,962	49,634	47,516
61,710	43,514	60,622	30,600	56,477	60,747
54,275	54,266	54,367	48,420	40,549	65,291
64,842	52,435	49,179	48,042	63,648	49,142

(a) Construct a 95% confidence interval for the population mean annual earnings for locksmiths.

(b) A researcher claims that the mean annual earnings for locksmiths is $53,000. At $\alpha = 0.05$, can you reject the researcher's claim? Interpret the decision in the context of the original claim.

15. A medical research team studied the use of a marijuana extract to treat children with an epilepsy disorder. Of the 52 children who were given the extract, the number of convulsive seizures was reduced from 12 to 6 per month. Of the 56 children who were given a placebo, the number of convulsive seizures was reduced from 15 to 14 per month. At $\alpha = 0.10$, can you support the claim that the proportion of monthly convulsive seizure reduction is greater for the group that received the extract than for the group that received the placebo? Interpret the decision in the context of the original claim. *(Adapted from the New England Journal of Medicine)*

16. A random sample of 40 ostrich eggs has a mean incubation period of 42 days. Assume the population standard deviation is 1.6 days.

(a) Construct a 95% confidence interval for the population mean incubation period.

(b) A zoologist claims that the mean incubation period for ostriches is at least 45 days. At $\alpha = 0.05$, can you reject the zoologist's claim? Interpret the decision in the context of the original claim.

17. A researcher claims that 5% of people who wear eyeglasses purchase their eyeglasses online. Describe type I and type II errors for a hypothesis test of the claim. *(Source: Consumer Reports)*

CHAPTER **9**

Correlation and Regression

In 2016, the Los Angeles Dodgers had the highest team salary in Major League Baseball at $231.3 million and the highest average attendance at 45,720. The Tampa Bay Rays had the lowest team salary at $48.2 million and the lowest average attendance at 15,879.

In Chapters 1–8, you studied descriptive statistics, probability, and inferential statistics. One of the techniques you learned in descriptive statistics was graphing paired data with a scatter plot (see Section 2.2). For instance, the salaries and average attendances at home games for the teams in Major League Baseball in 2016 are shown in the scatter plot at the right and in the table below.

Major League Baseball

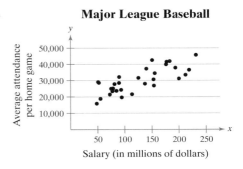

Salary (in millions of dollars)	78.4	75.0	153.7	218.7	176.1	113.4	77.3	94.5	89.7	199.9
Average attendance per home game	25,138	24,950	26,819	36,487	39,906	21,559	23,384	19,650	32,130	31,173

Salary (in millions of dollars)	89.5	125.1	139.7	231.3	72.5	52.1	93.3	155.2	193.2	55.0
Average attendance per home game	28,477	31,577	37,236	45,720	21,405	28,575	24,246	34,440	37,820	18,784

Salary (in millions of dollars)	84.8	81.2	50.7	137.2	177.0	150.4	48.2	212.1	182.7	153.0
Average attendance per home game	23,644	27,768	29,030	27,999	41,546	42,525	15,879	33,462	41,878	30,641

In this chapter, you will study how to describe and test the significance of relationships between two variables when data are presented as ordered pairs. For instance, in the scatter plot above, it appears that higher team salaries tend to correspond to higher average attendances and lower team salaries tend to correspond to lower average attendances. This relationship is described by saying that the team salaries are positively correlated to the average attendances. Graphically, the relationship can be described by drawing a line, called a regression line, that fits the points as closely as possible, as shown below. The second scatter plot below shows the salaries and wins for the teams in Major League Baseball in 2016. From the scatter plot, it appears that there is a positive correlation between the team salaries and wins.

Major League Baseball

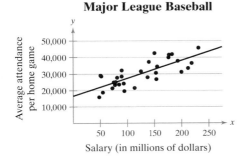

Major League Baseball

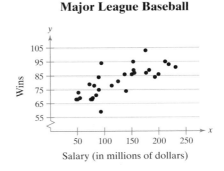

9.1 Correlation

What You Should Learn

▶ An introduction to linear correlation, independent and dependent variables, and the types of correlation

▶ How to find a correlation coefficient

▶ How to test a population correlation coefficient ρ using a table

▶ How to perform a hypothesis test for a population correlation coefficient ρ

▶ How to distinguish between correlation and causation

An Overview of Correlation ■ Correlation Coefficient ■ Using a Table to Test a Population Correlation Coefficient ρ ■ Hypothesis Testing for a Population Correlation Coefficient ρ ■ Correlation and Causation

An Overview of Correlation

Suppose a safety inspector wants to determine whether a relationship exists between the number of hours of training for an employee and the number of accidents involving that employee. Or suppose a psychologist wants to know whether a relationship exists between the number of hours a person sleeps each night and that person's reaction time. How would he or she determine if any relationship exists?

In this section, you will study how to describe what type of relationship, or correlation, exists between two quantitative variables and how to determine whether the correlation is significant.

DEFINITION

A **correlation** is a relationship between two variables. The data can be represented by the ordered pairs (x, y), where x is the **independent** (or **explanatory**) **variable** and y is the **dependent** (or **response**) **variable.**

In Section 2.2, you learned that the graph of ordered pairs (x, y) is called a *scatter plot*. In a scatter plot, the ordered pairs (x, y) are graphed as points in a coordinate plane. The independent (explanatory) variable x is measured on the horizontal axis, and the dependent (response) variable y is measured on the vertical axis. A scatter plot can be used to determine whether a linear (straight line) correlation exists between two variables. The scatter plots below show several types of correlation.

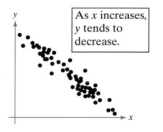

Negative Linear Correlation

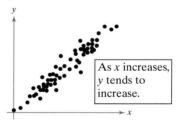

Positive Linear Correlation

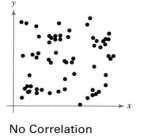

No Correlation

Nonlinear Correlation

GDP (in trillions of dollars), x	CO_2 emissions (in millions of metric tons), y
1.8	604.4
1.3	434.2
2.4	544.0
1.5	370.4
3.9	742.3
2.1	340.5
0.9	232.0
1.4	262.3
3.0	441.9
4.6	1157.7

Tech Tip

Remember that all data sets containing 20 or more entries are available electronically. Also, some of the data sets in this section are used throughout the chapter, so save any data that you enter. For instance, the data used in Example 1 is also used later in this section and in Sections 9.2 and 9.3.

Note to Instructor

Have students try to picture a line going through the points. If the line clearly has a positive slope, then the correlation is positive. If the slope is clearly negative, then the correlation is negative. If they have trouble picturing a line through the points, then there is probably no linear correlation.

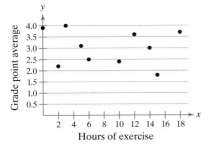

EXAMPLE 1

Constructing a Scatter Plot

An economist wants to determine whether there is a linear relationship between a country's gross domestic product (GDP) and carbon dioxide (CO_2) emissions. The data are shown in the table at the left. Display the data in a scatter plot and describe the type of correlation. *(Source: World Bank and U.S. Energy Information Administration)*

SOLUTION

The scatter plot is shown below. From the scatter plot, it appears that there is a positive linear correlation between the variables.

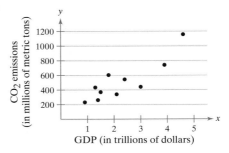

Interpretation Reading from left to right, as the gross domestic products increase, the carbon dioxide emissions tend to increase.

TRY IT YOURSELF 1

A director of alumni affairs at a small college wants to determine whether there is a linear relationship between the number of years alumni have been out of school and their annual contributions (in thousands of dollars). The data are shown in the table below. Display the data in a scatter plot and describe the type of correlation.

Number of years out of school, x	1	10	5	15	3	24	30
Annual contribution (in 1000s of $), y	12.5	8.7	14.6	5.2	9.9	3.1	2.7

Answer: Page A38

EXAMPLE 2

Constructing a Scatter Plot

A student conducts a study to determine whether there is a linear relationship between the number of hours a student exercises each week and the student's grade point average (GPA). The data are shown in the table below. Display the data in a scatter plot and describe the type of correlation.

Hours of exercise, x	12	3	0	6	10	2	18	14	15	5
GPA, y	3.6	4.0	3.9	2.5	2.4	2.2	3.7	3.0	1.8	3.1

SOLUTION

The scatter plot is shown at the left. From the scatter plot, it appears that there is no linear correlation between the variables.

Interpretation The number of hours a student exercises each week does not appear to be related to the student's grade point average.

TRY IT YOURSELF 2

A researcher conducts a study to determine whether there is a linear relationship between a person's height (in inches) and pulse rate (in beats per minute). The data are shown in the table below. Display the data in a scatter plot and describe the type of correlation.

Height, x	68	72	65	70	62	75	78	64	68
Pulse rate, y	90	85	88	100	105	98	70	65	72

Answer: Page A38

Duration, x	Time, y	Duration, x	Time, y
1.80	56	3.78	79
1.82	58	3.83	85
1.90	62	3.88	80
1.93	56	4.10	89
1.98	57	4.27	90
2.05	57	4.30	89
2.13	60	4.43	89
2.30	57	4.47	86
2.37	61	4.53	89
2.82	73	4.55	86
3.13	76	4.60	92
3.27	77	4.63	91
3.65	77		

EXAMPLE 3

Constructing a Scatter Plot Using Technology

Old Faithful, located in Yellowstone National Park, is the world's most famous geyser. The durations (in minutes) of several of Old Faithful's eruptions and the times (in minutes) until the next eruption are shown in the table at the left. Use technology to display the data in a scatter plot. Describe the type of correlation.

SOLUTION

MINITAB, Excel, the TI-84 Plus, and StatCrunch each have features for graphing scatter plots. Try using this technology to draw the scatter plots shown. From the scatter plots, it appears that the variables have a positive linear correlation.

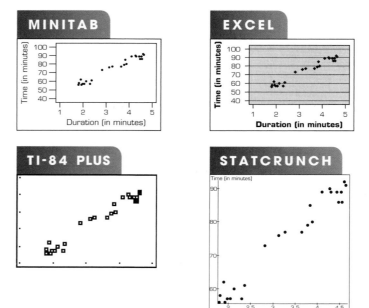

Interpretation Reading from left to right, as the durations of the eruptions increase, the times until the next eruption tend to increase.

TRY IT YOURSELF 3

Consider the data on page 469 on the salaries and average attendances at home games for the teams in Major League Baseball. Use technology to display the data in a scatter plot. Describe the type of correlation.

Answer: Page A38

Correlation Coefficient

Interpreting correlation using a scatter plot is subjective. A precise measure of the type and strength of a linear correlation between two variables is to calculate the **correlation coefficient.** A formula for the sample correlation coefficient is given, but it is more convenient to use technology to calculate this value.

> **DEFINITION**
>
> The **correlation coefficient** is a measure of the strength and the direction of a linear relationship between two variables. The symbol r represents the sample correlation coefficient. A formula for r is
>
> $$r = \frac{n\Sigma xy - (\Sigma x)(\Sigma y)}{\sqrt{n\Sigma x^2 - (\Sigma x)^2}\sqrt{n\Sigma y^2 - (\Sigma y)^2}}$$ Sample correlation coefficient
>
> where n is the number of pairs of data. The **population correlation coefficient** is represented by ρ (the lowercase Greek letter rho, pronounced "row").

The range of the correlation coefficient is -1 to 1, inclusive. When x and y have a strong positive linear correlation, r is close to 1. When x and y have a strong negative linear correlation, r is close to -1. When x and y have perfect positive linear correlation or perfect negative linear correlation, r is equal to 1 or -1, respectively. When there is no linear correlation, r is close to 0. It is important to remember that when r is close to 0, it does not mean that there is no relation between x and y, just that there is no *linear* relation. Several examples are shown below.

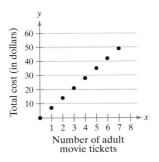

Perfect positive correlation
$r = 1$

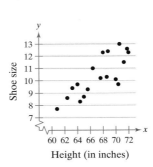

Strong positive correlation
$r = 0.81$

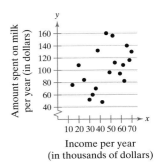

Weak positive correlation
$r = 0.45$

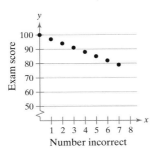

Perfect negative correlation
$r = -1$

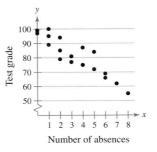

Strong negative correlation
$r = -0.92$

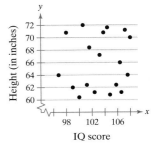

No correlation
$r = 0.04$

To use a correlation coefficient r to make an inference about a population, it is required that (1) the sample paired data (x, y) is random and (2) x and y have a *bivariate normal distribution* (you will learn more about this distribution in Section 9.3). In this text, unless stated otherwise, you can assume that these requirements are met.

GUIDELINES

Calculating a Correlation Coefficient

In Words	In Symbols
1. Find the sum of the x-values.	Σx
2. Find the sum of the y-values.	Σy
3. Multiply each x-value by its corresponding y-value and find the sum.	Σxy
4. Square each x-value and find the sum.	Σx^2
5. Square each y-value and find the sum.	Σy^2
6. Use these five sums to calculate the correlation coefficient.	$r = \dfrac{n\Sigma xy - (\Sigma x)(\Sigma y)}{\sqrt{n\Sigma x^2 - (\Sigma x)^2}\sqrt{n\Sigma y^2 - (\Sigma y)^2}}$

EXAMPLE 4

Calculating a Correlation Coefficient

Calculate the correlation coefficient for the gross domestic products and carbon dioxide emissions data in Example 1. Interpret the result in the context of the data.

SOLUTION Use a table to help calculate the correlation coefficient.

GDP (in trillions of dollars), x	CO₂ emissions (in millions of metric tons), y	xy	x^2	y^2
1.8	604.4	1087.92	3.24	365,299.36
1.3	434.2	564.46	1.69	188,529.64
2.4	544.0	1305.6	5.76	295,936
1.5	370.4	555.6	2.25	137,196.16
3.9	742.3	2894.97	15.21	551,009.29
2.1	340.5	715.05	4.41	115,940.25
0.9	232.0	208.8	0.81	53,824
1.4	262.3	367.22	1.96	68,801.29
3.0	441.9	1325.7	9	195,275.61
4.6	1157.7	5325.42	21.16	1,340,269.29
$\Sigma x = 22.9$	$\Sigma y = 5129.7$	$\Sigma xy = 14{,}350.74$	$\Sigma x^2 = 65.49$	$\Sigma y^2 = 3{,}312{,}080.89$

With these sums and $n = 10$, the correlation coefficient is

$$r = \frac{n\Sigma xy - (\Sigma x)(\Sigma y)}{\sqrt{n\Sigma x^2 - (\Sigma x)^2}\sqrt{n\Sigma y^2 - (\Sigma y)^2}}$$

$$= \frac{10(14{,}350.74) - (22.9)(5129.7)}{\sqrt{10(65.49) - (22.9)^2}\sqrt{10(3{,}312{,}080.89) - (5129.7)^2}}$$

$$= \frac{26{,}037.27}{\sqrt{130.49}\sqrt{6{,}806{,}986.81}}$$

$$\approx 0.874. \qquad \text{Round to three decimal places.}$$

The result $r \approx 0.874$ suggests a strong positive linear correlation.

Interpretation As the gross domestic product increases, the carbon dioxide emissions tend to increase.

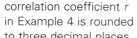
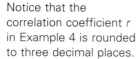

Number of years out of school, x	Annual contribution (in 1000s of $), y
1	12.5
10	8.7
5	14.6
15	5.2
3	9.9
24	3.1
30	2.7

9.1 To explore this topic further, see **Activity 9.1** on page 485.

Tech Tip

Before using the TI-84 Plus to calculate *r*, make sure the *diagnostics* feature is on. To turn on this feature, from the home screen, press 2nd CATALOG and cursor to *DiagnosticOn*. Then press ENTER twice.

TRY IT YOURSELF 4

Calculate the correlation coefficient for the number of years out of school and annual contribution data in Try It Yourself 1. Interpret the result in the context of the data.

Answer: Page A38

EXAMPLE 5

Using Technology to Calculate a Correlation Coefficient

Use technology to calculate the correlation coefficient for the Old Faithful data in Example 3. Interpret the result in the context of the data.

SOLUTION

Minitab, Excel, the TI-84 Plus, and StatCrunch each have features that allow you to calculate a correlation coefficient for paired data sets. Try using this technology to find *r*. You should obtain results similar to the displays shown.

MINITAB

Correlations: Duration, Time

Pearson correlation of Duration and Time = 0.979 ← Correlation coefficient

EXCEL

	A	B	C
26	CORREL(A1:A25,B1:B25)		
27			0.978659213

← Correlation coefficient

TI-84 PLUS

LinReg
y=ax+b
a=12.48094391
b=33.68290034
r²=.9577738551
r=.9786592129 ← Correlation coefficient

STATCRUNCH

Correlation between Duration and Time is:
0.97865921 ← Correlation coefficient

Rounded to three decimal places, the correlation coefficient is

$r \approx 0.979.$ Round to three decimal places.

This value of *r* suggests a strong positive linear correlation.

Interpretation As the duration of the eruptions increases, the time until the next eruption tends to increase.

TRY IT YOURSELF 5

Use technology to calculate the correlation coefficient for the data on page 469 on the salaries and average attendances at home games for the teams in Major League Baseball. Interpret the result in the context of the data.

Answer: Page A38

Note to Instructor

Another way to determine whether the population correlation coefficient ρ is significant is to perform a hypothesis test (which is explained later in this section).

Using a Table to Test a Population Correlation Coefficient ρ

Once you have calculated r, the sample correlation coefficient, you will want to determine whether there is enough evidence to decide that the population correlation coefficient ρ is significant. In other words, based on a few pairs of data, can you make an inference about the population of all such data pairs? Remember that you are using sample data to make a decision about population data, so it is always possible that your inference may be wrong. In correlation studies, the small percentage of times when you decide that the correlation is significant when it is really not is called the *level of significance.* It is typically set at $\alpha = 0.01$ or 0.05. When $\alpha = 0.05$, you will probably decide that the population correlation coefficient is significant when it is really not 5% of the time. (Of course, 95% of the time, you will correctly determine that a correlation coefficient is significant.) When $\alpha = 0.01$, you will make this type of error only 1% of the time. When using a lower level of significance, however, you may fail to identify some significant correlations.

In order for a correlation coefficient to be significant, its absolute value must be close to 1. To determine whether the population correlation coefficient ρ is significant, use the critical values given in Table 11 in Appendix B. A portion of the table is shown below. If $|r|$ is greater than the critical value, then there is enough evidence to decide that the correlation is significant. Otherwise, there is *not* enough evidence to say that the correlation is significant. For instance, to determine whether ρ is significant for five pairs of data ($n = 5$) at a level of significance of $\alpha = 0.01$, you need to compare $|r|$ with a critical value of 0.959, as shown in the table.

Number n of pairs of data in sample ⟶   Critical values for $\alpha = 0.05$ and $\alpha = 0.01$

n	$\alpha = 0.05$	$\alpha = 0.01$
4	0.950	0.990
5	0.878	(0.959)
6	0.811	0.917

If $|r| > 0.959$, then the correlation is significant. Otherwise, there is *not* enough evidence to conclude that the correlation is significant. Here are the guidelines for this process.

GUIDELINES

Using Table 11 for the Correlation Coefficient ρ

In Words	In Symbols
1. Determine the number of pairs of data in the sample.	Determine n.
2. Specify the level of significance.	Identify α.
3. Find the critical value.	Use Table 11 in Appendix B.
4. Decide whether the correlation is significant.	If $|r|$ is greater than the critical value, then the correlation is significant. Otherwise, there is *not* enough evidence to conclude that the correlation is significant.
5. Interpret the decision in the context of the original claim.	

To use Table 11 to test a correlation coefficient, note that the requirements for calculating a correlation coefficient given on page 474 also apply to the test. In this text, unless stated otherwise, you can assume that these requirements are met.

EXAMPLE 6

Using Table 11 for a Correlation Coefficient

In Example 5, you used 25 pairs of data to find $r \approx 0.979$. Is the correlation coefficient significant? Use $\alpha = 0.05$.

SOLUTION

The number of pairs of data is 25, so $n = 25$. The level of significance is $\alpha = 0.05$. Using Table 11, find the critical value in the $\alpha = 0.05$ column that corresponds to the row with $n = 25$. The number in that column and row is 0.396.

Critical values
for $\alpha = 0.05$
↓

n	$\alpha = 0.05$	$\alpha = 0.01$
4	0.950	0.990
5	0.878	0.959
6	0.811	0.917
7	0.754	0.875
8	0.707	0.834
9	0.666	0.798
10	0.632	0.765
11	0.602	0.735
12	0.576	0.708
13	0.553	0.684
14	0.532	0.661
15	0.514	0.641
16	0.497	0.623
17	0.482	0.606
18	0.468	0.590
19	0.456	0.575
20	0.444	0.561
21	0.433	0.549
22	0.423	0.537
23	0.413	0.526
24	0.404	0.515
$n = 25 \rightarrow$ 25	(0.396)	0.505
26	0.388	0.496
27	0.381	0.487
28	0.374	0.479
29	0.367	0.471

Because $|r| \approx 0.979 > 0.396$, you can decide that the population correlation is significant.

Interpretation   There is enough evidence at the 5% level of significance to conclude that there is a significant linear correlation between the duration of Old Faithful's eruptions and the time between eruptions.

TRY IT YOURSELF 6

In Try It Yourself 4, you calculated the correlation coefficient of the number of years out of school and annual contribution data to be $r \approx -0.908$. Is the correlation coefficient significant? Use $\alpha = 0.01$.

Answer: Page A38

In Table 11, notice that for fewer data pairs (smaller values of n), stronger evidence is needed to conclude that the correlation coefficient is significant.

Note to Instructor

The material on hypothesis testing requires previous coverage of Chapter 7. This material can be omitted if you cover correlation and regression before covering Chapters 7 and 8.

Hypothesis Testing for a Population Correlation Coefficient ρ

You can also use a hypothesis test to determine whether the sample correlation coefficient r provides enough evidence to conclude that the population correlation coefficient ρ is significant. A hypothesis test for ρ can be one-tailed or two-tailed. The null and alternative hypotheses for these tests are listed below.

$$\begin{cases} H_0: \rho \geq 0 \text{ (no significant negative correlation)} \\ H_a: \rho < 0 \text{ (significant negative correlation)} \end{cases}$$ Left-tailed test

$$\begin{cases} H_0: \rho \leq 0 \text{ (no significant positive correlation)} \\ H_a: \rho > 0 \text{ (significant positive correlation)} \end{cases}$$ Right-tailed test

$$\begin{cases} H_0: \rho = 0 \text{ (no significant correlation)} \\ H_a: \rho \neq 0 \text{ (significant correlation)} \end{cases}$$ Two-tailed test

In this text, you will consider only two-tailed hypothesis tests for ρ.

Note to Instructor

Remind students that this is the same family of distributions that was used for hypothesis testing of means with unknown population standard deviations and with dependent samples. There are $n - 2$ degrees of freedom because one degree of freedom is lost for each variable.

The *t*-Test for the Correlation Coefficient

A **t-test** can be used to test whether the correlation between two variables is significant. The **test statistic** is r and the **standardized test statistic**

$$t = \frac{r}{\sigma_r} = \frac{r}{\sqrt{\dfrac{1 - r^2}{n - 2}}}$$

follows a *t*-distribution with $n - 2$ degrees of freedom, where n is the number of pairs of data. (Note that there are $n - 2$ degrees of freedom because one degree of freedom is lost for each variable.)

GUIDELINES

Using the *t*-Test for the Correlation Coefficient ρ

In Words	In Symbols
1. Identify the null and alternative hypotheses.	State H_0 and H_a.
2. Specify the level of significance.	Identify α.
3. Identify the degrees of freedom.	d.f. $= n - 2$
4. Determine the critical value(s) and the rejection region(s).	Use Table 5 in Appendix B.
5. Find the standardized test statistic.	$t = \dfrac{r}{\sqrt{\dfrac{1 - r^2}{n - 2}}}$
6. Make a decision to reject or fail to reject the null hypothesis.	If t is in the rejection region, then reject H_0. Otherwise, fail to reject H_0.
7. Interpret the decision in the context of the original claim.	

To use the *t*-test for a correlation coefficient, note that the requirements for calculating a correlation coefficient given on page 474 also apply to the test. In this text, unless stated otherwise, you can assume that these requirements are met.

EXAMPLE 7

The *t*-Test for a Correlation Coefficient

In Example 4, you used 10 pairs of data to find $r \approx 0.874$. Test the significance of this correlation coefficient. Use $\alpha = 0.05$.

SOLUTION

The null and alternative hypotheses are

$$H_0: \rho = 0 \text{ (no correlation)} \quad \text{and} \quad H_a: \rho \neq 0 \text{ (significant correlation)}.$$

Because there are 10 pairs of data in the sample, there are $10 - 2 = 8$ degrees of freedom. Because the test is a two-tailed test, $\alpha = 0.05$, and d.f. = 8, the critical values are $-t_0 = -2.306$ and $t_0 = 2.306$. The rejection regions are $t < -2.306$ and $t > 2.306$. Using the *t*-test, the standardized test statistic is

$$t = \frac{r}{\sqrt{\dfrac{1 - r^2}{n - 2}}} \qquad \text{Use the } t\text{-test for } \rho.$$

$$\approx \frac{0.874}{\sqrt{\dfrac{1 - (0.874)^2}{10 - 2}}} \qquad \text{Substitute 0.874 for } r \text{ and 10 for } n.$$

$$\approx 5.087. \qquad \text{Round to three decimal places.}$$

TI-84 PLUS

```
LinRegTTest
y=a+bx
β≠0 and ρ≠0
t=5.078276982
p=9.5516561ε-4
df=8
↓a=56.03576519
```

You can check this result using technology. For instance, using a TI-84 Plus, you can find the standardized test statistic, as shown at the left. (Note that the result differs slightly due to rounding.) The figure below shows the location of the rejection regions and the standardized test statistic.

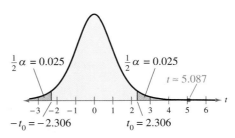

Because *t* is in the rejection region, you reject the null hypothesis.

Interpretation There is enough evidence at the 5% level of significance to conclude that there is a significant linear correlation between gross domestic products and carbon dioxide emissions.

TRY IT YOURSELF 7

In Try It Yourself 5, you calculated the correlation coefficient of the salaries and average attendances at home games for the teams in Major League Baseball to be $r \approx 0.775$. Test the significance of this correlation coefficient. Use $\alpha = 0.01$.

Answer: Page A38

In Example 7, you can use Table 11 in Appendix B to test the population correlation coefficient ρ. Given $n = 10$ and $\alpha = 0.05$, the critical value from Table 11 is 0.632. Because $|r| \approx 0.874 > 0.632$, the correlation is significant. Note that this is the same result you obtained using a *t*-test for the population correlation coefficient ρ.

Picturing the World

The scatter plot shows the results of a survey conducted as a group project by students in a high school statistics class in the San Francisco area. In the survey, 125 high school students were asked their grade point average (GPA) and the number of caffeine drinks they consumed each day.

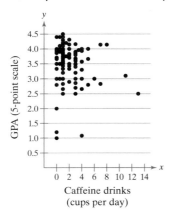

What type of correlation, if any, does the scatter plot show between caffeine consumption and GPA?

The scatter plot shows no correlation between caffeine consumption and GPA.

Correlation and Causation

The fact that two variables are strongly correlated does not in itself imply a cause-and-effect relationship between the variables. More in-depth study is usually needed to determine whether there is a causal relationship between the variables.

When there is a significant correlation between two variables, a researcher should consider these possibilities.

1. Is there a direct cause-and-effect relationship between the variables?

That is, does x cause y? For instance, consider the relationship between gross domestic products and carbon dioxide emissions that has been discussed throughout this section. It is reasonable to conclude that an increase in a country's gross domestic product will result in higher carbon dioxide emissions.

2. Is there a reverse cause-and-effect relationship between the variables?

That is, does y cause x? For instance, consider the Old Faithful data that have been discussed throughout this section. These variables have a positive linear correlation, and it is possible to conclude that the duration of an eruption affects the time before the next eruption. However, it is also possible that the time between eruptions affects the duration of the next eruption.

3. Is it possible that the relationship between the variables can be caused by a third variable or perhaps a combination of several other variables?

For instance, consider the salaries and average attendances per home game for the teams in Major League Baseball listed on page 469. Although these variables have a positive linear correlation, it is doubtful that just because a team's salary decreases, the average attendance per home game will also decrease. The relationship is probably due to other variables, such as the economy, the players on the team, and whether or not the team is winning games. Variables that have an effect on the variables being studied but are not included in the study are called **lurking variables.**

4. Is it possible that the relationship between two variables may be a coincidence?

For instance, although it may be possible to find a significant correlation between the number of animal species living in certain regions and the number of people who own more than two cars in those regions, it is highly unlikely that the variables are directly related. The relationship is probably due to coincidence.

Determining which of the cases above is valid for a data set can be difficult. For instance, consider this example. A person breaks out in a rash after eating shrimp at a certain restaurant. This happens every time the person eats shrimp at the restaurant. The natural conclusion is that the person is allergic to shrimp. However, upon further study by an allergist, it is found that the person is not allergic to shrimp, but to a type of seasoning the chef is putting into the shrimp.

9.1 EXERCISES

For Extra Help: MyLab Statistics

1. Increase; Decrease

2. The range of values for the correlation coefficient is −1 to 1, inclusive.

3. The sample correlation coefficient r measures the strength and direction of a linear relationship between two variables; $r = -0.932$ indicates a stronger correlation because $|-0.932| = 0.932$ is closer to 1 than $|0.918| = 0.918$.

4. *Sample answer:* Perfect positive linear correlation: price per gallon of gasoline and total cost of gasoline

Perfect negative linear correlation: distance from door and height of wheelchair ramp

5. A table can be used to compare r with a critical value, or a hypothesis test can be performed using a t-test.

6. r is the sample correlation coefficient, while ρ is the population correlation coefficient.

7. H_0: $\rho = 0$ (no significant correlation)

H_a: $\rho \neq 0$ (significant correlation)

Reject the null hypothesis if t is in the rejection region.

8. The fact that two variables have a linear relationship does not necessarily imply that one variable is the cause of the other; Answers will vary.

9. Strong negative linear correlation

10. No linear correlation

11. No linear correlation

12. Perfect positive linear correlation

13. Explanatory variable: Amount of water consumed

Response variable: Weight loss

14. Explanatory variable: Hours of safety classes

Response variable: Number of driving accidents

Building Basic Skills and Vocabulary

1. Two variables have a positive linear correlation. Does the dependent variable increase or decrease as the independent variable increases? What if the variables have a negative linear correlation?

2. Describe the range of values for the correlation coefficient.

3. What does the sample correlation coefficient r measure? Which value indicates a stronger correlation: $r = 0.918$ or $r = -0.932$? Explain your reasoning.

4. Give examples of two variables that have perfect positive linear correlation and two variables that have perfect negative linear correlation.

5. Explain how to determine whether a sample correlation coefficient indicates that the population correlation coefficient is significant.

6. Discuss the difference between r and ρ.

7. What are the null and alternate hypotheses for a two-tailed t-test for the population correlation coefficient ρ? When do you reject the null hypothesis?

8. In your own words, what does it mean to say "correlation does not imply causation"? List a pair of variables that have correlation but no cause-and-effect relationship.

Graphical Analysis *In Exercises 9–12, determine whether there is a perfect positive linear correlation, a strong positive linear correlation, a perfect negative linear correlation, a strong negative linear correlation, or no linear correlation between the variables.*

9.
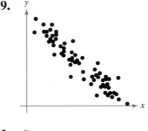

10.
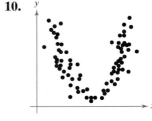

11.
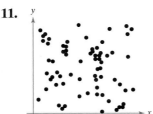

12.
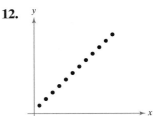

In Exercises 13 and 14, identify the explanatory variable and the response variable.

13. A nutritionist wants to determine whether the amounts of water consumed each day by persons of the same weight and on the same diet can be used to predict individual weight loss.

14. An actuary at an insurance company wants to determine whether the number of hours of safety driving classes can be used to predict the number of driving accidents for each driver.

15. c; You would expect a positive linear correlation between age and income.

16. d; You would not expect age and height to be correlated.

17. b; You would expect a negative linear correlation between age and balance on student loans.

18. a; You would expect the relationship between age and body temperature to be fairly constant.

19. *Sample answer:* People who can afford more valuable homes will live longer because they have more money to take care of themselves.

20. *Sample answer:* People who use alcohol are more likely to use tobacco because they are less concerned about their health than the general population.

21. *Sample answer:* Ice cream sales are higher when the weather is warm and people are outside more often. This is when homicides rates go up as well.

22. *Sample answer:* The correlation is a coincidence.

23. (a)

(b) 0.979

(c) Strong positive correlation; As age increases, the number of words in children's vocabulary tends to increase.

(d) There is enough evidence at the 1% level of significance to conclude that there is a significant linear correlation between children's ages and number of words in their vocabulary.

24. See Selected Answers, page A105.

Graphical Analysis *In Exercises 15–18, the scatter plots show the results of a survey of 20 randomly selected males ages 24–35. Using age as the explanatory variable, match each graph with the appropriate description. Explain your reasoning.*

(a) *Age and body temperature* (b) *Age and balance on student loans*

(c) *Age and income* (d) *Age and height*

15.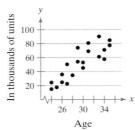

16.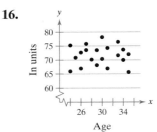

17.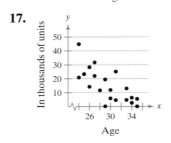

18.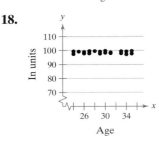

In Exercises 19–22, two variables are given that have been shown to have correlation but no cause-and-effect relationship. Describe at least one possible reason for the correlation.

19. Value of home and life span

20. Alcohol use and tobacco use

21. Ice cream sales and homicide rates

22. Marriage rate in Kentucky and number of deaths caused by falling out of a fishing boat

Using and Interpreting Concepts

Constructing a Scatter Plot and Determining Correlation *In Exercises 23–28, (a) display the data in a scatter plot, (b) calculate the sample correlation coefficient r, (c) describe the type of correlation, if any, and interpret the correlation in the context of the data, and (d) use Table 11 in Appendix B to make a conclusion about the correlation coefficient. If convenient, use technology. Let α = 0.01.*

23. **Age and Vocabulary** The ages (in years) of 11 children and the numbers of words in their vocabulary

Age, x	1	2	3	4	5	6	3	5	2	4	6
Vocabulary size, y	3	220	540	1100	2100	2600	730	2200	260	1200	2500

24. **Height and IQ** The height (in inches) of 8 high school girls and their scores on an IQ test

Height, x	62	58	65	67	59	64	65	57
IQ score, y	109	102	107	114	96	110	116	128

25. (a)

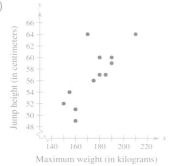

(b) 0.756

(c) Strong positive linear correlation; As the maximum weight for one repetition of a half squat increases, the jump height tends to increase.

(d) There is enough evidence at the 1% level of significance to conclude that there is a significant linear correlation between maximum weight for one repetition of a half squat and jump height.

26. (a)

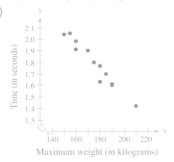

(b) −0.975

(c) Strong negative linear correlation; As the maximum weight for one repetition of a half squat increases, the time to run a 10-meter sprint tends to decrease.

(d) There is enough evidence at the 1% level of significance to conclude that there is a significant linear correlation between maximum weight for one repetition of a half squat and time to run a 10-meter sprint.

27. See Odd Answers, page A79.

28. See Selected Answers, page A105.

29. See Odd Answers, page A79.

30. See Selected Answers, page A105.

31. See Odd Answers, page A79.

32. See Selected Answers, page A105.

25. Maximal Strength and Jump Height The maximum weights (in kilograms) for which one repetition of a half squat can be performed and the jump heights (in centimeters) for 12 international soccer players *(Adapted from British Journal of Sports Medicine)*

Maximum weight, x	190	185	155	180	175	170
Jump height, y	60	57	54	60	56	64

Maximum weight, x	150	160	160	180	190	210
Jump height, y	52	51	49	57	59	64

26. Maximal Strength and Sprint Performance The maximum weights (in kilograms) for which one repetition of a half squat can be performed and the times (in seconds) to run a 10-meter sprint for 12 international soccer players *(Adapted from British Journal of Sports Medicine)*

Maximum weight, x	175	180	155	210	150	190
Time, y	1.80	1.77	2.05	1.42	2.04	1.61

Maximum weight, x	185	160	190	180	160	170
Time, y	1.70	1.91	1.60	1.63	1.98	1.90

27. Earnings and Dividends The earnings per share (in dollars) and the dividends per share (in dollars) for 6 companies in a recent year *(Source: The Value Line Investment Survey)*

Earnings per share, x	1.22	4.00	3.53	8.21	1.74	3.14
Dividends per share, y	0.90	0.31	2.10	1.00	0.55	1.48

28. Speed of Sound Eleven altitudes (in thousands of feet) and the speeds of sound (in feet per second) at these altitudes

Altitude, x	0	5	10	15	20	25
Speed of sound, y	1116.3	1096.9	1077.3	1057.2	1036.8	1015.8

Altitude, x	30	35	40	45	50
Speed of sound, y	994.5	969.0	967.7	967.7	967.7

29. In Exercise 23, add data for a child who is 6 years old and has a vocabulary size of 900 words to the data set. Describe how this affects the correlation coefficient r.

30. In Exercise 24, remove the data for the girl who is 57 inches tall and scored 128 on the IQ test from the data set. Describe how this affects the correlation coefficient r.

31. In Exercise 25, remove the data for the international soccer player with a maximum weight of 170 kilograms and a jump height of 64 centimeters from the data set. Describe how this affects the correlation coefficient r.

32. In Exercise 26, add data for an international soccer player who can perform the half squat with a maximum of 210 kilograms and can sprint 10 meters in 2.00 seconds to the data set. Describe how this affects the correlation coefficient r.

33. There is not enough evidence at the 1% level of significance to conclude that there is a significant linear correlation between vehicle weight and the variability in braking distance on a dry surface.

34. There is enough evidence at the 5% level of significance to conclude that there is a significant linear correlation between vehicle weight and the variability in braking distance on a wet surface.

35. There is enough evidence at the 5% level of significance to conclude that there is a significant linear correlation between the maximum weight for one repetition of a half squat and the jump height.

36. There is enough evidence at the 1% level of significance to conclude that there is a significant linear correlation between the maximum weight for one repetition of a half squat and the time to run a 10-meter sprint.

37. $r \approx -0.975$; The correlation coefficient remains unchanged when the x-values and y-values are switched.

38. Answers will vary.

The t-Test for Correlation Coefficients   *In Exercises 33–36, perform a hypothesis test using Table 5 in Appendix B to make a conclusion about the correlation coefficient.*

33. **Braking Distances: Dry Surface**   The weights (in pounds) of eight vehicles and the variabilities of their braking distances (in feet) when stopping on a dry surface are shown in the table. At $\alpha = 0.01$, is there enough evidence to conclude that there is a significant linear correlation between vehicle weight and variability in braking distance on a dry surface? *(Adapted from National Highway Traffic Safety Administration)*

Weight, x	5940	5340	6500	5100	5850	4800	5600	5890
Variability, y	1.78	1.93	1.91	1.59	1.66	1.50	1.61	1.70

34. **Braking Distances: Wet Surface**   The weights (in pounds) of eight vehicles and the variabilities of their braking distances (in feet) when stopping on a wet surface are shown in the table. At $\alpha = 0.05$, is there enough evidence to conclude that there is a significant linear correlation between vehicle weight and variability in braking distance on a wet surface? *(Adapted from National Highway Traffic Safety Administration)*

Weight, x	5890	5340	6500	4800	5940	5600	5100	5850
Variability, y	2.92	2.40	4.09	1.72	2.88	2.53	2.32	2.78

35. **Maximal Strength and Jump Height**   The table in Exercise 25 shows the maximum weights (in kilograms) for which one repetition of a half squat can be performed and the jump heights (in centimeters) for 12 international soccer players. At $\alpha = 0.05$, is there enough evidence to conclude that there is a significant linear correlation between the data? (Use the value of r found in Exercise 25.)

36. **Maximal Strength and Sprint Performance**   The table in Exercise 26 shows the maximum weights (in kilograms) for which one repetition of a half squat can be performed and the times (in seconds) to run a 10-meter sprint for 12 international soccer players. At $\alpha = 0.01$, is there enough evidence to conclude that there is a significant linear correlation between the data? (Use the value of r found in Exercise 26.)

Extending Concepts

37. **Interchanging x and y**   In Exercise 26, let the time (in seconds) to sprint 10 meters represent the x-values and the maximum weight (in kilograms) for which one repetition of a half squat can be performed represent the y-values. Calculate the correlation coefficient r. What effect does switching the explanatory and response variables have on the correlation coefficient?

38. **Writing**   Use your school's library, the Internet, or some other reference source to find a real-life data set with the indicated cause-and-effect relationship. Write a paragraph describing each variable and explain why you think the variables have the indicated cause-and-effect relationship.

    (a) *Direct Cause-and-Effect:* Changes in one variable cause changes in the other variable.

    (b) *Other Factors:* The relationship between the variables is caused by a third variable.

    (c) *Coincidence:* The relationship between the variables is a coincidence.

9.1 ACTIVITY

Correlation by Eye

APPLET

You can find the interactive applet for this activity within MyLab Statistics or at *www.pearsonhighered.com/ mathstatsresources.*

The *correlation by eye* applet allows you to guess the sample correlation coefficient r for a data set. When the applet loads, a data set consisting of 20 points is displayed. Points can be added to the plot by clicking the mouse. Points on the plot can be removed by clicking on the point and then dragging the point into the trash can. All of the points on the plot can be removed by simply clicking inside the trash can. You can enter your guess for r in the "Guess" field, and then click SHOW R! to see whether you are within 0.1 of the true value. When you click NEW DATA, a new data set is generated.

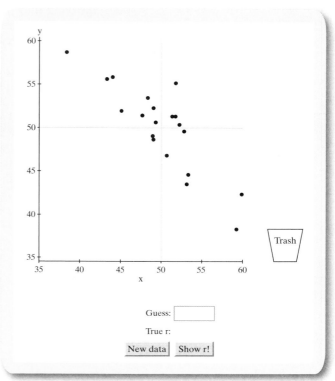

EXPLORE

Step 1   Add five points to the plot.
Step 2   Enter a guess for r.
Step 3   Click SHOW R!.
Step 4   Click NEW DATA.
Step 5   Remove five points from the plot.
Step 6   Enter a guess for r.
Step 7   Click SHOW R!.

DRAW CONCLUSIONS

APPLET

1. Generate a new data set. Using your knowledge of correlation, try to guess the value of r for the data set. Repeat this 10 times. How many times were you correct? Describe how you chose each r value.

2. Describe how to create a data set with a value of r that is approximately 1.

3. Describe how to create a data set with a value of r that is approximately 0.

4. Try to create a data set with a value of r that is approximately -0.9. Then try to create a data set with a value of r that is approximately 0.9. What did you do differently to create the two data sets?

SECTION 9.1   Correlation   **485**

9.2 Linear Regression

What You Should Learn

▶ How to find the equation of a regression line

▶ How to predict *y*-values using a regression equation

Regression Lines ■ Applications of Regression Lines

Regression Lines

After verifying that the linear correlation between two variables is significant, the next step is to determine the equation of the line that best models the data. This line is called a **regression line,** and its equation can be used to predict the value of *y* for a given value of *x*. Although many lines can be drawn through a set of points, a regression line is determined by specific criteria.

Consider the scatter plot and the line shown below. For each data point, d_i represents the difference between the observed *y*-value and the predicted *y*-value for a given *x*-value. These differences are called **residuals** and can be positive, negative, or zero. When the point is above the line, d_i is positive. When the point is below the line, d_i is negative. When the observed *y*-value equals the predicted *y*-value, $d_i = 0$. Of all possible lines that can be drawn through a set of points, the regression line is the line for which the sum of the squares of all the residuals

$$\Sigma d_i^2 \qquad \text{Sum of the squares of the residuals}$$

is a minimum.

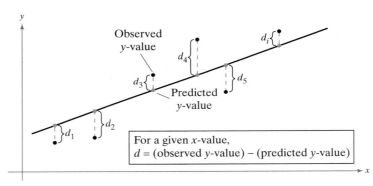

For a given *x*-value,
$d = $ (observed *y*-value) − (predicted *y*-value)

DEFINITION

A **regression line,** also called a **line of best fit,** is the line for which the sum of the squares of the residuals is a minimum.

Study Tip

When determining the equation of a regression line, it is helpful to construct a scatter plot of the data to check for outliers, which can greatly influence a regression line. You should also check for gaps and clusters in the data.

In algebra, you learned that you can write an equation of a line by finding its slope *m* and *y*-intercept *b*. The equation has the form

$$y = mx + b.$$

Recall that the slope of a line is the ratio of its rise over its run and the *y*-intercept is the *y*-value of the point at which the line crosses the *y*-axis. It is the *y*-value when $x = 0$. For instance, the graph of $y = 2x + 1$ is shown in the figure at the right. The slope of the line is 2 and the *y*-intercept is 1.

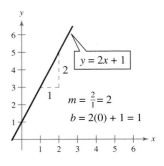

In algebra, you used two points to determine the equation of a line. In statistics, you will use every point in the data set to determine the equation of the regression line.

The equation of a regression line allows you to use the independent (explanatory) variable x to make predictions for the dependent (response) variable y.

Tech Tip

Although formulas for the slope and y-intercept are given, it is more convenient to use technology to calculate the equation of a regression line (see Example 2).

The Equation of a Regression Line

The equation of a regression line for an independent variable x and a dependent variable y is

$$\hat{y} = mx + b$$

where $\hat{y}$ is the predicted y-value for a given x-value. The slope m and y-intercept b are given by

$$m = \frac{n\Sigma xy - (\Sigma x)(\Sigma y)}{n\Sigma x^2 - (\Sigma x)^2} \quad \text{and} \quad b = \bar{y} - m\bar{x} = \frac{\Sigma y}{n} - m\frac{\Sigma x}{n}$$

where $\bar{y}$ is the mean of the y-values in the data set, $\bar{x}$ is the mean of the x-values, and n is the number of pairs of data. The regression line always passes through the point $(\bar{x}, \bar{y})$.

EXAMPLE 1

Finding the Equation of a Regression Line

Find the equation of the regression line for the gross domestic products and carbon dioxide emissions data used in Section 9.1. (See table at the left.)

GDP (in trillions of dollars), x	CO_2 emissions (in millions of metric tons), y
1.8	604.4
1.3	434.2
2.4	544.0
1.5	370.4
3.9	742.3
2.1	340.5
0.9	232.0
1.4	262.3
3.0	441.9
4.6	1157.7

SOLUTION

Recall from Example 7 of Section 9.1 that there is a significant linear correlation between gross domestic products and carbon dioxide emissions. Also, in Example 4 of Section 9.1, you found that $n = 10$, $\Sigma x = 22.9$, $\Sigma y = 5129.7$, $\Sigma xy = 14,350.74$, and $\Sigma x^2 = 65.49$. You can use these values to calculate the slope m of the regression line

$$m = \frac{n\Sigma xy - (\Sigma x)(\Sigma y)}{n\Sigma x^2 - (\Sigma x)^2} = \frac{10(14,350.74) - (22.9)(5129.7)}{10(65.49) - (22.9)^2} \approx 199.534600$$

and its y-intercept b.

$$b = \bar{y} - m\bar{x}$$
$$\approx \frac{5129.7}{10} - (199.534600)\left(\frac{22.9}{10}\right)$$
$$\approx 56.036$$

So, the equation of the regression line is

$$\hat{y} = 199.535x + 56.036.$$

Study Tip

When writing the equation of a regression line, the slope m and the y-intercept b are rounded to three decimal places, as shown in Example 1. This *round-off rule* will be used throughout the text.

To sketch the regression line, first choose two x-values between the least and greatest x-values in the data set. Next, calculate the corresponding y-values using the regression equation. Then draw a line through the two points. The regression line and scatter plot of the data are shown at the right. Notice that the line passes through the point $(\bar{x}, \bar{y}) = (2.29, 512.97)$.

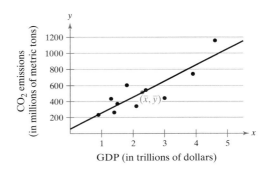

Duration, x	Time, y	Duration, x	Time, y
1.80	56	3.78	79
1.82	58	3.83	85
1.90	62	3.88	80
1.93	56	4.10	89
1.98	57	4.27	90
2.05	57	4.30	89
2.13	60	4.43	89
2.30	57	4.47	86
2.37	61	4.53	89
2.82	73	4.55	86
3.13	76	4.60	92
3.27	77	4.63	91
3.65	77		

Note to Instructor

Point out the different notation used by Minitab, Excel, and the TI-84 Plus.

9.2 To explore this topic further, see **Activity 9.2** on page 496.

TRY IT YOURSELF 1

Find the equation of the regression line for the number of years out of school and annual contribution data used in Try It Yourself 4 in Section 9.1.

Answer: Page A38

EXAMPLE 2

Using Technology to Find a Regression Equation

Use technology to find the equation of the regression line for the Old Faithful data used in Section 9.1. (See table at the left.)

SOLUTION

Recall from Example 6 of Section 9.1 that there is a significant linear correlation between the duration of Old Faithful's eruptions and the time between eruptions. Minitab, Excel, and the TI-84 Plus each have features that calculate a regression equation. Try using this technology to find the regression equation. You should obtain results similar to the displays shown below.

MINITAB

Regression Analysis: Time versus Duration

Coefficients

Term	Coef	SE Coef	T-Value	P-Value
Constant	33.68	1.89	17.79	0.000
Duration	12.481	0.546	22.84	0.000

Regression Equation

Time = 33.68 + 12.481 Duration

EXCEL

	A	B	C	D
26	Slope:			
27	SLOPE(B1:B25, A1:A25)			
28				12.48094
29				
30	Y-intercept:			
31	INTERCEPT(B1:B25, A1:A25)			
32				33.6829

TI-84 PLUS

LinReg
y=ax+b
a=12.48094391
b=33.68290034
r^2=.9577738551
r=.9786592129

From the displays, you can see that the regression equation is

$$\hat{y} = 12.481x + 33.683.$$

The TI-84 Plus display at the right shows the regression line and a scatter plot of the data in the same viewing window. To do this, use the *Stat Plot* feature to construct the scatter plot and enter the regression equation as y_1.

TI-84 PLUS

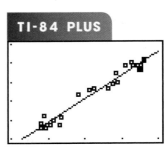

TRY IT YOURSELF 2

Use technology to find the equation of the regression line for the salaries and average attendances at home games for the teams in Major League Baseball listed on page 469.

Answer: Page A38

Applications of Regression Lines

When the correlation between x and y is *significant* (see Section 9.1), the equation of a regression line can be used to predict y-values for certain x-values. Prediction values are meaningful only for x-values in (or close to) the range of the observed x-values in the data. For instance, in Example 1 the observed x-values in the data range from $0.9 trillion to $4.6 trillion. So, it would not be appropriate to use the regression equation found in Example 1 to predict carbon dioxide emissions for gross domestic products such as $0.2 trillion or $14.5 trillion.

To predict y-values, substitute an x-value into the regression equation, then calculate $\hat{y}$, the predicted y-value. This process is shown in the next example.

EXAMPLE 3

Predicting y-Values Using Regression Equations

The regression equation for the gross domestic products (in trillions of dollars) and carbon dioxide emissions (in millions of metric tons) data is

$$\hat{y} = 199.535x + 56.036. \qquad \text{See Example 1.}$$

Use this equation to predict the *expected* carbon dioxide emissions for each gross domestic product.

1. $1.2 trillion

2. $2.0 trillion

3. $2.6 trillion

SOLUTION

Recall from Section 9.1, Example 7, that x and y have a significant linear correlation. So, you can use the regression equation to predict y-values. Note that the given gross domestic products are in the range ($0.9 trillion to $4.6 trillion) of the observed x-values. To predict the expected carbon dioxide emissions, substitute each gross domestic product for x in the regression equation. Then calculate $\hat{y}$.

1. $\hat{y} = 199.535x + 56.036$
  $= 199.535(1.2) + 56.036$
  $= 295.478$

Interpretation When the gross domestic product is $1.2 trillion, the predicted CO_2 emissions are 295.478 million metric tons.

2. $\hat{y} = 199.535x + 56.036$
  $= 199.535(2.0) + 56.036$
  $= 455.106$

Interpretation When the gross domestic product is $2.0 trillion, the predicted CO_2 emissions are 455.106 million metric tons.

3. $\hat{y} = 199.535x + 56.036$
  $= 199.535(2.6) + 56.036$
  $= 574.827$

Interpretation When the gross domestic product is $2.6 trillion, the predicted CO_2 emissions are 574.827 million metric tons.

TRY IT YOURSELF 3

The regression equation for the Old Faithful data is $\hat{y} = 12.481x + 33.683$. Use this to predict the time until the next eruption for each eruption duration. (Recall from Section 9.1, Example 6, that x and y have a significant linear correlation.)

1. 2 minutes

2. 3.32 minutes

Answer: Page A38

When the correlation between x and y is *not* significant, the best predicted y-value is $\bar{y}$, the mean of the y-values in the data.

Picturing the World

The scatter plot shows the relationship between the number of farms (in thousands) in a state and the total value of the farms (in billions of dollars). (Source: U.S. Department of Agriculture, National Agriculture Statistics Service)

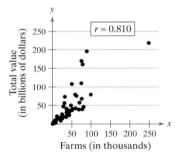

Describe the correlation between these two variables in words. Use the scatter plot to predict the total value of farms in a state that has 150,000 farms. The regression line for this scatter plot is $\hat{y} = 1.014x + 2.611$. Use this equation to predict the total value in a state that has 150,000 farms ($x = 150$). (Assume x and y have a significant linear correlation.) How does your algebraic prediction compare with your graphical one?

$150 billion; 154.711 billion; The predictions are very close to each other.

9.2 EXERCISES

1. A residual is the difference between the observed y-value of a data point and the predicted y-value on the regression line for the x-coordinate of the data point. A residual is positive when the data point is above the line, negative when the point is below the line, and zero when the observed y-value equals the predicted y-value.

2. Positive

3. Substitute a value of x into the equation of a regression line and solve for $\hat{y}$.

4. Prediction values are meaningful only for x-values in (or close to) the range of the original data.

5. The correlation between variables must be significant.

6. *Sample answer:* Because the regression line models the trend of the given data, and it is not known if the trend continues beyond the range of those data.

7. b

8. a

9. e

10. c

11. f

12. d

13. b

14. c

15. d

16. a

Building Basic Skills and Vocabulary

1. What is a residual? Explain when a residual is positive, negative, and zero.

2. Two variables have a positive linear correlation. Is the slope of the regression line for the variables positive or negative?

3. Explain how to predict y-values using the equation of a regression line.

4. For a set of data and a corresponding regression line, describe all values of x that provide meaningful predictions for y.

5. In order to predict y-values using the equation of a regression line, what must be true about the correlation coefficient of the variables?

6. Why is it not appropriate to use a regression line to predict y-values for x-values that are not in (or close to) the range of x-values found in the data?

In Exercises 7–12, match the description in the left column with its symbol(s) in the right column.

7. The y-value of a data point corresponding to x_i **a.** $\hat{y}_i$

8. The y-value for a point on the regression line corresponding to x_i **b.** y_i

                                                              c. b

9. Slope   **d.** $(\bar{x}, \bar{y})$

10. y-intercept

11. The mean of the y-values   **e.** m

                                        f. $\bar{y}$

12. The point a regression line always passes through

Graphical Analysis   *In Exercises 13–16, match the regression equation with the appropriate graph.*

13. $\hat{y} = -1.361x + 21.952$

14. $\hat{y} = 2.115x + 21.958$

15. $\hat{y} = 2.125x + 9.588$

16. $\hat{y} = -0.705x + 27.214$

a.
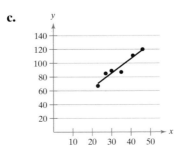

b.
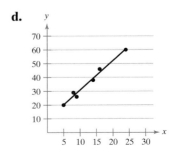

c.
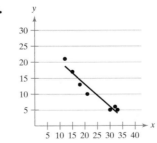

d.
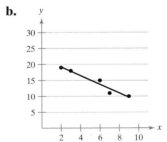

17. See Odd Answers, page A79.

18. See Selected Answers, page A105.

19. See Odd Answers, page A79.

20. See Selected Answers, page A105.

21. See Odd Answers, page A79.

Square footage, x	Sale price, y
1490	145.0
1122	84.9
1112	119.9
720	44.9
2509	289.0
2448	229.0
1786	189.9

TABLE FOR EXERCISE 18

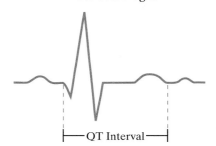

Electrocardiogram

QT Interval

The QT interval is a measure of electrical waves of the heart. A lengthened QT interval can indicate heart health problems.

FIGURE FOR EXERCISE 21

Using and Interpreting Concepts

Finding the Equation of a Regression Line   *In Exercises 17–26, find the equation of the regression line for the data. Then construct a scatter plot of the data and draw the regression line. (Each pair of variables has a significant correlation.) Then use the regression equation to predict the value of y for each of the x-values, if meaningful. If the x-value is not meaningful to predict the value of y, explain why not. If convenient, use technology.*

17. **Height and Number of Stories**   The heights (in feet) and the numbers of stories of the nine tallest buildings in Houston, Texas *(Source: Emporis Corporation)*

Height, x	1002	992	901	780	762	756	752	741	732
Stories, y	75	71	64	56	53	55	48	47	53

(a) $x = 950$ feet   (b) $x = 850$ feet
(c) $x = 800$ feet   (d) $x = 700$ feet

18. **Square Footage and Home Sale Price**   The square footages and sale prices (in thousands of dollars) of seven homes in Akron, Ohio, are shown in the table at the left. *(Source: Howard Hanna)*

(a) $x = 1450$ square feet   (b) $x = 2720$ square feet
(c) $x = 2175$ square feet   (d) $x = 890$ square feet

19. **Hours Studying and Test Scores**   The number of hours 9 students spent studying for a test and their scores on that test

Hours spent studying, x	0	2	4	5	5	5	6	7	8
Test scores, y	40	51	64	69	73	75	93	90	95

(a) $x = 3$ hours   (b) $x = 6.5$ hours
(c) $x = 13$ hours   (d) $x = 4.5$ hours

20. **Goals and Wins**   The number of goals scored and the number of wins for the top 10 teams in the 2016–2017 English Premier League season *(Source: Premier League)*

Goals, x	85	86	80	78	77	54	62	41	55	43
Wins, y	30	26	23	22	23	18	17	12	12	12

(a) $x = 50$ goals   (b) $x = 70$ goals
(c) $x = 75$ goals   (d) $x = 95$ goals

21. **Heart Rate and QT Interval**   The heart rates (in beats per minute) and QT intervals (in milliseconds) for 13 males (The figure at the left shows the QT interval of a heartbeat in an electrocardiogram.) *(Adapted from Chest)*

Heart rate, x	60	75	62	68	84	97	66
QT interval, y	403	363	381	367	341	317	401

Heart rate, x	65	86	78	93	75	88
QT interval, y	384	342	377	329	377	349

(a) $x = 120$ beats per minute   (b) $x = 67$ beats per minute
(c) $x = 90$ beats per minute   (d) $x = 83$ beats per minute

22. $\hat{y} = 0.682x + 11.872$

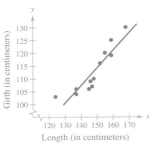

(a) 107 centimeters

(b) 129 centimeters

(c) 124 centimeters

(d) 120 centimeters

23. $\hat{y} = 2.979x + 52.476$

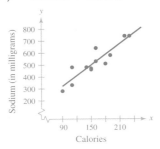

(a) 559 milligrams

(b) 350 milligrams

(c) It is not meaningful to predict the value of y for $x = 260$ because $x = 260$ is outside the range of the original data.

(d) 678 milligrams

24. $\hat{y} = 0.140x + 398.579$

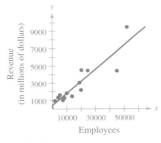

(a) 4949 employees

(b) 1239 employees

(c) It is not meaningful to predict the value of y for $x = 1350$ because $x = 1350$ is outside the range of the original data.

(d) 6979 employees

25. See Odd Answers, page A80.

 22. Length and Girth of Harbor Seals   The lengths (in centimeters) and girths (in centimeters) of 12 harbor seals *(Adapted from Moss Landing Marine Laboratories)*

Length, x	137	168	152	145	159	159
Girth, y	106	130	116	106	125	119

Length, x	124	137	155	148	147	146
Girth, y	103	104	120	110	107	109

(a) $x = 140$ centimeters   (b) $x = 172$ centimeters

(c) $x = 164$ centimeters   (d) $x = 158$ centimeters

 23. Hot Dogs: Caloric and Sodium Content   The caloric contents and the sodium contents (in milligrams) of 12 brands of beef hot dogs *(Source: Walmart)*

Calories, x	180	220	230	90	160	190
Sodium, y	510	740	740	280	530	580

Calories, x	150	110	110	160	140	150
Sodium, y	490	480	330	640	480	460

(a) $x = 170$ calories   (b) $x = 100$ calories

(c) $x = 260$ calories   (d) $x = 210$ calories

 24. Employees and Revenue   The number of employees and the 2016 revenue (in millions of dollars) of 13 hotel and gaming companies *(Source: Value Line)*

Employees, x	1800	8300	45,000	7300	52,000	10,000	20,200
Revenue, y	925	1271	4429	1006	9455	1811	4519

Employees, x	13,700	5200	24,600	19,900	4000	18,800
Revenue, y	1452	1601	4466	2184	1309	3034

(a) $x = 32,500$ employees   (b) $x = 6000$ employees

(c) $x = 1350$ employees   (d) $x = 47,000$ employees

 25. Shoe Size and Height   The shoe sizes and heights (in inches) of 14 men

Shoe size, x	8.5	9.0	9.0	9.5	10.0	10.0	10.5
Height, y	66.0	68.5	67.5	70.0	70.0	72.0	71.5

Shoe size, x	10.5	11.0	11.0	11.0	12.0	12.0	12.5
Height, y	69.5	71.5	72.0	73.0	73.5	74.0	74.0

(a) $x = $ size 11.5   (b) $x = $ size 8.0

(c) $x = $ size 15.5   (d) $x = $ size 10.0

26. $\hat{y} = -0.780x + 14.421$

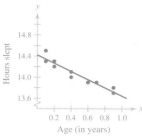

(a) 14.187 hours

(b) It is not meaningful to predict the value of y for x = 3.9 because x = 3.9 is outside the range of the original data.

(c) 13.953 hours

(d) 13.797 hours

27. Strong positive linear correlation; As the years of experience of the registered nurses increase, their salaries tend to increase.

28. $\hat{y} = 1.008x + 51.453$

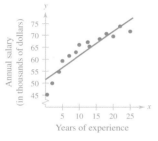

29. No, it is not meaningful to predict a salary for a registered nurse with 28 years of experience because x = 28 is outside the range of the original data.

30. $|r| \approx 0.920 > 0.661$ (the critical value), so the population has a significant correlation.

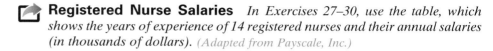

26. Age and Hours Slept The ages (in years) of 10 infants and the numbers of hours each slept in a day

Age, x	0.1	0.2	0.4	0.7	0.6	0.9
Hours slept, y	14.5	14.3	14.1	13.9	13.9	13.7

Age, x	0.1	0.2	0.4	0.9
Hours slept, y	14.3	14.2	14.0	13.8

(a) x = 0.3 year   (b) x = 3.9 years

(c) x = 0.6 year   (d) x = 0.8 year

Registered Nurse Salaries *In Exercises 27–30, use the table, which shows the years of experience of 14 registered nurses and their annual salaries (in thousands of dollars).* (Adapted from Payscale, Inc.)

Years of experience, x	0.5	2	4	5	7	9	10
Annual salary (in thousands of dollars), y	45.2	49.9	54.7	59.3	61.4	62.9	66.0

Years of experience, x	12.5	13	16	18	20	22	25
Annual salary (in thousands of dollars), y	67.1	65.3	68.4	70.6	69.5	73.9	71.6

27. Correlation Using the scatter plot of the registered nurse salary data shown below, what type of correlation, if any, do you think the data have? Explain.

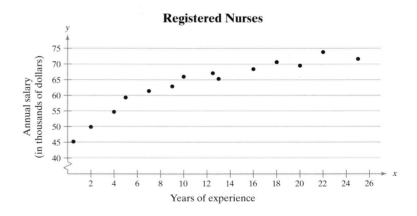

Registered Nurses

28. Regression Line Find an equation of the regression line for the data. Sketch a scatter plot of the data and draw the regression line.

29. Using the Regression Line An analyst used the regression line you found in Exercise 28 to predict the annual salary for a registered nurse with 28 years of experience. Is this a valid prediction? Explain your reasoning.

30. Significant Correlation? A salary analyst claims that the population has a significant correlation for $\alpha = 0.01$. Test this claim.

31. (a) $\hat{y} = -4.297x + 94.200$

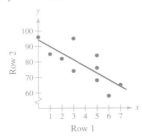

(b) $\hat{y} = -0.141x + 14.763$

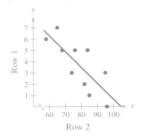

(c) The slope of the line keeps the same sign, but the values of m and b change.

32. (a) $\hat{y} = 1.724x + 79.733$

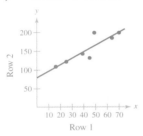

(b) $\hat{y} = 0.453x - 26.448$

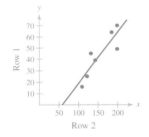

(c) The slope of the line keeps the same sign, but the values of m and b change.

33. See Odd Answers, page A80.

34. See Selected Answers, page A106.

35. See Odd Answers, page A80.

Extending Concepts

Interchanging x and y   *In Exercises 31 and 32, perform the steps below.*

(a) *Find the equation of the regression line for the data, letting Row 1 represent the x-values and Row 2 the y-values. Sketch a scatter plot of the data and draw the regression line.*

(b) *Find the equation of the regression line for the data, letting Row 2 represent the x-values and Row 1 the y-values. Sketch a scatter plot of the data and draw the regression line.*

(c) *Describe the effect of switching the explanatory and response variables on the regression line.*

31.

Row 1	0	1	2	3	3	5	5	5	6	7
Row 2	96	85	82	74	95	68	76	84	58	65

32.

Row 1	16	25	39	45	49	64	70
Row 2	109	122	143	132	199	185	199

Residual Plots   *A **residual plot** allows you to assess correlation data and check for possible problems with a regression model. To construct a residual plot, make a scatter plot of $(x, y - \hat{y})$, where $y - \hat{y}$ is the residual of each y-value. If the resulting plot shows any type of pattern, then the regression line is not a good representation of the relationship between the two variables. If it does not show a pattern—that is, if the residuals fluctuate about 0—then the regression line is a good representation. Be aware that if a point on the residual plot appears to be outside the pattern of the other points, then it may be an outlier.*

In Exercises 33 and 34, (a) find the equation of the regression line, (b) construct a scatter plot of the data and draw the regression line, (c) construct a residual plot, and (d) determine whether there are any patterns in the residual plot and explain what they suggest about the relationship between the variables.

33.

x	38	34	40	46	43	48	60	55	52
y	24	22	27	32	30	31	27	26	28

34.

x	8	4	15	7	6	3	12	10	5
y	18	11	29	18	14	8	25	20	12

Influential Points   *An **influential point** is a point in a data set that can greatly affect the graph of a regression line. An outlier may or may not be an influential point. To determine whether a point is influential, find two regression lines: one including all the points in the data set, and the other excluding the possible influential point. If the slope or y-intercept of the regression line shows significant changes, then the point can be considered influential. An influential point can be removed from a data set only when there is proper justification.*

In Exercises 35 and 36, (a) construct a scatter plot of the data, (b) identify any possible outliers, and (c) determine whether the point is influential. Explain your reasoning.

35.

x	5	6	9	10	14	17	19	44
y	32	33	28	26	25	23	23	8

36. See Selected Answers, page A106.

37. See Odd Answers, page A80.

38. See Selected Answers, page A106.

39. See Odd Answers, page A80.

40. The exponential equation is a much better model for the data. The graph of the exponential equation fits the data better than the regression line.

Number of hours, x	Number of bacteria, y
1	165
2	280
3	468
4	780
5	1310
6	1920
7	4900

TABLE FOR EXERCISES 37–40

41. See Odd Answers, page A80.

42. See Selected Answers, page A106.

43. See Odd Answers, page A80.

x	y
1	695
2	410
3	256
4	110
5	80
6	75
7	68
8	74

TABLE FOR EXERCISES 41–44

44. The power equation is a much better model for the data. The graph of the power equation fits the data better than the regression line.

45. See Odd Answers, page A81.

46. See Selected Answers, page A106.

47. The logarithmic equation is a better model for the data. The graph of the logarithmic equation fits the data better than the regression line.

48. The logarithmic equation is a better model for the data. The graph of the logarithmic equation fits the data better than the regression line.

36.

x	1	3	6	8	12	14
y	4	7	10	9	15	3

Transformations to Achieve Linearity *When a linear model is not appropriate for representing data, other models can be used. In some cases, the values of x or y must be transformed to find an appropriate model. In a **logarithmic transformation,** the logarithms of the variables are used instead of the original variables when creating a scatter plot and calculating the regression line.*

In Exercises 37–40, use the data shown in the table at the left, which shows the number of bacteria present after a certain number of hours.

37. Find the equation of the regression line for the data. Then construct a scatter plot of (x, y) and sketch the regression line with it.

38. Replace each y-value in the table with its logarithm, $\log y$. Find the equation of the regression line for the transformed data. Then construct a scatter plot of $(x, \log y)$ and sketch the regression line with it. What do you notice?

39. An **exponential equation** is a nonlinear regression equation of the form $y = ab^x$. Use technology to find and graph the exponential equation for the original data. Include the original data in your graph. Note that you can also find this model by solving the equation $\log y = mx + b$ from Exercise 38 for y.

40. Compare your results in Exercise 39 with the equation of the regression line and its graph in Exercise 37. Which equation is a better model for the data? Explain.

In Exercises 41–44, use the data shown in the table at the left.

41. Find the equation of the regression line for the data. Then construct a scatter plot of (x, y) and sketch the regression line with it.

42. Replace each x-value and y-value in the table with its logarithm. Find the equation of the regression line for the transformed data. Then construct a scatter plot of $(\log x, \log y)$ and sketch the regression line with it. What do you notice?

43. A **power equation** is a nonlinear regression equation of the form $y = ax^b$. Use technology to find and graph the power equation for the original data. Include a scatter plot in your graph. Note that you can also find this model by solving the equation $\log y = m(\log x) + b$ from Exercise 42 for y.

44. Compare your results in Exercise 43 with the equation of the regression line and its graph in Exercise 41. Which equation is a better model for the data? Explain.

Logarithmic Equation *The **logarithmic equation** is a nonlinear regression equation of the form $y = a + b \ln x$. In Exercises 45–48, use this information and technology.*

45. Find and graph the logarithmic equation for the data in Exercise 25.

46. Find and graph the logarithmic equation for the data in Exercise 26.

47. Compare your results in Exercise 45 with the equation of the regression line and its graph. Which equation is a better model for the data? Explain.

48. Compare your results in Exercise 46 with the equation of the regression line and its graph. Which equation is a better model for the data? Explain.

APPLET

You can find the interactive applet for this activity within MyLab Statistics or at *www.pearsonhighered.com/mathstatsresources.*

The *regression by eye* applet allows you to interactively estimate the regression line for a data set. When the applet loads, a data set consisting of 20 points is displayed. Points on the plot can be added to the plot by clicking the mouse. Points on the plot can be removed by clicking on the point and then dragging the point into the trash can. All of the points on the plot can be removed by simply clicking inside the trash can. You can move the green line on the plot by clicking and dragging the endpoints. You should try to move the line in order to minimize the sum of the squares of the residuals, also known as the sum of square error (SSE). Note that the regression line minimizes the SSE. The SSE for the green line and for the regression line are shown below the plot. The equations of each line are shown above the plot. Click SHOW REGRESSION LINE! to see the regression line in the plot. Click NEW DATA to generate a new data set.

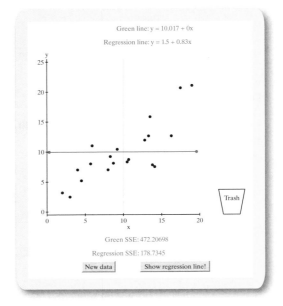

EXPLORE

Step 1 Move the endpoints of the green line to try to approximate the regression line.

Step 2 Click SHOW REGRESSION LINE!.

DRAW CONCLUSIONS

APPLET

1. Click NEW DATA to generate a new data set. Try to move the green line to where the regression line should be. Then click SHOW REGRESSION LINE!. Repeat this five times. Describe how you moved each green line.

2. On a blank plot, place 10 points so that they have a strong positive correlation. Record the equation of the regression line. Then, add a point in the upper left corner of the plot and record the equation of the regression line. How does the regression line change?

3. Remove the point from the upper-left corner of the plot. Add 10 more points so that there is still a strong positive correlation. Record the equation of the regression line. Add a point in the upper-left corner of the plot and record the equation of the regression line. How does the regression line change?

4. Use the results of Exercises 2 and 3 to describe what happens to the slope of the regression line when an outlier is added as the sample size increases.

Correlation of Body Measurements

In a study published in *Medicine and Science in Sports and Exercise* (volume 17, no. 2, page 189), the measurements of 252 men (ages 22–81) were taken. Of the 14 measurements taken of each man, some have significant correlations and others do not. For instance, the scatter plot at the right shows that the hip and abdomen circumferences of the men have a strong linear correlation ($r \approx 0.874$). The partial table shown here lists only the first nine rows of the data.

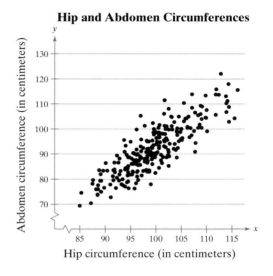

Hip and Abdomen Circumferences

Hip circumference (in centimeters)

Age (yr)	Weight (lb)	Height (in.)	Neck (cm)	Chest (cm)	Abdom. (cm)	Hip (cm)	Thigh (cm)	Knee (cm)	Ankle (cm)	Bicep (cm)	Forearm (cm)	Wrist (cm)	Body fat %
22	173.25	72.25	38.5	93.6	83.0	98.7	58.7	37.3	23.4	30.5	28.9	18.2	6.1
22	154.00	66.25	34.0	95.8	87.9	99.2	59.6	38.9	24.0	28.8	25.2	16.6	25.3
23	154.25	67.75	36.2	93.1	85.2	94.5	59.0	37.3	21.9	32.0	27.4	17.1	12.3
23	198.25	73.50	42.1	99.6	88.6	104.1	63.1	41.7	25.0	35.6	30.0	19.2	11.7
23	159.75	72.25	35.5	92.1	77.1	93.9	56.1	36.1	22.7	30.5	27.2	18.2	9.4
23	188.15	77.50	38.0	96.6	85.3	102.5	59.1	37.6	23.2	31.8	29.7	18.3	10.3
24	184.25	71.25	34.4	97.3	100.0	101.9	63.2	42.2	24.0	32.2	27.7	17.7	28.7
24	210.25	74.75	39.0	104.5	94.4	107.8	66.0	42.0	25.6	35.7	30.6	18.8	20.9
24	156.00	70.75	35.7	92.7	81.9	95.3	56.4	36.5	22.0	33.5	28.3	17.3	14.2

Source: "Generalized Body Composition Prediction Equation for Men Using Simple Measurement Techniques" by K.W. Penrose et al. (1985). MEDICINE AND SCIENCE IN SPORTS AND EXERCISE, vol. 17, no.2, p. 189.

EXERCISES

1. Using your intuition, classify each (x, y) pair as having a weak correlation $(0 < r < 0.5)$, a moderate correlation $(0.5 < r < 0.8)$, or a strong correlation $(0.8 < r < 1.0)$.

    (a) (weight, neck)   (b) (weight, height)
    (c) (age, body fat)   (d) (chest, hip)
    (e) (age, wrist)   (f) (ankle, wrist)
    (g) (forearm, height)   (h) (bicep, forearm)
    (i) (weight, body fat)   (j) (knee, thigh)
    (k) (hip, abdomen)   (l) (abdomen, hip)

2. Use technology to find the correlation coefficient for each pair in Exercise 1. Compare your results with those obtained by intuition.

3. Use technology to find the regression line for each pair in Exercise 1 that has a strong correlation.

4. Use the results of Exercise 3 to predict the following.

    (a) The hip circumference of a man whose chest circumference is 95 centimeters
    (b) The height of a man whose forearm circumference is 28 centimeters

5. Are there pairs of measurements that have stronger correlation coefficients than 0.85? Use technology and intuition to reach a conclusion.

9.3 Measures of Regression and Prediction Intervals

What You Should Learn

▶ How to interpret the three types of variation about a regression line

▶ How to find and interpret the coefficient of determination

▶ How to find and interpret the standard error of estimate for a regression line

▶ How to construct and interpret a prediction interval for y

Variation about a Regression Line ■ The Coefficient of Determination ■ The Standard Error of Estimate ■ Prediction Intervals

Variation About a Regression Line

In this section, you will study two measures used in correlation and regression studies—the coefficient of determination and the standard error of estimate. You will also learn how to construct a prediction interval for y using a regression equation and a given value of x. Before studying these concepts, you need to understand the three types of variation about a regression line.

To find the total variation, the explained variation, and the unexplained variation about a regression line, you must first calculate the **total deviation,** the **explained deviation,** and the **unexplained deviation** for each ordered pair (x_i, y_i) in a data set. These deviations are shown in the figure.

$$\text{Total deviation} = y_i - \bar{y}$$

$$\text{Explained deviation} = \hat{y}_i - \bar{y}$$

$$\text{Unexplained deviation} = y_i - \hat{y}_i$$

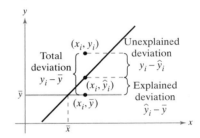

After calculating the deviations for each data point (x_i, y_i), you can find the **total variation,** the **explained variation,** and the **unexplained variation.**

DEFINITION

The **total variation** about a regression line is the sum of the squares of the differences between the y-value of each ordered pair and the mean of y.

$$\text{Total variation} = \Sigma(y_i - \bar{y})^2$$

The **explained variation** is the sum of the squares of the differences between each predicted y-value and the mean of y.

$$\text{Explained variation} = \Sigma(\hat{y}_i - \bar{y})^2$$

The **unexplained variation** is the sum of the squares of the differences between the y-value of each ordered pair and each corresponding predicted y-value.

$$\text{Unexplained variation} = \Sigma(y_i - \hat{y}_i)^2$$

The sum of the explained and unexplained variations is equal to the total variation.

$$\text{Total variation} = \text{Explained variation} + \text{Unexplained variation}$$

As its name implies, the *explained variation* can be explained by the relationship between x and y. The *unexplained variation* cannot be explained by the relationship between x and y and is due to other factors, such as sampling error, coincidence, or lurking variables. (Recall from Section 9.1 that lurking variables are variables that have an effect on the variables being studied but are not included in the study.)

Picturing the World

Janette Benson (Psychology Department, University of Denver) performed a study relating the age at which infants crawl (in weeks after birth) with the average monthly temperature six months after birth. Her results are based on a sample of 414 infants. Benson believes that the reason for the correlation of temperature and crawling age is that parents tend to bundle infants in more restrictive clothing and blankets during cold months. This bundling doesn't allow the infant as much opportunity to move and experiment with crawling.

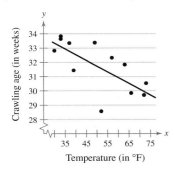

The correlation coefficient is $r \approx -0.701$. What percent of the variation in the data can be explained? What percent is due to other factors, such as sampling error, coincidence, or lurking variables?

About 49.1% of the variation in the crawling age can be explained by the variation in the temperature. About 50.9% of the variation is unexplained and is due to other factors, such as sampling error, coincidence, or lurking variables.

The Coefficient of Determination

You already know how to calculate the correlation coefficient r. The square of this coefficient is called the **coefficient of determination.** It can be shown that the coefficient of determination is equal to the ratio of the explained variation to the total variation.

DEFINITION

The **coefficient of determination r^2** is the ratio of the explained variation to the total variation. That is,

$$r^2 = \frac{\text{Explained variation}}{\text{Total variation}}.$$

It is important that you interpret the coefficient of determination correctly. For instance, if the correlation coefficient is $r = 0.900$, then the coefficient of determination is

$$r^2 = (0.900)^2$$
$$= 0.810.$$

This means that 81% of the variation in y can be explained by the relationship between x and y. The remaining 19% of the variation is unexplained and is due to other factors, such as sampling error, coincidence, or lurking variables.

EXAMPLE 1

Finding the Coefficient of Determination

The correlation coefficient for the gross domestic products and carbon dioxide emissions data is

$$r \approx 0.874. \qquad \text{See Example 4 in Section 9.1.}$$

Find the coefficient of determination. What does this tell you about the explained variation of the data about the regression line? about the unexplained variation?

SOLUTION

The coefficient of determination is

$$r^2 \approx (0.874)^2$$
$$\approx 0.764. \qquad \text{Round to three decimal places.}$$

Interpretation About 76.4% of the variation in the carbon dioxide emissions can be explained by the relationship between the gross domestic products and carbon dioxide emissions. About 23.6% of the variation is unexplained and is due to other factors, such as sampling error, coincidence, or lurking variables.

TRY IT YOURSELF 1

The correlation coefficient for the Old Faithful data is

$$r \approx 0.979. \qquad \text{See Example 5 in Section 9.1.}$$

Find the coefficient of determination. What does this tell you about the explained variation of the data about the regression line? about the unexplained variation?

Answer: Page A38

The Standard Error of Estimate

When a $\hat{y}$-value is predicted from an x-value, the prediction is a point estimate. You can construct an interval estimate for $\hat{y}$, but first you need to calculate the **standard error of estimate.**

DEFINITION

The **standard error of estimate** s_e is the standard deviation of the observed y_i-values about the predicted $\hat{y}$-value for a given x_i-value. It is given by

$$s_e = \sqrt{\frac{\Sigma (y_i - \hat{y}_i)^2}{n - 2}}$$

where n is the number of pairs of data.

From this formula, you can see that the standard error of estimate is the square root of the unexplained variation divided by $n - 2$. So, the closer the observed y-values are to the predicted $\hat{y}$-values, the smaller the standard error of estimate will be.

GUIDELINES

Finding the Standard Error of Estimate s_e

In Words	In Symbols
1. Make a table that includes the five column headings shown at the right.	$x_i, y_i, \hat{y}_i, (y_i - \hat{y}_i),$ $(y_i - \hat{y}_i)^2$
2. Use the regression equation to calculate the predicted y-values.	$\hat{y}_i = mx_i + b$
3. Calculate the sum of the squares of the differences between each observed y-value and the corresponding predicted y-value.	$\Sigma (y_i - \hat{y}_i)^2$
4. Find the standard error of estimate.	$s_e = \sqrt{\dfrac{\Sigma (y_i - \hat{y}_i)^2}{n - 2}}$

Instead of the formula used in Step 4, you can also find the standard error of estimate using the formula

$$s_e = \sqrt{\frac{\Sigma y^2 - b\Sigma y - m\Sigma xy}{n - 2}}.$$

This formula is easy to use if you have already calculated the slope m, the y-intercept b, and several of the sums. For instance, consider the gross domestic products and carbon dioxide emissions data (see Example 4 in Section 9.1 and Example 1 in Section 9.2). To use the alternative formula, note that the regression equation for these data is $\hat{y} = 199.535x + 56.036$ and the values of the sums are $\Sigma y^2 = 3{,}312{,}080.89$, $\Sigma y = 5129.7$, and $\Sigma xy = 14{,}350.74$. So, using the alternative formula, the standard error of estimate is

$$s_e = \sqrt{\frac{\Sigma y^2 - b\Sigma y - m\Sigma xy}{n - 2}}$$

$$= \sqrt{\frac{3{,}312{,}080.89 - (56.036)(5129.7) - (199.535)(14{,}350.74)}{10 - 2}}$$

$$\approx 141.932.$$

EXAMPLE 2

Finding the Standard Error of Estimate

The regression equation for the gross domestic products and carbon dioxide emissions data is

$$\hat{y} = 199.535x + 56.036. \qquad \text{See Example 1 in Section 9.2.}$$

Find the standard error of estimate.

SOLUTION

Use a table to calculate the sum of the squared differences of each observed y-value and the corresponding predicted y-value.

x_i	y_i	$\hat{y}_i$	$y_i - \hat{y}_i$	$(y_i - \hat{y}_i)^2$
1.8	604.4	415.199	189.201	35,797.0184
1.3	434.2	315.4315	118.7685	14,105.95659
2.4	544.0	534.92	9.08	82.4464
1.5	370.4	355.3385	15.0615	226.8487822
3.9	742.3	834.2225	−91.9225	8,449.746006
2.1	340.5	475.0595	−134.5595	18,106.25904
0.9	232.0	235.6175	−3.6175	13.08630625
1.4	262.3	335.385	−73.085	5,341.417225
3.0	441.9	654.641	−212.741	45,258.73308
4.6	1157.7	973.897	183.803	33,783.54281
				$\Sigma = 161,165.0546$

Unexplained variation

When $n = 10$ and $\Sigma (y_i - \hat{y}_i)^2 = 161,165.0546$ are used, the standard error of estimate is

$$s_e = \sqrt{\frac{\Sigma (y_i - \hat{y}_i)^2}{n - 2}}$$

$$= \sqrt{\frac{161,165.0546}{10 - 2}}$$

$$\approx 141.935.$$

Interpretation The standard error of estimate of the carbon dioxide emissions for a specific gross domestic product is about 141.935 million metric tons.

TRY IT YOURSELF 2

A researcher collects the data shown below and concludes that there is a significant relationship between the amount of radio advertising time (in minutes per week) and the weekly sales of a product (in hundreds of dollars).

Radio ad time, x	15	20	20	30	40	45	50	60
Weekly sales, y	26	32	38	56	54	78	80	88

Find the standard error of estimate. Use the regression equation

$$\hat{y} = 1.405x + 7.311.$$

Answer: Page A38

Prediction Intervals

Recall from Section 9.1 that one of the requirements for calculating a correlation coefficient is that the two variables x and y have a bivariate normal distribution. Two variables have a **bivariate normal distribution** when for any fixed values of x the corresponding values of y are normally distributed, and for any fixed values of y the corresponding values of x are normally distributed.

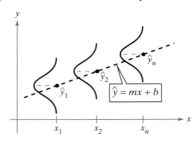

Bivariate Normal Distribution

Because regression equations are determined using random samples of paired data and because x and y are assumed to have a bivariate normal distribution, you can construct a **prediction interval** for the true value of y. To construct the prediction interval, use a t-distribution with $n - 2$ degrees of freedom.

DEFINITION

Given a linear regression equation $\hat{y} = mx + b$ and x_0, a specific value of x, a **c-prediction interval** for y is $\hat{y} - E < y < \hat{y} + E$ where

$$E = t_c s_e \sqrt{1 + \frac{1}{n} + \frac{n(x_0 - \bar{x})^2}{n\Sigma x^2 - (\Sigma x)^2}}.$$

The point estimate is $\hat{y}$ and the margin of error is E. The probability that the prediction interval contains y is c (the level of confidence), assuming that the estimation process is repeated a large number of times.

GUIDELINES

Constructing a Prediction Interval for y for a Specific Value of x

In Words	In Symbols
1. Identify the number n of pairs of data and the degrees of freedom.	d.f. $= n - 2$
2. Use the regression equation and the given x-value to find the point estimate $\hat{y}$.	$\hat{y}_i = mx_i + b$
3. Find the critical value t_c that corresponds to the given level of confidence c.	Use Table 5 in Appendix B.
4. Find the standard error of estimate s_e.	$s_e = \sqrt{\dfrac{\Sigma(y_i - \hat{y}_i)^2}{n - 2}}$
5. Find the margin of error E.	$E = t_c s_e \sqrt{1 + \dfrac{1}{n} + \dfrac{n(x_0 - \bar{x})^2}{n\Sigma x^2 - (\Sigma x)^2}}$
6. Find the left and right endpoints and form the prediction interval.	Left endpoint: $\hat{y} - E$ Right endpoint: $\hat{y} + E$ Interval: $\hat{y} - E < y < \hat{y} + E$

Study Tip

The formulas for s_e and E use the quantities $\Sigma(y_i - \hat{y}_i)^2$, $(\Sigma x)^2$, and Σx^2. Use a table to calculate these quantities.

EXAMPLE 3

Constructing a Prediction Interval

Using the results of Example 2, construct a 90% prediction interval for the carbon dioxide emissions when the gross domestic product is $2.8 trillion. What can you conclude?

SOLUTION

Because $n = 10$, there are d.f. $= 10 - 2 = 8$ degrees of freedom. Using the regression equation

$$\hat{y} = 199.535x + 56.036$$

and

$$x = 2.8$$

the point estimate is

$$\begin{aligned} \hat{y} &= 199.535x + 56.036 \\ &= 199.535(2.8) + 56.036 \\ &= 614.734. \end{aligned}$$

From Table 5, the critical value is $t_c = 1.860$ and from Example 2, $s_e \approx 141.935$. From Example 4 in Section 9.1, you found that $\Sigma x = 22.9$ and $\Sigma x^2 = 65.49$. Also, $\bar{x} = 2.29$. Using these values, the margin of error is

$$\begin{aligned} E &= t_c s_e \sqrt{1 + \frac{1}{n} + \frac{n(x_0 - \bar{x})^2}{n\Sigma x^2 - (\Sigma x)^2}} \\ &\approx (1.860)(141.935)\sqrt{1 + \frac{1}{10} + \frac{10(2.8 - 2.29)^2}{10(65.49) - (22.9)^2}} \\ &\approx 279.382. \end{aligned}$$

Using $\hat{y} = 614.734$ and $E \approx 279.382$, the prediction interval is constructed as shown.

Left Endpoint	Right Endpoint
$\hat{y} - E \approx 614.734 - 279.382$	$\hat{y} + E \approx 614.734 + 279.382$
$= 335.352$	$= 894.116$

$$335.352 < y < 894.116$$

Interpretation You can be 90% confident that when the gross domestic product is $2.8 trillion, the carbon dioxide emissions will be between 335.352 and 894.116 million metric tons.

TRY IT YOURSELF 3

Using the results of Example 2, construct a 95% prediction interval for the carbon dioxide emissions when the gross domestic product is $4 trillion. What can you conclude?

Answer: Page A38

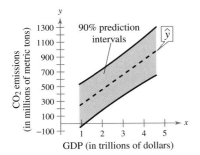

For x-values near $\bar{x}$, the prediction interval for y becomes narrower. For x-values further from $\bar{x}$, the prediction interval for y becomes wider. (This is one reason why the regression equation should not be used to predict y-values for x-values outside the range of the observed x-values in the data.) For instance, consider the 90% prediction intervals for y in Example 3 shown at the left. The range of the x-values is $0.9 \leq x \leq 4.6$. Notice how the confidence interval bands curve away from the regression line as x gets closer to 0.9 or to 4.6.

9.3 EXERCISES

Building Basic Skills and Vocabulary

1. The total variation is the sum of the squares of the differences between the y-values of each ordered pair and the mean of the y-values of the ordered pairs, or $\Sigma(y_i - \bar{y})^2$.

2. The explained variation is the sum of the squares of the differences between the predicted y-values and the mean of the y-values of the ordered pairs, or $\Sigma(\hat{y}_i - \bar{y})^2$.

3. The unexplained variation is the sum of the squares of the differences between the observed y-values and the predicted y-values, or $\Sigma(y_i - \hat{y}_i)^2$.

4. The coefficient of determination
$$r^2 = \frac{\Sigma(\hat{y}_i - \bar{y})^2}{\Sigma(y_i - \bar{y})^2}$$
is the ratio of the explained variation to the total variation and is the percent of variation of y that is explained by the relationship between x and y; $1 - r^2$ is the percent of the variation that is unexplained.

5. Two variables that have perfect positive or perfect negative linear correlation have a correlation coefficient of 1 or -1, respectively. In either case, the coefficient of determination is 1, which means that 100% of the variation in the response variable is explained by the variation in the explanatory variable.

6. Two variables have a bivariate normal distribution when, for any fixed values of either variable, the corresponding values of the other variable are normally distributed.

7. 0.216; About 21.6% of the variation is explained. About 78.4% of the variation is unexplained.

8. See Selected Answers, page A106.

9. See Odd Answers, page A81.

10. See Selected Answers, page A106.

11. See Odd Answers, page A81.

Graphical Analysis   *In Exercises 1–3, use the figure.*

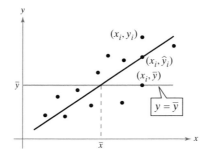

1. Describe the total variation about a regression line in words and in symbols.

2. Describe the explained variation about a regression line in words and in symbols.

3. Describe the unexplained variation about a regression line in words and in symbols.

4. The coefficient of determination r^2 is the ratio of which two types of variations? What does r^2 measure? What does $1 - r^2$ measure?

5. What is the coefficient of determination for two variables that have perfect positive linear correlation or perfect negative linear correlation? Interpret your answer.

6. Two variables have a bivariate normal distribution. Explain what this means.

In Exercises 7–10, use the value of the correlation coefficient r to calculate the coefficient of determination r^2. What does this tell you about the explained variation of the data about the regression line? about the unexplained variation?

7. $r = 0.465$

8. $r = -0.328$

9. $r = -0.957$

10. $r = 0.881$

Using and Interpreting Concepts

Finding the Coefficient of Determination and the Standard Error of Estimate   *In Exercises 11–20, use the data to (a) find the coefficient of determination r^2 and interpret the result, and (b) find the standard error of estimate s_e and interpret the result.*

 11. Stock Offerings   The numbers of initial public offerings of stock issued and the total proceeds of these offerings (in millions of dollars) for 12 years are shown in the table. The equation of the regression line is $\hat{y} = 104.965x + 14{,}093.666$. *(Source: University of Florida)*

Number of offerings, x	316	485	382	79	70	67
Proceeds, y	34,314	64,906	64,876	34,241	22,136	10,068

Number of offerings, x	183	168	162	162	21	43
Proceeds, y	31,927	28,593	30,648	35,762	22,762	13,307

12. (a) 0.948; About 94.8% of the variation in the median annual earnings of female workers can be explained by the relationship between the median annual earnings of male and female workers, and about 5.2% of the variation is unexplained.

    (b) 1396.779; The standard error of estimate of the median annual earnings of female workers for a specific median annual earnings of a male worker is about $1396.78.

13. (a) 0.729; About 72.9% of the variation in points earned can be explained by the relationship between the number of goals allowed and points earned, and about 27.1% of the variation is unexplained.

    (b) 9.438; The standard error of estimate of the points earned for a specific number of goals allowed is about 9.438.

14. (a) 0.671; About 67.1% of the variation in the trunk diameters can be explained by the relationship between the tree heights and trunk diameters, and about 32.9% of the variation is unexplained.

    (b) 1.780; The standard error of estimate of the trunk diameter for a specific tree height is about 1.780 inches.

15. (a) 0.651; About 65.1% of the variation in mean annual wages can be explained by the relationship between percentages of employment in STEM occupations and mean annual wages, and about 34.9% of the variation is unexplained.

    (b) 8.141; The standard error of estimate of the mean annual wages for a specific percentage of employment in STEM occupations is about $8141.

 12. Earnings of Men and Women The table shows the median annual earnings (in dollars) of male and female workers from 10 states in a recent year. The equation of the regression line is $\hat{y} = 1.005x - 10{,}770.313$. *(Source: U.S. Census Bureau)*

Median annual earnings of male workers, x	50,976	46,763	46,934	41,092	47,960
Median annual earnings of female workers, y	40,214	36,834	36,841	31,110	40,173

Median annual earnings of male workers, x	43,829	47,092	51,628	46,123	61,666
Median annual earnings of female workers, y	32,096	35,753	41,690	33,443	50,802

 13. Goals Allowed and Points The table shows the number of goals allowed and the total points earned (2 points for a win and 1 point for an overtime or shootout loss) by the 14 Western Conference teams in the 2016–2017 National Hockey League season. The equation of the regression line is $\hat{y} = -0.573x + 220.087$. *(Source: ESPN)*

Goals allowed, x	213	208	218	224	256	262	278
Points, y	109	106	99	94	87	79	48

Goals allowed, x	200	212	201	221	205	260	243
Points, y	105	103	99	94	86	70	69

14. Trees The table shows the heights (in feet) and trunk diameters (in inches) of eight trees. The equation of the regression line is $\hat{y} = 0.479x - 24.086$.

Height, x	70	72	75	76	85	78	77	82
Trunk diameter, y	8.3	10.5	11.0	11.4	14.9	14.0	16.3	15.8

 15. STEM Employment and Mean Wage The table shows the percentage of employment in STEM (science, technology, engineering, and math) occupations and mean annual wage (in thousands of dollars) for 16 industries. The equation of the regression line is $\hat{y} = 1.153x + 46.374$. *(Source: U.S. Bureau of Labor Statistics)*

Percentage of employment in STEM occupations, x	10.5	15.8	1.6	11.0	8.4	1.1	23.7	7.0
Mean annual wage, y	63.3	73.1	51.3	49.6	54.8	46.2	70.4	67.4

Percentage of employment in STEM occupations, x	1.0	34	16.7	3.7	4.8	1.0	1.4	8.4
Mean annual wage, y	45.0	77.6	79.6	36.7	52.4	51.0	39.2	57.4

16. (a) 0.823; About 82.3% of the variation in ballots cast in federal elections can be explained by the relationship between the voting age population and the ballots cast in federal elections, and about 17.7% of the variation is unexplained.

    (b) 3.694; The standard error of estimate of the ballots cast in federal elections for a specific voting age population is about 3,694,000 people.

17. (a) 0.885; About 88.5% of the variation in the amount of crude oil imported can be explained by the relationship between the amount of crude oil produced and the amount imported, and about 11.5% of the variation is unexplained.

    (b) 280.083; The standard error of estimate of the amount of crude oil imported for a specific amount of crude oil produced is about 280,083 barrels per day.

18. (a) 0.916; 91.6% of the variation in assets in federal pension plans can be explained by the relationship between IRA assets and federal DB plan assets, and 8.4% of the variation is unexplained.

    (b) 63.913; The standard error of estimate of assets in federal DB plans for a specific total of IRA assets is about $63,913,000,000.

19. (a) 0.816; About 81.6% of the variation in the new-vehicle sales of General Motors can be explained by the relationship between the new-vehicle sales of Ford and General Motors, and about 18.4% of the variation is unexplained.

    (b) 346.341; The standard error of estimate of the new-vehicle sales of General Motors for a specific amount of new-vehicle sales of Ford is about 346,341 new vehicles.

16. **Voter Turnout** The U.S. voting age populations (in millions) and the number of ballots cast (in millions) for the highest office in federal elections for nine nonpresidential election years are shown in the table. The equation of the regression line is $\hat{y} = 0.276x + 18.881$. *(Source: United States Elections Project)*

Voting age population, x	166.0	177.9	186.2	195.3	205.3
Ballots cast in federal elections, y	67.6	65.0	67.9	75.1	72.5

Voting age population, x	215.5	225.5	236.0	245.7
Ballots cast in federal elections, y	78.4	83.8	89.1	81.7

17. **Crude Oil** The table shows the amounts of crude oil (in thousands of barrels per day) produced by the United States and the amounts of crude oil (in thousands of barrels per day) imported by the United States for seven years. The equation of the regression line is $\hat{y} = -0.438x + 11,404.947$. *(Source: U.S. Energy Information Administration)*

Produced, x	5475	5646	6487	7468	8764	9415	8875
Imported, y	9213	8935	8527	7730	7344	7363	7877

18. **Fund Assets** The table shows the total assets (in billions of dollars) of individual retirement accounts (IRAs) and federal defined benefit (DB) plans for ten years. The equation of the regression line is $\hat{y} = 0.140x + 453.959$. *(Source: Investment Company Institute)*

IRAs, x	4748	3681	4488	5029	5153
Federal DB plans, y	978	1033	1095	1161	1230

IRAs, x	5785	6819	7292	7329	7850
Federal DB plans, y	1270	1370	1438	1512	1595

19. **New-Vehicle Sales** The table shows the numbers of new-vehicle sales (in thousands) in the United States for Ford and General Motors for 11 years. The equation of the regression line is $\hat{y} = 1.624x - 747.304$. *(Source: NADA Industry Analysis Division)*

New-vehicle sales (Ford), x	3107	2848	2502	1942	1656	1905
New-vehicle sales (General Motors), y	4457	4068	3825	2956	2072	2211

New-vehicle sales (Ford), x	2111	2206	2435	2418	2549
New-vehicle sales (General Motors), y	2504	2596	2786	2935	3082

20. (a) 0.894; About 89.4% of the variation in the new-vehicle sales of Honda can be explained by the relationship between the new-vehicle sales of Toyota and Honda, and 10.6% of the variation is unexplained.

  (b) 55.612; The standard error of the estimate of the new-vehicle sales of Honda for a specific amount of new-vehicle sales of Toyota is about 55,612 new vehicles.

21. $40{,}083.251 < y < 82{,}572.581$ You can be 95% confident that the proceeds will be between \$40,083,251,000 and \$82,572,581,000 when the number of initial offerings is 450 issues.

22. $31{,}674.779 < y < 38{,}514.965$ You can be 95% confident that the median annual earnings of female workers will be between \$31,674.78 and \$38,514.97 when the median annual earnings of male workers is \$45,637.

23. See Odd Answers, page A81.

24. See Selected Answers, page A106.

25. See Odd Answers, page A81.

26. See Selected Answers, page A106.

27. See Odd Answers, page A81.

28. See Selected Answers, page A106.

29. See Odd Answers, page A81.

30. See Selected Answers, page A106.

31. See Odd Answers, page A81.

32. See Selected Answers, page A106.

33. See Odd Answers, page A81.

34. See Selected Answers, page A106.

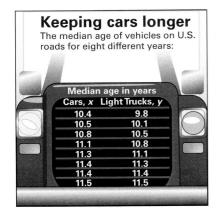

Keeping cars longer
The median age of vehicles on U.S. roads for eight different years:

Median age in years	
Cars, x	Light Trucks, y
10.4	9.8
10.5	10.1
10.8	10.5
11.1	10.8
11.3	11.1
11.4	11.3
11.4	11.4
11.5	11.5

(Source: Polk Co., IHS Automotive)

FIGURE FOR EXERCISES 31–34

 20. New-Vehicle Sales The table shows the numbers of new-vehicle sales (in thousands) in the United States for Toyota and Honda for 11 years. The equation of the regression line is $\hat{y} = 0.460x + 410.839$. *(Source: NADA Industry Analysis Division)*

New-vehicle sales (Toyota), x	2260	2543	2621	2218	1770	1764
New-vehicle sales (Honda), y	1463	1509	1552	1429	1151	1231

New-vehicle sales (Toyota), x	1645	2083	2236	2374	2499
New-vehicle sales (Honda), y	1147	1423	1525	1541	1587

Constructing and Interpreting a Prediction Interval   *In Exercises 21–30, construct the indicated prediction interval and interpret the results.*

21. **Proceeds**   Construct a 95% prediction interval for the proceeds from initial public offerings in Exercise 11 when the number of offerings is 450.

22. **Earnings of Women**   Construct a 95% prediction interval for the median annual earnings of female workers in Exercise 12 when the median annual earnings of male workers is \$45,637.

23. **Points**   Construct a 90% prediction interval for total points earned in Exercise 13 when the number of goals allowed by the team is 250.

24. **Trees**   Construct a 90% prediction interval for the trunk diameter of a tree in Exercise 14 when the height is 80 feet.

25. **Mean Wage**   Construct a 99% prediction interval for the mean annual wage in Exercise 15 when the percentage of employment in STEM occupations is 13% in the industry.

26. **Voter Turnout**   Construct a 99% prediction interval for number of ballots cast in Exercise 16 when the voting age population is 210 million.

27. **Crude Oil**   Construct a 95% prediction interval for the amount of crude oil imported by the United States in Exercise 17 when the amount of crude oil produced by the United States is 8 million barrels per day.

28. **Total Assets**   Construct a 90% prediction interval for the total assets in federal defined benefit plans in Exercise 18 when the total assets in IRAs is \$6200 billion.

29. **New-Vehicle Sales**   Construct a 95% prediction interval for new-vehicle sales for General Motors in Exercise 19 when the number of new vehicles sold by Ford is 2628 thousand.

30. **New-Vehicle Sales**   Construct a 99% prediction interval for new-vehicle sales for Honda in Exercise 20 when the number of new vehicles sold by Toyota is 2359 thousand.

Old Vehicles   *In Exercises 31–34, use the figure shown at the left.*

31. **Scatter Plot**   Construct a scatter plot of the data. Show $\bar{y}$ and $\bar{x}$ on the graph.

32. **Regression Line**   Find and draw the regression line.

33. **Coefficient of Determination**   Find the coefficient of determination r^2 and interpret the results.

34. **Error of Estimate**   Find the standard error of estimate s_e and interpret the results.

35. Fail to reject H_0. There is not enough evidence at the 1% level of significance to support the claim that there is a linear relationship between weight and number of hours slept.

36. Reject H_0. There is enough evidence at the 5% level of significance to support the claim that there is a linear relationship between age and salary.

37. $-175.836 < B < 287.908$; $108.928 < M < 290.142$

38. $-281.314 < B < 393.386$; $67.711 < M < 331.359$

Extending Concepts

Hypothesis Testing for Slope *When testing the slope M of the regression line for the population, you usually test that the slope is 0, or H_0: M = 0. A slope of 0 indicates that there is no linear relationship between x and y. To perform the t-test for the slope M, use the standardized test statistic*

$$t = \frac{m}{s_e}\sqrt{\Sigma x^2 - \frac{(\Sigma x)^2}{n}}$$

with n − 2 degrees of freedom. Then, using the critical values found in Table 5 in Appendix B, make a decision whether to reject or fail to reject the null hypothesis. You can also use the LinRegTTest feature on a TI-84 Plus to calculate the standardized test statistic as well as the corresponding P-value. If P ≤ α, then reject the null hypothesis. If P > α, then do not reject H_0.

In Exercises 35 and 36, test the claim and interpret the results in the context of the problem. If convenient, use technology.

35. The table shows the weights (in pounds) and the numbers of hours slept in a day by a random sample of infants. Test the claim that $M \neq 0$. Use $\alpha = 0.01$.

Weight, x	8.1	10.2	9.9	7.2	6.9	11.2	11	15
Hours slept, y	14.8	14.6	14.1	14.2	13.8	13.2	13.9	12.5

36. The table shows the ages (in years) and salaries (in thousands of dollars) of a random sample of engineers at a company. Test the claim that $M \neq 0$. Use $\alpha = 0.05$.

Age, x	25	34	29	30	42	38	49	52	35	40
Salary, y	57.5	61.2	59.9	58.7	87.5	67.4	89.2	85.3	69.5	75.1

Confidence Intervals for y-Intercept and Slope *You can construct confidence intervals for the y-intercept B and slope M of the regression line y = Mx + B for the population by using the inequalities below.*

y-intercept B: $\quad b - E < B < b + E$

$$\text{where } E = t_c s_e \sqrt{\frac{1}{n} + \frac{\bar{x}^2}{\Sigma x^2 - \frac{(\Sigma x)^2}{n}}} \text{ and}$$

slope M: $\quad m - E < M < m + E$

$$\text{where } E = \frac{t_c s_e}{\sqrt{\Sigma x^2 - \frac{(\Sigma x)^2}{n}}}$$

The values of m and b are obtained from the sample data, and the critical value t_c is found using Table 5 in Appendix B with n − 2 degrees of freedom.

In Exercises 37 and 38, construct the indicated confidence intervals for B and M using the gross domestic products and carbon dioxide emissions data found in Example 2.

37. 95% confidence interval   **38.** 99% confidence interval

9.4 Multiple Regression

What You Should Learn

▶ How to use technology to find and interpret a multiple regression equation, the standard error of estimate, and the coefficient of determination

▶ How to use a multiple regression equation to predict y-values

Finding a Multiple Regression Equation ■ Predicting y-Values

Finding a Multiple Regression Equation

In many instances, a better prediction model can be found for a dependent (response) variable by using more than one independent (explanatory) variable. For instance, a more accurate prediction for the carbon dioxide emissions discussed in previous sections might be made by considering the number of cars as well as the gross domestic product. Models that contain more than one independent variable are multiple regression models.

DEFINITION

A **multiple regression equation** for independent variables x_1, x_2, x_3, . . ., x_k and a dependent variable y has the form

$$\hat{y} = b + m_1x_1 + m_2x_2 + m_3x_3 + \cdots + m_kx_k$$

where $\hat{y}$ is the predicted y-value for given x_i values and b is the y-intercept. The y-intercept b is the value of $\hat{y}$ when all x_i are 0. Each coefficient m_i is the amount of change in $\hat{y}$ when the independent variable x_i is changed by one unit and all other independent variables are held constant.

Because the mathematics associated with multiple regression is complicated, this section focuses on how to use technology to find a multiple regression equation and how to interpret the results.

Tech Tip

Detailed instructions for using Minitab and Excel to find a multiple regression equation are shown in the technology manuals that accompany this text.

EXAMPLE 1

Finding a Multiple Regression Equation

A researcher wants to determine how employee salaries at a company are related to the length of employment, previous experience, and education. The researcher selects eight employees from the company and obtains the data shown in the table.

Employee	Salary (in dollars), y	Employment (in years), x_1	Experience (in years), x_2	Education (in years), x_3
A	57,310	10	2	16
B	57,380	5	6	16
C	54,135	3	1	12
D	56,985	6	5	14
E	58,715	8	8	16
F	60,620	20	0	12
G	59,200	8	4	18
H	60,320	14	6	17

Use Minitab to find a multiple regression equation that models the data.

SOLUTION

Enter the y-values in C1 and the x_1-, x_2-, and x_3-values in C2, C3, and C4, respectively. Select "Regression▶Regression▶Fit Regression Model" from the *Stat* menu. Using the salaries as the response variable and the remaining data as the predictors, you should obtain results similar to the display shown.

MINITAB

Regression Analysis: Salary, y versus x1, x2, x3

Model Summary

S	R-sq	R-sq(adj)
659.490	94.38%	90.17%

Coefficients

Term	Coef	SE Coef	T-Value	P-Value
Constant	49764 — b	1981	25.12	0.000
x1	364.4 — m_1	48.3	7.54	0.002
x2	228 — m_2	124	1.84	0.140
x3	267 — m_3	147	1.81	0.144

Regression Equation

Salary, y = 49764 + 364.4 x1 + 228 x2 + 267 x3

The regression equation is $\hat{y} = 49{,}764 + 364x_1 + 228x_2 + 267x_3$.

Study Tip

In Example 1, it is important that you interpret the coefficients m_1, m_2, and m_3 correctly. For instance, if x_2 and x_3 are held constant and x_1 increases by 1, then y increases by $364. Similarly, if x_1 and x_3 are held constant and x_2 increases by 1, then y increases by $228. If x_1 and x_2 are held constant and x_3 increases by 1, then y increases by $267.

TRY IT YOURSELF 1

A statistics professor wants to determine how students' final grades are related to the midterm exam grades and number of classes missed. The professor selects 10 students and obtains the data shown in the table.

Student	Final grade, y	Midterm exam, x_1	Classes missed, x_2
1	81	75	1
2	90	80	0
3	86	91	2
4	76	80	3
5	51	62	6
6	75	90	4
7	44	60	7
8	81	82	2
9	94	88	0
10	93	96	1

Use technology to find a multiple regression equation that models the data.

Answer: Page A38

Minitab displays much more than the regression equation and the coefficients of the independent variables. For instance, it also displays the standard error of estimate, denoted by S, and the coefficient of determination, denoted by *R-Sq*. In Example 1, $S = 659.490$ and *R-Sq* $= 94.38\%$. So, the standard error of estimate is $659.49. The coefficient of determination tells you that 94.38% of the variation in y can be explained by the multiple regression model. The remaining 5.62% is unexplained and is due to other factors, such as sampling error, coincidence, or lurking variables.

Picturing the World

In a lake in Finland, 159 fish of 7 species were caught and measured for weight G (in grams), length L (in centimeters), height H, and width W (H and W are percents of L). The regression equation for G and L is

$$G = -491 + 28.5L,$$
$$r \approx 0.925, r^2 \approx 0.855.$$

When all four variables are used, the regression equation is

$$G = -712 + 28.3L +$$
$$1.46H + 13.3W,$$
$$r \approx 0.930, r^2 \approx 0.865.$$

(Source: Journal of Statistics Education)

Predict the weight of a fish with the following measurements: $L = 40$, $H = 17$, and $W = 11$. How do your predictions vary when you use a single variable versus many variables? Which do you think is more accurate?

The predicted weight of the fish using the single variable model is 649 grams and the predicted weight of the fish using the model with many variables is 591.12 grams. These predictions vary greatly. Because the coefficient of determination is greater for the model with many variables, this model is most likely more accurate.

Predicting y-Values

After finding the equation of the multiple regression line, you can use the equation to predict y-values over the range of the data. To predict y-values, substitute the given value for each independent variable into the equation, then calculate $\hat{y}$.

EXAMPLE 2

Predicting y-Values Using Multiple Regression Equations

Use the regression equation

$$\hat{y} = 49{,}764 + 364x_1 + 228x_2 + 267x_3$$

found in Example 1 to predict an employee's salary for each set of conditions.

1. 12 years of current employment
  5 years of previous experience
  16 years of education

2. 4 years of current employment
  2 years of previous experience
  12 years of education

3. 8 years of current employment
  7 years of previous experience
  17 years of education

SOLUTION

To predict each employee's salary, substitute the values for x_1, x_2, and x_3 into the regression equation. Then calculate $\hat{y}$.

1. $\hat{y} = 49{,}764 + 364x_1 + 228x_2 + 267x_3$
   $= 49{,}764 + 364(12) + 228(5) + 267(16)$
   $= 59{,}544$

The employee's predicted salary is $59,544.

2. $\hat{y} = 49{,}764 + 364x_1 + 228x_2 + 267x_3$
   $= 49{,}764 + 364(4) + 228(2) + 267(12)$
   $= 54{,}880$

The employee's predicted salary is $54,880.

3. $\hat{y} = 49{,}764 + 364x_1 + 228x_2 + 267x_3$
   $= 49{,}764 + 364(8) + 228(7) + 267(17)$
   $= 58{,}811$

The employee's predicted salary is $58,811.

TRY IT YOURSELF 2

Use the regression equation found in Try It Yourself 1 to predict a student's final grade for each set of conditions.

1. A student has a midterm exam score of 89 and misses 1 class.

2. A student has a midterm exam score of 78 and misses 3 classes.

3. A student has a midterm exam score of 83 and misses 2 classes.

Answer: Page A38

9.4 EXERCISES

1. (a) 18,832.7 pounds per acre
   (b) 18,016.4 pounds per acre
   (c) 17,350.6 pounds per acre
   (d) 16,190.3 pounds per acre
2. (a) 51.68 bushels per acre
   (b) 67.2 bushels per acre
   (c) 69.64 bushels per acre
   (d) 55.58 bushels per acre
3. (a) 7.5 cubic feet
   (b) 16.8 cubic feet
   (c) 51.9 cubic feet
   (d) 62.1 cubic feet
4. (a) 4316.7 kilograms
   (b) 2126.55 kilograms
   (c) 3564.3 kilograms
   (d) 734.75 kilograms

Building Basic Skills and Vocabulary

Predicting y-Values   *In Exercises 1–4, use the multiple regression equation to predict the y-values for the values of the independent variables.*

1. **Cauliflower Yield**   The equation used to predict the annual cauliflower yield (in pounds per acre) is

   $$\hat{y} = 24{,}791 + 4.508x_1 - 4.723x_2$$

   where x_1 is the number of acres planted and x_2 is the number of acres harvested. *(Adapted from United States Department of Agriculture)*

   (a) $x_1 = 36{,}500, x_2 = 36{,}100$
   (b) $x_1 = 38{,}100, x_2 = 37{,}800$
   (c) $x_1 = 39{,}000, x_2 = 38{,}800$
   (d) $x_1 = 42{,}200, x_2 = 42{,}100$

2. **Sorghum Yield**   The equation used to predict the annual sorghum yield (in bushels per acre) is

   $$\hat{y} = 80.1 - 20.2x_1 + 21.2x_2$$

   where x_1 is the number of acres planted (in millions) and x_2 is the number of acres harvested (in millions). *(Adapted from United States Department of Agriculture)*

   (a) $x_1 = 5.5, x_2 = 3.9$
   (b) $x_1 = 8.3, x_2 = 7.3$
   (c) $x_1 = 6.5, x_2 = 5.7$
   (d) $x_1 = 9.4, x_2 = 7.8$

3. **Black Cherry Tree Volume**   The volume (in cubic feet) of a black cherry tree can be modeled by the equation

   $$\hat{y} = -52.2 + 0.3x_1 + 4.5x_2$$

   where x_1 is the tree's height (in feet) and x_2 is the tree's diameter (in inches). *(Source: Journal of the Royal Statistical Society)*

   (a) $x_1 = 70, x_2 = 8.6$
   (b) $x_1 = 65, x_2 = 11.0$
   (c) $x_1 = 83, x_2 = 17.6$
   (d) $x_1 = 87, x_2 = 19.6$

4. **Elephant Weight**   The equation used to predict the weight of an elephant (in kilograms) is

   $$\hat{y} = -4016 + 11.5x_1 + 7.55x_2 + 12.5x_3$$

   where x_1 represents the girth of the elephant (in centimeters), x_2 represents the length of the elephant (in centimeters), and x_3 represents the circumference of a footpad (in centimeters). *(Source: Field Trip Earth)*

   (a) $x_1 = 421, x_2 = 224, x_3 = 144$
   (b) $x_1 = 311, x_2 = 171, x_3 = 102$
   (c) $x_1 = 376, x_2 = 226, x_3 = 124$
   (d) $x_1 = 231, x_2 = 135, x_3 = 86$

5. (a)
$\hat{y} = 17,899 - 606.58x_1 - 52.9x_2$

(b) 564.314

(c) 0.966

The standard error of estimate of the predicted price given specific age and milage of pre-owned Honda Civic Sedans is about $564.31. The multiple regression model explains about 96.6% of the variation.

6. (a)
$\hat{y} = -29.69 + 0.229x_1 - 0.004x_2$

(b) 1.847

(c) 0.843

The standard error of estimate of the predicted shareholder's equity given specific net sales and total assets for Wal-Mart is about $1.847 billion. The multiple regression model explains about 84.3% of the variation.

7. 0.955; About 95.5% of the variation in y can be explained by the relationship between variables; $r^2_{adj} < r^2$.

8. 0.739; About 73.9% of the variation in y can be explained by the relationship between variables; $r^2_{adj} < r^2$.

Using and Interpreting Concepts

Finding a Multiple Regression Equation *In Exercises 5 and 6, use technology to find (a) the multiple regression equation for the data shown in the table, (b) the standard error of estimate, and (c) the coefficient of determination. Interpret the results.*

5. Used Cars The table shows the prices (in dollars), age (in years), and mileage (in thousands of miles) of eight pre-owned Honda Civic Sedans.

Price, y	Age, x_1	Mileage, x_2
9454	6	91.2
10,920	5	77.1
13,929	3	45.1
14,604	2	37.7
11,500	4	52.1
15,308	2	34.7
14,500	3	35.6
14,878	3	21.6
8000	9	87.9

6. Shareholder's Equity The table shows the net sales (in billions of dollars), total assets (in billions of dollars), and shareholder's equities (in billions of dollars) for Wal-Mart for six years. *(Adapted from Wal-Mart Stores, Inc.)*

Shareholder's equity, y	Net sales, x_1	Total assets, x_2
71.3	443.9	193.4
76.3	465.6	202.9
76.3	473.1	204.5
81.4	482.2	203.5
80.5	478.6	199.6
77.8	481.3	198.8

Extending Concepts

Adjusted r^2 *The calculation of the coefficient of determination r^2 depends on the number of data pairs and the number of independent variables. An adjusted value of r^2 based on the number of degrees of freedom is calculated using the formula*

$$r^2_{adj} = 1 - \left[\frac{(1 - r^2)(n - 1)}{n - k - 1} \right]$$

where n is the number of data pairs and k is the number of independent variables.

In Exercises 7 and 8, calculate r^2_{adj} and determine the percentage of the variation in y that can be explained by the relationships between variables according to r^2_{adj}. Compare this result with the one obtained using r^2.

7. Calculate r^2_{adj} for the data in Exercise 5.

8. Calculate r^2_{adj} for the data in Exercise 6.

Uses

Correlation and Regression Correlation and regression analysis can be used to determine whether there is a significant relationship between two variables. When there is, you can use one of the variables to predict the value of the other variable. For instance, educators have used correlation and regression analysis to determine that there is a significant correlation between a student's SAT score and the grade point average from a student's freshman year at college. Consequently, many colleges and universities use SAT scores of high school applicants as a predictor of the applicant's initial success at college.

Abuses

Confusing Correlation and Causation The most common abuse of correlation in studies is to confuse the concepts of correlation with those of causation (see page 480). Good SAT scores do not cause good college grades. Rather, there are other variables, such as good study habits and motivation, that contribute to both. When a strong correlation is found between two variables, look for other variables that are correlated with both.

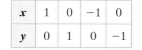

x	1	0	−1	0
y	0	1	0	−1

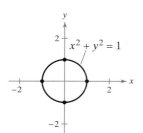

Considering Only Linear Correlation The correlation studied in this chapter is linear correlation. When the correlation coefficient is close to 1 or close to −1, the data points can be modeled by a straight line. It is possible that a correlation coefficient is close to 0 but there is still a strong correlation of a different type. Consider the data listed in the table at the left. The value of the correlation coefficient is 0. However, the data are perfectly correlated with the equation $x^2 + y^2 = 1$, as shown in the figure at the left.

Ethics

When data are collected, all of the data should be used when calculating statistics. In this chapter, you learned that before finding the equation of a regression line, it is helpful to construct a scatter plot of the data to check for outliers, gaps, and clusters in the data. Researchers cannot use only those data points that fit their hypotheses or those that show a significant correlation. Although eliminating outliers may help a data set coincide with predicted patterns or fit a regression line, it is unethical to amend data in such a way. An outlier or any other point that influences a regression model can be removed only when it is properly justified.

In most cases, the best and sometimes safest approach for presenting statistical measurements is with and without an outlier being included. By doing this, the decision as to whether or not to recognize the outlier is left to the reader.

EXERCISES

1. ***Confusing Correlation and Causation*** Find an example of an article that confuses correlation and causation. Discuss other variables that could contribute to the relationship between the variables.

2. ***Considering Only Linear Correlation*** Find an example of two real-life variables that have a nonlinear correlation.

9 | Chapter Summary

What Did You Learn?	Example(s)	Review Exercises

Section 9.1

▷ How to construct a scatter plot and how to find a correlation coefficient — Example(s) 1–5, Review Exercises 1–4

$$r = \frac{n\Sigma xy - (\Sigma x)(\Sigma y)}{\sqrt{n\Sigma x^2 - (\Sigma x)^2}\sqrt{n\Sigma y^2 - (\Sigma y)^2}}$$

▷ How to test a population correlation coefficient ρ using a table and how to perform a hypothesis test for a population correlation coefficient ρ — Example(s) 6, 7, Review Exercises 5–8

$$t = \frac{r}{\sqrt{\dfrac{1 - r^2}{n - 2}}}$$

Section 9.2

▷ How to find the equation of a regression line — Example(s) 1, 2, Review Exercises 9–12

$$\hat{y} = mx + b$$

$$m = \frac{n\Sigma xy - (\Sigma x)(\Sigma y)}{n\Sigma x^2 - (\Sigma x)^2}$$

$$b = \bar{y} - m\bar{x} = \frac{\Sigma y}{n} - m\frac{\Sigma x}{n}$$

▷ How to predict y-values using a regression equation — Example(s) 3, Review Exercises 9–12

Section 9.3

▷ How to find and interpret the coefficient of determination — Example(s) 1, Review Exercises 13–18

$$r^2 = \frac{\text{Explained variation}}{\text{Total variation}}$$

▷ How to find and interpret the standard error of estimate for a regression line — Example(s) 2, Review Exercises 17, 18

$$s_e = \sqrt{\frac{\Sigma(y_i - \hat{y}_i)^2}{n - 2}} = \sqrt{\frac{\Sigma y^2 - b\Sigma y - m\Sigma xy}{n - 2}}$$

▷ How to construct and interpret a prediction interval for y — Example(s) 3, Review Exercises 19–24

$$\hat{y} - E < y < \hat{y} + E, \quad E = t_c s_e \sqrt{1 + \frac{1}{n} + \frac{n(x_0 - \bar{x})^2}{n\Sigma x^2 - (\Sigma x)^2}}$$

Section 9.4

▷ How to use technology to find and interpret a multiple regression equation, the standard error of estimate, and the coefficient of determination — Example(s) 1, Review Exercises 25, 26

$$\hat{y} = b + m_1x_1 + m_2x_2 + m_3x_3 + \cdots + m_kx_k$$

▷ How to use a multiple regression equation to predict y-values — Example(s) 2, Review Exercises 27, 28

Review Exercises

1. (a)

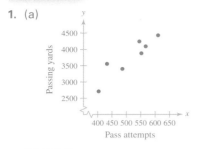

(b) 0.917

(c) Strong positive linear correlation; As the number of pass attempts increase, the number of passing yards tends to increase.

2. See Selected Answers, page A106.

3. See Odd Answers, page A82.

4. See Selected Answers, page A107.

5. There is enough evidence at the 5% level of significance to conclude that there is a significant linear correlation between a quarterback's pass attempts and passing yards.

6. There is not enough evidence at the 5% level of significance to conclude that there is a significant linear correlation between the number of wildland fires and the number of acres burned.

7. There is not enough evidence at the 1% level of significance to conclude that there is a significant linear correlation between IQ and brain size.

8. There is enough evidence at the 1% level of significance to conclude that there is a significant linear correlation between sugar consumption and number of cavities.

Section 9.1

In Exercises 1–4, (a) display the data in a scatter plot, (b) calculate the sample correlation coefficient r, and (c) describe the type of correlation and interpret the correlation in the context of the data.

1. The numbers of pass attempts and passing yards for seven professional quarterbacks for a recent regular season *(Source: National Football League)*

Pass attempts, x	610	545	567	552	432	486	403
Passing yards, y	4428	4240	4090	3877	3554	3401	2710

2. The numbers of wildland fires (in thousands) and wildland acres burned (in millions) in the United States for eight years *(Source: National Interagency Coordinate Center)*

Fires, x	78.8	72.0	74.1	67.8	47.6	63.3	68.2	67.7
Acres, y	5.9	3.4	8.7	9.3	4.3	3.6	10.1	5.5

3. The intelligence quotient (IQ) scores and brain sizes, as measured by the total pixel count (in thousands) from an MRI scan, for nine female college students *(Adapted from Intelligence)*

IQ score, x	138	140	96	83	101	135	85	77	88
Pixel count, y	991	856	879	865	808	791	799	794	894

4. The annual per capita sugar consumptions (in kilograms) and the average numbers of cavities of 11- and 12-year-old children in seven countries

Sugar consumption, x	2.1	5.0	6.3	6.5	7.7	8.7	11.6
Cavities, y	0.59	1.51	1.55	1.70	2.18	2.10	2.73

In Exercises 5–8, use Table 11 in Appendix B, or perform a hypothesis test using Table 5 in Appendix B to make a conclusion about the correlation coefficient.

5. Refer to the data in Exercise 1. At $\alpha = 0.05$, is there enough evidence to conclude that there is a significant linear correlation between the data? (Use the value of r found in Exercise 1.)

6. Refer to the data in Exercise 2. At $\alpha = 0.05$, is there enough evidence to conclude that there is a significant linear correlation between the data? (Use the value of r found in Exercise 2.)

7. Refer to the data in Exercise 3. At $\alpha = 0.01$, is there enough evidence to conclude that there is a significant linear correlation between the data? (Use the value of r found in Exercise 3.)

8. Refer to the data in Exercise 4. At $\alpha = 0.01$, is there enough evidence to conclude that there is a significant linear correlation between the data? (Use the value of r found in Exercise 4.)

9. $\hat{y} = 0.106x - 781.327$

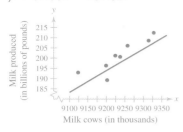

(a) It is not meaningful to predict the value of y for $x = 9080$ because $x = 9080$ is outside the range of the original data.

(b) 197.053 billions of pounds

(c) 199.173 billions of pounds

(d) 204.473 billions of pounds

10. $\hat{y} = 0.866x - 0.059$

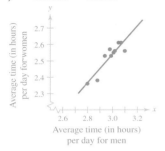

(a) 2.409 hours (b) 2.513 hours

(c) 2.574 hours

(d) It is not meaningful to predict the value of y when $x = 3.13$ because $x = 3.13$ is outside the range of the original data.

11. $\hat{y} = -0.086x + 10.450$

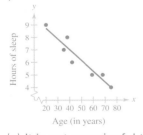

(a) It is not meaningful to predict the value of y when $x = 16$ because $x = 16$ is outside the range of the original data.

(b) 8.3 hours

(c) It is not meaningful to predict the value of y when $x = 85$ because $x = 85$ is outside the range of the original data.

(d) 6.15 hours

12. See Selected Answers, page A107.

Section 9.2

In Exercises 9–12, find the equation of the regression line for the data. Then construct a scatter plot of the data and draw the regression line. (Each pair of variables has a significant correlation.) Then use the regression equation to predict the value of y for each of the x-values, if meaningful. If the x-value is not meaningful to predict the value of y, explain why not. If convenient, use technology.

9. The average number (in thousands) of milk cows and the amounts (in billions of pounds) of milk produced in the United States for eight years *(Source: U.S. Department of Agriculture)*

Milk cows, x	9202	9123	9199	9237
Milk produced, y	189.2	192.9	196.3	200.6

Milk cows, x	9224	9257	9314	9328
Milk produced, y	201.2	206.1	208.6	212.4

(a) $x = 9080$ cows (b) $x = 9230$ cows

(c) $x = 9250$ cows (d) $x = 9300$ cows

10. The average times (in hours) per day spent watching television for men and women for 10 years *(Source: U.S. Bureau of Labor Statistics)*

Men, x	2.80	2.88	3.01	3.10	2.94
Women, y	2.36	2.38	2.55	2.56	2.53

Men, x	2.99	3.07	2.98	3.05	3.02
Women, y	2.53	2.61	2.57	2.61	2.56

(a) $x = 2.85$ hours (b) $x = 2.97$ hours

(c) $x = 3.04$ hours (d) $x = 3.13$ hours

11. The ages (in years) and the numbers of hours of sleep in one night for seven adults

Age, x	35	20	59	42	68	38	75
Hours of sleep, y	7	9	5	6	5	8	4

(a) $x = 16$ years (b) $x = 25$ years

(c) $x = 85$ years (d) $x = 50$ years

12. The engine displacements (in cubic inches) and the fuel efficiencies (in miles per gallon) of seven automobiles

Displacement, x	170	134	220	305	109	256	322
Fuel efficiency, y	29.5	34.5	23.0	17.0	33.5	23.0	15.5

(a) $x = 86$ cubic inches (b) $x = 198$ cubic inches

(c) $x = 289$ cubic inches (d) $x = 407$ cubic inches

13. 0.203; About 20.3% of the variation is explained. About 79.7% of the variation is unexplained.

14. 0.878; About 87.8% of the variation is explained. About 12.2% of the variation is unexplained.

15. 0.412; About 41.2% of the variation is explained. About 58.8% of the variation is unexplained.

16. 0.632; About 63.2% of the variation is explained. About 36.8% of the variation is unexplained.

17. (a) 0.690; About 69.0% of the variation in top speed for hybrid and electric cars can be explained by the relationship between their fuel efficiencies and top speeds, and about 31.0% of the variation is unexplained.

(b) 5.851; The standard error of estimate of the top speed for hybrid and electric cars for a specific fuel efficiency is about 5.851 miles per hour.

18. (a) 0.565; About 56.5% of the variation in the price of gas grills can be explained by the relationship between the cooking areas and their prices, and about 43.5% of the variation is unexplained.

(b) 360.348; The standard error of estimate of the price of a gas grill for a specific cooking area is about $360.35.

19. $193.364 < y < 210.282$
You can be 90% confident that the amount of milk produced will be between 193.364 billion pounds and 210.282 billion pounds when the average number of cows is 9275.

20. $2.528 < y < 2.688$
You can be 90% confident that the average time women spend per day watching television will be between 2.528 and 2.688 hours when the average time men spend per day watching television is 3.08 hours.

21. See Odd Answers, page A82.

22. See Selected Answers, page A107.

23. See Odd Answers, page A82.

24. See Selected Answers, page A107.

Section 9.3

In Exercises 13–16, use the value of the correlation coefficient r to calculate the coefficient of determination r^2. What does this tell you about the explained variation of the data about the regression line? about the unexplained variation?

13. $r = -0.450$

14. $r = -0.937$

15. $r = 0.642$

16. $r = 0.795$

In Exercises 17 and 18, use the data to (a) find the coefficient of determination r^2 and interpret the result, and (b) find the standard error of estimate s_e and interpret the result.

17. The table shows the combined city and highway fuel efficiency (in miles per gallon gasoline equivalent) and top speeds (in miles per hour) for nine hybrid and electric cars. The regression equation is $\hat{y} = -0.465x + 139.433$. *(Source: Car and Driver)*

Fuel efficiency, x	114	95	120	105	107	116	118	68	84
Top speed, y	80	103	78	85	92	88	92	105	101

 18. The table shows the cooking areas (in square inches) of 18 gas grills and their prices (in dollars). The regression equation is $\hat{y} = 2.335x - 853.278$. *(Source: Lowe's)*

Area, x	650	669	529	725	844	445	669	844	740
Price, y	149	699	499	374	1599	187	1299	899	374

Area, x	529	450	644	600	575	998	529	265	530
Price, y	599	399	499	269	299	1999	519	99	109

In Exercises 19–24, construct the indicated prediction interval and interpret the results.

19. Construct a 90% prediction interval for the amount of milk produced in Exercise 9 when there are an average of 9275 milk cows.

20. Construct a 90% prediction interval for the average time women spend per day watching television in Exercise 10 when the average time men spend per day watching television is 3.08 hours.

21. Construct a 95% prediction interval for the number of hours of sleep for an adult in Exercise 11 who is 45 years old.

22. Construct a 95% prediction interval for the fuel efficiency of an automobile in Exercise 12 that has an engine displacement of 265 cubic inches.

23. Construct a 99% prediction interval for the top speed of a hybrid or electric car in Exercise 17 that has a combined city and highway fuel economy of 90 miles per gallon equivalent.

24. Construct a 99% prediction interval for the price of a gas grill in Exercise 18 with a usable cooking area of 900 square inches.

25. (a)
$\hat{y} = 3.6738 + 1.2874x_1 - 7.531x_2$

(b) 0.710; The standard error of estimate of the predicted carbon monoxide content given specific tar and nicotine contents is about 0.710 milligram.

(c) 0.943; The multiple regression model explains about 94.3% of the variation in y.

26. (a)
$\hat{y} = 40.21 + 2.264x_1 - 1.9933x_2$

(b) 1205.22; The standard error of estimate of the annual yield of spinach given a specific acres planted and acres harvested is about 1205.22 pounds.

(c) 0.705; The multiple regression model explains about 70.5% of the variation in y.

27. (a) 21.705 miles per gallon

(b) 25.21 miles per gallon

(c) 30.1 miles per gallon

(d) 25.86 miles per gallon

28. (a) 11.276 milligrams

(b) 14.701 milligrams

(c) 12.879 milligrams

(d) 9.236 milligrams

Section 9.4

In Exercises 25 and 26, use technology to find (a) the multiple regression equation for the data shown in the table, (b) the standard error of estimate, and (c) the coefficient of determination. Interpret the result.

 25. The table shows the carbon monoxide, tar, and nicotine content, all in milligrams, of 14 brands of U.S. cigarettes. *(Source: Federal Trade Commission)*

Carbon monoxide, y	Tar, x_1	Nicotine, x_2
15	16	1.1
17	16	1.0
11	10	0.8
12	11	0.9
14	13	0.8
16	14	0.8
14	16	1.2
16	16	1.2
10	10	0.8
18	19	1.4
17	17	1.2
11	12	1.0
10	9	0.7
14	15	1.2

26. The table shows the numbers of acres planted, the numbers of acres harvested, and the annual yields (in pounds) of spinach for five years. *(Source: United States Department of Agriculture)*

Yield, y	Acres planted, x_1	Acres harvested, x_2
15,200	36,400	35,000
18,600	35,400	32,900
17,900	34,400	32,300
18,600	38,500	36,600
16,000	36,400	35,680

In Exercises 27 and 28, use the multiple regression equation to predict the y-values for the values of the independent variables.

27. An equation that can be used to predict fuel economy (in miles per gallon) for automobiles is

$$\hat{y} = 41.3 - 0.004x_1 - 0.0049x_2$$

where x_1 is the engine displacement (in cubic inches) and x_2 is the vehicle weight (in pounds).

(a) $x_1 = 305, x_2 = 3750$ (b) $x_1 = 225, x_2 = 3100$

(c) $x_1 = 105, x_2 = 2200$ (d) $x_1 = 185, x_2 = 3000$

28. Use the regression equation found in Exercise 25.

(a) $x_1 = 10, x_2 = 0.7$ (b) $x_1 = 15, x_2 = 1.1$

(c) $x_1 = 13, x_2 = 1.0$ (d) $x_1 = 9, x_2 = 0.8$

9 Chapter Quiz

1. See Odd Answers, page A83.

2. 0.992; Strong positive linear correlation; As the average annual salaries of secondary school teachers increase, the average annual salaries of elementary school teachers tend to increase.

3. Reject H_0. There is enough evidence at the 5% level of significance to conclude that there is a significant linear correlation between the average annual salaries of secondary school teachers and the average annual salaries of elementary school teachers.

4. See Odd Answers, page A83.

5. $50,382.50

6. 0.984; About 98.4% of the variation in the average annual salaries of elementary school teachers can be explained by the relationship between the average annual salaries of secondary school teachers and elementary school teachers, and about 1.6% of the variation is unexplained.

7. 0.422; The standard error of estimate of the average annual salaries of elementary school teachers for a specific average annual salary of secondary school teachers is about $422.

8. $49.311 < y < 51.455$

   You can be 95% confident that the average annual salary of elementary school teachers will be between $49,311 and $51,455 when the average annual salary of secondary school teachers is $52,500.

9. (a) $95.26
   (b) $70.28
   (c) $67.74
   (d) $59.46

Take this quiz as you would take a quiz in class. After you are done, check your work against the answers given in the back of the book.

 For Exercises 1–8, use the data in the table, which shows the average annual salaries (both in thousands of dollars) for secondary and elementary school teachers, excluding special and vocational education teachers, in the United States for 11 years. (Source: Bureau of Labor Statistics)

Secondary school teachers, x	Elementary school teachers, y
51.2	48.7
52.5	50.0
54.4	52.2
55.2	53.2
56.0	54.3
56.8	55.3
57.8	56.1
58.3	56.3
59.3	56.8
60.4	57.7
61.4	59.0

1. Construct a scatter plot for the data. Do the data appear to have a positive linear correlation, a negative linear correlation, or no linear correlation? Explain.

2. Calculate the correlation coefficient r and interpret the result.

3. Test the significance of the correlation coefficient r that you found in Exercise 2. Use $\alpha = 0.05$.

4. Find the equation of the regression line for the data. Draw the regression line on the scatter plot that you constructed in Exercise 1.

5. Use the regression equation that you found in Exercise 4 to predict the average annual salary of elementary school teachers when the average annual salary of secondary school teachers is $52,500.

6. Find the coefficient of determination r^2 and interpret the result.

7. Find the standard error of estimate s_e and interpret the result.

8. Construct a 95% prediction interval for the average annual salary of elementary school teachers when the average annual salary of secondary school teachers is $52,500. Interpret the results.

9. **Stock Price**   The equation used to predict the stock price (in dollars) at the end of the year for a restaurant chain is

   $$\hat{y} = -86 + 7.46x_1 - 1.61x_2$$

   where x_1 is the total revenue (in billions of dollars) and x_2 is the shareholders' equity (in billions of dollars). Use the multiple regression equation to predict the y-values for the values of the independent variables.

   (a) $x_1 = 27.6, x_2 = 15.3$ (b) $x_1 = 24.1, x_2 = 14.6$
   (c) $x_1 = 23.5, x_2 = 13.4$ (d) $x_1 = 22.8, x_2 = 15.3$

9 Chapter Test

1. (a) $2372.08 million

(b) $2286.97 million

(c) $2223.95 million

(d) $1933.98 million

2. See Selected Answers, page A107.

3. 0.949; Strong positive linear correlation; As the average annual salaries of librarians increase, the average annual salaries of library science teachers tend to increase.

4. Reject H_0. There is enough evidence at the 1% level of significance to conclude that there is a significant linear correlation between the average annual salaries of librarians and the average annual salaries of library science teachers.

5. See Selected Answers, page A107.

6. $67,116

7. 0.902; About 90.2% of the variation in the average annual salaries of library science teachers can be explained by the relationship between the average annual salaries of librarians and library science teachers, and about 9.8% of the variation is unexplained.

8. 2.082; The standard error of estimate of the average annual salary of library science teachers for a specific average annual salary of librarians is about $2082.

9. $60.246 < y < 73.986$

You can be 99% confident that the average annual salary of library science teachers is between $60,246 and $73,986 when the average annual salary of librarians is $56,000.

Take this test as you would take a test in class.

1. Net Sales   The equation used to predict the net sales (in millions of dollars) for a fiscal year for a clothing retailer is

$$\hat{y} = 23{,}769 + 9.18x_1 - 8.41x_2$$

where x_1 is the number of stores open at the end of the fiscal year and x_2 is the average square footage per store. Use the multiple regression equation to predict the y-values for the values of the independent variables.

(a) $x_1 = 1057, x_2 = 3698$

(b) $x_1 = 1012, x_2 = 3659$

(c) $x_1 = 952, x_2 = 3601$

(d) $x_1 = 914, x_2 = 3594$

 For Exercises 2–9, use the data in the table, which shows the average annual salaries (both in thousands of dollars) for librarians and postsecondary library science teachers in the United States for 12 years. (Source: Bureau of Labor Statistics)

Librarians, x	Library science teachers, y
49.1	56.6
50.9	57.6
52.9	59.7
54.7	61.6
55.7	64.3
56.4	67.0
57.0	70.0
57.2	70.8
57.6	73.3
58.1	72.4
58.9	73.0
59.9	72.3

2. Construct a scatter plot for the data. Do the data appear to have a positive linear correlation, a negative linear correlation, or no linear correlation? Explain.

3. Calculate the correlation coefficient r and interpret the result.

4. Test the significance of the correlation coefficient r that you found in Exercise 3. Use $\alpha = 0.01$.

5. Find the equation of the regression line for the data. Draw the regression line on the scatter plot that you constructed in Exercise 2.

6. Use the regression equation that you found in Exercise 5 to predict the average annual salary of postsecondary library science teachers when the average annual salary of librarians is $56,000.

7. Find the coefficient of determination r^2 and interpret the result.

8. Find the standard error of estimate s_e and interpret the result.

9. Construct a 99% prediction interval for the average annual salary of postsecondary library science teachers when the average annual salary of librarians is $56,000. Interpret the results.

Acid rain affects the environment by increasing the acidity of lakes and streams to dangerous levels, damaging trees and soil, accelerating the decay of building materials and paint, and destroying national monuments. The goal of the Environmental Protection Agency's (EPA) Acid Rain Program is to achieve environmental health benefits by reducing the emissions of the primary causes of acid rain: sulfur dioxide and nitrogen oxides.

You work for the EPA and you want to determine whether there is a significant correlation between the average concentrations of sulfur dioxide and nitrogen dioxide.

EXERCISES

1. **Analyzing the Data**

   (a) The data in the table show the annual averages of the daily maximum concentrations of sulfur dioxide (in parts per billion) and nitrogen dioxide (in parts per billion) for 12 years. Construct a scatter plot of the data and make a conclusion about the type of correlation between the average concentrations of sulfur dioxide and nitrogen dioxide.

   (b) Calculate the correlation coefficient r and verify your conclusion in part (a).

   (c) Test the significance of the correlation coefficient found in part (b). Use $\alpha = 0.05$.

   (d) Find the equation of the regression line for the average concentrations of sulfur dioxide and nitrogen dioxide. Add the graph of the regression line to your scatter plot in part (a). Does the regression line appear to be a good fit?

   (e) Can you use the equation of the regression line to predict the average concentration of nitrogen dioxide given the average concentration of sulfur dioxide? Why or why not?

   (f) Find the coefficient of determination r^2 and the standard error of estimate s_e. Interpret your results.

2. **Making Predictions**

   Construct a 95% prediction interval for the average concentration of nitrogen dioxide when the average concentration of sulfur dioxide is 28 parts per billion. Interpret the results.

Average sulfur dioxide concentration, x	Average nitrogen dioxide concentration, y
75.6	56.3
74.9	55.9
68.9	54.9
64.7	53.2
59.0	52.2
50.8	48.0
46.3	47.5
37.9	47.9
36.7	44.8
30.5	45.9
31.9	46.9
25.3	44.6

(Source: Environmental Protection Agency)

TECHNOLOGY

MINITAB   EXCEL   TI-84 PLUS

Nutrients in Breakfast Cereals

U.S. Food and Drug
Administration

C	S	F	R
100	12	0.5	25
130	11	1.5	29
100	1	2	20
130	15	2	31
130	13	1.5	29
120	3	0.5	26
100	2	0	24
120	10	0	29
150	16	1.5	31
110	4	0	25
110	12	1	25
150	15	0	36
160	15	1.5	35
150	12	2	29
150	15	1.5	29
110	6	1	23
190	19	1.5	45
100	3	0	23
120	4	0.5	23
120	11	1.5	28
130	5	0.5	29

The U.S. Food and Drug Administration (FDA) requires nutrition labeling for most foods. Under FDA regulations, manufacturers are required to list the amounts of certain nutrients in their foods, such as calories, sugar, fat, and carbohydrates. This nutritional information is displayed in the "Nutrition Facts" panel on the food's package.

The table shows the nutritional content below for one cup of each of 21 different breakfast cereals.

C = calories
S = sugar in grams
F = fat in grams
R = carbohydrates in grams

EXERCISES

1. Use technology to draw a scatter plot of the (x, y) pairs in each data set.

(a) (calories, sugar)

(b) (calories, fat)

(c) (calories, carbohydrates)

(d) (sugar, fat)

(e) (sugar, carbohydrates)

(f) (fat, carbohydrates)

2. From the scatter plots in Exercise 1, which pairs of variables appear to have a strong linear correlation?

3. Use technology to find the correlation coefficient for each pair of variables in Exercise 1. Which has the strongest linear correlation?

4. Use technology to find an equation of a regression line for each pair of variables.

(a) (calories, sugar)

(b) (calories, carbohydrates)

5. Use the results of Exercise 4 to predict each value.

(a) The sugar content of one cup of cereal that has 120 calories

(b) The carbohydrate content of one cup of cereal that has 120 calories

6. Use technology to find the multiple regression equations of each form.

(a) $C = b + m_1 S + m_2 F + m_3 R$

(b) $C = b + m_1 S + m_2 R$

7. Use the equations from Exercise 6 to predict the calories in 1 cup of cereal that has 7 grams of sugar, 0.5 gram of fat, and 31 grams of carbohydrates.

Extended solutions are given in the technology manuals that accompany this text. Technical instruction is provided for Minitab, Excel, and the TI-84 Plus.

Chi-Square Tests and the *F*-Distribution

Crash tests performed by the Insurance Institute for Highway Safety demonstrate how a vehicle will react when in a realistic collision. Tests are performed on the front, side, rear, and roof of the vehicles. Results of these tests are classified using the ratings *good, acceptable, marginal,* and *poor.*

In Chapter 8, you learned how to test a hypothesis that compares two populations by basing your decisions on sample statistics and their distributions. For instance, the Insurance Institute for Highway Safety buys new vehicles each year and crashes them into a barrier at 40 miles per hour to compare how different vehicles protect drivers in a frontal offset crash. In this test, 40% of the total width of the vehicle strikes the barrier on the driver side. The forces and impacts that occur during a crash test are measured by equipping dummies with special instruments and placing them in the car. The crash test results include data on head, chest, and leg injuries. For a low crash test number, the injury potential is low. If the crash test number is high, then the injury potential is high. Using the techniques of Chapter 8, you can determine whether the mean chest injury potential is the same for midsize SUVs and large pickups. (Assume the populations are normally distributed and the population variances are equal.) The table shows the sample statistics. *(Adapted from Insurance Institute for Highway Safety)*

Vehicle	Number	Mean chest injury	Standard deviation
Large Pickups	$n_1 = 12$	$\bar{x}_1 = 23.0$	$s_1 = 2.09$
Midsize SUVs	$n_2 = 19$	$\bar{x}_2 = 22.4$	$s_2 = 4.26$

For the means of chest injury, the P-value for the hypothesis that $\mu_1 = \mu_2$ is about 0.6655. At $\alpha = 0.01$, you fail to reject the null hypothesis. So, you do not have enough evidence to conclude that there is a significant difference in the means of the chest injury potential in a frontal offset crash at 40 miles per hour for large pickups and midsize SUVs.

In this chapter, you will learn how to test a hypothesis that compares three or more populations.

For instance, in addition to the crash tests for large pickups and midsize SUVs, a third group of vehicles was also tested. The table shows the results for all three types of vehicles.

Vehicle	Number	Mean chest injury	Standard deviation
Large Pickups	$n_1 = 12$	$\bar{x}_1 = 23.0$	$s_1 = 2.09$
Midsize SUVs	$n_2 = 19$	$\bar{x}_2 = 22.4$	$s_2 = 4.26$
Large Cars	$n_3 = 10$	$\bar{x}_3 = 27.2$	$s_3 = 6.65$

From these three samples, is there evidence of a difference in chest injury potential among large pickups, midsize SUVs, and large cars in a frontal offset crash at 40 miles per hour?

You can answer this question by testing the hypothesis that the three means are equal. For the means of chest injury, the P-value for the hypothesis that $\mu_1 = \mu_2 = \mu_3$ is about 0.0283. At $\alpha = 0.01$, you fail to reject the null hypothesis. So, there is not enough evidence at the 1% level of significance to conclude that at least one of the means is different from the others.

10.1 Goodness-of-Fit Test

The Chi-Square Goodness-of-Fit Test

The Chi-Square Goodness-of-Fit Test

A tax preparation company wants to determine the proportions of people who used different methods to prepare their taxes. To determine these proportions, the company can perform a multinomial experiment. A **multinomial experiment** is a probability experiment consisting of a fixed number of independent trials in which there are more than two possible outcomes for each trial. The probability of each outcome is fixed, and each outcome is classified into **categories.** (Remember from Section 4.2 that a binomial experiment has only two possible outcomes.)

The company wants to test a retail trade association's claim concerning the expected distribution of proportions of people who used different methods to prepare their taxes. To do so, the company could compare the distribution of proportions obtained in the multinomial experiment with the association's expected distribution. To compare the distributions, the company can perform a **chi-square goodness-of-fit test.**

Study Tip

The hypothesis tests described in Sections 10.1 and 10.2 can be used for qualitative data.

> **DEFINITION**
>
> A **chi-square goodness-of-fit test** is used to test whether a frequency distribution fits an expected distribution.

To begin a goodness-of-fit test, you must first state a null and an alternative hypothesis. Generally, the null hypothesis states that the frequency distribution fits an expected distribution and the alternative hypothesis states that the frequency distribution does not fit the expected distribution.

For instance, the association claims that the expected distribution of people who used different methods to prepare their taxes is as shown below.

Distribution of tax preparation methods	
Accountant	24%
By hand	20%
Computer software	35%
Friend/family	6%
Tax preparation service	15%

To test the association's claim, the company can perform a chi-square goodness-of-fit test using these null and alternative hypotheses.

H_0: The expected distribution of tax preparation methods is 24% by accountant, 20% by hand, 35% by computer software, 6% by friend or family, and 15% by tax preparation service. (Claim)

H_a: The distribution of tax preparation methods differs from the expected distribution.

To calculate the test statistic for the chi-square goodness-of-fit test, you can use **observed frequencies** and **expected frequencies.** To calculate the expected frequencies, you must assume the null hypothesis is true.

DEFINITION

The **observed frequency** O of a category is the frequency for the category observed in the sample data.

The **expected frequency** E of a category is the *calculated* frequency for the category. Expected frequencies are found by using the expected (or hypothesized) distribution and the sample size. The expected frequency for the ith category is

$$E_i = np_i$$

where n is the number of trials (the sample size) and p_i is the assumed probability of the ith category.

EXAMPLE 1

Finding Observed Frequencies and Expected Frequencies

A tax preparation company randomly selects 300 adults and asks them how they prepare their taxes. The results are shown at the right. Find the observed frequency and the expected frequency (using the distribution on the preceding page) for each tax preparation method. *(Adapted from National Retail Federation)*

Survey results ($n = 300$)	
Accountant	63
By hand	40
Computer software	115
Friend/family	29
Tax preparation service	53

SOLUTION

The observed frequency for each tax preparation method is the number of adults in the survey naming a particular tax preparation method. The expected frequency for each tax preparation method is the product of the number of adults in the survey and the assumed probability that an adult will name a particular tax preparation method. The observed frequencies and expected frequencies are shown in the table below.

Tax preparation method	% of people	Observed frequency	Expected frequency
Accountant	24%	63	$300(0.24) = 72$
By hand	20%	40	$300(0.20) = 60$
Computer software	35%	115	$300(0.35) = 105$
Friend/family	6%	29	$300(0.06) = 18$
Tax preparation service	15%	53	$300(0.15) = 45$

TRY IT YOURSELF 1

The tax preparation company in Example 1 decides it wants a larger sample size, so it randomly selects 500 adults. Find the expected frequency for each tax preparation method for $n = 500$.

Answer: Page A38

The sum of the expected frequencies always equals the sum of the observed frequencies. For instance, in Example 1 the sum of the observed frequencies and the sum of the expected frequencies are both 300.

Before performing a chi-square goodness-of-fit test, you must verify that (1) the observed frequencies were obtained from a random sample and (2) each expected frequency is at least 5. Note that when the expected frequency of a category is less than 5, it may be possible to combine the category with another one to meet the second requirement.

Study Tip

Remember that a chi-square distribution is positively skewed and its shape is determined by the degrees of freedom. Its graph is not symmetric, but it appears to become more symmetric as the degrees of freedom increase, as shown in Section 6.4.

The Chi-Square Goodness-of-Fit Test

To perform a chi-square goodness-of-fit test, these conditions must be met.

1. The observed frequencies must be obtained using a random sample.

2. Each expected frequency must be greater than or equal to 5.

If these conditions are met, then the sampling distribution for the test is approximated by a chi-square distribution with $k - 1$ degrees of freedom, where k is the number of categories. The **test statistic** is

$$\chi^2 = \Sigma \frac{(O - E)^2}{E}$$

where O represents the observed frequency of each category and E represents the expected frequency of each category.

When the observed frequencies closely match the expected frequencies, the differences between O and E will be small and the chi-square test statistic will be close to 0. As such, the null hypothesis is unlikely to be rejected. However, when there are large discrepancies between the observed frequencies and the expected frequencies, the differences between O and E will be large, resulting in a large chi-square test statistic. A large chi-square test statistic is evidence for rejecting the null hypothesis. So, the chi-square goodness-of-fit test is always a right-tailed test.

GUIDELINES

Performing a Chi-Square Goodness-of-Fit Test

In Words	In Symbols
1. Verify that the observed frequencies were obtained from a random sample and each expected frequency is at least 5.	$E_i = np_i \geq 5$
2. Identify the claim. State the null and alternative hypotheses.	State H_0 and H_a.
3. Specify the level of significance.	Identify α.
4. Identify the degrees of freedom.	d.f. $= k - 1$
5. Determine the critical value.	Use Table 6 in Appendix B.
6. Determine the rejection region.	
7. Find the test statistic and sketch the sampling distribution.	$\chi^2 = \Sigma \dfrac{(O - E)^2}{E}$
8. Make a decision to reject or fail to reject the null hypothesis.	If χ^2 is in the rejection region, then reject H_0. Otherwise, fail to reject H_0.
9. Interpret the decision in the context of the original claim.	

EXAMPLE 2

Performing a Chi-Square Goodness-of-Fit Test

A retail trade association claims that the tax preparation methods of adults are distributed as shown in the table at the left below. A tax preparation company randomly selects 300 adults and asks them how they prepare their taxes. The results are shown in the table at the right below. At $\alpha = 0.01$, test the association's claim. *(Adapted from National Retail Federation)*

Distribution of tax preparation methods	
Accountant	24%
By hand	20%
Computer software	35%
Friend/family	6%
Tax preparation service	15%

Survey results ($n = 300$)	
Accountant	63
By hand	40
Computer software	115
Friend/family	29
Tax preparation service	53

SOLUTION

The observed and expected frequencies are shown in the table at the left. The expected frequencies were calculated in Example 1. Because the observed frequencies were obtained using a random sample and each expected frequency is at least 5, you can use the chi-square goodness-of-fit test to test the proposed distribution. Here are the null and alternative hypotheses.

Tax preparation method	Observed frequency	Expected frequency
Accountant	63	72
By hand	40	60
Computer software	115	105
Friend/ family	29	18
Tax preparation service	53	45

H_0: The expected distribution of tax preparation methods is 24% by accountant, 20% by hand, 35% by computer software, 6% by friend or family, and 15% by tax preparation service. (Claim)

H_a: The distribution of tax preparation methods differs from the expected distribution.

Because there are 5 categories, the chi-square distribution has

$$\text{d.f.} = k - 1 = 5 - 1 = 4$$

degrees of freedom. With d.f. $= 4$ and $\alpha = 0.01$, the critical value is $\chi_0^2 = 13.277$. The rejection region is

$$\chi^2 > 13.277. \qquad \text{Rejection region}$$

With the observed and expected frequencies, the chi-square test statistic is

$$\begin{aligned} \chi^2 &= \Sigma \frac{(O - E)^2}{E} \\ &= \frac{(63 - 72)^2}{72} + \frac{(40 - 60)^2}{60} + \frac{(115 - 105)^2}{105} \\ &\quad + \frac{(29 - 18)^2}{18} + \frac{(53 - 45)^2}{45} \\ &\approx 16.888. \end{aligned}$$

The figure at the left shows the location of the rejection region and the chi-square test statistic. Because χ^2 is in the rejection region, you reject the null hypothesis.

Interpretation   There is enough evidence at the 1% level of significance to reject the claim that the distribution of tax preparation methods and the association's expected distribution are the same.

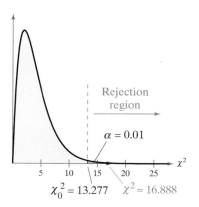

Rejection region

$\alpha = 0.01$

$\chi_0^2 = 13.277$ $\chi^2 \approx 16.888$

Ages	Previous age distribution	Survey results
0–9	16%	76
10–19	20%	84
20–29	8%	30
30–39	14%	60
40–49	15%	54
50–59	12%	40
60–69	10%	42
70+	5%	14

TRY IT YOURSELF 2

A sociologist claims that the age distribution for the residents of a city is different from the distribution 10 years ago. The distribution of ages 10 years ago is shown in the table at the left. You randomly select 400 residents and record the age of each. The survey results are shown in the table. At $\alpha = 0.05$, perform a chi-square goodness-of-fit test to test whether the distribution has changed.

Answer: Page A38

The chi-square goodness-of-fit test is often used to determine whether a distribution is uniform. For such tests, the expected frequencies of the categories are equal. When testing a uniform distribution, you can find the expected frequency of each category by dividing the sample size by the number of categories. For instance, suppose a company believes that the number of sales made by its sales force is uniform throughout a five-day workweek. If the sample consists of 1000 sales, then the expected value of the sales for each day will be $1000/5 = 200$.

EXAMPLE 3

Performing a Chi-Square Goodness-of-Fit Test

A researcher claims that the number of different-colored candies in bags of dark chocolate M&M's® is uniformly distributed. To test this claim, you randomly select a bag that contains 500 dark chocolate M&M's®. The results are shown in the table below. At $\alpha = 0.10$, test the researcher's claim. *(Adapted from Mars, Incorporated)*

Color	Frequency, f
Brown	80
Yellow	95
Red	88
Blue	83
Orange	76
Green	78

SOLUTION

The claim is that the distribution is uniform, so the expected frequencies of the colors are equal. To find each expected frequency, divide the sample size by the number of colors. So, for each color, $E = 500/6 \approx 83.333$. Because each expected frequency is at least 5 and the M&M's® were randomly selected, you can use the chi-square goodness-of-fit test to test the expected distribution. Here are the null and alternative hypotheses.

H_0: The expected distribution of the different-colored candies in bags of dark chocolate M&M's® is uniform. (Claim)

H_a: The distribution of the different-colored candies in bags of dark chocolate M&M's® is not uniform.

Because there are 6 categories, the chi-square distribution has

d.f. $= k - 1 = 6 - 1 = 5$

degrees of freedom. Using d.f. $= 5$ and $\alpha = 0.10$, the critical value is $\chi_0^2 = 9.236$. The rejection region is $\chi^2 > 9.236$. To find the chi-square test statistic using a table, use the observed and expected frequencies, as shown on the next page.

O	E	$O - E$	$(O - E)^2$	$\dfrac{(O - E)^2}{E}$
80	83.333	−3.333	11.108889	0.133307201
95	83.333	11.667	136.118889	1.633433202
88	83.333	4.667	21.780889	0.261371713
83	83.333	−0.333	0.110889	0.001330673
76	83.333	−7.333	53.772889	0.645277249
78	83.333	−5.333	28.440889	0.341292033
				$\chi^2 = \Sigma\dfrac{(O - E)^2}{E} \approx 3.016$

The figure shows the location of the rejection region and the chi-square test statistic. Because χ^2 is not in the rejection region, you fail to reject the null hypothesis.

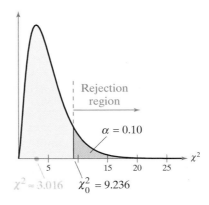

Interpretation   There is not enough evidence at the 10% level of significance to reject the claim that the distribution of the different-colored candies in bags of dark chocolate M&M's® is uniform.

TRY IT YOURSELF 3

A researcher claims that the number of different-colored candies in bags of peanut M&M's® is uniformly distributed. To test this claim, you randomly select a bag that contains 180 peanut M&M's®. The results are shown in the table below. Using $\alpha = 0.05$, test the researcher's claim. *(Adapted from Mars, Incorporated)*

Color	Frequency, f
Brown	22
Yellow	27
Red	22
Blue	41
Orange	41
Green	27

Answer: Page A38

TI-84 PLUS

```
χ²GOF-Test
χ²=3.016012072
p=.6975171071
df=5
CNTRB={.133307...
```

You can use technology and a *P*-value to perform a chi-square goodness-of-fit test. For instance, using a TI-84 Plus and the data in Example 3, you obtain $P = 0.6975171071$, as shown at the left. Because $P > \alpha$, you fail to reject the null hypothesis.

10.1 EXERCISES

For Extra Help: **MyLab Statistics**

Building Basic Skills and Vocabulary

1. What is a multinomial experiment?

2. What conditions are necessary to use the chi-square goodness-of-fit test?

Finding Expected Frequencies *In Exercises 3–6, find the expected frequency for the values of n and* p_i*.*

3. $n = 150, p_i = 0.3$

4. $n = 500, p_i = 0.9$

5. $n = 230, p_i = 0.25$

6. $n = 415, p_i = 0.08$

Using and Interpreting Concepts

Performing a Chi-Square Goodness-of-Fit Test *In Exercises 7–16, (a) identify the claim and state H_0 and H_a, (b) find the critical value and identify the rejection region, (c) find the chi-square test statistic, (d) decide whether to reject or fail to reject the null hypothesis, and (e) interpret the decision in the context of the original claim.*

7. Ages of Moviegoers A researcher claims that the ages of people who go to movies at least once a month are distributed as shown in the figure. You randomly select 1000 people who go to movies at least once a month and record the age of each. The table shows the results. At $\alpha = 0.10$, test the researcher's claim. *(Source: Motion Picture Association of America)*

Survey results	
Age	**Frequency, *f***
2–17	240
18–24	209
25–39	203
40–49	106
50+	242

8. Coffee A researcher claims that the numbers of cups of coffee U.S. adults drink per day are distributed as shown in the figure. You randomly select 1600 U.S. adults and ask them how many cups of coffee they drink per day. The table shows the results. At $\alpha = 0.05$, test the researcher's claim. *(Source: Gallup)*

Survey results	
Response	**Frequency, *f***
0 cups	570
1 cup	432
2 cups	282
3 cups	152
4 or more cups	164

Left margin answers:

1. A multinomial experiment is a probability experiment consisting of a fixed number of independent trials in which there are more than two possible outcomes for each trial. The probability of each outcome is fixed, and each outcome is classified into categories.

2. The observed frequencies must be obtained using a random sample, and each expected frequency must be greater than or equal to 5.

3. 45   **4.** 450   **5.** 57.5   **6.** 33.2

7. (a) H_0: The distribution of the ages of moviegoers is 23% ages 2–17, 20% ages 18–24, 22% ages 25–39, 9% ages 40–49, and 26% ages 50+. (claim)

H_a: The distribution of ages differs from the expected distribution.

(b) $\chi_0^2 = 7.779$
Rejection region: $\chi^2 > 7.779$

(c) 6.244   (d) Fail to reject H_0.

(e) There is not enough evidence at the 10% level of significance to reject the claim that the distribution of the ages of moviegoers and the expected distribution are the same.

8. (a) H_0: The distribution of the number of cups of coffee U.S. adults drink per day is 36% 0 cups, 26% 1 cup, 19% 2 cups, 8% 3 cups, and 11% 4 or more cups. (claim)

H_a: The distribution of amounts differs from the expected distribution.

(b) $\chi_0^2 = 9.488$
Rejection region: $\chi_0^2 > 9.488$

(c) 7.588   (d) Fail to reject H_0.

(e) There is not enough evidence at the 5% level of significance to reject the claim that the distribution of the number of cups of coffee U.S. adults drink per day and the expected distribution are the same.

9. (a) H_0: The distribution of the days people order food for delivery is 7% Sunday, 4% Monday, 6% Tuesday, 13% Wednesday, 10% Thursday, 36% Friday, and 24% Saturday.

  H_a: The distribution of days differs from the expected distribution. (claim)

  (b) $\chi_0^2 = 16.812$
  Rejection region: $\chi^2 > 16.812$

  (c) 17.595   (d) Reject H_0.

  (e) There is enough evidence at the 1% level of significance to conclude that the distribution of days differs from the expected distribution.

10. (a) H_0: The distribution of people who use cash to make their purchases is 19% all purchases, 17% most purchases, 20% half of purchases, 33% some purchases, and 11% no purchases.

  H_a: The distribution of people who use cash to make their purchases differs from the expected distribution. (claim)

  (b) $\chi_0^2 = 13.277$
  Rejection region: $\chi^2 > 13.277$

  (c) 45.228   (d) Reject H_0.

  (e) There is enough evidence at the 1% level of significance to conclude that the distribution of people who use cash to make purchases differs from the expected distribution.

11. (a) H_0: The distribution of the number of homicide crimes in California by county is uniform. (claim)

  H_a: The distribution of homicides by county is not uniform.

  (b) $\chi_0^2 = 30.578$
  Rejection region: $\chi^2 > 30.578$

  (c) 143.904   (d) Reject H_0.

  (e) There is enough evidence at the 1% level of significance to reject the claim that the distribution of the number of homicide crimes in California by county is uniform.

9. **Ordering Delivery** A research firm claims that the distribution of the days of the week that people are most likely to order food for delivery is different from the distribution shown in the figure. You randomly select 500 people and record which day of the week each is most likely to order food for delivery. The table shows the results. At $\alpha = 0.01$, test the research firm's claim. *(Source: Technomic, Inc.)*

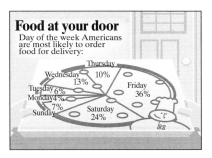

Survey results	
Day	Frequency, f
Sunday	43
Monday	16
Tuesday	25
Wednesday	49
Thursday	46
Friday	168
Saturday	153

10. **Going Cashless** A financial analyst claims that the distribution of people who use cash to make their purchases is different from the distribution shown in the figure. You randomly select 600 people and record the way they make purchases. The table shows the results. At $\alpha = 0.01$, test the financial analyst's claim. *(Adapted from Gallup)*

Making purchases

All purchases with cash 19%
Most purchases with cash 17%
Half of purchases with cash 20%
Some purchases with cash 33%
No purchases with cash 11%

Survey results	
Response	Frequency, f
All purchases with cash	60
Most purchases with cash	84
Half of purchases with cash	132
Some purchases with cash	252
No purchases with cash	72

11. **Homicides by County** A researcher claims that the number of homicide crimes in California by county is uniformly distributed. To test this claim, you randomly select 1000 homicides from a recent year and record the county in which each happened. The table shows the results. At $\alpha = 0.01$, test the researcher's claim. *(Adapted from California Department of Justice)*

County	Frequency, f	County	Frequency, f
Alameda	116	Sacramento	90
Contra Costa	55	San Bernardino	89
Fresno	57	San Diego	45
Kern	62	San Francisco	51
Los Angeles	101	San Joaquin	62
Monterey	58	Santa Clara	39
Orange	30	Stanislaus	37
Riverside	65	Tulare	43

12. (a) H_0: The distribution of the number of homicide crimes in California by month is uniform. (claim)

H_a: The distribution of homicides by month is not uniform.

(b) $\chi_0^2 = 17.275$
Rejection region: $\chi^2 > 17.275$

(c) 23.947   (d) Reject H_0.

(e) There is enough evidence at the 10% level of significance to reject the claim that the distribution of the number of homicide crimes in California by month is uniform.

13. (a) H_0: The distribution of the opinions of U.S. parents on whether a college education is worth the expense is 55% strongly agree, 30% somewhat agree, 5% neither agree nor disagree, 6% somewhat disagree, and 4% strongly disagree.

H_a: The distribution of opinions differs from the expected distribution. (claim)

(b) $\chi_0^2 = 9.488$
Rejection region: $\chi^2 > 9.488$

(c) 65.236   (d) Reject H_0.

(e) There is enough evidence at the 5% level of significance to conclude that the distribution of the opinions of U.S. parents on whether a college education is worth the expense differs from the expected distribution.

14. (a) See Selected Answers, page A107.

(b) $\chi_0^2 = 6.251$
Rejection region: $\chi^2 > 6.251$

(c) 29.057   (d) Reject H_0.

(e) There is enough evidence at the 10% level of significance to reject the claim that the distribution of how much married U.S. female adults trust their spouses to manage their finances is the same as the distribution of how much married U.S. male adults trust their spouses to manage their finances.

12. Homicides by Month   A researcher claims that the number of homicide crimes in California by month is uniformly distributed. To test this claim, you randomly select 1800 homicides from a recent year and record the month when each happened. The table shows the results. At $\alpha = 0.10$, test the researcher's claim. *(Adapted from California Department of Justice)*

Month	Frequency, f	Month	Frequency, f
January	135	July	164
February	112	August	161
March	141	September	168
April	132	October	162
May	141	November	148
June	168	December	168

13. College Education   The pie chart shows the results of a survey in which U.S. parents were asked their opinions on whether a college education is worth the expense. An economist claims that the distribution of the opinions of U.S. teenagers is different from the distribution given for U.S. parents. To test this claim, you randomly select 200 U.S. teenagers and ask each whether a college education is worth the expense. The table shows the results. At $\alpha = 0.05$, test the economist's claim. *(Adapted from Upromise, Inc.)*

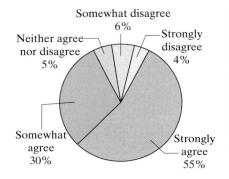

Survey results	
Response	**Frequency, f**
Strongly agree	86
Somewhat agree	62
Neither agree nor disagree	34
Somewhat disagree	14
Strongly disagree	4

14. Money Management   The pie chart shows the results of a survey in which married U.S. male adults were asked how much they trust their spouses to manage their finances. A financial services company claims that the distribution of how much married U.S. female adults trust their spouses to manage their finances is the same as the distribution given for married U.S. male adults. To test this claim, you randomly select 400 married U.S. female adults and ask each how much she trusts her spouse to manage their finances. The table shows the results. At $\alpha = 0.10$, test the company's claim. *(Adapted from Country Financial)*

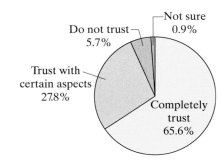

Survey results	
Response	**Frequency, f**
Completely trust	243
Trust with certain aspects	108
Do not trust	36
Not sure	13

Response	Frequency, f
Larger	285
Same size	224
Smaller	291

TABLE FOR EXERCISE 15

15. (a) H_0: The distribution of prospective home buyers by the size they want their next house to be is uniform.
    H_a: The distribution of prospective home buyers by the size they want their next house to be is not uniform. (claim)
    (b) $\chi_0^2 = 5.991$
        Rejection region: $\chi^2 > 5.991$
    (c) 10.308   (d) Reject H_0.
    (e) There is enough evidence at the 5% level of significance to conclude that the distribution of prospective home buyers by the size they want their next house to be is not uniform.

16. (a) H_0: The distribution of the number of births by day of the week is uniform. (claim)
    H_a: The distribution of the number of births is not uniform.
    (b) $\chi_0^2 = 10.645$
        Rejection region: $\chi^2 > 10.645$
    (c) 21.960   (d) Reject H_0.
    (e) There is enough evidence at the 10% level of significance to reject the claim that the distribution of the number of births by day of the week is uniform.

17. (a) The expected frequencies are 17, 63, 79, 34, and 5.
    (b) $\chi_0^2 = 13.277$
        Rejection region: $\chi^2 > 13.277$
    (c) 0.613   (d) Fail to reject H_0.
    (e) There is not enough evidence at the 1% level of significance to reject the claim that the test scores are normally distributed.

18. (a) The expected frequencies are 27, 103, 156, 90, and 20.
    (b) $\chi_0^2 = 9.488$
        Rejection region: $\chi^2 > 9.488$
    (c) 1.029   (d) Fail to reject H_0.
    (e) See Selected Answers, page A107.

15. **Home Sizes**   An organization claims that the number of prospective home buyers who want their next house to be larger, smaller, or the same size as their current house is not uniformly distributed. To test this claim, you randomly select 800 prospective home buyers and ask them what size they want their next house to be. The table at the left shows the results. At $\alpha = 0.05$, test the organization's claim. *(Adapted from Better Homes and Gardens)*

16. **Births by Day of the Week**   A doctor claims that the number of births by day of the week is uniformly distributed. To test this claim, you randomly select 700 births from a recent year and record the day of the week on which each takes place. The table shows the results. At $\alpha = 0.10$, test the doctor's claim. *(Adapted from National Center for Health Statistics)*

Day	Frequency, f
Sunday	68
Monday	108
Tuesday	115
Wednesday	113
Thursday	111
Friday	108
Saturday	77

Extending Concepts

Testing for Normality   *Using a chi-square goodness-of-fit test, you can decide, with some degree of certainty, whether a variable is normally distributed. In all chi-square tests for normality, the null and alternative hypotheses are as listed below.*

H_0: *The variable has a normal distribution.*

H_a: *The variable does not have a normal distribution.*

To determine the expected frequencies when performing a chi-square test for normality, first find the mean and standard deviation of the frequency distribution. Then, use the mean and standard deviation to compute the z-score for each class boundary. Then, use the z-scores to calculate the area under the standard normal curve for each class. Multiplying the resulting class areas by the sample size yields the expected frequency for each class.

In Exercises 17 and 18, (a) find the expected frequencies, (b) find the critical value and identify the rejection region, (c) find the chi-square test statistic, (d) decide whether to reject or fail to reject the null hypothesis, and (e) interpret the decision in the context of the original claim.

17. **Test Scores**   At $\alpha = 0.01$, test the claim that the 200 test scores shown in the frequency distribution are normally distributed.

Class boundaries	49.5–58.5	58.5–67.5	67.5–76.5	76.5–85.5	85.5–94.5
Frequency, f	19	61	82	34	4

18. **Test Scores**   At $\alpha = 0.05$, test the claim that the 400 test scores shown in the frequency distribution are normally distributed.

Class boundaries	50.5–60.5	60.5–70.5	70.5–80.5	80.5–90.5	90.5–100.5
Frequency, f	28	106	151	97	18

10.2 Independence

Contingency Tables ■ The Chi-Square Independence Test

What You Should Learn

▶ How to use a contingency table to find expected frequencies

▶ How to use a chi-square distribution to test whether two variables are independent

Contingency Tables

In Section 3.2, you learned that two events are *independent* when the occurrence of one event does not affect the probability of the occurrence of the other event. For instance, the outcomes of a roll of a die and a toss of a coin are independent. But, suppose a medical researcher wants to determine whether there is a relationship between caffeine consumption and heart attack risk. Are these variables independent or are they dependent? In this section, you will learn how to use the chi-square test for independence to answer such a question. To perform a chi-square test for independence, you will use sample data that are organized in a **contingency table.**

DEFINITION

An *r* × *c* **contingency table** shows the observed frequencies for two variables. The observed frequencies are arranged in *r* rows and *c* columns. The intersection of a row and a column is called a **cell.**

Study Tip

Note that "2 × 5" is read as "two-by-five."

A 2 × 5 contingency table is shown below. It has two rows and five columns and shows the results of a random sample of 2197 adults classified by two variables, *favorite way to eat ice cream* and *gender*. From the table, you can see that 182 of the adults who prefer ice cream in a sundae are males, and 158 of the adults who prefer ice cream in a sundae are females.

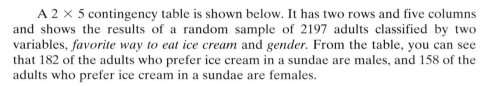

Gender	Cup	Cone	Sundae	Sandwich	Other
Male	504	287	182	43	53
Female	474	401	158	45	50

(Adapted from The Harris Poll)

Assuming two variables are independent, you can use a contingency table to find the expected frequency for each cell, as shown in the next definition.

Study Tip

In a contingency table, the notation $E_{r,c}$ represents the expected frequency for the cell in row *r*, column *c*. For instance, in the table above, $E_{1,4}$ represents the expected frequency for the cell in row 1, column 4.

Finding the Expected Frequency for Contingency Table Cells

The expected frequency for a cell $E_{r,c}$ in a contingency table is

$$\text{Expected frequency } E_{r,c} = \frac{(\text{Sum of row } r) \cdot (\text{Sum of column } c)}{\text{Sample size}}.$$

When you find the sum of each row and column in a contingency table, you are calculating the **marginal frequencies.** A marginal frequency is the frequency that an entire category of one of the variables occurs. For instance, in the table above, the marginal frequency for adults who prefer ice cream in a cone is 287 + 401 = 688. The observed frequencies in the interior of a contingency table are called **joint frequencies.**

In Example 1, notice that the marginal frequencies for the contingency table have already been calculated.

EXAMPLE 1

Finding Expected Frequencies

Find the expected frequency for each cell in the contingency table. Assume that the variables *favorite way to eat ice cream* and *gender* are independent.

Gender	\multicolumn{6}{c}{**Favorite way to eat ice cream**}
	Cup	**Cone**	**Sundae**	**Sandwich**	**Other**	**Total**
Male	504	287	182	43	53	1069
Female	474	401	158	45	50	1128
Total	978	688	340	88	103	2197

SOLUTION

After calculating the marginal frequencies, you can use the formula

$$\text{Expected frequency } E_{r,c} = \frac{(\text{Sum of row } r)\cdot(\text{Sum of column } c)}{\text{Sample size}}$$

to find each expected frequency. The expected frequencies for the first row are

$$E_{1,1} = \frac{1069 \cdot 978}{2197} \approx 475.868 \qquad E_{1,2} = \frac{1069 \cdot 688}{2197} \approx 334.762$$

$$E_{1,3} = \frac{1069 \cdot 340}{2197} \approx 165.435 \qquad E_{1,4} = \frac{1069 \cdot 88}{2197} \approx 42.818$$

$$E_{1,5} = \frac{1069 \cdot 103}{2197} \approx 50.117$$

and the expected frequencies for the second row are

$$E_{2,1} = \frac{1128 \cdot 978}{2197} \approx 502.132 \qquad E_{2,2} = \frac{1128 \cdot 688}{2197} \approx 353.238$$

$$E_{2,3} = \frac{1128 \cdot 340}{2197} \approx 174.565 \qquad E_{2,4} = \frac{1128 \cdot 88}{2197} \approx 45.182$$

$$E_{2,5} = \frac{1128 \cdot 103}{2197} \approx 52.883.$$

Study Tip

In Example 1, after finding $E_{1,1} \approx 475.868$, you can find $E_{2,1}$ by subtracting 475.868 from the first column's total, 978. So, $E_{2,1} \approx 978 - 475.868 = 502.132$. In general, you can find the expected value for the last cell in a column by subtracting the expected values for the other cells in that column from the column's total. Similarly, you can do this for the last cell in a row using the row's total.

TRY IT YOURSELF 1

The marketing consultant for a travel agency wants to determine whether certain travel concerns are related to travel purpose. The contingency table shows the results of a random sample of 300 travelers classified by their primary travel concern and travel purpose. Assume that the variables *travel concern* and *travel purpose* are independent. Find the expected frequency for each cell. *(Adapted from NPD Group for Embassy Suites)*

Travel purpose	\multicolumn{4}{c}{**Travel concern**}
	Hotel room	**Leg room on plane**	**Rental car size**	**Other**
Business	36	108	14	22
Leisure	38	54	14	14

Answer: Page A38

The Chi-Square Independence Test

After finding the expected frequencies, you can test whether the variables are independent using a **chi-square independence test.**

> ### DEFINITION
>
> A **chi-square independence test** is used to test the independence of two variables. Using this test, you can determine whether the occurrence of one variable affects the probability of the occurrence of the other variable.

Before performing a chi-square independence test, you must verify that (1) the observed frequencies were obtained from a random sample and (2) each expected frequency is at least 5.

> ### The Chi-Square Independence Test
>
> To perform a chi-square independence test, these conditions must be met.
>
> **1.** The observed frequencies must be obtained using a random sample.
>
> **2.** Each expected frequency must be greater than or equal to 5.
>
> If these conditions are met, then the sampling distribution for the test is approximated by a chi-square distribution with
>
> $$\text{d.f.} = (r - 1)(c - 1)$$
>
> degrees of freedom, where r and c are the number of rows and columns, respectively, of a contingency table. The **test statistic** is
>
> $$\chi^2 = \Sigma \frac{(O - E)^2}{E}$$
>
> where O represents the observed frequencies and E represents the expected frequencies.

To begin the independence test, you must first state a null hypothesis and an alternative hypothesis. For a chi-square independence test, the null and alternative hypotheses are always some variation of these statements.

H_0: The variables are independent.

H_a: The variables are dependent.

The expected frequencies are calculated on the assumption that the two variables are independent. If the variables are independent, then you can expect little difference between the observed frequencies and the expected frequencies. When the observed frequencies closely match the expected frequencies, the differences between O and E will be small and the chi-square test statistic will be close to 0. As such, the null hypothesis is unlikely to be rejected.

For dependent variables, however, there will be large discrepancies between the observed frequencies and the expected frequencies. When the differences between O and E are large, the chi-square test statistic is also large. A large chi-square test statistic is evidence for rejecting the null hypothesis. So, the chi-square independence test is always a right-tailed test.

Picturing the World

A researcher wants to determine whether a relationship exists between where people work (workplace or home) and their educational attainment. The results of a random sample of 925 employed persons are shown in the contingency table. (Source: U.S. Bureau of Labor Statistics)

	Where they work	
Educational attainment	**Workplace**	**Home**
Less than high school	35	2
High school diploma	250	21
Some college	226	30
BA degree or higher	293	68

Can the researcher use this sample to test for independence using a chi-square independence test? Why or why not?

No, because one of the expected frequencies, $E_{1,2} = 4.84$, is less than 5.

Study Tip

A contingency table with three rows and four columns will have

$$(3 - 1)(4 - 1) = (2)(3)$$
$$= 6 \text{ d.f.}$$

GUIDELINES

Performing a Chi-Square Independence Test

In Words	In Symbols
1. Verify that the observed frequencies were obtained from a random sample and each expected frequency is at least 5.	
2. Identify the claim. State the null and alternative hypotheses.	State H_0 and H_a.
3. Specify the level of significance.	Identify α.
4. Determine the degrees of freedom.	d.f. $= (r - 1)(c - 1)$
5. Determine the critical value.	Use Table 6 in Appendix B.
6. Determine the rejection region.	
7. Find the test statistic and sketch the sampling distribution.	$\chi^2 = \Sigma \dfrac{(O - E)^2}{E}$
8. Make a decision to reject or fail to reject the null hypothesis.	If χ^2 is in the rejection region, then reject H_0. Otherwise, fail to reject H_0.
9. Interpret the decision in the context of the original claim.	

EXAMPLE 2

Performing a Chi-Square Independence Test

The contingency table shows the results of a random sample of 2197 adults classified by their favorite way to eat ice cream and gender. The expected frequencies are displayed in parentheses. At $\alpha = 0.01$, can you conclude that the variables *favorite way to eat ice cream* and *gender* are related?

Gender	Favorite way to eat ice cream					Total
	Cup	**Cone**	**Sundae**	**Sandwich**	**Other**	**Total**
Male	504 (475.868)	287 (334.762)	182 (165.435)	43 (42.818)	53 (50.117)	1069
Female	474 (502.132)	401 (353.238)	158 (174.565)	45 (45.182)	50 (52.883)	1128
Total	978	688	340	88	103	2197

SOLUTION

The expected frequencies were calculated in Example 1. Because each expected frequency is at least 5 and the adults were randomly selected, you can use the chi-square independence test to test whether the variables are independent. Here are the null and alternative hypotheses.

H_0: The variables *favorite way to eat ice cream* and *gender* are independent.

H_a: The variables *favorite way to eat ice cream* and *gender* are dependent. (Claim)

The contingency table has two rows and five columns, so the chi-square distribution has

$$\text{d.f.} = (r - 1)(c - 1) = (2 - 1)(5 - 1) = 4$$

degrees of freedom. Because d.f. = 4 and $\alpha = 0.01$, the critical value is $\chi_0^2 = 13.277$. The rejection region is $\chi^2 > 13.277$. You can use a table to find the chi-square test statistic, as shown below.

O	E	$O - E$	$(O - E)^2$	$\dfrac{(O - E)^2}{E}$
504	475.868	28.132	791.409424	1.663086032
287	334.762	−47.762	2281.208644	6.814419331
182	165.435	16.565	274.399225	1.658652794
43	42.818	0.182	0.033124	0.0007736
53	50.117	2.883	8.311689	0.165845701
474	502.132	−28.132	791.409424	1.576098365
401	353.238	47.762	2281.208644	6.457993319
158	174.565	−16.565	274.399225	1.571902873
45	45.182	−0.182	0.033124	0.000733124
50	52.883	−2.883	8.311689	0.157171284
				$\chi^2 = \Sigma \dfrac{(O - E)^2}{E} \approx 20.067$

The figure at the right shows the location of the rejection region and the chi-square test statistic. Because

$$\chi^2 \approx 20.067$$

is in the rejection region, you reject the null hypothesis.

Interpretation There is enough evidence at the 1% level of significance to conclude that the variables *favorite way to eat ice cream* and *gender* are dependent.

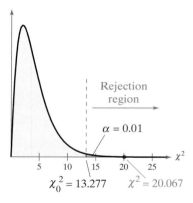

TRY IT YOURSELF 2

The marketing consultant for a travel agency wants to determine whether travel concerns are related to travel purpose. The contingency table shows the results of a random sample of 300 travelers classified by their primary travel concern and travel purpose. At $\alpha = 0.01$, can the consultant conclude that the variables *travel concern* and *travel purpose* are related? (The expected frequencies are displayed in parentheses.) *(Adapted from NPD Group for Embassy Suites)*

Travel purpose	Travel concern				Total
	Hotel room	Leg room on plane	Rental car size	Other	
Business	36 (44.4)	108 (97.2)	14 (16.8)	22 (21.6)	180
Leisure	38 (29.6)	54 (64.8)	14 (11.2)	14 (14.4)	120
Total	74	162	28	36	300

Answer: Page A38

EXAMPLE 3

Using Technology for a Chi-Square Independence Test

A health club manager wants to determine whether the number of days per week that college students exercise is related to gender. A random sample of 275 college students is selected and the results are classified as shown in the table. At $\alpha = 0.05$, is there enough evidence to conclude that the *number of days a student exercises per week* is related to *gender?*

Gender	\multicolumn{5}{c} Number of days of exercise per week
	0–1	2–3	4–5	6–7	Total
Male	40	53	26	6	125
Female	34	68	37	11	150
Total	74	121	63	17	275

SOLUTION   Here are the null and alternative hypotheses.

H_0: The *number of days of exercise per week* is independent of *gender.*

H_a: The *number of days of exercise per week* depends on *gender.*   (Claim)

Because d.f. = 3 and $\alpha = 0.05$, the critical value is $\chi_0^2 = 7.815$. So, the rejection region is $\chi^2 > 7.815$. Using Minitab (see below), the test statistic is $\chi^2 \approx 3.493$. Because $\chi^2 \approx 3.493$ is not in the rejection region, you fail to reject the null hypothesis.

Study Tip

You can also use a *P*-value to perform a chi-square independence test. For instance, in Example 3, note that Minitab displays $P = 0.322$. Because $P > \alpha$, you fail to reject the null hypothesis.

MINITAB

Tabulated Statistics: Gender, Number of days of exercise

Rows: Gender   Columns: Number of days of exercise

	0 to 1	2 to 3	4 to 5	6 to 7	All
Male	40	53	26	6	125
Female	34	68	37	11	150
All	74	121	63	17	275

Cell Contents: Count

Pearson Chi-Square = 3.493,   DF = 3,   P-Value = 0.322

Interpretation   There is not enough evidence to conclude that the number of days a student exercises per week is related to gender.

TRY IT YOURSELF 3

A researcher wants to determine whether age is related to whether or not a tax credit would influence an adult to purchase a hybrid vehicle. A random sample of 1250 adults is selected and the results are classified as shown in the table. At $\alpha = 0.01$, is there enough evidence to conclude that *age* is related to the *response?*   *(Adapted from HNTB)*

Response	\multicolumn{3}{c} Age	
	18–34	35–54	55 and older	Total
Yes	257	189	143	589
No	218	261	182	661
Total	475	450	325	1250

Answer: Page A38

10.2 EXERCISES

For Extra Help: **MyLab Statistics**

Building Basic Skills and Vocabulary

1. Explain how to find the expected frequency for a cell in a contingency table.

2. Explain the difference between marginal frequencies and joint frequencies in a contingency table.

3. Explain how the chi-square independence test and the chi-square goodness-of-fit test are similar. How are they different?

4. Explain why the chi-square independence test is always a right-tailed test.

True or False? *In Exercises 5 and 6, determine whether the statement is true or false. If it is false, rewrite it as a true statement.*

5. If the two variables in a chi-square independence test are dependent, then you can expect little difference between the observed frequencies and the expected frequencies.

6. When the test statistic for the chi-square independence test is large, you will, in most cases, reject the null hypothesis.

Finding Expected Frequencies *In Exercises 7–12, (a) calculate the marginal frequencies and (b) find the expected frequency for each cell in the contingency table. Assume that the variables are independent.*

7.

Result	Athlete has	
	Stretched	**Not stretched**
Injury	18	22
No injury	211	189

8.

Result	Treatment	
	Drug	**Placebo**
Nausea	36	13
No nausea	254	262

9.

Bank employee	Preference		
	New procedure	**Old procedure**	**No preference**
Teller	92	351	50
Customer service representative	76	42	8

10.

Size of restaurant	Rating		
	Excellent	**Fair**	**Poor**
Seats 100 or fewer	182	203	165
Seats over 100	180	311	159

11.

Gender	Type of car			
	Compact	**Full-size**	**SUV**	**Truck/van**
Male	28	39	21	22
Female	24	32	20	14

Sidebar answers

1. Find the sum of the row and the sum of the column in which the cell is located. Find the product of these sums. Divide the product by the sample size.

2. In a contingency table, a marginal frequency is the frequency that an entire category of a variable occurs, whereas a joint frequency is a frequency from a cell in the interior of a contingency table.

3. *Sample answer:* For both the chi-square independence test and the chi-square goodness-of-fit test, you are testing a claim about data that are in categories. However, the chi-square goodness-of-fit test has only one data value per category, while the chi-square independence test has multiple data values per category.

Both tests compare observed and expected frequencies. However, the chi-square goodness-of-fit test simply compares the distributions, whereas the chi-square independence test compares them and then draws a conclusion about the dependence or independence of the variables.

4. A chi-square independence test is always a right-tailed test because if the variables are dependent, then the chi-square test statistic will be large, which is evidence for rejecting the null hypothesis.

5. False. If the two variables of a chi-square independence test are dependent, then you can expect a large difference between the observed frequencies and the expected frequencies.

6. True

7. See Odd Answers, page A84.

8. See Selected Answers, page A107.

9. See Odd Answers, page A84.

10. See Selected Answers, page A107.

11. See Odd Answers, page A85.

12. See Selected Answers, page A108.

13. (a) H_0: An athlete's injury result is independent of whether or not the athlete has stretched. (claim)

H_a: An athlete's injury result is dependent on whether or not an athlete has stretched.

(b) d.f. $= 1$; $\chi_0^2 = 6.635$; Rejection region: $\chi^2 > 6.635$

(c) 0.875   (d) Fail to reject H_0.

(e) There is not enough evidence at the 1% level of significance to reject the claim that an athlete's injury result is independent of whether or not the athlete has stretched.

14. (a) H_0: Nausea result is independent of treatment.

H_a: Nausea result is dependent on treatment. (claim)

(b) d.f. $= 1$; $\chi_0^2 = 3.841$; Rejection region: $\chi^2 > 3.841$

(c) 10.530   (d) Reject H_0.

(e) There is enough evidence at the 5% level of significance to conclude that nausea result is dependent on treatment.

15. (a) H_0: The result is independent of the type of treatment.

H_a: The result is dependent on the type of treatment. (claim)

(b) d.f. $= 1$; $\chi_0^2 = 2.706$; Rejection region: $\chi^2 > 2.706$

(c) 12.478   (d) Reject H_0.

(e) There is enough evidence at the 10% level of significance to conclude that the result is dependent on the type of treatment.

16. (a) H_0: Attitudes about safety are independent of the type of school.

H_a: Attitudes about safety are dependent on the type of school. (claim)

(b) d.f. $= 1$; $\chi_0^2 = 6.635$; Rejection region: $\chi^2 > 6.635$

(c) 8.691   (d) Reject H_0.

(e) There is enough evidence at the 1% level of significance to conclude that attitudes about the safety steps taken by the school staff are dependent on the type of school.

12.

Type of movie rented	Age				
	18–24	**25–34**	**35–44**	**45–64**	**65 and older**
Comedy	38	30	24	10	8
Action	15	17	16	9	5
Drama	12	11	19	25	13

Using and Interpreting Concepts

Performing a Chi-Square Independence Test *In Exercises 13–28, perform the indicated chi-square independence test by performing the steps below.*

(a) *Identify the claim and state H_0 and H_a.*

(b) *Determine the degrees of freedom, find the critical value, and identify the rejection region.*

(c) *Find the chi-square test statistic.*

(d) *Decide whether to reject or fail to reject the null hypothesis.*

(e) *Interpret the decision in the context of the original claim.*

13. Use the contingency table and expected frequencies from Exercise 7. At $\alpha = 0.01$, test the hypothesis that the variables are independent.

14. Use the contingency table and expected frequencies from Exercise 8. At $\alpha = 0.05$, test the hypothesis that the variables are dependent.

15. Musculoskeletal Injury The contingency table shows the results of a random sample of patients with pain from musculoskeletal injuries treated with acetaminophen or ibuprofen. At $\alpha = 0.10$, can you conclude that the treatment is related to the result? *(Adapted from American Academy of Pediatrics)*

Result	Treatment	
	Acetaminophen	**Ibuprofen**
Significant improvement	58	81
Slight improvement	42	19

16. Attitudes about Safety The contingency table shows the results of a random sample of students by type of school and their attitudes on safety steps taken by the school staff. At $\alpha = 0.01$, can you conclude that attitudes about the safety steps taken by the school staff are related to the type of school? *(Adapted from Horatio Alger Association)*

Type of school	School staff has	
	Taken all steps necessary for student safety	**Taken some steps toward student safety**
Public	40	51
Private	64	34

17. (a) H_0: The number of times former smokers tried to quit is independent of gender.

H_a: The number of times former smokers tried to quit is dependent on gender. (claim)

(b) d.f. $= 2$; $\chi_0^2 = 5.991$; Rejection region: $\chi^2 > 5.991$

(c) 0.002   (d) Fail to reject H_0.

(e) There is not enough evidence at the 5% level of significance to conclude that the number of times former smokers tried to quit is dependent on gender.

18. (a) H_0: Skill level in a subject is independent of location. (claim)

H_a: Skill level in a subject is dependent on location.

(b) d.f. $= 2$; $\chi_0^2 = 9.210$; Rejection region: $\chi^2 > 9.210$

(c) 0.297   (d) Fail to reject H_0.

(e) There is not enough evidence at the 1% level of significance to reject the claim that skill level in a subject is independent of location.

19. (a) H_0: Reasons are independent of the type of worker.

H_a: Reasons are dependent on the type of worker. (claim)

(b) d.f. $= 2$; $\chi_0^2 = 9.210$; Rejection region: $\chi^2 > 9.210$

(c) 7.326   (d) Fail to reject H_0.

(e) There is not enough evidence at the 1% level of significance to conclude that reasons for continuing education are dependent on the type of worker.

20. (a) See Selected Answers, page A108.

(b) d.f. $= 4$; $\chi_0^2 = 7.779$; Rejection region: $\chi^2 > 7.779$

(c) 5.757   (d) Fail to reject H_0.

(e) There is not enough evidence at the 10% level of significance to conclude that the aspect of career development that is considered to be most important is dependent on age.

17. **Trying to Quit Smoking** The contingency table shows the results of a random sample of former smokers by the number of times they tried to quit smoking before they were habit-free and gender. At $\alpha = 0.05$, can you conclude that the number of times they tried to quit before they were habit-free is related to gender? *(Adapted from Porter Novelli HealthStyles for the American Lung Association)*

Gender	Number of times tried to quit before habit-free		
	1	2–3	4 or more
Male	271	257	149
Female	146	139	80

18. **Achievement and School Location** The contingency table shows the results of a random sample of students by the location of school and the number of those students achieving basic skill levels in three subjects. At $\alpha = 0.01$, test the hypothesis that the variables are independent. *(Adapted from HUD State of the Cities Report)*

Location of school	Subject		
	Reading	Math	Science
Urban	43	42	38
Suburban	63	66	65

19. **Continuing Education** You work for a college's continuing education department and want to determine whether the reasons given by workers for continuing their education are related to job type. In your study, you randomly collect the data shown in the contingency table. At $\alpha = 0.01$, can you conclude that the reason and the type of worker are dependent? *(Adapted from Market Research Institute for George Mason University)*

Type of worker	Reason for continuing education		
	Professional	Personal	Professional and personal
Technical	30	36	41
Other	47	25	30

20. **Ages and Goals** You are investigating the relationship between the ages of U.S. adults and what aspect of career development they consider to be the most important. You randomly collect the data shown in the contingency table. At $\alpha = 0.10$, is there enough evidence to conclude that age is related to which aspect of career development is considered to be most important? *(Adapted from The Harris Poll)*

Age	Career development aspect		
	Learning new skills	Pay increases	Career path
18–26 years	31	22	21
27–41 years	27	31	33
42–61 years	19	14	8

21. (a) H_0: A family borrowing money for college is independent of race.

    H_a: A family borrowing money for college is dependent on race. (claim)

    (b) d.f. $= 2$; $\chi_0^2 = 9.210$; Rejection region: $\chi^2 > 9.210$

    (c) 5.994   (d) Fail to reject H_0.

    (e) There is not enough evidence at the 1% level of significance to conclude that a family borrowing money for college is dependent on race.

22. (a) H_0: Who borrows money for college in a family is independent of the family's income.

    H_a: Who borrows money for college in a family is dependent on the family's income. (claim)

    (b) d.f. $= 6$; $\chi_0^2 = 16.812$; Rejection region: $\chi_0^2 > 16.812$

    (c) 38.746   (d) Reject H_0.

    (e) There is enough evidence at the 1% level of significance to conclude that who borrows money for college in a family is dependent on the family's income.

23. (a) H_0: Type of crash is independent of the type of vehicle.

    H_a: Type of crash is dependent on the type of vehicle. (claim)

    (b) d.f. $= 2$; $\chi_0^2 = 5.991$; Rejection region: $\chi^2 > 5.991$

    (c) 103.568   (d) Reject H_0.

    (e) There is enough evidence at the 5% level of significance to conclude that the type of crash is dependent on the type of vehicle.

24. (a) H_0: Age and gender are independent.

    H_a: Age and gender are dependent. (claim)

    (b) d.f. $= 5$; $\chi_0^2 = 11.071$; Rejection region: $\chi^2 > 11.071$

    (c) 0.417   (d) Fail to reject H_0.

    (e) There is not enough evidence at the 5% level of significance to conclude that age and gender are dependent.

21. **Borrowing for College** The contingency table shows a random sample of white, black, and Hispanic college students based on whether their family borrowed money to pay for their college education. At $\alpha = 0.01$, can you conclude that borrowing money for college and race are related? *(Adapted from Sallie Mae)*

Race	Family borrowed money?	
	Yes	No
White	49	64
Black	85	123
Hispanic	85	180

22. **Borrowing for College** A financial aid officer is studying the relationship between who borrows money to pay for college in a family and the income of the family. As part of the study, 1593 families are randomly selected and the resulting data are organized as shown in the contingency table. At $\alpha = 0.01$, can you conclude that who borrows money for college in a family is related to the income of the family? *(Adapted from Sallie Mae)*

Family income	Who borrowed money			
	Student only	Parent only	Both	No one
Less than $35,000	149	34	10	311
$35,000–$100,000	181	68	58	421
Greater than $100,000	69	40	14	238

23. **Vehicles and Crashes** You work for an insurance company and are studying the relationship between types of crashes and the vehicles involved in passenger vehicle occupant deaths. As part of your study, you randomly select 4270 vehicle crashes and organize the resulting data as shown in the contingency table. At $\alpha = 0.05$, can you conclude that the type of crash depends on the type of vehicle? *(Adapted from Insurance Institute for Highway Safety)*

Type of crash	Vehicle		
	Car	Pickup	Sport utility
Single-vehicle	1059	507	491
Multiple-vehicle	1476	354	383

24. **Alcohol-Related Accidents** The contingency table shows the results of a random sample of fatally injured passenger vehicle drivers (with blood alcohol concentrations greater than or equal to 0.08) by age and gender. At $\alpha = 0.05$, can you conclude that age is related to gender in such alcohol-related accidents? *(Adapted from Insurance Institute for Highway Safety)*

Gender	Age					
	16–20	21–30	31–40	41–50	51–60	61 and older
Male	31	147	95	67	57	42
Female	9	36	25	17	15	9

25. (a) H_0: Procedure preference is independent of bank employee.

    H_a: Procedure preference is dependent on bank employee. (claim)

    (b) d.f. = 2; $\chi_0^2 = 5.991$; Rejection region: $\chi^2 > 5.991$

    (c) 88.361   (d) Reject H_0.

    (e) There is enough evidence at the 5% level of significance to conclude that procedure preference is dependent on bank employee.

26. (a) H_0: Restaurant rating is independent of size of restaurant.

    H_a: Restaurant rating is dependent on size of restaurant. (claim)

    (b) d.f. = 2; $\chi_0^2 = 9.210$; Rejection region: $\chi^2 > 9.210$

    (c) 14.583   (d) Reject H_0.

    (e) There is enough evidence at the 1% level of significance to conclude that restaurant rating is dependent on size of restaurant.

27. (a) H_0: Type of car is independent of gender. (claim)

    H_a: Type of car is dependent on gender.

    (b) d.f. = 3; $\chi_0^2 = 6.251$; Rejection region: $\chi^2 > 6.251$

    (c) 0.808   (d) Fail to reject H_0.

    (e) There is not enough evidence at the 10% level of significance to conclude that type of car is dependent on gender.

28. See Selected Answers, page A108.

29. See Odd Answers, page A85.

30. See Selected Answers, page A108.

	Treatment	
Result	**Drug**	**Placebo**
Improvement	39	25
No change	54	70

TABLE FOR EXERCISE 30

25. Use the contingency table and expected frequencies from Exercise 9. At $\alpha = 0.05$, test the hypothesis that the variables are dependent.

26. Use the contingency table and expected frequencies from Exercise 10. At $\alpha = 0.01$, test the hypothesis that the variables are dependent.

27. Use the contingency table and expected frequencies from Exercise 11. At $\alpha = 0.10$, test the hypothesis that the variables are independent.

28. Use the contingency table and expected frequencies from Exercise 12. At $\alpha = 0.10$, test the hypothesis that the variables are dependent.

Extending Concepts

Homogeneity of Proportions Test *In Exercises 29–32, use this information about the homogeneity of proportions test. Another chi-square test that involves a contingency table is the **homogeneity of proportions test**. This test is used to determine whether several proportions are equal when samples are taken from different populations. Before the populations are sampled and the contingency table is made, the sample sizes are determined. After randomly sampling different populations, you can test whether the proportion of elements in a category is the same for each population using the same guidelines as the chi-square independence test. The null and alternative hypotheses are always some variation of these statements.*

H_0: *The proportions are equal.*

H_a: *At least one of the proportions is different from the others.*

Performing a homogeneity of proportions test requires that the observed frequencies be obtained using a random sample, and each expected frequency must be greater than or equal to 5.

29. **Motor Vehicle Crash Deaths** The contingency table shows the results of a random sample of motor vehicle crash deaths by age and gender. At $\alpha = 0.05$, perform a homogeneity of proportions test on the claim that the proportions of motor vehicle crash deaths involving males or females are the same for each age group. *(Adapted from Insurance Institute for Highway Safety)*

	Age			
Gender	**16–24**	**25–34**	**35–44**	**45–54**
Male	96	98	72	80
Female	39	33	25	29

	Age			
Gender	**55–64**	**65–74**	**75–84**	**85 and older**
Male	74	44	25	12
Female	26	21	16	10

30. **Obsessive-Compulsive Disorder** The contingency table at the left shows the results of a random sample of patients with obsessive-compulsive disorder after being treated with a drug or with a placebo. At $\alpha = 0.10$, perform a homogeneity of proportions test on the claim that the proportions of the results for drug and placebo treatments are the same. *(Adapted from The Journal of the American Medical Association)*

31. Right-tailed

32. *Sample answer:* Both tests are very similar, but the chi-square test for independence determines whether the occurrence of one variable affects the probability of the occurrence of another variable, while the chi-square homogeneity of proportions test determines whether the proportions for categories from a population follow the same distribution as another population.

33. See Odd Answers, page A85.

34. Several of the expected frequencies are less than 5.

35. (a) 0.9%
(b) 6.1%

36. (a) 15.6%
(b) 35.7%
(c) 10.9%

37. See Odd Answers, page A85.

38. 50.7%

39. 17.2%

40. See Selected Answers, page A108.

41. 26.3%

42. 3.8%

31. Is the chi-square homogeneity of proportions test a left-tailed, right-tailed, or two-tailed test?

32. Explain how the chi-square independence test is different from the chi-square homogeneity of proportions test.

Contingency Tables and Relative Frequencies   *In Exercises 33–36, use the information below.*

The frequencies in a contingency table can be written as relative frequencies by dividing each frequency by the sample size. The contingency table below shows the number of U.S. adults (in millions) ages 25 and over by employment status and educational attainment. *(Adapted from U.S. Census Bureau)*

Status	Educational attainment			
	Not a high school graduate	High school graduate	Some college, no degree	Associate's, bachelor's, or advanced degree
Employed	10.0	33.5	21.7	67.1
Unemployed	0.9	2.1	1.1	1.9
Not in the labor force	12.6	26.4	13.2	24.6

33. Rewrite the contingency table using relative frequencies.

34. Explain why you cannot perform the chi-square independence test on these data.

35. What percent of U.S. adults ages 25 and over (a) have a degree and are unemployed and (b) have some college education, but no degree, and are not in the labor force?

36. What percent of U.S. adults ages 25 and over (a) are employed and are only high school graduates, (b) are not in the labor force, and (c) are not high school graduates?

Conditional Relative Frequencies   *In Exercises 37–42, use the contingency table from Exercises 33–36, and the information below.*

Relative frequencies can also be calculated based on the row totals (by dividing each row entry by the row's total) or the column totals (by dividing each column entry by the column's total). These frequencies are **conditional relative frequencies** and can be used to determine whether an association exists between two categories in a contingency table.

37. Calculate the conditional relative frequencies in the contingency table based on the row totals.

38. What percent of U.S. adults ages 25 and over who are employed have a degree?

39. What percent of U.S. adults ages 25 and over who are not in the labor force have some college education, but no degree?

40. Calculate the conditional relative frequencies in the contingency table based on the column totals.

41. What percent of U.S. adults ages 25 and over who have a degree are not in the labor force?

42. What percent of U.S. adults ages 25 and over who are not high school graduates are unemployed?

Food Safety Survey

In your opinion, how safe is the food you buy? CBS News polled 1048 U.S. adults and asked them the question below.

> *Overall, how confident are you that the food you buy is safe to eat: very confident, somewhat confident, not too confident, not at all confident?*

The pie chart shows the responses to the question. You conduct a survey using the same question. The contingency table shows the results of your survey classified by gender.

How Confident Are You That the Food You Buy is Safe to Eat?

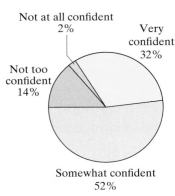

	Gender	
Response	**Female**	**Male**
Very confident	96	160
Somewhat confident	232	180
Not too confident	56	52
Not at all confident	12	4

EXERCISES

1. Assuming the variables *gender* and *response* are independent, did the number of female respondents or male respondents exceed the expected number of "very confident" responses?

2. Assuming the variables *gender* and *response* are independent, did the number of female respondents or male respondents exceed the expected number of "somewhat confident" responses?

3. At $\alpha = 0.01$, perform a chi-square independence test to determine whether the variables *response* and *gender* are independent. What can you conclude?

In Exercises 4 and 5, perform a chi-square goodness-of-fit test to compare the distribution of responses shown in the pie chart with the distribution of your survey results for each gender. Use the distribution shown in the pie chart as the expected distribution. Use $\alpha = 0.05$.

4. Compare the distribution of responses by females with the expected distribution. What can you conclude?

5. Compare the distribution of responses by males with the expected distribution. What can you conclude?

6. In addition to the variables used in the Case Study, what other variables do you think are important to consider when studying the distribution of U.S. consumers' attitudes about food safety?

10.3 Comparing Two Variances

What You Should Learn

▷ How to interpret the *F*-distribution and use an *F*-table to find critical values

▷ How to perform a two-sample *F*-test to compare two variances

Note to Instructor

If you prefer, you can cover this section before Section 8.2 so students can test whether variances of two populations are equal.

The *F*-Distribution ■ The Two-Sample *F*-Test for Variances

The *F*-Distribution

In Chapter 8, you learned how to perform hypothesis tests to compare population means and population proportions. Recall from Section 8.2 that the *t*-test for the difference between two population means depends on whether the population variances are equal. To determine whether the population variances are equal, you can perform a two-sample *F*-test.

In this section, you will learn about the **F-distribution** and how it can be used to compare two variances. As you read the next definition, recall that the sample variance s^2 is the square of the sample standard deviation s.

DEFINITION

Let s_1^2 and s_2^2 represent the sample variances of two different populations. If both populations are normal and the population variances σ_1^2 and σ_2^2 are equal, then the sampling distribution of

$$F = \frac{s_1^2}{s_2^2}$$

is an **F-distribution.** Here are several properties of the *F*-distribution.

1. The *F*-distribution is a family of curves, each of which is determined by two types of degrees of freedom: the degrees of freedom corresponding to the variance in the numerator, denoted by **d.f.$_N$**, and the degrees of freedom corresponding to the variance in the denominator, denoted by **d.f.$_D$**.

2. The *F*-distribution is positively skewed and therefore the distribution is not symmetric (see figure below).

3. The total area under each *F*-distribution curve is equal to 1.

4. All values of F are greater than or equal to 0.

5. For all *F*-distributions, the mean value of F is approximately equal to 1.

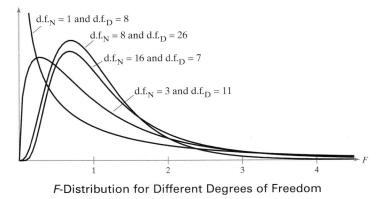

F-Distribution for Different Degrees of Freedom

For unequal variances, designate the greater sample variance as s_1^2. So, in the sampling distribution of $F = s_1^2 / s_2^2$, the variance in the numerator is greater than or equal to the variance in the denominator. This means that F is always greater than or equal to 1. As such, all one-tailed tests are right-tailed tests, and for all two-tailed tests, you need only to find the right-tailed critical value.

Table 7 in Appendix B lists the critical values for the *F*-distribution for selected levels of significance α and degrees of freedom d.f.$_N$ and d.f.$_D$.

GUIDELINES

Finding Critical Values for the *F*-Distribution

1. Specify the level of significance α.

2. Determine the degrees of freedom for the numerator d.f.$_N$.

3. Determine the degrees of freedom for the denominator d.f.$_D$.

4. Use Table 7 in Appendix B to find the critical value. When the hypothesis test is

    a. one-tailed, use the α *F*-table.

    b. two-tailed, use the $\frac{1}{2}\alpha$ *F*-table.

Note that because *F* is always greater than or equal to 1, all one-tailed tests are right-tailed tests. For two-tailed tests, you need only to find the right-tailed critical value.

In Examples 1 and 2, the values of d.f.$_N$ and d.f.$_D$ are given. You will learn how to determine these values on page 552.

EXAMPLE 1

Finding a Critical *F*-Value for a Right-Tailed Test

Find the critical *F*-value for a right-tailed test when $\alpha = 0.10$, d.f.$_N = 5$, and d.f.$_D = 28$.

SOLUTION

A portion of Table 7 is shown below. Using the $\alpha = 0.10$ *F*-table with d.f.$_N = 5$ and d.f.$_D = 28$, you can find the critical value, as shown by the highlighted areas in the table.

d.f.$_D$: Degrees of freedom, denominator	$\alpha = 0.10$ d.f.$_N$: Degrees of freedom, numerator							
	1	**2**	**3**	**4**	**5**	**6**	**7**	**8**
1	39.86	49.50	53.59	55.83	57.24	58.20	58.91	59.44
2	8.53	9.00	9.16	9.24	9.29	9.33	9.35	9.37
26	2.91	2.52	2.31	2.17	2.08	2.01	1.96	1.92
27	2.90	2.51	2.30	2.17	2.07	2.00	1.95	1.91
28	2.89	2.50	2.29	2.16	2.06	2.00	1.94	1.90
29	2.89	2.50	2.28	2.15	2.06	1.99	1.93	1.89
30	2.88	2.49	2.28	2.14	2.05	1.98	1.93	1.88

From the table, you can see that the critical value is

    $F_0 = 2.06.$ Critical value

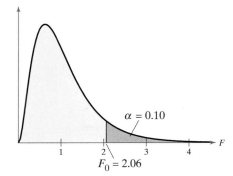

The figure at the left shows the *F*-distribution for $\alpha = 0.10$, d.f.$_N = 5$, d.f.$_D = 28$, and $F_0 = 2.06$.

TRY IT YOURSELF 1

Find the critical *F*-value for a right-tailed test when $\alpha = 0.05$, d.f.$_N = 8$, and d.f.$_D = 20$.

Answer: Page A39

When performing a two-tailed hypothesis test using the F-distribution, you need only to find the right-tailed critical value. You must, however, remember to use the $\frac{1}{2}\alpha$ F-table.

EXAMPLE 2

Finding a Critical F-Value for a Two-Tailed Test

Find the critical F-value for a two-tailed test when $\alpha = 0.05$, d.f.$_N = 4$, and d.f.$_D = 8$.

SOLUTION

A portion of Table 7 is shown below. Using the

$$\frac{1}{2}\alpha = \frac{1}{2}(0.05) = 0.025$$

F-table with d.f.$_N = 4$ and d.f.$_D = 8$, you can find the critical value, as shown by the highlighted areas in the table.

d.f.$_D$: Degrees of freedom, denominator	$\alpha = 0.025$							
	d.f.$_N$: Degrees of freedom, numerator							
	1	2	3	4	5	6	7	8
1	647.8	799.5	864.2	899.6	921.8	937.1	948.2	956.7
2	38.51	39.00	39.17	39.25	39.30	39.33	39.36	39.37
3	17.44	16.04	15.44	15.10	14.88	14.73	14.62	14.54
4	12.22	10.65	9.98	9.60	9.36	9.20	9.07	8.98
5	10.01	8.43	7.76	7.39	7.15	6.98	6.85	6.76
6	8.81	7.26	6.60	6.23	5.99	5.82	5.70	5.60
7	8.07	6.54	5.89	5.52	5.29	5.12	4.99	4.90
8	7.57	6.06	5.42	5.05	4.82	4.65	4.53	4.43
9	7.21	5.71	5.08	4.72	4.48	4.32	4.20	4.10

From the table, the critical value is

$$F_0 = 5.05. \qquad \text{Critical value}$$

The figure shows the F-distribution for $\frac{1}{2}\alpha = 0.025$, d.f.$_N = 4$, d.f.$_D = 8$, and $F_0 = 5.05$.

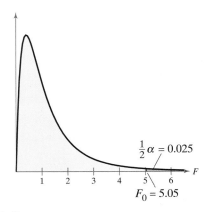

TRY IT YOURSELF 2

Find the critical F-value for a two-tailed test when $\alpha = 0.01$, d.f.$_N = 2$, and d.f.$_D = 5$.

Answer: Page A39

The Two-Sample *F*-Test for Variances

In the remainder of this section, you will learn how to perform a two-sample *F*-test for comparing two population variances using a sample from each population.

Two-Sample *F*-Test for Variances

A **two-sample *F*-test** is used to compare two population variances σ_1^2 and σ_2^2. To perform this test, these conditions must be met.

1. The samples must be random.
2. The samples must be independent.
3. Each population must have a normal distribution.

The **test statistic** is

$$F = \frac{s_1^2}{s_2^2}$$

where s_1^2 and s_2^2 represent the sample variances with $s_1^2 \geq s_2^2$. The numerator has d.f.$_N = n_1 - 1$ degrees of freedom and the denominator has d.f.$_D = n_2 - 1$ degrees of freedom, where n_1 is the size of the sample having variance s_1^2 and n_2 is the size of the sample having variance s_2^2.

GUIDELINES

Using a Two-Sample *F*-Test to Compare σ_1^2 and σ_2^2

In Words	In Symbols
1. Verify that the samples are random and independent, and the populations have normal distributions.	
2. Identify the claim. State the null and alternative hypotheses.	State H_0 and H_a.
3. Specify the level of significance.	Identify α.
4. Identify the degrees of freedom for the numerator and the denominator.	d.f.$_N = n_1 - 1$ d.f.$_D = n_2 - 1$
5. Determine the critical value.	Use Table 7 in Appendix B.
6. Determine the rejection region.	
7. Find the test statistic and sketch the sampling distribution.	$F = \dfrac{s_1^2}{s_2^2}$
8. Make a decision to reject or fail to reject the null hypothesis.	If F is in the rejection region, then reject H_0. Otherwise, fail to reject H_0.
9. Interpret the decision in the context of the original claim.	

In some cases, you will be given the sample standard deviations s_1 and s_2. Remember to square both standard deviations to calculate the sample variances s_1^2 and s_2^2 before using a two-sample *F*-test to compare variances.

Picturing the World

Does location have an effect on the variance of real estate selling prices? A random sample of selling prices (in thousands of dollars) of existing homes sold in the California counties of Los Angeles and San Diego is shown in the table. (Adapted from California Association of Realtors)

Los Angeles	San Diego
440	634
342	378
494	652
598	659
590	695
643	776
252	425
447	594
580	645
361	546

Assuming each population of selling prices is normally distributed, is it possible to use a two-sample F-test to compare the population variances?

Yes, because the samples are randomly selected and independent, and each population is normally distributed.

Normal solution	Treated solution
$n = 25$	$n = 20$
$s^2 = 180$	$s^2 = 56$

EXAMPLE 3

Performing a Two-Sample F-Test

A restaurant manager is designing a system that is intended to decrease the variance of the time customers wait before their meals are served. Under the old system, a random sample of 10 customers had a variance of 400. Under the new system, a random sample of 21 customers had a variance of 256. At $\alpha = 0.10$, is there enough evidence to convince the manager to switch to the new system? Assume both populations are normally distributed.

SOLUTION

Because $400 > 256$, $s_1^2 = 400$ and $s_2^2 = 256$. Therefore, s_1^2 and σ_1^2 represent the sample and population variances for the old system, respectively. With the claim "the variance of the waiting times under the new system is less than the variance of the waiting times under the old system," the null and alternative hypotheses are

$$H_0: \sigma_1^2 \leq \sigma_2^2 \quad \text{and} \quad H_a: \sigma_1^2 > \sigma_2^2. \quad \text{(Claim)}$$

Note that the test is a right-tailed test with $\alpha = 0.10$, and the degrees of freedom are

$$\text{d.f.}_N = n_1 - 1 = 10 - 1 = 9$$

and

$$\text{d.f.}_D = n_2 - 1 = 21 - 1 = 20.$$

So, the critical value is $F_0 = 1.96$ and the rejection region is $F > 1.96$. The test statistic is

$$F = \frac{s_1^2}{s_2^2} = \frac{400}{256} \approx 1.56.$$

The figure shows the location of the rejection region and the test statistic F. Because F is not in the rejection region, you fail to reject the null hypothesis.

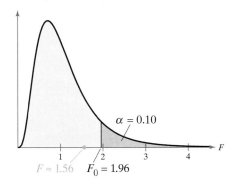

$\alpha = 0.10$

$F \approx 1.56 \quad F_0 = 1.96$

Interpretation There is not enough evidence at the 10% level of significance to convince the manager to switch to the new system.

TRY IT YOURSELF 3

A medical researcher claims that a specially treated intravenous solution decreases the variance of the time required for nutrients to enter the bloodstream. Independent samples from each type of solution are randomly selected, and the results are shown in the table at the left. At $\alpha = 0.01$, is there enough evidence to support the researcher's claim? Assume the populations are normally distributed.

Answer: Page A39

Stock A	Stock B
$n_2 = 30$	$n_1 = 31$
$s_2 = 3.5$	$s_1 = 5.7$

EXAMPLE 4

Using Technology for a Two-Sample *F*-Test

You want to purchase stock in a company and are deciding between two different stocks. Because a stock's risk can be associated with the standard deviation of its daily closing prices, you randomly select samples of the daily closing prices for each stock to obtain the results shown at the left. At $\alpha = 0.05$, can you conclude that one of the two stocks is a riskier investment? Assume the stock closing prices are normally distributed.

SOLUTION

Because $5.7^2 > 3.5^2$, $s_1^2 = 5.7^2$ and $s_2^2 = 3.5^2$. Therefore, s_1^2 and σ_1^2 represent the sample and population variances for Stock B, respectively. With the claim "one of the two stocks is a riskier investment," the null and alternative hypotheses are

$$H_0: \sigma_1^2 = \sigma_2^2 \quad \text{and} \quad H_a: \sigma_1^2 \neq \sigma_2^2. \quad \text{(Claim)}$$

Note that the test is a two-tailed test with $\frac{1}{2}\alpha = \frac{1}{2}(0.05) = 0.025$, and the degrees of freedom are d.f._N $= n_1 - 1 = 31 - 1 = 30$ and d.f._D $= n_2 - 1 = 30 - 1 = 29$. So, the critical value is $F_0 = 2.09$ and the rejection region is $F > 2.09$.

To perform a two-sample *F*-test using a TI-84 Plus, begin with the STAT keystroke. Choose the TESTS menu and select *E:2–SampFTest*. Then set up the two-sample *F*-test as shown in the first screen below. Because you are entering the descriptive statistics, select the *Stats* input option. When entering the original data, select the *Data* input option. The other displays below show the results of selecting *Calculate* or *Draw*.

TI-84 PLUS

```
            2-SampFTest
Inpt: Data Stats
Sx1:5.7
n1:31
Sx2:3.5
n2:30
σ1:≠σ2 <σ2 >σ2
Calculate Draw
```

TI-84 PLUS

```
            2-SampFTest
σ₁≠σ₂
  F=2.652244898
  p=.0102172459
  Sx₁=5.7
  Sx₂=3.5
↓n₁=31
```

TI-84 PLUS

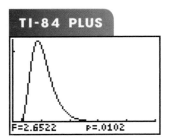

The test statistic $F \approx 2.65$ is in the rejection region, so you reject the null hypothesis.

Interpretation   There is enough evidence at the 5% level of significance to support the claim that one of the two stocks is a riskier investment.

TRY IT YOURSELF 4

Location A	Location B
$n = 16$	$n = 22$
$s = 0.95$	$s = 0.78$

A biologist claims that the pH levels of the soil in two geographic locations have equal standard deviations. Independent samples from each location are randomly selected, and the results are shown at the left. At $\alpha = 0.01$, is there enough evidence to reject the biologist's claim? Assume the pH levels are normally distributed.

Answer: Page A39

You can also use a *P*-value to perform a two-sample *F*-test. For instance, in Example 4, note that the TI-84 Plus displays $P = .0102172459$. Because $P < \alpha$, you reject the null hypothesis.

10.3 EXERCISES

For Extra Help: **MyLab Statistics**

Building Basic Skills and Vocabulary

1. Explain how to find the critical value for an F-test.

2. List five properties of the F-distribution.

3. List the three conditions that must be met in order to use a two-sample F-test.

4. Explain how to determine the values of d.f.$_N$ and d.f.$_D$ when performing a two-sample F-test.

Finding a Critical F-Value for a Right-Tailed Test *In Exercises 5–8, find the critical F-value for a right-tailed test using the level of significance α and degrees of freedom d.f.$_N$ and d.f.$_D$.*

5. $\alpha = 0.05$, d.f.$_N = 9$, d.f.$_D = 16$ **6.** $\alpha = 0.01$, d.f.$_N = 2$, d.f.$_D = 11$

7. $\alpha = 0.10$, d.f.$_N = 10$, d.f.$_D = 15$ **8.** $\alpha = 0.025$, d.f.$_N = 7$, d.f.$_D = 3$

Finding a Critical F-Value for a Two-Tailed Test *In Exercises 9–12, find the critical F-value for a two-tailed test using the level of significance α and degrees of freedom d.f.$_N$ and d.f.$_D$.*

9. $\alpha = 0.01$, d.f.$_N = 6$, d.f.$_D = 7$ **10.** $\alpha = 0.10$, d.f.$_N = 24$, d.f.$_D = 28$

11. $\alpha = 0.05$, d.f.$_N = 60$, d.f.$_D = 40$ **12.** $\alpha = 0.05$, d.f.$_N = 27$, d.f.$_D = 19$

In Exercises 13–18, test the claim about the difference between two population variances σ_1^2 and σ_2^2 at the level of significance α. Assume the samples are random and independent, and the populations are normally distributed.

13. Claim: $\sigma_1^2 > \sigma_2^2$; $\alpha = 0.10$.
Sample statistics: $s_1^2 = 773$,
$n_1 = 5$ and $s_2^2 = 765$, $n_2 = 6$

14. Claim: $\sigma_1^2 = \sigma_2^2$; $\alpha = 0.05$.
Sample statistics: $s_1^2 = 310$,
$n_1 = 7$ and $s_2^2 = 297$, $n_2 = 8$

15. Claim: $\sigma_1^2 \leq \sigma_2^2$; $\alpha = 0.01$.
Sample statistics: $s_1^2 = 842$,
$n_1 = 11$ and $s_2^2 = 836$, $n_2 = 10$

16. Claim: $\sigma_1^2 \neq \sigma_2^2$; $\alpha = 0.05$.
Sample statistics: $s_1^2 = 245$,
$n_1 = 31$ and $s_2^2 = 112$, $n_2 = 28$

17. Claim: $\sigma_1^2 = \sigma_2^2$; $\alpha = 0.01$.
Sample statistics: $s_1^2 = 9.8$,
$n_1 = 13$ and $s_2^2 = 2.5$, $n_2 = 20$

18. Claim: $\sigma_1^2 > \sigma_2^2$; $\alpha = 0.05$.
Sample statistics: $s_1^2 = 44.6$,
$n_1 = 16$ and $s_2^2 = 39.3$, $n_2 = 12$

Using and Interpreting Concepts

Performing a Two-Sample F-Test *In Exercises 19–26, (a) identify the claim and state H_0 and H_a, (b) find the critical value and identify the rejection region, (c) find the test statistic F, (d) decide whether to reject or fail to reject the null hypothesis, and (e) interpret the decision in the context of the original claim. Assume the samples are random and independent, and the populations are normally distributed.*

19. Life of Appliances Company A claims that the variance of the lives of its appliances is less than the variance of the lives of Company B's appliances. A sample of the lives of 20 of Company A's appliances has a variance of 1.8. A sample of the lives of 25 of Company B's appliances has a variance of 3.9. At $\alpha = 0.05$, can you support Company A's claim?

Answers to odd-numbered exercises (sidebar)

1. Specify the level of significance α. Determine the degrees of freedom for the numerator and denominator. Use Table 7 in Appendix B to find the critical value F.

2. (1) The F-distribution is a family of curves determined by two types of degrees of freedom, d.f.$_N$ and d.f.$_D$.
   (2) The F-distribution is positively skewed and therefore the distribution is not symmetric.
   (3)–(5) See Selected Answers, page A108.

3. (1) The samples must be random, (2) the samples must be independent, and (3) each population must have a normal distribution.

4. Determine the sample whose variance is greater. Use the size of this sample n_1 to find d.f.$_N = n_1 - 1$. Use the size of the other sample n_2 to find d.f.$_D = n_2 - 1$.

5. 2.54   **6.** 7.21   **7.** 2.06

8. 14.62   **9.** 9.16   **10.** 1.91

11. 1.80   **12.** 2.42

13. Fail to reject H_0. There is not enough evidence at the 10% level of significance to support the claim.

14. Fail to reject H_0. There is not enough evidence at the 5% level of significance to reject the claim.

15. Fail to reject H_0. There is not enough evidence at the 1% level of significance to reject the claim.

16. Reject H_0. There is enough evidence at the 5% level of significance to support the claim.

17. Reject H_0. There is enough evidence at the 1% level of significance to reject the claim.

18. Fail to reject H_0. There is not enough evidence at the 5% level of significance to support the claim.

19. See Odd Answers, page A86.

20. (a) $H_0: \sigma_1^2 \leq \sigma_2^2$

$H_a: \sigma_1^2 > \sigma_2^2$ (claim)

(b) $F_0 = 1.84$

Rejection region: $F > 1.84$

(c) 1.62   (d) Fail to reject H_0.

18–34		35–49		
208	210	229	217	218
229	213	245	222	256
223	168	232	236	244

TABLE FOR EXERCISE 21

Golfer 1			Golfer 2		
227	234	235	262	257	258
246	223	268	269	253	262
231	235	245	258	265	255
248			262		

TABLE FOR EXERCISE 22

20. (e) There is not enough evidence at the 10% level of significance to conclude that the variance of fuel consumption for the company's hybrid vehicles is less than that of the competitor's hybrid vehicles.

21. (a) $H_0: \sigma_1^2 = \sigma_2^2$

$H_a: \sigma_1^2 \neq \sigma_2^2$ (claim)

(b) $F_0 = 4.82$

Rejection region: $F > 4.82$

(c) 2:58   (d) Fail to reject H_0.

(e) There is not enough evidence at the 5% level of significance to conclude that the variances of the waiting times differ between the two age groups.

22. (a) $H_0: \sigma_1^2 = \sigma_2^2$

$H_a: \sigma_1^2 \neq \sigma_2^2$ (claim)

(b) $F_0 = 3.18$

Rejection region: $F > 3.18$

(c) 7.31   (d) Reject H_0.

(e) There is enough evidence at the 10% level of significance to conclude that the variances of the driving distances differ between the two golfers.

23. See Odd Answers, page A86.

24. See Selected Answers, page A108.

25. See Odd Answers, page A86.

26. See Selected Answers, page A108.

20. Fuel Consumption An automobile manufacturer claims that the variance of the fuel consumptions for its hybrid vehicles is less than the variance of the fuel consumptions for the hybrid vehicles of a top competitor. A sample of the fuel consumptions of 19 of the manufacturer's hybrids has a variance of 0.21. A sample of the fuel consumptions of 21 of its competitor's hybrids has a variance of 0.34. At $\alpha = 0.10$, can you support the manufacturer's claim? *(Adapted from GreenHybrid)*

21. Heart Transplant Waiting Times The table at the left shows a sample of the waiting times (in days) for a heart transplant for two age groups. At $\alpha = 0.05$, can you conclude that the variances of the waiting times differ between the two age groups? *(Adapted from Organ Procurement and Transplantation Network)*

22. Golf The table at the left shows a sample of the driving distances (in yards) for two golfers. At $\alpha = 0.10$, can you conclude that the variances of the driving distances differ between the two golfers?

23. Science Assessment Tests A state school administrator claims that the standard deviations of science assessment test scores for eighth-grade students are the same in Districts 1 and 2. A sample of 12 test scores from District 1 has a standard deviation of 36.8 points, and a sample of 14 test scores from District 2 has a standard deviation of 32.5 points. At $\alpha = 0.10$, can you reject the administrator's claim? *(Adapted from National Center for Education Statistics)*

24. U.S. History Assessment Tests A state school administrator claims that the standard deviations of U.S. history assessment test scores for eighth-grade students are the same in Districts 1 and 2. A sample of 10 test scores from District 1 has a standard deviation of 30.9 points, and a sample of 13 test scores from District 2 has a standard deviation of 27.2 points. At $\alpha = 0.01$, can you reject the administrator's claim? *(Adapted from National Center for Education Statistics)*

25. Annual Salaries An employment information service claims that the standard deviation of the annual salaries for actuaries is less in California than in New York. You select a sample of actuaries from each state. The results of each survey are shown in the figure. At $\alpha = 0.05$, can you support the service's claim? *(Adapted from America's Career InfoNet)*

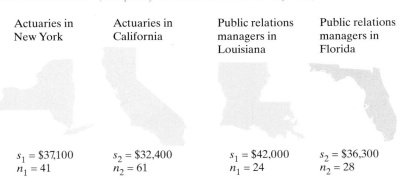

Actuaries in New York	Actuaries in California	Public relations managers in Louisiana	Public relations managers in Florida
$s_1 = \$37,100$	$s_2 = \$32,400$	$s_1 = \$42,000$	$s_2 = \$36,300$
$n_1 = 41$	$n_2 = 61$	$n_1 = 24$	$n_2 = 28$

FIGURE FOR EXERCISE 25   **FIGURE FOR EXERCISE 26**

26. Annual Salaries An employment information service claims that the standard deviation of the annual salaries for public relations managers is less in Florida than in Louisiana. You select a sample of public relations managers from each state. The results of each survey are shown in the figure. At $\alpha = 0.05$, can you support the service's claim? *(Adapted from America's Career InfoNet)*

Extending Concepts

Finding Left-Tailed Critical *F*-Values *In this section, you only needed to calculate the right-tailed critical F-value for a two-tailed test. For other applications of the F-distribution, you will need to calculate the left-tailed critical F-value. To calculate the left-tailed critical F-value, perform the steps below.*

(1) Interchange the values for d.f.$_N$ and d.f.$_D$.

(2) Find the corresponding F-value in Table 7.

(3) Calculate the reciprocal of the F-value to obtain the left-tailed critical F-value.

In Exercises 27 and 28, find the right- and left-tailed critical F-values for a two-tailed test using the level of significance α and degrees of freedom d.f.$_N$ and d.f.$_D$.

27. $\alpha = 0.05$, d.f.$_N = 6$, d.f.$_D = 3$ **28.** $\alpha = 0.10$, d.f.$_N = 20$, d.f.$_D = 15$

Confidence Interval for σ_1^2/σ_2^2 *When s_1^2 and s_2^2 are the variances of randomly selected, independent samples from normally distributed populations, then a confidence interval for σ_1^2/σ_2^2 is*

$$\frac{s_1^2}{s_2^2} \cdot \frac{1}{F_R} < \frac{\sigma_1^2}{\sigma_2^2} < \frac{s_1^2}{s_2^2} \cdot \frac{1}{F_L}$$

where F_R is the right-tailed critical F-value and F_L is the left-tailed critical F-value.

In Exercises 29 and 30, construct the confidence interval for σ_1^2/σ_2^2. Assume the samples are random and independent, and the populations are normally distributed.

29. Cholesterol Contents In a recent study of the cholesterol contents of grilled chicken sandwiches served at fast food restaurants, a nutritionist found that random samples of sandwiches from Restaurant A and from Restaurant B had the sample statistics shown in the table. Construct a 95% confidence interval for σ_1^2/σ_2^2, where σ_1^2 and σ_2^2 are the variances of the cholesterol contents of grilled chicken sandwiches from Restaurant A and Restaurant B, respectively.

| Cholesterol contents of grilled chicken sandwiches ||
Restaurant A	Restaurant B
$s_1^2 = 10.89$	$s_2^2 = 9.61$
$n_1 = 16$	$n_2 = 12$

30. Carbohydrate Contents In a recent study of the carbohydrate contents of grilled chicken sandwiches served at fast food restaurants, a nutritionist found that random samples of sandwiches from Restaurant A and from Restaurant B had the sample statistics shown in the table. Construct a 95% confidence interval for σ_1^2/σ_2^2, where σ_1^2 and σ_2^2 are the variances of the carbohydrate contents of grilled chicken sandwiches from Restaurant A and Restaurant B, respectively.

| Carbohydrate contents of grilled chicken sandwiches ||
Restaurant A	Restaurant B
$s_1^2 = 5.29$	$s_2^2 = 3.61$
$n_1 = 16$	$n_2 = 12$

10.4 Analysis of Variance

One-Way ANOVA ▪ Two-Way ANOVA

One-Way ANOVA

Suppose a medical researcher is analyzing the effectiveness of three types of pain relievers and wants to determine whether there is a difference in the mean lengths of time it takes the three medications to provide relief. To determine whether such a difference exists, the researcher can use the *F*-distribution together with a technique called **analysis of variance.** Because one independent variable is being studied, the process is called **one-way analysis of variance.**

> **DEFINITION**
>
> **One-way analysis of variance** is a hypothesis-testing technique that is used to compare the means of three or more populations. Analysis of variance is usually abbreviated as **ANOVA.**

To begin a one-way analysis of variance test, you should first state the null and alternative hypotheses. For a one-way ANOVA test, the null and alternative hypotheses are always similar to these statements.

H_0: $\mu_1 = \mu_2 = \mu_3 = \cdots = \mu_k$ (All population means are equal.)

H_a: At least one mean is different from the others.

When you reject the null hypothesis in a one-way ANOVA test, you can conclude that at least one of the means is different from the others. Without performing more statistical tests, however, you cannot determine which of the means is different.

Before performing a one-way ANOVA test, you must check that these conditions are satisfied.

1. Each sample must be randomly selected from a normal, or approximately normal, population.
2. The samples must be independent of each other.
3. Each population must have the same variance.

The test statistic for a one-way ANOVA test is the ratio of two variances: the variance between samples and the variance within samples.

$$\text{Test statistic} = \frac{\text{Variance between samples}}{\text{Variance within samples}}$$

1. The variance between samples measures the differences related to the treatment given to each sample. This variance, sometimes called the **mean square between,** is denoted by MS_B.
2. The variance within samples measures the differences related to entries within the same sample and is usually due to sampling error. This variance, sometimes called the **mean square within,** is denoted by MS_W.

One-Way Analysis of Variance Test

To perform a one-way ANOVA test, these conditions must be met.

1. Each of the k samples, $k \geq 3$, must be randomly selected from a normal, or approximately normal, population.

2. The samples must be independent of each other.

3. Each population must have the same variance.

If these conditions are met, then the sampling distribution for the test is approximated by the F-distribution. The **test statistic** is

$$F = \frac{MS_B}{MS_W}.$$

The degrees of freedom are

$$\text{d.f.}_N = k - 1 \qquad \text{Degrees of freedom for numerator}$$

and

$$\text{d.f.}_D = N - k \qquad \text{Degrees of freedom for denominator}$$

where k is the number of samples and N is the sum of the sample sizes.

If there is little or no difference between the means, then MS_B will be approximately equal to MS_W and the test statistic will be approximately 1. Values of F close to 1 suggest that you should fail to reject the null hypothesis. However, if one of the means differs significantly from the others, then MS_B will be greater than MS_W and the test statistic will be greater than 1. Values of F significantly greater than 1 suggest that you should reject the null hypothesis. So, all one-way ANOVA tests are right-tailed tests. That is, if the test statistic is greater than the critical value, then H_0 will be rejected.

Study Tip

The notations n_i, $\bar{x}_i$, and s_i^2 represent the sample size, mean, and variance of the ith sample, respectively. Also, note that $\bar{\bar{x}}$ is sometimes called the *grand mean*.

GUIDELINES

Finding the Test Statistic for a One-Way ANOVA Test

In Words	In Symbols
1. Find the mean and variance of each sample.	$\bar{x}_i = \dfrac{\Sigma x}{n}, \quad s_i^2 = \dfrac{\Sigma(x - \bar{x}_i)^2}{n - 1}$
2. Find the mean of all entries in all samples (the grand mean).	$\bar{\bar{x}} = \dfrac{\Sigma x}{N}$
3. Find the sum of squares between the samples.	$SS_B = \Sigma n_i(\bar{x}_i - \bar{\bar{x}})^2$
4. Find the sum of squares within the samples.	$SS_W = \Sigma(n_i - 1)s_i^2$
5. Find the variance between the samples.	$MS_B = \dfrac{SS_B}{\text{d.f.}_N} = \dfrac{\Sigma n_i(\bar{x}_i - \bar{\bar{x}})^2}{k - 1}$
6. Find the variance within the samples.	$MS_W = \dfrac{SS_W}{\text{d.f.}_D} = \dfrac{\Sigma(n_i - 1)s_i^2}{N - k}$
7. Find the test statistic.	$F = \dfrac{MS_B}{MS_W}$

Note that in Step 1 of the guidelines above, you are summing the values from just one sample. In Step 2, you are summing the values from all of the samples. The sums SS_B and SS_W are explained on the next page.

In the guidelines for finding the test statistic for a one-way ANOVA test, the notation SS_B represents the sum of squares between the samples.

$$SS_B = n_1(\bar{x}_1 - \bar{\bar{x}})^2 + n_2(\bar{x}_2 - \bar{\bar{x}})^2 + \cdots + n_k(\bar{x}_k - \bar{\bar{x}})^2$$

$$= \Sigma n_i(\bar{x}_i - \bar{\bar{x}})^2$$

Also, the notation SS_W represents the sum of squares within the samples.

$$SS_W = (n_1 - 1)s_i^2 + (n_2 - 1)s_2^2 + \cdots + (n_k - 1)s_k^2$$

$$= \Sigma(n_i - 1)s_i^2$$

GUIDELINES

Performing a One-Way Analysis of Variance Test

In Words	In Symbols
1. Verify that the samples are random and independent, the populations have normal distributions, and the population variances are equal.	
2. Identify the claim. State the null and alternative hypotheses.	State H_0 and H_a.
3. Specify the level of significance.	Identify α.
4. Determine the degrees of freedom for the numerator and the denominator.	d.f.$_N = k - 1$ d.f.$_D = N - k$
5. Determine the critical value.	Use Table 7 in Appendix B.
6. Determine the rejection region.	
7. Find the test statistic and sketch the sampling distribution.	$F = \dfrac{MS_B}{MS_W}$
8. Make a decision to reject or fail to reject the null hypothesis.	If F is in the rejection region, then reject H_0. Otherwise, fail to reject H_0.
9. Interpret the decision in the context of the original claim.	

Tables are a convenient way to summarize the results of a one-way analysis of variance test. ANOVA summary tables are set up as shown below.

ANOVA Summary Table

Variation	Sum of squares	Degrees of freedom	Mean squares	F
Between	SS_B	d.f.$_N = k - 1$	$MS_B = \dfrac{SS_B}{\text{d.f.}_N}$	$\dfrac{MS_B}{MS_W}$
Within	SS_W	d.f.$_D = N - k$	$MS_W = \dfrac{SS_W}{\text{d.f.}_D}$	

EXAMPLE 1

Performing a One-Way ANOVA Test

A medical researcher wants to determine whether there is a difference in the mean lengths of time it takes three types of pain relievers to provide relief from headache pain. Several headache sufferers are randomly selected and given one of the three medications. Each headache sufferer records the time (in minutes) it takes the medication to begin working. The results are shown in the table. At $\alpha = 0.01$, can you conclude that at least one mean time is different from the others? Assume that each population of relief times is normally distributed and that the population variances are equal.

Medication 1	Medication 2	Medication 3
12	16	14
15	14	17
17	21	20
12	15	15
	19	
$n_1 = 4$	$n_2 = 5$	$n_3 = 4$
$\overline{x}_1 = \frac{56}{4} = 14$	$\overline{x}_2 = \frac{85}{5} = 17$	$\overline{x}_3 = \frac{66}{4} = 16.5$
$s_1^2 = 6$	$s_2^2 = 8.5$	$s_3^2 = 7$

SOLUTION

The null and alternative hypotheses are as follows.

H_0: $\mu_1 = \mu_2 = \mu_3$

H_a: At least one mean is different from the others.   (Claim)

Because there are $k = 3$ samples, d.f.$_N = k - 1 = 3 - 1 = 2$. The sum of the sample sizes is $N = n_1 + n_2 + n_3 = 4 + 5 + 4 = 13$. So,

$$\text{d.f.}_D = N - k = 13 - 3 = 10.$$

Using d.f.$_N = 2$, d.f.$_D = 10$, and $\alpha = 0.01$, the critical value is $F_0 = 7.56$. The rejection region is $F > 7.56$. To find the test statistic, first calculate $\overline{\overline{x}}$, MS_B, and MS_W.

$$\overline{\overline{x}} = \frac{\Sigma x}{N} = \frac{56 + 85 + 66}{13} \approx 15.92$$

$$\begin{aligned} MS_B = \frac{SS_B}{\text{d.f.}_N} &= \frac{\Sigma n_i (\overline{x}_i - \overline{\overline{x}})^2}{k - 1} \\ &\approx \frac{4(14 - 15.92)^2 + 5(17 - 15.92)^2 + 4(16.5 - 15.92)^2}{3 - 1} \\ &= \frac{21.9232}{2} \\ &= 10.9616 \end{aligned}$$

$$\begin{aligned} MS_W = \frac{SS_W}{\text{d.f.}_D} &= \frac{\Sigma (n_i - 1)s_i^2}{N - k} \\ &= \frac{(4 - 1)(6) + (5 - 1)(8.5) + (4 - 1)(7)}{13 - 3} \\ &= \frac{73}{10} \\ &= 7.3 \end{aligned}$$

Picturing the World

A researcher wants to determine whether there is a difference in the mean lengths of time wasted at work for people in California, Georgia, and Pennsylvania. Several people from each state who work 8-hour days are randomly selected and they are asked how much time (in hours) they waste at work each day. The results are shown in the table. (Adapted from Salary.com)

CA	GA	PA
2	2	1.75
1.75	2.5	3
2.5	1.25	2.75
3	2.25	2
2.75	1.5	3
3.25	3	2.5
1.25	2.75	2.75
2	2.25	3.25
2.5	2	3
1.75	1	2.75
1.5		2.25
2.25		

At $\alpha = 0.10$, can the researcher conclude that there is a difference in the mean lengths of time wasted at work among the states? Assume that each population is normally distributed and that the population variances are equal.

There is enough evidence at the 10% level of significance to conclude that there is a difference in the mean lengths of time wasted at work among the states.

Using $MS_B \approx 10.9616$ and $MS_W = 7.3$, the test statistic is

$$F = \frac{MS_B}{MS_W}$$

$$\approx \frac{10.9616}{7.3}$$

$$\approx 1.50.$$

The figure shows the location of the rejection region and the test statistic F. Because F is not in the rejection region, you fail to reject the null hypothesis.

Interpretation There is not enough evidence at the 1% level of significance to conclude that there is a difference in the mean length of time it takes the three pain relievers to provide relief from headache pain.

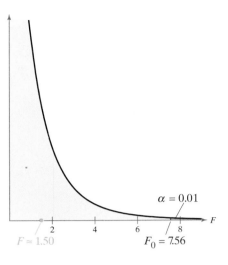

The ANOVA summary table for Example 1 is shown below.

Variation	Sum of squares	Degrees of freedom	Mean squares	*F*
Between	21.9232	2	10.9616	1.50
Within	73	10	7.3	

TRY IT YOURSELF 1

A sales analyst wants to determine whether there is a difference in the mean monthly sales of a company's four sales regions. Several salespersons from each region are randomly selected and they provide their sales amounts (in thousands of dollars) for the previous month. The results are shown in the table. At $\alpha = 0.05$, can the analyst conclude that there is a difference in the mean monthly sales among the sales regions? Assume that each population of sales is normally distributed and that the population variances are equal.

North	East	South	West
34	47	40	21
28	36	30	30
18	30	41	24
24	38	29	37
	44		23
$n_1 = 4$	$n_2 = 5$	$n_3 = 4$	$n_4 = 5$
$\bar{x}_1 = 26$	$\bar{x}_2 = 39$	$\bar{x}_3 = 35$	$\bar{x}_4 = 27$
$s_1^2 \approx 45.33$	$s_2^2 = 45$	$s_3^2 \approx 40.67$	$s_4^2 = 42.5$

Answer: Page A39

Using technology greatly simplifies the one-way ANOVA process. When using technology such as Minitab, Excel, StatCrunch, or the TI-84 Plus to perform a one-way analysis of variance test, you can use *P*-values to decide whether to reject the null hypothesis. If the *P*-value is less than α, then reject H_0.

Compact	Midsize	Large
12	21	18
23	23	17
17	19	14
20	14	17
25	14	20
18	21	17
24	26	17
27	18	13
29	25	
17	16	
31	27	
24	22	
24	21	
	25	
	21	

Tech Tip

Here are instructions for performing a one-way analysis of variance test on a TI-84 Plus. Begin by storing the data in L1, L2, and so on.

STAT

Choose the TESTS menu.

H: ANOVA(

Then enter L1, L2, and so on, separated by commas.

EXAMPLE 2

Using Technology to Perform a One-Way ANOVA Test

A researcher believes that for city driving, the fuel economy of compact, midsize, and large cars are the same. The gas mileages (in miles per gallon) for city driving for several randomly selected cars from each category are shown in the table at the left. Assume that the populations are normally distributed, the samples are independent, and the population variances are equal. At $\alpha = 0.05$, can you reject the claim that the mean gas mileages for city driving are the same for the three categories? Use technology to test the claim. *(Source: Fueleconomy.gov)*

SOLUTION

Here are the null and alternative hypotheses.

$H_0: \mu_1 = \mu_2 = \mu_3$ (Claim)

H_a: At least one mean is different from the others.

The results obtained by performing the test using Excel are shown below. From the results, you can see that $P \approx 0.02$. Because $P < \alpha$, you reject the null hypothesis.

EXCEL

	A	B	C	D	E	F
1	Anova: Single factor					
2						
3	SUMMARY					
4	*Groups*	*Count*	*Sum*	*Average*	*Variance*	
5	compact	13	291	22.38462	28.75641	
6	midsize	15	313	20.86667	16.69524	
7	large	8	133	16.625	4.839286	
8						
9						
10	ANOVA					
11	*Source of Variation*	*SS*	*df*	*MS*	*F*	*P-value*
12	Between Groups	168.287	2	84.14348	4.532074	0.018236
13	Within Groups	612.6853	33	18.56622		
14						
15	Total	780.9722	35			
16						

Interpretation   There is enough evidence at the 5% level of significance to reject the claim that the mean gas mileages for city driving are the same.

TRY IT YOURSELF 2

The data shown in the table represent the GPAs of randomly selected freshmen, sophomores, juniors, and seniors. At $\alpha = 0.05$, can you conclude that there is a difference in the means of the GPAs? Assume that the populations of GPAs are normally distributed and that the population variances are equal. Use technology to test the claim.

Freshmen	2.34	2.38	3.31	2.39	3.40	2.70	2.34			
Sophomores	3.26	2.22	3.26	3.29	2.95	3.01	3.13	3.59	2.84	3.00
Juniors	2.80	2.60	2.49	2.83	2.34	3.23	3.49	3.03	2.87	
Seniors	3.31	2.35	3.27	2.86	2.78	2.75	3.05	3.31		

Answer: Page A39

Two-Way ANOVA

When you want to test the effect of *two* independent variables, or factors, on one dependent variable, you can use a **two-way analysis of variance test.** For instance, suppose a medical researcher wants to test the effect of gender *and* type of medication on the mean length of time it takes pain relievers to provide relief. To perform such an experiment, the researcher can use the two-way ANOVA block design shown below.

Gender

		M	F
	I	Males taking type I	Females taking type I
Type of medication	**II**	Males taking type II	Females taking type II
	III	Males taking type III	Females taking type III

A two-way ANOVA test has three null hypotheses—one for each main effect and one for the interaction effect. A **main effect** is the effect of one independent variable on the dependent variable, and the **interaction effect** is the effect of both independent variables on the dependent variable. For instance, the hypotheses for the pain reliever experiment are listed below.

Hypotheses for main effects:

H_0: Gender has no effect on the mean length of time it takes a pain reliever to provide relief.

H_a: Gender has an effect on the mean length of time it takes a pain reliever to provide relief.

H_0: The type of medication has no effect on the mean length of time it takes a pain reliever to provide relief.

H_a: The type of medication has an effect on the mean length of time it takes a pain reliever to provide relief.

Hypotheses for interaction effect:

H_0: There is no interaction effect between gender and type of medication on the mean length of time it takes a pain reliever to provide relief.

H_a: There is an interaction effect between gender and type of medication on the mean length of time it takes a pain reliever to provide relief.

To test these hypotheses, you can perform a two-way ANOVA test. Note that the conditions for a two-way ANOVA test are the same as those for a one-way ANOVA test with the additional condition that all samples must be of equal size. Using the *F*-distribution, a two-way ANOVA test calculates an *F*-test statistic for each hypothesis. As a result, it is possible to reject none, one, two, or all of the null hypotheses.

The statistics involved with a two-way ANOVA test is beyond the scope of this course. You can, however, use technology such as Minitab to perform a two-way ANOVA test.

Study Tip

If gender and type of medication have no effect on the length of time it takes a pain reliever to provide relief, then there will be no significant difference in the means of the relief times.

10.4 EXERCISES

1. H_0: $\mu_1 = \mu_2 = \mu_3 = \cdots = \mu_k$
H_a: At least one of the means is different from the others.

2. Each sample must be randomly selected from a normal, or approximately normal, population. The samples must be independent of each other. Each population must have the same variance.

3. The MS_B measures the differences related to the treatment given to each sample. The MS_W measures the differences related to entries within the same sample.

4. There are three null hypotheses for a two-way ANOVA test. There is one for each main effect and one for the interaction effect.

5. (a) H_0: $\mu_1 = \mu_2 = \mu_3$
H_a: At least one mean is different from the others. (claim)

(b) $F_0 = 3.89$
Rejection region: $F > 3.89$

(c) 4.80 (d) Reject H_0.

(e) There is enough evidence at the 5% level of significance to conclude that at least one mean cost per ounce is different from the others.

6. (a) H_0: $\mu_1 = \mu_2 = \mu_3$
H_a: At least one mean is different from the others. (claim)

(b) $F_0 = 3.89$
Rejection region: $F > 3.89$

(c) 3.35 (d) Fail to reject H_0.

(e) There is not enough evidence at the 5% level of significance to conclude that at least one mean battery price is different from the others.

7. (a) H_0: $\mu_1 = \mu_2 = \mu_3$
H_a: At least one mean is different from the others. (claim)

(b)–(e) See Odd Answers, page A86.

Building Basic Skills and Vocabulary

1. State the null and alternative hypotheses for a one-way ANOVA test.

2. What conditions are necessary in order to use a one-way ANOVA test?

3. Describe the difference between the variance between samples MS_B and the variance within samples MS_W.

4. Describe the hypotheses for a two-way ANOVA test.

Using and Interpreting Concepts

Performing a One-Way ANOVA Test *In Exercises 5–14, (a) identify the claim and state H_0 and H_a, (b) find the critical value and identify the rejection region, (c) find the test statistic F, (d) decide whether to reject or fail to reject the null hypothesis, and (e) interpret the decision in the context of the original claim. Assume the samples are random and independent, the populations are normally distributed, and the population variances are equal.*

5. Toothpaste The table shows the costs per ounce (in dollars) for a sample of toothpastes exhibiting very good stain removal, good stain removal, and fair stain removal. At $\alpha = 0.05$, can you conclude that at least one mean cost per ounce is different from the others? *(Source: Consumer Reports)*

Very good	0.47	0.49	0.41	0.37	0.48	0.51
Good	0.60	0.64	0.58	0.75	0.46	
Fair	0.34	0.46	0.44	0.60		

6. Automobile Batteries The table shows the prices (in dollars) for a sample of automobile batteries. The prices are classified according to battery type. At $\alpha = 0.05$, is there enough evidence to conclude that at least one mean battery price is different from the others? *(Adapted from Consumer Reports)*

Group size 35	110	100	125	90	120
Group size 65	280	145	180	175	90
Group size 24/24F	140	125	85	140	80

7. Vacuum Cleaners The table shows the weights (in pounds) for a sample of vacuum cleaners. The weights are classified according to vacuum cleaner type. At $\alpha = 0.01$, can you conclude that at least one mean vacuum cleaner weight is different from the others? *(Source: Consumer Reports)*

Bagged upright	21	22	23	21	17	19
Bagless upright	16	18	19	18	17	20
Bagged canister	26	24	23	25	27	21

8. (a) H_0: $\mu_1 = \mu_2 = \mu_3$
   H_a: At least one mean is different from the others. (claim)
   (b) $F_0 = 5.49$
   Rejection region: $F > 5.49$
   (c) 34.33   (d) Reject H_0.
   (e) There is enough evidence at the 1% level of significance to conclude that at least one mean salary is different from the others.

9. (a) H_0: $\mu_1 = \mu_2 = \mu_3 = \mu_4$
   H_a: At least one mean is different from the others. (claim)
   (b) $F_0 = 2.84$
   Rejection region: $F > 2.84$
   (c) 0.62   (d) Fail to reject H_0.
   (e) There is not enough evidence at the 5% level of significance to conclude that at least one mean age is different from the others.

10. (a) H_0: $\mu_1 = \mu_2 = \mu_3 = \mu_4 = \mu_5$
    H_a: At least one mean is different from the others. (claim)
    (b) $F_0 = 4.37$
    Rejection region: $F > 4.37$
    (c) 9.82   (d) Reject H_0.
    (e) There is enough evidence at the 1% level of significance to conclude that at least one mean cost per mile is different from the others.

 8. Government Salaries The table shows the salaries (in thousands of dollars) for a sample of individuals from the federal, state, and local levels of government. At $\alpha = 0.01$, can you conclude that at least one mean salary is different from the others? *(Adapted from Bureau of Labor Statistics)*

Federal	State	Local
75.2	57.9	52.5
67.9	42.0	42.0
79.3	59.0	46.3
87.1	59.5	44.7
86.4	61.7	55.3
90.5	66.8	49.4
61.1	44.9	64.0
76.0	55.4	44.5
85.7	58.6	40.9
69.4	52.4	37.1

 9. Ages of Professional Athletes The table shows the ages (in years) for a sample of professional athletes from several sports. At $\alpha = 0.05$, can you conclude that at least one mean age is different from the others? *(Source: ESPN)*

MLB	NBA	NFL	NHL
30	28	26	29
25	27	28	23
26	29	27	26
31	30	26	30
27	24	29	27
29	27	27	25
27	28	26	24
25	33	26	26
27	26	27	29
23	28	27	32
26	27	29	28
34	28	25	25
29	26	24	27

10. Cost Per Mile The table shows the costs per mile (in cents) for a sample of automobiles. At $\alpha = 0.01$, can you conclude that at least one mean cost per mile is different from the others? *(Adapted from American Automobile Association)*

Small sedan	Medium sedan	Large sedan	SUV 4WD	Minivan
41	65	60	79	64
39	47	69	58	74
47	61	79	67	57
52	57	71	70	49
44	62	76		68
	50	68		

11. (a) H_0: $\mu_1 = \mu_2 = \mu_3 = \mu_4$
(claim)

H_a: At least one mean is different from the others.

(b) $F_0 = 2.28$
Rejection region: $F > 2.28$

(c) 3.67 (d) Reject H_0.

(e) There is enough evidence at the 10% level of significance to reject the claim that the mean scores are the same for all regions.

12. (a) H_0: $\mu_1 = \mu_2 = \mu_3 = \mu_4$ (claim)

H_a: At least one mean is different from the others.

(b) $F_0 = 4.54$
Rejection region: $F > 4.54$

(c) 0.56 (d) Fail to reject H_0.

(e) There is not enough evidence at the 1% level of significance for the company to reject the claim that the mean number of days patients spend at the hospital is the same for all four regions.

13. (a)
H_0: $\mu_1 = \mu_2 = \mu_3 = \mu_4 = \mu_5 = \mu_6$

H_a: At least one mean is different from the others. (claim)

(b) $F_0 = 2.53$
Rejection region: $F > 2.53$

(c) 2.06 (d) Fail to reject H_0.

(e) There is not enough evidence at the 5% level of significance to conclude that the mean salary is different in at least one of the areas.

 11. Well-Being Index The well-being index is a way to measure how people are faring physically, emotionally, socially, and professionally, as well as to rate the overall quality of their lives and their outlooks for the future. The table shows the well-being index scores for a sample of states from four regions of the United States. At $\alpha = 0.10$, can you reject the claim that the mean score is the same for all regions? *(Adapted from Gallup and Healthways)*

Northeast	Midwest	South	West
61.7	61.6	61.0	64.0
63.6	61.8	60.8	63.0
62.6	61.4	63.1	63.5
61.8	63.2	62.3	63.2
62.1	61.7	60.5	61.8
63.5	62.9	62.0	62.6
	62.9	61.3	62.5
	63.7	60.5	62.8
		62.3	62.5
		61.5	
		63.1	

 12. Days Spent at the Hospital In a recent study, a health insurance company investigated the number of days patients spent at the hospital. In part of the study, the company selected a sample of patients from four regions of the United States and recorded the number of days each patient spent at the hospital. The table shows the results of the study. At $\alpha = 0.01$, can the company reject the claim that the mean number of days patients spend at the hospital is the same for all four regions? *(Adapted from National Center for Health Statistics)*

Northeast	Midwest	South	West
6	6	3	3
4	6	5	4
7	7	6	6
2	3	6	4
3	5	3	6
4	4	7	6
6	4	4	5
8	3		2
9	2		

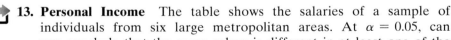 **13. Personal Income** The table shows the salaries of a sample of individuals from six large metropolitan areas. At $\alpha = 0.05$, can you conclude that the mean salary is different in at least one of the areas? *(Adapted from U.S. Bureau of Economic Analysis)*

Chicago	Dallas	Miami	Denver	San Diego	Seattle
48,581	42,524	49,357	48,790	53,370	63,678
42,731	39,709	53,207	49,970	50,470	54,043
51,831	46,209	40,557	53,990	48,920	51,943
58,031	57,704	52,357	57,290	59,670	58,543
57,551	46,909	44,907	60,565	46,770	63,418
47,131	59,259	48,757	51,390		
	53,269	53,557			

14. (a) $H_0: \mu_1 = \mu_2 = \mu_3$
H_a: At least one mean is different from the others. (claim)

(b) $F_0 = 2.50$
Rejection region: $F > 2.50$

(c) 0.93   (d) Fail to reject H_0.

(e) There is not enough evidence at the 10% level of significance to conclude that at least one mean sale price is different from the others.

15. Fail to reject all null hypotheses. The interaction between the advertising medium and the length of the ad has no effect on the rating and therefore there is no significant difference in the means of the ratings.

16. Type of vehicle has an effect on the mean number of vehicles sold in a month. There is an interaction effect between type of vehicle and gender on the number of vehicles sold in a month.

 14. Housing Prices The table shows the sale prices (in thousands of dollars) of a sample of one-family houses in three cities. At $\alpha = 0.10$, can you conclude that at least one mean sale price is different from the others? *(Adapted from National Association of Realtors)*

Gainesville	Orlando	Tampa
173.0	243.9	230.7
145.5	201.1	115.7
190.6	185.3	211.0
186.3	187.5	203.5
248.7	207.9	149.9
206.4	234.8	166.8
86.8	253.2	134.1
204.6	144.7	214.2
174.5	163.9	105.5
220.0	173.3	216.2
173.0		

Extending Concepts

Using Technology to Perform a Two-Way ANOVA Test *In Exercises 15–18, use technology and the block design to perform a two-way ANOVA test. Use $\alpha = 0.10$. Interpret the results. Assume the samples are random and independent, the populations are normally distributed, and the population variances are equal.*

 15. Advertising A study was conducted in which a sample of 20 adults was asked to rate the effectiveness of advertisements. Each adult rated a radio or television advertisement that lasted 30 or 60 seconds. The block design shows these ratings (on a scale of 1 to 5, with 5 being extremely effective).

	Advertising medium	
Length of ad	Radio	Television
30 sec	2, 3, 5, 1, 3	3, 5, 4, 1, 2
60 sec	1, 4, 2, 2, 5	2, 5, 3, 4, 4

 16. Vehicle Sales The owner of a car dealership wants to determine whether the gender of a salesperson and the type of vehicle sold affect the number of vehicles sold in a month. The block design shows the numbers of vehicles, listed by type, sold in a month by a sample of eight salespeople.

	Type of vehicle		
Gender	Car	Truck	Van/SUV
Male	6, 5, 4, 5	2, 2, 1, 3	4, 3, 4, 2
Female	5, 7, 8, 7	1, 0, 1, 2	4, 2, 0, 1

17. Fail to reject all null hypotheses. The interaction between age and gender has no effect on GPA and therefore there is no significant difference in the means of the GPAs.

18. There is an interaction effect between technician and brand on the mean time it takes to repair a laptop.

19. $CV_{\text{Scheffé}} = 7.78$
    (1, 2) → 8.05 → Significant difference
    (1, 3) → 0.01 → No difference
    (2, 3) → 6.13 → No difference

20. $CV_{\text{Scheffé}} = 12.72$
    (1, 2) → 4.95 → No difference
    (1, 3) → 11.61 → No difference
    (2, 3) → 31.73 → Significant difference

21. $CV_{\text{Scheffé}} = 10.98$
    (1, 2) → 34.18 → Significant difference
    (1, 3) → 64.14 → Significant difference
    (2, 3) → 4.67 → No difference

22. $CV_{\text{Scheffé}} = 6.84$
    (1, 2) → 0.11 → No difference
    (1, 3) → 4.23 → No difference
    (1, 4) → 0.55 → No difference
    (2, 3) → 3.46 → No difference
    (2, 4) → 1.37 → No difference
    (3, 4) → 10.17 → Significant difference

 17. Grade Point Average A study was conducted in which a sample of 24 high school students was asked to give their grade point average (GPA). The block design shows the GPAs of male and female students from four different age groups.

		Age			
		15	16	17	18
Gender	Male	2.5, 2.1, 3.8	4.0, 1.4, 2.0	3.5, 2.2, 2.0	3.1, 0.7, 2.8
	Female	4.0, 2.1, 1.9	3.5, 3.0, 2.1	4.0, 2.2, 1.7	1.6, 2.5, 3.6

18. Laptop Repairs The manager of a computer repair service wants to determine whether there is a difference in the time it takes four technicians to repair different brands of laptops. The block design shows the times (in minutes) it took for each technician to repair three laptops of each brand.

		Technician			
		Technician 1	Technician 2	Technician 3	Technician 4
Brand	Brand A	67, 82, 64	42, 56, 39	69, 47, 38	70, 44, 50
	Brand B	44, 62, 55	47, 58, 62	55, 45, 66	47, 29, 40
	Brand C	47, 36, 68	39, 74, 51	74, 80, 70	45, 62, 59

The Scheffé Test *If the null hypothesis is rejected in a one-way ANOVA test of three or more means, then a **Scheffé Test** can be performed to find which means have a significant difference. In a Scheffé Test, the means are compared two at a time. For instance, with three means you would have these comparisons: $\bar{x}_1$ versus $\bar{x}_2$, $\bar{x}_1$ versus $\bar{x}_3$, and $\bar{x}_2$ versus $\bar{x}_3$. For each comparison, calculate*

$$\frac{(\bar{x}_a - \bar{x}_b)^2}{\dfrac{SS_W}{\Sigma(n_i - 1)}\left(\dfrac{1}{n_a} + \dfrac{1}{n_b}\right)}$$

where $\bar{x}_a$ and $\bar{x}_b$ are the means being compared and n_a and n_b are the corresponding sample sizes. Calculate the critical value by multiplying the critical value of the one-way ANOVA test by $k - 1$. Then compare the value that is calculated using the formula above with the critical value. The means have a significant difference when the value calculated using the formula above is greater than the critical value.

Use the information above to solve Exercises 19–22.

19. Refer to the data in Exercise 5. At $\alpha = 0.05$, perform a Scheffé Test to determine which means have a significant difference.

20. Refer to the data in Exercise 7. At $\alpha = 0.01$, perform a Scheffé Test to determine which means have a significant difference.

21. Refer to the data in Exercise 8. At $\alpha = 0.01$, perform a Scheffé Test to determine which means have a significant difference.

22. Refer to the data in Exercise 11. At $\alpha = 0.10$, perform a Scheffé Test to determine which means have a significant difference.

Uses

One-Way Analysis of Variance (ANOVA) ANOVA can help you make important decisions about the allocation of resources. For instance, suppose you work for a large manufacturing company and part of your responsibility is to determine the distribution of the company's sales throughout the world and decide where to focus the company's efforts. Because wrong decisions will cost your company money, you want to make sure that you make the right decisions.

Abuses

Preconceived Notions There are several ways that the tests presented in this chapter can be abused. For instance, it is easy to allow preconceived notions to affect the results of a chi-square goodness-of-fit test and a chi-square independence test. When testing to see whether a distribution has changed, do not let the existing distribution "cloud" the study results. Similarly, when determining whether two variables are independent, do not let your intuition "get in the way." As with any hypothesis test, you must properly gather appropriate data and perform the corresponding test before you can reach a logical conclusion.

Incorrect Interpretation of Rejection of Null Hypothesis It is important to remember that when you reject the null hypothesis of an ANOVA test, you are simply stating that you have enough evidence to determine that at least one of the population means is different from the others. You are not finding them all to be different. One way to further test which of the population means differs from the others is explained in Extending Concepts in Section 10.4 Exercises.

EXERCISES

1. ***Preconceived Notions*** ANOVA depends on having independent variables. Describe an abuse that might occur by having dependent variables. Then describe how the abuse could be avoided.

2. ***Incorrect Interpretation of Rejection of Null Hypothesis*** Find an example of the use of ANOVA. In that use, describe what would be meant by "rejection of the null hypothesis." How should rejection of the null hypothesis be correctly interpreted?

10 Chapter Summary

What Did You Learn?	Example(s)	Review Exercises
Section 10.1		
▷ How to use the chi-square distribution to test whether a frequency distribution fits an expected distribution $$\chi^2 = \Sigma \frac{(O-E)^2}{E}$$	1–3	1–4
Section 10.2		
▷ How to use a contingency table to find expected frequencies $$E_{r,\,c} = \frac{(\text{Sum of row } r) \cdot (\text{Sum of column } c)}{\text{Sample size}}$$	1	5–8
▷ How to use a chi-square distribution to test whether two variables are independent	2, 3	5–8
Section 10.3		
▷ How to interpret the F-distribution and use an F-table to find critical values $$F = \frac{s_1^2}{s_2^2}$$	1, 2	9–16
▷ How to perform a two-sample F-test to compare two variances	3, 4	17–20
Section 10.4		
▷ How to use one-way analysis of variance to test claims involving three or more means $$F = \frac{MS_B}{MS_W}$$	1, 2	21, 22

10 Review Exercises

1. (a) H_0: The distribution of the lengths of office visits is 4% less than 9 minutes, 24% 10–12 minutes, 26% 13–16 minutes, 22% 17–20 minutes, 6% 21–24 minutes, and 18% 25 or more minutes.

   H_a: The distribution of the lengths differs from the expected distribution. (claim)

   (b) $\chi_0^2 = 15.086$
   Rejection region: $\chi^2 > 15.086$

   (c) 18.770   (d) Reject H_0.

   (e) There is enough evidence at the 1% level of significance to conclude that the distribution of the lengths differs from the expected distribution.

2. See Selected Answers, page A108.

3. See Odd Answers, page A87.

FIGURE FOR EXERCISE 2

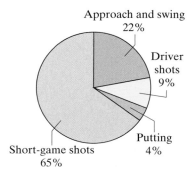

FIGURE FOR EXERCISE 3

Section 10.1

In Exercises 1–4, (a) identify the claim and state H_0 and H_a, (b) find the critical value and identify the rejection region, (c) find the chi-square test statistic, (d) decide whether to reject or fail to reject the null hypothesis, and (e) interpret the decision in the context of the original claim.

1. A researcher claims that the distribution of the lengths of visits at physician offices is different from the distribution shown in the pie chart. You randomly select 400 people and ask them how long their office visits with a physician were. The table shows the results. At $\alpha = 0.01$, test the researcher's claim. *(Adapted from Medscape)*

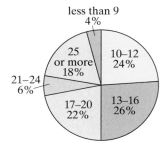

Survey results	
Minutes	**Frequency, f**
less than 9	20
10–12	80
13–16	113
17–20	91
21–24	40
25 or more	56

2. A researcher claims that the distribution of the amounts that parents give for an allowance is different from the distribution shown in the pie chart. You randomly select 1103 parents and ask them how much they give for an allowance. The table shows the results. At $\alpha = 0.10$, test the researcher's claim. *(Adapted from Echo Research)*

Survey results	
Response	**Frequency, f**
Less than $10	353
$10 to $20	167
More than $21	94
Don't give one/other	489

3. A sports magazine claims that the opinions of golf students about what they need the most help with in golf are distributed as shown in the pie chart. You randomly select 435 golf students and ask them what they need the most help with in golf. The table shows the results. At $\alpha = 0.05$, test the sports magazine's claim. *(Adapted from PGA of America)*

Survey results	
Response	**Frequency, f**
Short-game shots	276
Approach and swing	99
Driver shots	42
Putting	18

4. (a) H_0: The distribution of charges for tuition, fees, room, and board at 4-year degree-granting postsecondary institutions is uniform. (claim)

H_a: The distribution of charges for tuition, fees, room, and board is not uniform.

(b) $\chi_0^2 = 9.488$
Rejection region: $\chi^2 > 9.488$

(c) 78.038　　(d) Reject H_0.

(e) There is enough evidence at the 5% level of significance to reject the claim that the distribution of charges for tuition, fees, room, and board at 4-year degree-granting postsecondary institutions is uniform.

5. (a) $E_{1,1} \approx 95.4$, $E_{1,2} \approx 349.2$, $E_{1,3} \approx 383.4$, $E_{1,4} \approx 222.0$, $E_{2,1} \approx 222.6$, $E_{2,2} \approx 814.8$, $E_{2,3} \approx 894.6$, $E_{2,4} \approx 518.0$

(b) H_0: The years of full-time teaching experience is independent of gender.

H_a: The years of full-time teaching experience is dependent on gender. (claim)

(c) d.f. = 3; $\chi_0^2 = 11.345$;
Rejection region: $\chi^2 > 11.345$

(d) 3.815　　(e) Fail to reject H_0.

(f) There is not enough evidence at the 1% level of significance to conclude that the years of full-time teaching experience is dependent on gender.

6. (a) $E_{1,1} = 94.25$, $E_{1,2} = 81.2$, $E_{1,3} = 50.75$, $E_{1,4} = 5.8$, $E_{2,1} = 100.75$, $E_{2,2} = 86.8$, $E_{2,3} = 54.25$, $E_{2,4} = 6.2$

(b) H_0: The type of vehicle owned is independent of gender.

H_a: The type of vehicle owned is dependent on gender. (claim)

(c) d.f. = 3; $\chi_0^2 = 11.345$;
Rejection region: $\chi^2 > 11.345$

(d) 9.649　　(e) Fail to reject H_0.

(f) See Selected Answers, page A108.

7. See Odd Answers, page A87.

---

4. An education researcher claims that the charges for tuition, fees, room, and board at 4-year degree-granting postsecondary institutions are uniformly distributed. To test this claim, you randomly select 800 4-year degree-granting postsecondary institutions and determine the charges for tuition, fees, room, and board at each. The table shows the results. At $\alpha = 0.05$, test the education researcher's claim. *(Adapted from National Center for Education Statistics)*

Cost	Frequency, f
$15,000–$17,499	138
$17,500–$19,999	154
$20,000–$22,499	246
$22,500–$24,499	169
$25,000 or more	93

Section 10.2

In Exercises 5–8, (a) find the expected frequency for each cell in the contingency table, (b) identify the claim and state H_0 and H_a, (c) determine the degrees of freedom, find the critical value, and identify the rejection region, (d) find the chi-square test statistic, (e) decide whether to reject or fail to reject the null hypothesis, and (f) interpret the decision in the context of the original claim.

5. The contingency table shows the results of a random sample of public elementary and secondary school teachers by gender and years of full-time teaching experience. At $\alpha = 0.01$, can you conclude that gender is related to the years of full-time teaching experience? *(Adapted from U.S. National Center for Education Statistics)*

Gender	Years of full-time teaching experience			
	Less than 3 years	**3–9 years**	**10–20 years**	**20 years or more**
Male	102	339	402	207
Female	216	825	876	533

6. The contingency table shows the results of a random sample of individuals by gender and type of vehicle owned. At $\alpha = 0.01$, can you conclude that gender is related to the type of vehicle owned?

Gender	Type of vehicle owned			
	Car	**Truck**	**SUV**	**Van**
Male	85	95	44	8
Female	110	73	61	4

7. The contingency table shows the results of a random sample of endangered and threatened species by status and vertebrate group. At $\alpha = 0.05$, test the hypothesis that the variables are independent. *(Adapted from U.S. Fish and Wildlife Service)*

Status	Vertebrate group				
	Mammals	**Birds**	**Reptiles**	**Amphibians**	**Fish**
Endangered	151	137	37	18	50
Threatened	23	17	22	12	33

8. (a) $E_{1,1} \approx 552.6$, $E_{1,2} \approx 602.1$,
$E_{1,3} \approx 918.1$, $E_{1,4} \approx 928.2$,
$E_{2,1} \approx 318.4$, $E_{2,2} \approx 346.9$,
$E_{2,3} \approx 528.9$, $E_{2,4} \approx 534.8$

(b) H_0: The time of day of a collision is independent of gender.

H_a: The time of day of a collision is dependent on gender. (claim)

(c) d.f. $= 3$; $\chi_0^2 = 6.251$
Rejection region: $\chi^2 > 6.251$

(d) 21.436 **(e)** Reject H_0.

(f) There is enough evidence at the 10% level of significance to conclude that the time of day of a collision is dependent on gender.

9. 2.295 **10.** 4.71 **11.** 2.39

12. 2.01 **13.** 2.06 **14.** 4.36

15. 2.08 **16.** 4.73

17. (a) H_0: $\sigma_1 = \sigma_2$ (claim)

H_a: $\sigma_1 \neq \sigma_2$

(b) $F_0 = 2.575$
Rejection region: $F > 2.575$

(c) 2.905 **(d)** Reject H_0.

(e) There is enough evidence at the 1% level of significance to reject the claim that the standard deviations of hotel room rates for San Francisco, CA, and Sacramento, CA, are the same.

18. (a) H_0: $\sigma_1^2 \leq \sigma_2^2$

H_a: $\sigma_1^2 > \sigma_2^2$ (claim)

(b) $F_0 = 1.92$
Rejection region: $F > 1.92$

(c) 1.72 **(d)** Fail to reject H_0.

(e) There is not enough evidence at the 10% level of significance to support the claim that the variation in wheat production is greater in Garfield County than in Kay County.

19. (a) H_0: $\sigma_1^2 = \sigma_2^2$

H_a: $\sigma_1^2 \neq \sigma_2^2$ (claim)

(b) $F_0 = 5.32$
Rejection region: $F > 5.32$

(c) 1.37 **(d)** Fail to reject H_0.

(e) There is not enough evidence at the 1% level of significance to support the claim that the test score variance for females is different from that for males.

---

8. The contingency table shows the distribution of a random sample of fatal pedestrian motor vehicle collisions by time of day and gender in a recent year. At $\alpha = 0.10$, can you conclude that time of day and gender are related? *(Adapted from National Highway Traffic Safety Administration)*

	Time of day			
Gender	**12 A.M.–5:59 A.M.**	**6 A.M.–11:59 A.M.**	**12 P.M.–5:59 P.M.**	**6 P.M.–11:59 P.M.**
Male	611	595	884	911
Female	260	354	563	552

Section 10.3

In Exercises 9–12, find the critical F-value for a right-tailed test using the level of significance α and degrees of freedom d.f.$_N$ and d.f.$_D$.

9. $\alpha = 0.05$, d.f.$_N = 6$, d.f.$_D = 50$

10. $\alpha = 0.01$, d.f.$_N = 12$, d.f.$_D = 10$

11. $\alpha = 0.10$, d.f.$_N = 5$, d.f.$_D = 12$

12. $\alpha = 0.05$, d.f.$_N = 20$, d.f.$_D = 25$

In Exercises 13–16, find the critical F-value for a two-tailed test using the level of significance α and degrees of freedom d.f.$_N$ and d.f.$_D$.

13. $\alpha = 0.10$, d.f.$_N = 15$, d.f.$_D = 27$

14. $\alpha = 0.05$, d.f.$_N = 9$, d.f.$_D = 8$

15. $\alpha = 0.01$, d.f.$_N = 40$, d.f.$_D = 60$

16. $\alpha = 0.01$, d.f.$_N = 11$, d.f.$_D = 13$

In Exercises 17–20, (a) identify the claim and state H_0 and H_a, (b) find the critical value and identify the rejection region, (c) find the test statistic F, (d) decide whether to reject or fail to reject the null hypothesis, and (e) interpret the decision in the context of the original claim. Assume the samples are random and independent, and the populations are normally distributed.

17. A travel consultant claims that the standard deviations of hotel room rates for San Francisco, CA, and Sacramento, CA, are the same. A sample of 36 hotel room rates in San Francisco has a standard deviation of \$75 and a sample of 31 hotel room rates in Sacramento has a standard deviation of \$44. At $\alpha = 0.01$, can you reject the travel consultant's claim? *(Adapted from I-Map Data Systems LLC)*

18. An agricultural analyst is comparing the wheat production in Oklahoma counties. The analyst claims that the variation in wheat production is greater in Garfield County than in Kay County. A sample of 21 Garfield County farms has a standard deviation of 0.76 bushel per acre. A sample of 16 Kay County farms has a standard deviation of 0.58 bushel per acre. At $\alpha = 0.10$, can you support the analyst's claim? *(Adapted from Environmental Verification and Analysis Center — University of Oklahoma)*

 19. An instructor claims that the variance of SAT critical reading scores for females is different than the variance of SAT critical reading scores for males. The table shows the SAT critical reading scores for 12 randomly selected female students and 12 randomly selected male students. At $\alpha = 0.01$, can you support the instructor's claim?

Female		Male	
480	600	560	310
610	800	680	730
340	540	360	740
630	750	530	520
520	650	380	560
690	630	460	400

20. (a) H_0: $\sigma_1^2 \le \sigma_2^2$
H_a: $\sigma_1^2 > \sigma_2^2$ (claim)

(b) $F_0 = 2.82$
Rejection region: $F > 2.82$

(c) 2.91   (d) Reject H_0.

(e) There is enough evidence at the 5% level of significance to support the claim that the new mold produces inserts that are less variable in diameter than those produced with the current mold.

21. (a) H_0: $\mu_1 = \mu_2 = \mu_3 = \mu_4$
H_a: At least one mean is different from the others. (claim)

(b) $F_0 = 2.29$
Rejection region: $F > 2.29$

(c) 6.19   (d) Reject H_0.

(e) There is enough evidence at the 10% level of significance to conclude that at least one mean amount spent on energy is different from the others.

22. (a) H_0: $\mu_1 = \mu_2 = \mu_3 = \mu_4$
H_a: At least one mean is different from the others. (claim)

(b) $F_0 = 3.10$
Rejection region: $F > 3.10$

(c) 1.85   (d) Fail to reject H_0.

(e) There is not enough evidence at the 5% level of significance to conclude that at least one of the mean incomes is different from the others.

 20. A quality technician claims that the variance of the insert diameters produced by a plastic company's new injection mold for automobile dashboard inserts is less than the variance of the insert diameters produced by the company's current mold. The table shows samples of insert diameters (in centimeters) for both the current and new molds. At $\alpha = 0.05$, can you support the technician's claim?

New	9.611	9.618	9.594	9.580	9.611	9.597
Current	9.571	9.642	9.650	9.651	9.596	9.636

New	9.638	9.568	9.605	9.603	9.647	9.590
Current	9.570	9.537	9.641	9.625	9.626	9.579

Section 10.4

In Exercises 21 and 22, (a) identify the claim and state H_0 and H_a, (b) find the critical value and identify the rejection region, (c) find the test statistic F, (d) decide whether to reject or fail to reject the null hypothesis, and (e) interpret the decision in the context of the original claim. Assume the samples are random and independent, the populations are normally distributed, and the population variances are equal.

 21. The table shows the amounts spent (in dollars) on energy in one year for a sample of households from four regions of the United States. At $\alpha = 0.10$, can you conclude that the mean amount spent on energy in one year is different in at least one of the regions? *(Adapted from U.S. Energy Information Administration)*

Northeast	**Midwest**	**South**	**West**
1896	1712	1689	1455
2606	2096	2256	1164
1649	1923	1834	1851
2436	2281	2365	1776
2811	2703	1958	2030
2384	2092	1947	1640
2840	1499	2433	1678
2445	2146	1578	1547

 22. The table shows the annual incomes (in dollars) for a sample of families from four regions of the United States. At $\alpha = 0.05$, can you conclude that the mean annual income of families is different in at least one of the regions? *(Adapted from U.S. Census Bureau)*

Northeast	**Midwest**	**South**	**West**
78,123	54,930	52,623	70,496
69,388	78,543	76,365	62,904
78,251	76,602	50,668	59,113
54,379	57,357	50,373	57,191
75,210	54,907	48,536	60,668
	70,119	63,073	60,415
	46,833		

10 | Chapter Quiz

1. (a) H_0: The distribution of educational attainment for people in the United States ages 30–34 is 4.7% none–8th grade, 6.9% 9th–11th grade, 29.5% high school graduates, 16.6% some college, no degree, 9.8% associate's degree, 20.5% bachelor's degree, 8.7% master's degree, and 3.3% professional/doctoral degree.

   H_a: The distribution of educational attainment for people in the United States ages 30–34 differs from the distribution for people ages 25 and older. (claim)

   (b) $\chi_0^2 = 14.067$
   Rejection region: $\chi^2 > 14.067$

   (c) 6.026   (d) Fail to reject H_0.

   (e) There is not enough evidence at the 5% level of significance to conclude that the distribution for people in the United States ages 30–34 differs from the distribution for people ages 25 and older.

2. (a) H_0: Age and educational attainment are independent.

   H_a: Age and educational attainment are dependent. (claim)

   (b) $\chi_0^2 = 18.475$
   Rejection region: $\chi^2 > 18.475$

   (c) 8.187   (d) Fail to reject H_0.

   (e) There is not enough evidence at the 1% level of significance to conclude that educational attainment is dependent on age.

3. (a) H_0: $\sigma_1^2 = \sigma_2^2$
   H_a: $\sigma_1^2 \neq \sigma_2^2$ (claim)

   (b) $F_0 = 4.43$
   Rejection region: $F > 4.43$

   (c) 1.38   (d) Fail to reject H_0.

   (e) See Odd Answers, page A87.

4. See Odd Answers, page A87.

Take this quiz as you would take a quiz in class. After you are done, check your work against the answers given in the back of the book.

In each exercise,

(a) identify the claim and state H_0 and H_a,

(b) find the critical value and identify the rejection region,

(c) find the test statistic,

(d) decide whether to reject or fail to reject the null hypothesis, and

(e) interpret the decision in the context of the original claim.

In Exercises 1 and 2, use the table, which lists the distribution of educational achievement for people in the United States ages 25 and older. It also lists the results of a random survey for two additional age groups. (Adapted from U.S. Census Bureau)

Educational attainment	Ages		
	25 and older	30–34	65–69
None–8th grade	4.7%	10	23
9th–11th grade	6.9%	23	26
High school graduate	29.5%	80	136
Some college, no degree	16.6%	56	77
Associate's degree	9.8%	34	41
Bachelor's degree	20.5%	75	78
Master's degree	8.7%	31	41
Professional/doctoral degree	3.3%	11	18

1. Does the distribution for people in the United States ages 25 and older differ from the distribution for people in the United States ages 30–34? Use $\alpha = 0.05$.

2. Use the data for 30- to 34-year-olds and 65- to 69-year-olds to test whether age and educational attainment are related. Use $\alpha = 0.01$.

 In Exercises 3 and 4, use the data, which list the annual wages (in thousands of dollars) for randomly selected individuals from three metropolitan areas. Assume the wages are normally distributed and that the samples are independent. (Adapted from U.S. Bureau of Economic Analysis)

   Ithaca, NY: 44.2, 51.5, 25.8, 28.3, 37.8, 38.0, 32.6, 41.8, 42.0, 40.6, 26.2, 27.9, 48.3

   Little Rock, AR: 45.1, 38.1, 47.8, 34.4, 39.6, 47.1, 29.6, 54.8, 34.4, 40.3, 40.1, 41.7, 40.9, 38.9, 25.9

   Madison, WI: 50.3, 41.8, 55.5, 40.8, 55.6, 38.6, 50.0, 46.8, 49.0, 52.9, 48.3, 47.5, 39.2, 32.7, 54.1

3. At $\alpha = 0.01$, is there enough evidence to conclude that the variances of the annual wages for Ithaca, NY, and Little Rock, AR, are different?

4. Are the mean annual wages the same for all three cities? Use $\alpha = 0.10$. Assume that the population variances are equal.

10 Chapter Test

1. (a) H_0: $\sigma_1^2 = \sigma_2^2$ (claim)
   H_a: $\sigma_1^2 \neq \sigma_2^2$

   (b) $F_0 = 3.66$
   Rejection region: $F > 3.66$

   (c) 1.03   (d) Fail to reject H_0.

   (e) There is not enough evidence at the 5% level of significance to reject the claim that the variances of the hourly wages are the same.

2. (a) H_0: $\sigma_1^2 \leq \sigma_2^2$
   H_a: $\sigma_1^2 > \sigma_2^2$ (claim)

   (b) $F_0 = 4.63$
   Rejection region: $F > 4.63$

   (c) 1.09   (d) Fail to reject H_0.

   (e) There is not enough evidence at the 1% level of significance to support the claim that the variance of the hourly wages in Oklahoma is greater than the variance of the hourly wages in Massachusetts.

3. (a) H_0: $\mu_1 = \mu_2 = \mu_3$ (claim)
   H_a: At least one mean is different from the others.

   (b) $F_0 = 5.39$
   Rejection region: $F > 5.39$

   (c) 13.12   (d) Reject H_0.

   (e) There is enough evidence at the 1% level of significance to reject the claim that the mean hourly wages are the same for all three states.

4. (a) H_0: The distribution of ages of workers in Oklahoma is 6.1% ages 16–19, 11.8% ages 20–24, 41.8% ages 25–44, 21.2% ages 45–54, 10.1% ages 55–59, and 9.0% ages 60+.

   H_a: The distribution of ages of workers in Oklahoma differs from the distribution of ages of workers in Maine. (claim)

   (b)–(e) See Selected Answers, page A108.

5. See Selected Answers, page A108.

6. See Selected Answers, page A108.

Take this test as you would take a test in class.

In each exercise,

(a) identify the claim and state H_0 and H_a,

(b) find the critical value and identify the rejection region,

(c) find the test statistic,

(d) decide whether to reject or fail to reject the null hypothesis, and

(e) interpret the decision in the context of the original claim.

In Exercises 1–3, use the data, which list the hourly wages (in dollars) for randomly selected respiratory therapy technicians from three states. Assume the wages are normally distributed and that the samples are independent. (Adapted from U.S. Bureau of Labor Statistics)

Maine: 20.92, 25.37, 23.06, 15.64, 27.72, 24.90, 19.26, 23.46, 18.49, 21.76, 22.36

Oklahoma: 22.70, 19.95, 17.85, 16.76, 21.32, 18.96, 17.99, 28.35, 25.30, 21.93

Massachusetts: 25.43, 23.21, 30.81, 26.62, 31.42, 31.34, 34.58, 28.22, 27.25, 22.83, 27.45, 27.71

1. At $\alpha = 0.05$, is there enough evidence to conclude that the variances of the hourly wages for respiratory therapy technicians in Maine and Massachusetts are the same?

2. At $\alpha = 0.01$, is there enough evidence to conclude that the variance of the hourly wages for respiratory therapy technicians in Oklahoma is greater than the variance of the hourly wages for respiratory therapy technicians in Massachusetts?

3. Are the mean hourly wages of respiratory therapist technicians the same for all three states? Use $\alpha = 0.01$. Assume that the population variances are equal.

In Exercises 4–6, use the table, which lists the distribution of the ages of workers who carpool in Maine. It also lists the results of a random survey for two additional states. (Adapted from U.S. Census Bureau)

		State	
Ages	**Maine**	**Oklahoma**	**Massachusetts**
16–19	6.1%	11	15
20–24	11.8%	27	22
25–44	41.8%	96	86
45–54	21.2%	37	42
55–59	10.1%	14	17
60+	9.0%	15	18

4. Does the distribution of the ages of workers who carpool in Maine differ from the distribution of the ages of workers who carpool in Oklahoma? Use $\alpha = 0.10$.

5. Is the distribution of the ages of workers who carpool in Maine the same as the distribution of the ages of workers who carpool in Massachusetts? Use $\alpha = 0.01$.

6. Use the data for Oklahoma and Massachusetts to test whether state and age are independent. Use $\alpha = 0.05$.

Fraud.org was created by the National Consumers League (NCL) to combat the growing problem of telemarketing and Internet fraud by improving prevention and enforcement. NCL works to protect and promote social and economic justice for consumers and workers in the United States and abroad.

You work for the NCL as a statistical analyst. You are studying data on fraud. Part of your analysis involves testing the goodness-of-fit, testing for independence, comparing variances, and performing ANOVA.

EXERCISES

1. Goodness-of-Fit

The table at the right shows an expected distribution of the ages of fraud victims. The table also shows the results of a survey of 1000 randomly selected fraud victims. Using $\alpha = 0.01$, perform a chi-square goodness-of-fit test. What can you conclude?

2. Independence

The contingency table below shows the results of a random sample of 2000 fraud victims classified by age and type of fraud. The frauds were committed using bogus sweepstakes or credit card offers.

(a) Calculate the expected frequency for each cell in the contingency table. Assume the variables *age* and *type of fraud* are independent.

(b) Can you conclude that the ages of the victims are related to the type of fraud? Use $\alpha = 0.01$.

Age	Expected distribution	Survey results
Under 18	0.66%	8
18–25	14.92%	148
26–35	22.09%	206
36–45	17.29%	171
46–55	17.05%	175
56–65	14.71%	153
Over 65	13.28%	139

TABLE FOR EXERCISE 1

Type of Fraud	Age								
	Under 20	20–29	30–39	40–49	50–59	60–69	70–79	80+	Total
Sweepstakes	10	60	70	130	90	160	280	200	1000
Credit cards	20	180	260	240	180	70	30	20	1000
Total	30	240	330	370	270	230	310	220	2000

TECHNOLOGY

Teacher Salaries

The Illinois State Board of Education conducts an annual study of the salaries of Illinois teachers. The study looks at how teachers' salaries are distributed based on factors such as degree and experience level, district size, and geographic region.

The table shows the beginning salaries of a random sample of Illinois teachers from different-sized districts. District size is measured by the number of students enrolled.

Teacher salaries

Under 500 students	1000–2999 students	At least 12,000 students
36,462	41,862	40,726
40,877	40,482	39,640
33,937	38,292	40,686
32,957	38,264	37,347
38,313	43,385	46,239
30,313	36,195	44,064
33,490	44,117	41,855
41,357	31,188	39,476
29,892	38,746	44,136
35,237	32,760	44,966
29,760	41,527	46,992
36,580	40,814	39,257
33,547	29,997	39,572

EXERCISES

In Exercises 1–3, refer to the samples listed below. Use $\alpha = 0.05$.

(a) Under 500 students

(b) 1000–2999 students

(c) At least 12,000 students

1. Are the samples independent of each other? Explain.

2. Use technology to determine whether each sample is from a normal, or approximately normal, population.

3. Use technology to determine whether the samples were selected from populations having equal variances.

4. Using the results of Exercises 1–3, discuss whether the three conditions for a one-way ANOVA test are satisfied. If so, use technology to test the claim that teachers from districts of the three sizes have the same mean salary. Use $\alpha = 0.05$.

5. Repeat Exercises 1–4 using the data in the table below. The table displays the beginning salaries of a random sample of Illinois teachers from different geographic regions of Illinois.

Teacher salaries

Northeast	Northwest	Southwest
43,652	32,569	37,176
39,169	24,888	29,454
37,836	29,265	39,337
39,883	36,363	36,502
42,228	32,495	37,233
35,647	32,032	29,984
39,221	42,818	30,694
33,503	31,126	38,303
44,346	31,525	43,313
43,038	32,867	46,053
40,813	33,380	31,086
36,581	33,341	29,364
47,039	38,142	33,607

Extended solutions are given in the technology manuals that accompany this text. Technical instruction is provided for Minitab, Excel, and the TI-84 Plus.

Sections 9.1 and 9.2 **1.** The table below shows the winning times (in seconds) for the men's and women's 100-meter runs in the Summer Olympics from 1928 to 2016. *(Source: The International Association of Athletics Federations)*

Men, x	10.80	10.38	10.30	10.30	10.79	10.62	10.32
Women, y	12.20	11.90	11.50	12.20	11.67	11.82	11.18

Men, x	10.06	9.95	10.14	10.06	10.25	9.99	9.92
Women, y	11.49	11.08	11.07	11.08	11.06	10.97	10.54

Men, x	9.96	9.84	9.87	9.85	9.69	9.63	9.81
Women, y	10.82	10.94	10.75	10.93	10.78	10.75	10.90

(a) Display the data in a scatter plot, calculate the correlation coefficient r, and describe the type of correlation.

(b) At $\alpha = 0.05$, is there enough evidence to conclude that there is a significant linear correlation between the winning times for the men's and women's 100-meter runs?

(c) Find the equation of the regression line for the data. Draw the regression line on the scatter plot.

(d) Use the regression equation to predict the women's 100-meter time when the men's 100-meter time is 9.90 seconds.

Section 10.4 **2.** The table at the right shows the residential natural gas expenditures (in dollars) in one year for a random sample of households in four regions of the United States. Assume that the populations are normally distributed and the population variances are equal. At $\alpha = 0.10$, can you reject the claim that the mean expenditures are the same for all four regions? *(Adapted from U.S. Energy Information Administration)*

Northeast	Midwest	South	West
1608	449	509	591
779	1036	394	504
964	665	769	1011
1303	1213	753	463
1143	921	931	271
1695	1393	574	324
785	926	526	515
778	866	1096	599

Section 9.4 **3.** The equation used to predict the annual sweet potato yield (in pounds per acre) is $\hat{y} = 16{,}212 - 0.227x_1 + 0.212x_2$, where x_1 is the number of acres planted and x_2 is the number of acres harvested. Use the multiple regression equation to predict the annual sweet potato yields for the values of the independent variables. *(Adapted from U.S. Department of Agriculture)*

(a) $x_1 = 110{,}000$, $x_2 = 100{,}000$ (b) $x_1 = 125{,}000$, $x_2 = 115{,}000$

4. A school administrator claims that the standard deviations of reading test scores for eighth-grade students are the same in Colorado and Utah. A random sample of 16 test scores from Colorado has a standard deviation of 34.6 points, and a random sample of 15 test scores from Utah has a standard deviation of 33.2 points. At $\alpha = 0.10$, can you reject the administrator's claim? Assume the samples are independent and each population has a normal distribution. *(Adapted from National Center for Education Statistics)*

5. A researcher claims that the credit card debts of college students are distributed as shown in the pie chart. You randomly select 900 college students and record the credit card debt of each. The table shows the results. At $\alpha = 0.05$, test the researcher's claim. *(Adapted from Sallie Mae, Inc.)*

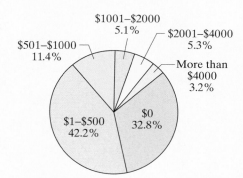

Survey results	
Response	**Frequency, f**
$0	290
$1–$500	397
$501–$1000	97
$1001–$2000	54
$2001–$4000	40
More than $4000	22

6. Reviewing a Movie The contingency table shows how a random sample of adults rated a newly released movie and gender. At $\alpha = 0.05$, can you conclude that the adults' ratings are related to gender?

	Rating			
Gender	**Excellent**	**Good**	**Fair**	**Poor**
Male	97	42	26	5
Female	101	33	25	11

7. The table shows the metacarpal bone lengths (in centimeters) and the heights (in centimeters) of 12 adults. The equation of the regression line is $\hat{y} = 1.707x + 94.380$. *(Adapted from the American Journal of Physical Anthropology)*

Metacarpal bone length, x	45	51	39	41	48	49
Height, y	171	178	157	163	172	183

Metacarpal bone length, x	46	43	47	42	40	44
Height, y	173	175	173	169	160	172

(a) Find the coefficient of determination r^2 and interpret the results.

(b) Find the standard error of estimate s_e and interpret the results.

(c) Construct a 95% prediction interval for the height of an adult whose metacarpal bone length is 50 centimeters. Interpret the results.

APPENDIX A

In this appendix, we use a 0-to-z table as an alternative development of the standard normal distribution. It is intended that this appendix be used after completion of the "Properties of a Normal Distribution" subsection of Section 5.1 in the text. If used, this appendix should replace the material in the "Standard Normal Distribution" subsection of Section 5.1 except for the exercises.

Standard Normal Distribution (0-to-z)

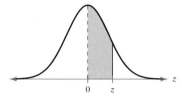

z	.00	.01	.02	.03	.04	.05	.06	.07	.08	.09
0.0	.0000	.0040	.0080	.0120	.0160	.0199	.0239	.0279	.0319	.0359
0.1	.0398	.0438	.0478	.0517	.0557	.0596	.0636	.0675	.0714	.0753
0.2	.0793	.0832	.0871	.0910	.0948	.0987	.1026	.1064	.1103	.1141
0.3	.1179	.1217	.1255	.1293	.1331	.1368	.1406	.1443	.1480	.1517
0.4	.1554	.1591	.1628	.1664	.1700	.1736	.1772	.1808	.1844	.1879
0.5	.1915	.1950	.1985	.2019	.2054	.2088	.2123	.2157	.2190	.2224
0.6	.2257	.2291	.2324	.2357	.2389	.2422	.2454	.2486	.2517	.2549
0.7	.2580	.2611	.2642	.2673	.2704	.2734	.2764	.2794	.2823	.2852
0.8	.2881	.2910	.2939	.2967	.2995	.3023	.3051	.3078	.3106	.3133
0.9	.3159	.3186	.3212	.3238	.3264	.3289	.3315	.3340	.3365	.3389
1.0	.3413	.3438	.3461	.3485	.3508	.3531	.3554	.3577	.3599	.3621
1.1	.3643	.3665	.3686	.3708	.3729	.3749	.3770	.3790	.3810	.3830
1.2	.3849	.3869	.3888	.3907	.3925	.3944	.3962	.3980	.3997	.4015
1.3	.4032	.4049	.4066	.4082	.4099	.4115	.4131	.4147	.4162	.4177
1.4	.4192	.4207	.4222	.4236	.4251	.4265	.4279	.4292	.4306	.4319
1.5	.4332	.4345	.4357	.4370	.4382	.4394	.4406	.4418	.4429	.4441
1.6	.4452	.4463	.4474	.4484	.4495	.4505	.4515	.4525	.4535	.4545
1.7	.4554	.4564	.4573	.4582	.4591	.4599	.4608	.4616	.4625	.4633
1.8	.4641	.4649	.4656	.4664	.4671	.4678	.4686	.4693	.4699	.4706
1.9	.4713	.4719	.4726	.4732	.4738	.4744	.4750	.4756	.4761	.4767
2.0	.4772	.4778	.4783	.4788	.4793	.4798	.4803	.4808	.4812	.4817
2.1	.4821	.4826	.4830	.4834	.4838	.4842	.4846	.4850	.4854	.4857
2.2	.4861	.4864	.4868	.4871	.4875	.4878	.4881	.4884	.4887	.4890
2.3	.4893	.4896	.4898	.4901	.4904	.4906	.4909	.4911	.4913	.4916
2.4	.4918	.4920	.4922	.4925	.4927	.4929	.4931	.4932	.4934	.4936
2.5	.4938	.4940	.4941	.4943	.4945	.4946	.4948	.4949	.4951	.4952
2.6	.4953	.4955	.4956	.4957	.4959	.4960	.4961	.4962	.4963	.4964
2.7	.4965	.4966	.4967	.4968	.4969	.4970	.4971	.4972	.4973	.4974
2.8	.4974	.4975	.4976	.4977	.4977	.4978	.4979	.4979	.4980	.4981
2.9	.4981	.4982	.4982	.4983	.4984	.4984	.4985	.4985	.4986	.4986
3.0	.4987	.4987	.4987	.4988	.4988	.4989	.4989	.4989	.4990	.4990
3.1	.4990	.4991	.4991	.4991	.4992	.4992	.4992	.4992	.4993	.4993
3.2	.4993	.4993	.4994	.4994	.4994	.4994	.4994	.4995	.4995	.4995
3.3	.4995	.4995	.4995	.4996	.4996	.4996	.4996	.4996	.4996	.4997
3.4	.4997	.4997	.4997	.4997	.4997	.4997	.4997	.4997	.4997	.4998

Reprinted with permission of Frederick Mosteller

A   Alternative Presentation of the Standard Normal Distribution

What You Should Learn

▶ How to find areas under the standard normal curve

Study Tip

Because every normal distribution can be transformed to the standard normal distribution, you can use z-scores and the standard normal curve to find areas (and therefore probabilities) under any normal curve.

Note to Instructor

Mention that the formula for a normal probability density function on page 234 is greatly simplified when $\mu = 0$ and $\sigma = 1$.

$$y = \frac{e^{-x^2/2}}{\sqrt{2\pi}}$$

Study Tip

It is important that you know the difference between x and z. The random variable x is sometimes called a raw score and represents values in a nonstandard normal distribution, whereas z represents values in the standard normal distribution.

The Standard Normal Distribution

There are infinitely many normal distributions, each with its own mean and standard deviation. The normal distribution with a mean of 0 and a standard deviation of 1 is called the **standard normal distribution.** The horizontal scale of the graph of the standard normal distribution corresponds to z-scores. In Section 2.5, you learned that a z-score is a measure of position that indicates the number of standard deviations a value lies from the mean. Recall that you can transform an x-value to a z-score using the formula

$$z = \frac{\text{Value} - \text{Mean}}{\text{Standard deviation}} = \frac{x - \mu}{\sigma}.$$ Round to the nearest hundredth.

DEFINITION

The **standard normal distribution** is a normal distribution with a mean of 0 and a standard deviation of 1. The total area under its normal curve is 1.

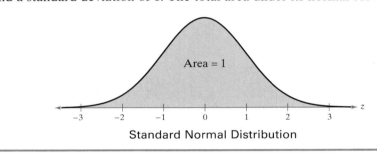

Standard Normal Distribution

When each data value of a normally distributed random variable x is transformed into a z-score, the result will be the standard normal distribution. After this transformation takes place, the area that falls in the interval under the nonstandard normal curve is the *same* as that under the standard normal curve within the corresponding z-boundaries.

In Section 2.4, you learned to use the Empirical Rule to approximate areas under a normal curve when the values of the random variable x corresponded to −3, −2, −1, 0, 1, 2, or 3 standard deviations from the mean. Now, you will learn to calculate areas corresponding to other x-values. After you use the formula above to transform an x-value to a z-score, you can use the Standard Normal Table (0-to-z) on page A1. The table lists the area under the standard normal curve between 0 and the given z-score. As you examine the table, notice the following.

Properties of the Standard Normal Distribution

1. The distribution is symmetric about the mean $(z = 0)$.
2. The area under the standard normal curve to the left of $z = 0$ is 0.5 and the area to the right of $z = 0$ is 0.5.
3. The area under the standard normal curve increases as the distance between 0 and z increases.

At first glance, the table on page A1 appears to give areas for positive z-scores only. However, because of the symmetry of the standard normal curve, the table also gives areas for negative z-scores (see Example 1).

EXAMPLE 1

Using the Standard Normal Table (0-to-z)

1. Find the area under the standard normal curve between $z = 0$ and $z = 1.15$.

2. Find the z-scores that correspond to an area of 0.0948.

SOLUTION

1. Find the area that corresponds to $z = 1.15$ by finding 1.1 in the left column and then moving across the row to the column under 0.05. The number in that row and column is 0.3749. So, the area between $z = 0$ and $z = 1.15$ is 0.3749, as shown in the figure at the left.

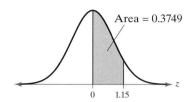
Area = 0.3749

z	.00	.01	.02	.03	.04	.05	.06
0.0	.0000	.0040	.0080	.0120	.0160	.0199	.0239
0.1	.0398	.0438	.0478	.0517	.0557	.0596	.0636
0.2	.0793	.0832	.0871	.0910	.0948	.0987	.1026
0.3	.1179	.1217	.1255	.1293	.1331	.1368	.1406

z	.00	.01	.02	.03	.04	.05	.06
0.9	.3159	.3186	.3212	.3238	.3264	.3289	.3315
1.0	.3413	.3438	.3461	.3485	.3508	.3531	.3554
1.1	.3643	.3665	.3686	.3708	.3729	.3749	.3770
1.2	.3849	.3869	.3888	.3907	.3925	.3944	.3962
1.3	.4032	.4049	.4066	.4082	.4099	.4115	.4131
1.4	.4192	.4207	.4222	.4236	.4251	.4265	.4279

2. Find the z-scores that correspond to an area of 0.0948 by locating 0.0948 in the table. The values at the beginning of the corresponding row and at the top of the corresponding column give the z-score. For an area of 0.0948, the row value is 0.2 and the column value is 0.04. So, the z-scores are $z = -0.24$ and $z = 0.24$, as shown in the figures at the left.

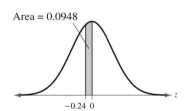
Area = 0.0948

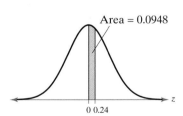
Area = 0.0948

z	.00	.01	.02	.03	.04	.05	.06
0.0	.0000	.0040	.0080	.0120	.0160	.0199	.0239
0.1	.0398	.0438	.0478	.0517	.0557	.0596	.0636
0.2	.0793	.0832	.0871	.0910	.0948	.0987	.1026
0.3	.1179	.1217	.1255	.1293	.1331	.1368	.1406
0.4	.1554	.1591	.1628	.1664	.1700	.1736	.1772
0.5	.1915	.1950	.1985	.2019	.2054	.2088	.2123

TRY IT YOURSELF 1

1. Find the area under the standard normal curve between $z = 0$ and $z = 2.19$.

2. Find the z-scores that correspond to an area of 0.4850.

Answer: Page A39

When the z-score is not in the table, use the entry closest to it. When the z-score is exactly midway between two z-scores, use the area midway between the corresponding areas. In addition to using the table, you can use technology to find the area under the standard normal curve that corresponds to a z-score, as shown at the left using a TI-84 Plus for Part 1 of Example 1.

TI-84 PLUS

```
normalcdf(0,1.15
)
        .3749280109
```

You can use the following guidelines to find various types of areas under the standard normal curve.

GUIDELINES

Finding Areas Under the Standard Normal Curve

1. Sketch the standard normal curve and shade the appropriate area under the curve.

2. Use the Standard Normal Table (0-to-z) on page A1 to find the area that corresponds to the z-score(s).

3. Find the area by following the directions for each case shown.

   a. Area to the left of z

   i. When $z < 0$, *subtract* the area from 0.5.

   2. Subtract to find the area to the left of $z = -1.23$; $0.5 - 0.3907 = 0.1093$.

   1. The area between $z = 0$ and $z = -1.23$ is 0.3907.

   ii. When $z > 0$, *add* 0.5 to the area.

   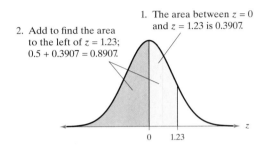

   2. Add to find the area to the left of $z = 1.23$; $0.5 + 0.3907 = 0.8907$.

   1. The area between $z = 0$ and $z = 1.23$ is 0.3907.

   b. Area to the right of z

   i. When $z < 0$, *add* 0.5 to the area.

   1. The area between $z = 0$ and $z = -1.23$ is 0.3907.   2. Add to find the area to the right of $z = -1.23$; $0.5 + 0.3907 = 0.8907$.

   ii. When $z > 0$, *subtract* the area from 0.5.

   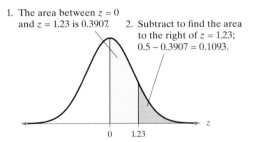

   1. The area between $z = 0$ and $z = 1.23$ is 0.3907.   2. Subtract to find the area to the right of $z = 1.23$; $0.5 - 0.3907 = 0.1093$.

   c. Area between two z-scores

   i. When the two z-scores have the same sign (both positive or both negative), *subtract* the smaller area from the larger area.

   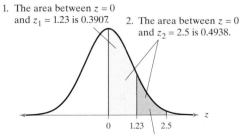

   1. The area between $z = 0$ and $z_1 = 1.23$ is 0.3907.   2. The area between $z = 0$ and $z_2 = 2.5$ is 0.4938.

   3. Subtract to find the area between $z_1 = 1.23$ and $z_2 = 2.5$; $0.4938 - 0.3907 = 0.1031$.

   ii. When the two z-scores have opposite signs (one negative and one positive), *add* the areas.

   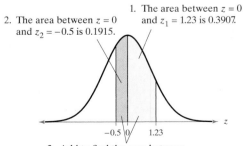

   2. The area between $z = 0$ and $z_2 = -0.5$ is 0.1915.   1. The area between $z = 0$ and $z_1 = 1.23$ is 0.3907.

   3. Add to find the area between $z_1 = 1.23$ and $z_2 = -0.5$; $0.3907 + 0.1915 = 0.5822$.

EXAMPLE 2

Finding Area Under the Standard Normal Curve

Find the area under the standard normal curve to the left of $z = -0.99$.

SOLUTION

The area under the standard normal curve to the left of $z = -0.99$ is shown.

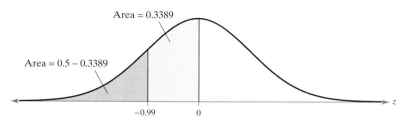

Area = 0.3389

Area = 0.5 − 0.3389

−0.99   0 z

From the Standard Normal Table (0-to-z), the area corresponding to $z = -0.99$ is 0.3389. Because the area to the left of $z = 0$ is 0.5, the area to the left of $z = -0.99$ is

Area $= 0.5 - 0.3389 = 0.1611$.

You can use technology to find the area to the left of $z = -0.99$, as shown at the left.

TI-84 PLUS

.5−normalcdf(0,.
99)
         .1610870617

TRY IT YOURSELF 2

Find the area under the standard normal curve to the left of $z = 2.13$.

Answer: Page A39

EXAMPLE 3

Finding Area Under the Standard Normal Curve

Find the area under the standard normal curve to the right of $z = 1.06$.

SOLUTION

The area under the standard normal curve to the right of $z = 1.06$ is shown.

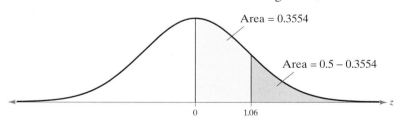

Area = 0.3554

Area = 0.5 − 0.3554

0   1.06 z

From the Standard Normal Table (0-to-z), the area corresponding to $z = 1.06$ is 0.3554. Because the area to the right of $z = 0$ is 0.5, the area to the right of $z = 1.06$ is

Area $= 0.5 - 0.3554 = 0.1446$.

You can use technology to find the area to the right of $z = 1.06$, as shown at the left.

TI-84 PLUS

.5−normalcdf(0,1
.06)
       .1445723279

TRY IT YOURSELF 3

Find the area under the standard normal curve to the right of $z = -2.16$.

Answer: Page A39

EXAMPLE 4

Finding Area Under the Standard Normal Curve

Find the area under the standard normal curve between $z = -1.5$ and $z = 1.25$.

SOLUTION

The area under the standard normal curve between $z = -1.5$ and $z = 1.25$ is shown.

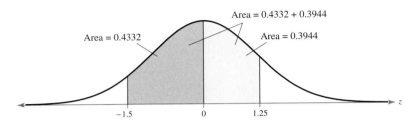

Area = 0.4332

Area = 0.4332 + 0.3944

Area = 0.3944

-1.5 0   1.25 z

From the Standard Normal Table (0-to-z), the area corresponding to $z = -1.5$ is 0.4332 and the area corresponding to $z = 1.25$ is 0.3944. To find the area between these two z-scores, add the resulting areas.

Area $= 0.4332 + 0.3944 = 0.8276$

Note that when you use technology, your answers may differ slightly from those found using the Standard Normal Table. For instance, when finding the area between $z = -1.5$ and $z = 1.25$ on a TI-84 Plus, you get the result shown at the left.

Interpretation   So, 82.76% of the area under the curve falls between $z = -1.5$ and $z = 1.25$.

TRY IT YOURSELF 4

Find the area under the standard normal curve between $z = -2.165$ and $z = -1.35$.

Answer: Page A39

TI-84 PLUS

normalcdf(-1.5,1
.25)
       .8275429323

Because the normal distribution is a continuous probability distribution, the area under the standard normal curve to the left of a z-score gives the probability that z is less than that z-score. For instance, in Example 2, the area to the left of $z = -0.99$ is 0.1611. So, $P(z < -0.99) = 0.1611$, which is read as "the probability that z is less than -0.99 is 0.1611." The table shows the probabilities for Examples 3 and 4.

	Area	**Probability**
Example 3	To the right of $z = 1.06$: 0.1446	$P(z > 1.06) = 0.1446$
Example 4	Between $z = -1.5$ and $z = 1.25$: 0.8276	$P(-1.5 < z < 1.25) = 0.8276$

Recall from Section 2.4 that values lying more than two standard deviations from the mean are considered unusual. Values lying more than three standard deviations from the mean are considered *very* unusual. So, a z-score greater than 2 or less than -2 is unusual. A z-score greater than 3 or less than -3 is *very* unusual.

You are now ready to continue Section 5.1 on page 242 with the section exercises.

Table 1—Random Numbers

92630	78240	19267	95457	53497	23894	37708	79862	76471	66418
79445	78735	71549	44843	26104	67318	00701	34986	66751	99723
59654	71966	27386	50004	05358	94031	29281	18544	52429	06080
31524	49587	76612	39789	13537	48086	59483	60680	84675	53014
06348	76938	90379	51392	55887	71015	09209	79157	24440	30244
28703	51709	94456	48396	73780	06436	86641	69239	57662	80181
68108	89266	94730	95761	75023	48464	65544	96583	18911	16391
99938	90704	93621	66330	33393	95261	95349	51769	91616	33238
91543	73196	34449	63513	83834	99411	58826	40456	69268	48562
42103	02781	73920	56297	72678	12249	25270	36678	21313	75767
17138	27584	25296	28387	51350	61664	37893	05363	44143	42677
28297	14280	54524	21618	95320	38174	60579	08089	94999	78460
09331	56712	51333	06289	75345	08811	82711	57392	25252	30333
31295	04204	93712	51287	05754	79396	87399	51773	33075	97061
36146	15560	27592	42089	99281	59640	15221	96079	09961	05371
29553	18432	13630	05529	02791	81017	49027	79031	50912	09399
23501	22642	63081	08191	89420	67800	55137	54707	32945	64522
57888	85846	67967	07835	11314	01545	48535	17142	08552	67457
55336	71264	88472	04334	63919	36394	11196	92470	70543	29776
10087	10072	55980	64688	68239	20461	89381	93809	00796	95945
34101	81277	66090	88872	37818	72142	67140	50785	21380	16703
53362	44940	60430	22834	14130	96593	23298	56203	92671	15925
82975	66158	84731	19436	55790	69229	28661	13675	99318	76873
54827	84673	22898	08094	14326	87038	42892	21127	30712	48489
25464	59098	27436	89421	80754	89924	19097	67737	80368	08795
67609	60214	41475	84950	40133	02546	09570	45682	50165	15609
44921	70924	61295	51137	47596	86735	35561	76649	18217	63446
33170	30972	98130	95828	49786	13301	36081	80761	33985	68621
84687	85445	06208	17654	51333	02878	35010	67578	61574	20749
71886	56450	36567	09395	96951	35507	17555	35212	69106	01679
00475	02224	74722	14721	40215	21351	08596	45625	83981	63748
25993	38881	68361	59560	41274	69742	40703	37993	03435	18873
92882	53178	99195	93803	56985	53089	15305	50522	55900	43026
25138	26810	07093	15677	60688	04410	24505	37890	67186	62829
84631	71882	12991	83028	82484	90339	91950	74579	03539	90122
34003	92326	12793	61453	48121	74271	28363	66561	75220	35908
53775	45749	05734	86169	42762	70175	97310	73894	88606	19994
59316	97885	72807	54966	60859	11932	35265	71601	55577	67715
20479	66557	50705	26999	09854	52591	14063	30214	19890	19292
86180	84931	25455	26044	02227	52015	21820	50599	51671	65411
21451	68001	72710	40261	61281	13172	63819	48970	51732	54113
98062	68375	80089	24135	72355	95428	11808	29740	81644	86610
01788	64429	14430	94575	75153	94576	61393	96192	03227	32258
62465	04841	43272	68702	01274	05437	22953	18946	99053	41690
94324	31089	84159	92933	99989	89500	91586	02802	69471	68274
05797	43984	21575	09908	70221	19791	51578	36432	33494	79888
10395	14289	52185	09721	25789	38562	54794	04897	59012	89251
35177	56986	25549	59730	64718	52630	31100	62384	49483	11409
25633	89619	75882	98256	02126	72099	57183	55887	09320	73463
16464	48280	94254	45777	45150	68865	11382	11782	22695	41988

A Million Random Digits with 100,000 Normal Deviates by the Rand Corporation (New York: The Free Press, 1955).

Table 2—Binomial Distribution

This table shows the probability of x successes in n independent trials, each with probability of success p.

n	x	.01	.05	.10	.15	.20	.25	.30	.35	.40	.45	.50	.55	.60	.65	.70	.75	.80	.85	.90	.95
2	0	.980	.902	.810	.723	.640	.563	.490	.423	.360	.303	.250	.203	.160	.123	.090	.063	.040	.023	.010	.002
	1	.020	.095	.180	.255	.320	.375	.420	.455	.480	.495	.500	.495	.480	.455	.420	.375	.320	.255	.180	.095
	2	.000	.002	.010	.023	.040	.063	.090	.123	.160	.203	.250	.303	.360	.423	.490	.563	.640	.723	.810	.902
3	0	.970	.857	.729	.614	.512	.422	.343	.275	.216	.166	.125	.091	.064	.043	.027	.016	.008	.003	.001	.000
	1	.029	.135	.243	.325	.384	.422	.441	.444	.432	.408	.375	.334	.288	.239	.189	.141	.096	.057	.027	.007
	2	.000	.007	.027	.057	.096	.141	.189	.239	.288	.334	.375	.408	.432	.444	.441	.422	.384	.325	.243	.135
	3	.000	.000	.001	.003	.008	.016	.027	.043	.064	.091	.125	.166	.216	.275	.343	.422	.512	.614	.729	.857
4	0	.961	.815	.656	.522	.410	.316	.240	.179	.130	.092	.062	.041	.026	.015	.008	.004	.002	.001	.000	.000
	1	.039	.171	.292	.368	.410	.422	.412	.384	.346	.300	.250	.200	.154	.112	.076	.047	.026	.011	.004	.000
	2	.001	.014	.049	.098	.154	.211	.265	.311	.346	.368	.375	.368	.346	.311	.265	.211	.154	.098	.049	.014
	3	.000	.000	.004	.011	.026	.047	.076	.112	.154	.200	.250	.300	.346	.384	.412	.422	.410	.368	.292	.171
	4	.000	.000	.000	.001	.002	.004	.008	.015	.026	.041	.062	.092	.130	.179	.240	.316	.410	.522	.656	.815
5	0	.951	.774	.590	.444	.328	.237	.168	.116	.078	.050	.031	.019	.010	.005	.002	.001	.000	.000	.000	.000
	1	.048	.204	.328	.392	.410	.396	.360	.312	.259	.206	.156	.113	.077	.049	.028	.015	.006	.002	.000	.000
	2	.001	.021	.073	.138	.205	.264	.309	.336	.346	.337	.312	.276	.230	.181	.132	.088	.051	.024	.008	.001
	3	.000	.001	.008	.024	.051	.088	.132	.181	.230	.276	.312	.337	.346	.336	.309	.264	.205	.138	.073	.021
	4	.000	.000	.000	.002	.006	.015	.028	.049	.077	.113	.156	.206	.259	.312	.360	.396	.410	.392	.328	.204
	5	.000	.000	.000	.000	.000	.001	.002	.005	.010	.019	.031	.050	.078	.116	.168	.237	.328	.444	.590	.774
6	0	.941	.735	.531	.377	.262	.178	.118	.075	.047	.028	.016	.008	.004	.002	.001	.000	.000	.000	.000	.000
	1	.057	.232	.354	.399	.393	.356	.303	.244	.187	.136	.094	.061	.037	.020	.010	.004	.002	.000	.000	.000
	2	.001	.031	.098	.176	.246	.297	.324	.328	.311	.278	.234	.186	.138	.095	.060	.033	.015	.006	.001	.000
	3	.000	.002	.015	.042	.082	.132	.185	.236	.276	.303	.312	.303	.276	.236	.185	.132	.082	.042	.015	.002
	4	.000	.000	.001	.006	.015	.033	.060	.095	.138	.186	.234	.278	.311	.328	.324	.297	.246	.176	.098	.031
	5	.000	.000	.000	.000	.002	.004	.010	.020	.037	.061	.094	.136	.187	.244	.303	.356	.393	.399	.354	.232
	6	.000	.000	.000	.000	.000	.000	.001	.002	.004	.008	.016	.028	.047	.075	.118	.178	.262	.377	.531	.735
7	0	.932	.698	.478	.321	.210	.133	.082	.049	.028	.015	.008	.004	.002	.001	.000	.000	.000	.000	.000	.000
	1	.066	.257	.372	.396	.367	.311	.247	.185	.131	.087	.055	.032	.017	.008	.004	.001	.000	.000	.000	.000
	2	.002	.041	.124	.210	.275	.311	.318	.299	.261	.214	.164	.117	.077	.047	.025	.012	.004	.001	.000	.000
	3	.000	.004	.023	.062	.115	.173	.227	.268	.290	.292	.273	.239	.194	.144	.097	.058	.029	.011	.003	.000
	4	.000	.000	.003	.011	.029	.058	.097	.144	.194	.239	.273	.292	.290	.268	.227	.173	.115	.062	.023	.004
	5	.000	.000	.000	.001	.004	.012	.025	.047	.077	.117	.164	.214	.261	.299	.318	.311	.275	.210	.124	.041
	6	.000	.000	.000	.000	.000	.001	.004	.008	.017	.032	.055	.087	.131	.185	.247	.311	.367	.396	.372	.257
	7	.000	.000	.000	.000	.000	.000	.000	.001	.002	.004	.008	.015	.028	.049	.082	.133	.210	.321	.478	.698
8	0	.923	.663	.430	.272	.168	.100	.058	.032	.017	.008	.004	.002	.001	.000	.000	.000	.000	.000	.000	.000
	1	.075	.279	.383	.385	.336	.267	.198	.137	.090	.055	.031	.016	.008	.003	.001	.000	.000	.000	.000	.000
	2	.003	.051	.149	.238	.294	.311	.296	.259	.209	.157	.109	.070	.041	.022	.010	.004	.001	.000	.000	.000
	3	.000	.005	.033	.084	.147	.208	.254	.279	.279	.257	.219	.172	.124	.081	.047	.023	.009	.003	.000	.000
	4	.000	.000	.005	.018	.046	.087	.136	.188	.232	.263	.273	.263	.232	.188	.136	.087	.046	.018	.005	.000
	5	.000	.000	.000	.003	.009	.023	.047	.081	.124	.172	.219	.257	.279	.279	.254	.208	.147	.084	.033	.005
	6	.000	.000	.000	.000	.001	.004	.010	.022	.041	.070	.109	.157	.209	.259	.296	.311	.294	.238	.149	.051
	7	.000	.000	.000	.000	.000	.000	.001	.003	.008	.016	.031	.055	.090	.137	.198	.267	.336	.385	.383	.279
	8	.000	.000	.000	.000	.000	.000	.000	.000	.001	.002	.004	.008	.017	.032	.058	.100	.168	.272	.430	.663
9	0	.914	.630	.387	.232	.134	.075	.040	.021	.010	.005	.002	.001	.000	.000	.000	.000	.000	.000	.000	.000
	1	.083	.299	.387	.368	.302	.225	.156	.100	.060	.034	.018	.008	.004	.001	.000	.000	.000	.000	.000	.000
	2	.003	.063	.172	.260	.302	.300	.267	.216	.161	.111	.070	.041	.021	.010	.004	.001	.000	.000	.000	.000
	3	.000	.008	.045	.107	.176	.234	.267	.272	.251	.212	.164	.116	.074	.042	.021	.009	.003	.001	.000	.000
	4	.000	.001	.007	.028	.066	.117	.172	.219	.251	.260	.246	.213	.167	.118	.074	.039	.017	.005	.001	.000
	5	.000	.000	.001	.005	.017	.039	.074	.118	.167	.213	.246	.260	.251	.219	.172	.117	.066	.028	.007	.001
	6	.000	.000	.000	.001	.003	.009	.021	.042	.074	.116	.164	.212	.251	.272	.267	.234	.176	.107	.045	.008
	7	.000	.000	.000	.000	.000	.001	.004	.010	.021	.041	.070	.111	.161	.216	.267	.300	.302	.260	.172	.063
	8	.000	.000	.000	.000	.000	.000	.000	.001	.004	.008	.018	.034	.060	.100	.156	.225	.302	.368	.387	.299
	9	.000	.000	.000	.000	.000	.000	.000	.000	.000	.001	.002	.005	.010	.021	.040	.075	.134	.232	.387	.630

From Brase/Brase, *Understandable Statistics*, Sixth Edition.

Table 2—Binomial Distribution *(continued)*

n	x	.01	.05	.10	.15	.20	.25	.30	.35	.40	.45	.50	.55	.60	.65	.70	.75	.80	.85	.90	.95
10	0	.904	.599	.349	.197	.107	.056	.028	.014	.006	.003	.001	.000	.000	.000	.000	.000	.000	.000	.000	.000
	1	.091	.315	.387	.347	.268	.188	.121	.072	.040	.021	.010	.004	.002	.000	.000	.000	.000	.000	.000	.000
	2	.004	.075	.194	.276	.302	.282	.233	.176	.121	.076	.044	.023	.011	.004	.001	.000	.000	.000	.000	.000
	3	.000	.010	.057	.130	.201	.250	.267	.252	.215	.166	.117	.075	.042	.021	.009	.003	.001	.000	.000	.000
	4	.000	.001	.011	.040	.088	.146	.200	.238	.251	.238	.205	.160	.111	.069	.037	.016	.006	.001	.000	.000
	5	.000	.000	.001	.008	.026	.058	.103	.154	.201	.234	.246	.234	.201	.154	.103	.058	.026	.008	.001	.000
	6	.000	.000	.000	.001	.006	.016	.037	.069	.111	.160	.205	.238	.251	.238	.200	.146	.088	.040	.011	.001
	7	.000	.000	.000	.000	.001	.003	.009	.021	.042	.075	.117	.166	.215	.252	.267	.250	.201	.130	.057	.010
	8	.000	.000	.000	.000	.000	.000	.001	.004	.011	.023	.044	.076	.121	.176	.233	.282	.302	.276	.194	.075
	9	.000	.000	.000	.000	.000	.000	.000	.000	.002	.004	.010	.021	.040	.072	.121	.188	.268	.347	.387	.315
	10	.000	.000	.000	.000	.000	.000	.000	.000	.000	.000	.001	.003	.006	.014	.028	.056	.107	.197	.349	.599
11	0	.895	.569	.314	.167	.086	.042	.020	.009	.004	.001	.000	.000	.000	.000	.000	.000	.000	.000	.000	.000
	1	.099	.329	.384	.325	.236	.155	.093	.052	.027	.013	.005	.002	.001	.000	.000	.000	.000	.000	.000	.000
	2	.005	.087	.213	.287	.295	.258	.200	.140	.089	.051	.027	.013	.005	.002	.001	.000	.000	.000	.000	.000
	3	.000	.014	.071	.152	.221	.258	.257	.225	.177	.126	.081	.046	.023	.010	.004	.001	.000	.000	.000	.000
	4	.000	.001	.016	.054	.111	.172	.220	.243	.236	.206	.161	.113	.070	.038	.017	.006	.002	.000	.000	.000
	5	.000	.000	.002	.013	.039	.080	.132	.183	.221	.236	.226	.193	.147	.099	.057	.027	.010	.002	.000	.000
	6	.000	.000	.000	.002	.010	.027	.057	.099	.147	.193	.226	.236	.221	.183	.132	.080	.039	.013	.002	.000
	7	.000	.000	.000	.000	.002	.006	.017	.038	.070	.113	.161	.206	.236	.243	.220	.172	.111	.054	.016	.001
	8	.000	.000	.000	.000	.000	.001	.004	.010	.023	.046	.081	.126	.177	.225	.257	.258	.221	.152	.071	.014
	9	.000	.000	.000	.000	.000	.000	.001	.002	.005	.013	.027	.051	.089	.140	.200	.258	.295	.287	.213	.087
	10	.000	.000	.000	.000	.000	.000	.000	.000	.001	.002	.005	.013	.027	.052	.093	.155	.236	.325	.384	.329
	11	.000	.000	.000	.000	.000	.000	.000	.000	.000	.000	.000	.001	.004	.009	.020	.042	.086	.167	.314	.569
12	0	.886	.540	.282	.142	.069	.032	.014	.006	.002	.001	.000	.000	.000	.000	.000	.000	.000	.000	.000	.000
	1	.107	.341	.377	.301	.206	.127	.071	.037	.017	.008	.003	.001	.000	.000	.000	.000	.000	.000	.000	.000
	2	.006	.099	.230	.292	.283	.232	.168	.109	.064	.034	.016	.007	.002	.001	.000	.000	.000	.000	.000	.000
	3	.000	.017	.085	.172	.236	.258	.240	.195	.142	.092	.054	.028	.012	.005	.001	.000	.000	.000	.000	.000
	4	.000	.002	.021	.068	.133	.194	.231	.237	.213	.170	.121	.076	.042	.020	.008	.002	.001	.000	.000	.000
	5	.000	.000	.004	.019	.053	.103	.158	.204	.227	.223	.193	.149	.101	.059	.029	.011	.003	.001	.000	.000
	6	.000	.000	.000	.004	.016	.040	.079	.128	.177	.212	.226	.212	.177	.128	.079	.040	.016	.004	.000	.000
	7	.000	.000	.000	.001	.003	.011	.029	.059	.101	.149	.193	.223	.227	.204	.158	.103	.053	.019	.004	.000
	8	.000	.000	.000	.000	.001	.002	.008	.020	.042	.076	.121	.170	.213	.237	.231	.194	.133	.068	.021	.002
	9	.000	.000	.000	.000	.000	.000	.001	.005	.012	.028	.054	.092	.142	.195	.240	.258	.236	.172	.085	.017
	10	.000	.000	.000	.000	.000	.000	.000	.001	.002	.007	.016	.034	.064	.109	.168	.232	.283	.292	.230	.099
	11	.000	.000	.000	.000	.000	.000	.000	.000	.000	.001	.003	.008	.017	.037	.071	.127	.206	.301	.377	.341
	12	.000	.000	.000	.000	.000	.000	.000	.000	.000	.000	.000	.001	.002	.006	.014	.032	.069	.142	.282	.540
15	0	.860	.463	.206	.087	.035	.013	.005	.002	.000	.000	.000	.000	.000	.000	.000	.000	.000	.000	.000	.000
	1	.130	.366	.343	.231	.132	.067	.031	.013	.005	.002	.000	.000	.000	.000	.000	.000	.000	.000	.000	.000
	2	.009	.135	.267	.286	.231	.156	.092	.048	.022	.009	.003	.001	.000	.000	.000	.000	.000	.000	.000	.000
	3	.000	.031	.129	.218	.250	.225	.170	.111	.063	.032	.014	.005	.002	.000	.000	.000	.000	.000	.000	.000
	4	.000	.005	.043	.116	.188	.225	.219	.179	.127	.078	.042	.019	.007	.002	.001	.000	.000	.000	.000	.000
	5	.000	.001	.010	.045	.103	.165	.206	.212	.186	.140	.092	.051	.024	.010	.003	.001	.000	.000	.000	.000
	6	.000	.000	.002	.013	.043	.092	.147	.191	.207	.191	.153	.105	.061	.030	.012	.003	.001	.000	.000	.000
	7	.000	.000	.000	.003	.014	.039	.081	.132	.177	.201	.196	.165	.118	.071	.035	.013	.003	.001	.000	.000
	8	.000	.000	.000	.001	.003	.013	.035	.071	.118	.165	.196	.201	.177	.132	.081	.039	.014	.003	.000	.000
	9	.000	.000	.000	.000	.001	.003	.012	.030	.061	.105	.153	.191	.207	.191	.147	.092	.043	.013	.002	.000
	10	.000	.000	.000	.000	.000	.001	.003	.010	.024	.051	.092	.140	.186	.212	.206	.165	.103	.045	.010	.001
	11	.000	.000	.000	.000	.000	.000	.001	.002	.007	.019	.042	.078	.127	.179	.219	.225	.188	.116	.043	.005
	12	.000	.000	.000	.000	.000	.000	.000	.000	.002	.005	.014	.032	.063	.111	.170	.225	.250	.218	.129	.031
	13	.000	.000	.000	.000	.000	.000	.000	.000	.000	.001	.003	.009	.022	.048	.092	.156	.231	.286	.267	.135
	14	.000	.000	.000	.000	.000	.000	.000	.000	.000	.000	.000	.002	.005	.013	.031	.067	.132	.231	.343	.366
	15	.000	.000	.000	.000	.000	.000	.000	.000	.000	.000	.000	.000	.000	.002	.005	.013	.035	.087	.206	.463

Table 2—Binomial Distribution *(continued)*

		.01	.05	.10	.15	.20	.25	.30	.35	.40	.45	.50	.55	.60	.65	.70	.75	.80	.85	.90	.95
n	**x**																				
16	0	.851	.440	.185	.074	.028	.010	.003	.001	.000	.000	.000	.000	.000	.000	.000	.000	.000	.000	.000	.000
	1	.138	.371	.329	.210	.113	.053	.023	.009	.003	.001	.000	.000	.000	.000	.000	.000	.000	.000	.000	.000
	2	.010	.146	.275	.277	.211	.134	.073	.035	.015	.006	.002	.001	.000	.000	.000	.000	.000	.000	.000	.000
	3	.000	.036	.142	.229	.246	.208	.146	.089	.047	.022	.009	.003	.001	.000	.000	.000	.000	.000	.000	.000
	4	.000	.006	.051	.131	.200	.225	.204	.155	.101	.057	.028	.011	.004	.001	.000	.000	.000	.000	.000	.000
	5	.000	.001	.014	.056	.120	.180	.210	.201	.162	.112	.067	.034	.014	.005	.001	.000	.000	.000	.000	.000
	6	.000	.000	.003	.018	.055	.110	.165	.198	.198	.168	.122	.075	.039	.017	.006	.001	.000	.000	.000	.000
	7	.000	.000	.000	.005	.020	.052	.101	.152	.189	.197	.175	.132	.084	.044	.019	.006	.001	.000	.000	.000
	8	.000	.000	.000	.001	.006	.020	.049	.092	.142	.181	.196	.181	.142	.092	.049	.020	.006	.001	.000	.000
	9	.000	.000	.000	.000	.001	.006	.019	.044	.084	.132	.175	.197	.189	.152	.101	.052	.020	.005	.000	.000
	10	.000	.000	.000	.000	.000	.001	.006	.017	.039	.075	.122	.168	.198	.198	.165	.110	.055	.018	.003	.000
	11	.000	.000	.000	.000	.000	.000	.001	.005	.014	.034	.067	.112	.162	.201	.210	.180	.120	.056	.014	.001
	12	.000	.000	.000	.000	.000	.000	.000	.001	.004	.011	.028	.057	.101	.155	.204	.225	.200	.131	.051	.006
	13	.000	.000	.000	.000	.000	.000	.000	.000	.001	.003	.009	.022	.047	.089	.146	.208	.246	.229	.142	.036
	14	.000	.000	.000	.000	.000	.000	.000	.000	.000	.001	.002	.006	.015	.035	.073	.134	.211	.277	.275	.146
	15	.000	.000	.000	.000	.000	.000	.000	.000	.000	.000	.000	.001	.003	.009	.023	.053	.113	.210	.329	.371
	16	.000	.000	.000	.000	.000	.000	.000	.000	.000	.000	.000	.000	.001	.003	.010	.028	.074	.185	.440	
20	0	.818	.358	.122	.039	.012	.003	.001	.000	.000	.000	.000	.000	.000	.000	.000	.000	.000	.000	.000	.000
	1	.165	.377	.270	.137	.058	.021	.007	.002	.000	.000	.000	.000	.000	.000	.000	.000	.000	.000	.000	.000
	2	.016	.189	.285	.229	.137	.067	.028	.010	.003	.001	.000	.000	.000	.000	.000	.000	.000	.000	.000	.000
	3	.001	.060	.190	.243	.205	.134	.072	.032	.012	.004	.001	.000	.000	.000	.000	.000	.000	.000	.000	.000
	4	.000	.013	.090	.182	.218	.190	.130	.074	.035	.014	.005	.001	.000	.000	.000	.000	.000	.000	.000	.000
	5	.000	.002	.032	.103	.175	.202	.179	.127	.075	.036	.015	.005	.001	.000	.000	.000	.000	.000	.000	.000
	6	.000	.000	.009	.045	.109	.169	.192	.171	.124	.075	.036	.015	.005	.001	.000	.000	.000	.000	.000	.000
	7	.000	.000	.002	.016	.055	.112	.164	.184	.166	.122	.074	.037	.015	.005	.001	.000	.000	.000	.000	.000
	8	.000	.000	.000	.005	.022	.061	.114	.161	.180	.162	.120	.073	.035	.014	.004	.001	.000	.000	.000	.000
	9	.000	.000	.000	.001	.007	.027	.065	.116	.160	.177	.160	.119	.071	.034	.012	.003	.000	.000	.000	.000
	10	.000	.000	.000	.000	.002	.010	.031	.069	.117	.159	.176	.159	.117	.069	.031	.010	.002	.000	.000	.000
	11	.000	.000	.000	.000	.000	.003	.012	.034	.071	.119	.160	.177	.160	.116	.065	.027	.007	.001	.000	.000
	12	.000	.000	.000	.000	.000	.001	.004	.014	.035	.073	.120	.162	.180	.161	.114	.061	.022	.005	.000	.000
	13	.000	.000	.000	.000	.000	.000	.001	.005	.015	.037	.074	.122	.166	.184	.164	.112	.055	.016	.002	.000
	14	.000	.000	.000	.000	.000	.000	.000	.001	.005	.015	.037	.075	.124	.171	.192	.169	.109	.045	.009	.000
	15	.000	.000	.000	.000	.000	.000	.000	.000	.001	.005	.015	.036	.075	.127	.179	.202	.175	.103	.032	.002
	16	.000	.000	.000	.000	.000	.000	.000	.000	.000	.001	.005	.014	.035	.074	.130	.190	.218	.182	.090	.013
	17	.000	.000	.000	.000	.000	.000	.000	.000	.000	.000	.001	.004	.012	.032	.072	.134	.205	.243	.190	.060
	18	.000	.000	.000	.000	.000	.000	.000	.000	.000	.000	.000	.001	.003	.010	.028	.067	.137	.229	.285	.189
	19	.000	.000	.000	.000	.000	.000	.000	.000	.000	.000	.000	.000	.002	.007	.021	.058	.137	.270	.377	
	20	.000	.000	.000	.000	.000	.000	.000	.000	.000	.000	.000	.000	.000	.001	.003	.012	.039	.122	.358	

Table 3—Poisson Distribution

x	μ 0.1	0.2	0.3	0.4	0.5	0.6	0.7	0.8	0.9	1.0
0	.9048	.8187	.7408	.6703	.6065	.5488	.4966	.4493	.4066	.3679
1	.0905	.1637	.2222	.2681	.3033	.3293	.3476	.3595	.3659	.3679
2	.0045	.0164	.0333	.0536	.0758	.0988	.1217	.1438	.1647	.1839
3	.0002	.0011	.0033	.0072	.0126	.0198	.0284	.0383	.0494	.0613
4	.0000	.0001	.0003	.0007	.0016	.0030	.0050	.0077	.0111	.0153
5	.0000	.0000	.0000	.0001	.0002	.0004	.0007	.0012	.0020	.0031
6	.0000	.0000	.0000	.0000	.0000	.0000	.0001	.0002	.0003	.0005
7	.0000	.0000	.0000	.0000	.0000	.0000	.0000	.0000	.0000	.0001

x	μ 1.1	1.2	1.3	1.4	1.5	1.6	1.7	1.8	1.9	2.0
0	.3329	.3012	.2725	.2466	.2231	.2019	.1827	.1653	.1496	.1353
1	.3662	.3614	.3543	.3452	.3347	.3230	.3106	.2975	.2842	.2707
2	.2014	.2169	.2303	.2417	.2510	.2584	.2640	.2678	.2700	.2707
3	.0738	.0867	.0998	.1128	.1255	.1378	.1496	.1607	.1710	.1804
4	.0203	.0260	.0324	.0395	.0471	.0551	.0636	.0723	.0812	.0902
5	.0045	.0062	.0084	.0111	.0141	.0176	.0216	.0260	.0309	.0361
6	.0008	.0012	.0018	.0026	.0035	.0047	.0061	.0078	.0098	.0120
7	.0001	.0002	.0003	.0005	.0008	.0011	.0015	.0020	.0027	.0034
8	.0000	.0000	.0001	.0001	.0001	.0002	.0003	.0005	.0006	.0009
9	.0000	.0000	.0000	.0000	.0000	.0000	.0001	.0001	.0001	.0002

x	μ 2.1	2.2	2.3	2.4	2.5	2.6	2.7	2.8	2.9	3.0
0	.1225	.1108	.1003	.0907	.0821	.0743	.0672	.0608	.0550	.0498
1	.2572	.2438	.2306	.2177	.2052	.1931	.1815	.1703	.1596	.1494
2	.2700	.2681	.2652	.2613	.2565	.2510	.2450	.2384	.2314	.2240
3	.1890	.1966	.2033	.2090	.2138	.2176	.2205	.2225	.2237	.2240
4	.0992	.1082	.1169	.1254	.1336	.1414	.1488	.1557	.1622	.1680
5	.0417	.0476	.0538	.0602	.0668	.0735	.0804	.0872	.0940	.1008
6	.0146	.0174	.0206	.0241	.0278	.0319	.0362	.0407	.0455	.0504
7	.0044	.0055	.0068	.0083	.0099	.0118	.0139	.0163	.0188	.0216
8	.0011	.0015	.0019	.0025	.0031	.0038	.0047	.0057	.0068	.0081
9	.0003	.0004	.0005	.0007	.0009	.0011	.0014	.0018	.0022	.0027
10	.0001	.0001	.0001	.0002	.0002	.0003	.0004	.0005	.0006	.0008
11	.0000	.0000	.0000	.0000	.0000	.0001	.0001	.0001	.0002	.0002
12	.0000	.0000	.0000	.0000	.0000	.0000	.0000	.0000	.0000	.0001

x	μ 3.1	3.2	3.3	3.4	3.5	3.6	3.7	3.8	3.9	4.0
0	.0450	.0408	.0369	.0334	.0302	.0273	.0247	.0224	.0202	.0183
1	.1397	.1304	.1217	.1135	.1057	.0984	.0915	.0850	.0789	.0733
2	.2165	.2087	.2008	.1929	.1850	.1771	.1692	.1615	.1539	.1465
3	.2237	.2226	.2209	.2186	.2158	.2125	.2087	.2046	.2001	.1954
4	.1734	.1781	.1823	.1858	.1888	.1912	.1931	.1944	.1951	.1954
5	.1075	.1140	.1203	.1264	.1322	.1377	.1429	.1477	.1522	.1563
6	.0555	.0608	.0662	.0716	.0771	.0826	.0881	.0936	.0989	.1042
7	.0246	.0278	.0312	.0348	.0385	.0425	.0466	.0508	.0551	.0595
8	.0095	.0111	.0129	.0148	.0169	.0191	.0215	.0241	.0269	.0298
9	.0033	.0040	.0047	.0056	.0066	.0076	.0089	.0102	.0116	.0132
10	.0010	.0013	.0016	.0019	.0023	.0028	.0033	.0039	.0045	.0053
11	.0003	.0004	.0005	.0006	.0007	.0009	.0011	.0013	.0016	.0019
12	.0001	.0001	.0001	.0002	.0002	.0003	.0003	.0004	.0005	.0006
13	.0000	.0000	.0000	.0000	.0001	.0001	.0001	.0001	.0002	.0002
14	.0000	.0000	.0000	.0000	.0000	.0000	.0000	.0000	.0000	.0001

W. H. Beyer, *Handbook of Tables for Probability and Statistics*, 2e, CRC Press, Boca Raton, Florida, 1986.

Table 3—Poisson Distribution *(continued)*

x	μ 4.1	4.2	4.3	4.4	4.5	4.6	4.7	4.8	4.9	5.0
0	.0166	.0150	.0136	.0123	.0111	.0101	.0091	.0082	.0074	.0067
1	.0679	.0630	.0583	.0540	.0500	.0462	.0427	.0395	.0365	.0337
2	.1393	.1323	.1254	.1188	.1125	.1063	.1005	.0948	.0894	.0842
3	.1904	.1852	.1798	.1743	.1687	.1631	.1574	.1517	.1460	.1404
4	.1951	.1944	.1933	.1917	.1898	.1875	.1849	.1820	.1789	.1755
5	.1600	.1633	.1662	.1687	.1708	.1725	.1738	.1747	.1753	.1755
6	.1093	.1143	.1191	.1237	.1281	.1323	.1362	.1398	.1432	.1462
7	.0640	.0686	.0732	.0778	.0824	.0869	.0914	.0959	.1002	.1044
8	.0328	.0360	.0393	.0428	.0463	.0500	.0537	.0575	.0614	.0653
9	.0150	.0168	.0188	.0209	.0232	.0255	.0280	.0307	.0334	.0363
10	.0061	.0071	.0081	.0092	.0104	.0118	.0132	.0147	.0164	.0181
11	.0023	.0027	.0032	.0037	.0043	.0049	.0056	.0064	.0073	.0082
12	.0008	.0009	.0011	.0014	.0016	.0019	.0022	.0026	.0030	.0034
13	.0002	.0003	.0004	.0005	.0006	.0007	.0008	.0009	.0011	.0013
14	.0001	.0001	.0001	.0001	.0002	.0002	.0003	.0003	.0004	.0005
15	.0000	.0000	.0000	.0000	.0001	.0001	.0001	.0001	.0001	.0002

x	μ 5.1	5.2	5.3	5.4	5.5	5.6	5.7	5.8	5.9	6.0
0	.0061	.0055	.0050	.0045	.0041	.0037	.0033	.0030	.0027	.0025
1	.0311	.0287	.0265	.0244	.0225	.0207	.0191	.0176	.0162	.0149
2	.0793	.0746	.0701	.0659	.0618	.0580	.0544	.0509	.0477	.0446
3	.1348	.1293	.1239	.1185	.1133	.1082	.1033	.0985	.0938	.0892
4	.1719	.1681	.1641	.1600	.1558	.1515	.1472	.1428	.1383	.1339
5	.1753	.1748	.1740	.1728	.1714	.1697	.1678	.1656	.1632	.1606
6	.1490	.1515	.1537	.1555	.1571	.1584	.1594	.1601	.1605	.1606
7	.1086	.1125	.1163	.1200	.1234	.1267	.1298	.1326	.1353	.1377
8	.0692	.0731	.0771	.0810	.0849	.0887	.0925	.0962	.0998	.1033
9	.0392	.0423	.0454	.0486	.0519	.0552	.0586	.0620	.0654	.0688
10	.0200	.0220	.0241	.0262	.0285	.0309	.0334	.0359	.0386	.0413
11	.0093	.0104	.0116	.0129	.0143	.0157	.0173	.0190	.0207	.0225
12	.0039	.0045	.0051	.0058	.0065	.0073	.0082	.0092	.0102	.0113
13	.0015	.0018	.0021	.0024	.0028	.0032	.0036	.0041	.0046	.0052
14	.0006	.0007	.0008	.0009	.0011	.0013	.0015	.0017	.0019	.0022
15	.0002	.0002	.0003	.0003	.0004	.0005	.0006	.0007	.0008	.0009
16	.0001	.0001	.0001	.0001	.0001	.0002	.0002	.0002	.0003	.0003
17	.0000	.0000	.0000	.0000	.0000	.0000	.0001	.0001	.0001	.0001

Table 3—Poisson Distribution *(continued)*

x	μ 6.1	6.2	6.3	6.4	6.5	6.6	6.7	6.8	6.9	7.0
0	.0022	.0020	.0018	.0017	.0015	.0014	.0012	.0011	.0010	.0009
1	.0137	.0126	.0116	.0106	.0098	.0090	.0082	.0076	.0070	.0064
2	.0417	.0390	.0364	.0340	.0318	.0296	.0276	.0258	.0240	.0223
3	.0848	.0806	.0765	.0726	.0688	.0652	.0617	.0584	.0552	.0521
4	.1294	.1249	.1205	.1162	.1118	.1076	.1034	.0992	.0952	.0912
5	.1579	.1549	.1519	.1487	.1454	.1420	.1385	.1349	.1314	.1277
6	.1605	.1601	.1595	.1586	.1575	.1562	.1546	.1529	.1511	.1490
7	.1399	.1418	.1435	.1450	.1462	.1472	.1480	.1486	.1489	.1490
8	.1066	.1099	.1130	.1160	.1188	.1215	.1240	.1263	.1284	.1304
9	.0723	.0757	.0791	.0825	.0858	.0891	.0923	.0954	.0985	.1014
10	.0441	.0469	.0498	.0528	.0558	.0588	.0618	.0649	.0679	.0710
11	.0245	.0265	.0285	.0307	.0330	.0353	.0377	.0401	.0426	.0452
12	.0124	.0137	.0150	.0164	.0179	.0194	.0210	.0227	.0245	.0264
13	.0058	.0065	.0073	.0081	.0089	.0098	.0108	.0119	.0130	.0142
14	.0025	.0029	.0033	.0037	.0041	.0046	.0052	.0058	.0064	.0071
15	.0010	.0012	.0014	.0016	.0018	.0020	.0023	.0026	.0029	.0033
16	.0004	.0005	.0005	.0006	.0007	.0008	.0010	.0011	.0013	.0014
17	.0001	.0002	.0002	.0002	.0003	.0003	.0004	.0004	.0005	.0006
18	.0000	.0001	.0001	.0001	.0001	.0001	.0001	.0002	.0002	.0002
19	.0000	.0000	.0000	.0000	.0000	.0000	.0000	.0001	.0001	.0001

x	μ 7.1	7.2	7.3	7.4	7.5	7.6	7.7	7.8	7.9	8.0
0	.0008	.0007	.0007	.0006	.0006	.0005	.0005	.0004	.0004	.0003
1	.0059	.0054	.0049	.0045	.0041	.0038	.0035	.0032	.0029	.0027
2	.0208	.0194	.0180	.0167	.0156	.0145	.0134	.0125	.0116	.0107
3	.0492	.0464	.0438	.0413	.0389	.0366	.0345	.0324	.0305	.0286
4	.0874	.0836	.0799	.0764	.0729	.0696	.0663	.0632	.0602	.0573
5	.1241	.1204	.1167	.1130	.1094	.1057	.1021	.0986	.0951	.0916
6	.1468	.1445	.1420	.1394	.1367	.1339	.1311	.1282	.1252	.1221
7	.1489	.1486	.1481	.1474	.1465	.1454	.1442	.1428	.1413	.1396
8	.1321	.1337	.1351	.1363	.1373	.1382	.1388	.1392	.1395	.1396
9	.1042	.1070	.1096	.1121	.1144	.1167	.1187	.1207	.1224	.1241
10	.0740	.0770	.0800	.0829	.0858	.0887	.0914	.0941	.0967	.0993
11	.0478	.0504	.0531	.0558	.0585	.0613	.0640	.0667	.0695	.0722
12	.0283	.0303	.0323	.0344	.0366	.0388	.0411	.0434	.0457	.0481
13	.0154	.0168	.0181	.0196	.0211	.0227	.0243	.0260	.0278	.0296
14	.0078	.0086	.0095	.0104	.0113	.0123	.0134	.0145	.0157	.0169
15	.0037	.0041	.0046	.0051	.0057	.0062	.0069	.0075	.0083	.0090
16	.0016	.0019	.0021	.0024	.0026	.0030	.0033	.0037	.0041	.0045
17	.0007	.0008	.0009	.0010	.0012	.0013	.0015	.0017	.0019	.0021
18	.0003	.0003	.0004	.0004	.0005	.0006	.0006	.0007	.0008	.0009
19	.0001	.0001	.0001	.0002	.0002	.0002	.0003	.0003	.0003	.0004
20	.0000	.0000	.0001	.0001	.0001	.0001	.0001	.0001	.0001	.0002
21	.0000	.0000	.0000	.0000	.0000	.0000	.0000	.0000	.0001	.0001

Table 3—Poisson Distribution *(continued)*

x	8.1	8.2	8.3	8.4	8.5	8.6	8.7	8.8	8.9	9.0
0	.0003	.0003	.0002	.0002	.0002	.0002	.0002	.0002	.0001	.0001
1	.0025	.0023	.0021	.0019	.0017	.0016	.0014	.0013	.0012	.0011
2	.0100	.0092	.0086	.0079	.0074	.0068	.0063	.0058	.0054	.0050
3	.0269	.0252	.0237	.0222	.0208	.0195	.0183	.0171	.0160	.0150
4	.0544	.0517	.0491	.0466	.0443	.0420	.0398	.0377	.0357	.0337
5	.0882	.0849	.0816	.0784	.0752	.0722	.0692	.0663	.0635	.0607
6	.1191	.1160	.1128	.1097	.1066	.1034	.1003	.0972	.0941	.0911
7	.1378	.1358	.1338	.1317	.1294	.1271	.1247	.1222	.1197	.1171
8	.1395	.1392	.1388	.1382	.1375	.1366	.1356	.1344	.1332	.1318
9	.1256	.1269	.1280	.1290	.1299	.1306	.1311	.1315	.1317	.1318
10	.1017	.1040	.1063	.1084	.1104	.1123	.1140	.1157	.1172	.1186
11	.0749	.0776	.0802	.0828	.0853	.0878	.0902	.0925	.0948	.0970
12	.0505	.0530	.0555	.0579	.0604	.0629	.0654	.0679	.0703	.0728
13	.0315	.0334	.0354	.0374	.0395	.0416	.0438	.0459	.0481	.0504
14	.0182	.0196	.0210	.0225	.0240	.0256	.0272	.0289	.0306	.0324
15	.0098	.0107	.0116	.0126	.0136	.0147	.0158	.0169	.0182	.0194
16	.0050	.0055	.0060	.0066	.0072	.0079	.0086	.0093	.0101	.0109
17	.0024	.0026	.0029	.0033	.0036	.0040	.0044	.0048	.0053	.0058
18	.0011	.0012	.0014	.0015	.0017	.0019	.0021	.0024	.0026	.0029
19	.0005	.0005	.0006	.0007	.0008	.0009	.0010	.0011	.0012	.0014
20	.0002	.0002	.0002	.0003	.0003	.0004	.0004	.0005	.0005	.0006
21	.0001	.0001	.0001	.0001	.0001	.0002	.0002	.0002	.0002	.0003
22	.0000	.0000	.0000	.0000	.0001	.0001	.0001	.0001	.0001	.0001

x	9.1	9.2	9.3	9.4	9.5	9.6	9.7	9.8	9.9	10.0
0	.0001	.0001	.0001	.0001	.0001	.0001	.0001	.0001	.0001	.0000
1	.0010	.0009	.0009	.0008	.0007	.0007	.0006	.0005	.0005	.0005
2	.0046	.0043	.0040	.0037	.0034	.0031	.0029	.0027	.0025	.0023
3	.0140	.0131	.0123	.0115	.0107	.0100	.0093	.0087	.0081	.0076
4	.0319	.0302	.0285	.0269	.0254	.0240	.0226	.0213	.0201	.0189
5	.0581	.0555	.0530	.0506	.0483	.0460	.0439	.0418	.0398	.0378
6	.0881	.0851	.0822	.0793	.0764	.0736	.0709	.0682	.0656	.0631
7	.1145	.1118	.1091	.1064	.1037	.1010	.0982	.0955	.0928	.0901
8	.1302	.1286	.1269	.1251	.1232	.1212	.1191	.1170	.1148	.1126
9	.1317	.1315	.1311	.1306	.1300	.1293	.1284	.1274	.1263	.1251
10	.1198	.1210	.1219	.1228	.1235	.1241	.1245	.1249	.1250	.1251
11	.0991	.1012	.1031	.1049	.1067	.1083	.1098	.1112	.1125	.1137
12	.0752	.0776	.0799	.0822	.0844	.0866	.0888	.0908	.0928	.0948
13	.0526	.0549	.0572	.0594	.0617	.0640	.0662	.0685	.0707	.0729
14	.0342	.0361	.0380	.0399	.0419	.0439	.0459	.0479	.0500	.0521
15	.0208	.0221	.0235	.0250	.0265	.0281	.0297	.0313	.0330	.0347
16	.0118	.0127	.0137	.0147	.0157	.0168	.0180	.0192	.0204	.0217
17	.0063	.0069	.0075	.0081	.0088	.0095	.0103	.0111	.0119	.0128
18	.0032	.0035	.0039	.0042	.0046	.0051	.0055	.0060	.0065	.0071
19	.0015	.0017	.0019	.0021	.0023	.0026	.0028	.0031	.0034	.0037
20	.0007	.0008	.0009	.0010	.0011	.0012	.0014	.0015	.0017	.0019
21	.0003	.0003	.0004	.0004	.0005	.0006	.0006	.0007	.0008	.0009
22	.0001	.0001	.0002	.0002	.0002	.0002	.0003	.0003	.0004	.0004
23	.0000	.0001	.0001	.0001	.0001	.0001	.0001	.0001	.0002	.0002
24	.0000	.0000	.0000	.0000	.0000	.0000	.0000	.0001	.0001	.0001

Table 3—Poisson Distribution *(continued)*

x	11	12	13	14	15	16	17	18	19	20
					μ					
0	.0000	.0000	.0000	.0000	.0000	.0000	.0000	.0000	.0000	.0000
1	.0002	.0001	.0000	.0000	.0000	.0000	.0000	.0000	.0000	.0000
2	.0010	.0004	.0002	.0001	.0000	.0000	.0000	.0000	.0000	.0000
3	.0037	.0018	.0008	.0004	.0002	.0001	.0000	.0000	.0000	.0000
4	.0102	.0053	.0027	.0013	.0006	.0003	.0001	.0001	.0000	.0000
5	.0224	.0127	.0070	.0037	.0019	.0010	.0005	.0002	.0001	.0001
6	.0411	.0255	.0152	.0087	.0048	.0026	.0014	.0007	.0004	.0002
7	.0646	.0437	.0281	.0174	.0104	.0060	.0034	.0018	.0010	.0005
8	.0888	.0655	.0457	.0304	.0194	.0120	.0072	.0042	.0024	.0013
9	.1085	.0874	.0661	.0473	.0324	.0213	.0135	.0083	.0050	.0029
10	.1194	.1048	.0859	.0663	.0486	.0341	.0230	.0150	.0095	.0058
11	.1194	.1144	.1015	.0844	.0663	.0496	.0355	.0245	.0164	.0106
12	.1094	.1144	.1099	.0984	.0829	.0661	.0504	.0368	.0259	.0176
13	.0926	.1056	.1099	.1060	.0956	.0814	.0658	.0509	.0378	.0271
14	.0728	.0905	.1021	.1060	.1024	.0930	.0800	.0655	.0514	.0387
15	.0534	.0724	.0885	.0989	.1024	.0992	.0906	.0786	.0650	.0516
16	.0367	.0543	.0719	.0866	.0960	.0992	.0963	.0884	.0772	.0646
17	.0237	.0383	.0550	.0713	.0847	.0934	.0963	.0936	.0863	.0760
18	.0145	.0256	.0397	.0554	.0706	.0830	.0909	.0936	.0911	.0844
19	.0084	.0161	.0272	.0409	.0557	.0699	.0814	.0887	.0911	.0888
20	.0046	.0097	.0177	.0286	.0418	.0559	.0692	.0798	.0866	.0888

x	11	12	13	14	15	16	17	18	19	20
					μ					
21	.0024	.0055	.0109	.0191	.0299	.0426	.0560	.0684	.0783	.0846
22	.0012	.0030	.0065	.0121	.0204	.0310	.0433	.0560	.0676	.0769
23	.0006	.0016	.0037	.0074	.0133	.0216	.0320	.0438	.0559	.0669
24	.0003	.0008	.0020	.0043	.0083	.0144	.0226	.0328	.0442	.0557
25	.0001	.0004	.0010	.0024	.0050	.0092	.0154	.0237	.0336	.0446
26	.0000	.0002	.0005	.0013	.0029	.0057	.0101	.0164	.0246	.0343
27	.0000	.0001	.0002	.0007	.0016	.0034	.0063	.0109	.0173	.0254
28	.0000	.0000	.0001	.0003	.0009	.0019	.0038	.0070	.0117	.0181
29	.0000	.0000	.0001	.0002	.0004	.0011	.0023	.0044	.0077	.0125
30	.0000	.0000	.0000	.0001	.0002	.0006	.0013	.0026	.0049	.0083
31	.0000	.0000	.0000	.0000	.0001	.0003	.0007	.0015	.0030	.0054
32	.0000	.0000	.0000	.0000	.0001	.0001	.0004	.0009	.0018	.0034
33	.0000	.0000	.0000	.0000	.0000	.0001	.0002	.0005	.0010	.0020
34	.0000	.0000	.0000	.0000	.0000	.0000	.0001	.0002	.0006	.0012
35	.0000	.0000	.0000	.0000	.0000	.0000	.0000	.0001	.0003	.0007
36	.0000	.0000	.0000	.0000	.0000	.0000	.0000	.0001	.0002	.0004
37	.0000	.0000	.0000	.0000	.0000	.0000	.0000	.0000	.0001	.0002
38	.0000	.0000	.0000	.0000	.0000	.0000	.0000	.0000	.0000	.0001
39	.0000	.0000	.0000	.0000	.0000	.0000	.0000	.0000	.0000	.0001

Table 4—Standard Normal Distribution

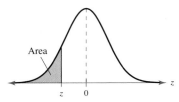

z	.09	.08	.07	.06	.05	.04	.03	.02	.01	.00
−3.4	.0002	.0003	.0003	.0003	.0003	.0003	.0003	.0003	.0003	.0003
−3.3	.0003	.0004	.0004	.0004	.0004	.0004	.0004	.0005	.0005	.0005
−3.2	.0005	.0005	.0005	.0006	.0006	.0006	.0006	.0006	.0007	.0007
−3.1	.0007	.0007	.0008	.0008	.0008	.0008	.0009	.0009	.0009	.0010
−3.0	.0010	.0010	.0011	.0011	.0011	.0012	.0012	.0013	.0013	.0013
−2.9	.0014	.0014	.0015	.0015	.0016	.0016	.0017	.0018	.0018	.0019
−2.8	.0019	.0020	.0021	.0021	.0022	.0023	.0023	.0024	.0025	.0026
−2.7	.0026	.0027	.0028	.0029	.0030	.0031	.0032	.0033	.0034	.0035
−2.6	.0036	.0037	.0038	.0039	.0040	.0041	.0043	.0044	.0045	.0047
−2.5	.0048	.0049	.0051	.0052	.0054	.0055	.0057	.0059	.0060	.0062
−2.4	.0064	.0066	.0068	.0069	.0071	.0073	.0075	.0078	.0080	.0082
−2.3	.0084	.0087	.0089	.0091	.0094	.0096	.0099	.0102	.0104	.0107
−2.2	.0110	.0113	.0116	.0119	.0122	.0125	.0129	.0132	.0136	.0139
−2.1	.0143	.0146	.0150	.0154	.0158	.0162	.0166	.0170	.0174	.0179
−2.0	.0183	.0188	.0192	.0197	.0202	.0207	.0212	.0217	.0222	.0228
−1.9	.0233	.0239	.0244	.0250	.0256	.0262	.0268	.0274	.0281	.0287
−1.8	.0294	.0301	.0307	.0314	.0322	.0329	.0336	.0344	.0351	.0359
−1.7	.0367	.0375	.0384	.0392	.0401	.0409	.0418	.0427	.0436	.0446
−1.6	.0455	.0465	.0475	.0485	.0495	.0505	.0516	.0526	.0537	.0548
−1.5	.0559	.0571	.0582	.0594	.0606	.0618	.0630	.0643	.0655	.0668
−1.4	.0681	.0694	.0708	.0721	.0735	.0749	.0764	.0778	.0793	.0808
−1.3	.0823	.0838	.0853	.0869	.0885	.0901	.0918	.0934	.0951	.0968
−1.2	.0985	.1003	.1020	.1038	.1056	.1075	.1093	.1112	.1131	.1151
−1.1	.1170	.1190	.1210	.1230	.1251	.1271	.1292	.1314	.1335	.1357
−1.0	.1379	.1401	.1423	.1446	.1469	.1492	.1515	.1539	.1562	.1587
−0.9	.1611	.1635	.1660	.1685	.1711	.1736	.1762	.1788	.1814	.1841
−0.8	.1867	.1894	.1922	.1949	.1977	.2005	.2033	.2061	.2090	.2119
−0.7	.2148	.2177	.2206	.2236	.2266	.2296	.2327	.2358	.2389	.2420
−0.6	.2451	.2483	.2514	.2546	.2578	.2611	.2643	.2676	.2709	.2743
−0.5	.2776	.2810	.2843	.2877	.2912	.2946	.2981	.3015	.3050	.3085
−0.4	.3121	.3156	.3192	.3228	.3264	.3300	.3336	.3372	.3409	.3446
−0.3	.3483	.3520	.3557	.3594	.3632	.3669	.3707	.3745	.3783	.3821
−0.2	.3859	.3897	.3936	.3974	.4013	.4052	.4090	.4129	.4168	.4207
−0.1	.4247	.4286	.4325	.4364	.4404	.4443	.4483	.4522	.4562	.4602
−0.0	.4641	.4681	.4721	.4761	.4801	.4840	.4880	.4920	.4960	.5000

Critical Values

Level of Confidence c	z_c
0.80	1.28
0.90	1.645
0.95	1.96
0.99	2.575

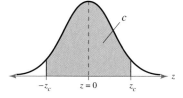

Table A-3, pp. 681–682 from *Probability and Statistics for Engineers and Scientists,* 6e by Walpole, Myers, and Myers. Copyright 1997. Pearson Prentice Hall, Upper Saddle River, N.J.

Table 4—Standard Normal Distribution *(continued)*

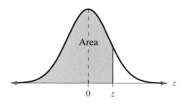

z	.00	.01	.02	.03	.04	.05	.06	.07	.08	.09
0.0	.5000	.5040	.5080	.5120	.5160	.5199	.5239	.5279	.5319	.5359
0.1	.5398	.5438	.5478	.5517	.5557	.5596	.5636	.5675	.5714	.5753
0.2	.5793	.5832	.5871	.5910	.5948	.5987	.6026	.6064	.6103	.6141
0.3	.6179	.6217	.6255	.6293	.6331	.6368	.6406	.6443	.6480	.6517
0.4	.6554	.6591	.6628	.6664	.6700	.6736	.6772	.6808	.6844	.6879
0.5	.6915	.6950	.6985	.7019	.7054	.7088	.7123	.7157	.7190	.7224
0.6	.7257	.7291	.7324	.7357	.7389	.7422	.7454	.7486	.7517	.7549
0.7	.7580	.7611	.7642	.7673	.7704	.7734	.7764	.7794	.7823	.7852
0.8	.7881	.7910	.7939	.7967	.7995	.8023	.8051	.8078	.8106	.8133
0.9	.8159	.8186	.8212	.8238	.8264	.8289	.8315	.8340	.8365	.8389
1.0	.8413	.8438	.8461	.8485	.8508	.8531	.8554	.8577	.8599	.8621
1.1	.8643	.8665	.8686	.8708	.8729	.8749	.8770	.8790	.8810	.8830
1.2	.8849	.8869	.8888	.8907	.8925	.8944	.8962	.8980	.8997	.9015
1.3	.9032	.9049	.9066	.9082	.9099	.9115	.9131	.9147	.9162	.9177
1.4	.9192	.9207	.9222	.9236	.9251	.9265	.9279	.9292	.9306	.9319
1.5	.9332	.9345	.9357	.9370	.9382	.9394	.9406	.9418	.9429	.9441
1.6	.9452	.9463	.9474	.9484	.9495	.9505	.9515	.9525	.9535	.9545
1.7	.9554	.9564	.9573	.9582	.9591	.9599	.9608	.9616	.9625	.9633
1.8	.9641	.9649	.9656	.9664	.9671	.9678	.9686	.9693	.9699	.9706
1.9	.9713	.9719	.9726	.9732	.9738	.9744	.9750	.9756	.9761	.9767
2.0	.9772	.9778	.9783	.9788	.9793	.9798	.9803	.9808	.9812	.9817
2.1	.9821	.9826	.9830	.9834	.9838	.9842	.9846	.9850	.9854	.9857
2.2	.9861	.9864	.9868	.9871	.9875	.9878	.9881	.9884	.9887	.9890
2.3	.9893	.9896	.9898	.9901	.9904	.9906	.9909	.9911	.9913	.9916
2.4	.9918	.9920	.9922	.9925	.9927	.9929	.9931	.9932	.9934	.9936
2.5	.9938	.9940	.9941	.9943	.9945	.9946	.9948	.9949	.9951	.9952
2.6	.9953	.9955	.9956	.9957	.9959	.9960	.9961	.9962	.9963	.9964
2.7	.9965	.9966	.9967	.9968	.9969	.9970	.9971	.9972	.9973	.9974
2.8	.9974	.9975	.9976	.9977	.9977	.9978	.9979	.9979	.9980	.9981
2.9	.9981	.9982	.9982	.9983	.9984	.9984	.9985	.9985	.9986	.9986
3.0	.9987	.9987	.9987	.9988	.9988	.9989	.9989	.9989	.9990	.9990
3.1	.9990	.9991	.9991	.9991	.9992	.9992	.9992	.9992	.9993	.9993
3.2	.9993	.9993	.9994	.9994	.9994	.9994	.9994	.9995	.9995	.9995
3.3	.9995	.9995	.9995	.9996	.9996	.9996	.9996	.9996	.9996	.9997
3.4	.9997	.9997	.9997	.9997	.9997	.9997	.9997	.9997	.9997	.9998

Table 5—*t*-Distribution

d.f.	Level of confidence, *c*	0.80	0.90	0.95	0.98	0.99
	One tail, α	0.10	0.05	0.025	0.01	0.005
	Two tails, α	0.20	0.10	0.05	0.02	0.01
1		3.078	6.314	12.706	31.821	63.657
2		1.886	2.920	4.303	6.965	9.925
3		1.638	2.353	3.182	4.541	5.841
4		1.533	2.132	2.776	3.747	4.604
5		1.476	2.015	2.571	3.365	4.032
6		1.440	1.943	2.447	3.143	3.707
7		1.415	1.895	2.365	2.998	3.499
8		1.397	1.860	2.306	2.896	3.355
9		1.383	1.833	2.262	2.821	3.250
10		1.372	1.812	2.228	2.764	3.169
11		1.363	1.796	2.201	2.718	3.106
12		1.356	1.782	2.179	2.681	3.055
13		1.350	1.771	2.160	2.650	3.012
14		1.345	1.761	2.145	2.624	2.977
15		1.341	1.753	2.131	2.602	2.947
16		1.337	1.746	2.120	2.583	2.921
17		1.333	1.740	2.110	2.567	2.898
18		1.330	1.734	2.101	2.552	2.878
19		1.328	1.729	2.093	2.539	2.861
20		1.325	1.725	2.086	2.528	2.845
21		1.323	1.721	2.080	2.518	2.831
22		1.321	1.717	2.074	2.508	2.819
23		1.319	1.714	2.069	2.500	2.807
24		1.318	1.711	2.064	2.492	2.797
25		1.316	1.708	2.060	2.485	2.787
26		1.315	1.706	2.056	2.479	2.779
27		1.314	1.703	2.052	2.473	2.771
28		1.313	1.701	2.048	2.467	2.763
29		1.311	1.699	2.045	2.462	2.756
30		1.310	1.697	2.042	2.457	2.750
31		1.309	1.696	2.040	2.453	2.744
32		1.309	1.694	2.037	2.449	2.738
33		1.308	1.692	2.035	2.445	2.733
34		1.307	1.691	2.032	2.441	2.728
35		1.306	1.690	2.030	2.438	2.724
36		1.306	1.688	2.028	2.434	2.719
37		1.305	1.687	2.026	2.431	2.715
38		1.304	1.686	2.024	2.429	2.712
39		1.304	1.685	2.023	2.426	2.708
40		1.303	1.684	2.021	2.423	2.704
45		1.301	1.679	2.014	2.412	2.690
50		1.299	1.676	2.009	2.403	2.678
60		1.296	1.671	2.000	2.390	2.660
70		1.294	1.667	1.994	2.381	2.648
80		1.292	1.664	1.990	2.374	2.639
90		1.291	1.662	1.987	2.368	2.632
100		1.290	1.660	1.984	2.364	2.626
500		1.283	1.648	1.965	2.334	2.586
1000		1.282	1.646	1.962	2.330	2.581
∞		1.282	1.645	1.960	2.326	2.576

The critical values in Table 5 were generated using Excel.

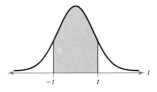

c-confidence interval

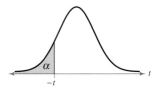

Left-tailed test

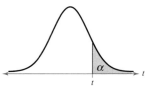

Right-tailed test

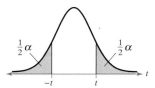

Two-tailed test

Table 6—Chi-Square Distribution

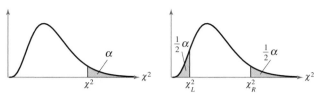

Right tail   Two tails

Degrees of freedom	α									
	0.995	0.99	0.975	0.95	0.90	0.10	0.05	0.025	0.01	0.005
1	—	—	0.001	0.004	0.016	2.706	3.841	5.024	6.635	7.879
2	0.010	0.020	0.051	0.103	0.211	4.605	5.991	7.378	9.210	10.597
3	0.072	0.115	0.216	0.352	0.584	6.251	7.815	9.348	11.345	12.838
4	0.207	0.297	0.484	0.711	1.064	7.779	9.488	11.143	13.277	14.860
5	0.412	0.554	0.831	1.145	1.610	9.236	11.071	12.833	15.086	16.750
6	0.676	0.872	1.237	1.635	2.204	10.645	12.592	14.449	16.812	18.548
7	0.989	1.239	1.690	2.167	2.833	12.017	14.067	16.013	18.475	20.278
8	1.344	1.646	2.180	2.733	3.490	13.362	15.507	17.535	20.090	21.955
9	1.735	2.088	2.700	3.325	4.168	14.684	16.919	19.023	21.666	23.589
10	2.156	2.558	3.247	3.940	4.865	15.987	18.307	20.483	23.209	25.188
11	2.603	3.053	3.816	4.575	5.578	17.275	19.675	21.920	24.725	26.757
12	3.074	3.571	4.404	5.226	6.304	18.549	21.026	23.337	26.217	28.299
13	3.565	4.107	5.009	5.892	7.042	19.812	22.362	24.736	27.688	29.819
14	4.075	4.660	5.629	6.571	7.790	21.064	23.685	26.119	29.141	31.319
15	4.601	5.229	6.262	7.261	8.547	22.307	24.996	27.488	30.578	32.801
16	5.142	5.812	6.908	7.962	9.312	23.542	26.296	28.845	32.000	34.267
17	5.697	6.408	7.564	8.672	10.085	24.769	27.587	30.191	33.409	35.718
18	6.265	7.015	8.231	9.390	10.865	25.989	28.869	31.526	34.805	37.156
19	6.844	7.633	8.907	10.117	11.651	27.204	30.144	32.852	36.191	38.582
20	7.434	8.260	9.591	10.851	12.443	28.412	31.410	34.170	37.566	39.997
21	8.034	8.897	10.283	11.591	13.240	29.615	32.671	35.479	38.932	41.401
22	8.643	9.542	10.982	12.338	14.042	30.813	33.924	36.781	40.289	42.796
23	9.260	10.196	11.689	13.091	14.848	32.007	35.172	38.076	41.638	44.181
24	9.886	10.856	12.401	13.848	15.659	33.196	36.415	39.364	42.980	45.559
25	10.520	11.524	13.120	14.611	16.473	34.382	37.652	40.646	44.314	46.928
26	11.160	12.198	13.844	15.379	17.292	35.563	38.885	41.923	45.642	48.290
27	11.808	12.879	14.573	16.151	18.114	36.741	40.113	43.194	46.963	49.645
28	12.461	13.565	15.308	16.928	18.939	37.916	41.337	44.461	48.278	50.993
29	13.121	14.257	16.047	17.708	19.768	39.087	42.557	45.722	49.588	52.336
30	13.787	14.954	16.791	18.493	20.599	40.256	43.773	46.979	50.892	53.672
40	20.707	22.164	24.433	26.509	29.051	51.805	55.758	59.342	63.691	66.766
50	27.991	29.707	32.357	34.764	37.689	63.167	67.505	71.420	76.154	79.490
60	35.534	37.485	40.482	43.188	46.459	74.397	79.082	83.298	88.379	91.952
70	43.275	45.442	48.758	51.739	55.329	85.527	90.531	95.023	100.425	104.215
80	51.172	53.540	57.153	60.391	64.278	96.578	101.879	106.629	112.329	116.321
90	59.196	61.754	65.647	69.126	73.291	107.565	113.145	118.136	124.116	128.299
100	67.328	70.065	74.222	77.929	82.358	118.498	124.342	129.561	135.807	140.169

D. B. Owen, HANDBOOK OF STATISTICAL TABLES, A.5, Published by
Addison Wesley Longman, Inc.

Table 7—F-Distribution

$\alpha = 0.005$

d.f.$_N$: Degrees of freedom, numerator

d.f.$_D$: Degrees of freedom, denominator	1	2	3	4	5	6	7	8	9	10	12	15	20	24	30	40	60	120	∞
1	16211	20000	21615	22500	23056	23437	23715	23925	24091	24224	24426	24630	24836	24940	25044	25148	25253	25359	25465
2	198.5	199.0	199.2	199.2	199.3	199.3	199.4	199.4	199.4	199.4	199.4	199.4	199.4	199.5	199.5	199.5	199.5	199.5	199.5
3	55.55	49.80	47.47	46.19	45.39	44.84	44.43	44.13	43.88	43.69	43.39	43.08	42.78	42.62	42.47	42.31	42.15	41.99	41.83
4	31.33	26.28	24.26	23.15	22.46	21.97	21.62	21.35	21.14	20.97	20.70	20.44	20.17	20.03	19.89	19.75	19.61	19.47	19.32
5	22.78	18.31	16.53	15.56	14.94	14.51	14.20	13.96	13.77	13.62	13.38	13.15	12.90	12.78	12.66	12.53	12.40	12.27	12.14
6	18.63	14.54	12.92	12.03	11.46	11.07	10.79	10.57	10.39	10.25	10.03	9.81	9.59	9.47	9.36	9.24	9.12	9.00	8.88
7	16.24	12.40	10.88	10.05	9.52	9.16	8.89	8.68	8.51	8.38	8.18	7.97	7.75	7.65	7.53	7.42	7.31	7.19	7.08
8	14.69	11.04	9.60	8.81	8.30	7.95	7.69	7.50	7.34	7.21	7.01	6.81	6.61	6.50	6.40	6.29	6.18	6.06	5.95
9	13.61	10.11	8.72	7.96	7.47	7.13	6.88	6.69	6.54	6.42	6.23	6.03	5.83	5.73	5.62	5.52	5.41	5.30	5.19
10	12.83	9.43	8.08	7.34	6.87	6.54	6.30	6.12	5.97	5.85	5.66	5.47	5.27	5.17	5.07	4.97	4.86	4.75	4.64
11	12.23	8.91	7.60	6.88	6.42	6.10	5.86	5.68	5.54	5.42	5.24	5.05	4.86	4.76	4.65	4.55	4.44	4.34	4.23
12	11.75	8.51	7.23	6.52	6.07	5.76	5.52	5.35	5.20	5.09	4.91	4.72	4.53	4.43	4.33	4.23	4.12	4.01	3.90
13	11.37	8.19	6.93	6.23	5.79	5.48	5.25	5.08	4.94	4.82	4.64	4.46	4.27	4.17	4.07	3.97	3.87	3.76	3.65
14	11.06	7.92	6.68	6.00	5.56	5.26	5.03	4.86	4.72	4.60	4.43	4.25	4.06	3.96	3.86	3.76	3.66	3.55	3.44
15	10.80	7.70	6.48	5.80	5.37	5.07	4.85	4.67	4.54	4.42	4.25	4.07	3.88	3.79	3.69	3.58	3.48	3.37	3.26
16	10.58	7.51	6.30	5.64	5.21	4.91	4.69	4.52	4.38	4.27	4.10	3.92	3.73	3.64	3.54	3.44	3.33	3.22	3.11
17	10.38	7.35	6.16	5.50	5.07	4.78	4.56	4.39	4.25	4.14	3.97	3.79	3.61	3.51	3.41	3.31	3.21	3.10	2.98
18	10.22	7.21	6.03	5.37	4.96	4.66	4.44	4.28	4.14	4.03	3.86	3.68	3.50	3.40	3.30	3.20	3.10	2.99	2.87
19	10.07	7.09	5.92	5.27	4.85	4.56	4.34	4.18	4.04	3.93	3.76	3.59	3.40	3.31	3.21	3.11	3.00	2.89	2.78
20	9.94	6.99	5.82	5.17	4.76	4.47	4.26	4.09	3.96	3.85	3.68	3.50	3.32	3.22	3.12	3.02	2.92	2.81	2.69
21	9.83	6.89	5.73	5.09	4.68	4.39	4.18	4.01	3.88	3.77	3.60	3.43	3.24	3.15	3.05	2.95	2.84	2.73	2.61
22	9.73	6.81	5.65	5.02	4.61	4.32	4.11	3.94	3.81	3.70	3.54	3.36	3.18	3.08	2.98	2.88	2.77	2.66	2.55
23	9.63	6.73	5.58	4.95	4.54	4.26	4.05	3.88	3.75	3.64	3.47	3.30	3.12	3.02	2.92	2.82	2.71	2.60	2.48
24	9.55	6.66	5.52	4.89	4.49	4.20	3.99	3.83	3.69	3.59	3.42	3.25	3.06	2.97	2.87	2.77	2.66	2.55	2.43
25	9.48	6.60	5.46	4.84	4.43	4.15	3.94	3.78	3.64	3.54	3.37	3.20	3.01	2.92	2.82	2.72	2.61	2.50	2.38
26	9.41	6.54	5.41	4.79	4.38	4.10	3.89	3.73	3.60	3.49	3.33	3.15	2.97	2.87	2.77	2.67	2.56	2.45	2.33
27	9.34	6.49	5.36	4.74	4.34	4.06	3.85	3.69	3.56	3.45	3.28	3.11	2.93	2.83	2.73	2.63	2.52	2.41	2.29
28	9.28	6.44	5.32	4.70	4.30	4.02	3.81	3.65	3.52	3.41	3.25	3.07	2.89	2.79	2.69	2.59	2.48	2.37	2.25
29	9.23	6.40	5.28	4.66	4.26	3.98	3.77	3.61	3.48	3.38	3.21	3.04	2.86	2.76	2.66	2.56	2.45	2.33	2.24
30	9.18	6.35	5.24	4.62	4.23	3.95	3.74	3.58	3.45	3.34	3.18	3.01	2.82	2.73	2.63	2.52	2.42	2.30	2.18
40	8.83	6.07	4.98	4.37	3.99	3.71	3.51	3.35	3.22	3.12	2.95	2.78	2.60	2.50	2.40	2.30	2.18	2.06	1.93
60	8.49	5.79	4.73	4.14	3.76	3.49	3.29	3.13	3.01	2.90	2.74	2.57	2.39	2.29	2.19	2.08	1.96	1.83	1.69
120	8.18	5.54	4.50	3.92	3.55	3.28	3.09	2.93	2.81	2.71	2.54	2.37	2.19	2.09	1.98	1.87	1.75	1.61	1.43
∞	7.88	5.30	4.28	3.72	3.35	3.09	2.90	2.74	2.62	2.52	2.36	2.19	2.00	1.90	1.79	1.67	1.53	1.36	1.00

Table 7—*F*-Distribution (continued)

$\alpha = 0.01$

d.f._D: Degrees of freedom, denominator	d.f._N: Degrees of freedom, numerator																		
	1	**2**	**3**	**4**	**5**	**6**	**7**	**8**	**9**	**10**	**12**	**15**	**20**	**24**	**30**	**40**	**60**	**120**	**∞**
1	4052	4999.5	5403	5625	5764	5859	5928	5982	6022	6056	6106	6157	6209	6235	6261	6287	6313	6339	6366
2	98.50	99.00	99.17	99.25	99.30	99.33	99.36	99.37	99.39	99.40	99.42	99.43	99.45	99.46	99.47	99.47	99.48	99.49	99.50
3	34.12	30.82	29.46	28.71	28.24	27.91	27.67	27.49	27.35	27.23	27.05	26.87	26.69	26.60	26.50	26.41	26.32	26.22	26.13
4	21.20	18.00	16.69	15.98	15.52	15.21	14.98	14.80	14.66	14.55	14.37	14.20	14.02	13.93	13.84	13.75	13.65	13.56	13.46
5	16.26	13.27	12.06	11.39	10.97	10.67	10.46	10.29	10.16	10.05	9.89	9.72	9.55	9.47	9.38	9.29	9.20	9.11	9.02
6	13.75	10.92	9.78	9.15	8.75	8.47	8.26	8.10	7.98	7.87	7.72	7.56	7.40	7.31	7.23	7.14	7.06	6.97	6.88
7	12.25	9.55	8.45	7.85	7.46	7.19	6.99	6.84	6.72	6.62	6.47	6.31	6.16	6.07	5.99	5.91	5.82	5.74	5.65
8	11.26	8.65	7.59	7.01	6.63	6.37	6.18	6.03	5.91	5.81	5.67	5.52	5.36	5.28	5.20	5.12	5.03	4.95	4.86
9	10.56	8.02	6.99	6.42	6.06	5.80	5.61	5.47	5.35	5.26	5.11	4.96	4.81	4.73	4.65	4.57	4.48	4.40	4.31
10	10.04	7.56	6.55	5.99	5.64	5.39	5.20	5.06	4.94	4.85	4.71	4.56	4.41	4.33	4.25	4.17	4.08	4.00	3.91
11	9.65	7.21	6.22	5.67	5.32	5.07	4.89	4.74	4.63	4.54	4.40	4.25	4.10	4.02	3.94	3.86	3.78	3.69	3.60
12	9.33	6.93	5.95	5.41	5.06	4.82	4.64	4.50	4.39	4.30	4.16	4.01	3.86	3.78	3.70	3.62	3.54	3.45	3.36
13	9.07	6.70	5.74	5.21	4.86	4.62	4.44	4.30	4.19	4.10	3.96	3.82	3.66	3.59	3.51	3.43	3.34	3.25	3.17
14	8.86	6.51	5.56	5.04	4.69	4.46	4.28	4.14	4.03	3.94	3.80	3.66	3.51	3.43	3.35	3.27	3.18	3.09	3.00
15	8.68	6.36	5.42	4.89	4.56	4.32	4.14	4.00	3.89	3.80	3.67	3.52	3.37	3.29	3.21	3.13	3.05	2.96	2.87
16	8.53	6.23	5.29	4.77	4.44	4.20	4.03	3.89	3.78	3.69	3.55	3.41	3.26	3.18	3.10	3.02	2.93	2.84	2.75
17	8.40	6.11	5.18	4.67	4.34	4.10	3.93	3.79	3.68	3.59	3.46	3.31	3.16	3.08	3.00	2.92	2.83	2.75	2.65
18	8.29	6.01	5.09	4.58	4.25	4.01	3.84	3.71	3.60	3.51	3.37	3.23	3.08	3.00	2.92	2.84	2.75	2.66	2.57
19	8.18	5.93	5.01	4.50	4.17	3.94	3.77	3.63	3.52	3.43	3.30	3.15	3.00	2.92	2.84	2.76	2.67	2.58	2.49
20	8.10	5.85	4.94	4.43	4.10	3.87	3.70	3.56	3.46	3.37	3.23	3.09	2.94	2.86	2.78	2.69	2.61	2.52	2.42
21	8.02	5.78	4.87	4.37	4.04	3.81	3.64	3.51	3.40	3.31	3.17	3.03	2.88	2.80	2.72	2.64	2.55	2.46	2.36
22	7.95	5.72	4.82	4.31	3.99	3.76	3.59	3.45	3.35	3.26	3.12	2.98	2.83	2.75	2.67	2.58	2.50	2.40	2.31
23	7.88	5.66	4.76	4.26	3.94	3.71	3.54	3.41	3.30	3.21	3.07	2.93	2.78	2.70	2.62	2.54	2.45	2.35	2.26
24	7.82	5.61	4.72	4.22	3.90	3.67	3.50	3.36	3.26	3.17	3.03	2.89	2.74	2.66	2.58	2.49	2.40	2.31	2.21
25	7.77	5.57	4.68	4.18	3.85	3.63	3.46	3.32	3.22	3.13	2.99	2.85	2.70	2.62	2.54	2.45	2.36	2.27	2.17
26	7.72	5.53	4.64	4.14	3.82	3.59	3.42	3.29	3.18	3.09	2.96	2.81	2.66	2.58	2.50	2.42	2.33	2.23	2.13
27	7.68	5.49	4.60	4.11	3.78	3.56	3.39	3.26	3.15	3.06	2.93	2.78	2.63	2.55	2.47	2.38	2.29	2.20	2.10
28	7.64	5.45	4.57	4.07	3.75	3.53	3.36	3.23	3.12	3.03	2.90	2.75	2.60	2.52	2.44	2.35	2.26	2.17	2.06
29	7.60	5.42	4.54	4.04	3.73	3.50	3.33	3.20	3.09	3.00	2.87	2.73	2.57	2.49	2.41	2.33	2.23	2.14	2.03
30	7.56	5.39	4.51	4.02	3.70	3.47	3.30	3.17	3.07	2.98	2.84	2.70	2.55	2.47	2.39	2.30	2.21	2.11	2.01
40	7.31	5.18	4.31	3.83	3.51	3.29	3.12	2.99	2.89	2.80	2.66	2.52	2.37	2.29	2.20	2.11	2.02	1.92	1.80
60	7.08	4.98	4.13	3.65	3.34	3.12	2.95	2.82	2.72	2.63	2.50	2.35	2.20	2.12	2.03	1.94	1.84	1.73	1.60
120	6.85	4.79	3.95	3.48	3.17	2.96	2.79	2.66	2.56	2.47	2.34	2.19	2.03	1.95	1.86	1.76	1.66	1.53	1.38
∞	6.63	4.61	3.78	3.32	3.02	2.80	2.64	2.51	2.41	2.32	2.18	2.04	1.88	1.79	1.70	1.59	1.47	1.32	1.00

Table 7—*F*-Distribution (continued)

α = 0.025

d.f._D: Degrees of freedom, denominator	d.f._N: Degrees of freedom, numerator																		
	1	2	3	4	5	6	7	8	9	10	12	15	20	24	30	40	60	120	∞
1	647.8	799.5	864.2	899.6	921.8	937.1	948.2	956.7	963.3	968.6	976.7	984.9	993.1	997.2	1001	1006	1010	1014	1018
2	38.51	39.00	39.17	39.25	39.30	39.33	39.36	39.37	39.39	39.40	39.41	39.43	39.45	39.46	39.46	39.47	39.48	39.49	39.50
3	17.44	16.04	15.44	15.10	14.88	14.73	14.62	14.54	14.47	14.42	14.34	14.25	14.17	14.12	14.08	14.04	13.99	13.95	13.90
4	12.22	10.65	9.98	9.60	9.36	9.20	9.07	8.98	8.90	8.84	8.75	8.66	8.56	8.51	8.46	8.41	8.36	8.31	8.26
5	10.01	8.43	7.76	7.39	7.15	6.98	6.85	6.76	6.68	6.62	6.52	6.43	6.33	6.28	6.23	6.18	6.12	6.07	6.02
6	8.81	7.26	6.60	6.23	5.99	5.82	5.70	5.60	5.52	5.46	5.37	5.27	5.17	5.12	5.07	5.01	4.96	4.90	4.85
7	8.07	6.54	5.89	5.52	5.29	5.12	4.99	4.90	4.82	4.76	4.67	4.57	4.47	4.42	4.36	4.31	4.25	4.20	4.14
8	7.57	6.06	5.42	5.05	4.82	4.65	4.53	4.43	4.36	4.30	4.20	4.10	4.00	3.95	3.89	3.84	3.78	3.73	3.67
9	7.21	5.71	5.08	4.72	4.48	4.32	4.20	4.10	4.03	3.96	3.87	3.77	3.67	3.61	3.56	3.51	3.45	3.39	3.33
10	6.94	5.46	4.83	4.47	4.24	4.07	3.95	3.85	3.78	3.72	3.62	3.52	3.42	3.37	3.31	3.26	3.20	3.14	3.08
11	6.72	5.26	4.63	4.28	4.04	3.88	3.76	3.66	3.59	3.53	3.43	3.33	3.23	3.17	3.12	3.06	3.00	2.94	2.88
12	6.55	5.10	4.47	4.12	3.89	3.73	3.61	3.51	3.44	3.37	3.28	3.18	3.07	3.02	2.96	2.91	2.85	2.79	2.72
13	6.41	4.97	4.35	4.00	3.77	3.60	3.48	3.39	3.31	3.25	3.15	3.05	2.95	2.89	2.84	2.78	2.72	2.66	2.60
14	6.30	4.86	4.24	3.89	3.66	3.50	3.38	3.29	3.21	3.15	3.05	2.95	2.84	2.79	2.73	2.67	2.61	2.55	2.49
15	6.20	4.77	4.15	3.80	3.58	3.41	3.29	3.20	3.12	3.06	2.96	2.86	2.76	2.70	2.64	2.59	2.52	2.46	2.40
16	6.12	4.69	4.08	3.73	3.50	3.34	3.22	3.12	3.05	2.99	2.89	2.79	2.68	2.63	2.57	2.51	2.45	2.38	2.32
17	6.04	4.62	4.01	3.66	3.44	3.28	3.16	3.06	2.98	2.92	2.82	2.72	2.62	2.56	2.50	2.44	2.38	2.32	2.25
18	5.98	4.56	3.95	3.61	3.38	3.22	3.10	3.01	2.93	2.87	2.77	2.67	2.56	2.50	2.44	2.38	2.32	2.26	2.19
19	5.92	4.51	3.90	3.56	3.33	3.17	3.05	2.96	2.88	2.82	2.72	2.62	2.51	2.45	2.39	2.33	2.27	2.20	2.13
20	5.87	4.46	3.86	3.51	3.29	3.13	3.01	2.91	2.84	2.77	2.68	2.57	2.46	2.41	2.35	2.29	2.22	2.16	2.09
21	5.83	4.42	3.82	3.48	3.25	3.09	2.97	2.87	2.80	2.73	2.64	2.53	2.42	2.37	2.31	2.25	2.18	2.11	2.04
22	5.79	4.38	3.78	3.44	3.22	3.05	2.93	2.84	2.76	2.70	2.60	2.50	2.39	2.33	2.27	2.21	2.14	2.08	2.00
23	5.75	4.35	3.75	3.41	3.18	3.02	2.90	2.81	2.73	2.67	2.57	2.47	2.36	2.30	2.24	2.18	2.11	2.04	1.97
24	5.72	4.32	3.72	3.38	3.15	2.99	2.87	2.78	2.70	2.64	2.54	2.44	2.33	2.27	2.21	2.15	2.08	2.01	1.94
25	5.69	4.29	3.69	3.35	3.13	2.97	2.85	2.75	2.68	2.61	2.51	2.41	2.30	2.24	2.18	2.12	2.05	1.98	1.91
26	5.66	4.27	3.67	3.33	3.10	2.94	2.82	2.73	2.65	2.59	2.49	2.39	2.28	2.22	2.16	2.09	2.03	1.95	1.88
27	5.63	4.24	3.65	3.31	3.08	2.92	2.80	2.71	2.63	2.57	2.47	2.36	2.25	2.19	2.13	2.07	2.00	1.93	1.85
28	5.61	4.22	3.63	3.29	3.06	2.90	2.78	2.69	2.61	2.55	2.45	2.34	2.23	2.17	2.11	2.05	1.98	1.91	1.83
29	5.59	4.20	3.61	3.27	3.04	2.88	2.76	2.67	2.59	2.53	2.43	2.32	2.21	2.15	2.09	2.03	1.96	1.89	1.81
30	5.57	4.18	3.59	3.25	3.03	2.87	2.75	2.65	2.57	2.51	2.41	2.31	2.20	2.14	2.07	2.01	1.94	1.87	1.79
40	5.42	4.05	3.46	3.13	2.90	2.74	2.62	2.53	2.45	2.39	2.29	2.18	2.07	2.01	1.94	1.88	1.80	1.72	1.64
60	5.29	3.93	3.34	3.01	2.79	2.63	2.51	2.41	2.33	2.27	2.17	2.06	1.94	1.88	1.82	1.74	1.67	1.58	1.48
120	5.15	3.80	3.23	2.89	2.67	2.52	2.39	2.30	2.22	2.16	2.05	1.94	1.82	1.76	1.69	1.61	1.53	1.43	1.31
∞	5.02	3.69	3.12	2.79	2.57	2.41	2.29	2.19	2.11	2.05	1.94	1.83	1.71	1.64	1.57	1.48	1.39	1.27	1.00

Table 7—F-Distribution (continued)

$\alpha = 0.05$

d.f._D: Degrees of freedom, denominator	d.f._N: Degrees of freedom, numerator																		
	1	2	3	4	5	6	7	8	9	10	12	15	20	24	30	40	60	120	∞
1	161.4	199.5	215.7	224.6	230.2	234.0	236.8	238.9	240.5	241.9	243.9	245.9	248.0	249.1	250.1	251.1	252.2	253.3	254.3
2	18.51	19.00	19.16	19.25	19.30	19.33	19.35	19.37	19.38	19.40	19.41	19.43	19.45	19.45	19.46	19.47	19.48	19.49	19.50
3	10.13	9.55	9.28	9.12	9.01	8.94	8.89	8.85	8.81	8.79	8.74	8.70	8.66	8.64	8.62	8.59	8.57	8.55	8.53
4	7.71	6.94	6.59	6.39	6.26	6.16	6.09	6.04	6.00	5.96	5.91	5.86	5.80	5.77	5.75	5.72	5.69	5.66	5.63
5	6.61	5.79	5.41	5.19	5.05	4.95	4.88	4.82	4.77	4.74	4.68	4.62	4.56	4.53	4.50	4.46	4.43	4.40	4.36
6	5.99	5.14	4.76	4.53	4.39	4.28	4.21	4.15	4.10	4.06	4.00	3.94	3.87	3.84	3.81	3.77	3.74	3.70	3.67
7	5.59	4.74	4.35	4.12	3.97	3.87	3.79	3.73	3.68	3.64	3.57	3.51	3.44	3.41	3.38	3.34	3.30	3.27	3.23
8	5.32	4.46	4.07	3.84	3.69	3.58	3.50	3.44	3.39	3.35	3.28	3.22	3.15	3.12	3.08	3.04	3.01	2.97	2.93
9	5.12	4.26	3.86	3.63	3.48	3.37	3.29	3.23	3.18	3.14	3.07	3.01	2.94	2.90	2.86	2.83	2.79	2.75	2.71
10	4.96	4.10	3.71	3.48	3.33	3.22	3.14	3.07	3.02	2.98	2.91	2.85	2.77	2.74	2.70	2.66	2.62	2.58	2.54
11	4.84	3.98	3.59	3.36	3.20	3.09	3.01	2.95	2.90	2.85	2.79	2.72	2.65	2.61	2.57	2.53	2.49	2.45	2.40
12	4.75	3.89	3.49	3.26	3.11	3.00	2.91	2.85	2.80	2.75	2.69	2.62	2.54	2.51	2.47	2.43	2.38	2.34	2.30
13	4.67	3.81	3.41	3.18	3.03	2.92	2.83	2.77	2.71	2.67	2.60	2.53	2.46	2.42	2.38	2.34	2.30	2.25	2.21
14	4.60	3.74	3.34	3.11	2.96	2.85	2.76	2.70	2.65	2.60	2.53	2.46	2.39	2.35	2.31	2.27	2.22	2.18	2.13
15	4.54	3.68	3.29	3.06	2.90	2.79	2.71	2.64	2.59	2.54	2.48	2.40	2.33	2.29	2.25	2.20	2.16	2.11	2.07
16	4.49	3.63	3.24	3.01	2.85	2.74	2.66	2.59	2.54	2.49	2.42	2.35	2.28	2.24	2.19	2.15	2.11	2.06	2.01
17	4.45	3.59	3.20	2.96	2.81	2.70	2.61	2.55	2.49	2.45	2.38	2.31	2.23	2.19	2.15	2.10	2.06	2.01	1.96
18	4.41	3.55	3.16	2.93	2.77	2.66	2.58	2.51	2.46	2.41	2.34	2.27	2.19	2.15	2.11	2.06	2.02	1.97	1.92
19	4.38	3.52	3.13	2.90	2.74	2.63	2.54	2.48	2.42	2.38	2.31	2.23	2.16	2.11	2.07	2.03	1.98	1.93	1.88
20	4.35	3.49	3.10	2.87	2.71	2.60	2.51	2.45	2.39	2.35	2.28	2.20	2.12	2.08	2.04	1.99	1.95	1.90	1.84
21	4.32	3.47	3.07	2.84	2.68	2.57	2.49	2.42	2.37	2.32	2.25	2.18	2.10	2.05	2.01	1.96	1.92	1.87	1.81
22	4.30	3.44	3.05	2.82	2.66	2.55	2.46	2.40	2.34	2.30	2.23	2.15	2.07	2.03	1.98	1.94	1.89	1.84	1.78
23	4.28	3.42	3.03	2.80	2.64	2.53	2.44	2.37	2.32	2.27	2.20	2.13	2.05	2.01	1.96	1.91	1.86	1.81	1.76
24	4.26	3.40	3.01	2.78	2.62	2.51	2.42	2.36	2.30	2.25	2.18	2.11	2.03	1.98	1.94	1.89	1.84	1.79	1.73
25	4.24	3.39	2.99	2.76	2.60	2.49	2.40	2.34	2.28	2.24	2.16	2.09	2.01	1.96	1.92	1.87	1.82	1.77	1.71
26	4.23	3.37	2.98	2.74	2.59	2.47	2.39	2.32	2.27	2.22	2.15	2.07	1.99	1.95	1.90	1.85	1.80	1.75	1.69
27	4.21	3.35	2.96	2.73	2.57	2.46	2.37	2.31	2.25	2.20	2.13	2.06	1.97	1.93	1.88	1.84	1.79	1.73	1.67
28	4.20	3.34	2.95	2.71	2.56	2.45	2.36	2.29	2.24	2.19	2.12	2.04	1.96	1.91	1.87	1.82	1.77	1.71	1.65
29	4.18	3.33	2.93	2.70	2.55	2.43	2.35	2.28	2.22	2.18	2.10	2.03	1.94	1.90	1.85	1.81	1.75	1.70	1.64
30	4.17	3.32	2.92	2.69	2.53	2.42	2.33	2.27	2.21	2.16	2.09	2.01	1.93	1.89	1.84	1.79	1.74	1.68	1.62
40	4.08	3.23	2.84	2.61	2.45	2.34	2.25	2.18	2.12	2.08	2.00	1.92	1.84	1.79	1.74	1.69	1.64	1.58	1.51
60	4.00	3.15	2.76	2.53	2.37	2.25	2.17	2.10	2.04	1.99	1.92	1.84	1.75	1.70	1.65	1.59	1.53	1.47	1.39
120	3.92	3.07	2.68	2.45	2.29	2.17	2.09	2.02	1.96	1.91	1.83	1.75	1.66	1.61	1.55	1.50	1.43	1.35	1.25
∞	3.84	3.00	2.60	2.37	2.21	2.10	2.01	1.94	1.88	1.83	1.75	1.67	1.57	1.52	1.46	1.39	1.32	1.22	1.00

Table 7—F-Distribution (continued)

$\alpha = 0.10$

d.f._D: Degrees of freedom, denominator	d.f._N: Degrees of freedom, numerator																		
	1	2	3	4	5	6	7	8	9	10	12	15	20	24	30	40	60	120	∞
1	39.86	49.50	53.59	55.83	57.24	58.20	58.91	59.44	59.86	60.19	60.71	61.22	61.74	62.00	62.26	62.53	62.79	63.06	63.33
2	8.53	9.00	9.16	9.24	9.29	9.33	9.35	9.37	9.38	9.39	9.41	9.42	9.44	9.45	9.46	9.47	9.47	9.48	9.49
3	5.54	5.46	5.39	5.34	5.31	5.28	5.27	5.25	5.24	5.23	5.22	5.20	5.18	5.18	5.17	5.16	5.15	5.14	5.13
4	4.54	4.32	4.19	4.11	4.05	4.01	3.98	3.95	3.94	3.92	3.90	3.87	3.84	3.83	3.82	3.80	3.79	3.78	3.76
5	4.06	3.78	3.62	3.52	3.45	3.40	3.37	3.34	3.32	3.30	3.27	3.24	3.21	3.19	3.17	3.16	3.14	3.12	3.10
6	3.78	3.46	3.29	3.18	3.11	3.05	3.01	2.98	2.96	2.94	2.90	2.87	2.84	2.82	2.80	2.78	2.76	2.74	2.72
7	3.59	3.26	3.07	2.96	2.88	2.83	2.78	2.75	2.72	2.70	2.67	2.63	2.59	2.58	2.56	2.54	2.51	2.49	2.47
8	3.46	3.11	2.92	2.81	2.73	2.67	2.62	2.59	2.56	2.54	2.50	2.46	2.42	2.40	2.38	2.36	2.34	2.32	2.29
9	3.36	3.01	2.81	2.69	2.61	2.55	2.51	2.47	2.44	2.42	2.38	2.34	2.30	2.28	2.25	2.23	2.21	2.18	2.16
10	3.29	2.92	2.73	2.61	2.52	2.46	2.41	2.38	2.35	2.32	2.28	2.24	2.20	2.18	2.16	2.13	2.11	2.08	2.06
11	3.23	2.86	2.66	2.54	2.45	2.39	2.34	2.30	2.27	2.25	2.21	2.17	2.12	2.10	2.08	2.05	2.03	2.00	1.97
12	3.18	2.81	2.61	2.48	2.39	2.33	2.28	2.24	2.21	2.19	2.15	2.10	2.06	2.04	2.01	1.99	1.96	1.93	1.90
13	3.14	2.76	2.56	2.43	2.35	2.28	2.23	2.20	2.16	2.14	2.10	2.05	2.01	1.98	1.96	1.93	1.90	1.88	1.85
14	3.10	2.73	2.52	2.39	2.31	2.24	2.19	2.15	2.12	2.10	2.05	2.01	1.96	1.94	1.91	1.89	1.86	1.83	1.80
15	3.07	2.70	2.49	2.36	2.27	2.21	2.16	2.12	2.09	2.06	2.02	1.97	1.92	1.90	1.87	1.85	1.82	1.79	1.76
16	3.05	2.67	2.46	2.33	2.24	2.18	2.13	2.09	2.06	2.03	1.99	1.94	1.89	1.87	1.84	1.81	1.78	1.75	1.72
17	3.03	2.64	2.44	2.31	2.22	2.15	2.10	2.06	2.03	2.00	1.96	1.91	1.86	1.84	1.81	1.78	1.75	1.72	1.69
18	3.01	2.62	2.42	2.29	2.20	2.13	2.08	2.04	2.00	1.98	1.93	1.89	1.84	1.81	1.78	1.75	1.72	1.69	1.66
19	2.99	2.61	2.40	2.27	2.18	2.11	2.06	2.02	1.98	1.96	1.91	1.86	1.81	1.79	1.76	1.73	1.70	1.67	1.63
20	2.97	2.59	2.38	2.25	2.16	2.09	2.04	2.00	1.96	1.94	1.89	1.84	1.79	1.77	1.74	1.71	1.68	1.64	1.61
21	2.96	2.57	2.36	2.23	2.14	2.08	2.02	1.98	1.95	1.92	1.87	1.83	1.78	1.75	1.72	1.69	1.66	1.62	1.59
22	2.95	2.56	2.35	2.22	2.13	2.06	2.01	1.97	1.93	1.90	1.86	1.81	1.76	1.73	1.70	1.67	1.64	1.60	1.57
23	2.94	2.55	2.34	2.21	2.11	2.05	1.99	1.95	1.92	1.89	1.84	1.80	1.74	1.72	1.69	1.66	1.62	1.59	1.55
24	2.93	2.54	2.33	2.19	2.10	2.04	1.98	1.94	1.91	1.88	1.83	1.78	1.73	1.70	1.67	1.64	1.61	1.57	1.53
25	2.92	2.53	2.32	2.18	2.09	2.02	1.97	1.93	1.89	1.87	1.82	1.77	1.72	1.69	1.66	1.63	1.59	1.56	1.52
26	2.91	2.52	2.31	2.17	2.08	2.01	1.96	1.92	1.88	1.86	1.81	1.76	1.71	1.68	1.65	1.61	1.58	1.54	1.50
27	2.90	2.51	2.30	2.17	2.07	2.00	1.95	1.91	1.87	1.85	1.80	1.75	1.70	1.67	1.64	1.60	1.57	1.53	1.49
28	2.89	2.50	2.29	2.16	2.06	2.00	1.94	1.90	1.87	1.84	1.79	1.74	1.69	1.66	1.63	1.59	1.56	1.52	1.48
29	2.89	2.50	2.28	2.15	2.06	1.99	1.93	1.89	1.86	1.83	1.78	1.73	1.68	1.65	1.62	1.58	1.55	1.51	1.47
30	2.88	2.49	2.28	2.14	2.05	1.98	1.93	1.88	1.85	1.82	1.77	1.72	1.67	1.64	1.61	1.57	1.54	1.50	1.46
40	2.84	2.44	2.23	2.09	2.00	1.93	1.87	1.83	1.79	1.76	1.71	1.66	1.61	1.57	1.54	1.51	1.47	1.42	1.38
60	2.79	2.39	2.18	2.04	1.95	1.87	1.82	1.77	1.74	1.71	1.66	1.60	1.54	1.51	1.48	1.44	1.40	1.35	1.29
120	2.75	2.35	2.13	1.99	1.90	1.82	1.77	1.72	1.68	1.65	1.60	1.55	1.48	1.45	1.41	1.37	1.32	1.26	1.19
∞	2.71	2.30	2.08	1.94	1.85	1.77	1.72	1.67	1.63	1.60	1.55	1.49	1.42	1.38	1.34	1.30	1.24	1.17	1.00

From M. Merrington and C.M. Thompson, "Table of Percentage Points of the Inverted Beta (F) Distribution," Biometrika 33 (1943), pp. 74-87, Oxford University Press.

Table 8—Critical Values for the Sign Test

Reject the null hypothesis when the test statistic x is less than or equal to the value in the table.

n	One-tailed, $\alpha = 0.005$ Two-tailed, $\alpha = 0.01$	$\alpha = 0.01$ $\alpha = 0.02$	$\alpha = 0.025$ $\alpha = 0.05$	$\alpha = 0.05$ $\alpha = 0.10$
8	0	0	0	1
9	0	0	1	1
10	0	0	1	1
11	0	1	1	2
12	1	1	2	2
13	1	1	2	3
14	1	2	3	3
15	2	2	3	3
16	2	2	3	4
17	2	3	4	4
18	3	3	4	5
19	3	4	4	5
20	3	4	5	5
21	4	4	5	6
22	4	5	5	6
23	4	5	6	7
24	5	5	6	7
25	5	6	6	7

Note: Table 8 is for one-tailed or two-tailed tests. The sample size n represents the total number of + and − signs. The test value is the smaller number of + or − signs.

From *Journal of American Statistical Association* Vol. 41 (1946), pp. 557– 566. W. J. Dixon and A. M. Mood.

Table 9—Critical Values for the Wilcoxon Signed-Rank Test

Reject the null hypothesis when the test statistic w_s is less than or equal to the value in the table.

n	One-tailed, $\alpha = 0.05$ Two-tailed, $\alpha = 0.10$	$\alpha = 0.025$ $\alpha = 0.05$	$\alpha = 0.01$ $\alpha = 0.02$	$\alpha = 0.005$ $\alpha = 0.01$
5	1	—	—	—
6	2	1	—	—
7	4	2	0	—
8	6	4	2	0
9	8	6	3	2
10	11	8	5	3
11	14	11	7	5
12	17	14	10	7
13	21	17	13	10
14	26	21	16	13
15	30	25	20	16
16	36	30	24	19
17	41	35	28	23
18	47	40	33	28
19	54	46	38	32
20	60	52	43	37
21	68	59	49	43
22	75	66	56	49
23	83	73	62	55
24	92	81	69	61
25	101	90	77	68
26	110	98	85	76
27	120	107	93	84
28	130	117	102	92
29	141	127	111	100
30	152	137	120	109

From *Some Rapid Approximate Statistical Procedures.* Copyright 1949, 1964 Lederle Laboratories, American Cyanamid Co., Wayne, N.J.

Table 10—Critical Values for the Spearman Rank Correlation Coefficient

Reject H_0: $\rho_s = 0$ when the absolute value of r_s is greater than the value in the table.

n	α = 0.10	α = 0.05	α = 0.01
5	0.900	—	—
6	0.829	0.886	—
7	0.714	0.786	0.929
8	0.643	0.738	0.881
9	0.600	0.700	0.833
10	0.564	0.648	0.794
11	0.536	0.618	0.818
12	0.497	0.591	0.780
13	0.475	0.566	0.745
14	0.457	0.545	0.716
15	0.441	0.525	0.689
16	0.425	0.507	0.666
17	0.412	0.490	0.645
18	0.399	0.476	0.625
19	0.388	0.462	0.608
20	0.377	0.450	0.591
21	0.368	0.438	0.576
22	0.359	0.428	0.562
23	0.351	0.418	0.549
24	0.343	0.409	0.537
25	0.336	0.400	0.526
26	0.329	0.392	0.515
27	0.323	0.385	0.505
28	0.317	0.377	0.496
29	0.311	0.370	0.487
30	0.305	0.364	0.478

From the Institute of Mathematical Statistics.

Table 11—Critical Values for the Pearson Correlation Coefficient

The correlation is significant when the absolute value of r is greater than the value in the table.

n	α = 0.05	α = 0.01
4	0.950	0.990
5	0.878	0.959
6	0.811	0.917
7	0.754	0.875
8	0.707	0.834
9	0.666	0.798
10	0.632	0.765
11	0.602	0.735
12	0.576	0.708
13	0.553	0.684
14	0.532	0.661
15	0.514	0.641
16	0.497	0.623
17	0.482	0.606
18	0.468	0.590
19	0.456	0.575
20	0.444	0.561
21	0.433	0.549
22	0.423	0.537
23	0.413	0.526
24	0.404	0.515
25	0.396	0.505
26	0.388	0.496
27	0.381	0.487
28	0.374	0.479
29	0.367	0.471
30	0.361	0.463
35	0.334	0.430
40	0.312	0.403
45	0.294	0.380
50	0.279	0.361
55	0.266	0.345
60	0.254	0.330
65	0.244	0.317
70	0.235	0.306
75	0.227	0.296
80	0.220	0.286
85	0.213	0.278
90	0.207	0.270
95	0.202	0.263
100	0.197	0.256

The critical values in Table 11 were generated using Excel.

Table 12—Critical Values for the Number of Runs

Reject the null hypothesis when the test statistic G is less than or equal to the smaller entry or greater than or equal to the larger entry.

		Value of n_2																	
		2	**3**	**4**	**5**	**6**	**7**	**8**	**9**	**10**	**11**	**12**	**13**	**14**	**15**	**16**	**17**	**18**	**19**	**20**
Value of n_1	**2**	1	1	1	1	1	1	1	1	1	1	2	2	2	2	2	2	2	2	2
		6	6	6	6	6	6	6	6	6	6	6	6	6	6	6	6	6	6	6
	3	1	1	1	1	2	2	2	2	2	2	2	2	2	3	3	3	3	3	3
		6	8	8	8	8	8	8	8	8	8	8	8	8	8	8	8	8	8	8
	4	1	1	1	2	2	2	3	3	3	3	3	3	3	3	4	4	4	4	4
		6	8	9	9	9	10	10	10	10	10	10	10	10	10	10	10	10	10	10
	5	1	1	2	2	3	3	3	3	3	4	4	4	4	4	4	4	5	5	5
		6	8	9	10	10	11	11	12	12	12	12	12	12	12	12	12	12	12	12
	6	1	2	2	3	3	3	3	4	4	4	4	5	5	5	5	5	5	6	6
		6	8	9	10	11	12	12	13	13	13	13	14	14	14	14	14	14	14	14
	7	1	2	2	3	3	3	4	4	5	5	5	5	5	6	6	6	6	6	6
		6	8	10	11	12	13	13	14	14	14	14	15	15	15	16	16	16	16	16
	8	1	2	3	3	3	4	4	5	5	5	6	6	6	6	6	7	7	7	7
		6	8	10	11	12	13	14	14	15	15	16	16	16	16	17	17	17	17	17
	9	1	2	3	3	4	4	5	5	5	6	6	6	7	7	7	7	8	8	8
		6	8	10	12	13	14	14	15	16	16	16	17	17	18	18	18	18	18	18
	10	1	2	3	3	4	5	5	5	6	6	7	7	7	7	8	8	8	8	9
		6	8	10	12	13	14	15	16	16	17	17	18	18	18	19	19	19	20	20
	11	1	2	3	4	4	5	5	6	6	7	7	7	8	8	8	9	9	9	9
		6	8	10	12	13	14	15	16	17	17	18	19	19	19	20	20	20	21	21
	12	2	2	3	4	4	5	6	6	7	7	7	8	8	8	9	9	9	10	10
		6	8	10	12	13	14	16	16	17	18	19	19	20	20	21	21	21	22	22
	13	2	2	3	4	5	5	6	6	7	7	8	8	9	9	9	10	10	10	10
		6	8	10	12	14	15	16	17	18	19	19	20	20	21	21	22	22	23	23
	14	2	2	3	4	5	5	6	7	7	8	8	9	9	9	10	10	10	11	11
		6	8	10	12	14	15	16	17	18	19	20	20	21	22	22	23	23	23	24
	15	2	3	3	4	5	6	6	7	7	8	8	9	9	10	10	11	11	11	12
		6	8	10	12	14	15	16	18	18	19	20	21	22	22	23	23	24	24	25
	16	2	3	4	4	5	6	6	7	8	8	9	9	10	10	11	11	11	12	12
		6	8	10	12	14	16	17	18	19	20	21	21	22	23	23	24	25	25	25
	17	2	3	4	4	5	6	7	7	8	9	9	10	10	11	11	11	12	12	13
		6	8	10	12	14	16	17	18	19	20	21	22	23	23	24	25	25	26	26
	18	2	3	4	5	5	6	7	8	8	9	9	10	10	11	11	12	12	13	13
		6	8	10	12	14	16	17	18	19	20	21	22	23	24	25	25	26	26	27
	19	2	3	4	5	6	6	7	8	8	9	10	10	11	11	12	12	13	13	13
		6	8	10	12	14	16	17	18	20	21	22	23	23	24	25	26	26	27	27
	20	2	3	4	5	6	6	7	8	9	9	10	10	11	12	12	13	13	13	14
		6	8	10	12	14	16	17	18	20	21	22	23	24	25	25	26	27	27	28

Note: Table 12 is for a two-tailed test with $\alpha = 0.05$.

From the Institute of Mathematical Statistics.

APPENDIX C

C Normal Probability Plots

What You Should Learn

▶ How to construct and interpret a normal probability plot

Normal Probability Plots

For many of the examples and exercises in this text, it has been assumed that a random sample is selected from a population that has a normal distribution. After selecting a random sample from a population with an unknown distribution, how can you determine whether the sample was selected from a population that has a normal distribution?

You have already learned that a histogram or stem-and-leaf plot can reveal the shape of a distribution and any outliers, clusters, or gaps in a distribution (see Sections 2.1, 2.2, and 2.3). These data displays are useful for assessing large sets of data, but assessing small data sets in this manner can be difficult and unreliable. A reliable method for assessing normality in *any* data set is to use a **normal probability plot.**

DEFINITION

A **normal probability plot** (also called a **normal quantile plot**) is a graph that plots each observed value from the data set along with its expected z-score. The observed values are usually plotted along the horizontal axis while the expected z-scores are plotted along the vertical axis.

The guidelines below can help you determine whether data come from a population that has a normal distribution.

1. If the plotted points in a normal probability plot are approximately linear, then you can conclude that the data come from a normal distribution.

2. If the plotted points are not approximately linear or follow some type of pattern that is not linear, then you can conclude that the data come from a distribution that is not normal.

3. Multiple outliers or clusters of points indicate a distribution that is not normal.

Two normal probability plots are shown below. The normal probability plot on the left is approximately linear. So, you can conclude that the data come from a population that has a normal distribution. The normal probability plot on the right follows a nonlinear pattern. So, you can conclude that the data do not come from a population that has a normal distribution.

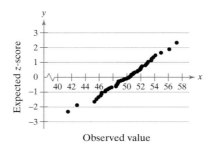

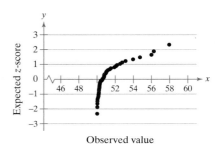

Constructing a normal probability plot by hand can be rather tedious. You can use technology such as Minitab, Excel, StatCrunch, or the TI-84 Plus to construct a normal probability plot, as shown in Example 1.

EXAMPLE 1

Constructing a Normal Probability Plot

The heights (in inches) of 12 randomly selected current National Basketball Association players are listed. Use technology to construct a normal probability plot to determine whether the data come from a population that has a normal distribution.

74   69   78   75   73   71   80   82   81   76   86   77

SOLUTION

Using Minitab, enter the heights into column C1. From the *Graph* menu, select "Probability Plot," choose the option "Single," and click OK. Next, select column C1 as the graph variable. Then click "Distribution" and choose "Normal" from the drop-down menu. Click the *Data Display* tab, select "Symbols only," and click OK. After clicking "Scale," click the *Y-Scale Type* tab, select "Score," and click OK. Click OK to construct the normal probability plot. Your result should be similar to the one shown below. (To construct a normal probability plot using a TI-84 Plus, follow the instructions in the Tech Tip at the left.)

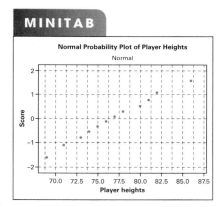

Interpretation   Because the points are approximately linear, you can conclude that the sample data come from a population that has a normal distribution.

TRY IT YOURSELF 1

The balances (in dollars) on student loans for 18 randomly selected college seniors are listed. Use technology to construct a normal probability plot to determine whether the data come from a population that has a normal distribution.

| 29,150 | 16,980 | 12,470 | 19,235 | 15,875 | 8,960 | 16,105 | 14,575 | 39,860 |
| 20,170 | 9,710 | 19,650 | 21,590 | 8,200 | 18,100 | 25,530 | 9,285 | 10,075 |

Answer: Page A39

To see that the points are approximately linear, you can graph the regression line for the observed values from the data set and their expected z-scores. The regression line for the heights and expected z-scores from Example 1 is shown in the graph at the left. From the graph, you can see that the points lie along the regression line. You can also approximate the mean of the data set by determining where the line crosses the x-axis.

Tech Tip

Here are instructions for constructing a normal probability plot using a TI-84 Plus. First, enter the data into List 1. Then use *Stat Plot* to construct the normal probability plot, as shown below.

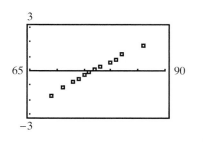

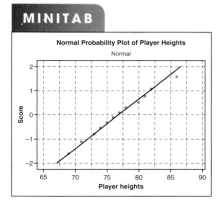

C EXERCISES

1. The observed values are usually plotted along the horizontal axis. The expected *z*-scores are plotted along the vertical axis.

3. Because the points appear to follow a nonlinear pattern, you can conclude that the data do not come from a population that has a normal distribution.

5.

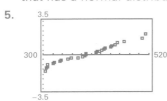

Because the points are approximately linear, you can conclude that the data come from a population that has a normal distribution.

2. If the plotted points in a normal probability plot are approximately linear, then you can conclude that the data come from a normal distribution. If the plotted points are not approximately linear or follow some type of pattern that is not linear, then you can conclude that the data come from a distribution that is not normal. Multiple outliers or clusters of points indicate a distribution that is not normal.

4. Because the points are approximately linear, you can conclude that the data come from a population that has a normal distribution.

6.

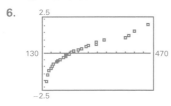

Because the points appear to follow a nonlinear pattern, you can conclude that the data do not come from a population that has a normal distribution.

1. In a normal probability plot, what is usually plotted along the horizontal axis? What is usually plotted along the vertical axis?

2. Describe how you can use a normal probability plot to determine whether data come from a normal distribution.

Graphical Analysis   *In Exercises 3 and 4, use the histogram and normal probability plot to determine whether the data come from a normal distribution. Explain your reasoning.*

3.

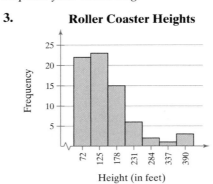

Roller Coaster Heights

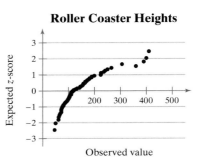

Roller Coaster Heights

4.

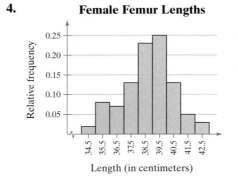

Female Femur Lengths

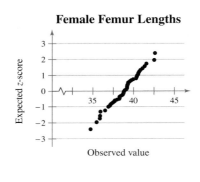

Female Femur Lengths

Constructing a Normal Probability Plot   *In Exercises 5 and 6, use technology to construct a normal probability plot to determine whether the data come from a population that has a normal distribution.*

 5. Reaction Times   The reaction times (in milliseconds) of 30 randomly selected adult females to an auditory stimulus

507	389	305	291	336	310	514	442
373	428	387	454	323	441	388	426
411	382	320	450	309	416	359	388
307	337	469	351	422	413		

 6. Triglyceride Levels   The triglyceride levels (in milligrams per deciliter of blood) of 26 randomly selected patients

209	140	155	170	265	138	180
295	250	320	270	225	215	390
420	462	150	200	400	295	240
200	190	145	160	175		

TRY IT YOURSELF ANSWERS

Chapter 1

Section 1.1

1. The population consists of the responses of all ninth to twelfth graders in the United States. The sample consists of the responses of the 1501 ninth to twelfth graders in the survey. The sample data set consists of 1215 ninth to twelfth graders who said leaders today are more concerned with their own agenda than with achieving the overall goals of the organization they serve and 286 ninth to twelfth graders who did not say that.

2. **a.** Population parameter, because the total spent on employees' salaries, $5,150,694, is based on the entire company.
   b. Sample statistic, because 43% is based on a subset of the population.

3. **a.** The population consists of the responses of all U.S. adults, and the sample consists of the responses of the 1000 U.S. adults in the study.
   b. The part of this study that represents the descriptive branch of statistics involves the statement "three out of four adults will consult with their physician or pharmacist and only 8% visit a medication-specific website [when they have a question about their medication]."
   c. A possible inference drawn from the study is that most adults consult with their physician or pharmacist when they have a question about their medication.

Section 1.2

1. The city names are nonnumerical entries, so these are qualitative data. The city populations are numerical entries, so these are quantitative data.

2. (1) Ordinal, because the data can be put in order.
   (2) Nominal, because no mathematical computations can be made.

3. (1) Interval, because the data can be ordered and meaningful differences can be calculated, but it does not make sense to write a ratio using the temperatures.
   (2) Ratio, because the data can be ordered, meaningful differences can be calculated, the data can be written as a ratio, and the data set contains an inherent zero.

Section 1.3

1. This is an observational study.

2. There is no way to tell why the people quit smoking. They could have quit smoking as a result of either chewing the gum or watching the DVD. The gum and the DVD could be confounding variables. To improve the study, two experiments could be done, one using the gum and the other using the DVD. Or just conduct one experiment using either the gum or the DVD.

3. *Sample answer:* Assign numbers 1 to 79 to the employees of the company. Use the table of random numbers and obtain 63, 7, 40, 19, and 26. The employees assigned these numbers will make up the sample.

4. (1) The sample was selected by using the students in a randomly chosen class. This is cluster sampling.
   (2) The sample was selected by numbering each student in the school, randomly choosing a starting number, and selecting students at regular intervals from the starting number. This is systematic sampling.

Chapter 2

Section 2.1

1.

Class	Frequency, f
14–20	8
21–27	15
28–34	14
35–41	7
42–48	4
49–55	3

2.

Class	Frequency, f	Midpoint	Relative frequency	Cumulative frequency
14–20	8	17	0.1569	8
21–27	15	24	0.2941	23
28–34	14	31	0.2745	37
35–41	7	38	0.1373	44
42–48	4	45	0.0784	48
49–55	3	52	0.0588	51
	$\Sigma f = 51$		$\Sigma \dfrac{f}{n} = 1$	

Sample answer: The most common range of points scored by winning teams is 21 to 27. About 14% of the winning teams scored more than 41 points.

3.

Points Scored by Winning Super Bowl Teams

Sample answer: The most common range of points scored by winning teams is 21 to 27. About 14% of the winning teams scored more than 41 points.

4.

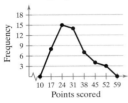

Points Scored by Winning Super Bowl Teams

Sample answer: The frequency of points scored increases up to 24 points and then decreases.

5.

Points Scored by Winning Super Bowl Teams

6.

Points Scored by Winning Super Bowl Teams

7.

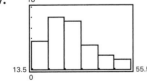

Section 2.2

1.

```
1 | 4 6 6 6 7                            Key: 1|4 = 14
2 | 0 0 0 1 1 1 3 3 4 4 4 4 6 7 7 7 7 7 8 9
3 | 0 1 1 1 1 2 2 3 4 4 4 4 5 5 5 7 8 8 9
4 | 2 3 6 8 9
5 | 2 5
```

Sample answer: Most of the winning teams scored between 20 and 39 points.

2.

```
1 | 4                    Key: 1|4 = 14
1 | 6 6 6 7
2 | 0 0 0 1 1 1 3 3 4 4 4 4
2 | 6 7 7 7 7 7 8 9
3 | 0 1 1 1 1 2 2 3 4 4 4 4
3 | 5 5 5 7 8 8 9
4 | 2 3
4 | 6 8 9
5 | 2
5 | 5
```

Sample answer: Most of the winning teams scored from 20 to 35 points.

3.   **Points Scored by Winning Super Bowl Teams**

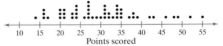

Sample answer: Most of the points scored by winning teams cluster between 20 and 40.

4.   **Earned Degrees Conferred in 1990**

From 1990 to 2014, as percentages of the total degrees conferred, associate's degrees increased by 2.9%, bachelor's degrees decreased by 5.1%, master's degrees increased by 2.8%, and doctoral degrees decreased by 0.7%.

5.   **Causes of BBB Complaints**

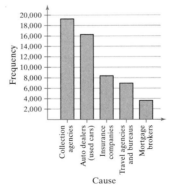

Collection agencies are the greatest cause of complaints.

6.

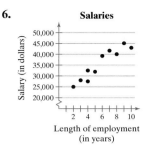

It appears that the longer an employee is with the company, the greater the employee's salary.

7.

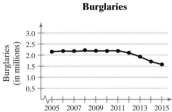

Sample answer: The number of burglaries remained about the same until 2012 and then decreased through 2015.

Section 2.3

1. About 30.2   **2.** 30   **3.** 28.5   **4.** 27   **5.** "some"
6. $\bar{x} \approx 21.6$; median = 21; mode = 20
The mean in Example 6 ($\bar{x} \approx 23.8$) was heavily influenced by the entry 65. Neither the median nor the mode was affected as much by the entry 65.
7. About 2.6
8. About 30.0; This is very close to the mean found using the original data set.

Section 2.4

1. 35, or $35,000; The range of the starting salaries for Corporation B, which is $35,000, is much larger than the range of Corporation A.
2. $\sigma^2 \approx 110.3$; $\sigma \approx 10.5$, or $10,500
3. $s^2 \approx 177.1$; $s \approx 13.3$ **4.** $\bar{x} \approx 19.8$; $s \approx 7.8$
5. *Sample answer:* 7, 7, 7, 7, 7, 13, 13, 13, 13, 13   **6.** 34.13%
7. At least 75% of Iowa's population is between 0 and 86.3 years old. Because $80 < 86.3$, an age of 80 lies within two standard deviations of the mean. So, the age is not unusual.
8. $\bar{x} = 1.7$; $s \approx 1.5$
Both the mean and sample standard deviation decreased slightly.
9. $\bar{x} \approx 195.5$; $s \approx 169.5$
Both the mean and sample standard deviation increased.
10. Los Angeles: $CV \approx 47.2\%$
Dallas: $CV \approx 39.4\%$
The office rental rates are more variable in Los Angeles than in Dallas.

Section 2.5

1. $Q_1 = 23$, $Q_2 = 30$, $Q_3 = 35$
About one-quarter of the winning scores were 23 points or less, about one-half were 30 points or less, and about three-quarters were 35 points or less.
2. $Q_1 = 23.5$, $Q_2 = 30$, $Q_3 = 41$
About one-quarter of these universities charge tuition of $23,500 or less, about one-half charge $30,000 or less, and about three-quarters charge $41,000 or less.
3. IQR = 12; 55 is an outlier.
4.

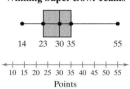

About 50% of the winning scores were between 23 and 35 points. About 25% of the winning scores were less than 23 points. About 25% of the winning scores were greater than 35 points.
5. 19.5; About 10% of the winning scores were 19 points or less.
6. 28th percentile
7. For $60, $z = -1.25$.
For $71, $z = 0.125$.
For $92, $z = 2.75$.
8. Man: $z = -3.3$; Woman: $z \approx -1.7$
The z-score for the 5-foot-tall man is 3.3 standard deviations below the mean. This is a very unusual height for a man. The z-score for the 5-foot-tall woman is 1.7 standard deviations below the mean. This is among the typical heights for a woman.

Chapter 3

Section 3.1

1. (1)

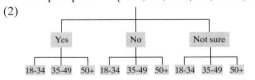

6 outcomes
Let Y = Yes, N = No, NS = Not sure, M = Male, F = Female.
Sample space = {YM, YF, NM, NF, NSM, NSF}

(2)

9 outcomes
Let Y = Yes, N = No, NS = Not sure,
50+ = 50 and older.
Sample space = {Y18–34, Y35–49, Y50+, N18–34, N35–49, N50+, NS18–34, NS35–49, NS50+}

(3)

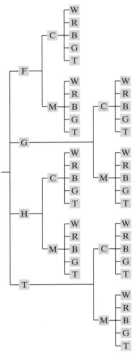

12 outcomes
Let Y = Yes, N = No, NS = Not sure,
NE = Northeast, S = South, MW = Midwest,
W = West.
Sample space = {YNE, YS, YMW, YW, NNE, NS,
NMW, NW, NSNE, NSS, NSMW, NSW}

2. (1) 6; Not a simple event because it is an event that consists of more than a single outcome.
(2) 1; Simple event because it is an event that consists of a single outcome.

3. 40

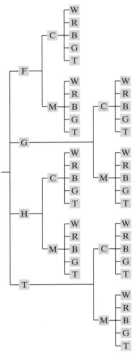

4. (1) 308,915,776 (2) 165,765,600 (3) 261,390,272
(4) 106,932,384
5. (1) 0.019 (2) 0.25 (3) 1 **6.** 0.061 **7.** 0.261
8. Empirical probability **9.** 0.84 **10.** 0.313

11. $\dfrac{1}{10,000,000}$

Section 3.2

1. 0.488
2. (1) Dependent (2) Independent
3. (1) 0.723 (2) 0.059
4. (1) 0.729 (2) 0.001 (3) 0.999
5. (1) 0.163 (2) 0.488
Both of the events are not unusual because their probabilities are not less than or equal to 0.05.

Section 3.3

1. (1) Not mutually exclusive; The events can occur at the same time.
(2) Mutually exclusive; The events cannot occur at the same time.
2. (1) 0.667 (2) 0.423 **3.** 0.222
4. (1) 0.149 (2) 0.149 (3) 0.910 (4) 0.499 **5.** 0.806

Section 3.4

1. 3,628,800 **2.** 336 **3.** 11,880 **4.** 77,597,520
5. 1140 **6.** 0.003 **7.** 0.0009 **8.** 0.045

Chapter 4

Section 4.1

1. (1) The random variable is continuous because x can be any speed up to the maximum speed of a rocket.
(2) The random variable is discrete because the number of calves born on a farm in one year is countable.
(3) The random variable is discrete because the number of days of rain for the next three days is countable.

2.

x	f	$P(x)$
0	16	0.16
1	19	0.19
2	15	0.15
3	21	0.21
4	9	0.09
5	10	0.10
6	8	0.08
7	2	0.02
	$n = 100$	$\Sigma P(x) = 1$

New Employee Sales

3. Each $P(x)$ is between 0 and 1 and $\Sigma P(x) = 1$. Because both conditions are met, the distribution is a probability distribution.
4. (1) Probability distribution; The probability of each outcome is between 0 and 1, and the sum of all the probabilities is 1.
(2) Not a probability distribution; The sum of all the probabilities is not 1.
5. $\mu = 2.6$; On average, a new employee makes 2.6 sales per day.
6. $\sigma^2 \approx 3.7$; $\sigma \approx 1.9$
7. −$3.08; Because the expected value is negative, you can expect to lose an average of $3.08 for each ticket you buy.

Section 4.2

1. Binomial experiment
$n = 10, p = 0.25, q = 0.75, x = 0, 1, 2, 3, 4, 5, 6, 7, 8, 9, 10$
2. 0.088

3.

x	P(x)
0	0.193
1	0.376
2	0.293
3	0.114
4	0.022
5	0.002

4. 0.007

5. (1) 0.284   (2) 0.409   (3) 0.591   **6.** 0.031

7.

x	P(x)
0	0.269
1	0.418
2	0.244
3	0.063
4	0.006

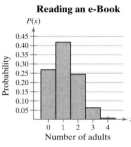

Reading an e-Book

8. $\mu \approx 13.6$; $\sigma^2 \approx 7.6$; $\sigma \approx 2.8$; On average, there are about 14 clear days during the month of May. A May with fewer than 8 clear days or more than 19 clear days would be unusual.

Section 4.3

1. 0.066   **2.** 0.185   **3.** 0.0002

Chapter 5

Section 5.1

1. (1) Curve B has the greatest mean.
(2) Curve C is more spread out, so curve C has the greatest standard deviation.
2. $\mu = 300$; $\sigma = 37$ **3.** (1) 0.0143   (2) 0.9850
4. 0.9834   **5.** 0.9846   **6.** 0.0733

Section 5.2

1. 0.1957   **2.** 0.7357; 147   **3.** 0.4352

Section 5.3

1. (1) −1.77   (2) 1.96
2. (1) −1.28   (2) −0.84   (3) 2.33
3. (1) 17.05 pounds   (2) 97 pounds   (3) 60.7 pounds
17.05 pounds is to the left of the mean, 97 pounds is to the right of the mean, and 60.7 pounds is to the right of the mean.
4. The longest braking distance one of these cars could have and still be in the bottom 1% is about 117 feet.
5. The maximum length of time an employee could have worked and still be laid off is about 8.5 years.

Section 5.4

1.

Sample	Mean	Sample	Mean
1, 1, 1	1	3, 3, 5	3.67
1, 1, 3	1.67	3, 5, 1	3
1, 1, 5	2.33	3, 5, 3	3.67
1, 3, 1	1.67	3, 5, 5	4.33
1, 3, 3	2.33	5, 1, 1	2.33
1, 3, 5	3	5, 1, 3	3
1, 5, 1	2.33	5, 1, 5	3.67
1, 5, 3	3	5, 3, 1	3
1, 5, 5	3.67	5, 3, 3	3.67
3, 1, 1	1.67	5, 3, 5	4.33
3, 1, 3	2.33	5, 5, 1	3.67
3, 1, 5	3	5, 5, 3	4.33
3, 3, 1	2.33	5, 5, 5	5
3, 3, 3	3		

$\mu_{\bar{x}} = 3$; $(\sigma_{\bar{x}})^2 \approx 0.889$; $\sigma_{\bar{x}} \approx 0.943$
$\mu_{\bar{x}} = \mu = 3$
$(\sigma_{\bar{x}})^2 = \dfrac{\sigma^2}{n} = \dfrac{8/3}{3} = \dfrac{8}{9} \approx 0.889$

$\sigma_{\bar{x}} = \dfrac{\sigma}{\sqrt{n}} = \dfrac{\sqrt{8/3}}{\sqrt{3}} = \dfrac{\sqrt{8}}{3} \approx 0.943$

2. $\mu_{\bar{x}} = 6.8$, $\sigma_{\bar{x}} \approx 0.18$

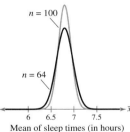

With a smaller sample size, the mean stays the same but the standard deviation increases.

3. $\mu_{\bar{x}} = 3.5$, $\sigma_{\bar{x}} = 0.05$

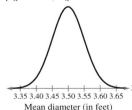

4. 0.9744   **5.** 0.7673
6. 0.5832; 0.7454
There is about a 58% chance that an LCD computer monitor will cost less than $200. There is about a 75% chance that the mean of a sample of 10 LCD computer monitors is less than $200.

Section 5.5

1. Because np and nq are greater than 5, a normal distribution can be used.
2. (1) $56.5 < x < 83.5$ (2) $x < 54.5$ 3. 0.3707
4. 0.0281   5. 0.0083

Chapter 6

Section 6.1

1. 20.9   2. 0.8 hour
3. (20.1, 21.7); This confidence interval is wider than the one found in Example 3.
4. (20.6, 21.5); (20.5, 21.6); (20.4, 21.7); As the confidence level increases, so does the width of the interval.
5. (22.4, 23.4) [*Tech:* (22.5, 23.4)]; Because of the larger sample size, the confidence interval is slightly narrower.
6. 37; Because of the larger margin of error, the sample size needed is smaller.

Section 6.2

1. 1.721   2. (157.6, 166.4); (154.6, 169.4)
3. (9.08, 10.42); (8.94, 10.56); The 90% confidence interval is slightly narrower.
4. Use a t-distribution because σ is not known and the population is normally distributed.

Section 6.3

1. 15%   2. (0.14, 0.16)   3. (0.454, 0.546)
4. (1) 1692   (2) 400

Section 6.4

1. 42.557, 17.708
2. Population variance: (0.98, 2.36), (0.91, 2.60)
   Standard deviation: (0.99, 1.54), (0.96, 1.61)

Chapter 7

Section 7.1

1. (1) The mean is not 74 months.
   $\mu \neq 74$
   $H_0: \mu = 74$; $H_a: \mu \neq 74$ (claim)
   (2) The variance is less than or equal to 2.7.
   $\sigma^2 \leq 2.7$
   $H_0: \sigma^2 \leq 2.7$ (claim); $H_a: \sigma^2 > 2.7$
   (3) The proportion is more than 24%.
   $p > 0.24$
   $H_0: p \leq 0.24$; $H_a: p > 0.24$ (claim)

2. A type I error will occur when the actual proportion is less than or equal to 0.01, but you reject H_0.
   A type II error will occur when the actual proportion is greater than 0.01, but you fail to reject H_0.
   A type II error is more serious because you would be misleading the consumer, possibly causing serious injury or death.
3. (1) H_0: The mean life of a certain type of automobile battery is 74 months.
   H_a: The mean life of a certain type of automobile battery is not 74 months.
   $H_0: \mu = 74$; $H_a: \mu \neq 74$
   Two-tailed

   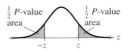

   (2) H_0: The variance of the life of a manufacturer's home theater systems is less than or equal to 2.7.
   H_a: The variance of the life of a manufacturer's home theater systems is greater than 2.7.
   $H_0: \sigma^2 \leq 2.7$; $H_a: \sigma^2 > 2.7$
   Right-tailed

   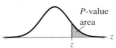

   (3) H_0: The proportion of homeowners who feel their house is too small for their family is less than or equal to 24%.
   H_a: The proportion of homeowners who feel their house is too small for their family is greater than 24%.
   $H_0: p \leq 0.24$; $H_a: p > 0.24$
   Right-tailed

   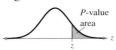

4. (1) There is enough evidence to support the claim that the mean life of a certain type of automobile battery is not 74 months.
   There is not enough evidence to support the claim that the mean life of a certain type of automobile battery is not 74 months.
   (2) There is enough evidence to support the realtor's claim that the proportion of homeowners who feel their house is too small for their family is more than 24%.
   There is not enough evidence to support the realtor's claim that the proportion of homeowners who feel their house is too small for their family is more than 24%.
5. (1) $H_0: \mu \geq 650$; $H_a: \mu < 650$ (claim)
   If you reject H_0, then you will support the claim that the mean repair cost per automobile is less than $650.
   (2) $H_0: \mu = 98.6$ (claim); $H_a: \mu \neq 98.6$
   If you reject H_0, then you will reject the claim that the mean temperature is about 98.6°F.

Section 7.2

1. (1) Fail to reject H_0.   (2) Reject H_0.
2. 0.0436; Reject H_0 because 0.0436 < 0.05.
3. 0.1010; Fail to reject H_0 because 0.1010 > 0.01.
4. There is enough evidence at the 5% level of significance to support the claim that the average speed is greater than 35 miles per hour.
5. There is not enough evidence at the 1% level of significance to support the claim that the mean number of workdays missed due to illness or injury in the past 12 months is 3.5 days.
6. Fail to reject H_0.
7. $z_0 = -1.28$; Rejection region: $z < -1.28$
8. $-z_0 = -1.75$, $z_0 = 1.75$
   Rejection regions: $z < -1.75$, $z > 1.75$
9. There is enough evidence at the 1% level of significance to support the claim that the mean workday is less than 8.5 hours.
10. There is not enough evidence at the 1% level of significance to reject the claim that the mean cost of raising a child (age 2 and under) by married-couple families in the United States is $14,050.

Section 7.3

1. -2.650 2. 1.397 3. $-2.131, 2.131$
4. There is enough evidence at the 10% level of significance to support the claim that the mean age of a used car sold in the last 12 months is less than 4.1 years.
5. There is enough evidence at the 1% level of significance to reject the company's claim that the mean conductivity of the river is 1890 milligrams per liter.
6. There is not enough evidence at the 5% level of significance to reject the office's claim that the mean wait time is at most 18 minutes.

Section 7.4

1. There is not enough evidence at the 1% level of significance to support the claim that more than 90% of U.S. adults have access to a smartphone.
2. There is enough evidence at the 10% level of significance to reject the claim that 67% of U.S. adults believe that doctors prescribing antibiotics for viral infections for which antibiotics are not effective is a significant cause of drug-resistant superbugs.

Section 7.5

1. $\chi^2 = 33.409$ 2. $\chi^2 = 17.708$
3. $\chi_L^2 = 27.991$, $\chi_R^2 = 79.490$
4. There is enough evidence at the 1% level of significance to reject the bottling company's claim that the variance of the amount of sports drink in a 12-ounce bottle is no more than 0.40.

5. There is not enough evidence at the 5% level of significance to support the police chief's claim that the standard deviation of the lengths of response times is less than 3.7 minutes.
6. There is enough evidence at the 10% level of significance to reject the company's claim that the variance of the weight losses of the users is 25.5.

Chapter 8

Section 8.1

1. (1) Independent   (2) Dependent
2. There is not enough evidence at the 10% level of significance to support the claim that there is a difference in the mean annual wages for forensic science technicians working for local and state governments.
3. There is not enough evidence at the 5% level of significance to support the travel agency's claim that the average daily cost of meals and lodging for vacationing in Alaska is greater than the average daily cost in Colorado.

Section 8.2

1. There is enough evidence at the 5% level of significance to support the claim that there is a difference in the mean annual earnings based on level of education.
2. There is not enough evidence at the 10% level of significance to support the manufacturer's claim that the mean driving cost per mile of its minivans is less than that of its leading competitor.

Section 8.3

1. There is not enough evidence at the 5% level of significance to support the claim that athletes can decrease their times in the 40-yard dash.
2. There is not enough evidence at the 5% level of significance to support the claim that the drug changes the body's temperature.

Section 8.4

1. There is not enough evidence at the 5% level of significance to support the claim that there is a difference between the proportion of 40- to 49-year-olds who are yoga users and the proportion of 40- to 49-year-olds who are non-yoga users.
2. There is enough evidence at the 5% level of significance to support the claim that the proportion of yoga users with incomes of $20,000 to $34,499 is less than the proportion of non-yoga users with incomes of $20,000 to $34,499.

Chapter 9

Section 9.1

1.

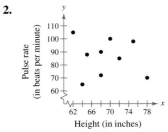

It appears that there is a negative linear correlation. As the number of years out of school increases, the annual contribution tends to decrease.

2.

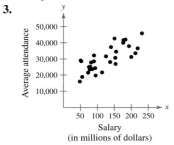

It appears that there is no linear correlation between height and pulse rate.

3.

It appears that there is a positive linear correlation. As the team salary increases, the average attendance per home game tends to increase.

4. Because r is close to -1, this suggests a strong negative linear correlation. As the number of years out of school increases, the annual contribution tends to decrease.

5. 0.775; Because r is close to 1, this suggests a strong positive linear correlation. As the team salaries increase, the average attendance per home game tends to increase.

6. $|r| \approx 0.908 > 0.875$; The correlation is significant.
There is enough evidence at the 1% level of significance to conclude that there is a significant linear correlation between the number of years out of school and the annual contribution.

7. There is enough evidence at the 1% level of significance to conclude that there is a significant linear correlation between the salaries and average attendances per home game for the teams in Major League Baseball.

Section 9.2

1. $\hat{y} = -0.380x + 12.876$
2. $\hat{y} = 108.022x + 16{,}586.282$
3. (1) 58.645 minutes   (2) 75.120 minutes

Section 9.3

1. 0.958; About 95.8% of the variation in the times is explained. About 4.2% of the variation is unexplained.
2. 6.218
3. $477.553 < y < 1230.799$
You can be 95% confident that when the gross domestic product is $4 trillion, the carbon dioxide emissions will be between 477.553 and 1230.799 million metric tons.

Section 9.4

1. $\hat{y} = 46.385 + 0.540x_1 - 4.897x_2$
2. (1) 90   (2) 74   (3) 81

Chapter 10

Section 10.1

1.

Tax preparation method	% of people	Expected frequency
Accountant	24%	120
By hand	20%	100
Computer software	35%	175
Friend/family	6%	30
Tax preparation service	15%	75

2. There is not enough evidence at the 5% level of significance to support the sociologist's claim that the age distribution differs from the age distribution 10 years ago.
3. There is enough evidence at the 5% level of significance to reject the claim that the distribution of different-colored candies in bags of peanut M&M's is uniform.

Section 10.2

1. $E_{1,1} = 44.4$, $E_{1,2} = 97.2$, $E_{1,3} = 16.8$, $E_{1,4} = 21.6$, $E_{2,1} = 29.6$, $E_{2,2} = 64.8$, $E_{2,3} = 11.2$, $E_{2,4} = 14.4$
2. There is not enough evidence at the 1% level of significance for the consultant to conclude that travel concern is dependent on travel purpose.
3. There is enough evidence at the 1% level of significance to conclude that whether or not a tax credit would influence an adult to purchase a hybrid vehicle is dependent on age.

Section 10.3

1. 2.45   **2.** 18.31

3. There is enough evidence at the 1% level of significance to support the researcher's claim that a specially treated intravenous solution decreases the variance of the time required for nutrients to enter the bloodstream.

4. There is not enough evidence at the 1% level of significance to reject the biologist's claim that the pH levels of the soil in the two geographic locations have equal standard deviations.

Section 10.4

1. There is enough evidence at the 5% level of significance for the analyst to conclude that there is a difference in the mean monthly sales among the sales regions.

2. There is not enough evidence at the 5% level of significance to conclude that there is a difference in the means of the GPAs.

Appendix A

1. (1) 0.4857   (2) $z = \pm 2.17$

2. 0.9834   **3.** 0.9846   **4.** 0.0733

Appendix C

1.

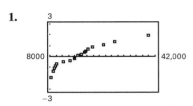

Because the points do not appear to be approximately linear and there is an outlier, you can conclude that the sample data do not come from a population that has a normal distribution.

ODD ANSWERS

Chapter 1

Section 1.1 (page 6)

1. A sample is a subset of a population.
3. A parameter is a numerical description of a population characteristic. A statistic is a numerical description of a sample characteristic.
5. False. A statistic is a numerical description of a sample characteristic.
7. True
9. False. A population is the collection of *all* outcomes, responses, measurements, or counts that are of interest.
11. Population, because it is a collection of the salaries of each member of a Major League Baseball team.
13. Sample, because the collection of the 300 people is a subset of the population of 13,000 people in the auditorium.
15. Sample, because the collection of the 10 patients is a subset of the population of 50 patients at the clinic.
17. Population, because it is a collection of all the gamers' scores in the tournament.
19. Population, because it is a collection of all the U.S. senators' political parties.
21. Population: Parties of registered voters
    Sample: Parties of registered voters who respond to a survey
23. Population: Ages of adults in the United States who own automobiles
    Sample: Ages of adults in the United States who own Honda automobiles
25. Population: Collections of the responses of all U.S. adults
    Sample: Collection of the responses of the 1020 U.S. adults surveyed
    Sample data set: 42% of adults who said they trust their political leaders and 58% who said they did not
27. Population: Collection of the influenza immunization status of all adults in the United States
    Sample: Collection of the influenza immunization status of the 3301 U.S. adults surveyed
    Sample data set: 39% of U.S. adults who received an influenza vaccine and 61% who did not
29. Population: Collection of the average hourly billing rates of all U.S. law firms
    Sample: Collection of the average hourly billing rates for partners of the 159 U.S. law firms surveyed
    Sample data set: The average hourly billing rate for partners of 159 U.S. law firms is $604.
31. Population: Collection of all U.S. adults
    Sample: Collection of the responses of those suffering with chronic pain of the 1029 U.S. adults surveyed
    Sample data set: 23% of respondents suffering with chronic pain who were diagnosed with a sleeping disorder and 77% who were not

33. Population: Collection of all companies listed in the Standard & Poor's 500
    Sample: Collection of the responses of the 54 Standard & Poor's 500 companies surveyed
    Sample data set: Starting salaries of the 54 companies surveyed
35. Sample statistic. The value $72,000 is a numerical description of a sample of average salaries
37. Population parameter. The 62 surviving passengers out of 97 total passengers is a numerical description of all of the passengers of the Hindenburg that survived.
39. Sample statistic. The value 7% is a numerical description of a sample of computer users.
41. Sample statistic. The value 80% is a numerical description of a sample of U.S. adults.
43. The statement "23% of those suffering with chronic pain had been diagnosed with a sleep disorder" is an example of descriptive statistics. Using inferential statistics, you may conclude that an association exists between chronic pain and sleep disorders.
45. Answers will vary.
47. The inference may incorrectly imply that exercise increases a person's cognitive ability. The study shows a slower decline in cognitive ability, not an increase.
49. (a) The sample is the results on the standardized test by the participants in the study.
    (b) The population is the collection of all the results of the standardized test.
    (c) The statement "the closer that participants were to an optimal sleep duration target, the better they performed on a standardized test" is an example of descriptive statistics.
    (d) Individuals who obtain optimal sleep will be more likely to perform better on a standardized test then they would without optimal sleep.

Section 1.2 (page 13)

1. Nominal and ordinal
3. False. Data at the ordinal level can be qualitative or quantitative.
5. False. More types of calculations can be performed with data at the interval level than with data at the nominal level.
7. Quantitative, because dog weights are numerical measurements.
9. Qualitative, because hair colors are attributes.
11. Quantitative, because infant heights are numerical measurements.
13. Qualitative, because the poll responses are attributes.
15. Interval. Data can be ordered and meaningful differences can be calculated, but it does not make sense to say that one year is a multiple of another.

17. Nominal. No mathematical computations can be made, and data are categorized using numbers.

19. Ordinal. Data can be arranged in order, but the differences between data entries are not meaningful.

21. Horizontal: Nominal; Vertical: Ratio

23. Horizontal: Nominal; Vertical: Ratio

25. (a) Interval (b) Nominal (c) Ratio (d) Ordinal

27. Qualitative. Ordinal. Data can be arranged in order, but the differences between data entries make no sense.

29. Qualitative. Nominal. No mathematical computations can be made and data are categorized by region.

31. Qualitative. Ordinal. Data can be arranged in order, but the differences between data entries are not meaningful.

33. An inherent zero is a zero that implies "none." Answers will vary.

Section 1.3 (page 24)

1. In an experiment, a treatment is applied to part of a population and responses are observed. In an observational study, a researcher measures characteristics of interest of a part of a population but does not change existing conditions.

3. In a random sample, every member of the population has an equal chance of being selected. In a simple random sample, every possible sample of the same size has an equal chance of being selected.

5. False. A placebo is a fake treatment.

7. False. Using stratified sampling guarantees that members of each group within a population will be sampled.

9. False. A systematic sample is selected by ordering a population in some way and then selecting members of the population at regular intervals.

11. Observational study. The study does not apply a treatment to the adults.

13. Experiment. The study applies a treatment (different photographs) to the subjects.

15. Answers will vary.   **17.** Answers will vary.

19. (a) The experimental units are the 500 females ages 25 to 45 years old who suffer from migraine headaches. The treatment is the new drug used to treat migraine headaches.

(b) A problem with the design is that the sample is not representative of the entire population because only females ages 25 to 45 were used. To increase validity, use a stratified sample.

(c) For the experiment to be double-blind, neither the subjects nor the company would know whether the subjects are receiving the drug or the placebo.

21. *Sample answer:* Treatment group: Jake, Maria, Lucy, Adam, Bridget, Vanessa, Rick, Dan, and Mary. Control group: Mike, Ron, Carlos, Steve, Susan, Kate, Pete, Judy, and Connie. A random number table was used.

23. Simple random sampling is used because each employee has an equal chance of being contacted, and all samples of 300 people have an equal chance of being selected. A possible source of bias is that the random sample may contain a much greater percentage of employees from one department than from others.

25. Cluster sampling is used because the disaster area is divided into grids, and 30 grids are then entirely selected. A possible source of bias is that certain grids may have been much more severely damaged than others.

27. Stratified sampling is used because a sample is taken from each one-acre subplot.

29. Census, because it is relatively easy to obtain the ages of the 115 residents.

31. The question is biased because it already suggests that eating whole-grain foods improves your health. The question could be rewritten as "How does eating whole-grain foods affect your health?"

33. The survey question is unbiased.

35. Answers will vary.

37. Open Question

Advantage: Allows respondent to express some depth and shades of meaning in the answer. Allows for new solutions to be introduced.

Disadvantage: Not easily quantified and difficult to compare surveys.

Closed Question

Advantage: Easy to analyze results.

Disadvantage: May not provide appropriate alternatives and may influence the opinion of the respondent.

Section 1.3 Activity (page 27)

1. Answers will vary. The list contains one number at least twice.

2. The minimum is 1, the maximum is 731, and the number of samples is 8. Answers will vary.

Uses and Abuses for Chapter 1 (page 28)

1. Answers will vary.   **2.** Answers will vary.

Review Exercises for Chapter 1 (page 30)

1. Population: Collection of the responses of all U.S. adults
Sample: Collection of the responses of the 4787 U.S. adults who were sampled
Sample data set: 15% of adults who use ride-hailing applications and 85% who do not

3. Population: Collection of the responses of all U.S. adults
Sample: Collection of the responses of the 2223 U.S. adults who were sampled
Sample data set: 62% of adults who would encourage a child to pursue a career as a video game developer or designer and 38% who would not

5. Population parameter. The value $22.7 million is a numerical description of the total infrastructure-strengthening investments.

7. Population parameter. The 10 students minoring in physics is a numerical description of all math majors at a university.

9. The statement "62% would encourage a child to pursue a career as a video game developer or designer" is an example of descriptive statistics. An inference drawn from the sample is that a majority of people encourage children to pursue a career as a video game developer or designer.

11. Quantitative, because ages are numerical measurements.

13. Quantitative, because revenues are numerical measurements.

15. Interval. The data can be ordered and meaningful differences can be calculated, but it does not make sense to say that 84 degrees is 1.05 times as hot as 80 degrees.

17. Nominal. The data are qualitative and cannot be arranged in a meaningful order.

19. Experiment. The study applies a treatment (drug to treat hypertension in patients with obstructive sleep apnea) to the subjects.

21. *Sample answer:* The subjects could be split into male and female and then be randomly assigned to each of the five treatment groups.

23. Simple random sampling is used because random telephone numbers were generated and called. A potential source of bias is that telephone sampling only samples individuals who have telephones, who are available, and who are willing to respond.

25. Cluster sampling is used because each district is considered a cluster and every pregnant woman in a selected district is surveyed. A potential source of bias is that the selected districts may not be representative of the entire area.

27. Stratified sampling is used because the population is divided by grade level and then 25 students are randomly selected from each grade level.

29. Sampling, because the population of students at the university is too large for their favorite spring break destinations to be easily recorded. Random sampling would be advised because it would be easy to select students randomly and then record their favorite spring break destinations.

Quiz for Chapter 1 (page 32)

1. Population: Collection of the school performance of all Korean adolescents
Sample: Collection of the school performance of the 359,264 Korean adolescents in the study

2. (a) Sample statistic. The value 52% is a numerical description of a sample of U.S. adults.

(b) Population parameter. The 90% of members that approved the contract of the new president is a numerical description of all Board of Trustees members.

(c) Sample statistic. The value 25% is a numerical description of a sample of small business owners.

3. (a) Qualitative, because debit card personal identification numbers are labels and it does not make sense to find differences between numbers.

(b) Quantitative, because final scores are numerical measurements.

4. (a) Ordinal, because badge numbers can be ordered and often indicate seniority of service, but no meaningful mathematical computation can be performed.

(b) Ratio, because one data entry can be expressed as a multiple of another.

(c) Ordinal, because data can be arranged in order, but the differences between data entries make no sense.

(d) Interval, because meaningful differences between entries can be calculated but a zero entry is not an inherent zero.

5. (a) Observational study. The study does not attempt to influence the responses of the subjects and there is no treatment.

(b) Experiment. The study applies a treatment (multivitamin) to the subjects.

6. Randomized block design

7. (a) Convenience sampling, because all of the people sampled are in one convenient location.

(b) Systematic sampling, because every tenth machine part is sampled.

(c) Stratified sampling, because the population is first stratified and then a sample is collected from each stratum.

8. Convenience sampling. People at campgrounds may be strongly against air pollution because they are at an outdoor location.

Real Statistics—Real Decisions for Chapter 1 (page 34)

1. (a)–(b) Answers will vary.
(c) *Sample answer:* Use surveys.
(d) *Sample answer:* You may take too large a percentage of your sample from a subgroup of the population that is relatively small.

2. (a) *Sample answer:* Qualitative, because questions will ask for demographics and the sample questions have nonnumerical categories.
(b) *Sample answer:* Nominal and ordinal, because the results can be put in categories and the categories can be ranked.
(c) Sample (d) Statistics

3. (a) *Sample answer:* Sample includes only members of the population with access to the Internet.
(b) Answers will vary.

Chapter 2

Section 2.1   (page 49)

1. Organizing the data into a frequency distribution may make patterns within the data more evident. Sometimes it is easier to identify patterns of a data set by looking at a graph of the frequency distribution.

3. Class limits determine which numbers can belong to each class. Class boundaries are the numbers that separate classes without forming gaps between them.

5. The sum of the relative frequencies must be 1 or 100% because it is the sum of all portions or percentages of the data.

7. False. Class width is the difference between lower or upper limits of consecutive classes.

9. False. An ogive is a graph that displays cumulative frequencies.

11. Class width = 8; Lower class limits: 9, 17, 25, 33, 41, 49, 57; Upper class limits: 16, 24, 32, 40, 48, 56, 64

13. Class width = 15; Lower class limits: 17, 32, 47, 62, 77, 92, 107, 122; Upper class limits: 31, 46, 61, 76, 91, 106, 121, 136

15. (a) 11
    (b) and (c)

Class	Midpoint	Class boundaries
0–10	5	−0.5–10.5
11–21	16	10.5–21.5
22–32	27	21.5–32.5
33–43	38	32.5–43.5
44–54	49	43.5–54.5
55–65	60	54.5–65.5
66–76	71	65.5–76.5

17.

Class	Frequency, f	Midpoint	Relative frequency	Cumulative frequency
0–10	188	5	0.15	188
11–21	372	16	0.30	560
22–32	264	27	0.22	824
33–43	205	38	0.17	1029
44–54	83	49	0.07	1112
55–65	76	60	0.06	1188
66–76	32	71	0.03	1220
	$\Sigma f = 1220$		$\Sigma \dfrac{f}{n} = 1$	

19. (a) 7
    (b) Greatest frequency: about 300
        Least frequency: about 10
    (c) 10
    (d) *Sample answer:* About half of the employee salaries are between $50,000 and $69,000.

21. Class with greatest frequency: 506–510
    Classes with least frequency: 474–478

23. (a) Class with greatest relative frequency: 35–36 centimeters
        Class with least relative frequency: 39–40 centimeters
    (b) Greatest relative frequency ≈ 0.25
        Least relative frequency ≈ 0.01
    (c) *Sample answer:* From the graph, 0.25 or 25% of females have a fibula length between 35 and 36 centimeters.

25. (a) 75   (b) 158.5–201.5 pounds

27. (a) 47   (b) 287.5 pounds   (c) 40   (d) 6

29.

Class	Frequency, f	Midpoint	Relative frequency	Cumulative frequency
0–7	8	3.5	0.33	8
8–15	7	11.5	0.29	15
16–23	3	19.5	0.13	18
24–31	3	27.5	0.13	21
32–39	3	35.5	0.13	24
	$\Sigma f = 24$		$\Sigma \dfrac{f}{n} \approx 1$	

Class with greatest frequency: 0–7
Classes with least frequency: 16–23, 24–31, 32–39

31.

Class	Frequency, f	Mid-point	Relative frequency	Cumulative frequency
1000–2019	11	1509.5	0.52	11
2020–3039	3	2529.5	0.14	14
3040–4059	2	3549.5	0.10	16
4060–5079	3	4569.5	0.14	19
5080–6099	1	5589.5	0.05	20
6100–7119	1	6609.5	0.05	21
	$\Sigma f = 21$		$\Sigma \dfrac{f}{n} = 1$	

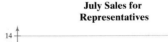

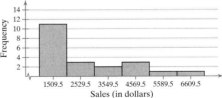

Sample answer: The graph shows that most of the sales representatives at the company sold from $1000 to $2019.

33.

Class	Frequency, f	Mid-point	Relative frequency	Cumulative frequency
291–318	5	304.5	0.1667	5
319–346	4	332.5	0.1333	9
347–374	3	360.5	0.1000	12
375–402	5	388.5	0.1667	17
403–430	6	416.5	0.2000	23
431–458	4	444.5	0.1333	27
459–486	1	472.5	0.0333	28
487–514	2	500.5	0.0667	30
	$\Sigma f = 30$		$\Sigma \dfrac{f}{n} = 1$	

Reaction Times for Females

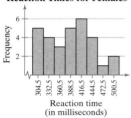

Sample answer: The graph shows that the most frequent reaction times were from 403 to 430 milliseconds.

35.

Class	Frequency, f	Midpoint	Relative frequency	Cumulative frequency
42–46	4	44	0.0889	4
47–51	11	49	0.2444	15
52–56	14	54	0.3111	29
57–61	9	59	0.2000	38
62–66	4	64	0.0889	42
67–71	3	69	0.0667	45
	$\Sigma f = 45$		$\Sigma \dfrac{f}{n} = 1$	

Ages of U.S. Presidents at Inauguration

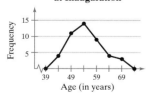

Sample answer: The graph shows that the number of U.S. presidents who were 52 or older at inauguration was twice as many as those who were 51 and younger.

37.

Class	Frequency, f	Midpoint	Relative frequency	Cumulative frequency
1–2	7	1.5	0.19	7
3–4	8	3.5	0.22	15
5–6	10	5.5	0.28	25
7–8	2	7.5	0.06	27
9–10	9	9.5	0.25	36
	$\Sigma f = 36$		$\Sigma \dfrac{f}{n} = 1$	

Taste Test Ratings

Class with greatest relative frequency: 5–6
Class with least relative frequency: 7–8

39.

Class	Frequency, f	Midpoint	Relative frequency	Cumulative frequency
60–63	6	61.5	0.2143	6
64–67	7	65.5	0.2500	13
68–71	9	69.5	0.3214	22
72–75	6	73.5	0.2143	28
76–79	0	77.5	0.0000	28
	$\Sigma f = 28$		$\Sigma \dfrac{f}{n} = 1$	

Lengths of Fijian Banded Iguanas

Class with greatest relative frequency: 68–71
Class with least relative frequency: 76–79

41.

Class	Frequency, f	Relative frequency	Cumulative frequency
52–55	6	0.1714	6
56–59	4	0.1143	10
60–63	6	0.1714	16
64–67	10	0.2857	26
68–71	5	0.1429	31
72–75	4	0.1143	35
	$\Sigma f = 35$	$\Sigma \dfrac{f}{n} = 1$	

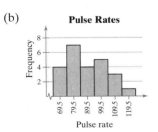

Retirement Ages

Location of the greatest increase in frequency: 64–67

43. (a)

Class	Frequency, f	Mid-point	Relative frequency	Cumulative frequency
65–74	4	69.5	0.1667	4
75–84	7	79.5	0.2917	11
85–94	4	89.5	0.1667	15
95–104	5	99.5	0.2083	20
105–114	3	109.5	0.1250	23
115–124	1	119.5	0.0417	24
	$\Sigma f = 24$		$\Sigma \dfrac{f}{n} \approx 1$	

(b)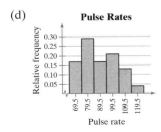
Pulse Rates

(c)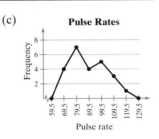
Pulse Rates

(d)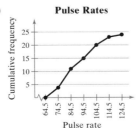
Pulse Rates

(e)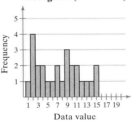
Pulse Rates

45. (a)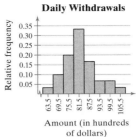
Daily Withdrawals

(b) 16.7%, because the sum of the relative frequencies for the last three classes is 0.167.

(c) $9700, because the sum of the relative frequencies for the last two classes is 0.10.

47.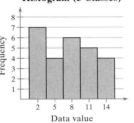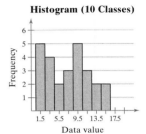
Histogram (5 Classes)   Histogram (10 Classes)

Histogram (20 Classes)

In general, a greater number of classes better preserves the actual values of the data set but is not as helpful for observing general trends and making conclusions. In choosing the number of classes, an important consideration is the size of the data set. For instance, you would not want to use 20 classes if your data set contained 20 entries. In this particular example, as the number of classes increases, the histogram shows more fluctuation. The histograms with 10 and 20 classes have classes with zero frequencies. Not much is gained by using more than five classes. Therefore, it appears that five classes would be best.

Section 2.2 (page 62)

1. Quantitative: stem-and-leaf plot, dot plot, histogram, scatter plot, time series chart
Qualitative: pie chart, Pareto chart

3. Both the stem-and-leaf plot and the dot plot allow you to see how data are distributed, to determine specific data entries, and to identify unusual data values.

5. b   **6.** d   **7.** a   **8.** c

9. 27, 32, 41, 43, 43, 44, 47, 47, 48, 50, 51, 51, 52, 53, 53, 53, 54, 54, 54, 54, 55, 56, 56, 58, 59, 68, 68, 68, 73, 78, 78, 85
Max: 85; Min: 27

11. 13, 13, 14, 14, 14, 15, 15, 15, 15, 15, 16, 17, 17, 18, 19
Max: 19; Min: 13

13. *Sample answer:* Facebook has the most users, and Pinterest has the least. Tumblr and Instagram have about the same number of users.

15. *Sample answer:* The Texter is the least popular driver. The Left-Lane Hog is tolerated more than the Tailgater. The Speedster and the Drifter have the same popularity.

17. Exam Scores

```
6 | 7 8        Key: 6|7 = 67
7 | 3 5 5 6 9
8 | 0 0 2 3 5 5 7 7 8
9 | 0 1 1 1 2 4 5 5
```

Sample answer: Most grades for the biology midterm were in the 80s or 90s.

19. Ice Thickness (in centimeters)

```
4 | 3 9        Key: 4|3 = 4.3
5 | 1 8 8 8 9
6 | 4 8 9 9 9
7 | 0 0 2 2 2 5
8 | 0 1
```

Sample answer: Most of the ice had a thickness of 5.8 centimeters to 7.2 centimeters.

21. Incomes (in millions) of Highest Paid Athletes

```
3 | 3 4 4 4        Key: 3|3 = 33
3 | 5 6 7 7 8 8 8
4 | 1 2 3 4 4
4 | 5 5 5 6
5 | 0 3 3 3
5 | 6 6
6 |
6 | 8
7 |
7 | 7
8 | 1
8 | 8
```

Sample answer: Most of the highest-paid athletes have an income of $33 million to $56 million.

23.

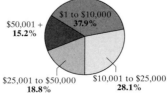

Sample answer: Systolic blood pressure tends to be from 120 to 150 millimeters of mercury.

25. Student Loan Borrowers by Balance Owed in Fourth Quarter 2015

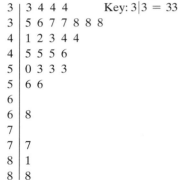

Sample answer: The majority of student loan borrowers owe $25,000 or less.

27.

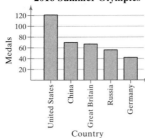

Sample answer: The United States won the most medals out of the five countries and Germany won the least.

29.

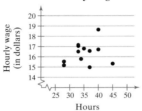

Sample answer: It appears that there is no relation between hourly wages and hours worked.

31.

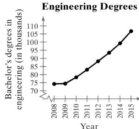

Sample answer: The number of bachelor's degrees in engineering conferred in the U.S. has increased from 2008 to 2015.

33. Heights (in inches)

```
7 | 2 2 4        Key: 7|2 = 72
7 | 5 5 5 5 6
8 | 1 1 2 2 2 4
8 |
```

The dot plot helps you see that the data are clustered from 72 to 76 and 81 to 84, with 75 being the most frequent value. The stem-and-leaf plot helps you see that most values are 75 or greater.

35.

Favorite Season of U.S.
Adults Ages 18 and Older

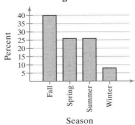

The pie chart helps you to see the percentages as parts of a whole, with fall being the largest. It also shows that while fall is the largest percentage, it makes up less than half of the pie chart. That means that a majority of U.S. adults ages 18 and older prefer a season other than fall. This means it would not be a fair statement to say that most U.S. adults ages 18 and older prefer fall. The Pareto chart helps you to see the rankings of the seasons.

37. (a) The graph is misleading because the large gap from 0 to 90 makes it appear that the sales for the 3rd quarter are disproportionately larger than the other quarters.

(b)

Sales for Company A

39. (a) The graph is misleading because the angle makes it appear as though the 3rd quarter had a larger percent of sales than the others, when the 1st and 3rd quarters have the same percent.

(b) **Sales for Company B**

41. (a) At Law Firm A, the lowest salary was $90,000 and the highest salary was $203,000. At Law Firm B, the lowest salary was $90,000 and the highest salary was $190,000. There are 30 lawyers at Law Firm A and 32 lawyers at Law Firm B.

(b) At Law Firm A, the salaries are clustered at the far ends of the distribution range. At Law Firm B, the salaries are spread out.

43. (a)

```
2 | 6          Key: 2|6 = 26
3 | 1
4 | 0 4 4 5 6 7 9 9
5 | 5 5 5
6 | 3 4 4
7 | 0 1 2 2
```

(b)

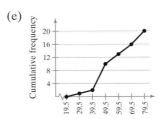

(c)

(d)

(e)

Sample answer: The stem-and-leaf plot, dot plot, frequency histogram, and ogive display the data best because the data is quantitative.

Section 2.3 (page 74)

1. True   **3.** True   **5.** *Sample answer:* 1, 2, 2, 2, 3

7. *Sample answer:* 2, 5, 7, 9, 35

9. The shape of the distribution is skewed right because the bars have a "tail" to the right.

11. The shape of the distribution is uniform because the bars are approximately the same height.

13. (11), because the distribution of values ranges from 1 to 12 and has (approximately) equal frequencies.

14. (9), because the distribution has values in the thousands and is skewed right due to the few vehicles that have much higher mileages than the majority of the vehicles.

15. (12), because the distribution has a maximum value of 90 and is skewed left due to a few students scoring much lower than the majority of the students.

16. (10), because the distribution is approximately symmetric and the weights range from 80 to 160 pounds.

17. $\bar{x} \approx 14.9$; median = 15; mode = 16

19. $\bar{x} \approx 902.3$; median = 788; mode = none; The mode cannot be found because no data entry is repeated. The mean does not represent the center of data because it is influenced by the outliers of 1242 and 1462.

21. $\bar{x} \approx 49.8$; median = 50.5; mode = 51

23. $\bar{x} \approx 7.4$; median = 6; mode = 6

25. $\bar{x} = 14.3$; median = 9; mode = none; The mode cannot be found because no data entry is repeated. The mean does not represent the center of the data set because it is influenced by the outlier of 42.

27. $\bar{x}$ is not possible; median is not possible; mode = "Search and buy online"; The mean and median cannot be found because the data are at the nominal level of measurement.

29. $\bar{x}$ is not possible; median is not possible; mode = "Junior"; The mean and median cannot be found because the data are at the nominal level of measurement.

31. $\bar{x} \approx 29.2$; median = 30.5; mode = 23, 34

33. $\bar{x} \approx 19.5$; median = 20; mode = 15

35. Cluster around 275–425

37. Mode, because the data are at the nominal level of measurement.

39. Mean, because the distribution is symmetric and there are no outliers.

41. 90.5   **43.** $612.73   **45.** 84   **47.** 87

49. 36.2 miles per gallon   **51.** 42.3 years old

53.

Class	Frequency, f	Midpoint
127–161	7	144
162–196	6	179
197–231	3	214
232–266	3	249
267–301	1	284

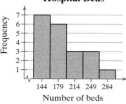

Hospital Beds

Positively skewed

55.

Class	Frequency, f	Midpoint
62–64	3	63
65–67	7	66
68–70	9	69
71–73	8	72
74–76	3	75

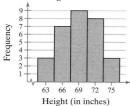

Heights of Males

Symmetric

57. (a) $\bar{x} \approx 1518.2$, median = 1520.5
(b) $\bar{x} \approx 1521.2$ median = 1522.5
(c) Mean

59. The data are skewed right.
A = mode, because it is the data entry that occurred most often.
B = median, because the median is to the left of the mean in a skewed right distribution.
C = mean, because the mean is to the right of the median in a skewed right distribution.

61. Increase one of the three-credit B classes to an A. The three-credit class is weighted more than the two-credit classes, so it will have a greater effect on the grade point average.

63. (a) Mean, because Car A has the highest mean of the three.
(b) Median, because Car B has the highest median of the three.
(c) Mode, because Car C has the highest mode of the three.

65. (a) $\bar{x} \approx 49.2$; median = 46.5
(b) **Test Scores**

```
1 | 1 3      Key: 3|6 = 36
2 | 2 8
3 | 6 6 6 7 7 7 8
4 | 1 3 4 6 7——mean
5 | 1 1 1 3
6 | 1 2 3 4     median
7 | 2 2 4 6
8 | 5
9 | 0
```

(c) Positively skewed

Section 2.3 Activity   (page 81)

1. The distribution is symmetric. The mean and median both decrease slightly. Over time, the median will decrease dramatically and the mean will also decrease, but to a lesser degree.

2. Neither the mean nor the median can be any of the points that were plotted. Because there are 10 points in each region, the mean will fall somewhere between the two regions. By the same logic, the median will be the average of the greatest point between 0 and 0.75 and the least point between 20 and 25.

Section 2.4   (page 93)

1. The range is the difference between the maximum and minimum values of a data set. The advantage of the range is that it is easy to calculate. The disadvantage is that it uses only two entries from the data set.

3. The units of variance are squared. Its units are meaningless (example: dollars^2). The units of standard deviation are the same as the data.

5. When calculating the population standard deviation, you divide the sum of the squared deviations by N, then take the square root of that value. When calculating the sample standard deviation, you divide the sum of the squared deviations by $n - 1$, then take the square root of that value.

7. Similarity: Both estimate proportions of the data contained within k standard deviations of the mean.
Difference: The Empirical Rule assumes the distribution is approximately symmetric and bell-shaped and Chebychev's Theorem makes no such assumption.

9. Approximately 35, or \$35,000   **11.** (a) 17.8   (b) 39.8

13. Range $= 4$; $\mu \approx 11.2$; $\sigma^2 \approx 2.1$; $\sigma \approx 1.5$

15. Range $= 6$; $\bar{x} = 19$; $s^2 \approx 3.5$; $s \approx 1.9$

17. The data set in (a) has a standard deviation of 2.4 and the data set in (b) has a standard deviation of 5 because the data in (b) have more variability.

19. Company B; An offer of \$43,000 is two standard deviations from the mean of Company A's starting salaries, which makes it unlikely. The same offer is within one standard deviation of the mean of Company B's starting salaries, which makes the offer likely.

21. (a) Greatest sample standard deviation: (ii)
Data set (ii) has more entries that are farther away from the mean.
Least sample standard deviation: (iii)
Data set (iii) has more entries that are close to the mean.
(b) The three data sets have the same mean, median, and mode, but have a different standard deviation.
(c) Estimates will vary; (i) $s \approx 1.1$; (ii) $s \approx 1.3$; (iii) $s \approx 0.8$

23. (a) Greatest sample standard deviation: (i)
Data set (i) has more entries that are farther away from the mean.
Least sample standard deviation: (iii)
Data set (iii) has more entries that are close to the mean.
(b) The three data sets have the same mean, median, and mode, but have a different standard deviation.
(c) Estimates will vary; (i) $s \approx 9.6$; (ii) $s \approx 9.0$; (iii) $s \approx 5.1$

25. *Sample answer:* 3, 3, 3, 7, 7, 7

27. *Sample answer:* 9, 9, 9, 9, 9, 9, 9

29. 68%   **31.** (a) 51   (b) 17

33. 78, 76, and 82 are unusual; 82 is very unusual because it is more than 3 standard deviations from the mean.

35. 30

37. At least 93.75% of the exam scores are from 70 to 94.

39.

x	f	xf	$x - \bar{x}$	$(x - \bar{x})^2$	$(x - \bar{x})^2 f$
0	3	0	-5	25	75
1	4	4	-4	16	64
2	3	6	-3	9	27
3	9	27	-2	4	36
4	3	12	-1	1	3
5	3	15	0	0	0
6	8	48	1	1	8
7	5	35	2	4	20
8	6	48	3	9	54
9	6	54	4	16	96
	$\Sigma = 50$	$\Sigma = 249$			$\Sigma = 383$

$\bar{x} \approx 5.0$, $s \approx 2.8$

41.

Class	x	f	xf
15,000–17,499	16,249.5	9	146,245.5
17,500–19,999	18,749.5	10	187,495
20,000–22,499	21,249.5	16	339,992
22,500–24,999	23,749.5	11	261,244.5
25,000 or more	26,249.5	6	157,497
		$n = 52$	$\Sigma xf = 1,092,474$

$x - \bar{x}$	$(x - \bar{x})^2$	$(x - \bar{x})^2 f$
-4759.62	22,653,982.54	203,885,842.9
-2259.62	5,105,882.54	51,058,825.4
240.38	57,782.54	924,520.64
2740.38	7,509,682.54	82,606,507.94
5240.38	27,461,582.54	164,769,495.20
	$\Sigma (x - \bar{x})^2 f = 503,245,192.1$	

$\bar{x} \approx \$21,009.12$; $s \approx \$3141.27$

43.

x	f	xf	$x - \bar{x}$	$(x - \bar{x})^2$	$(x - \bar{x})^2 f$
1	2	2	-1.9	3.61	7.22
2	18	36	-0.9	0.81	14.58
3	24	72	0.1	0.01	0.24
4	16	64	1.1	1.21	19.36
	$n = 60$	$\Sigma xf = 174$		$\Sigma (x - \bar{x})^2 f = 41.4$	

$\bar{x} \approx 2.9$; $s \approx 0.8$

45. $CV_{\text{Denver}} \approx 9.7\%$, $CV_{\text{LA}} \approx 8.8\%$
Salaries for entry level architects are more variable in Denver than in Los Angeles.

47. $CV_{\text{ages}} \approx 13.3\%$, $CV_{\text{heights}} \approx 3.5\%$
Ages are more variable than heights for all members of the 2016 Women's U.S. Olympic swimming team.

49. $CV_{\text{males}} \approx 20.7\%$, $CV_{\text{females}} \approx 17.6\%$
SAT scores are more variable for males than for females.

51. (a) Answers will vary.   (b) $s \approx 1.9$
(c) They are the same.

53. (a) $\bar{x} \approx 42.1$; $s \approx 5.6$ (b) $\bar{x} \approx 44.3$; $s \approx 5.9$
(c) 3.5, 3, 3, 4, 4, 2.75, 4.25, 3.25, 3.25, 3.5, 3.25, 3.75, 3.5, 4.17
$\bar{x} \approx 3.5$; $s \approx 0.47$
(d) When each entry is multiplied by a constant k, the new sample mean is $k \cdot \bar{x}$, and the new sample standard deviation is $k \cdot s$.

55. (a) $P \approx -2.61$
The data are skewed left.
(b) $P \approx 4.12$
The data are skewed right.
(c) $P = 0$
The data are symmetric.
(d) $P = 1$
The data are skewed right.
(e) $P = -3$
The data are skewed left.

Section 2.4 Activity   (page 100)

1. When a point with a value of 15 is added, the mean remains constant or changes very little, and the standard deviation decreases. When a point with a value of 20 is added, the mean is raised and the standard deviation increases.

2. To get the largest standard deviation, plot four of the points at 30 and four of the points at 40.

To get the smallest standard deviation, plot all of the points at the same number. That way, each $x - \bar{x}$ is 0, so the standard deviation will be 0.

Section 2.5   (page 109)

1. The talk is longer in length than 75% of the lectures in the series.

3. The student scored higher than 89% of the students who took the Fundamentals of Engineering exam.

5. The interquartile range of a data set can be used to identify outliers because data entries that are greater than $Q_3 + 1.5(\text{IQR})$ or less than $Q_1 - 1.5(\text{IQR})$ are considered outliers.

7. True

9. False; An outlier is any number above $Q_3 + 1.5(\text{IQR})$ or below $Q_1 - 1.5(\text{IQR})$.

11. (a) $Q_1 = 57$, $Q_2 = 60$, $Q_3 = 63$ (b) IQR $= 6$ (c) 80

13. Min $= 0$, $Q_1 = 2$, $Q_2 = 5$, $Q_3 = 8$, Max $= 10$

15. (a) Min $= 24$, $Q_1 = 28$, $Q_2 = 35$, $Q_3 = 41$, Max $= 60$
(b)

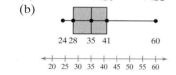

17. (a) Min $= 1$, $Q_1 = 4.5$, $Q_2 = 6$, $Q_3 = 7.5$, Max $= 9$
(b)

19. None. The data are not skewed or symmetric.

21. Skewed left. Most of the data lie to the right on the box plot.

23.   **Studying**

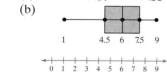

25.   **Commuting Distances**

27. (a) 6.5 hours   (b) about 50%   (c) about 25%

29. About 158; About 70% of quantitative reasoning scores on the Graduate Record Examination are less than 158.

31. About 8th percentile; About 8% of quantitative reasoning scores on the Graduate Record Examination are less that 140.

33. 10th percentile   **35.** 57, 57, 61, 61, 65, 66

37.

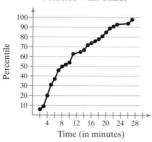

Depatment of Motor Vehicles Wait Times

39. About 85th percentile

41. $A \rightarrow z = -1.43$
$B \rightarrow z = 0$
$C \rightarrow z = 2.14$
A z-score of 2.14 would be unusual.

43. Not unusual; The z-score is 0.94, so the age of 31 is about 0.94 standard deviation above the mean.

45. Not unusual; The z-score is -0.27, so the age of 27 is about 0.27 standard deviation below the mean.

47. Unusual; The z-score is -2.39, so the age of 20 is about 2.39 standard deviations below the mean.

49. (a) For 34,000, $z \approx -0.44$; For 37,000, $z \approx 0.89$; For 30,000, $z \approx -2.22$
The tire with a life span of 30,000 miles has an unusually short life span.
(b) For 30,500, about 2.5th percentile;
For 37,250, about 84th percentile;
For 35,000, about 50th percentile

51. Robert Duvall: $z \approx 1.07$; Jack Nicholson: $z \approx -0.32$; The age of Robert Duvall was about 1 standard deviation above the mean age of Best Actor winners, and the age of Jack Nicholson was less than 1 standard deviation below the mean age of Best Supporting Actor winners. Neither actor's age is unusual.

53. John Wayne: $z \approx 2.10$; Gig Young: $z \approx 0.41$; The age of John Wayne was more than 2 standard deviations above the mean age of Best Actor winners, which is unusual. The age of Gig Young was less than 1 standard deviation above the mean age of Best Supporting Actor winners, which is not unusual.

55. 5

57. (a) The distribution of Concert 1 is symmetric. The distribution of Concert 2 is skewed right. Concert 1 has less variation.
(b) Concert 2 is more likely to have outliers because it has more variation.
(c) Concert 1, because 68% of the data should be between ± 16.3 of the mean.
(d) No, you do not know the number of songs played at either concert or the actual lengths of the songs.

59. (a) 24, 2   (b)

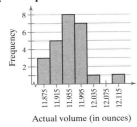

61. (a) 1   (b)

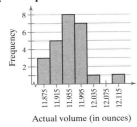

63. Answers will vary.

Uses and Abuses for Chapter 2   (page 114)

1. Answers will vary.

2. No, it is not ethical because it misleads the consumer to believe that drinking red wine is more effective at preventing heart disease than it may actually be.

Review Exercises for Chapter 2   (page 116)

1.

Class	Midpoint	Class boundaries
26–31	28.5	25.5–31.5
32–37	34.5	31.5–37.5
38–43	40.5	37.5–43.5
44–49	46.5	43.5–49.5
50–55	52.5	49.5–55.5

Frequency, f	Relative frequency	Cumulative frequency
5	0.25	5
4	0.20	9
6	0.30	15
3	0.15	18
2	0.10	20
$\Sigma f = 20$	$\Sigma \dfrac{f}{n} = 1$	

3.

Liquid Volume 12-oz Cans

Actual volume (in ounces)

5.

Class	Midpoint	Frequency, f
79–93	86	9
94–108	101	12
109–123	116	5
124–138	131	4
139–153	146	2
154–168	161	1
		$\Sigma f = 33$

Rooms Reserved

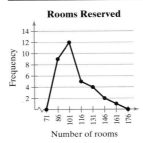

Number of rooms

7. Pollution Indices of U.S. Cities

```
2 | 2 3 8 8          Key: 2|2 = 22
3 | 2 3 6 8 8 9 9
4 | 1 1 3 6 9
5 | 0 3 4 6 7
6 | 3 5 5
```

Sample answer: Most U.S. cities have a pollution index from 32 to 57.

9.

College Students' Activities and Time Use

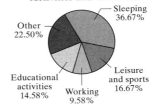

Sample answer: Full-time university and college students spend the least amount of time working.

11.

Heights of Buildings

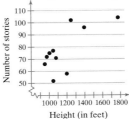

Height (in feet)

Sample answer: The number of stories appears to increase with height.

13. $\bar{x} = 29.5$; median $= 29.5$; mode $= 29.5$

15. 82.1   **17.** 38.4   **19.** Skewed right   **21.** Skewed right

23. Mean; When a distribution is skewed right, the mean is to the right of the median.

25. Range $= 14$; $\mu \approx 6.9$; $\sigma^2 \approx 21.1$; $\sigma \approx 4.6$

27. Range $= \$2044$; $\bar{x} \approx \$6266.81$; $s^2 \approx 455{,}944.30$; $s \approx \$675.24$

29. \$75 and \$145   **31.** 30 customers
33. $\bar{x} \approx 2.5$; $s \approx 1.2$
35. $CV_{freshmen} \approx 41.3\%$; $CV_{seniors} \approx 24.2\%$
Grade point averages are more variable for freshmen than seniors.
37. Min $= 16$, $Q_1 = 25$, $Q_2 = 35$, $Q_3 = 56$, Max $= 136$
39. **Model 2017 Vehicle Fuel Economies**

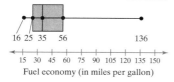

Fuel economy (in miles per gallon)

41. 7 inches   **43.** 35%
45. Not unusual; The *z*-score is 1.97, so a towing capacity of 16,500 pounds is about 1.97 standard deviations above the mean.
47. Unusual; The *z*-score is 2.60, so a towing capacity of 18,000 pounds is about 2.60 standard deviations above the mean.

Quiz for Chapter 2   (page 120)

1. (a)

Class	Midpoint	Class boundaries
101–112	106.5	100.5–112.5
113–124	118.5	112.5–124.5
125–136	130.5	124.5–136.5
137–148	142.5	136.5–148.5
149–160	154.5	148.5–160.5

Frequency, *f*	Relative frequency	Cumulative frequency
3	0.11	3
11	0.41	14
8	0.30	22
3	0.11	25
2	0.07	27

(b) **Weekly Exercise**

Number of minutes

(c) **Weekly Exercise**

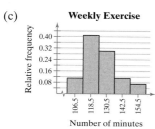

Number of minutes

(d) Skewed right

(e) **Weekly Exercise**

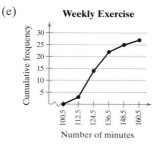

Number of minutes

(f) **Weekly Exercise (in minutes)**

```
10 | 1 8       Key: 10|8 = 108
11 | 1 4 6 7 8 9 9
12 | 0 0 3 3 4 7 7 8
13 | 0 1 1 2 5 9 9
14 | 2
15 | 0 7
```

(g) **Weekly Exercise**

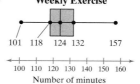

Number of minutes

2. $\bar{x} \approx 126.1$; $s \approx 13.0$
3. (a) **Elements with Known Properties**

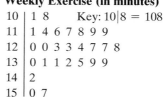

(b) **Elements with Known Properties**

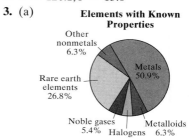

Element

4. (a) $\bar{x} \approx 1016.4$; median $= 1019$; mode $= 1100$; The mean or median best describes a typical salary because there are no outliers.
(b) Range $= 666$; $s^2 \approx 47{,}120.9$; $s \approx 217.1$
(c) $CV \approx 21.4\%$
5. \$150,000 and \$210,000

6. (a) Unusual; The z-score is 3, so a new home price of $225,000 is about 3 standard deviations above the mean.

(b) Unusual; The z-score is -6.67, so a new home price of $80,000 is about 6.67 standard deviations below the mean.

(c) Not unusual; The z-score is 1.33, so a new home price of $200,000 is about 1.33 standard deviations above the mean.

(d) Unusual; The z-score is -2.2, so a new home price of $147,000 is about 2.2 standard deviations below the mean.

7. **Wins for Each MLB Team**

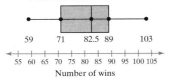

Number of wins

Real Statistics—Real Decisions for Chapter 2   (page 122)

1. (a) Find the average cost of renting an apartment for each area and do a comparison.

(b) The mean would best represent the data sets for the four areas of the city.

(c) Area A: $\bar{x} = \$1131.58$
Area B: $\bar{x} = \$998.33$
Area C: $\bar{x} = \$991.58$
Area D: $\bar{x} = \$1064.17$

2. (a) Construct a Pareto chart, because the data are quantitative and a Pareto chart positions data in order of decreasing height, with the tallest bar positioned at the left.

(b)   **Cost of Monthly Rent per Area**

Area

(c) Yes. From the Pareto chart, you can see that Area A has the highest average cost of monthly rent, followed by Area D, Area B, and Area C.

3. *Sample answer:*

(a) You could use the range and sample standard deviation for each area.

(b)

Area A	**Area B**
range $= \$467$	range $= \$474$
$s \approx \$138.45$	$s \approx \$163.11$

Area C	**Area D**
range $= \$518$	range $= \$560$
$s \approx \$164.51$	$s \approx \$156.26$

(c) No. Area A has the lowest range and standard deviation, so the rents in Areas B–D are more spread out. There could be one or two inexpensive rents that lower the means for these areas. It is possible that the population means of Areas B–D are close to the populations mean of Area A.

4. (a) Answers will vary.

(b) Location, weather, population

Cumulative Review for Chapters 1–2   (page 126)

1. Systematic sampling is used because every fortieth toothbrush from each assembly line is tested. It is possible for bias to enter into the sample if, for some reason, an assembly line makes a consistent error.

2. Simple random sampling is used because each telephone number has an equal chance of being dialed, and all samples of 1090 phone numbers have an equal chance of being selected. The sample may be biased because telephone sampling only samples those individuals who have telephones, who are available, and who are willing to respond.

3.   **Workplace Fraud**

Fraud detection

4. Parameter. The median salary is based on all marketing account executives.

5. Statistic. The percent, 88%, is based on a subset of the population.

6. (a) 95%

(b) For $93,500, $z \approx 4.67$; For $85,600, $z \approx -0.6$; For $82,750, $z \approx -2.5$.
The salaries of $93,500 and $82,750 are unusual.

7. Population: Collection of opinions of all college and university admissions directors and enrollment officers.
Sample: Collection of opinions of the 339 college and university admission directors and enrollment officers surveyed.

8. Population: Reasons for pain reliever use of all Americans ages 12 or older
Sample: Reasons for pain reliever use of the 67,901 Americans ages 12 or older surveyed

9. Experiment. The study applies a treatment (digital device) to the subjects.

10. Observational study. The study does not attempt to influence the responses of the subjects.

11. Quantitative; Ratio

12. Qualitative; Nominal

13. (a)   **Tornadoes by State**

Number of tornadoes

(b) Skewed right

14. 88.9

15. (a) $\bar{x} \approx 5.49$; median = 5.4; mode = none; Both the mean and the median accurately describe a typical American alligator tail length.
(b) Range = 4.1; $s^2 \approx 2.34$; $s \approx 1.53$

16. (a) An inference drawn from the study is that the life expectancies for Americans will continue to increase or remain stable.
(b) This inference may incorrectly imply that Americans will have higher life expectancies in the future.

17.

Class	Midpoint	Class boundaries
0–8	4	−0.5–8.5
9–17	13	8.5–17.5
18–26	22	17.5–26.5
27–35	31	26.5–35.5
36–44	40	35.5–44.5
45–53	49	44.5–53.5
54–62	58	53.5–62.5
63–71	67	62.5–71.5

Frequency, f	Relative frequency	Cumulative frequency
20	0.500	20
7	0.175	27
6	0.150	33
1	0.025	34
2	0.050	36
1	0.025	37
2	0.050	39
1	0.025	40

18. Skewed right

19.

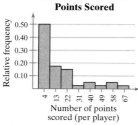

Montreal Canadiens Points Scored

Class with greatest frequency: 0–8
Classes with least frequency: 27–35, 45–53, and 63–71

Chapter 3

Section 3.1 (page 140)

1. An outcome is the result of a single trial in a probability experiment, whereas an event is a set of one or more outcomes.

3. The probability of an event cannot exceed 100%.

5. The law of large numbers states that as an experiment is repeated over and over, the probabilities found in the experiment will approach the actual probabilities of the event. Examples will vary.

7. False. The event "choosing false on a true or false question and choosing A or B on a multiple choice question" is not simple because it consists of two possible outcomes and can be represented as $A = \{FA, FB\}$.

9. False. A probability of less than $\frac{1}{20} = 0.05$ indicates an unusual event.

11. d   **12.** f   **13.** b   **14.** c   **15.** a   **16.** e

17. $\frac{4}{23}$ **19.** 0.97   **21.** 0.05   **23.** $\frac{1}{4}$

25. {A, B, C, D, E, F, G, H, I, J, K, L, M, N, O, P, Q, R, S, T, U, V, W, X, Y, Z}; 26

27. {A♥, K♥, Q♥, J♥, 10♥, 9♥, 8♥, 7♥, 6♥, 5♥, 4♥, 3♥, 2♥, A♦, K♦, Q♦, J♦, 10♦, 9♦, 8♦, 7♦, 6♦, 5♦, 4♦, 3♦, 2♦, A♠, K♠, Q♠, J♠, 10♠, 9♠, 8♠, 7♠, 6♠, 5♠, 4♠, 3♠, 2♠, A♣, K♣, Q♣, J♣, 10♣, 9♣, 8♣, 7♣, 6♣, 5♣, 4♣, 3♣, 2♣}; 52

29.

{HH, HT, TH, TT}; 4

31.

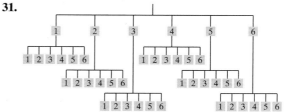

{(1, 1), (1, 2), (1, 3), (1, 4), (1, 5), (1, 6), (2, 1), (2, 2), (2, 3), (2, 4), (2, 5), (2, 6), (3, 1), (3, 2), (3, 3), (3, 4), (3, 5), (3, 6), (4, 1), (4, 2), (4, 3), (4, 4), (4, 5), (4, 6), (5, 1), (5, 2), (5, 3), (5, 4), (5, 5), (5, 6), (6, 1), (6, 2), (6, 3), (6, 4), (6, 5), (6, 6)}; 36

33. 1; Simple event because it is an event that consists of a single outcome.

35. 13; Not a simple event because it is an event that consists of more than a single outcome.

37. 576   **39.** 4500   **41.** 0.083   **43.** 0.667   **45.** 0.333

47. 0.712   **49.** 0.216   **51.** 0.344

53. Empirical probability because company records were used to calculate the frequency of a washing machine breaking down.

55. Subjective probability because it is most likely based on an educated guess.

57. Classical probability because each outcome in the sample space is equally likely to occur.

59. 0.842   **61.** 0.777   **63.** 0.042; unusual

65. 0.208; not unusual

67. (a) 0.001   (b) 0.999

69. 0.125   **71.** 0.375   **73.** 0.450   **75.** 0.033   **77.** 0.275

79. No; None of the events have a probability of 0.05 or less.

81. (a) 0.5   (b) 0.25   (c) 0.25   **83.** 0.808   **85.** 0.103

87. (a) 0.313   (b) 0.078
    (c) 0.031; This event is unusual because its probability is less than or equal to 0.05.
89. The probability of randomly choosing a tea drinker who does not have a college degree
91. No. The odds of winning a prize are 1 : 6 (one winning cap and six losing caps). So, the statement should read, "one in seven game pieces wins a prize."
93. (a) 0.444   (b) 0.556   **95.** 39 : 13 = 3 : 1
97. (a)

Sum	Probability
2	0.028
3	0.056
4	0.083
5	0.111
6	0.139
7	0.167
8	0.139
9	0.111
10	0.083
11	0.056
12	0.028

    (b) Answers will vary.
    (c) Answers will vary.

Section 3.1 Activity   (page 146)

1–2. Answers will vary.

Section 3.2   (page 152)

1. Two events are independent when the occurrence of one of the events does not affect the probability of the occurrence of the other event, whereas two events are dependent when the occurrence of one of the events does affect the probability of the occurrence of the other event.
3. The notation $P(B|A)$ means the probability of event B occurring, given that event A has occurred.
5. False. If two events are independent, then $P(A|B) = P(A)$.
7. (a) 0.526   (b) 0.159
9. Independent. The outcome of the first draw does not affect the outcome of the second draw.
11. Dependent. The outcome of returning a movie after its due date affects the outcome of receiving a late fee.
13. Dependent. The sum of the rolls depends on which numbers came up on the first and second rolls.
15. Events: obstructive sleep apnea, heart disease; Dependent. People with obstructive sleep apnea are more likely to have heart disease.
17. Events: playing violent video games, aggressive or bullying behavior; Independent. Playing violent video games does not cause aggressive or bullying behavior in teens.
19. 0.063   **21.** 0.001
23. (a) 0.040   (b) 0.640   (c) 0.360

25. (a) 0.022   (b) 0.722   (c) 0.278
    (d) The event in part (a) is unusual because its probability is less than or equal to 0.05.
27. (a) 0.011   (b) 0.022   (c) 0.978
    (d) The events in parts (a) and (b) are unusual because their probabilities are less than or equal to 0.05.
29. (a) 0.007   (b) 0.589
    (c) Yes, this is unusual because the probability is less than or equal to 0.05.
31. 0.32   **33.** 0.444   **35.** 0.167   **37.** 0.792
39. (a) 0.074   (b) 0.999   **41.** 0.954

Section 3.3   (page 162)

1. $P(A \text{ and } B) = 0$ because A and B cannot occur at the same time.
3. True
5. False. The probability that event A or event B will occur is $P(A \text{ or } B) = P(A) + P(B) - P(A \text{ and } B)$.
7. Not mutually exclusive. A presidential candidate can lose the popular vote and win the election.
9. Not mutually exclusive. A psychology major can be male and 20 years old.
11. Mutually exclusive. A voter cannot be both a Republican and a Democrat.
13. 0.625   **15.** 0.126
17. (a) 0.308   (b) 0.538   (c) 0.308
19. (a) 0.06   (b) 0.426   (c) 0.81   (d) 0.201
21. (a) 0.780   (b) 0.410   (c) 0.590   (d) 0.400
23. (a) 0.520   (b) 0.899   (c) 0.909
25. (a) 0.589   (b) 0.762   (c) 0.461   (d) 0.922
27. 0.63
29. If events A, B, and C are not mutually exclusive, then $P(A \text{ and } B \text{ and } C)$ must be added because $P(A) + P(B) + P(C)$ counts the intersection of all three events three times and $-P(A \text{ and } B) - P(A \text{ and } C) - P(B \text{ and } C)$ subtracts the intersection of all three events three times. So, if $P(A \text{ and } B \text{ and } C)$ is not added at the end, then it will not be counted.

Section 3.3 Activity   (page 166)

1. Answers will vary.
2. The theoretical probability is 0.5, so the green line should be placed there.

Section 3.4   (page 174)

1. The number of ordered arrangements of n objects taken r at a time.
    Sample answer: An example of a permutation is the number of seating arrangements of you and three of your friends.
3. False. A permutation is an ordered arrangement of objects.
5. True   **7.** 15,120   **9.** 56   **11.** 0.076   **13.** 0.462
15. Permutation. The order of the 16 floats in line matters.
17. Combination. The order does not matter because the position of one captain is the same as the other.
19. 5040   **21.** 720   **23.** 117,600   **25.** 96,909,120

27. 2,042,040 **29.** 50,400 **31.** 4845 **33.** 9880
35. 6240 **37.** 86,296,950 **39.** 0.005 **41.** 0.005
43. (a) 0.016 (b) 0.385 **45.** 0.0009
47. 0.242 **49.** 0.0000015
51. 0.166 **53.** 0.070 **55.** 0.933 **57.** 0.086
59. 0.066 **61.** 0.001

Uses and Abuses for Chapter 3 (page 178)

1. (a) 0.000001 **2.** 0.001 **3.** 0.001

Review Exercises for Chapter 3 (page 180)

1.

{HHHH, HHHT, HHTH, HHTT, HTHH, HTHT, HTTH,
HTTT, THHH, THHT, THTH, THTT, TTHH, TTHT, TTTH,
TTTT}; 4
3. {January, February, March, April, May, June, July, August,
September, October, November, December}; 3
5. 84
7. Empirical probability because prior counts were used to
calculate the frequency of a part being defective.
9. Subjective probability because it is based on opinion.
11. Classical probability because all of the outcomes in the
event and the sample space can be counted.
13. 0.258 **15.** 1.25×10^{-7} **17.** 0.555
19. Independent. The outcomes of the first four coin tosses do
not affect the outcome of the fifth coin toss.
21. Dependent. The outcome of taking a driver's education course
affects the outcome of passing the driver's license exam.
23. 0.025; Yes, the event is unusual because its probability is less
than or equal to 0.05.
25. Mutually exclusive. A jelly bean cannot be both completely
red and completely yellow.
27. 0.9 **29.** 0.538 **31.** 0.583 **33.** 0.576 **35.** 0.584
37. 0.568
39. No; You do not know whether events A and B are mutually
exclusive.
41. 110 **43.** 35 **45.** 2730 **47.** 2380 **49.** 0.000009
51. (a) 0.955 (b) 0.0000008 (c) 0.045 (d) 0.9999992
53. (a) 0.071 (b) 0.005 (c) 0.429 (d) 0.114

Quiz for Chapter 3 (page 184)

1. 450,000
2. (a) 0.713 (b) 0.662 (c) 0.778 (d) 0.937
(e) 0.049 (f) 0.606 (g) 0.346 (h) 0.515
3. The event in part (e) is unusual because its probability is less
than or equal to 0.05.
4. Not mutually exclusive. A bowler can have the highest game
in a 40-game tournament and still lose the tournament.
Dependent. One event can affect the occurrence of the
second event.
5. 657,720
6. (a) 2,481,115 (b) 1 (c) 2,572,999
7. (a) 0.964 (b) 0.0000004 (c) 0.9999996

Real Statistics—Real Decisions for Chapter 3 (page 186)

1. (a) *Sample answer:* Investigate the number of possible
passwords when different sets of characters, such as
lowercase and capital letters, numbers, and special
characters.
(b) You could use the definition of theoretical probability,
the Fundamental Counting Principal, and the
Multiplication Rule.
2. (a) *Sample answer:* Allow lowercase letters, uppercase
letters, and numerical digits.
(b) *Sample answer:* Because there are 26 lowercase letters,
26 uppercase letters, and 10 numerical digits, there are
$26 + 26 + 10 = 62$ choices for each digit. So, there are
62^8 8-digit passwords and the probability of guessing a
password correctly on one try is $\frac{1}{62^8}$, which is less than $\frac{1}{60^8}$.
3. (a) Without the requirement, the number of possible PINs
is $10^5 = 100,000$. With the requirement, the number of
possible PINs is $_{10}P_5 = 10 \cdot 9 \cdot 8 \cdot 7 \cdot 6 = 30,240$.
(b) *Sample answer:* No, although the requirement would
likely discourage customers from choosing predictable
PINs, the numbers of possible PINs would significantly
decrease, and the most popular PIN, 12345, would still
be allowed.

Chapter 4

Section 4.1 (page 197)

1. A random variable represents a value associated with each
outcome of a probability experiment.
Examples: Answers will vary.
3. No; The expected value may not be a possible value of x for
one trial, but it represents the average value of x over a large
number of trials.
5. False. In most applications, discrete random variables
represent counted data, while continuous random variables
represent measured data.
7. False. The mean of the random variable of a probability
distribution describes a typical outcome. The variance and
standard deviation of the random variable of a probability
distribution describe how the outcomes vary.

9. Discrete; Attendance is a random variable that is countable.
11. Continuous; Distance traveled is a random variable that must be measured.
13. Discrete; The number of cars in a university parking lot is a random variable that is countable.
15. Continuous; The volume of blood drawn for a blood test is a random variable that must be measured.
17. Discrete; The number of texts a student sends in one day is a random variable that is countable.
19. (a)

x	$P(x)$
0	0.01
1	0.17
2	0.28
3	0.54

(b)

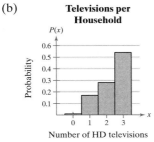

Televisions per Household

Skewed left
21. (a) 0.45   (b) 0.82   (c) 0.99   (d) 0.46
23. Yes, because the probability is less than 0.05.
25. 0.34   27. Yes
29. (a) $\mu \approx 0.5$; $\sigma^2 \approx 0.8$; $\sigma \approx 0.9$
    (b) The mean is 0.5, so the average number of dogs per household is about 0 or 1 dog. The standard deviation is 0.9, so most of the households differ from the mean by no more than about 1 dog.
31. (a) $\mu \approx 1.5$; $\sigma^2 \approx 1.5$; $\sigma \approx 1.2$
    (b) The mean is 1.5, so the average batch of 1000 machine parts has 1 or 2 defects. The standard deviation is 1.2, so most of the batches of 1000 differ from the mean by no more than about 1 defect.
33. (a) $\mu \approx 2.0$; $\sigma^2 \approx 1.0$; $\sigma \approx 1.0$
    (b) The mean is 2.0, so the average hurricane that hits the U.S. mainland is a category 2 hurricane. The standard deviation is 1.0, so most of the hurricanes differ from the mean by no more than 1 category level.
35. An expected value of 0 means that the money gained is equal to the money spent, representing the break-even point.
37. $-\$0.05$ 39. $47,980   41. 1018; 30

Section 4.2   (page 210)

1. Each trial is independent of the other trials when the outcome of one trial does not affect the outcome of any of the other trials.
3. c; Because the probability is greater than 0.5, the distribution is skewed left.
4. b; Because the probability is 0.5, the distribution is symmetric.
5. a; Because the probability is less than 0.5, the distribution is skewed right.
6. c; The histogram shows probabilities for 12 trials.
7. a; The histogram shows probabilities for 4 trials.
8. b; The histogram shows probabilities for 8 trials.
    As n increases, the distribution becomes more symmetric.
9. (3) 0, 1   (4) 0, 5   (5) 4, 5
11. $\mu = 20$; $\sigma^2 = 12$; $\sigma \approx 3.5$

13. $\mu \approx 32.2$; $\sigma^2 \approx 23.9$; $\sigma \approx 4.9$
15. Binomial experiment
    Success: frequent gamer who plays video games on smartphone
    $n = 10$; $p = 0.36$; $q = 0.64$; $x = 0, 1, 2, 3, 4, 5, 6, 7, 8, 9, 10$
17. Not a binomial experiment because the probability of a success is not the same for each trial.
19. (a) 0.019   (b) 0.272   (c) 0.905
21. (a) 0.150   (b) 0.759   (c) 0.712
23. (a) 0.221   (b) 0.247   (c) 0.753
25. (a) 0.089   (b) 0.017   (c) 0.106
27. (a)

x	$P(x)$
0	0.008974
1	0.060355
2	0.173965
3	0.278572
4	0.267647
5	0.154291
6	0.049413
7	0.006782

(b)

Health Insurance Deductibles

Approximately symmetric
(c) The values 0, 6, and 7 are unusual because their probabilities are less than 0.05.
29. (a)

x	$P(x)$
0	0.00064
1	0.01077
2	0.07214
3	0.24151
4	0.40426
5	0.27068

(b)

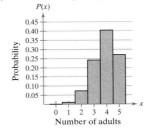

Living to Age 100

Skewed left
(c) The values 0 and 1 are unusual because their probabilities are less than 0.05.

31. $\mu \approx 5.0$; $\sigma^2 \approx 1.4$; $\sigma \approx 1.2$; On average, 5 out of every 7 U.S. adults think that political correctness is a problem in America today. The standard deviation is 1.2, so most samples of 7 U.S. adults would differ from the mean by at most 1.2 U.S. adults.

33. $\mu \approx 6.3$; $\sigma^2 \approx 1.3$; $\sigma \approx 1.2$; On average, 6.3 out of every 8 adults believe that life on other planets is possible. The standard deviation is 1.2, so most samples of 8 adults would differ from the mean by at most 1.2 adults.

35. $\mu \approx 1.9$; $\sigma^2 \approx 1.3$; $\sigma \approx 1.1$; On average, 1.9 out of every 6 U.S. employees who are late for work blame oversleeping. The standard deviation is 1.1, so most samples of 6 U.S. employees who are late would differ from the mean by at most 1.1 U.S. employees.

37. 0.033

39. (a) 0.107 (b) 0.107 (c) The results are the same.

Section 4.2 Activity (page 214)

1–3. Answers will vary.

Section 4.3 (page 220)

1. 0.080   **3.** 0.062   **5.** 0.175   **7.** 0.251

9. In a binomial distribution, the value of x represents the number of successes in n trials. In a geometric distribution, the value of x represents the first trial that results in a success.

11. (a) 0.082 (b) 0.469 (c) 0.531

13. (a) 0.036; unusual (b) 0.053 (c) 0.017; unusual

15. (a) 0.230 (b) 0.871 (c) 0.129

17. (a) 0.002; unusual (b) 0.006; unusual (c) 0.980

19. (a) 0.311 (b) 0.493 (c) 0.507

21. (a) 0.273 (b) 0.615 (c) 0.868

23. (a) 0.322 (b) 0.513 (c) 0.809

25. (a) 0.071 (b) 0.827 (c) 0.173

27. (a) 0.12542
(b) 0.12541; The results are approximately the same.

29. (a) $\mu = 1000$; $\sigma^2 = 999{,}000$; $\sigma \approx 999.5$ (b) 1000 times
(c) Lose money. On average, you would win $500 once in every 1000 times you play the lottery. So, the net gain would be −$500.

31. (a) $\sigma^2 = 4.1$; $\sigma \approx 2.0$; The standard deviation is 2.0 strokes, so most of Steven's scores per hole differ from the mean by no more than 2 strokes.
(b) 0.553

Uses and Abuses for Chapter 4 (page 223)

1. At least 20; The probability of at least 20 incidents is about 0.125, whereas there is about a 0.102 chance of 15 incidents.

2. Less than 14; The probability of less than 14 incidents is about 0.363, whereas there is about a 0.301 chance of 14 to 16 incidents.

3. Yes. The probability of 21 incidents is about 0.030, which is less than 0.05.

Review Exercises for Chapter 4 (page 225)

1. Discrete; The number of pumps in use at a gas station is a random variable that is countable.

3. (a)

x	$P(x)$
0	0.207
1	0.443
2	0.236
3	0.086
4	0.021
5	0.007

(b)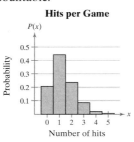
Skewed right

5. Yes

7. (a) $\mu \approx 2.8$; $\sigma^2 \approx 1.7$; $\sigma \approx 1.3$
(b) The mean is 2.8, so the average number of cell phones per household is about 3. The standard deviation is 1.3, so most of the households differ from the mean by no more than about 1 cell phone.

9. −$3.13

11. Binomial experiment
Success: a green candy is selected
$n = 12$, $p = 0.16$, $q = 0.84$,
$x = 0, 1, 2, 3, 4, 5, 6, 7, 8, 9, 10, 11, 12$

13. (a) 0.191 (b) 0.891 (c) 0.700

15. (a) 0.067 (b) 0.984 (c) 0.917

17. (a)

x	$P(x)$
0	0.0008
1	0.0126
2	0.0798
3	0.2529
4	0.4003
5	0.2536

(b)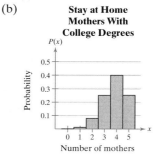
Skewed left

(c) The values 0 and 1 are unusual because their probabilities are less than 0.05.

19. $\mu \approx 1.0$; $\sigma^2 \approx 0.9$; $\sigma \approx 1.0$; On average, 1 out of every 8 drivers is uninsured. The standard deviation is 1.0, so most samples of 8 drivers would differ from the mean by at most 1 driver.

21. (a) 0.148 (b) 0.006; unusual (c) 0.820

23. (a) 0.154 (b) 0.217 (c) 0.011; unusual

25. (a) 0.085 (b) 0.410 (c) 0.430

Quiz for Chapter 4   (page 228)

1. (a) Discrete; The number of lightning strikes that occur in Wyoming during the month of June is a random variable that is countable.

(b) Continuous; The fuel (in gallons) used by a jet during takeoff is a random variable that has an infinite number of possible outcomes and cannot be counted.

(c) Discrete; The number of die rolls required for an individual to roll a five is a random variable that is countable.

2. (a)

x	$P(x)$
0	0.238
1	0.405
2	0.209
3	0.090
4	0.040
5	0.019

(b)

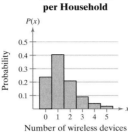

Wireless Devices per Household

Skewed right

(c) $\mu \approx 1.3$; $\sigma^2 \approx 1.4$; $\sigma \approx 1.2$; The mean is 1.3, so the average number of wireless devices per household is 1.3. The standard deviation is 1.2, so most households will differ from the mean by no more than 1.2 wireless devices.

(d) 0.058

3. (a) 0.269   (b) 0.811   (c) 0.061

4. (a)

x	$P(x)$
0	0.000008
1	0.000278
2	0.004262
3	0.034907
4	0.160820
5	0.395159
6	0.404567

(b)

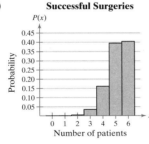

Successful Surgeries

Skewed left

(c) $\mu \approx 5.2$; $\sigma^2 \approx 0.7$; $\sigma \approx 0.8$; On average, 5.2 out of every 6 patients have a successful surgery. The standard deviation is 0.8, so most samples of 6 surgeries would differ from the mean by at most 0.8 surgery.

5. (a) 0.175   (b) 0.440   (c) 0.007

6. (a) 0.048   (b) 0.355   (c) 0.085

7. Event (a) is unusual because its probability is less than 0.05.

Real Statistics—Real Decisions for Chapter 4   (page 230)

1. (a) *Sample answer:* Calculate the probability of obtaining 0 clinical pregnancies out of 10 randomly selected ART cycles.

(b) Binomial. The distribution is discrete because the number of clinical pregnancies is countable.

2. $n = 10$, $p = 0.33$

x	$P(x)$
0	0.01823
1	0.08978
2	0.19899
3	0.26136
4	0.22528
5	0.13315
6	0.05465
7	0.01538
8	0.00284
9	0.00031
10	0.00002

Sample answer: Because $P(0) \approx 0.018$, this event is unusual but not impossible.

3. (a) Suspicious, because the probability is less than 0.05.

(b) Not suspicious, because the probability is greater than 0.05.

Chapter 5

Section 5.1   (page 242)

1. Answers will vary.   **3.** 1

5. Answers will vary.

Similarities: The two curves will have the same line of symmetry.

Differences: The curve with the larger standard deviation will be more spread out than the curve with the smaller standard deviation.

7. $\mu = 0$, $\sigma = 1$

9. "The" standard normal distribution is used to describe one specific normal distribution ($\mu = 0, \sigma = 1$). "A" normal distribution is used to describe a normal distribution with any mean and standard deviation.

11. No, the graph is skewed left.

13. No, the graph crosses the x-axis.

15. Yes, the graph fulfills the properties of the normal distribution.

$\mu \approx 11.5$, $\sigma \approx 1.5$

17. 0.9032   **19.** 0.0228   **21.** 0.6429   **23.** 0.5675

25. 0.0050   **27.** 0.7422   **29.** 0.6387   **31.** 0.4979

33. 0.8788   **35.** 0.2006 (*Tech:* 0.2005)

37. (a)

Life Spans of Tires

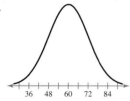

It is reasonable to assume that the life spans are normally distributed because the histogram is symmetric and bell-shaped.

(b) 37,234.7, 6259.2

(c) The sample mean of 37,234.7 hours is less than the claimed mean, so, on average, the tires in the sample lasted for a shorter time. The sample standard deviation of 6259.2 is greater than the claimed standard deviation, so the tires in the sample had a greater variation in life span than the manufacturer's claim.

39. (a) $x = 162 \rightarrow z \approx 1.37$
$x = 168 \rightarrow z \approx 2.06$
$x = 155 \rightarrow z \approx 0.57$
$x = 138 \rightarrow z \approx -1.37$

(b) $x = 168$ is unusual because its corresponding z-score (2.06) lies more than 2 standard deviations from the mean.

41. 0.9750   **43.** 0.9832   **45.** 0.6826 (*Tech:* 0.6827)
47. 0.8770   **49.** 0.0148   **51.** 0.3133
53. 0.9250 (*Tech:* 0.9249)   **55.** 0.0098 (*Tech:* 0.0099)
57.

The normal distribution curve is centered at its mean (60) and has 2 points of inflection (48 and 72) representing $\mu \pm \sigma$.

59. (1) The area under the curve is

$$(b - a)\left(\frac{1}{b - a}\right) = \frac{b - a}{b - a} = 1.$$

(Because $a < b$, you do not have to worry about division by 0.)

(2) All of the values of the probability density function are positive because $\frac{1}{b - a}$ is positive when $a < b$.

Section 5.2   (page 249)

1. 0.4207   **3.** 0.3446   **5.** 0.1787 (*Tech:* 0.1788)
7. (a) 0.2611 (*Tech:* 0.2623)
(b) 0.3453 (*Tech:* 0.3452)
(c) 0.1190 (*Tech:* 0.1186)
(d) No unusual events because all of the probabilities are greater than 0.05.

9. (a) 0.1492 (*Tech:* 0.1497)
(b) 0.4262 (*Tech:* 0.4269)
(c) 0.0188 (*Tech:* 0.0190)
(d) The event in part (c) is unusual because its probability is less than 0.05.

11. (a) 0.0062   (b) 0.7492 (*Tech:* 0.7499)   (c) 0.0004
13. 0.2918 (*Tech:* 0.2914)   **15.** 0.0324 (*Tech:* 0.0325)
17. (a) 86.86% (*Tech:* 86.96%)
(b) 464 scores (*Tech:* 465 scores)
19. (a) 98.93%   (b) 75.75% (*Tech:* 75.76%)
(c) 6 mothers
21. Out of control, because there is a point more than three standard deviations beyond the mean.
23. Out of control, because there are nine consecutive points below the mean, and two out of three consecutive points lie more than two standard deviations from the mean.

Section 5.3   (page 257)

1. −0.81   **3.** 0.45   **5.** −1.645   **7.** 1.555   **9.** −1.04
11. 1.175   **13.** −0.67   **15.** 1.34   **17.** −0.38
19. 1.99   **21.** −1.96, 1.96   **23.** −1.18   **25.** −0.35
27. −2.00   **29.** 1.28
31. (a) 68.97 inches   (b) 63.71 inches (*Tech:* 63.69 inches)
(c) 62.26 inches (*Tech:* 62.24 inches)
33. (a) 1315.99 kilowatt-hours (*Tech:* 1316.08 kilowatt-hours)
(b) 1719.67 kilowatt-hours (*Tech:* 1719.58 kilowatt-hours)
(c) 2671.34 kilowatt-hours (*Tech:* 2671.04 kilowatt-hours)
35. (a) 3.66   (b) 3.24 and 3.48
37. (a) 5.67 millions of cells per microliter
(b) 4.98 millions of cells per microliter (*Tech:* 4.99 millions of cells per microliter)
39. 32.61 ounces   **41.** 7.93 ounces

Section 5.4   (page 269)

1. 150, 3.536   **3.** 790, 3.036
5. False. As the size of a sample increases, the mean of the distribution of sample means does not change.
7. False. A sampling distribution is normal when either $n \geq 30$ or the population is normal.
9. (c), because $\mu_{\bar{x}} = 16.5$, $\sigma_{\bar{x}} = 1.19$, and the graph approximates a normal curve.

11. (a) $\mu = 53.2$, $\sigma \approx 19.9$

(b)

Sample	Mean
19, 19	19
19, 48	33.5
19, 56	37.5
19, 64	41.5
19, 79	49
48, 19	33.5
48, 48	48
48, 56	52
48, 64	56
48, 79	63.5
56, 19	37.5
56, 48	52
56, 56	56
56, 64	60
56, 79	67.5
64, 19	41.5
64, 48	56
64, 56	60
64, 64	64
64, 79	71.5
79, 19	49
79, 48	63.5
79, 56	67.5
79, 64	71.5
79, 79	79

(c) $\mu_{\bar{x}} = 53.2$, $\sigma_{\bar{x}} \approx 14.1$
The means are equal, but the standard deviation of the sampling distribution is smaller.

13. (a) $\mu = 389$, $\sigma \approx 28.65$

(b)

Sample	Mean
350, 350, 350	350
350, 350, 399	366.33
350, 350, 418	372.67
350, 399, 350	366.33
350, 399, 399	382.67
350, 399, 418	389
350, 418, 350	372.67
350, 418, 399	389
350, 418, 418	395.33
399, 350, 350	366.33
399, 350, 399	382.67
399, 350, 418	389
399, 399, 350	382.67
399, 399, 399	399
399, 399, 418	405.33
399, 418, 350	389
399, 418, 399	405.33
399, 418, 418	411.67
418, 350, 350	372.67
418, 350, 399	389
418, 350, 418	395.33
418, 399, 350	389
418, 399, 399	405.33
418, 399, 418	411.67
418, 418, 350	395.33
418, 418, 399	411.67
418, 418, 418	418

(c) $\mu_{\bar{x}} = 389$, $\sigma_{\bar{x}} \approx 16.54$
The means are equal, but the standard deviation of the sampling distribution is smaller.

15. 0.9726; not unusual

17. 0.0351 (*Tech:* 0.0349); unusual

19. $\mu_{\bar{x}} = 495$, $\sigma_{\bar{x}} \approx 26.83$

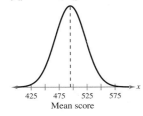

21. $\mu_{\bar{x}} = 23$, $\sigma_{\bar{x}} = 0.26$

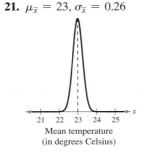

23. $\mu_{\bar{x}} = 1.64$, $\sigma_{\bar{x}} \approx 0.83$

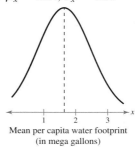

25. $\mu_{\bar{x}} = 132,000$, $\sigma_{\bar{x}} \approx 3042.56$

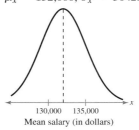

27. $n = 40$: $\mu_{\bar{x}} = 495$, $\sigma_{\bar{x}} \approx 18.97$
$n = 60$: $\mu_{\bar{x}} = 495$, $\sigma_{\bar{x}} \approx 15.49$

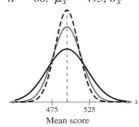

As the sample size increases, the standard deviation of the sample means decreases, while the mean of the sample means remains constant.

29. 0.4623 (*Tech:* 0.4645); About 46% of samples of 32 years will have a mean gain between 200 and 500.

31. 0.0708 (*Tech:* 0.0702); About 7% of samples of 30 Chinese cities will have a mean childhood asthma rate greater than 2.6%.

33. It is more likely to select a sample of 10 cities with a mean childhood asthma prevalence less than 3.2% because the sample of 10 has a higher probability.

35. Yes, it is very unlikely that you would have randomly sampled 40 cans with a mean equal to 127.9 ounces because it is more than 3 standard deviations from the mean of the sample means.

37. (a) 0.3085   (b) 0.0008

39. Yes, the finite correction factor should be used; 0.6772 (*Tech:* 0.6755)

41. 0.0446 (*Tech:* 0.0448); The probability that less than 55% of a sample of 105 residents are in favor of building a new high school is about 4.5%. Because the probability is less than 0.05, this is an unusual event.

Section 5.4 Activity   (page 274)

1–2. Answers will vary.

Section 5.5   (page 281)

1. Cannot use normal distribution
3. Cannot use normal distribution
5. a   6. d   7. c   8. b
9. The probability of getting fewer than 25 successes;
   $P(x < 24.5)$
11. The probability of getting exactly 33 successes;
   $P(32.5 < x < 33.5)$
13. The probability of getting at most 150 successes;
   $P(x < 150.5)$
15. Binomial: $P(5 \le x \le 7) \approx 0.549$
   Normal: $P(4.5 < x < 7.5) = 0.5463$ (*Tech:* 0.5466)
   The results are about the same.
17. Can use normal distribution; $\mu = 9.3$, $\sigma \approx 2.533$
19. Can use normal distribution
   (a) 0.0793

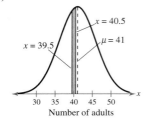

   (b) 0.6198

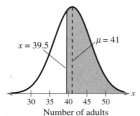

   (c) 0.3802

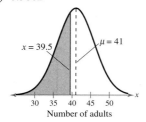

No unusual events because all of the probabilities are greater than 0.05.

21. Can use normal distribution
   (a) 0.3936 (*Tech:* 0.3925)

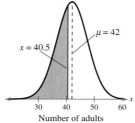

(b) 0.0606 (*Tech:* 0.0611)

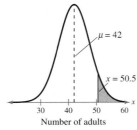

(c) 0.0182

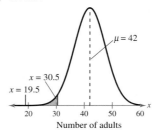

The event in part (c) is unusual because its probability is less than 0.05.

23. Cannot use normal distribution because $np < 5$.
   (a) 0.0382   (b) 0.4862   (c) 0.6994
   The event in part (a) is unusual because its probability is less than 0.05.
25. Can use normal distribution.
   (a) 0.4013 (*Tech:* 0.4001)

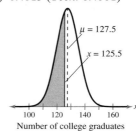

(b) 0.1867 (*Tech:* 0.1879)

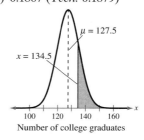

(c) 0.3999

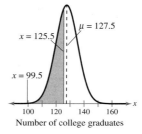

No unusual events because all of the probabilities are greater than 0.05.

27. (a) 0.0885 (*Tech:* 0.0878)   (b) 0.1660 (*Tech:* 0.1658)
     (c) 0.7324 (*Tech:* 0.7322)
29. Highly unlikely. Answers will vary.   **31.** 0.1020

Uses and Abuses for Chapter 5   (page 284)

1. (a) Not unusual; A sample mean of 112 is less than
     2 standard deviations from the population mean.
   (b) Unusual; A sample mean of 105 is more than 2 standard
     deviations from the population mean.
2. The ages of students at a high school may not be normally
   distributed.
3. Answers will vary.

Review Exercises for Chapter 5   (page 286)

1. $\mu = 15, \sigma = 3$
3. Curve *B* has the greatest mean because its line of symmetry
   occurs the farthest to the right.
5. 0.6772   **7.** 0.6293   **9.** 0.7157
11. 0.00235 (*Tech:* 0.00236)   **13.** 0.4495
15. 0.4365 (*Tech:* 0.4364)   **17.** 0.1336
19. $x = 17 \rightarrow z \approx -0.66$ **21.** 0.8997
     $x = 29 \rightarrow z \approx 1.18$
     $x = 8 \rightarrow z \approx -2.05$
     $x = 23 \rightarrow z \approx 0.26$
23. 0.9236 (*Tech:* 0.9237)   **25.** 0.0124   **27.** 0.8944
29. 0.2266   **31.** 0.2684 (*Tech:* 0.2685)
33. (a) 0.4168 (*Tech:* 0.4173)   (b) 0.1425 (*Tech:* 0.1407)
     (c) 0.3974 (*Tech:* 0.3971)
35. No unusual events because all of the probabilities are
   greater than 0.05.
37. −0.07   **39.** 2.455 (*Tech:* 2.457)   **41.** 1.04   **43.** 0.51
45. 119.54 feet   **47.** 137.80 feet (*Tech:* 137.81 feet)
49. 136.71 feet (*Tech:* 136.70 feet)
51. (a) $\mu = 1.5, \sigma \approx 1.118$
   (b)

Sample	Mean
0, 0	0
0, 1	0.5
0, 2	1
0, 3	1.5
1, 0	0.5
1, 1	1
1, 2	1.5
1, 3	2
2, 0	1
2, 1	1.5
2, 2	2
2, 3	2.5
3, 0	1.5
3, 1	2
3, 2	2.5
3, 3	3

(c) $\mu_{\bar{x}} = 1.5, \sigma_{\bar{x}} \approx 0.791$
The means are equal, but
the standard deviation of
the sampling distribution
is smaller.

53. $\mu_{\bar{x}} = 471.5, \sigma_{\bar{x}} \approx 31.761$

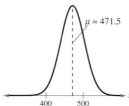
$\mu = 471.5$
Mean electric power consumption
(in kilowatt-hours)

55. (a) 0.3840 (*Tech:* 0.3839)   (b) 0.1898 (*Tech:* 0.1923)
     (c) 0.3557 (*Tech:* 0.3561)
   The probabilities in parts (a) and (c) are smaller, and the
   probability in part (b) is larger.
57. (a) 0.8051 (*Tech:* 0.8043)   (b) 0.8577 (*Tech:* 0.8580)
     (c) 0.3993 (*Tech:* 0.3994)
59. (a) 0.2709 (*Tech:* 0.2710)   (b) 0.1112 (*Tech:* 0.1113)
61. Can use normal distribution; $\mu = 15, \sigma \approx 1.936$
63. The probability of getting at least 25 successes; $P(x > 24.5)$
65. The probability of getting exactly 45 successes;
   $P(44.5 < x < 45.5)$
67. The probability of getting less than 60 successes; $P(x < 59.5)$
69. Can use normal distribution
   (a) 0.0384 (*Tech:* 0.0385)

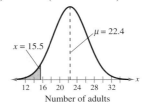
$\mu = 22.4$
$x = 15.5$
Number of adults

   (b) 0.0798 (*Tech:* 0.0818)

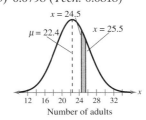
$x = 24.5$
$\mu = 22.4$
$x = 25.5$
Number of adults

   (c) 0.0188 (*Tech:* 0.0190)

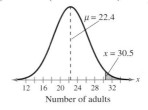
$\mu = 22.4$
$x = 30.5$
Number of adults

The events in parts (a) and (c) are unusual because their
probabilities are less than 0.05.

Quiz for Chapter 5   (page 290)

1. (a) 0.9535   (b) 0.9871   (c) 0.3616
   (d) 0.7703 (*Tech:* 0.7702)

2. (a) 0.0233 (*Tech:* 0.0231)   (b) 0.9929 (*Tech:* 0.9928)
   (c) 0.9198 (*Tech:* 0.9199)   (d) 0.3607 (*Tech:* 0.3610)
3. 0.0475 (*Tech:* 0.0478); Yes, the event is unusual because its probability is less than 0.05.
4. 0.2586 (*Tech:* 0.2611); No, the event is not unusual because its probability is greater than 0.05.
5. 21.19%   **6.** 503 people (*Tech:* 505 people)
7. 125   **8.** 80
9. 0.0049; About 0.5% of samples of 60 people will have a mean IQ score greater than 105. This is a very unusual event.
10. More likely to select one person with an IQ score greater than 105 because the standard error of the mean is less than the standard deviation.
11. Can use normal distribution; $\mu = 40$, $\sigma \approx 5.797$
12. (a) 0.5359 (*Tech:* 0.5344)   (b) 0.7823 (*Tech:* 0.7812)
   (c) 0.0277 (*Tech:* 0.0266)
   The event in part (c) is unusual because its probability is less than 0.05.

Real Statistics—Real Decisions for Chapter 5   (page 292)

1. (a) 0.4207   (b) 0.9988
2. (a) 0.3264 (*Tech:* 0.3274)   (b) 0.6944 (*Tech:* 0.6957)
   (c) randomly selected sample mean
3. Answers will vary.

Cumulative Review for Chapters 3–5   (page 294)

1. (a) $np = 18.3 \geq 5$, $nq = 11.7 \geq 5$
   (b) 0.0778 (*Tech:* 0.0775)
   (c) Yes, because the probability is less than 0.05.
2. (a) 2.46   (b) 1.95   (c) 1.40   (d) 2.46
   The size of a household on average is about 2.5 persons. The standard deviation is 1.4, so most households differ from the mean by no more than about 1 person.
3. (a) 2.45   (b) 2.29   (c) 1.51   (d) 2.45
   The number of fouls per game for Garrett Temple is about 2.45 fouls. The standard deviation is 1.5, so most of Temple's games differ from the mean by no more than about 1 or 2 fouls.
4. (a) 0.777   (b) 0.514   (c) 0.626
5. (a) 43,680   (b) 0.019
6. 0.7642   **7.** 0.0010   **8.** 0.7995   **9.** 0.4984
10. 0.2862   **11.** 0.5905
12. (a) 0.0367   (b) 0.3735   (c) 0.0029
   (d) The events in parts (a) and (c) are unusual because their probabilities are less than 0.05.
13. (a) 0.0049   (b) 0.0149   (c) 0.9046
14. (a) 0.277   (b) 0.886
   (c) Dependent.
   $P($Being a public school teacher $\mid$ having 20 years or more of full-time teaching experience$) \neq P($Being a public school teacher$)$
   (d) 0.413

15. (a) $\mu_{\bar{x}} = 70$, $\sigma_{\bar{x}} \approx 0.190$ (b) 0.0006

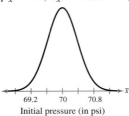

Initial pressure (in psi)

16. (a) 0.0548   (b) 0.6547   (c) 52.2 months
17. (a) 495   (b) 0.002
18. (a)

x	$P(x)$
0	0.000006
1	0.0001
2	0.0014
3	0.0090
4	0.0368
5	0.1029
6	0.2001
7	0.2668
8	0.2335
9	0.1211
10	0.0282

(b)

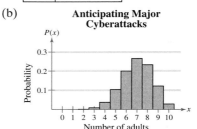

Anticipating Major Cyberattacks

Skewed left
(c) The values 0, 1, 2, 3, 4, and 10 are unusual because their probabilities are less than 0.05.

Chapter 6

Section 6.1   (page 305)

1. You are more likely to be correct using an interval estimate because it is unlikely that a point estimate will exactly equal the population mean.
3. d; As the level of confidence increases, z_c increases, causing wider intervals.
5. 1.28   **7.** 1.15   **9.** −0.47   **11.** 1.76   **13.** 1.861
15. 0.192   **17.** c   **18.** d   **19.** b   **20.** a
21. (12.0, 12.6)   **23.** (9.7, 11.3)   **25.** $E = 1.4$, $\bar{x} = 13.4$
27. $E = 0.17$, $\bar{x} = 1.88$ **29.** 126   **31.** 7
33. $E = 1.95$, $\bar{x} = 28.15$
35. (1320.4, 1416.6); (1311.2, 1425.8)
   With 90% confidence, you can say that the population mean price is between $1320.40 and $1416.60. With 95% confidence, you can say that the population mean price is between $1311.20 and $1425.80. The 95% CI is wider.

37. (81.14, 87.12); (80.56, 87.70)

With 90% confidence, you can say that the population mean temperature is between 81.14°F and 87.12°F. With 95% confidence, you can say that the population mean temperature is between 80.56°F and 87.70°F. The 95% CI is wider.

39. No; The margin of error is large ($E = 48.1$).

41. No; The right endpoint of the 95% CI is 87.70.

43. (a) An increase in the level of confidence will widen the confidence interval and the less certain you can be about a point estimate.

    (b) An increase in the sample size will narrow the confidence interval because it decreases the standard error.

    (c) An increase in the population standard deviation will widen the confidence interval because small standard deviations produce more precise intervals, which are smaller.

45. (20.9, 24.8); (19.8, 25.9)

With 90% confidence, you can say that the population mean length of time is between 20.9 and 24.8 minutes. With 99% confidence, you can say that the population mean length of time is between 19.8 and 25.9 minutes. The 99% CI is wider.

47. 89

49. (a) 66 servings

    (b) No; Yes; The 95% CI is (28.252, 29.748). If the population mean is within 3% of the sample mean, then it falls outside the CI. If the population mean is within 0.3% of the sample mean, then it falls within the CI.

51. (a) 7 cans

    (b) Yes; The 90% CI is (127.3, 128.2) and 128 ounces falls within that interval.

53. (a) 74 balls

    (b) Yes; The 99% CI is (27.360, 27.640) and there are amounts less than 27.6 inches that fall within that interval.

55. *Sample answer:* A 99% CI may not be practical to use in all situations. It may produce a CI so wide that it has no practical application.

57. (a) 0.707   (b) 0.949   (c) 0.962   (d) 0.975

    (e) 0.711   (f) 0.937   (g) 0.964   (h) 0.979

The finite population correction factor approaches 1 as the sample size decreases and the population size remains the same.

The finite population correction factor approaches 1 as the population size increases and the sample size remains the same.

59. *Sample answer:*

$E = \dfrac{z_c \sigma}{\sqrt{n}}$ Write the original equation.

$E\sqrt{n} = z_c \sigma$ Multiply each side by $\sqrt{n}$.

$\sqrt{n} = \dfrac{z_c \sigma}{E}$ Divide each side by E.

$n = \left(\dfrac{z_c \sigma}{E}\right)^2$ Square each side.

Section 6.2 (page 315)

1. 1.833   **3.** 2.947   **5.** 2.664   **7.** 0.686   **9.** (10.9, 14.1)

11. (4.1, 4.5)   **13.** $E = 3.7, \bar{x} = 18.4$ **15.** $E = 9.5, \bar{x} = 74.1$

17. 6.0; (29.5, 41.5); With 95% confidence, you can say that the population mean commute time is between 29.5 and 41.5 minutes.

19. 153.83; (372.67, 680.33); With 95% confidence, you can say that the population mean cell phone price is between $372.67 and $680.33.

21. 6.4; (29.1, 41.9); With 95% confidence, you can say that the population mean commute time is between 29.1 and 41.9 minutes. This confidence interval is slightly wider than the one found in Exercise 17.

23. Yes   **25.** (a) 1185   (b) 168.1   (c) (1034.3, 1335.7)

27. (a) 7.49   (b) 1.64   (c) (6.28, 8.70)   **29.** No

31. (a) 68,757.94   (b) 15,834.18   (c) (61,892.21, 75,623.67)

33. Yes

35. Use a t-distribution because σ is unknown and $n \geq 30$. (26.0, 29.4); With 95% confidence, you can say that the population mean BMI is between 26.0 and 29.4.

37. Neither distribution can be used because $n < 30$ and the mileages are not normally distributed.

39. No; Half the sample mean is 2.18%, which falls outside the confidence interval.

41. No; They are not making good tennis balls because the t-value for the sample is $t = 10$, which is not between $-t_{0.99} = -2.797$ and $t_{0.99} = 2.797$.

Section 6.2 Activity (page 318)

1–2. Answers will vary.

Section 6.3 (page 325)

1. False. To estimate the value of p, the population proportion of successes, use the point estimate $\hat{p} = x/n$.

3. 0.060, 0.940   **5.** 0.650, 0.350

7. $E = 0.014, \hat{p} = 0.919$ **9.** $E = 0.042, \hat{p} = 0.554$

11. (0.573, 0.607); (0.570, 0.610)

With 90% confidence, you can say that the population proportion of U.S. adults who say they have made a New Year's resolution is between 57.3% and 60.7%. With 95% confidence, you can say it is between 57.0% and 61.0%. The 95% confidence interval is slightly wider.

13. (0.663, 0.737)

With 99% confidence, you can say that the population proportion of U.S. adults who say they think police officers should be required to wear body cameras while on duty is between 66.3% and 73.7%.

15. (0.030, 0.031)

17. (a) 601 adults   (b) 451 adults

    (c) Having an estimate of the population proportion reduces the minimum sample size needed.

19. (a) 752 adults   (b) 295 adults

    (c) Having an estimate of the population proportion reduces the minimum sample size needed.

21. Yes; It falls within both confidence intervals.

23. No; The minimum sample size needed is 451 adults.

25. United States: $(0.282, 0.358)$
Canada: $(0.177, 0.243)$
France: $(0.215, 0.285)$
Japan: $(0.459, 0.541)$
Australia: $(0.103, 0.157)$

27. (a) Expect to stay at first employer for 3 or more years: $(0.670, 0.710)$
Completed an apprenticeship or internship: $(0.660, 0.700)$
Employed in field of study: $(0.629, 0.671)$
Feel underemployed: $(0.488, 0.532)$
Prefer to work for a large company: $(0.125, 0.155)$

(b) Expect to stay at first employer for 3 or more years: $(0.663, 0.717)$
Completed an apprenticeship or internship: $(0.653, 0.707)$
Employed in field of study: $(0.623, 0.677)$
Feel underemployed: $(0.481, 0.539)$
Prefer to work for a large company: $(0.120, 0.160)$

29. $(0.666, 0.734)$ is approximately a 98.1% CI.

31. $(0.68, 0.74)$ is approximately a 96.3% CI.

33. $(0.45, 0.49)$ is approximately a 98.3% CI.
$(0.51, 0.55)$ is approximately a 98.3% CI.

35. If $n\hat{p} < 5$ or $n\hat{q} < 5$, the sampling distribution of $\hat{p}$ may not be normally distributed, so z_c cannot be used to calculate the confidence interval.

37.

$\hat{p}$	$\hat{q} = 1 - \hat{p}$	$\hat{p}\hat{q}$	$\hat{p}$	$\hat{q} = 1 - \hat{p}$	$\hat{p}\hat{q}$
0.0	1.0	0.00	0.45	0.55	0.2475
0.1	0.9	0.09	0.46	0.54	0.2484
0.2	0.8	0.16	0.47	0.53	0.2491
0.3	0.7	0.21	0.48	0.52	0.2496
0.4	0.6	0.24	0.49	0.51	0.2499
0.5	0.5	0.25	0.50	0.50	0.2500
0.6	0.4	0.24	0.51	0.49	0.2499
0.7	0.3	0.21	0.52	0.48	0.2496
0.8	0.2	0.16	0.53	0.47	0.2491
0.9	0.1	0.09	0.54	0.46	0.2484
1.0	0.0	0.00	0.55	0.45	0.2475

$\hat{p} = 0.5$ gives the maximum value of $\hat{p}\hat{q}$.

Section 6.3 Activity (page 329)

1–2. Answers will vary.

Section 6.4 (page 334)

1. Yes **3.** $\chi_R^2 = 14.067, \chi_L^2 = 2.167$
5. $\chi_R^2 = 32.852, \chi_L^2 = 8.907$
7. $\chi_R^2 = 52.336, \chi_L^2 = 13.121$
9. (a) $(7.33, 20.89)$ (b) $(2.71, 4.57)$
11. (a) $(755, 2401)$ (b) $(27, 49)$

13. (a) $(0.0426, 0.1699)$ (b) $(0.2063, 0.4122)$
With 95% confidence, you can say that the population variance is between 0.0426 and 0.1699, and the population standard deviation is between 0.2063 and 0.4122 inch.

15. (a) $(181.50, 976.54)$ (b) $(13.47, 31.25)$
With 99% confidence, you can say that the population variance is between 181.50 and 976.54, and the population standard deviation is between 13.47 and 31.25 thousand dollars.

17. (a) $(5.46, 45.70)$ (b) $(2.34, 6.76)$
With 99% confidence, you can say that the population variance is between 5.46 and 45.70, and the population standard deviation is between 2.34 and 6.76 days.

19. (a) $(128, 492)$ (b) $(11, 22)$
With 95% confidence, you can say that the population variance is between 128 and 492, and the population standard deviation is between 11 and 22 grains per gallon.

21. (a) $(0.04, 0.11)$ (b) $(0.21, 0.32)$
With 80% confidence, you can say that the population variance is between 0.04 and 0.11, and the population standard deviation is between 0.21 and 0.32 day.

23. (a) $(7.0, 30.6)$ (b) $(2.6, 5.5)$
With 98% confidence, you can say that the population variance is between 7.0 and 30.6, and the population standard deviation is between 2.6 and 5.5 minutes.

25. Yes, because all of the values in the confidence interval are less than 0.5.

27. No, because 0.25 is contained in the confidence interval.

29. *Sample answer:* Unlike a confidence interval for a population mean or proportion, a confidence interval for a population variance does not have a margin of error. The left and right endpoints must be calculated separately.

Uses and Abuses for Chapter 6 (page 336)

1–2. Answers will vary. **3.** (a) No (b) Yes

Review Exercises for Chapter 6 (page 338)

1. (a) 103.5 (b) 11.7

3. (a) $(91.8, 115.2)$; With 90% confidence, you can say that the population mean waking time is between 91.8 and 115.2 minutes past 5:00 A.M.
(b) Yes; If the population mean is within 10% of the sample mean, then it falls inside the confidence interval.

5. $E = 1.675, \bar{x} = 22.425$ **7.** 78 people **9.** 1.383

11. 2.624 **13.** (a) 11.2 (b) $(60.9, 83.3)$

15. (a) 0.7 (b) $(6.1, 7.5)$

17. $(146, 184)$; With 90% confidence, you can say that the population mean height is between 146 and 184 feet.

19. (a) $0.720, 0.280$ (b) $(0.697, 0.743)$; $(0.692, 0.747)$
(c) With 90% confidence, you can say that the population proportion of U.S. adults who say they want the U.S. to play a leading or major role in global affairs is between 69.7% and 74.3%. With 95% confidence, you can say it is between 69.2% and 74.7%. The 95% confidence interval is slightly wider.

21. (a) 0.530, 0.470   (b) (0.513, 0.547); (0.509, 0.551)
   (c) With 90% confidence, you can say that the population proportion of U.S. adults who think antibiotics are effective against viral infections is between 51.3% and 54.7%. With 95% confidence, you can say it is between 50.9% and 55.1%. The 95% confidence interval is slightly wider.
23. No; It falls outside both confidence intervals.
25. (a) 385 adults   (b) 335 adults
   (c) Having an estimate of the population proportion reduces the minimum sample size needed.
27. $\chi_R^2 = 23.337$, $\chi_L^2 = 4.404$ **29.** $\chi_R^2 = 24.996$, $\chi_L^2 = 7.261$
31. (a) (185.1, 980.8)   (b) (13.6, 31.3)
   With 95% confidence, you can say that the population variance is between 185.1 and 980.8, and the population standard deviation is between 13.6 and 31.3 knots.

Quiz for Chapter 6   (page 340)

1. (a) 2.598   (b) 0.123
   (c) (2.475, 2.721); With 95% confidence, you can say that the population mean winning time is between 2.475 and 2.721 hours.
   (d) No; It falls outside the confidence interval.
2. 42 champions
3. (a) $\bar{x} = 6.61$, $s \approx 3.38$
   (b) (4.65, 8.57); With 90% confidence, you can say that the population mean amount of time is between 4.65 and 8.57 minutes.
   (c) (4.79, 8.43); With 90% confidence, you can say that the population mean amount of time is between 4.79 and 8.43 minutes. This confidence interval is narrower than the one in part (b).
4. (109,990, 156,662); With 95% confidence, you can say that the population mean annual earnings is between \$109,990 and \$156,662.
5. Yes
6. (a) 0.740
   (b) (0.717, 0.763); With 90% confidence, you can say that the population proportion of U.S. adults who say that the energy situation in the United States is very or fairly serious is between 71.7% and 76.3%.
   (c) No; The values fall outside the confidence interval.
   (d) 798 adults
7. (a) (5.41, 38.08)
   (b) (2.32, 6.17); With 95% confidence, you can say that the population standard deviation is between 2.32 and 6.17 minutes.

Real Statistics—Real Decisions for Chapter 6   (page 342)

1. (a) Yes, there has been a change in the mean concentration level because the confidence interval for Year 1 does not overlap the confidence interval for Year 2.
   (b) No, there has not been a change in the mean concentration level because the confidence interval for Year 2 overlaps the confidence interval for Year 3.

   (c) Yes, there has been a change in the mean concentration level because the confidence interval for Year 1 does not overlap the confidence interval for Year 3.
2. The concentrations of cyanide in the drinking water have increased over the three-year period.
3. The width of the confidence interval for Year 2 may have been caused by greater variation in the levels of cyanide than in other years, which may be the result of outliers.
4. Increase the sample size.
5. Answers will vary.
   (a) *Sample answer:* The sampling distribution of the sample means was used because the "mean concentration" was used. The sample mean is the most unbiased point estimate of the population mean.
   (b) *Sample answer:* No, because typically σ is unknown. They could have used the sample standard deviation.

Chapter 7

Section 7.1   (page 359)

1. The two types of hypotheses used in a hypothesis test are the null hypothesis and the alternative hypothesis.
   The alternative hypothesis is the complement of the null hypothesis.
3. You can reject the null hypothesis, or you can fail to reject the null hypothesis.
5. False. In a hypothesis test, you assume the null hypothesis is true.
7. True
9. False. A small P-value in a test will favor rejection of the null hypothesis.
11. H_0: $\mu \le 645$ (claim); H_a: $\mu > 645$
13. H_0: $\sigma = 5$; H_a: $\sigma \ne 5$ (claim)
15. H_0: $p \ge 0.45$; H_a: $p < 0.45$ (claim)
17. c; H_0: $\mu \le 3$ **18.** d; H_0: $\mu \ge 3$
19. b; H_0: $\mu = 3$ **20.** a; H_0: $\mu \le 2$
21. Right-tailed   **23.** Two-tailed
25. $\mu > 8$
   H_0: $\mu \le 8$; H_a: $\mu > 8$ (claim)
27. $\sigma \le 320$
   H_0: $\sigma \le 320$ (claim); H_a: $\sigma > 320$
29. $p = 0.73$
   H_0: $p = 0.73$ (claim); H_a: $p \ne 0.73$
31. A type I error will occur when the actual proportion of new customers who return to buy their next textbook is at least 0.60, but you reject H_0: $p \ge 0.60$.
   A type II error will occur when the actual proportion of new customers who return to buy their next textbook is less than 0.60, but you fail to reject H_0: $p \ge 0.60$.

33. A type I error will occur when the actual standard deviation of the length of time to play a game is less than or equal to 12 minutes, but you reject H_0: $\sigma \leq 12$.
A type II error will occur when the actual standard deviation of the length of time to play a game is greater than 12 minutes, but you fail to reject H_0: $\sigma \leq 12$.

35. A type I error will occur when the actual proportion of applicants who become campus security officers is at most 0.25, but you reject H_0: $p \leq 0.25$.
A type II error will occur when the actual proportion of applicants who become campus security officers is greater than 0.25, but you fail to reject H_0: $p \leq 0.25$.

37. H_0: The proportion of homeowners who have a home security alarm is greater than or equal to 14%.
H_a: The proportion of homeowners who have a home security alarm is less than 14%.
H_0: $p \geq 0.14$; H_a: $p < 0.14$
Left-tailed because the alternative hypothesis contains $<$.

39. H_0: The standard deviation of the 18-hole scores for a golfer is greater than or equal to 2.1 strokes.
H_a: The standard deviation of the 18-hole scores for a golfer is less than 2.1 strokes.
H_0: $\sigma \geq 2.1$; H_a: $\sigma < 2.1$
Left-tailed because the alternative hypothesis contains $<$.

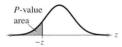

41. H_0: The mean graduation rate at a high school is less than or equal to 97%.
H_a: The mean graduation rate at a high school is greater than 97%.
H_0: $\mu \leq 97$; H_a: $\mu > 97$
Right-tailed because the alternative hypothesis contains $>$.

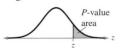

43. Alternative hypothesis
(a) There is enough evidence to support the scientist's claim that the mean incubation period for swan eggs is less than 40 days.
(b) There is not enough evidence to support the scientist's claim that the mean incubation period for swan eggs is less than 40 days.

45. Null hypothesis
(a) There is enough evidence to reject the researcher's claim that the standard deviation of the life of the lawn mower is at most 2.8 years.
(b) There is not enough evidence to reject the researcher's claim that the standard deviation of the life of the lawn mower is at most 2.8 years.

47. Null hypothesis
(a) There is enough evidence to reject the report's claim that at least 65% of individuals convicted of terrorism or terrorism-related offenses in the United States are foreign born.
(b) There is not enough evidence to reject the report's claim that at least 65% of individuals convicted of terrorism or terrorism-related offenses in the United States are foreign born.

49. H_0: $\mu \geq 60$; H_a: $\mu < 60$

51. (a) H_0: $\mu \geq 5$; H_a: $\mu < 5$
(b) H_0: $\mu \leq 5$; H_a: $\mu > 5$

53. If you decrease α, then you are decreasing the probability that you will reject H_0. Therefore, you are increasing the probability of failing to reject H_0. This could increase β, the probability of failing to reject H_0 when H_0 is false.

55. Yes; If the P-value is less than $\alpha = 0.05$, then it is also less than $\alpha = 0.10$.

57. (a) Fail to reject H_0 because the confidence interval includes values greater than 70.
(b) Reject H_0 because the confidence interval is located entirely to the left of 70.
(c) Fail to reject H_0 because the confidence interval includes values greater than 70.

59. (a) Reject H_0 because the confidence interval is located entirely to the right of 0.20.
(b) Fail to reject H_0 because the confidence interval includes values less than 0.20.
(c) Fail to reject H_0 because the confidence interval includes values less than 0.20.

Section 7.2   (page 373)

1. The z-test using a P-value compares the P-value with the level of significance α. In the z-test using rejection region(s), the test statistic is compared with critical values.

3. (a) Fail to reject H_0.   (b) Reject H_0.   (c) Reject H_0.

5. (a) Fail to reject H_0.   (b) Fail to reject H_0.
(c) Fail to reject H_0.

7. (a) Fail to reject H_0.   (b) Fail to reject H_0.
(c) Reject H_0.

9. $P = 0.0934$; Reject H_0.   **11.** $P = 0.0069$; Reject H_0.

13. $P = 0.0930$; Fail to reject H_0.

15. (a) $P = 0.0089$ (b) $P = 0.3050$
The larger P-value corresponds to the larger area.

17. Fail to reject H_0.

19. Critical value: $z_0 = -1.88$; Rejection region: $z < -1.88$

21. Critical value: $z_0 = 1.645$; Rejection region: $z > 1.645$

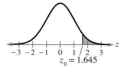

23. Critical values: $-z_0 = -2.33$, $z_0 = 2.33$
Rejection regions: $z < -2.33$, $z > 2.33$

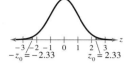

25. (a) Fail to reject H_0 because $z < 1.285$.
(b) Fail to reject H_0 because $z < 1.285$.
(c) Fail to reject H_0 because $z < 1.285$.
(d) Reject H_0 because $z > 1.285$.

27. Reject H_0. There is enough evidence at the 5% level of significance to reject the claim.

29. Fail to reject H_0. There is not enough evidence at the 3% level of significance to support the claim.

31. (a) The claim is "the mean total score for the school's applicants is more than 499."
$H_0: \mu \leq 499$; $H_a: \mu > 499$ (claim)
(b) 2.83   (c) 0.0023   (d) Reject H_0.
(e) There is enough evidence at the 1% level of significance to support the report's claim that the mean total score for the school's applicants is more than 499.

33. (a) The claim is "the mean winning times for Boston Marathon women's open division champions is at least 2.68 hours."
$H_0: \mu \geq 2.68$ (claim); $H_a: \mu < 2.68$
(b) -1.37 (c) 0.0853   (d) Fail to reject H_0.
(e) There is not enough evidence at the 5% level of significance to reject the statistician's claim that the mean winning times for Boston Marathon women's open division champions is at least 2.68 hours.

35. (a) The claim is "the mean height of top-rated roller coasters is 160 feet."
$H_0: \mu = 160$ (claim); $H_a: \mu \neq 160$
(b) 0.39   (c) 0.6992   (d) Fail to reject H_0.
(e) There is not enough evidence at the 5% level of significance to reject the claim that the mean height of top-rated roller coasters is 160 feet.

37. (a) The claim is "the mean caffeine content per 12-ounce bottle of a population of caffeinated soft drinks is 37.7 milligrams."
$H_0: \mu = 37.7$ (claim); $H_a: \mu \neq 37.7$
(b) $-z_0 = -2.575$, $z_0 = 2.575$
Rejection regions: $z < -2.575$, $z > 2.575$
(c) -0.72 (d) Fail to reject H_0.
(e) There is not enough evidence at the 1% level of significance to reject the consumer research organization's claim that the mean caffeine content per 12-ounce bottle of a population of caffeinated soft drinks is 37.7 milligrams.

39. (a) The claim is "the mean sodium content in a breakfast sandwich is no more than 920 milligrams."
$H_0: \mu \leq 920$ (claim); $H_a: \mu > 920$
(b) $z_0 = 1.28$; Rejection region: $z > 1.28$
(c) 1.84   (d) Reject H_0.
(e) There is enough evidence at the 10% level of significance to reject the restaurant's claim that the mean sodium content in one of their breakfast sandwiches is no more than 920 milligrams.

41. (a) The claim is "the mean life of a lamp is at least 10,000 hours."
$H_0: \mu \geq 10,000$ (claim); $H_a: \mu < 10,000$
(b) $z_0 = -1.23$; Rejection region: $z < -1.23$
(c) -1.28 (d) Reject H_0.
(e) There is enough evidence at the 11% level of significance to reject the lamp manufacturer's claim that the mean life of fluorescent lamps is at least 10,000 hours.

43. Outside; When the standardized test statistic is inside the rejection region, $P < \alpha$.

Section 7.3   (page 383)

1. Specify the level of significance α and the degrees of freedom, d.f. $= n - 1$. Find the critical value(s) using the t-distribution table in the row with $n - 1$ d.f. When the hypothesis test is
(1) left-tailed, use the "One Tail, α" column with a negative sign.
(2) right-tailed, use the "One Tail, α" column with a positive sign.
(3) two-tailed, use the "Two Tails, α" column with a negative and a positive sign.

3. Critical value: $t_0 = -1.328$; Rejection region: $t < -1.328$

5. Critical value: $t_0 = 1.717$; Rejection region: $t > 1.717$

7. Critical values: $-t_0 = -2.056$, $t_0 = 2.056$
Rejection regions: $t < -2.056$, $t > 2.056$

9. (a) Fail to reject H_0 because $t > -2.086$.
(b) Fail to reject H_0 because $t > -2.086$.
(c) Reject H_0 because $t < -2.086$.

11. (a) Reject H_0 because $t < -1.725$.
(b) Fail to reject H_0 because $-1.725 < t < 1.725$.
(c) Reject H_0 because $t > 1.725$.

13. Fail to reject H_0. There is not enough evidence at the 1% level of significance to reject the claim.

15. Reject H_0. There is enough evidence at the 1% level of significance to reject the claim.

17. Fail to reject H_0. There is not enough evidence at the 2% level of significance to reject the claim.

19. (a) The claim is "the mean price of a three-year-old sport utility vehicle (in good condition) is $20,000."
$H_0: \mu = 20,000$ (claim); $H_a: \mu \neq 20,000$
(b) $-t_0 = -2.080$, $t_0 = 2.080$
Rejection regions: $t < -2.080$, $t > 2.080$
(c) 1.51   (d) Fail to reject H_0.

(e) There is not enough evidence at the 5% level of significance to reject the claim that the mean price of a three-year-old sport utility vehicle (in good condition) is $20,000.

21. (a) The claim is "the mean credit card debt by state is greater than $5500 per person."
   $H_0: \mu \leq 5500$; $H_a: \mu > 5500$ (claim)
   (b) $t_0 = 1.699$; Rejection region: $t > 1.699$
   (c) 0.86   (d) Fail to reject H_0.
   (e) There is not enough evidence at the 5% level of significance to support the credit reporting agency's claim that the mean credit card debt by state is greater than $5500 per person.

23. (a) The claim is "the mean amount of carbon monoxide in the air in U.S. cities is less than 2.34 parts per million."
   $H_0: \mu \geq 2.34$; $H_a: \mu < 2.34$ (claim)
   (b) $t_0 = -1.295$; Rejection region: $t < -1.295$
   (c) 0.11   (d) Fail to reject H_0.
   (e) There is not enough evidence at the 10% level of significance to support the claim that the mean amount of carbon monoxide in the air in U.S. cities is less than 2.34 parts per million.

25. (a) The claim is "the mean annual salary for senior-level product engineers is $98,000."
   $H_0: \mu = 98,000$ (claim); $H_a: \mu \neq 98,000$
   (b) $-t_0 = -2.131$, $t_0 = 2.131$
   Rejection regions: $t < -2.131$, $t > 2.131$
   (c) -1.87 (d) Fail to reject H_0.
   (e) There is not enough evidence at the 5% level of significance to reject the employment information service's claim that the mean annual salary for senior-level product engineers is $98,000.

27. (a) The claim is "the mean minimum time it takes for a sedan to travel a quarter mile is greater than 14.7 seconds."
   $H_0: \mu \leq 14.7$; $H_a: \mu > 14.7$ (claim)
   (b) 0.0664   (c) Reject H_0.
   (d) There is enough evidence at the 10% level of significance to support the consumer group's claim that the mean minimum time it takes for a sedan to travel a quarter mile is greater than 14.7 seconds.

29. (a) The claim is "the mean class size for full-time faculty is fewer than 32 students."
   $H_0: \mu \geq 32$; $H_a: \mu < 32$ (claim)
   (b) 0.0344   (c) Reject H_0.
   (d) There is enough evidence at the 5% level of significance to support the brochure's claim that the mean class size for full-time faculty is fewer than 32 students.

31. Use the t-distribution because σ is unknown, the sample is random, and the population is normally distributed.
   Fail to reject H_0. There is not enough evidence at the 5% level of significance to reject the car company's claim that the mean gas mileage for the luxury sedan is at least 23 miles per gallon.

33. More likely; The tails of a t-distribution curve are thicker than those of a standard normal distribution curve. So, if you incorrectly use a standard normal sampling distribution instead of a t-sampling distribution, then the area under the curve at the tails will be smaller than what it would be for the t-test, meaning the critical value(s) will lie closer to the mean. This makes it more likely for the test statistic to be in the rejection region(s). This result is the same regardless of whether the test is left-tailed, right-tailed, or two-tailed; in each case, the tail thickness affects the location of the critical value(s).

Section 7.3 Activity   (page 386)

1–3. Answers will vary.

Section 7.4   (page 391)

1. If $np \geq 5$ and $nq \geq 5$, then the normal distribution can be used.
3. Cannot use normal distribution.
5. Can use normal distribution.
   Fail to reject H_0. There is not enough evidence at the 5% level of significance to support the claim.
7. (a) The claim is "less than 80% of U.S. adults think that healthy children should be required to be vaccinated."
   $H_0: p \geq 0.80$; $H_a: p < 0.80$ (claim)
   (b) $z_0 = -1.645$; Rejection region: $z < -1.645$
   (c) 0.707   (d) Fail to reject H_0.
   (e) There is not enough evidence at the 5% level of significance to support the medical researcher's claim that less than 80% of U.S. adults think that healthy children should be required to be vaccinated.
9. (a) The claim is "at most 3% of working college students are employed as teachers or teaching assistants."
   $H_0: p \leq 0.03$ (claim); $H_a: p > 0.03$
   (b) $z_0 = 2.33$; Rejection region: $z > 2.33$
   (c) 0.83   (d) Fail to reject H_0.
   (e) There is not enough evidence at the 1% level of significance to reject the education researcher's claim that at most 3% of working college students are employed as teachers or teaching assistants.
11. (a) The claim is "85% of Americans think they are unlikely to contract the Zika virus."
   $H_0: p = 0.85$ (claim); $H_a: p \neq 0.85$
   (b) $-z_0 = -1.96$, $z_0 = 1.96$
   Rejection region: $z < -1.96$, $z > 1.96$
   (c) 2.21   (d) Reject H_0.
   (e) There is enough evidence at the 5% level of significance to reject the medical researcher's claim that 85% of Americans think they are unlikely to contract the Zika virus.

13. (a) The claim is "27% of U.S. adults would travel into space on a commercial flight if they could afford it."
H_0: $p = 0.27$ (claim); H_a: $p \neq 0.27$
(b) 0.03   (c) Reject H_0.
(d) There is enough evidence at the 5% level of significance to reject the research center's claim that 27% of U.S. adults would travel into space on a commercial flight if they could afford it.

15. (a) The claim is "less than 67% of U.S. households own a pet."
H_0: $p \geq 0.67$; H_a: $p < 0.67$ (claim)
(b) 0.15   (c) Fail to reject H_0.
(d) There is not enough evidence at the 10% level of significance to support the humane society's claim that less than 67% of U.S. households own a pet.

17. Fail to reject H_0. There is not enough evidence at the 5% level of significance to reject the claim that at least 63% of adults make an effort to live in ways that help protect the environment some of the time.

19. (a) The claim is "less than 80% of U.S. adults think that healthy children should be required to be vaccinated."
H_0: $p \geq 0.80$; H_a: $p < 0.80$ (claim)
(b) $z_0 = -1.645$; Rejection region: $z < -1.645$
(c) 0.707   (d) Fail to reject H_0.
(e) There is not enough evidence at the 5% level of significance to support the medical researcher's claim that less than 80% of U.S. adults think that healthy children should be required to be vaccinated.
The results are the same.

Section 7.4 Activity   (page 393)

1–2. Answers will vary.

Section 7.5   (page 400)

1. Specify the level of significance α. Determine the degrees of freedom. Determine the critical values using the χ^2-distribution. For a right-tailed test, use the value that corresponds to d.f. and α; for a left-tailed test, use the value that corresponds to d.f. and $1 - \alpha$; for a two-tailed test, use the values that correspond to d.f. and $\frac{1}{2}\alpha$, and d.f. and $1 - \frac{1}{2}\alpha$.

3. The requirement of a normal distribution is more important when testing a standard deviation than when testing a mean. When the population is not normal, the results of a chi-square test can be misleading because the chi-square test is not as robust as the tests for the population mean.

5. Critical value: $\chi_0^2 = 38.885$; Rejection region: $\chi^2 > 38.885$

7. Critical value: $\chi_0^2 = 0.872$; Rejection region: $\chi^2 < 0.872$

9. Critical values: $\chi_L^2 = 60.391$, $\chi_R^2 = 101.879$
Rejection regions: $\chi^2 < 60.391$, $\chi^2 > 101.879$

11. Critical value: $\chi_0^2 = 49.588$; Rejection region: $\chi^2 > 49.588$

13. (a) Fail to reject H_0 because $\chi^2 < 6.251$.
(b) Fail to reject H_0 because $\chi^2 < 6.251$.
(c) Fail to reject H_0 because $\chi^2 < 6.251$.
(d) Reject H_0 because $\chi^2 > 6.251$.

15. Fail to reject H_0. There is not enough evidence at the 5% level of significance to reject the claim.

17. Reject H_0. There is enough evidence at the 1% level of significance to reject the claim.

19. Reject H_0. There is enough evidence at the 10% level of significance to support the claim.

21. Fail to reject H_0. There is not enough evidence at the 1% level of significance to support the claim.

23. (a) The claim is "the variance of the diameters in a certain tire model is 8.6."
H_0: $\sigma^2 = 8.6$ (claim); H_a: $\sigma^2 \neq 8.6$
(b) $\chi_L^2 = 1.735$, $\chi_R^2 = 23.589$
Rejection regions: $\chi^2 < 1.735$, $\chi^2 > 23.589$
(c) 4.5   (d) Fail to reject H_0.
(e) There is not enough evidence at the 1% level of significance to reject the tire manufacturer's claim that the variance of the diameters in a certain tire model is 8.6.

25. (a) The claim is "the standard deviation for grade 12 students on a mathematics assessment test is less than 35 points."
H_0: $\sigma \geq 35$; H_a: $\sigma < 35$ (claim)
(b) $\chi_0^2 = 18.114$; Rejection region: $\chi^2 < 18.114$
(c) 25.48   (d) Fail to reject H_0.
(e) There is not enough evidence at the 10% level of significance to support the school administrator's claim that the standard deviation for grade 12 students on a mathematics assessment test is less than 35 points.

27. (a) The claim is "the standard deviation of the waiting times for patients is no more than 0.5 minute."
H_0: $\sigma \leq 0.5$ (claim); H_a: $\sigma > 0.5$
(b) $\chi_0^2 = 33.196$; Rejection region: $\chi^2 > 33.196$
(c) 47.04   (d) Reject H_0.
(e) There is enough evidence at the 10% level of significance to reject the hospital's claim that the standard deviation of the waiting times for patients is no more than 0.5 minute.

29. (a) The claim is "the standard deviation of the annual salaries of senior-level graphic design specialists is different from $10,300."
H_0: $\sigma = 10,300$; H_a: $\sigma \neq 10,300$ (claim)
(b) $\chi_L^2 = 5.629$, $\chi_R^2 = 26.119$
Rejection regions: $\chi^2 < 5.629$, $\chi^2 > 26.119$
(c) 27.86   (d) Reject H_0.
(e) There is enough evidence at the 5% level of significance to support the claim that the standard deviation of the annual salaries of senior-level graphic design specialists is different from $10,300.

31. P-value $= 0.4524$; Fail to reject H_0.

33. P-value $= 0.0033$; Reject H_0.

Uses and Abuses for Chapter 7   (page 404)

1. H_0: $p = 0.57$; Answers will vary.   **2–3.** Answers will vary.

Review Exercises for Chapter 7 (page 406)

1. $H_0: \mu \le 375$ (claim); $H_a: \mu > 375$

3. $H_0: p \ge 0.205$; $H_a: p < 0.205$ (claim)

5. $H_0: \sigma \le 1.9$; $H_a: \sigma > 1.9$ (claim)

7. (a) $H_0: p = 0.65$ (claim); $H_a: p \ne 0.65$

   (b) A type I error will occur when the actual proportion of U.S. adults who have volunteered their time or donated money to help clean up the environment is 65%, but you reject $H_0: p = 0.65$.

   A type II error will occur when the actual proportion is not 65%, but you fail to reject $H_0: p = 0.65$.

   (c) Two-tailed because the alternative hypothesis contains $\ne$.

   (d) There is enough evidence to reject the polling organization's claim that the proportion of U.S. adults who have volunteered their time or donated money to help clean up the environment is 65%.

   (e) There is not enough evidence to reject the polling organization's claim that the proportion of U.S. adults who have volunteered their time or donated money to help clean up the environment is 65%.

9. (a) $H_0: \sigma \le 9.5$ (claim); $H_a: \sigma > 9.5$

   (b) A type I error will occur when the actual standard deviation of the fuel economies is no more than 9.5 miles per gallon, but you reject $H_0: \sigma \le 9.5$.

   A type II error will occur when the actual standard deviation of the fuel economies is more than 9.5 miles per gallon, but you fail to reject $H_0: \sigma \le 9.5$.

   (c) Right-tailed because the alternative hypothesis contains $>$.

   (d) There is enough evidence to reject the nonprofit consumer organization's claim that the standard deviation of the fuel economies of its top-rated vehicles for a recent year is no more than 9.5 miles per gallon.

   (e) There is not enough evidence to reject the nonprofit consumer organization's claim that the standard deviation of the fuel economies of its top-rated vehicles for a recent year is no more than 9.5 miles per gallon.

11. 0.1736; Fail to reject H_0.

13. Critical value: $z_0 = -2.05$; Rejection region: $z < -2.05$

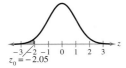

15. Critical value: $z_0 = 1.96$; Rejection region: $z > 1.96$

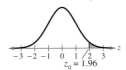

17. Fail to reject H_0 because $-1.645 < z < 1.645$.

19. Fail to reject H_0 because $-1.645 < z < 1.645$.

21. Fail to reject H_0. There is not enough evidence at the 5% level of significance to reject the claim.

23. Fail to reject H_0. There is not enough evidence at the 1% level of significance to support the claim.

25. (a) The claim is "the mean annual production of cotton is 3.5 million bales per country."

   $H_0: \mu = 3.5$ (claim); $H_a: \mu \ne 3.5$

   (b) -2.06 (c) 0.0394 (d) Reject H_0.

   (e) There is enough evidence at the 5% level of significance to reject the researcher's claim that the mean annual production of cotton is 3.5 million bales per country.

27. (a) The claim is "the mean amount of sulfur dioxide in the air in U.S. cities is 1.15 parts per billion."

   $H_0: \mu = 1.15$ (claim); $H_a: \mu \ne 1.15$

   (b) $-z_0 = -2.575$, $z_0 = 2.575$

   Rejection regions: $z < -2.575$, $z > 2.575$

   (c) -0.97 (d) Fail to reject H_0.

   (e) There is not enough evidence at the 1% level of significance to reject the environmental researcher's claim that the mean amount of sulfur dioxide in the air in U.S. cities is 1.15 parts per billion.

29. Critical values: $-t_0 = -2.093$, $t_0 = 2.093$

   Rejection regions: $t < -2.093$, $t > 2.093$

31. Critical value: $t_0 = 2.098$; Rejection region: $t > 2.098$

33. Critical value: $t_0 = -2.977$; Rejection region: $t < -2.977$

35. Reject H_0. There is enough evidence at the 0.5% level of significance to support the claim.

37. Reject H_0. There is enough evidence at the 1% level of significance to reject the claim.

39. Fail to reject H_0. There is not enough evidence at the 10% level of significance to reject the claim.

41. (a) The claim is "the mean monthly cost of joining a health club is $25."

   $H_0: \mu = 25$ (claim); $H_a: \mu \ne 25$

   (b) $-t_0 = -1.740$, $t_0 = 1.740$

   Rejection regions: $t < -1.740$, $t > 1.740$

   (c) 1.64 (d) Fail to reject H_0.

   (e) There is not enough evidence at the 10% level of significance to reject the advertisement's claim that the mean monthly cost of joining a health club is $25.

43. (a) The claim is "the mean score for grade 12 students on a science achievement test is more than 145."

   $H_0: \mu \le 145$; $H_a: \mu > 145$ (claim)

   (b) 0.0824 (c) Reject H_0.

   (d) There is enough evidence at the 10% level of significance to support the education publication's claim that the mean score for grade 12 students on a science achievement test is more than 145.

45. Can use normal distribution.

   Fail to reject H_0. There is not enough evidence at the 5% level of significance to reject the claim.

47. Can use normal distribution.

   Reject H_0. There is enough evidence at the 1% level of significance to support the claim.

49. (a) The claim is "over 40% of U.S. adults say they are less likely to travel to Europe in the next six months for fear of terrorist attacks."
H_0: $p \leq 0.40$; H_a: $p > 0.40$ (claim)
(b) $z_0 = 2.33$; Rejection region: $z > 2.33$ (c) 1.29
(d) Fail to reject H_0.
(e) There is not enough evidence at the 1% level of significance to support the polling agency's claim that over 40% of U.S. adults say they are less likely to travel to Europe in the next six months for fear of terrorist attacks.

51. Critical value: $\chi_0^2 = 30.144$; Rejection region: $\chi^2 > 30.144$

53. Critical values: $\chi_L^2 = 26.509$, $\chi_R^2 = 55.758$
Rejection regions: $\chi^2 < 26.509$, $\chi^2 > 55.758$

55. Reject H_0. There is enough evidence at the 10% level of significance to support the claim.

57. Fail to reject H_0. There is not enough evidence at the 5% level of significance to reject the claim.

59. (a) The claim is "the variance of the bolt widths is at most 0.01."
H_0: $\sigma^2 \leq 0.01$ (claim); H_a: $\sigma^2 > 0.01$
(b) $\chi_0^2 = 49.645$; Rejection region: $\chi^2 > 49.645$
(c) 172.8   (d) Reject H_0.
(e) There is enough evidence at the 0.5% level of significance to reject the bolt manufacturer's claim that the variance is at most 0.01.

61. You can reject H_0 at the 1% level of significance because $\chi^2 = 172.8 > 46.963$

Quiz for Chapter 7 (page 410)

1. (a) The claim is "the mean hat size for a male is at least 7.25."
H_0: $\mu \geq 7.25$ (claim); H_a: $\mu < 7.25$
(b) Left-tailed because the alternative hypothesis contains $<$; z-test because σ is known and the population is normally distributed.
(c) *Sample answer:* $z_0 = -2.33$;
Rejection region: $z < -2.33$; -1.28
(d) Fail to reject H_0.
(e) There is not enough evidence at the 1% level of significance to reject the company's claim that the mean hat size for a male is at least 7.25.

2. (a) The claim is "the mean daily base price for renting a full-size or less expensive vehicle in Vancouver, Washington, is more than $36."
H_0: $\mu \leq 36$; H_a: $\mu > 36$ (claim)
(b) Right-tailed because the alternative hypothesis contains $>$; z-test because σ is known and $n \geq 30$.
(c) *Sample answer:* $z_0 = 1.28$;
Rejection region: $z > 1.28$; 1.997
(d) Reject H_0.
(e) There is enough evidence at the 10% level of significance to support the travel analyst's claim that the mean daily base price for renting a full-size or less expensive vehicle in Vancouver, Washington, is more than $36.

3. (a) The claim is "the mean amount of earnings for full-time workers ages 18 to 24 with a bachelor's degree in a recent year is $47,254."
H_0: $\mu = 47{,}254$ (claim); H_a: $\mu \neq 47{,}254$
(b) Two-tailed because the alternative hypothesis contains $\neq$; t-test because σ is unknown and the population is normally distributed.
(c) *Sample answer:* $-t_0 = -2.145$, $t_0 = 2.145$;
Rejection regions: $t < -2.145$, $t > 2.145$; 2.58
(d) Reject H_0.
(e) There is not enough evidence at the 5% level of significance to support the government agency's claim that the mean amount of earnings for full-time workers ages 18 to 24 with a bachelor's degree is a recent year is $47,254.

4. (a) The claim is "program participants have a mean weight loss of at least 10.5 pounds after 1 month."
H_0: $\mu \geq 10.5$ (claim); H_a: $\mu < 10.5$
(b) Left-tailed because the alternative hypothesis contains $<$; t-test because σ is unknown and $n \geq 30$.
(c) *Sample answer:* $t_0 = -2.462$;
Rejection region: $t < -2.462$; -3.09
(d) Reject H_0.
(e) There is enough evidence at the 1% level of significance to reject the weight loss program's claim that program participants have a mean weight loss of at least 10.5 pounds after 1 month.

5. (a) The claim is "less than 18% of the vehicles a nonprofit consumer organization rated in a recent year have an overall score of 78 or more."
H_0: $p \geq 0.18$; H_a: $p < 0.18$ (claim)
(b) Left-tailed because the alternative hypothesis contains $<$; z-test because $np \geq 5$ and $nq \geq 5$.
(c) *Sample answer:* $z_0 = -1.645$;
Rejection region: $z < -1.645$; 0.49
(d) Fail to reject H_0.
(e) There is not enough evidence at the 5% level of significance to support the nonprofit consumer organization's claim that less than 18% of the vehicles a nonprofit consumer organization rated in a recent year have an overall score of 78 or more.

6. (a) The claim is "the standard deviation of vehicle rating scores is 11.90."
H_0: $\sigma = 11.90$ (claim); H_a: $\sigma \neq 11.90$
(b) Two-tailed because the alternative hypothesis contains $\neq$; chi-square test because the test is for a standard deviation and the population is normally distributed.
(c) *Sample answer:* $\chi_L^2 = 68.249$, $\chi_R^2 = 112.022$;
Rejection regions: $\chi^2 < 68.249$, $\chi^2 > 112.022$; 89.90
(d) Fail to reject H_0.
(e) There is not enough evidence at the 10% level of significance to reject the nonprofit consumer organization's claim that the standard deviation of vehicle rating scores is 11.90.

Real Statistics—Real Decisions for Chapter 7 (page 412)

1. Answers will vary.
2. (a) Not random; *Sample answer:* Randomly select students from the student directory.
   (b) Random   (c) Random
3. Alternative hypothesis because you cannot use a hypothesis test to support your claim if your claim is the null hypothesis.
4. No   5. No   6. Answers will vary.

Chapter 8

Section 8.1 (page 424)

1. Two samples are dependent when each member of one sample corresponds to a member of the other sample. Example: The weights of 22 people before starting an exercise program and the weights of the same 22 people 6 weeks after starting the exercise program.
   Two samples are independent when the sample selected from one population is not related to the sample selected from the other population. Example: The weights of 25 cats and the weights of 20 dogs.
3. Use P-values.
5. Dependent because the same football players were sampled.
7. Independent because different boats were sampled.
9. Reject H_0.
11. Reject H_0. There is enough evidence at the 1% level of significance to reject the claim.
13. Fail to reject H_0. There is not enough evidence at the 5% level of significance to support the claim.
15. (a) The claim is "the mean braking distances are different for the two makes of automobiles."
    H_0: $\mu_1 = \mu_2$; H_a: $\mu_1 \neq \mu_2$ (claim)
    (b) $-z_0 = -1.645$, $z_0 = 1.645$
    Rejection regions: $z < -1.645$, $z > 1.645$
    (c) 2.80   (d) Reject H_0.
    (e) There is enough evidence at the 10% level of significance to support the safety engineer's claim that the mean braking distances are different for the two makes of automobiles.
17. (a) The claim is "the wind speed in Region A is less than the wind speed in Region B."
    H_0: $\mu_1 \geq \mu_2$; H_a: $\mu_1 < \mu_2$ (claim)
    (b) $z_0 = -1.645$; Rejection region: $z < -1.645$
    (c) -1.94 (d) Reject H_0.
    (e) There is enough evidence at the 5% level of significance to support the claim that the wind speed in Region A is less than the wind speed in Region B.
19. (a) The claim is "ACT mathematics and science scores are equal."
    H_0: $\mu_1 = \mu_2$ (claim); H_a: $\mu_1 \neq \mu_2$
    (b) $-z_0 = -2.575$, $z_0 = 2.575$
    Rejection regions: $z < -2.575$, $z > 2.575$
    (c) -0.21 (d) Fail to reject H_0.

(e) There is not enough evidence at the 1% level of significance to reject the claim that ACT mathematics and science scores are equal.
21. (a) The claim is "the mean home sales price in Casper, Wyoming, is the same as in Cheyenne, Wyoming."
    H_0: $\mu_1 = \mu_2$ (claim); H_a: $\mu_1 \neq \mu_2$
    (b) $-z_0 = -2.575$, $z_0 = 2.575$
    Rejection regions: $z < -2.575$, $z > 2.575$
    (c) 0.15   (d) Fail to reject H_0.
    (e) There is not enough evidence at the 1% level of significance to reject the real estate agency's claim that the mean home sales price in Casper, Wyoming, is the same as in Cheyenne, Wyoming.
23. (a) The claim is "the precipitation in Seattle, Washington, was greater than in Birmingham, Alabama."
    H_0: $\mu_1 \leq \mu_2$; H_a: $\mu_1 > \mu_2$ (claim)
    (b) $z_0 = 1.645$; Rejection region: $z > 1.645$
    (c) 1.02   (d) Fail to reject H_0.
    (e) There is not enough evidence at the 5% level of significance to support the climatologist's claim that the precipitation is Seattle, Washington, was greater than in Birmingham, Alabama.
25. They are equivalent through algebraic manipulation of the equation.
    $\mu_1 = \mu_2 \Rightarrow \mu_1 - \mu_2 = 0$
27. H_0: $\mu_1 - \mu_2 \leq 2000$; H_a: $\mu_1 - \mu_2 > 2000$ (claim)
    Fail to reject H_0. There is not enough evidence at the 5% level of significance to support the claim that the difference between the mean annual salaries of entry-level software engineers in Raleigh, North Carolina, and Wichita, Kansas, is more than $2000.
29. $-\$3129 < \mu_1 - \mu_2 < \$6449$

Section 8.2 (page 432)

1. (1) The population standard deviations are unknown.
   (2) The samples are randomly selected.
   (3) The samples are independent.
   (4) The populations are normally distributed or each sample size is at least 30.
3. (a) $-t_0 = -1.714$, $t_0 = 1.714$
   (b) $-t_0 = -1.812$, $t_0 = 1.812$
5. (a) $t_0 = -1.746$ (b) $t_0 = -1.943$
7. (a) $t_0 = 1.729$ (b) $t_0 = 1.895$
9. Fail to reject H_0. There is not enough evidence at the 1% level of significance to reject the claim.
11. Reject H_0. There is enough evidence at the 5% level of significance to reject the claim.
13. (a) The claim is "the mean annual costs of food for dogs and cats are the same."
    H_0: $\mu_1 = \mu_2$ (claim); H_a: $\mu_1 \neq \mu_2$
    (b) $-t_0 = -1.694$, $t_0 = 1.694$
    Rejection regions: $t < -1.694$, $t > 1.694$
    (c) 8.19   (d) Reject H_0.
    (e) There is enough evidence at the 10% level of significance to reject the pet association's claim that the mean annual costs of food for dogs and cats are the same.

15. (a) The claim is "the stomachs of blue crabs from one location contain more fish than the stomachs of blue crabs from another location."
$H_0: \mu_1 \le \mu_2$; $H_a: \mu_1 > \mu_2$ (claim)
(b) $t_0 = 2.429$; Rejection region: $t > 2.429$
(c) 1.80   (d) Fail to reject H_0.
(e) There is not enough evidence at the 1% level of significance to support the claim that the stomachs of blue crabs from one location contain more fish than the stomachs of blue crabs from another location.

17. (a) The claim is "the mean household income in a recent year is greater in Cuyahoga County, Ohio, than it is in Wayne County, Michigan."
$H_0: \mu_1 \le \mu_2$; $H_a: \mu_1 > \mu_2$ (claim)
(b) $t_0 = 1.761$; Rejection region: $t > 1.761$
(c) 5.65   (d) Reject H_0.
(e) There is enough evidence at the 5% level of significance to support the demographics researcher's claim that the mean household income in a recent year is greater in Cuyahoga County, Ohio, than it is in Wayne County, Michigan.

19. (a) The claim is "an experimental method makes a difference in the tensile strength of steel bars."
$H_0: \mu_1 = \mu_2$; $H_a: \mu_1 \ne \mu_2$ (claim)
(b) $-t_0 = -2.819$, $t_0 = 2.819$
Rejection regions: $t < -2.819$, $t > 2.819$
(c) -2.64 (d) Fail to reject H_0.
(e) There is not enough evidence at the 1% level of significance to support the claim that an experimental method makes a difference in the tensile strength of steel bars.

21. (a) The claim is "the new method of teaching reading produces higher reading test scores than the old method."
$H_0: \mu_1 \ge \mu_2$; $H_a: \mu_1 < \mu_2$ (claim)
(b) $t_0 = -1.303$; Rejection region: $t < -1.303$
(c) -4.286 (d) Reject H_0.
(e) There is enough evidence at the 10% level of significance to support the claim that the new method of teaching reading produces higher reading test scores than the old method.

23. $-0.07 < \mu_1 - \mu_2 < 0.03$ **25.** $-20.8 < \mu_1 - \mu_2 < 24.8$

Section 8.3   (page 442)

1. (1) The samples are randomly selected.
(2) The samples are dependent.
(3) The populations are normally distributed or the number n of pairs of data is at least 30.

3. Fail to reject H_0. There is not enough evidence at the 5% level of significance to support the claim.

5. Reject H_0. There is enough evidence at the 10% level of significance to reject the claim.

7. Reject H_0. There is enough evidence at the 1% level of significance to reject the claim.

9. (a) The claim is "seven of the stocks that make up the Dow Jones Industrial Average lost value from one hour to the next on one business day."
$H_0: \mu_d \le 0$; $H_a: \mu_d > 0$ (claim)
(b) $t_0 = 3.143$; Rejection region: $t > 3.143$
(c) $\bar{d} \approx 0.087$; $s_d \approx 0.405$ (d) 0.569
(e) Fail to reject H_0.
(f) There is not enough evidence at the 1% level of significance to support the stock market analyst's claim that seven of the stocks that make up the Dow Jones Industrial Average lost value from one hour to the next on one business day.

11. (a) The claim is "caffeine ingestion improves repeated freestyle sprints in trained male swimmers."
$H_0: \mu_d \le 0$; $H_a: \mu_d > 0$ (claim)
(b) $t_0 = 3.365$; Rejection region: $t > 3.365$
(c) $\bar{d} \approx 0.533$; $s_d \approx 0.350$ (d) 3.730
(e) Reject H_0.
(f) There is enough evidence at the 1% level of significance to support the researcher's claim that caffeine ingestion improves repeated freestyle sprints in trained male swimmers.

13. (a) The claim is "soft tissue therapy helps to reduce the numbers of days per week patients suffer from headaches."
$H_0: \mu_d \le 0$; $H_a: \mu_d > 0$ (claim)
(b) $t_0 = 2.567$; Rejection region: $t > 2.567$
(c) $\bar{d} = 1.5$; $s_d \approx 1.249$ (d) 5.095
(e) Reject H_0.
(f) There is enough evidence at the 1% level of significance to support the physical therapist's claim that soft tissue therapy helps to reduce the numbers of days per week patients suffer from headaches.

15. (a) The claim is "student housing rates have increased from one academic year to the next."
$H_0: \mu_d \ge 0$; $H_a: \mu_d < 0$ (claim)
(b) $t_0 = -1.796$; Rejection region: $t < -1.796$
(c) $\bar{d} = -254.5$; $s_d \approx 291.767$ (d) -3.022
(e) Reject H_0.
(f) There is enough evidence at the 5% level of significance to support the college administrator's claim that student housing rates have increased from one academic year to the next.

17. (a) The claim is "the product ratings have changed from last year to this year."
$H_0: \mu_d = 0$; $H_a: \mu_d \ne 0$ (claim)
(b) $-t_0 = -2.365$, $t_0 = 2.365$
Rejection regions: $t < -2.365$, $t > 2.365$
(c) $\bar{d} = -1$; $s_d \approx 1.309$ (d) -2.161 (*Tech:* -2.160)
(e) Fail to reject H_0.
(f) There is not enough evidence at the 5% level of significance to support the claim that the product ratings have changed from last year to this year.

19. (a) The claim is "eating a new cereal as part of a daily diet lowers total blood cholesterol levels."
H_0: $\mu_d \le 0$; H_a: $\mu_d > 0$ (claim)
(b) $t_0 = 1.943$; Rejection region: $t > 1.943$
(c) $\bar{d} \approx 2.857$; $s_d \approx 4.451$ (d) 1.698
(e) Fail to reject H_0.
(f) There is not enough evidence at the 5% level of significance to support the claim that the new cereal lowers total blood cholesterol levels.

21. Yes; $P \approx 0.0058 < 0.05$, so you reject H_0.

23. $-1.76 < \mu_d < -1.29$

Section 8.4 (page 451)

1. (1) The samples are randomly selected.
(2) The samples are independent.
(3) $n_1\bar{p} \ge 5$, $n_1\bar{q} \ge 5$, $n_2\bar{p} \ge 5$, and $n_2\bar{q} \ge 5$

3. Can use normal sampling distribution; Fail to reject H_0. There is not enough evidence at the 1% level of significance to support the claim.

5. Can use normal sampling distribution; Reject H_0. There is enough evidence at the 10% level of significance to reject the claim.

7. (a) The claim is "there is a difference in the proportion of subjects who had no 12-week confirmed disability progression."
H_0: $p_1 = p_2$; H_a: $p_1 \ne p_2$ (claim)
(b) $-z_0 = -2.575$, $z_0 = 2.575$
Rejection regions: $z < -2.575$, $z > 2.575$
(c) -1.70 (d) Fail to reject H_0.
(e) There is not enough evidence at the 1% level of significance to support the claim that there is a difference in the proportion of subjects who had no 12-week confirmed disability progression.

9. (a) The claim is "there is a difference in the proportion of those employed between females ages 20 to 24 and males ages 20 to 24."
H_0: $p_1 = p_2$; H_a: $p_1 \ne p_2$ (claim)
(b) $-z_0 = -2.575$, $z_0 = 2.575$
Rejection regions: $z < -2.575$, $z > 2.575$
(c) -5.82 (d) Reject H_0.
(e) There is enough evidence at the 1% level of significance to support the claim that there is a difference in the proportion of those employed between females ages 20 to 24 and males ages 20 to 24.

11. (a) The claim is "the proportion of drivers who wear seat belts is greater in the West than in the Northeast."
H_0: $p_1 \le p_2$; H_a: $p_1 > p_2$ (claim)
(b) $z_0 = 1.645$; Rejection region: $z > 1.645$
(c) 2.08 (d) Reject H_0.
(e) There is enough evidence at the 5% level of significance to support the claim that the proportion of drivers who wear seat belts is greater in the West than in the Northeast.

13. No, there is not enough evidence at the 5% level of significance to reject the claim that the proportion of newlywed Asians who have a spouse of a different race or ethnicity is the same as the proportion of newlywed Hispanics who have a spouse of a different race or ethnicity.

15. Yes, there is enough evidence at the 1% level of significance to support the claim that the proportion of newlywed Asians who have a spouse of a different race or ethnicity is greater than the proportion of newlywed whites who have a spouse of a different race or ethnicity.

17. Yes, there is enough evidence at the 1% level of significance to support the claim that the proportion of newlywed whites who have a spouse of a different race or ethnicity is less than the proportion of newlywed blacks who have a spouse of a different race or ethnicity.

19. No, there is not enough evidence at the 1% level of significance to reject the claim that the proportion of men who work 40 hours per week is the same as the proportion of men who work more than 40 hours per week.

21. Yes, there is enough evidence at the 5% level of significance to support the claim that the U.S. workforce that works 40 hours per week is greater for women than for men.

23. $-0.028 < p_1 - p_2 < -0.012$

25. $0.011 < p_1 - p_2 < 0.069$; Answers will vary.

Uses and Abuses for Chapter 8 (page 454)

1. Answers will vary.

2. Blind: The patients do not know which group (medicine or placebo) they belong to.
Double-blind: Both the researcher and patient do not know which group (medicine or placebo) that the patient belongs to.

Review Exercises for Chapter 8 (page 456)

1. Dependent because the same adults were sampled.

3. Independent because different vehicles were sampled.

5. Fail to reject H_0. There is not enough evidence at the 5% level of significance to reject the claim.

7. Reject H_0. There is enough evidence at the 10% level of significance to support the claim.

9. (a) The claim is "the mean sodium content of chicken sandwiches at Restaurant A is less than the mean sodium content of chicken sandwiches at Restaurant B."
H_0: $\mu_1 \ge \mu_2$; H_a: $\mu_1 < \mu_2$ (claim)
(b) $z_0 = -1.645$; Rejection region: $z < -1.645$
(c) -2.82 (d) Reject H_0.
(e) There is enough evidence at the 5% level of significance to support the researcher's claim that the mean sodium content of chicken sandwiches at Restaurant A is less than the mean sodium content of chicken sandwiches at Restaurant B.

11. Reject H_0. There is enough evidence at the 5% level of significance to reject the claim.

13. Fail to reject H_0. There is not enough evidence at the 10% level of significance to reject the claim.

15. Reject H_0. There is enough evidence at the 1% level of significance to support the claim.

17. (a) The claim is "the new method of teaching mathematics produces higher mathematics test scores than the old method does."
H_0: $\mu_1 \leq \mu_2$; H_a: $\mu_1 > \mu_2$ (claim)
(b) $t_0 = 1.667$; Rejection region: $t > 1.667$
(c) 2.313   (d) Reject H_0.
(e) There is enough evidence at the 5% level of significance to support the claim that the new method of teaching mathematics produces higher mathematics test scores than the old method does.

19. Reject H_0. There is enough evidence at the 1% level of significance to reject the claim.

21. Reject H_0. There is enough evidence at the 10% level of significance to reject the claim.

23. (a) The claim is "the numbers of passing yards for college football quarterbacks change from their junior to their senior years."
H_0: $\mu_d = 0$; H_a: $\mu_d \neq 0$ (claim)
(b) $-t_0 = -2.262$, $t_0 = 2.262$
Rejection regions: $t < -2.262$, $t > 2.262$
(c) $\bar{d} = -12.3$; $s_d \approx 553.0877$ (d) -0.070
(e) Fail to reject H_0.
(f) There is not enough evidence at the 5% level of significance to support the sports statistician's claim that the numbers of passing yards for college football quarterbacks change from their junior to their senior years.

25. Can use normal sampling distribution; Fail to reject H_0. There is not enough evidence at the 5% level of significance to reject the claim.

27. Can use normal sampling distribution; Reject H_0. There is enough evidence at the 10% level of significance to support the claim.

29. (a) The claim is "the proportion of subjects who had at least 24 weeks of accrued remission is the same for the two groups."
H_0: $p_1 = p_2$ (claim); H_a: $p_1 \neq p_2$
(b) $-z_0 = -1.96$, $z_0 = 1.96$
Rejection regions: $z < -1.96$, $z > 1.96$
(c) 4.03   (d) Reject H_0.
(e) There is enough evidence at the 5% level of significance to reject the medical research team's claim that the proportion of subjects who had at least 24 weeks of accrued remission is the same for the two groups.

Quiz for Chapter 8   (page 460)

1. (a) The claim is "the mean score on the reading assessment test for male high school students is greater than the mean score for female high school students."
H_0: $\mu_1 \leq \mu_2$; H_a: $\mu_1 > \mu_2$ (claim)
(b) Right-tailed because H_a contains $>$; z-test because σ_1 and σ_2 are known, the samples are random samples, the samples are independent, and $n_1 \geq 30$ and $n_2 \geq 30$.
(c) $z_0 = 1.645$; Rejection region: $z > 1.645$

(d) 0.12   (e) Fail to reject H_0.
(f) There is not enough evidence at the 5% level of significance to support the claim that the mean score on the reading assessment test for the male high school students was higher than for the female high school students.

2. (a) The claim is "the mean scores on a music assessment test for eighth grade boys and girls are equal."
H_0: $\mu_1 = \mu_2$ (claim); H_a: $\mu_1 \neq \mu_2$
(b) Two-tailed because H_a contains $\neq$; t-test because σ_1 and σ_2 are unknown, the samples are random samples, the samples are independent, and the populations are normally distributed.
(c) $-t_0 = -1.706$, $t_0 = 1.706$
Rejection regions: $t < -1.706$, $t > 1.706$
(d) -0.814 (e) Fail to reject H_0.
(f) There is not enough evidence at the 10% level of significance to reject the teacher's claim that the mean scores on the music assessment test are the same for eighth grade boys and girls.

3. (a) The claim is "the seminar helps adults increase their credit scores."
H_0: $\mu_d \geq 0$; H_a: $\mu_d < 0$ (claim)
(b) Left-tailed because H_a contains $<$; t-test because both populations are normally distributed and the samples are dependent.
(c) $t_0 = -2.718$; Rejection region: $t < -2.718$
(d) -5.07 (e) Reject H_0.
(f) There is enough evidence at the 1% level of significance to support the claim that the seminar helps adults increase their credit scores.

4. (a) The claim is "the proportion of U.S. adults who approve of the job the Supreme Court is doing is less than it was 3 years prior."
H_0: $p_1 \geq p_2$; H_a: $p_1 < p_2$ (claim)
(b) Left-tailed because H_a contains $<$; z-test because you are testing proportions, the samples are random, the samples are independent, and the quantities $n_1\bar{p}$, $n_2\bar{p}$, $n_1\bar{q}$, and $n_2\bar{q}$ are at least 5.
(c) $z_0 = -1.645$; Rejection region: $z < -1.645$
(d) -0.48 (e) Fail to reject H_0.
(f) There is not enough evidence at the 5% level of significance to support the claim that the proportion of U.S. adults who approve of the job the Supreme Court is doing is less than it was 3 years prior.

Real Statistics—Real Decisions for Chapter 8   (page 462)

1. (a) *Sample answer:* Divide the records into groups according to the inpatients' ages, and then randomly select records from each group.
(b) *Sample answer:* Divide the records into groups according to geographic regions, and then randomly select records from each group.
(c) *Sample answer:* Assign a different number to each record, randomly choose a starting number, and then select every 50th record.

(d) *Sample answer:* Assign a different number to each record, and then use a table of random numbers to generate a sample of numbers.

2. (a)–(b) Answers will vary.

3. Use a *t*-test; independent; yes, you need to know if the population distributions are normal or not; yes, you need to know if the population variances are equal or not.

4. There is not enough evidence at the 5% level of significance to support the claim that there is a difference in the mean length of hospital stays for inpatients.
This decision does not support the claim.

Cumulative Review for Chapters 6−8 (page 466)

1. (a) (0.786, 0.814)
   (b) There is enough evidence at the 5% level of significance to support the researcher's claim that more than 75% of U.S. adults say their household contains a desktop or a laptop computer.

2. There is enough evidence at the 10% level of significance to support the claim that the fuel additive improved gas mileage.

3. (25.94, 28.00); *z*-distribution

4. (2.59, 4.33); *t*-distribution

5. (10.7, 13.5); *t*-distribution

6. (7.85, 8.57); *z*-distribution

7. H_0: $\mu \geq 33$; H_a: $\mu < 33$ (claim)

8. H_0: $p \geq 0.19$ (claim); H_a: $p < 0.19$

9. H_0: $\sigma = 0.63$ (claim); H_a: $\sigma \neq 0.63$

10. H_0: $\mu = 2.28$; H_a: $\mu \neq 2.28$ (claim)

11. There is enough evidence at the 10% level of significance to support the pediatrician's claim that the mean birth weight of a single-birth baby is greater than the mean birth weight of a baby that has a twin.

12. (a) (511.95, 2283.75)   (b) (22.63, 47.79)
    (c) There is not enough evidence at the 1% level of significance to reject the travel analyst's claim that the standard deviation of the mean room rate for two adults at three-star hotels in Cincinnati is at most $30.

13. There is enough evidence at the 5% level of significance to support the organization's claim that the mean SAT scores for male athletes and male non-athletes at a college are different.

14. (a) (49,996.92, 57,039.14)
    (b) There is not enough evidence at the 5% level of significance to reject the researcher's claim that the mean annual earnings for locksmiths is $53,000.

15. There is enough evidence at the 10% level of significance to support the medical research team's claim that the proportion of monthly convulsive seizure reduction is greater for the group that received the extract than for the group that received the placebo.

16. (a) (41.5, 42.5)
    (b) There is enough evidence at the 5% level of significance to reject the zoologist's claim that the mean incubation period for ostriches is at least 45 days.

17. A type I error will occur when the actual proportion of people who purchase their eyeglasses online is 0.05, but you reject H_0. A type II error will occur when the actual proportion of people who purchase their eyeglasses online is different from 0.05, but you fail to reject H_0.

Chapter 9

Section 9.1 (page 481)

1. Increase; Decrease

3. The sample correlation coefficient r measures the strength and direction of a linear relationship between two variables; $r = -0.932$ indicates a stronger correlation because $|-0.932| = 0.932$ is closer to 1 than $|0.918| = 0.918$.

5. A table can be used to compare r with a critical value, or a hypothesis test can be performed using a *t*-test.

7. H_0: $\rho = 0$ (no significant correlation)
   H_a: $\rho \neq 0$ (significant correlation)
   Reject the null hypothesis if t is in the rejection region.

9. Strong negative linear correlation

11. No linear correlation

13. Explanatory variable: Amount of water consumed
    Response variable: Weight loss

15. c; You would expect a positive linear correlation between age and income.

16. d; You would not expect age and height to be correlated.

17. b; You would expect a negative linear correlation between age and balance on student loans.

18. a; You would expect the relationship between age and body temperature to be fairly constant.

19. *Sample answer:* People who can afford more valuable homes will live longer because they have more money to take care of themselves.

21. *Sample answer:* Ice cream sales are higher when the weather is warm and people are outside more often. This is when homicides rates go up as well.

23. (a)

(b) 0.979
(c) Strong positive correlation; As age increases, the number of words in children's vocabulary tends to increase.
(d) There is enough evidence at the 1% level of significance to conclude that there is a significant linear correlation between children's ages and number of words in their vocabulary.

25. (a)

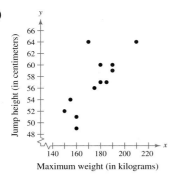

(b) 0.756

(c) Strong positive linear correlation; As the maximum weight for one repetition of a half squat increases, the jump height tends to increase.

(d) There is enough evidence at the 1% level of significance to conclude that there is a significant linear correlation between maximum weight for one repetition of a half squat and jump height.

27. (a)

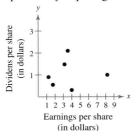

(b) 0.061

(c) No linear correlation; The earnings per share for the companies do not appear to be related to their dividends per share.

(d) There is not enough evidence at the 1% level of significance to conclude that there is a significant linear correlation between earning per share for the companies and their dividends per share.

29. The correlation coefficient gets weaker, going from $r \approx 0.979$ to $r \approx 0.863$.

31. The correlation coefficient gets stronger, going from $r \approx 0.756$ to $r \approx 0.908$.

33. There is not enough evidence at the 1% level of significance to conclude that there is a significant linear correlation between vehicle weight and the variability in braking distance on a dry surface.

35. There is enough evidence at the 5% level of significance to conclude that there is a significant linear correlation between the maximum weight for one repetition of a half squat and the jump height.

37. $r \approx -0.975$; The correlation coefficient remains unchanged when the x-values and y-values are switched.

Section 9.1 Activity   (page 485)

1–4. Answers will vary.

Section 9.2   (page 490)

1. A residual is the difference between the observed y-value of a data point and the predicted y-value on the regression line for the x-coordinate of the data point. A residual is positive when the data point is above the line, negative when the point is below the line, and zero when the observed y-value equals the predicted y-value.

3. Substitute a value of x into the equation of a regression line and solve for $\hat{y}$.

5. The correlation between variables must be significant.

7. b   **8.** a   **9.** e   **10.** c   **11.** f

12. d   **13.** b   **14.** c   **15.** d   **16.** a

17. $\hat{y} = 0.086x - 13.259$

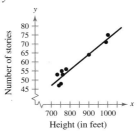

(a) 68 stories   (b) 60 stories
(c) 55 stories   (d) 47 stories

19. $\hat{y} = 7.451x + 37.449$

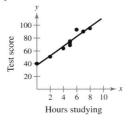

(a) 60   (b) 86
(c) It is not meaningful to predict the value of y for $x = 13$ because $x = 13$ is outside the range of the original data.
(d) 71

21. $\hat{y} = -2.044x + 520.668$

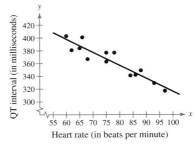

(a) It is not meaningful to predict the value of y for $x = 120$ because $x = 120$ is outside the range of the original data.
(b) 384 milliseconds   (c) 337 milliseconds
(d) 351 milliseconds

23. $\hat{y} = 2.979x + 52.476$

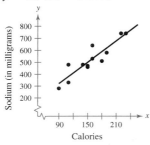

(a) 559 milligrams   (b) 350 milligrams
(c) It is not meaningful to predict the value of y for $x = 260$ because $x = 260$ is outside the range of the original data.
(d) 678 milligrams

25. $\hat{y} = 1.870x + 51.360$

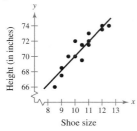

(a) 72.865 inches   (b) 66.320 inches
(c) It is not meaningful to predict the value of y for $x = 15.5$ because $x = 15.5$ is outside the range of the original data.
(d) 70.060 inches

27. Strong positive linear correlation; As the years of experience of the registered nurses increase, their salaries tend to increase.

29. No, it is not meaningful to predict a salary for a registered nurse with 28 years of experience because $x = 28$ is outside the range of the original data.

31. (a) $\hat{y} = -4.297x + 94.200$

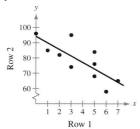

(b) $\hat{y} = -0.141x + 14.763$

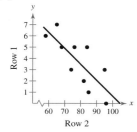

(c) The slope of the line keeps the same sign, but the values of m and b change.

33. (a) $\hat{y} = 0.139x + 21.024$
(b)

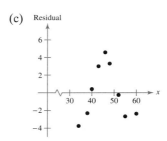

(c)   Residual

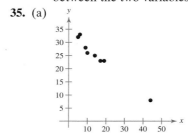

(d) The residual plot shows a pattern because the residuals do not fluctuate about 0. This implies that the regression line is not a good representation of the relationship between the two variables.

35. (a)

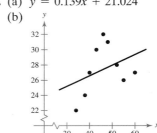

(b) The point $(44, 8)$ may be an outlier.
(c) The point $(44, 8)$ is not an influential point because the slopes and y-intercepts of the regression lines with the point included and without the point included are not significantly different.

37. $\hat{y} = 654.536x - 1214.857$

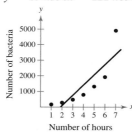

39. $y = 93.028(1.712)^x$

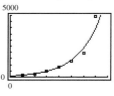

41. $\hat{y} = -78.929x + 576.179$

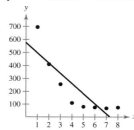

43. $y = 782.300x^{-1.251}$

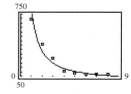

45. $y = 25.035 + 19.599 \ln x$

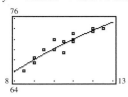

47. The logarithmic equation is a better model for the data. The graph of the logarithmic equation fits the data better than the regression line.

Section 9.2 Activity (page 496)

1–4. Answers will vary.

Section 9.3 (page 504)

1. The total variation is the sum of the squares of the differences between the y-values of each ordered pair and the mean of the y-values of the ordered pairs, or $\Sigma(y_i - \bar{y})^2$.

3. The unexplained variation is the sum of the squares of the differences between the observed y-values and the predicted y-values, or $\Sigma(y_i - \hat{y}_i)^2$.

5. Two variables that have perfect positive or perfect negative linear correlation have a correlation coefficient of 1 or -1, respectively. In either case, the coefficient of determination is 1, which means that 100% of the variation in the response variable is explained by the variation in the explanatory variable.

7. 0.216; About 21.6% of the variation is explained. About 78.4% of the variation is unexplained.

9. 0.916; About 91.6% of the variation is explained. About 8.4% of the variation is unexplained.

11. (a) 0.798; About 79.8% of the variation in proceeds can be explained by the relationship between the number of offerings and proceeds, and about 20.2% of the variation is unexplained.
(b) 8054.328; The standard error of estimate of the proceeds for a specific number of offerings is about 8,054,328,000.

13. (a) 0.729; About 72.9% of the variation in points earned can be explained by the relationship between the number of goals allowed and points earned, and about 27.1% of the variation is unexplained.
(b) 9.438; The standard error of estimate of the points earned for a specific number of goals allowed is about 9.438.

15. (a) 0.651; About 65.1% of the variation in mean annual wages can be explained by the relationship between percentages of employment in STEM occupations and mean annual wages, and about 34.9% of the variation is unexplained.
(b) 8.141; The standard error of estimate of the mean annual wages for a specific percentage of employment in STEM occupations is about $8141.

17. (a) 0.885; About 88.5% of the variation in the amount of crude oil imported can be explained by the relationship between the amount of crude oil produced and the amount imported, and about 11.5% of the variation is unexplained.
(b) 280.083; The standard error of estimate of the amount of crude oil imported for a specific amount of crude oil produced is about 280,083 barrels per day.

19. (a) 0.816; About 81.6% of the variation in the new-vehicle sales of General Motors can be explained by the relationship between the new-vehicle sales of Ford and General Motors, and about 18.4% of the variation is unexplained.
(b) 346.341; The standard error of estimate of the new-vehicle sales of General Motors for a specific amount of new-vehicle sales of Ford is about 346,341 new vehicles.

21. $40,083.251 < y < 82,572.581$
You can be 95% confident that the proceeds will be between $40,083,251,000 and $82,572,581,000 when the number of initial offerings is 450 issues.

23. $59.009 < y < 94.665$
You can be 90% confident that the total points earned will be between 59 and 95 when the number of goals allowed is 250.

25. $36.264 < y < 86.462$
You can be 99% confident that the mean annual wage will be between $36,264 and $86,462 when the percentage of employment in STEM occupations is 13% in the industry.

27. $7124.606 < y < 8677.288$
You can be 95% confident that the amount of crude oil imported will be between 7,124,606 and 8,677,288 barrels per day when the amount of crude oil produced is 8 million barrels per day.

29. $2684.712 < y < 4356.424$
You can be 95% confident that the new-vehicle sales of General Motors will be between 2,684,712 and 4,356,424 when the new-vehicle sales of Ford are 2,628,000.

31.

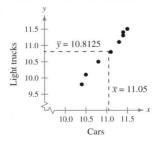

33. 0.987; About 98.7% of the variation in the median ages of light trucks can be explained by the relationship between the median ages of cars and light trucks, and about 1.3% of the variation is unexplained.

35. Fail to reject H_0. There is not enough evidence at the 1% level of significance to support the claim that there is a linear relationship between weight and number of hours slept.

37. $-175.836 < B < 287.908$; $108.928 < M < 290.142$

Section 9.4 (page 512)

1. (a) 18,832.7 pounds per acre
   (b) 18,016.4 pounds per acre
   (c) 17,350.6 pounds per acre
   (d) 16,190.3 pounds per acre
3. (a) 7.5 cubic feet
   (b) 16.8 cubic feet
   (c) 51.9 cubic feet
   (d) 62.1 cubic feet
5. (a) $\hat{y} = 17{,}899 - 606.58x_1 - 52.9x_2$
   (b) 564.314   (c) 0.966
   The standard error of estimate of the predicted price given specific age and milage of pre-owned Honda Civic Sedans is about \$564.31. The multiple regression model explains about 96.6% of the variation.
7. 0.955; About 95.5% of the variation in y can be explained by the relationship between variables; $r_{adj}^2 < r^2$.

Uses and Abuses for Chapter 9 (page 514)

1–2. Answers will vary.

Review Exercises for Chapter 9 (page 516)

1. (a)

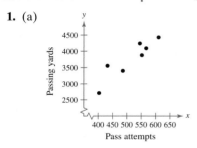

   (b) 0.917
   (c) Strong positive linear correlation; As the number of pass attempts increase, the number of passing yards tends to increase.
3. (a)

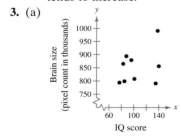

   (b) 0.338
   (c) Weak positive linear correlation; The IQ does not appear to be related to the brain size.
5. There is enough evidence at the 5% level of significance to conclude that there is a significant linear correlation between a quarterback's pass attempts and passing yards.
7. There is not enough evidence at the 1% level of significance to conclude that there is a significant linear correlation between IQ and brain size.

9. $\hat{y} = 0.106x - 781.327$

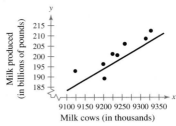

   (a) It is not meaningful to predict the value of y for $x = 9080$ because $x = 9080$ is outside the range of the original data.
   (b) 197.053 billions of pounds
   (c) 199.173 billions of pounds
   (d) 204.473 billions of pounds
11. $\hat{y} = -0.086x + 10.450$

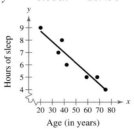

   (a) It is not meaningful to predict the value of y when $x = 16$ because $x = 16$ is outside the range of the original data.
   (b) 8.3 hours
   (c) It is not meaningful to predict the value of y when $x = 85$ because $x = 85$ is outside the range of the original data.
   (d) 6.15 hours
13. 0.203; About 20.3% of the variation is explained. About 79.7% of the variation is unexplained.
15. 0.412; About 41.2% of the variation is explained. About 58.8% of the variation is unexplained.
17. (a) 0.690; About 69.0% of the variation in top speed for hybrid and electric cars can be explained by the relationship between their fuel efficiencies and top speeds, and about 31.0% of the variation is unexplained.
   (b) 5.851; The standard error of estimate of the top speed for hybrid and electric cars for a specific fuel efficiency is about 5.851 miles per hour.
19. $193.364 < y < 210.282$
   You can be 90% confident that the amount of milk produced will be between 193.364 billion pounds and 210.282 billion pounds when the average number of cows is 9275.
21. $4.866 < y < 8.294$
   You can be 95% confident that the hours slept will be between 4.866 and 8.294 hours for an adult who is 45 years old.
23. $75.349 < y < 119.817$
   You can be 99% confident that the top speed of a hybrid or electric car will be between 75.349 and 119.817 miles per hour when the combined city and highway fuel efficiency is 90 miles per gallon equivalent.

25. (a) $\hat{y} = 3.6738 + 1.2874x_1 - 7.531x_2$

(b) 0.710; The standard error of estimate of the predicted carbon monoxide content given specific tar and nicotine contents is about 0.710 milligram.

(c) 0.943; The multiple regression model explains about 94.3% of the variation in y.

27. (a) 21.705 miles per gallon

(b) 25.21 miles per gallon

(c) 30.1 miles per gallon

(d) 25.86 miles per gallon

Quiz for Chapter 9   (page 520)

1.

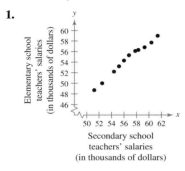

The data appear to have a positive linear correlation. As x increases, y tends to increase.

2. 0.992; Strong positive linear correlation; As the average annual salaries of secondary school teachers increase, the average annual salaries of elementary school teachers tend to increase.

3. Reject H_0. There is enough evidence at the 5% level of significance to conclude that there is a significant linear correlation between the average annual salaries of secondary school teachers and the average annual salaries of elementary school teachers.

4. $\hat{y} = 0.997x - 1.960$

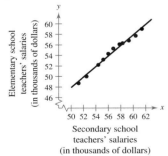

5. $50,382.50

6. 0.984; About 98.4% of the variation in the average annual salaries of elementary school teachers can be explained by the relationship between the average annual salaries of secondary school teachers and elementary school teachers, and about 1.6% of the variation is unexplained.

7. 0.422; The standard error of estimate of the average annual salaries of elementary school teachers for a specific average annual salary of secondary school teachers is about $422.

8. $49.311 < y < 51.455$

You can be 95% confident that the average annual salary of elementary school teachers will be between $49,311 and $51,455 when the average annual salary of secondary school teachers is $52,500.

9. (a) $95.26   (b) $70.28   (c) $67.74   (d) $59.46

Real Statistics—Real Decisions for Chapter 9   (page 522)

1. (a)

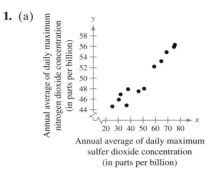

It appears that there is a positive linear correlation. As the annual average of the daily maximum sulfur dioxide concentration increases, the annual average of the daily maximum nitrogen dioxide concentration tends to increase.

(b) 0.966; There is a strong positive linear correlation.

(c) There is enough evidence at the 5% level of significance to conclude that there is a significant linear correlation between annual averages of the daily maximum concentrations of sulfur dioxide and nitrogen dioxide.

(d) $\hat{y} = 0.234x + 38.081$

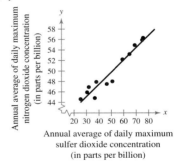

(e) Yes, for x-values that are within the range of the data set.

(f) $r^2 \approx 0.934$; About 93.4% of the variation in nitrogen dioxide concentrations can be explained by the variation in sulfur dioxide concentrations, and about 6.6% of the variation is unexplained.

$s_e \approx 1.178$; The standard error of estimate of the annual averages of the daily maximum concentration of nitrogen dioxide for a specific annual average of the daily maximum concentration of sulfur dioxide is about 1.178 parts per billion.

2. $41.733 < y < 47.533$

You can be 95% confident that the annual average of the daily maximum nitrogen dioxide concentration will be between 41.733 and 47.533 parts per billion when the annual average of the daily maximum sulfur dioxide concentration is 28 parts per billion.

Chapter 10

Section 10.1   (page 532)

1. A multinomial experiment is a probability experiment consisting of a fixed number of independent trials in which there are more than two possible outcomes for each trial. The probability of each outcome is fixed, and each outcome is classified into categories.

3. 45   5. 57.5

7. (a) H_0: The distribution of the ages of moviegoers is 23% ages 2–17, 20% ages 18–24, 22% ages 25–39, 9% ages 40–49, and 26% ages 50+. (claim)
   H_a: The distribution of ages differs from the expected distribution.
   (b) $\chi_0^2 = 7.779$; Rejection region: $\chi^2 > 7.779$
   (c) 6.244   (d) Fail to reject H_0.
   (e) There is not enough evidence at the 10% level of significance to reject the claim that the distribution of the ages of moviegoers and the expected distribution are the same.

9. (a) H_0: The distribution of the days people order food for delivery is 7% Sunday, 4% Monday, 6% Tuesday, 13% Wednesday, 10% Thursday, 36% Friday, and 24% Saturday.
   H_a: The distribution of days differs from the expected distribution. (claim)
   (b) $\chi_0^2 = 16.812$; Rejection region: $\chi^2 > 16.812$
   (c) 17.595   (d) Reject H_0.
   (e) There is enough evidence at the 1% level of significance to conclude that the distribution of days differs from the expected distribution.

11. (a) H_0: The distribution of the number of homicide crimes in California by county is uniform. (claim)
   H_a: The distribution of homicides by county is not uniform.
   (b) $\chi_0^2 = 30.578$; Rejection region: $\chi^2 > 30.578$
   (c) 143.904   (d) Reject H_0.
   (e) There is enough evidence at the 1% level of significance to reject the claim that the distribution of the number of homicide crimes in California by county is uniform.

13. (a) H_0: The distribution of the opinions of U.S. parents on whether a college education is worth the expense is 55% strongly agree, 30% somewhat agree, 5% neither agree nor disagree, 6% somewhat disagree, and 4% strongly disagree.
   H_a: The distribution of opinions differs from the expected distribution. (claim)
   (b) $\chi_0^2 = 9.488$; Rejection region: $\chi^2 > 9.488$
   (c) 65.236   (d) Reject H_0.
   (e) There is enough evidence at the 5% level of significance to conclude that the distribution of the opinions of U.S. parents on whether a college education is worth the expense differs from the expected distribution.

15. (a) H_0: The distribution of prospective home buyers by the size they want their next house to be is uniform.
   H_a: The distribution of prospective home buyers by the size they want their next house to be is not uniform. (claim)
   (b) $\chi_0^2 = 5.991$; Rejection region: $\chi^2 > 5.991$
   (c) 10.308   (d) Reject H_0.
   (e) There is enough evidence at the 5% level of significance to conclude that the distribution of prospective home buyers by the size they want their next house to be is not uniform.

17. (a) The expected frequencies are 17, 63, 79, 34, and 5.
   (b) $\chi_0^2 = 13.277$; Rejection region: $\chi^2 > 13.277$
   (c) 0.613   (d) Fail to reject H_0.
   (e) There is not enough evidence at the 1% level of significance to reject the claim that the test scores are normally distributed.

Section 10.2   (page 542)

1. Find the sum of the row and the sum of the column in which the cell is located. Find the product of these sums. Divide the product by the sample size.

3. *Sample answer:* For both the chi-square independence test and the chi-square goodness-of-fit test, you are testing a claim about data that are in categories. However, the chi-square goodness-of-fit test has only one data value per category, while the chi-square independence test has multiple data values per category.
   Both tests compare observed and expected frequencies. However, the chi-square goodness-of-fit test simply compares the distributions, whereas the chi-square independence test compares them and then draws a conclusion about the dependence or independence of the variables.

5. False. If the two variables of a chi-square independence test are dependent, then you can expect a large difference between the observed frequencies and the expected frequencies.

7. (a)–(b)

Result	Athlete has		
	Stretched	Not stretched	Total
Injury	18 (20.82)	22 (19.18)	40
No injury	211 (208.18)	189 (191.82)	400
Total	229	211	440

9. (a)–(b)

Bank employee	Preference			
	New procedure	Old procedure	No preference	Total
Teller	92 (133.80)	351 (313.00)	50 (46.19)	493
Customer service representative	76 (34.20)	42 (80.00)	8 (11.81)	126
Total	168	393	58	619

11. (a)–(b)

Gender	Type of car				
	Compact	**Full-size**	**SUV**	**Truck/van**	**Total**
Male	28 (28.6)	39 (39.05)	21 (22.55)	22 (19.8)	110
Female	24 (23.4)	32 (31.95)	20 (18.45)	14 (16.2)	90
Total	52	71	41	36	200

13. (a) H_0: An athlete's injury result is independent of whether or not the athlete has stretched. (claim)

H_a: An athlete's injury result is dependent on whether or not an athlete has stretched.

(b) d.f. = 1; $\chi_0^2 = 6.635$; Rejection region: $\chi^2 > 6.635$

(c) 0.875   (d) Fail to reject H_0.

(e) There is not enough evidence at the 1% level of significance to reject the claim that an athlete's injury result is independent of whether or not the athlete has stretched.

15. (a) H_0: The result is independent of the type of treatment.

H_a: The result is dependent on the type of treatment. (claim)

(b) d.f. = 1; $\chi_0^2 = 2.706$; Rejection region: $\chi^2 > 2.706$

(c) 12.478   (d) Reject H_0.

(e) There is enough evidence at the 10% level of significance to conclude that the result is dependent on the type of treatment.

17. (a) H_0: The number of times former smokers tried to quit is independent of gender.

H_a: The number of times former smokers tried to quit is dependent on gender. (claim)

(b) d.f. = 2; $\chi_0^2 = 5.991$; Rejection region: $\chi^2 > 5.991$

(c) 0.002   (d) Fail to reject H_0.

(e) There is not enough evidence at the 5% level of significance to conclude that the number of times former smokers tried to quit is dependent on gender.

19. (a) H_0: Reasons are independent of the type of worker.

H_a: Reasons are dependent on the type of worker. (claim)

(b) d.f. = 2; $\chi_0^2 = 9.210$; Rejection region: $\chi^2 > 9.210$

(c) 7.326   (d) Fail to reject H_0.

(e) There is not enough evidence at the 1% level of significance to conclude that reasons for continuing education are dependent on the type of worker.

21. (a) H_0: A family borrowing money for college is independent of race.

H_a: A family borrowing money for college is dependent on race. (claim)

(b) d.f. = 2; $\chi_0^2 = 9.210$; Rejection region: $\chi^2 > 9.210$

(c) 5.994   (d) Fail to reject H_0.

(e) There is not enough evidence at the 1% level of significance to conclude that a family borrowing money for college is dependent on race.

23. (a) H_0: Type of crash is independent of the type of vehicle.

H_a: Type of crash is dependent on the type of vehicle. (claim)

(b) d.f. = 2; $\chi_0^2 = 5.991$; Rejection region: $\chi^2 > 5.991$

(c) 103.568   (d) Reject H_0.

(e) There is enough evidence at the 5% level of significance to conclude that the type of crash is dependent on the type of vehicle.

25. (a) H_0: Procedure preference is independent of bank employee.

H_a: Procedure preference is dependent on bank employee. (claim)

(b) d.f. = 2; $\chi_0^2 = 5.991$; Rejection region: $\chi^2 > 5.991$

(c) 88.361   (d) Reject H_0.

(e) There is enough evidence at the 5% level of significance to conclude that procedure preference is dependent on bank employee.

27. (a) H_0: Type of car is independent of gender. (claim)

H_a: Type of car is dependent on gender.

(b) d.f. = 3; $\chi_0^2 = 6.251$; Rejection region: $\chi^2 > 6.251$

(c) 0.808   (d) Fail to reject H_0.

(e) There is not enough evidence at the 10% level of significance to conclude that type of car is dependent on gender.

29. Fail to reject H_0. There is not enough evidence at the 5% level of significance to reject the claim that the proportions of motor vehicle crash deaths involving males or females are the same for each age group.

31. Right-tailed

33.

Status	Educational Attainment			
	Not a high school graduate	**High school graduate**	**Some college, no degree**	**Associate's, bachelor's, or advanced degree**
Employed	0.046	0.156	0.101	0.312
Unemployed	0.004	0.010	0.005	0.009
Not in the labor force	0.059	0.123	0.061	0.114

35. (a) 0.9%   (b) 6.1%

37.

Status	Educational Attainment			
	Not a high school graduate	**High school graduate**	**Some college, no degree**	**Associate's, bachelor's, or advanced degree**
Employed	0.076	0.253	0.164	0.507
Unemployed	0.150	0.350	0.183	0.317
Not in the labor force	0.164	0.344	0.172	0.320

39. 17.2%   **41.** 26.3%

Section 10.3   (page 555)

1. Specify the level of significance α. Determine the degrees of freedom for the numerator and denominator. Use Table 7 in Appendix B to find the critical value F.
3. (1) The samples must be random, (2) the samples must be independent, and (3) each population must have a normal distribution.
5. 2.54   7. 2.06   9. 9.16   11. 1.80
13. Fail to reject H_0. There is not enough evidence at the 10% level of significance to support the claim.
15. Fail to reject H_0. There is not enough evidence at the 1% level of significance to reject the claim.
17. Reject H_0. There is enough evidence at the 1% level of significance to reject the claim.
19. (a) $H_0: \sigma_1^2 \le \sigma_2^2$; $H_a: \sigma_1^2 > \sigma_2^2$ (claim)
    (b) $F_0 = 2.11$; Rejection region: $F > 2.11$
    (c) 2.17   (d) Reject H_0.
    (e) There is enough evidence at the 5% level of significance to support Company A's claim that the variance of the life of its appliances is less than the variance of the life of Company B's appliances.
21. (a) $H_0: \sigma_1^2 = \sigma_2^2$; $H_a: \sigma_1^2 \ne \sigma_2^2$ (claim)
    (b) $F_0 = 4.82$; Rejection region: $F > 4.82$
    (c) 2.58   (d) Fail to reject H_0.
    (e) There is not enough evidence at the 5% level of significance to conclude that the variances of the waiting times differ between the two age groups.
23. (a) $H_0: \sigma_1^2 = \sigma_2^2$ (claim); $H_a: \sigma_1^2 \ne \sigma_2^2$
    (b) $F_0 = 2.635$; Rejection region: $F > 2.635$
    (c) 1.282   (d) Fail to reject H_0.
    (e) There is not enough evidence at the 10% level of significance to reject the administrator's claim that the standard deviations of science assessment test scores for eighth-grade students are the same in Districts 1 and 2.
25. (a) $H_0: \sigma_1^2 \le \sigma_2^2$; $H_a: \sigma_1^2 > \sigma_2^2$ (claim)
    (b) $F_0 = 1.59$; Rejection region: $F > 1.59$
    (c) 1.31   (d) Fail to reject H_0.
    (e) There is not enough evidence at the 5% level of significance to conclude that the standard deviation of the annual salaries for actuaries is less in California than in New York.
27. Right-tailed: 14.73; Left-tailed: 0.15   29. (0.340, 3.422)

Section 10.4   (page 565)

1. $H_0: \mu_1 = \mu_2 = \mu_3 = \cdots = \mu_k$
   H_a: At least one of the means is different from the others.
3. The MS_B measures the differences related to the treatment given to each sample. The MS_W measures the differences related to entries within the same sample.
5. (a) $H_0: \mu_1 = \mu_2 = \mu_3$
    H_a: At least one mean is different from the others. (claim)
    (b) $F_0 = 3.89$; Rejection region: $F > 3.89$
    (c) 4.80   (d) Reject H_0.

(e) There is enough evidence at the 5% level of significance to conclude that at least one mean cost per ounce is different from the others.
7. (a) $H_0: \mu_1 = \mu_2 = \mu_3$
    H_a: At least one mean is different from the others. (claim)
    (b) $F_0 = 6.36$; Rejection region: $F > 6.36$
    (c) 16.11   (d) Reject H_0.
    (e) There is enough evidence at the 1% level of significance to conclude that at least one mean vacuum cleaner weight is different from the others.
9. (a) $H_0: \mu_1 = \mu_2 = \mu_3 = \mu_4$
    H_a: At least one mean is different from the others. (claim)
    (b) $F_0 = 2.84$; Rejection region: $F > 2.84$
    (c) 0.62   (d) Fail to reject H_0.
    (e) There is not enough evidence at the 5% level of significance to conclude that at least one mean age is different from the others.
11. (a) $H_0: \mu_1 = \mu_2 = \mu_3 = \mu_4$ (claim)
    H_a: At least one mean is different from the others.
    (b) $F_0 = 2.28$; Rejection region: $F > 2.28$
    (c) 3.67   (d) Reject H_0.
    (e) There is enough evidence at the 10% level of significance to reject the claim that the mean scores are the same for all regions..
13. (a) $H_0: \mu_1 = \mu_2 = \mu_3 = \mu_4 = \mu_5 = \mu_6$
    H_a: At least one mean is different from the others. (claim)
    (b) $F_0 = 2.53$; Rejection region: $F > 2.53$
    (c) 2.06   (d) Fail to reject H_0.
    (e) There is not enough evidence at the 5% level of significance to conclude that the mean salary is different in at least one of the areas.
15. Fail to reject all null hypotheses. The interaction between the advertising medium and the length of the ad has no effect on the rating and therefore there is no significant difference in the means of the ratings.
17. Fail to reject all null hypotheses. The interaction between age and gender has no effect on GPA and therefore there is no significant difference in the means of the GPAs.
19. $CV_{\text{Scheffé}} = 7.78$
    $(1, 2) \rightarrow 8.05 \rightarrow$ Significant difference
    $(1, 3) \rightarrow 0.01 \rightarrow$ No difference
    $(2, 3) \rightarrow 6.13 \rightarrow$ No difference
21. $CV_{\text{Scheffé}} = 10.98$
    $(1, 2) \rightarrow 34.18 \rightarrow$ Significant difference
    $(1, 3) \rightarrow 64.14 \rightarrow$ Significant difference
    $(2, 3) \rightarrow 4.67 \rightarrow$ No difference

Uses and Abuses for Chapter 10   (page 570)

1–2. Answers will vary.

Review Exercises for Chapter 10   (page 572)

1. (a) H_0: The distribution of the lengths of office visits is 4% less than 9 minutes, 24% 10–12 minutes, 26% 13–16 minutes, 22% 17–20 minutes, 6% 21–24 minutes, and 18% 25 or more minutes.
   H_a: The distribution of the lengths differs from the expected distribution. (claim)
   (b) $\chi_0^2 = 15.086$; Rejection region: $\chi^2 > 15.086$
   (c) 18.770   (d) Reject H_0.
   (e) There is enough evidence at the 1% level of significance to conclude that the distribution of the lengths differs from the expected distribution.

3. (a) H_0: The distribution of responses from golf students about what they need the most help with is 22% approach and swing, 9% driver shots, 4% putting, and 65% short-game shots. (claim)
   H_a: The distribution of responses differs from the expected distribution.
   (b) $\chi_0^2 = 7.815$; Rejection region: $\chi^2 > 7.815$
   (c) 0.503   (d) Fail to reject H_0.
   (e) There is enough evidence at the 5% level of significance to conclude that the distribution of golf students' responses is the same as the expected distribution.

5. (a) $E_{1,1} \approx 95.4$, $E_{1,2} \approx 349.2$, $E_{1,3} \approx 383.4$,
   $E_{1,4} \approx 222.0$, $E_{2,1} \approx 222.6$, $E_{2,2} \approx 814.8$,
   $E_{2,3} \approx 894.6$, $E_{2,4} \approx 518.0$
   (b) H_0: The years of full-time teaching experience is independent of gender.
   H_a: The years of full-time teaching experience is dependent on gender. (claim)
   (c) d.f. = 3; $\chi_0^2 = 11.345$; Rejection region: $\chi^2 > 11.345$
   (d) 3.815   (e) Fail to reject H_0.
   (f) There is not enough evidence at the 1% level of significance to conclude that the years of full-time teaching experience is dependent on gender.

7. (a) $E_{1,1} \approx 136.8$, $E_{1,2} \approx 121.0$, $E_{1,3} \approx 46.4$, $E_{1,4} \approx 23.6$,
   $E_{1,5} \approx 65.2$, $E_{2,1} \approx 37.2$, $E_{2,2} \approx 33.0$, $E_{2,3} \approx 12.6$,
   $E_{2,4} \approx 6.4$, $E_{2,5} \approx 17.8$
   (b) H_0: A species' status is independent of vertebrate group. (claim).
   H_a: A species' status is dependent on vertebrate group.
   (c) d.f. = 4; $\chi_0^2 = 9.488$; Rejection region: $\chi^2 > 9.488$
   (d) 48.438   (e) Reject H_0.
   (f) There is enough evidence at the 5% level of significance to reject the claim that a species' status (endangered or threatened) is independent of vertebrate group.

9. 2.295   **11.** 2.39   **13.** 2.06   **15.** 2.08

17. (a) H_0: $\sigma_1 = \sigma_2$ (claim); H_a: $\sigma_1 \neq \sigma_2$
    (b) $F_0 = 2.575$; Rejection region: $F > 2.575$
    (c) 2.905   (d) Reject H_0.
    (e) There is enough evidence at the 1% level of significance to reject the claim that the standard deviations of hotel room rates for San Francisco, CA, and Sacramento, CA, are the same.

19. (a) H_0: $\sigma_1^2 = \sigma_2^2$
    H_a: $\sigma_1^2 \neq \sigma_2^2$ (claim)
    (b) $F_0 = 5.32$; Rejection region: $F > 5.32$
    (c) 1.37   (d) Fail to reject H_0.
    (e) There is not enough evidence at the 1% level of significance to support the claim that the test score variance for females is different from that for males.

21. (a) H_0: $\mu_1 = \mu_2 = \mu_3 = \mu_4$
    H_a: At least one mean is different from the others. (claim)
    (b) $F_0 = 2.29$; Rejection region: $F > 2.29$
    (c) 6.19   (d) Reject H_0.
    (e) There is enough evidence at the 10% level of significance to conclude that at least one mean amount spent on energy is different from the others.

Quiz for Chapter 10   (page 576)

1. (a) H_0: The distribution of educational attainment for people in the United States ages 30–34 is 4.7% none–8th grade, 6.9% 9th–11th grade, 29.5% high school graduates, 16.6% some college, no degree, 9.8% associate's degree, 20.5% bachelor's degree, 8.7% master's degree, and 3.3% professional/doctoral degree.
   H_a: The distribution of educational attainment for people in the United States ages 30–34 differs from the distribution for people ages 25 and older. (claim)
   (b) $\chi_0^2 = 14.067$; Rejection region: $\chi^2 > 14.067$
   (c) 6.026   (d) Fail to reject H_0.
   (e) There is not enough evidence at the 5% level of significance to conclude that the distribution for people in the United States ages 30–34 differs from the distribution for people ages 25 and older.

2. (a) H_0: Age and educational attainment are independent.
   H_a: Age and educational attainment are dependent. (claim)
   (b) $\chi_0^2 = 18.475$; Rejection region: $\chi^2 > 18.475$
   (c) 8.187   (d) Fail to reject H_0.
   (e) There is not enough evidence at the 1% level of significance to conclude that educational attainment is dependent on age.

3. (a) H_0: $\sigma_1^2 = \sigma_2^2$
   H_a: $\sigma_1^2 \neq \sigma_2^2$ (claim)
   (b) $F_0 = 4.43$; Rejection region: $F > 4.43$
   (c) 1.38   (d) Fail to reject H_0.
   (e) There is not enough evidence at the 1% level of significance to conclude that the variances in annual wages for Ithaca, NY, and Little Rock, AR, are different.

4. (a) H_0: $\mu_1 = \mu_2 = \mu_3$ (claim)
   H_a: At least one mean is different from the others.
   (b) $F_0 = 2.44$; Rejection region: $F > 2.44$
   (c) 6.18   (d) Reject H_0.
   (e) There is enough evidence at the 10% level of significance to reject the claim that the mean annual wages are the same for all three cities.

Real Statistics—Real Decisions for Chapter 10 (page 578)

1. Fail to reject H_0. There is not enough evidence at the 1% level of significance to conclude that the distribution of responses differs from the expected distribution.
2. (a) $E_{1,1} = 15$, $E_{1,2} = 120$, $E_{1,3} = 165$, $E_{1,4} = 185$, $E_{1,5} = 135$, $E_{1,6} = 115$, $E_{1,7} = 155$, $E_{1,8} = 110$, $E_{2,1} = 15$, $E_{2,2} = 120$, $E_{2,3} = 165$, $E_{2,4} = 185$, $E_{2,5} = 135$, $E_{2,6} = 115$, $E_{2,7} = 155$, $E_{2,8} = 110$
   (b) There is enough evidence at the 1% level of significance to conclude that the ages of the victims are related to the type of fraud.

Cumulative Review for Chapters 9 and 10 (page 580)

1. (a)

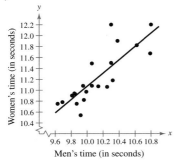

$r \approx 0.827$; strong positive linear correlation
   (b) There is enough evidence at the 5% level of significance to conclude that there is a significant linear correlation between the men's and women's winning 100-meter times.
   (c) $y = 1.216x - 1.088$

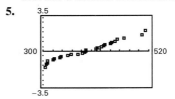

   (d) 10.95 seconds
2. There is enough evidence at the 10% level of significance to reject the claim that the mean expenditures are the same for all four regions.
3. (a) 12,442 pounds per acre
   (b) 12,217 pounds per acre
4. There is not enough evidence at the 10% level of significance to reject the administrator's claim that the standard deviations of reading test scores for eighth-grade students are the same in Colorado and Utah.
5. There is not enough evidence at the 5% level of significance to reject the claim that the distributions are the same.

6. There is not enough evidence at the 5% level of significance to conclude that the adults' ratings of the movie are dependent on gender.
7. (a) 0.751; About 75.1% of the variation in height can be explained by the relationship between metacarpal bone length and height, and about 24.9% of the variation is unexplained.
   (b) 3.87; The standard error of estimate of the height for a specific metacarpal bone length is about 3.87 centimeters.
   (c) $170.015 < y < 189.446$; You can be 95% confident that the height will be between 170.015 centimeters and 189.446 centimeters when the metacarpal bone length is 50 centimeters.

Appendix C

Appendix C (page A30)

1. The observed values are usually plotted along the horizontal axis. The expected z-scores are plotted along the vertical axis.
3. Because the points appear to follow a nonlinear pattern, you can conclude that the data do not come from a population that has a normal distribution.
5.

Because the points are approximately linear, you can conclude that the data come from a population that has a normal distribution.

SELECTED ANSWERS

Chapter 1

Section 1.1 (page 6)

16. Population, because it is a collection of the number of wireless devices in all U.S. households.

18. Sample, because only the age of every fourth person entering the grocery store is recorded.

20. Sample, because the collection of the 20 air contamination levels is a subset of the population.

22. Population: Student donations at a food drive
Sample: Student donations of canned goods

30. Population: Collection of plans after high school of all students at a high school
Sample: Collection of plans after high school of 496 students surveyed at a high school
Sample data set: 95% of those surveyed who are planning to go to college and 5% who are not

32. Population: Collection of the responses of all preowned automobile shoppers
Sample: Collection of the responses of the 1254 preowned automobile shoppers surveyed
Sample data set: 5% of respondents shopping for preowned automobiles who bought extended warranties and 95% who did not

34. Population: Collection of parents of 13- to 17-year-olds
Sample: Collection of responses of 1060 parents of 13- to 17-year-olds surveyed
Sample data set: 636 parents who said they check their teen's social media profile and 424 parents who did not

36. Sample statistic. The value 56.3% is a numerical description of a sample of college board members

Section 1.3 (page 24)

14. Observational study. The study does not apply a treatment to the motorists.

20. (a) The experimental units are the 31 patients with type 2 diabetes. The treatment is the dietary supplement designed to control metabolism in patients with type 2 diabetes.

(b) A problem with the design is that the sample size is small. The experiment could be replicated to increase validity.

(c) In a placebo-controlled, double-blind experiment, neither the subject nor the experimenter knows whether the subject is receiving a treatment or a placebo. The experimenter is informed after all the data have been collected.

(d) Divide the subjects into age categories and then, within each age group, randomly assign subjects to either the treatment group or the control group.

34. The question is biased because it already suggests that the media influences the opinions of voters. The question could be rewritten as "Does the media influence the opinions of voters?"

38. *Sample answer:* Observational studies may be referred to as natural experiments because they involve observing naturally occurring events that are not influenced by the study.

Review Exercises for Chapter 1 (page 30)

2. Population: Collection of the opinions on health care reform of all doctors in the St. Louis area
Sample: Collection of the opinions on health care reform of the 83 doctors in the St. Louis area who were sampled
Sample data set: Doctors in the St. Louis area and their opinions on health care reform

4. Population: Collection of the responses of all U.S. children and adults ages 16 years and older
Sample: Collection of the responses of the 1601 U.S. children and adults ages 16 and older who were sampled
Sample data set: 48% of children and adults who have visited a public library or a bookmobile over a recent span of 12 months and 52% who did not

6. Sample statistic. The value 29% is a numerical description of a sample of U.S. voters.

28. Convenience sampling is used because of the convenience of asking the people waiting for their baggage. A potential source of bias is that all the people just got off an airplane.

Test for Chapter 1 (page 33)

4. (a) Quantitative. Ratio. The number of employees are numerical measurements. A ratio of two data values can be formed, so it makes sense to say that 40 employees are twice as many as 20 employees.

(b) Quantitative. Interval. The grade point averages are numerical measurements. Data can be ordered and meaningful differences can be calculated, but it does not make sense to say that a person with a 3.8 GPA is twice as smart as a person with a 1.9 GPA.

5. (a) The survey question is unbiased.

(b) The question is biased because it already suggests that the town's ban on skateboarding in parks is unfair. The question could be rewritten as "What are your thoughts on the town's ban on skateboarding in parks?"

6. (a) Population: Collection of the responses of all U.S. physicians

Sample: Collection of the 19,183 U.S. physicians who were sampled.

(b) Both. Location, employment status, benefits received, and speciality are qualitative because they are attributes. Income and time spent seeing patients per week are quantitative because they are numerical measurements.

(c) Nominal: location, employment status, benefits received, specialty

Ratio: income, time spent seeing patients per week

(d) Observational study. The study does not attempt to influence the responses of the physicians and there is no treatment.

Chapter 2

Section 2.1   (page 49)

12. Class width = 13; Lower class limits: 12, 25, 38, 51, 64, 77; Upper class limits: 24, 37, 50, 63, 76, 89

14. Class width = 20; Lower class limits: 54, 74, 94, 114, 134, 154, 174, 194, 214, 234; Upper class limits: 73, 93, 113, 133, 153, 173, 193, 213, 233, 253

16. (a) 8

(b) and (c)

Class	Midpoint	Class boundaries
25–32	28.5	24.5–32.5
33–40	36.5	32.5–40.5
41–48	44.5	40.5–48.5
49–56	52.5	48.5–56.5
57–64	60.5	56.5–64.5
65–72	68.5	64.5–72.5
73–80	76.5	72.5–80.5

18.

Class	Frequency, f	Midpoint	Relative frequency	Cumulative frequency
25–32	86	28.5	0.24	86
33–40	39	36.5	0.11	125
41–48	41	44.5	0.11	166
49–56	48	52.5	0.13	214
57–64	43	60.5	0.12	257
65–72	68	68.5	0.19	325
73–80	40	76.5	0.11	365
	$\Sigma f = 365$		$\Sigma \dfrac{f}{n} \approx 1$	

30.

Class	Frequency, f	Midpoint	Relative frequency	Cumulative frequency
30–113	5	71.5	0.17	5
114–197	7	155.5	0.23	12
198–281	8	239.5	0.27	20
282–365	3	323.5	0.10	23
366–449	3	407.5	0.10	26
450–533	4	491.5	0.13	30
	$\Sigma f = 30$		$\Sigma \dfrac{f}{n} = 1$	

Class with greatest frequency: 198–281
Classes with least frequency: 282–365, 366–449

32.

Class	Frequency, f	Midpoint	Relative frequency	Cumulative frequency
32–35	3	33.5	0.1250	3
36–39	9	37.5	0.3750	12
40–43	8	41.5	0.3333	20
44–47	3	45.5	0.1250	23
48–51	1	49.5	0.0417	24
	$\Sigma f = 24$		$\Sigma \dfrac{f}{n} = 1$	

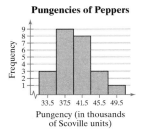

Pungencies of Peppers

Sample answer: The graph shows that most of the pungencies of the peppers were from 36,000 to 43,000 Scoville units.

34.

Class	Frequency, f	Mid-point	Relative frequency	Cumulative frequency
2243–2588	5	2415.5	0.2381	5
2589–2934	4	2761.5	0.1905	9
2935–3280	2	3107.5	0.0952	11
3281–3626	2	3453.5	0.0952	13
3627–3972	4	3799.5	0.1905	17
3973–4318	3	4145.5	0.1429	20
4319–4664	0	4491.5	0.0000	20
4665–5010	1	4837.5	0.0476	21
	$\Sigma f = 21$		$\Sigma \dfrac{f}{n} = 1$	

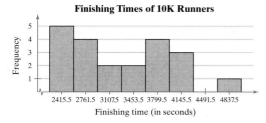

Finishing Times of 10K Runners

Sample answer: The graph shows that the most frequent finishing times were from 2243 to 2588 seconds.

36.

Class	Frequency, f	Mid-point	Relative frequency	Cumulative frequency
0–3	21	1.5	0.3750	21
4–7	18	5.5	0.3214	39
8–11	9	9.5	0.1607	48
12–15	6	13.5	0.1071	54
16–19	2	17.5	0.0357	56
	$\Sigma f = 56$		$\Sigma \dfrac{f}{n} \approx 1$	

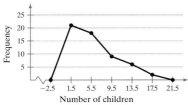

Number of Children of the Declaration of Independence Signers

Sample answer: The graph shows that most of the signers of the Declaration of Independence had 7 or fewer children.

38.

Class	Frequency, f	Midpoint	Relative frequency	Cumulative frequency
7–9	8	8	0.2857	8
10–12	11	11	0.3929	19
13–15	8	14	0.2857	27
16–18	0	17	0.0000	27
19–21	1	20	0.0357	28
	$\Sigma f = 28$		$\Sigma \dfrac{f}{n} = 1$	

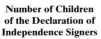

Years of Service

Class with greatest relative frequency: 10–12
Class with least relative frequency: 16–18

40.

Class	Frequency, f	Mid-point	Relative frequency	Cumulative frequency
138–202	13	170	0.46	13
203–267	7	235	0.25	20
268–332	4	300	0.14	24
333–397	1	365	0.04	25
398–462	3	430	0.11	28
	$\Sigma f = 28$		$\Sigma \dfrac{f}{n} = 1$	

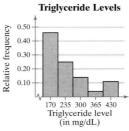

Triglyceride Levels

Class with greatest relative frequency: 138–202
Class with least relative frequency: 333–397

42.

Class	Frequency, f	Relative frequency	Cumulative frequency
7–10	2	0.0714	2
11–14	11	0.3929	13
15–18	7	0.2500	20
19–22	5	0.1786	25
23–26	3	0.1071	28
27–30	0	0.0000	28
	$\Sigma f = 28$	$\Sigma \dfrac{f}{n} \approx 1$	

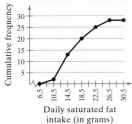

Saturated Fat Intakes

Location of the greatest increase in frequency: 11–14

44. (a)

Class	Frequency, f	Mid-point	Relative frequency	Cumulative frequency
7–53	22	30	0.44	22
54–100	16	77	0.32	38
101–147	6	124	0.12	44
148–194	2	171	0.04	46
195–241	2	218	0.04	48
242–288	0	265	0.00	48
289–335	0	312	0.00	48
336–382	2	359	0.04	50
	$\Sigma f = 50$		$\Sigma \dfrac{f}{n} = 1$	

(b)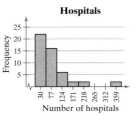

(c)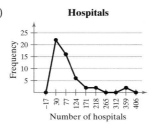

(d)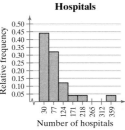

(e)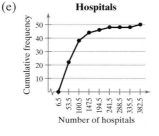

Section 2.2 (page 62)

20. Tomato Prices (in dollars per pound)

```
15 | 4 7        Key: 15|4 = 1.54
16 | 0 1 1 3 4 4 6 8
17 | 1 4 7 8 8 9
18 | 2 3 6 7 9
19 | 1 3 7 8
20 | 7 7 8
21 | 1 3
```

Sample answer: Most retail outlets charge $1.60 to $1.79 per pound of tomatoes.

22. Electoral Votes for the 50 States

```
0 | 3 3 3 3 3 3 3 4 4 4 4 4       Key: 0|3 = 3
0 | 5 5 5 6 6 6 6 6 6 7 7 7 8 8 9 9 9
1 | 0 0 0 0 1 1 1 1 2 3 4
1 | 5 6 6 8
2 | 0 0
2 | 9 9
3 |
3 | 8
4 |
4 |
5 |
5 | 5
```

Sample answer: Over half the states have less than 10 electoral votes.

26.

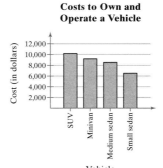

Marathon Winners' Countries of Origin

Sample answer: Most of the New York City Marathon winners are from the United States and Kenya.

28.

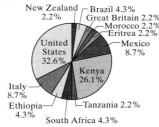

Costs to Own and Operate a Vehicle

Sample answer: It costs the least to own and operate a small sedan.

32.

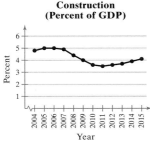
Construction (Percent of GDP)

Sample answer: The percentage of the U.S. gross domestic product that comes from the construction sector decreased from 2007 to 2011 but then increased from 2012 to 2015.

34.

Phone Screen Sizes (in inches)

The stem-and-leaf plot helps you see that most values are from 60 to 69. The dot plot helps you see that the values 55 and 60 occur most frequently.

36. **Favorite Day of The Week**

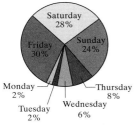

The Pareto chart helps you see the order from most favorite to least favorite day. The pie chart helps you visualize the data as parts of a whole and see that about 80% of people say their favorite day is Friday, Saturday, or Sunday.

38. (b)

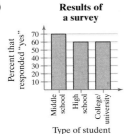

40. (a) The graph is misleading because the "non-OPEC countries" bar is wider than the "OPEC countries" bar.

(b)

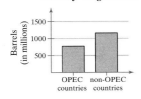

42. (a)

	3:00 P.M. Class		8:00 P.M. Class

```
                1 | 8 8 8 8 8 9 9 9 9 9
                2 | 0 0 0 2 3 4 4 5 5 8 9 9
          8 5   3 | 1 1 9
            0   4 | 3 4 4
      9 7 5 3 1   5 | 6
  9 8 8 8 8 4 2 0   6 |
  7 7 6 5 5 5 3 3   7 | 1
          5 4   8 |
```

Key: 5|3|1 = 35-year-old in 3:00 P.M. class and 31-year-old in 8:00 P.M. class

(b) In the 3:00 P.M. class, the lowest age is 35 years old and the highest age is 85 years old. In the 8:00 P.M. class, the lowest age is 18 years old and the highest age is 71 years old. There are 26 participants in the 3:00 P.M. class and there are 30 participants in the 8:00 P.M. class.

(c) *Sample answer:* The participants in each class are clustered at one of the ends of their distribution range. The 3:00 P.M. class mostly has participants over 50 years old and the 8:00 P.M. class mostly has participants under 50 years old.

Section 2.3   (page 74)

32. $\bar{x} \approx 2.49$; median = 2.35; mode = 4.0; The mode does not represent the center of the data set because it is the largest entry in the data set.

34. $\bar{x} \approx 197.5$; median = 200; mode = 160

36. Cluster around 450–1050, gap between 1950 and 2850, outlier at 3000

54.

Class	Frequency, f	Midpoint
104–150	3	127
151–197	4	174
198–244	3	221
245–291	2	268
292–338	1	315
339–385	1	362

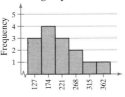
Emergency Room Visits

Number of patients

Positively skewed

56.

Number	Frequency, f
1	6
2	5
3	4
4	6
5	4
6	5

Results of Rolling Six-Sided Die

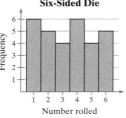

Uniform

Section 2.4   (page 93)

24. (a) Greatest sample standard deviation: (iii)
Data set (iii) has more entries that are farther away from the mean.
Least sample standard deviation: (i)
Data set (i) has more entries that are close to the mean.
(b) The three data sets have the same mean and median but have a different mode and standard deviation.
(c) Estimates will vary; (i) $s \approx 1.5$; (ii) $s \approx 1.8$; (iii) $s \approx 2.5$

40.

x	f	xf
0	30	0
1	20	20
	$\Sigma = 50$	$\Sigma = 20$

$x - \bar{x}$	$(x - \bar{x})^2$	$(x - \bar{x})^2 f$
−0.4	0.16	4.8
0.6	0.36	7.2
		$\Sigma = 12$

$\bar{x} \approx 0.4$, $s \approx 0.5$

42.

Class	x	f	xf
0–4	2	5	10
5–9	7	12	84
10–14	12	24	288
15–19	17	17	289
20–24	22	16	352
25–29	27	11	297
30+	32	5	160
		$n = 90$	$\Sigma xf = 1480$

$x - \bar{x}$	$(x - \bar{x})^2$	$(x - \bar{x})^2 f$
−14.44	208.5136	1042.5680
−9.44	89.1136	1069.3632
−4.44	19.7136	473.1264
0.56	0.3136	5.3312
5.56	30.9136	494.6176
10.56	111.5136	1226.6496
15.56	242.1136	1210.5680
		$\Sigma (x - \bar{x})^2 f = 5522.224$

$\bar{x} \approx 16.4$
$s \approx 7.9$

44.

Midpoint, x	f	xf
70.5	1	70.5
92.5	12	1110.0
114.5	25	2862.5
136.5	10	1365.0
158.5	2	317.0
	$n = 50$	$\Sigma xf = 5725$

$x - \bar{x}$	$(x - \bar{x})^2$	$(x - \bar{x})^2 f$
−44	1936	1936
−22	484	5808
0	0	0
22	484	4840
44	1936	3872
	$\Sigma (x - \bar{x})^2 f = 16{,}456$	

$\bar{x} = 114.5$
$s \approx 18.33$

Section 2.5   (page 109)

16. (a) Min $= 150$, $Q_1 = 172$, $Q_2 = 177$, $Q_3 = 180$, Max $= 182$
(b)

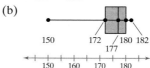

18. (a) Min $= 1$, $Q_1 = 3$, $Q_2 = 5$, $Q_3 = 8$, Max $= 9$

(b)

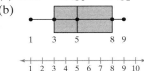

42. $A \rightarrow z = -1.54$

$B \rightarrow z = 0.77$

$C \rightarrow z = 1.54$

None of the z-scores are unusual.

52. Jamie Foxx: $z \approx -0.77$; Morgan Freeman: $z \approx 1.20$; The age of Jamie Foxx was less than 1 standard deviation below the mean age of Best Actor winners, and the age of Morgan Freeman was between 1 and 2 standard deviations above the mean age of Best Supporting Actor winners. Neither actor's age is unusual.

54. Henry Fonda: $z \approx 3.71$; John Gielgud: $z \approx 1.93$; The age of Henry Fonda was more than 3 standard deviations above the mean age of Best Actor winners, which is very unusual. The age of John Gielgud was less than 2 standard deviations above the mean age of Best Supporting Actor winners, which is not unusual.

Review Exercises for Chapter 2   (page 116)

2.

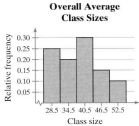

Overall Average
Class Sizes

Class with greatest relative frequency: 38–43

Class with least relative frequency: 50–55

8.   Pollution Indices of U.S. Cities

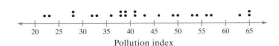

Sample answer: Most U.S. cities have a pollution index from 32 to 57.

10.   College Students'
Activities and Time Use

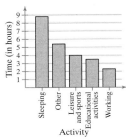

Sample answer: Full-time university and college students spend the most amount of time sleeping.

Test for Chapter 2   (page 121)

2. (a)

Class	Midpoint	Class boundaries
18–51	34.5	17.5–51.5
52–85	68.5	51.5–85.5
86–119	102.5	85.5–119.5
120–153	136.5	119.5–153.5
154–187	170.5	153.5–187.5
188–221	204.5	187.5–221.5

Frequency, f	Relative frequency	Cumulative frequency
2	0.10	2
4	0.20	6
5	0.25	11
5	0.25	16
3	0.15	19
1	0.05	20

(b)   Movies Watched in a Year

(c)   Movies Watched in a Year

(d) Symmetric

(e)   Movies Watched
in a Year

5. (a) The Beatles' Albums

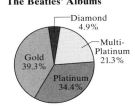

(b)   The Beatles' Albums

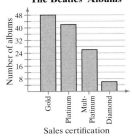

Chapter 3

Section 3.1   (page 140)

2. (c) Could not represent the probability of an event. The probability of an event occurring cannot be greater than 1.

    (d) Could not represent the probability of an event. The probability of an event occurring cannot be less than 0.

    (e) Could represent the probability of an event. The probability of an event occurring must be contained in the interval $[0, 1]$ or $[0\%, 100\%]$.

    (f) Could not represent the probability of an event. The probability of an event occurring cannot be greater than 1.

30.

```
              ┌─ H
        ┌─ H ─┤
        │     └─ T
   ┌─ H ─┤
   │     │     ┌─ H
   │     └─ T ─┤
   │           └─ T
───┤
   │           ┌─ H
   │     ┌─ H ─┤
   │     │     └─ T
   └─ T ─┤
         │     ┌─ H
         └─ T ─┤
               └─ T
```

{HHH, HHT, HTH, HTT, THH, THT, TTH, TTT}; 8

58. Subjective probability because it is most likely based on an educated guess.

64. 0.125; not unusual   **66.** 0.042; unusual

68. (a) 0.0000000245   (b) 0.99999998

80. Yes; The events in Exercises 75 and 78 can be considered unusual because their probabilities are less than or equal to 0.05.

Review Exercises for Chapter 3   (page 180)

2.

```
   ┌─1
   │ ┌─2
 1─┤ ├─3
   │ ├─4
   │ ├─5
   │ └─6
 2─┤
   │
 3─┤
   │
 4─┤
   │
 5─┤
   │
 6─┘
```

{(1, 1), (1, 2), (1, 3), (1, 4), (1, 5), (1, 6), (2, 1), (2, 2), (2, 3), (2, 4), (2, 5), (2, 6), (3, 1), (3, 2), (3, 3), (3, 4), (3, 5), (3, 6), (4, 1), (4, 2), (4, 3), (4, 4), (4, 5), (4, 6), (5, 1), (5, 2), (5, 3), (5, 4), (5, 5), (5, 6), (6, 1), (6, 2), (6, 3), (6, 4), (6, 5), (6, 6)}; 7

Chapter 4

Section 4.1   (page 197)

10. Continuous; The length of time is a random variable that has an infinite number of possible outcomes and cannot be counted.

12. Discrete; The number of arrests is a random variable that is countable.

14. Continuous; The length of time it takes to complete an exam is a random variable that cannot be counted.

16. Discrete; The number of tornadoes in the month of May in Oklahoma is a random variable that is countable.

18. Continuous; The amount of snow in Nome, Alaska, last winter is a random variable that cannot be counted.

36. A "fair bet" in a game of chance has an expected value of 0, which means that the chances of losing are equal to the chances of winning.

Section 4.2   (page 210)

28. (a)

x	P(x)
0	0.000216
1	0.002865
2	0.017089
3	0.060407
4	0.140129
5	0.222904
6	0.246231
7	0.186514
8	0.092715
9	0.027311
10	0.003620

(b)

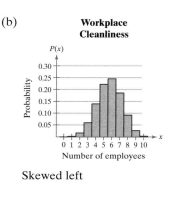

Skewed left

(c) The values 0, 1, 2, 9, and 10 are unusual because their probabilities are less than 0.05.

30. (a)

x	P(x)
0	0.00117
1	0.01239
2	0.05751
3	0.15246
4	0.25262
5	0.26790
6	0.17756
7	0.06725
8	0.01114

(b)

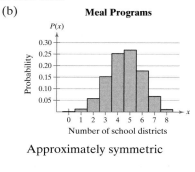

Approximately symmetric

(c) The values 0, 1, and 8 are unusual because their probabilities are less than 0.05.

Review Exercises for Chapter 4 (page 225)

18. (a)

x	P(x)
0	0.000003
1	0.000131
2	0.002409
3	0.023552
4	0.129534
5	0.379967
6	0.464404

(b)

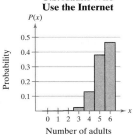

U.S. Adults Who Use the Internet

Skewed left

(c) The values 0, 1, 2, and 3 are unusual because their probabilities are less than 0.05.

Test for Chapter 4 (page 229)

6. (a)

x	P(x)
0	0.0006
1	0.0108
2	0.0721
3	0.2415
4	0.4043
5	0.2707

(b)

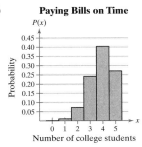

Paying Bills on Time

Skewed left

Chapter 5

Section 5.4 (page 269)

12. (a) $\mu = 1.002$, $\sigma \approx 0.0016$

(b)

Sample	Mean
1.000, 1.000	1.000
1.000, 1.001	1.0005
1.000, 1.003	1.0015
1.000, 1.004	1.002
1.001, 1.000	1.0005
1.001, 1.001	1.001
1.001, 1.003	1.002
1.001, 1.004	1.0025
1.003, 1.000	1.0015
1.003, 1.001	1.002
1.003, 1.003	1.003
1.003, 1.004	1.0035
1.004, 1.000	1.002
1.004, 1.001	1.0025
1.004, 1.003	1.0035
1.004, 1.004	1.004

(c) $\mu_{\bar{x}} = 1.002$, $\sigma_{\bar{x}} \approx 0.0011$
The means are equal, but the standard deviation of the sampling distribution is smaller.

14. (a) $\mu = 75.75$, $\sigma \approx 7.46$

(b)

Sample	Mean	Sample	Mean	Sample	Mean
67, 67, 67	67	70, 70, 81	73.67	81, 85, 67	77.67
67, 67, 70	68	70, 70, 85	75	81, 85, 70	78.67
67, 67, 81	71.67	70, 81, 67	72.67	81, 85, 81	82.33
67, 67, 85	73	70, 81, 70	73.67	81, 85, 85	83.67
67, 70, 67	68	70, 81, 81	77.33	85, 67, 67	73
67, 70, 70	69	70, 81, 85	78.67	85, 67, 70	74
67, 70, 81	72.67	70, 85, 67	74	85, 67, 81	77.67
67, 70, 85	74	70, 85, 70	75	85, 67, 85	79
67, 81, 67	71.67	70, 85, 81	78.67	85, 70, 67	74
67, 81, 70	72.67	70, 85, 85	80	85, 70, 70	75
67, 81, 81	76.33	81, 67, 67	71.67	85, 70, 81	78.67
67, 81, 85	77.67	81, 67, 70	72.67	85, 70, 85	80
67, 85, 67	73	81, 67, 81	76.33	85, 81, 67	77.67
67, 85, 70	74	81, 67, 85	77.67	85, 81, 70	78.67
67, 85, 81	77.67	81, 70, 67	72.67	85, 81, 81	82.33
67, 85, 85	79	81, 70, 70	73.67	85, 81, 85	83.67
70, 67, 67	68	81, 70, 81	77.33	85, 85, 67	79
70, 67, 70	69	81, 70, 85	78.67	85, 85, 70	80
70, 67, 81	72.67	81, 81, 67	76.33	85, 85, 81	83.67
70, 67, 85	74	81, 81, 70	77.33	85, 85, 85	85
70, 70, 67	69	81, 81, 81	81		
70, 70, 70	70	81, 81, 85	82.33		

(c) $\mu_{\bar{x}} = 75.75$, $\sigma_{\bar{x}} \approx 4.31$
The means are equal, but the standard deviation of the sampling distribution is smaller.

24. $\mu_{\bar{x}} = 196$, $\sigma_{\bar{x}} \approx 18.07$

26. $\mu_{\bar{x}} = 111{,}000$, $\sigma_{\bar{x}} \approx 1876.39$

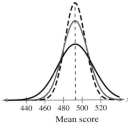

28. $n = 72$: $\mu_{\bar{x}} = 493$, $\sigma_{\bar{x}} \approx 13.44$
$n = 108$: $\mu_{\bar{x}} = 493$, $\sigma_{\bar{x}} \approx 10.97$

As the sample size increases, the standard deviation of the sample means decreases, while the mean of the sample means remains constant.

Section 5.5   (page 281)

20. (b) 0.1841 (*Tech:* 0.1833)

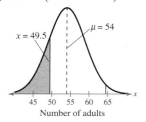

(c) 0.0968 (*Tech:* 0.0961)

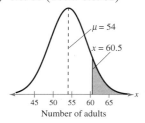

22. Can use normal distribution
(a) 0.0066 (*Tech:* 0.0068)

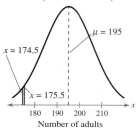

(b) 0.9974

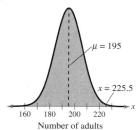

(c) 0.6915 (*Tech:* 0.6930)

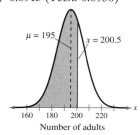

The event in part (a) is unusual because its probability is less than 0.05.

24. Can use normal distribution
(a) 0.5000

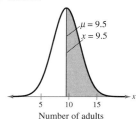

(b) 0.1075 (*Tech:* 0.1082)

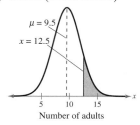

(c) 0.2914 (*Tech:* 0.2905)

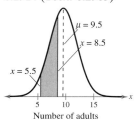

No unusual event because all of the probabilities are greater than 0.05.

26. Can use normal distribution
(a) 0.0010   (b) 0.8749 (*Tech:* 0.8756)

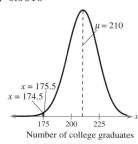 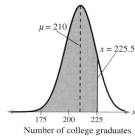

(c) 0.2389 (*Tech:* 0.2398)

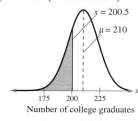

The event in part (a) is unusual because its probability is less than 0.05.

Review Exercises for Chapter 5   (page 286)

52. (b)

Sample	Mean		Sample	Mean
90, 90, 90	90		120, 120, 210	150
90, 90, 120	100		120, 210, 90	140
90, 90, 210	130		120, 210, 120	150
90, 120, 90	100		120, 210, 210	180
90, 120, 120	110		210, 90, 90	130
90, 120, 210	140		210, 90, 120	140
90, 210, 90	130		210, 90, 210	170
90, 210, 120	140		210, 120, 90	140
90, 210, 210	170		210, 120, 120	150
120, 90, 90	100		210, 120, 210	180
120, 90, 120	110		210, 210, 90	170
120, 90, 210	140		210, 210, 120	180
120, 120, 90	110		210, 210, 210	210
120, 120, 120	120			

54. $\mu_{\bar{x}} = 155.69$, $\sigma_{\bar{x}} \approx 0.798$

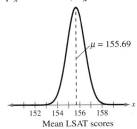

70. Can use normal distribution
(a) 0.1210 (*Tech:* 0.1204)   (b) 0.0505 (*Tech:* 0.0504)

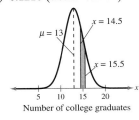

(c) 0.000219

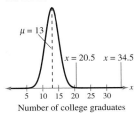

The event in part (c) is unusual because its probability is less than 0.05.

Test for Chapter 5   (page 291)

5. (b) 0.0013   (c) 0.1446

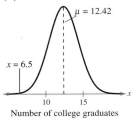
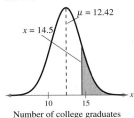

Chapter 6

Section 6.4   (page 334)

16. (a) (153.4, 499.6)   (b) (12.4, 22.4)
With 95% confidence, you can say that the population variance is between 153.4 and 499.6, and the population standard deviation is between 12.4 and 22.4.

22. (a) (193.3, 446.1)   (b) (13.9, 21.1)
With 98% confidence, you can say that the population variance is between 193.3 and 446.1, and the population standard deviation is between 13.9°F and 21.1°F.

24. (a) (9,586,982, 28,564,792)   (b) (3096, 5345)
With 90% confidence, you can say that the population variance is between 9,586,982 and 28,564,792, and the population standard deviation is between $3096 and $5345.

Review Exercises for Chapter 6   (page 338)

20. (a) 0.450, 0.550   (b) (0.424, 0.476); (0.419, 0.481)
(c) With 90% confidence, you can say that the population proportion of U.S. adults who believe that for a person to be considered truly American, it is very important that he or she share American customs and traditions is between 42.4% and 47.6%. With 95% confidence, you can say it is between 41.9% and 48.1%. The 95% confidence interval is slightly wider.

Chapter 7

Section 7.1   (page 359)

34. A type I error will occur when the actual proportion of U.S. adults who own a video game system is 0.26, but you reject H_0: $p = 0.26$.
A type II error will occur when the actual proportion of U.S. adults who own a video game system is not 0.26, but you fail to reject H_0: $p = 0.26$.

36. A type I error will occur when the actual mean cost of repairing a phone screen is greater than or equal to $75, but you reject H_0: $\mu \geq 75$.
A type II error will occur when the actual mean cost of repairing a phone screen is less than $75, but you fail to reject H_0: $\mu \geq 75$.

38. H_0: The mean time that the manufacturer's clocks lose is less than or equal to 0.02 second per day.

H_a: The mean time that the manufacturer's clocks lose is greater than 0.02 second per day.

H_0: $\mu \leq 0.02$; H_a: $\mu > 0.02$

Right-tailed because the alternative hypothesis contains $>$.

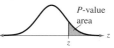

40. H_0: The percentage of cancer diagnoses attributable to lung cancer is 25%.

H_a: The percentage of cancer diagnoses attributable to lung cancer is not 25%.

H_0: $p = 0.25$; H_a: $p \neq 0.25$

Two-tailed because the alternative hypothesis contains $\neq$.

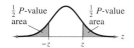

42. H_0: The percentage of responses to a survey is 100%.

H_a: The percentage of responses to a survey is not 100%.

H_0: $p = 1$; H_a: $p \neq 1$

Two-tailed because the alternative hypothesis contains $\neq$.

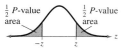

44. Alternative hypothesis

(a) There is enough evidence to support the report's claim that more than 40% of households in a New York county struggle to afford basic necessities.

(b) There is not enough evidence to support the report's claim that more than 40% of households in a New York county struggle to afford basic necessities.

Section 7.2   (page 373)

22. Critical value: $z_0 = 1.41$; Rejection region: $z > 1.41$

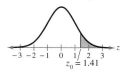

24. Critical values: $-z_0 = -1.555$, $z_0 = 1.555$

Rejection regions: $z < -1.555$, $z > 1.555$

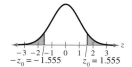

34. (a) The claim is "the mean acceleration time from 0 to 60 miles per hour for a sedan is 6.3 seconds."

H_0: $\mu = 6.3$ (claim); H_a: $\mu \neq 6.3$

(b) 2.07   (c) 0.0384   (d) Reject H_0.

(e) There is enough evidence at the 5% level of significance to reject the consumer group's claim that the mean acceleration time from 0 to 60 miles per hour for a sedan is 6.3 seconds.

36. (a) The claim is "the mean annual salary for intermediate level architects in Wichita, Kansas, is more than the national mean, $52,000."

H_0: $\mu \leq 52{,}000$; H_a: $\mu > 52{,}000$ (claim)

(b) -1.11 (c) 0.8665   (d) Fail to reject H_0.

(e) There is not enough evidence at the 9% level of significance to support the analyst's claim that the mean annual salary for intermediate level architects in Wichita, Kansas, is more than the national mean, $52,000.

40. (a) The claim is "the mean life of a light bulb is at least 750 hours."

H_0: $\mu \geq 750$ (claim); H_a: $\mu < 750$

(b) $z_0 = -2.05$; Rejection region: $z < -2.05$

(c) -0.42 (d) Fail to reject H_0.

(e) There is not enough evidence at the 2% level of significance to reject the light bulb manufacturer's claim that the mean life of the bulb is at least 750 hours.

42. (a) The claim is "the mean carbon dioxide emissions per country in a recent year are greater than 150 megatons."

H_0: $\mu \leq 150$; H_a: $\mu > 150$ (claim)

(b) $z_0 = 1.555$; Rejection region: $z < 1.555$

(c) -0.577 (d) Fail to reject H_0.

(e) There is not enough evidence at the 6% level of significance to support the scientist's claim that the mean carbon dioxide emissions per country in a recent year are greater than 150 megatons.

44. To the right; When $P < \alpha$, the standardized test statistic is in the rejection region.

Section 7.3   (page 383)

2. Identify the claim. State H_0 and H_a. Specify the level of significance. Identify the degrees of freedom. Determine the critical value(s) and rejection region(s). Find the standardized test statistic. Make a decision and interpret it in the context of the original claim.

The sample must be a random sample, and either the population must be normally distributed or n must be greater than or equal to 30.

14. Reject H_0. There is enough evidence at the 5% level of significance to support the claim.

16. Fail to reject H_0. There is not enough evidence at the 2% level of significance to reject the claim.

18. Reject H_0. There is enough evidence at the 5% level of significance to support the claim.

22. (a) The claim is "the mean battery life of an MP3 player is at least 30 hours."

H_0: $\mu \geq 30$ (claim); H_a: $\mu < 30$

(b) $t_0 = -2.567$; Rejection region: $t < -2.567$

(c) -3.74 (d) Reject H_0.

(e) There is enough evidence at the 1% level of significance to reject the company's claim that the mean battery life of their MP3 player is at least 30 hours.

24. (a) The claim is "the mean amount of lead in the air in U.S. cities is less than 0.036 microgram per cubic meter."
$H_0: \mu \geq 0.036$; $H_a: \mu < 0.036$ (claim)
(b) $t_0 = -2.396$; Rejection region: $t < -2.396$
(c) 0.33   (d) Fail to reject H_0.
(e) There is not enough evidence at the 1% level of significance to support the claim that the mean amount of lead in the air in U.S. cities is less than 0.036 microgram per cubic meter.

26. (a) The claim is "the mean annual salary for home care physical therapists is more than $80,000."
$H_0: \mu \leq 80,000$; $H_a: \mu > 80,000$ (claim)
(b) $t_0 = 1.363$; Rejection region: $t > 1.363$
(c) 0.62   (d) Fail to reject H_0.
(d) There is not enough evidence at the 10% level of significance to support the employment information service's claim that the mean annual salary for home care physical therapists is more than $80,000.

30. (a) The claim is "the mean number of classroom hours per week for full-time faculty is 11.0."
$H_0: \mu = 11.0$ (claim); $H_a: \mu \neq 11.0$
(b) 0.3155   (c) Fail to reject H_0.
(d) There is not enough evidence at the 1% level of significance to reject the dean's claim that the mean number of classroom hours per week for full-time faculty is 11.0.

32. Use the standard normal distribution because σ is known, the sample is random, and $n \geq 30$.
Fail to reject H_0. There is not enough evidence at the 1% level of significance to support the education publication's claim that the mean in-state tuition and fees at public four-year institutions by state is more than $9000.

Section 7.4   (page 391)

8. (a) The claim is "at least 27% of U.S. adults think that the IRS will audit their taxes."
$H_0: p \geq 0.27$ (claim); $H_a: p < 0.27$
(b) $z_0 = -2.33$; Rejection region: $z < -2.33$
(c) -2.85 (d) Reject H_0.
(e) There is enough evidence at the 1% level of significance to reject the research center's claim that at least 27% of U.S. adults think that the IRS will audit their taxes.

10. (a) The claim is "57% of college students work year-round."
$H_0: p = 0.57$ (claim); $H_a: p \neq 0.57$
(b) $-z_0 = -1.645$, $z_0 = 1.645$
Rejection region: $z < -1.645$, $z > 1.645$
(c) 0   (d) Fail to reject H_0.
(e) There is not enough evidence at the 10% level of significance to reject the researcher's claim that 57% of college students work year-round.

12. (a) The claim is "more than 29% of U.S. employees have changed jobs in the past three years."
$H_0: p \leq 0.29$; $H_a: p > 0.29$ (claim)

(b) $z_0 = 1.28$; Rejection region: $z > 1.28$
(c) 1.77   (d) Reject H_0.
(e) There is enough evidence at the 10% level of significance to support the research center's claim that more than 29% of U.S. employees have changed jobs in the past three years.

14. (a) The claim is "at most 18% of U.S. adults' online food purchases are for snacks."
$H_0: p \leq 0.18$ (claim); $H_a: p > 0.18$
(b) 0.01   (c) Reject H_0.
(d) There is enough evidence at the 10% level of significance to reject the research center's claim that at most 18% of U.S. adults' online food purchases are for snacks.

16. (a) The claim is "5% of U.S. households have taken in a stray dog."
$H_0: p = 0.05$ (claim); $H_a: p \neq 0.05$
(b) 0.52   (c) Fail to reject H_0.
(d) There is not enough evidence at the 5% level of significance to reject the humane society's claim that 5% of U.S. households have taken in a stray dog.

20. *Sample answer:*

$$z = \frac{\hat{p} - p}{\sqrt{pq/n}} \qquad \text{Write the original equation.}$$

$$= \frac{(x/n) - p}{\sqrt{pq/n}} \qquad \text{Substitute } \frac{x}{n} \text{ for } \hat{p}.$$

$$= \frac{x - np}{n\sqrt{\dfrac{pq}{n}}} \qquad \text{Multiply by } \frac{n}{n}.$$

$$= \frac{x - np}{\sqrt{\dfrac{n^2 pq}{n}}} \qquad \begin{array}{l}\text{Use the fact that} \\ n = \sqrt{n^2} \text{ and} \\ \text{combine radicals.}\end{array}$$

$$= \frac{x - np}{\sqrt{npq}} \qquad \text{Simplify.}$$

Section 7.5   (page 400)

2. No; In a χ^2-distribution, all χ^2-values are greater than or equal to 0 because anything squared is greater than or equal to 0.

4. Verify that the sample is random and the population is normally distributed. State H_0 and H_a, and identify the claim. Specify the level of significance. Determine the degrees of freedom. Determine the critical value(s) and rejection region(s). Find the standardized test statistic. Make a decision and interpret it in the context of the original claim.

16. Fail to reject H_0. There is not enough evidence at the 5% level of significance to reject the claim.

18. Reject H_0. There is enough evidence at the 10% level of significance to reject the claim.

20. Fail to reject H_0. There is not enough evidence at the 1% level of significance to reject the claim.

24. (a) The claim is "the variance of the gas mileages in a model of a hybrid vehicle is 0.16."
$H_0: \sigma^2 = 0.16$ (claim); $H_a: \sigma^2 \neq 0.16$
(b) $\chi_L^2 = 16.047$, $\chi_R^2 = 45.722$
Rejection regions: $\chi^2 < 16.047$, $\chi^2 > 45.722$
(c) 47.125 **(d)** Reject H_0.
(e) There is enough evidence at the 5% level of significance to reject the auto manufacture's claim that the variance of the gas mileages in a model of a hybrid vehicle is 0.16.

26. (a) The claim is "the standard deviation for grade 12 students on a vocabulary assessment test is greater than 45 points."
$H_0: \sigma \leq 45$; $H_a: \sigma > 45$ (claim)
(b) $\chi_0^2 = 42.98$; Rejection region: $\chi^2 > 42.98$
(c) 25.08 **(d)** Fail to reject H_0.
(e) There is not enough evidence at the 1% level of significance to support the school administrator's claim that the standard deviation for grade 12 students on a vocabulary assessment test is greater than 45 points.

28. (a) The claim is "the standard deviation of the room rates for two adults at three-star hotels in Denver is at least $68."
$H_0: \sigma \geq 68$ (claim); $H_a: \sigma < 68$
(b) $\chi_0^2 = 6.408$; Rejection region: $\chi^2 < 6.408$
(c) 5.88 **(d)** Reject H_0.
(e) There is enough evidence at the 1% level of significance to reject the travel analyst's claim that the standard deviation of the room rates for two adults at three-star hotels in Denver is at least $68.

30. (a) The claim is "the standard deviation of the annual salaries of nursing supervisors is $16,500."
$H_0: \sigma = 16,500$ (claim); $H_a: \sigma \neq 16,500$
(b) $\chi_L^2 = 4.575$, $\chi_R^2 = 19.675$
Rejection regions: $\chi^2 < 4.575$, $\chi^2 > 19.675$
(c) 8.51 **(d)** Fail to reject H_0.
(e) There is not enough evidence at the 10% level of significance to reject the claim that the standard deviation of the annual salaries of nursing supervisors is $16,500.

Review Exercises for Chapter 7 (page 406)

8. (a) $H_0: \mu \geq 400$ (claim); $H_a: \mu < 400$
(b) A type I error will occur when the actual mean shelf life of the dried fruit is at least 400 days, but you reject $H_0: \mu \geq 400$.
A type II error will occur when the actual mean shelf life of the dried fruit is less than 400 days, but you fail to reject $H_0: \mu \geq 400$.
(c) Left-tailed because the alternative hypothesis contains $<$.
(d) There is enough evidence to reject the agricultural cooperative's claim that the mean shelf life of the dried fruit is at least 400 days.
(e) There is not enough evidence to reject the agricultural cooperative's claim that the mean shelf life of the dried fruit is at least 400 days.

10. (a) $H_0: \mu \geq 25$; $H_a: \mu < 25$ (claim)
(b) A type I error will occur when the actual mean number of grams of carbohydrates in one bar is at least 25, but you reject $H_0: \mu \geq 25$.
A type II error will occur when the actual mean number of grams of carbohydrates in one bar is less than 25, but you fail to reject $H_0: \mu \geq 25$.
(c) Left-tailed because the alternative hypothesis contains $<$.
(d) There is enough evidence to support the energy bar maker's claim that the mean number of grams of carbohydrates in one bar is less than 25.
(e) There is not enough evidence to support the energy bar maker's claim that the mean number of grams of carbohydrates in one bar is less than 25.

14. Critical values: $-z_0 = -2.81$, $z_0 = 2.81$
Rejection regions: $z < -2.81$, $z > 2.81$

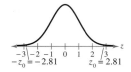

16. Critical values: $-z_0 = -2.17$, $z_0 = 2.17$
Rejection regions: $z < -2.17$, $z > 2.17$

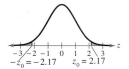

24. Fail to reject H_0. There is not enough evidence at the 10% level of significance to reject the claim.

26. (a) The claim is "the mean annual consumption of cotton is greater than 1.1 million bales per country."
$H_0: \mu \leq 1.1$; $H_a: \mu > 1.1$ (claim)
(b) -0.19 **(c)** 0.5753 **(d)** Fail to reject H_0.
(e) There is not enough evidence at the 1% level of significance to support the researcher's claim that the mean annual consumption of cotton is greater than 1.1 million bales per country.

28. (a) The claim is "the mean price of a round-trip flight from New York City to Los Angeles is less than $507."
$H_0: \mu \geq 507$; $H_a: \mu < 507$ (claim)
(b) $-z_0 = -1.645$; Rejection region: $z < -1.645$
(c) -0.33 **(d)** Fail to reject H_0.
(e) There is not enough evidence at the 5% level of significance to support the travel analyst's claim that the mean price of a round-trip flight from New York City to Los Angeles is less than $507.

44. (a) The claim is "the overall average score of 15-year-old students on an international mathematics literacy test is 494."
$H_0: \mu = 494$ (claim); $H_a: \mu \neq 494$
(b) 0.0086 **(c)** Reject H_0.
(d) There is enough evidence at the 5% level of significance to reject the education researcher's claim that the overall average score of 15-year-old students on an international mathematics literacy test is 494.

50. (a) The claim is "6% of U.S. employees say it is likely they will be laid off in the next year."
H_0: $p = 0.06$ (claim); H_a: $p \neq 0.06$
(b) $-z_0 = -1.96$, $z_0 = 1.96$
Rejection regions: $z < -1.96$, $z > 1.96$
(c) 2.013   (d) Reject H_0.
(e) There is enough evidence at the 5% level of significance to reject the labor researcher's claim that 6% of U.S. employees say it is likely they will be laid off in the next year.

60. (a) The claim is "the standard deviation of the lengths of serving times is 3 minutes."
H_0: $\sigma = 3$ (claim); H_a: $\sigma \neq 3$
(b) $\chi_L^2 = 11.160$, $\chi_R^2 = 48.290$
Rejection regions: $\chi^2 < 11.160$, $\chi^2 > 48.290$
(c) 43.94   (d) Fail to reject H_0.
(e) There is not enough evidence at the 1% level of significance to reject the restaurant's claim that the standard deviation of the lengths of serving times is 3 minutes.

Test for Chapter 7   (page 411)

3. (a) The claim is "the mean price of a meal for a family of 4 in a resort restaurant is at most $100."
H_0: $\mu \leq 100$ (claim); H_a: $\mu > 100$
(b) Right-tailed because the alternative hypothesis contains $>$; t-test because σ is unknown and $n \geq 30$.
(c) *Sample answer:* $t_0 = 2.449$;
Rejection region: $t > 2.449$; 3.02
(d) Reject H_0.
(e) There is enough evidence at the 1% level of significance to reject the travel analyst's claim that the mean price of a meal for a family of 4 in a resort restaurant is at most $100."

4. (a) The claim is "more than 80% of U.S. adults think that mothers should have paid maternity leave."
H_0: $p \leq 0.80$; H_a: $p > 0.80$ (claim)
(b) Right-tailed because the alternative hypothesis contains $>$; z-test because $np \geq 5$ and $nq \geq 5$.
(c) *Sample answer:* $z_0 = 1.645$;
Rejection region: $z > 1.645$; 0.35
(d) Fail to reject H_0.
(e) There is not enough evidence at the 5% level of significance to support the researcher's center's claim that more than 80% of U.S. adults think that mothers should have paid maternity leave.

5. (a) The claim is "the standard deviation of the number of grams of carbohydrates in a bar is 1.11 grams."
H_0: $\sigma = 1.11$ (claim); H_a: $\sigma \neq 1.11$
(b) Two-tailed because the alternative hypothesis contains $\neq$; chi-square test because the test is for a standard deviation and the population is normally distributed.
(c) *Sample answer:* $\chi_L^2 = 13.120$, $\chi_R^2 = 40.646$;
Rejection region: $\chi^2 < 13.120$, $\chi^2 > 40.646$; 28.733
(d) Fail to reject H_0.

(e) There is not enough evidence at the 5% level of significance to reject the nutrition bar manufacturer's claim that the standard deviation of the number of grams of carbohydrates in a bar is 1.11 grams.

6. (a) The claim is "the mean price of the vehicles the nonprofit consumer organization rated in a recent year is at least $41,000."
H_0: $\mu \geq 41{,}000$ (claim); H_a: $\mu < 41{,}000$
(b) Left-tailed because the alternative hypothesis contains $<$; t-test because σ is unknown and $n \geq 30$.
(c) *Sample answer:* $t_0 = -2.352$;
Rejection region: $t < -2.352$; -0.28
(d) Fail to reject H_0.
(e) There is not enough evidence at the 1% level of significance to reject the nonprofit consumer organization's claim that the mean price of the vehicles the organization rated in a recent year is at least $41,000.

7. (a) The claim is "the mean age of the residents of a small town is more than 38 years."
H_0: $\mu \leq 38$; H_a: $\mu > 38$ (claim)
(b) Right-tailed because the alternative hypothesis contains $>$; z-test because σ is known and $n \geq 30$.
(c) *Sample answer:* $z_0 = 1.28$;
Rejection region: $z > 1.28$; 2.47
(d) Reject H_0.
(e) There is enough evidence at the 10% level of significance to support the researcher's claim that the mean age of the residents of a small town is more than 38 years.

Chapter 8

Section 8.1   (page 424)

18. (a) The claim is "the mean repair costs for Model A and Model B are equal."
H_0: $\mu_1 = \mu_2$ (claim); H_a: $\mu_1 \neq \mu_2$
(b) $z_0 = -2.575$, $z_0 = 2.575$
Rejection regions: $z < -2.575$, $z > 2.575$
(c) -2.29 (d) Fail to reject H_0.
(e) There is not enough evidence at the 1% level of significance to reject the claim that the mean repair costs for Model A and Model B are equal.

22. (a) The claim is "the mean home sales price in Casper, Wyoming, is the same as in Cheynne, Wyoming."
H_0: $\mu_1 = \mu_2$ (claim); H_a: $\mu_1 \neq \mu_2$
(b) $-z_0 = -2.575$, $z_0 = 2.575$
Rejection regions: $z < -2.575$, $z > 2.575$
(c) -2.92 (d) Reject H_0.
(e) There is enough evidence at the 1% level of significance to reject the real estate agency's claim that the mean home sales price in Casper, Wyoming, is the same as in Cheynne, Wyoming.
Yes, the new samples lead to a different conclusion.

24. (a) The claim is "the temperature in Seattle, Washington, was lower than in Birmingham, Alabama."
H_0: $\mu_1 \geq \mu_2$; H_a: $\mu_1 < \mu_2$ (claim)
(b) $z_0 = -2.33$; Rejection region: $z < -2.33$
(c) -3.72 (d) Reject H_0.
(e) There is enough evidence at the 1% level of significance to support the climatologist's claim that the temperature in Seattle, Washington, was lower than in Birmingham, Alabama.

Section 8.2   (page 432)

16. (a) The claim is "the mean fork length of yellowfin tuna is different in two zones in the eastern tropical Pacific Ocean."
H_0: $\mu_1 = \mu_2$; H_a: $\mu_1 \neq \mu_2$ (claim)
(b) $-t_0 = -2.678$, $t_0 = 2.678$
Rejection regions: $t < -2.678$, $t > 2.678$
(c) -0.84 (d) Fail to reject H_0.
(e) There is not enough evidence at the 1% level of significance to support the marine biologist's claim that the mean fork length of yellowfin tuna is different in two zones in the eastern tropical Pacific Ocean.

18. (a) The claim is "the mean household income in a recent year is the same in Ada County, Idaho, and Cameron Parish, Louisiana."
H_0: $\mu_1 = \mu_2$ (claim); H_a: $\mu_1 \neq \mu_2$
(b) $-t_0 = -1.740$, $t_0 = 1.740$
Rejection regions: $t < -1.740$, $t > 1.740$
(c) 0.42 (d) Fail to reject H_0.
(e) There is not enough evidence at the 10% level of significance to reject the demographics researcher's claim that the mean household income in a recent year is the same in Ada County, Idaho, and Cameron Parish, Louisiana.

Section 8.3   (page 442)

10. (a) The claim is "an SAT preparation course will improve the test scores of students."
H_0: $\mu_d \geq 0$; H_a: $\mu_d < 0$ (claim)
(b) $t_0 = -2.821$; Rejection region: $t < -2.821$
(c) $\bar{d} = -60$; $s_d \approx 27.889$ (d) -6.803 (e) Reject H_0.
(f) There is enough evidence at the 1% level of significance to support the SAT preparation course's claim that its course will improve the test scores of students.

20. (a) The claim is "the contestants' times have changed."
H_0: $\mu_d = 0$; H_a: $\mu_d \neq 0$ (claim)
(b) $t_0 = -3.499$, $t_0 = 3.499$
Rejection regions: $t < -3.499$, $t > 3.499$
(c) $\bar{d} = 8.2375$; $s_d \approx 5.222$
(d) 4.462 (*Tech*: 4.461)   (e) Reject H_0.
(f) There is enough evidence at the 1% level of significance to support the claim that the contestants' times have changed.

Section 8.4   (page 451)

2. State the hypotheses and identify the claim. Specify the level of significance. Find the critical value(s) and rejection region(s). Find $\bar{p}$ and $\bar{q}$. Find the standardized test statistic. Make a decision and interpret it in the context of the claim. The test can also be performed by calculating the P-value and comparing it to α.

8. (a) The claim is "the proportion of 12-year survivors is greater for subjects who were given the chemotherapy than for subjects who were given the placebo."
H_0: $p_1 \leq p_2$; H_a: $p_1 > p_2$ (claim)
(b) $z_0 = 1.28$; Rejection region: $z > 1.28$
(c) 1.58 (d) Reject H_0.
(e) There is enough evidence at the 10% level of significance to support the claim that the proportion of 12-year survivors is greater for subjects who were given the chemotherapy than for subjects who were given the placebo.

10. (a) The claim is "the proportion of males ages 20 to 24 who were neither in school nor working is less than the proportion of females ages 20 to 24 who were neither in school nor working."
H_0: $p_1 \geq p_2$; H_a: $p_1 < p_2$ (claim)
(b) $z_0 = -1.645$; Rejection region: $z < -1.645$
(c) -0.85 (d) Fail to reject H_0.
(e) There is not enough evidence at the 5% level of significance to support the claim that the proportion of males ages 20 to 24 who were neither in school nor working is less than the proportion of females ages 20 to 24 who were neither in school nor working.

12. (a) The claim is "the proportion of drivers who wear seat belts in the Midwest is less than the proportion of drivers who wear seat belts in the South."
H_0: $p_1 \geq p_2$; H_a: $p_1 < p_2$ (claim)
(b) $z_0 = -1.28$; Rejection region: $z < -1.28$
(c) -3.74 (d) Reject H_0.
(e) There is enough evidence at the 10% level of significance to support the claim that the proportion of drivers who wear seat belts in the Midwest is less than the proportion of drivers who wear seat belts in the South.

Review Exercises for Chapter 8   (page 456)

10. (a) The claim is "the mean annual salary of entry-level paralegals in Peoria, Illinois, and Gary, Indiana, is the same."
H_0: $\mu_1 = \mu_2$ (claim); H_a: $\mu_1 \neq \mu_2$
(b) $-z_0 = -1.645$, $z_0 = 1.645$
Rejection regions: $z < -1.645$, $z > 1.645$
(c) 1.42 (d) Fail to reject H_0.
(e) There is not enough evidence at the 10% level of significance to reject the career counselor's claim that the mean annual salary of entry-level paralegals in Peoria, Illinois, and Gary, Indiana, is the same.

24. (a) The claim is "the weight loss supplement will help users lose weight after two weeks."
$H_0: \mu_d \leq 0$; $H_a: \mu_d > 0$ (claim)
(b) $t_0 = 1.397$;   Rejection region: $t > 1.397$
(c) $\bar{d} \approx 1.444$; $s_d \approx 2.744$ (d) 1.579   (e) Reject H_0.
(f) There is enough evidence at the 10% level of significance to support the physical fitness instructor's claim that the weight loss supplement will help users lose weight after two weeks.

Test for Chapter 8   (page 461)

3. (a) The claim is "soft tissue massage therapy helps to reduce the lengths of time patients suffer from headaches."
$H_0: \mu_d \leq 0$; $H_a: \mu_d > 0$ (claim)
(b) Right-tailed because H_a contains >; t-test because the samples are dependent and the populations are normally distributed.
(c) $t_0 = 1.740$; Rejection region: $t > 1.740$
(d) 4.34   (e) Reject H_0.
(f) There is enough evidence at the 5% level of significance to support the physical therapist's claim that soft tissue massage therapy helps to reduce the lengths of time patients suffer from headaches.

4. (a) The claim is "the mean household income in a recent year is different in Polk County, Iowa, than it is in Woodward County, Oklahoma."
$H_0: \mu_1 = \mu_2$; $H_a: \mu_1 \neq \mu_2$ (claim)
(b) Two-tailed because H_a contains $\neq$; t-test because σ_1 and σ_2 are unknown, the samples are random, the samples are independent, and the populations are normally distributed.
(c) $-t_0 = -3.055$, $t_0 = 3.055$
Rejection regions: $t < -3.055$, $t > 3.055$
(d) 0.68   (e) Fail to reject H_0.
(f) There is not enough evidence at the 1% level of significance to support the demographics researcher's claim that the mean household income in a recent year is different in Polk County, Iowa, than it is in Woodward County, Oklahoma.

Chapter 9

Section 9.1   (page 481)

24. (a)

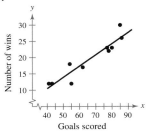

(b) 0.051
(c) No linear correlation; The heights of high school girls do not appear to be related to their IQ scores.

(d) There is not enough evidence at the 1% level of significance to conclude that there is a significant linear correlation between high school girls' heights and their IQ scores.

28. (a)

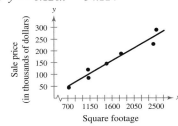

(b) -0.972
(c) Strong negative linear correlation; As the altitude increases, the speed of sound tends to decrease.
(d) There is enough evidence at the 1% level of significance to conclude that there is a significant linear correlation between altitude and speed of sound.

30. The correlation coefficient gets stronger, going from $r \approx 0.051$ to $r \approx 0.845$.

32. The correlation coefficient gets weaker, going from $r \approx -0.975$ to $r \approx -0.655$.

Section 9.2   (page 490)

18. $\hat{y} = 0.120x - 34.114$

(a) \$139,886
(b) It is not meaningful to predict the value of y for $x = 2720$ because $x = 2720$ is outside the range of the original data.
(c) \$226,886   (d) \$72,686

20. $\hat{y} = 0.345x - 3.326$

(a) 14 wins   (b) 21 wins   (c) 23 wins
(d) It is not meaningful to predict the value of y for $x = 95$ because $x = 95$ is outside the range of the original data.

34. (a) $\hat{y} = 1.711x + 3.912$

(b)

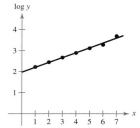

(c)

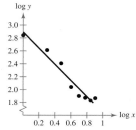

(d) The residual plot shows no pattern in the residuals because the residuals fluctuate about 0. This suggests that the regression line is a good representation of the data.

36. (a)

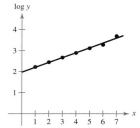

(b) The point $(14, 3)$ may be an outlier.

(c) The point $(14, 3)$ is an influential point because the slopes and y-intercepts of the regression lines with the point included and without the point included are significantly different.

38. $\log y = 0.233x + 1.969$

A linear model is more appropriate for the transformed data.

42. $\log y = -1.251 \log x + 2.893$

A linear model is more appropriate for the transformed data.

46. $y = 13.758 - 0.291 \ln x$

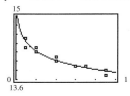

Section 9.3   (page 504)

8. 0.108; About 10.8% of the variation is explained. About 89.2% of the variation is unexplained.

10. 0.776; About 77.6% of the variation is explained. About 22.4% of the variation is unexplained.

24. $10.473 < y < 17.995$

You can be 90% confident that the trunk diameter will be between 10.473 and 17.995 inches when the tree height is 80 feet.

26. $63.199 < y < 90.483$

You can be 99% confident that the number of ballots cast will be between 63,199,000 and 90,483,000 when the voting age population is 210 million.

28. $1196.822 < y < 1447.096$

You can be 90% confident that the total assets of defined benefit plans will be between $1,196,822,000,000 and $1,447,096,000,000 when the total assets of IRAs is $6200 billion.

30. $1304.811 < y < 1687.147$

You can be 99% confident that the new-vehicle sales of Honda will be between 1,304,811 and 1,687,147 when the new-vehicle sales of Toyota are 2,359,000.

32. $\hat{y} = 1.45x - 5.21$

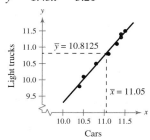

34. 0.077; The standard error of estimate of the median age of light trucks for a specific median age of cars is about 0.077 year.

Review Exercises for Chapter 9   (page 516)

2. (a)

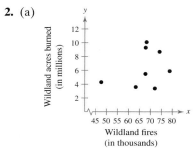

(b) 0.305

(c) Weak positive linear correlation; The number of wildland fires does not appear to be related to the number of wildland acres burned.

4. (a)

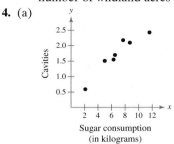

Sugar consumption
(in kilograms)

(b) 0.979

(c) Strong positive linear correlation; As sugar consumption increases, the number of cavities tends to increase.

12. $\hat{y} = -0.090x + 44.675$

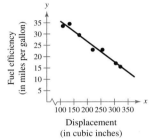

Displacement
(in cubic inches)

(a) It is not meaningful to predict the value of y when $x = 86$ because $x = 86$ is outside the range of the original data.

(b) 26.855 miles per gallon   (c) 18.665 miles per gallon

(d) It is not meaningful to predict the value of y when $x = 407$ because $x = 407$ is outside the range of the original data.

22. $16.668 < y < 24.982$

You can be 95% confident that the fuel efficiency will be between 16.668 and 24.982 miles per gallon when the engine displacement is 265 cubic inches.

24. $90.822 < y < 2405.622$

You can be 99% confident that the price of a gas grill will be between \$90.82 and \$2405.62 when the cooking area is 900 square inches.

Test for Chapter 9   (page 521)

2.

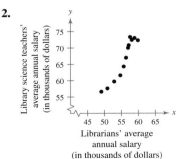

Librarians' average
annual salary
(in thousands of dollars)

The data appear to have a positive linear correlation. As x increases, y tends to increase.

5. $\hat{y} = 1.841x - 35.980$

Librarians' average
annual salary
(in thousands of dollars)

Chapter 10

Section 10.1   (page 532)

14. (a) H_0: The distribution of how much married U.S. female adults trust their spouses to manage their finances is 65.6% completely trust, 27.8% trust with certain aspects, 5.7% do not trust, and 0.9% not sure. (claim)

H_a: The distribution of how much married U.S. female adults trust their spouses to manage their finances differs from the distribution of how much married U.S. male adults trust their spouses to manage their finances.

18. (e) There is not enough evidence at the 5% level of significance to reject the claim that the test scores are normally distributed.

Section 10.2   (page 542)

8. (a)–(b)

Result	Treatment		Total
	Drug	**Placebo**	**Total**
Nausea	36 (25.15)	13 (23.85)	49
No nausea	254 (264.85)	262 (251.15)	516
Total	290	275	565

10. (a)–(b)

Size of restaurant	Rating			Total
	Excellent	**Fair**	**Poor**	**Total**
Seats 100 or fewer	182 (165.92)	203 (235.58)	165 (148.5)	550
Seats over 100	180 (196.08)	311 (278.42)	159 (175.5)	650
Total	362	514	324	1200

12. (a)–(b)

Type of movie rented	Age		
	18–24	**25–34**	**35–44**
Comedy	38 (28.37)	30 (25.32)	24 (25.75)
Action	15 (15.99)	17 (14.27)	16 (14.52)
Drama	12 (20.63)	11 (18.41)	19 (18.73)
Total	65	58	59

Type of movie rented	Age		
	45–64	**65 and older**	**Total**
Comedy	10 (19.21)	8 (11.35)	110
Action	9 (10.83)	5 (6.40)	62
Drama	25 (13.97)	13 (8.25)	80
Total	44	26	252

20. (a) H_0: The aspect of career development that is considered to be most important is independent of age.
H_a: The aspect of career development that is considered to be most important is dependent on age. (claim)

28. (a) H_0: Type of movie rental is independent of age.
H_a: Type of movie rental is dependent on age. (claim)
(b) d.f. = 8; $\chi_0^2 = 13.362$; Rejection region: $\chi^2 > 13.362$
(c) 29.045 (d) Reject H_0.
(e) There is enough evidence at the 10% level of significance to conclude that type of movie rental is dependent on age.

30. Reject H_0. There is enough evidence at the 10% level of significance to reject the claim that the proportions of the results for drug and placebo treatments are the same.

40.

Status	Educational Attainment			
	Not a high school graduate	**High school graduate**	**Some college, no degree**	**Associate's, bachelor's, or advanced degree**
Employed	0.426	0.540	0.603	0.717
Unemployed	0.038	0.034	0.031	0.020
Not in the labor force	0.536	0.426	0.367	0.263

Section 10.3 (page 555)

2. (3) The area under the F-distribution curve is equal to 1.
(4) All values of F are greater than or equal to 0.
(5) For all F-distributions, the mean value of F is approximately equal to 1.

24. (a) H_0: $\sigma_1^2 = \sigma_2^2$ (claim); H_a: $\sigma_1^2 \neq \sigma_2^2$
(b) $F_0 = 5.20$; Rejection region: $F > 5.20$
(c) 1.29 (d) Fail to reject H_0.
(e) There is not enough evidence at the 1% level of significance to reject the administrator's claim that the standard deviations of U.S. history assessment test scores for eighth-grade students are the same in Districts 1 and 2.

26. (a) H_0: $\sigma_1^2 \leq \sigma_2^2$; H_a: $\sigma_1^2 > \sigma_2^2$ (claim)
(b) $F_0 = 1.985$; Rejection region: $F > 1.985$
(c) 1.34 (d) Fail to reject H_0.
(e) There is not enough evidence at the 5% level of significance to conclude that the standard deviation of the annual salaries for public relations managers is less in Florida than in Louisiana.

Review Exercises for Chapter 10 (page 572)

2. (a) H_0: The distribution of the allowance amounts is 29% less than $10, 16% $10 to $20, 9% more than $21, and 46% don't give one/other.
H_a: The distribution of amounts differs from the expected distribution. (claim)
(b) $\chi_0^2 = 6.251$; Rejection region: $\chi^2 > 6.251$
(c) 4.886 (d) Fail to reject H_0.
(e) There is not enough evidence at the 10% level of significance to conclude that the distribution of the amounts that parents give for an allowance differs from the expected distribution.

6. (f) There is not enough evidence at the 1% level of significance to conclude that the type of vehicle owned is dependent on gender.

Test for Chapter 10 (page 577)

4. (b) $\chi_0^2 = 9.236$; Rejection region: $\chi^2 > 9.236$
(c) 5.55 (d) Fail to reject H_0.
(e) There is not enough evidence at the 10% level of significance to support the claim that the distribution of ages of workers in Oklahoma differs from the distribution of ages of workers in Maine.

5. (a) H_0: The distribution of ages of workers in Massachusetts is 6.1% ages 16–19, 11.8% ages 20–24, 41.8% ages 25–44, 21.2% ages 45–54, 10.1% ages 55–59, and 9.0% ages 60+. (claim)
H_a: The distribution of ages of workers in Massachusetts differs from the distribution of ages of workers in Maine.
(b) $\chi_0^2 = 15.086$; Rejection region: $\chi^2 > 15.086$
(c) 1.33 (d) Fail to reject H_0.
(e) There is not enough evidence at the 1% level of significance to reject the claim that the distribution of ages of workers in Massachusetts is the same as the distribution of ages of workers in Maine.

6. (a) H_0: State and age are independent. (claim)
H_a: State and age are dependent.
(b) $\chi_0^2 = 11.071$; Rejection region: $\chi^2 > 11.071$
(c) 2.555 (d) Fail to reject H_0.
(e) There is not enough evidence at the 5% level of significance to reject the claim that state and age are independent.

INDEX

PHOTO CREDITS

Cover Credits: Russian dolls, Prague, Czech Republic: Martin Child/DigitalVision/Getty Images; Great White Shark: Fuse/Corbis/Getty Images; Parked cars, Delhi, India: WIN-Initiative/Photodisc/Getty Images; Dim Sum, Taiwan: AME/a.collectionRF/Getty Images; India, Uttar Pradesh, Agra, Taj Mahal: Tetra Images - Bryan Mullennix/Brand X Pictures/Getty Images; Nyhavn, colorful harbour of Copenhagen (Denmark): Yoann JEZEQUEL Photography/Moment/Getty Images; Camel in dahnaa: sultanalsweed/Moment/Getty Images; Tasiilaq Greenland: Christine Zenino Travel Photography/Moment/Getty Images; African elephant (Loxodonta africana): Lost Horizon Images/Cultura/Getty Images; Old Faithful at Yellowstone National Park: Daniel A. Leifheit/Moment Open/Getty Images; Iceberg stacking over lake baikal: coolbiere photograph/Moment/Getty Images; Eiffel Tower, Paris, France: Thanapol Tontinikorn/Moment/Getty Images; High Angle View Of Panama Canal: Marian Stoev/EyeEm/Getty Images; St Basils Cathedral Against Clear Blue Sky: Philippe Jacquemart/EyeEm/Getty Images; Tegallalang Rice Terraces in Ubud: ©Daniela White Images/Moment/Getty Images; Within Liard River Hotsprings Provincial Park: DESPITE STRAIGHT LINES (Paul Williams)/Moment/Getty Images; Uganda, Bwindi Impenetrable National Park, Bwindi Impenetrable Forest, mountain gorilla: Westend61/Getty Images; Breaching humpback whale: Betty Wiley/Moment/Getty Images; Aurora Borealis on Iceland: Sascha Kilmer/Moment/Getty Images; High Angle View Of Red Chili Peppers On Field: Ana Soledad Guido Y Spano/EyeEm/Getty Images; Cuernos of Torres del Paine: Eric Hanson/Moment/Getty Images; Manhattan-New York: YUBO/Moment/Getty Images; Multi Colored Baskets For Sale At Market Stall: Nicolas Ayer/EyeEm/Getty Images; Occupation: CP Photo Art/Photographer's Choice/Getty Images; Goalkeeper catching ball, close-up: DAJ/amana images/Getty Images; World map concept infographic template with map made out of puzzle pieces: madpixblue/Shutterstock; Dogon women carrying millet to the village - Mali: Philippe Marion/Moment/Getty Images; Sheep Herd in New Zealand: Clickhere/E+/Getty Images

Multiple Uses: Female in pink shirt with books and backpack: Sze Fei/Shutterstock; Male in gray shirt sitting with laptop: Susan Kim/Fotolia; Female in reddish orange shirt with books: Hugo Félix/Ftolia; Female in plaid shirt with books and backpack: Elnur/Fotolia; Male in blue shirt with books and backpack: Odua Images/Shutterstock; Female in light gray shirt holding books: Kurhan/Fotolia; Male in gray shirt holding laptop with backpack: Odua Images/Shutterstock; Male in plaid shirt with backpack: Michael Jung/Shutterstock; Male in light blue shirt holding paper: Antonio Diaz/Fotolia; Female in tan jacket sitting with open book: Lithian/Shutterstock; Male in green shirt with backpack: GVS/Fotolia; Female in white shirt with scarf holding laptop: Lenets_Tan/Fotolia; Abstract 3d vector sphere with glossy mosaic design (Picturing the World icon): Red shine studio/Shutterstock; Internet icon (Applet icon): Icojam/Shutterstock

Chapter 1 p. xxii B Brown/Shutterstock; **p. 20** Olinchuk/Shutterstock; **p. 35** Pearson Education, Inc.

Chapter 2 p. 38 Mike Segar/Reuters/Alamy Stock Photo; **p. 101** Michaeljung/Fotolia

Chapter 3 p. 128 Steve Powell/Staff/Getty Images; **p. 150** DNY59/E+/Getty Images

Chapter 4 p. 188 Beeboys/Shutterstock; **p. 215** Mark Lomoglio/Icon Smi Ccx/Newscom

Chapter 5 p. 232 Smereka/Shutterstock

Chapter 6 p. 296 WavebreakMediaMicro/Fotolia; **p. 313** Pearson Education, Inc.; **p. 319** Maridav/Shutterstock; **p. 342** Environmental Protection Agency

Chapter 7 p. 346 Corbis/Getty Images; **p. 371** Nitinut380/Shutterstock

Chapter 8 p. 416 Lucky Images/Fotolia; **p. 429** Emily2k/iStock/Getty Images; **p. 433** Kletr/Shutterstock; **p. 436** Andres Rodriguez/Fotolia

Chapter 9 p. 468 Robert Beck/Sports Illustrated/Getty Images; **p. 522** Smileus/Shutterstock; **p. 523** U.S. Food and Drug Administration

Chapter 10 p. 524 Lisa S/Shutterstock; **p. 570** Bogdan Vasilescu/Shutterstock

Table 5— *t*-Distribution

d.f.	Level of confidence, *c*	0.80	0.90	0.95	0.98	0.99
	One tail, α	0.10	0.05	0.025	0.01	0.005
	Two tails, α	0.20	0.10	0.05	0.02	0.01
1		3.078	6.314	12.706	31.821	63.657
2		1.886	2.920	4.303	6.965	9.925
3		1.638	2.353	3.182	4.541	5.841
4		1.533	2.132	2.776	3.747	4.604
5		1.476	2.015	2.571	3.365	4.032
6		1.440	1.943	2.447	3.143	3.707
7		1.415	1.895	2.365	2.998	3.499
8		1.397	1.860	2.306	2.896	3.355
9		1.383	1.833	2.262	2.821	3.250
10		1.372	1.812	2.228	2.764	3.169
11		1.363	1.796	2.201	2.718	3.106
12		1.356	1.782	2.179	2.681	3.055
13		1.350	1.771	2.160	2.650	3.012
14		1.345	1.761	2.145	2.624	2.977
15		1.341	1.753	2.131	2.602	2.947
16		1.337	1.746	2.120	2.583	2.921
17		1.333	1.740	2.110	2.567	2.898
18		1.330	1.734	2.101	2.552	2.878
19		1.328	1.729	2.093	2.539	2.861
20		1.325	1.725	2.086	2.528	2.845
21		1.323	1.721	2.080	2.518	2.831
22		1.321	1.717	2.074	2.508	2.819
23		1.319	1.714	2.069	2.500	2.807
24		1.318	1.711	2.064	2.492	2.797
25		1.316	1.708	2.060	2.485	2.787
26		1.315	1.706	2.056	2.479	2.779
27		1.314	1.703	2.052	2.473	2.771
28		1.313	1.701	2.048	2.467	2.763
29		1.311	1.699	2.045	2.462	2.756
30		1.310	1.697	2.042	2.457	2.750
31		1.309	1.696	2.040	2.453	2.744
32		1.309	1.694	2.037	2.449	2.738
33		1.308	1.692	2.035	2.445	2.733
34		1.307	1.691	2.032	2.441	2.728
35		1.306	1.690	2.030	2.438	2.724
36		1.306	1.688	2.028	2.434	2.719
37		1.305	1.687	2.026	2.431	2.715
38		1.304	1.686	2.024	2.429	2.712
39		1.304	1.685	2.023	2.426	2.708
40		1.303	1.684	2.021	2.423	2.704
45		1.301	1.679	2.014	2.412	2.690
50		1.299	1.676	2.009	2.403	2.678
60		1.296	1.671	2.000	2.390	2.660
70		1.294	1.667	1.994	2.381	2.648
80		1.292	1.664	1.990	2.374	2.639
90		1.291	1.662	1.987	2.368	2.632
100		1.290	1.660	1.984	2.364	2.626
500		1.283	1.648	1.965	2.334	2.586
1000		1.282	1.646	1.962	2.330	2.581
∞		1.282	1.645	1.960	2.326	2.576

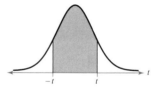

c-confidence interval

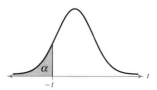

Left-tailed test

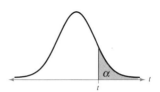

Right-tailed test

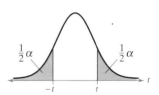

Two-tailed test

Table 6— Chi-Square Distribution

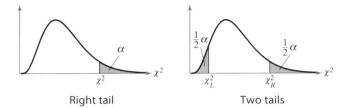

Right tail   Two tails

Degrees of freedom	α									
	0.995	0.99	0.975	0.95	0.90	0.10	0.05	0.025	0.01	0.005
1	—	—	0.001	0.004	0.016	2.706	3.841	5.024	6.635	7.879
2	0.010	0.020	0.051	0.103	0.211	4.605	5.991	7.378	9.210	10.597
3	0.072	0.115	0.216	0.352	0.584	6.251	7.815	9.348	11.345	12.838
4	0.207	0.297	0.484	0.711	1.064	7.779	9.488	11.143	13.277	14.860
5	0.412	0.554	0.831	1.145	1.610	9.236	11.071	12.833	15.086	16.750
6	0.676	0.872	1.237	1.635	2.204	10.645	12.592	14.449	16.812	18.548
7	0.989	1.239	1.690	2.167	2.833	12.017	14.067	16.013	18.475	20.278
8	1.344	1.646	2.180	2.733	3.490	13.362	15.507	17.535	20.090	21.955
9	1.735	2.088	2.700	3.325	4.168	14.684	16.919	19.023	21.666	23.589
10	2.156	2.558	3.247	3.940	4.865	15.987	18.307	20.483	23.209	25.188
11	2.603	3.053	3.816	4.575	5.578	17.275	19.675	21.920	24.725	26.757
12	3.074	3.571	4.404	5.226	6.304	18.549	21.026	23.337	26.217	28.299
13	3.565	4.107	5.009	5.892	7.042	19.812	22.362	24.736	27.688	29.819
14	4.075	4.660	5.629	6.571	7.790	21.064	23.685	26.119	29.141	31.319
15	4.601	5.229	6.262	7.261	8.547	22.307	24.996	27.488	30.578	32.801
16	5.142	5.812	6.908	7.962	9.312	23.542	26.296	28.845	32.000	34.267
17	5.697	6.408	7.564	8.672	10.085	24.769	27.587	30.191	33.409	35.718
18	6.265	7.015	8.231	9.390	10.865	25.989	28.869	31.526	34.805	37.156
19	6.844	7.633	8.907	10.117	11.651	27.204	30.144	32.852	36.191	38.582
20	7.434	8.260	9.591	10.851	12.443	28.412	31.410	34.170	37.566	39.997
21	8.034	8.897	10.283	11.591	13.240	29.615	32.671	35.479	38.932	41.401
22	8.643	9.542	10.982	12.338	14.042	30.813	33.924	36.781	40.289	42.796
23	9.260	10.196	11.689	13.091	14.848	32.007	35.172	38.076	41.638	44.181
24	9.886	10.856	12.401	13.848	15.659	33.196	36.415	39.364	42.980	45.559
25	10.520	11.524	13.120	14.611	16.473	34.382	37.652	40.646	44.314	46.928
26	11.160	12.198	13.844	15.379	17.292	35.563	38.885	41.923	45.642	48.290
27	11.808	12.879	14.573	16.151	18.114	36.741	40.113	43.194	46.963	49.645
28	12.461	13.565	15.308	16.928	18.939	37.916	41.337	44.461	48.278	50.993
29	13.121	14.257	16.047	17.708	19.768	39.087	42.557	45.722	49.588	52.336
30	13.787	14.954	16.791	18.493	20.599	40.256	43.773	46.979	50.892	53.672
40	20.707	22.164	24.433	26.509	29.051	51.805	55.758	59.342	63.691	66.766
50	27.991	29.707	32.357	34.764	37.689	63.167	67.505	71.420	76.154	79.490
60	35.534	37.485	40.482	43.188	46.459	74.397	79.082	83.298	88.379	91.952
70	43.275	45.442	48.758	51.739	55.329	85.527	90.531	95.023	100.425	104.215
80	51.172	53.540	57.153	60.391	64.278	96.578	101.879	106.629	112.329	116.321
90	59.196	61.754	65.647	69.126	73.291	107.565	113.145	118.136	124.116	128.299
100	67.328	70.065	74.222	77.929	82.358	118.498	124.342	129.561	135.807	140.169

Applet Correlation

Applet	Concept Illustrated	Descriptor	Applet Activity
Random numbers	This applet simulates selecting a random sample from a population by first assigning a unique integer to each experimental unit and then using the random numbers generated to determine the experimental units that will be included in the sample.	This applet generates random numbers from a range of integers specified by the user.	1.3
Mean versus median	The mean and the median of a data set respond differently to changes in the data. This applet investigates how skewedness and outliers affect measures of central tendency.	This applet allows the user to visualize the relationship between the mean and median of a data set. The user may easily add and delete data points. The applet automatically updates the mean and median for each change in the data.	2.3
Standard deviation	Standard deviation measures the spread of a data set. Use this applet to investigate how the shape and spread of a distribution affect the standard deviation.	This applet allows the user to visualize the relationship between the mean and standard deviation of a data set. The user may easily add and delete data points. The applet automatically updates the mean and standard deviation for each change in the data.	2.4
Simulating the stock market	Theoretical probabilities are long run experimental probabilities.	This applet simulates fluctuation in the stock market, where on any given day going up is equally likely as going down. The user specifies the number of days and the applet reports whether the stock market goes up or down each day and creates a bar graph for the outcomes. It also calculates and plots the proportion of days that the stock market goes up during the simulation.	3.1
Simulating the probability of rolling a 3 or 4	Theoretical probabilities are long run experimental probabilities. Use this applet to investigate the relationship between the theoretical and experimental probabilities of rolling a 3 or a 4 as the number of times the die is rolled increases.	This applet simulates rolling a fair die. The user specifies the number of rolls and the applet reports the outcome of each roll and creates a frequency histogram for the outcomes. It also calculates and plots the proportion of 3s and 4s rolled during the simulation.	3.3
Binomial distribution	As the number of samples increases, the estimated probability gets closer to the true value.	This applet simulates values from a binomial distribution. The user specifies the parameters for the binomial distribution (n and p) and the number of values to be simulated (N). The applet plots N values from the specified binomial distribution in a bar graph and reports the frequency of each outcome.	4.2
Sampling distributions	The mean and standard deviation of the distribution of sample means are unbiased estimators of the mean and standard deviation of the population distribution. This applet compares the means and standard deviations of the distributions and assesses the effect of sample size.	This applet simulates repeatedly choosing samples of a fixed size n from a population. The user specifies the size of the sample, the number of samples to be chosen, and the shape of the population distribution. The applet reports the means, medians, and standard deviations of both the sample means and the sample medians and creates plots for both.	5.4

(continued on next page)

Applet	Concept Illustrated	Descriptor	Applet Activity
Confidence intervals for a mean (the impact of not knowing the standard deviation)	Confidence intervals obtained using the sample standard deviation are different from those obtained using the population standard deviation. This applet investigates the effect of not knowing the population standard deviation.	This applet generates confidence intervals for a population mean. The user specifies the sample size, the shape of the distribution, the population mean, and the population standard deviation. The applet simulates selecting 100 random samples from the population and finds the 95% z-interval and 95% t-interval for each sample. The confidence intervals are plotted and the number and proportion containing the true mean are reported.	6.2
Confidence intervals for a proportion	Not all confidence intervals contain the population mean. This applet investigates the meaning of 95% and 99% confidence.	This applet generates confidence intervals for a population proportion. The user specifies the population proportion and the sample size. The applet simulates selecting 100 random samples from the population and finds the 95% and 99% confidence intervals for each sample. The confidence intervals are plotted and the number and proportion containing the true proportion are reported.	6.3
Hypothesis tests for a mean	Not all tests of hypotheses lead correctly to either rejecting or failing to reject the null hypothesis. This applet investigates the relationship between the level of confidence and the probabilities of making Type I and Type II errors.	This applet performs hypotheses tests for a population mean. The user specifies the shape of the population distribution, the population mean and standard deviation, the sample size, and the null and alternative hypotheses. The applet simulates selecting 100 random samples from the population and calculates and plots the t statistic and P-value for each sample. The applet reports the number and proportion of times the null hypothesis is rejected at both the 0.05 level and the 0.01 level.	7.3
Hypothesis tests for a proportion	Not all tests of hypotheses lead correctly to either rejecting or failing to reject the null hypothesis. This applet investigates the relationship between the level of confidence and the probabilities of making Type I and Type II errors.	This applet performs hypotheses tests for a population proportion. The user specifies the population proportion, the sample size, and the null and alternative hypotheses. The applet simulates selecting 100 random samples from the population and calculates and plots the z statistic and P-value for each sample. The applet reports the number and proportion of times the null hypothesis is rejected at both the 0.05 level and the 0.01 level.	7.4
Correlation by eye	The correlation coefficient measures the strength of a linear relationship between two variables. This applet teaches the user how to assess the strength of a linear relationship from a scatter plot.	This applet computes the correlation coefficient r for a set of bivariate data plotted on a scatter plot. The user can easily add or delete points and guess the value of r. The applet then compares the guess to its calculated value.	9.1
Regression by eye	The least squares regression line has a smaller SSE than any other line that might approximate a set of bivariate data. This applet teaches the user how to approximate the location of a regression line on a scatter plot.	This applet computes the least squares regression line for a set of bivariate data plotted on a scatter plot. The user can easily add or delete points and guess the location of the regression line by manipulating a line provided on the scatter plot. The applet will then plot the least squares line. It displays the equations and the SSEs for both lines.	9.2